Computer Vision

Katsushi Ikeuchi
Editor

Computer Vision

A Reference Guide

Volume 1

A–I

With 433 Figures and 16 Tables

Editor
Katsushi Ikeuchi
Institute of Industrial Science
The University of Tokyo
Tokyo, Japan

ISBN 978-0-387-30771-8 ISBN 978-0-387-31439-6 (eBook)
ISBN 978-0-387-49138-7 (print and electronic bundle)
DOI 10.1007/978-0-387-31439-6
Springer New York Heidelberg Dordrecht London

Library of Congress Control Number: 2014936288

Printed on acid-free paper

Springer is part of Springer Science+Business Media (www.springer.com)

Preface

Computer vision is a field of computer science and engineering; its goal is to make a computer that can see its outer world and understand what is happening. As David Marr defined, computer vision is "an information processing task that constructs efficient symbolic descriptions of the world from images." Computer vision aims to create an alternative for human visual systems on computers.

Takeo Kanade says, "computer vision looks easy, but is difficult. But, it is fun." Computer vision looks easy because each human uses vision in daily life without any effort. Even a new-born baby uses its vision capability to recognize the mother. It is computationally difficult, however, because the original outer world is made up of three dimensional objects, while those projected on the retina or an image plane, are of only two dimensional images. This dimensional reduction from 3D to 2D occurs along the projection from the outer world to images. "Common sense" needs to be used to augment the descriptions of the original 3D world from the 2D images. Computer vision is fun, because we have to discover this common sense. This search for common sense attracts the interest of vision researchers.

The origin of computer vision can be traced back to Lawrence Roberts' research, "Machine Perception of Three-Dimensional Solids." Later, this line of research has been extended through Project MAC of MIT. Professor Marvin Minski, the then director of Project MAC, initially believed that computer vision could be solved as a summer project of an MIT graduate student. His original estimation was wrong, and for more than 40 years we have been investigating various aspects of computer vision.

This 40-year effort proved that computer vision is one of the fundamental sciences, and the field is rich enough for researchers to devote their entire research lives to it. This period also reveals that the field contains a wide variety of topics from low-level optics to high-level recognition problems. This richness and diversity were an important motivation for us to decide to compile a reference book on computer vision.

Lawrence Roberts' research contains all of the essential components of the computer vision paradigm, which modern computer vision still follows: homogeneous coordinate system to define the coordinates, cross operator for edge detection, and object models represented as a combination of edges. David Marr defines his paradigm of computer vision: shape-from-x low-level vision, interpolation and fusion of such fragmental representations, 2-1/2D viewer-centered representation as the result of interpolation and fusion, and 3D object-centered representation. Roughly, this reference guide follows these paradigms, and defines the sections accordingly.

The online version of the reference guide is intended to be developed continuously, both by the updates of existing entries and by the addition of new entries. In this way, it will provide the resources to help both vision researchers and newcomers to the field be on the same page with the continuing and exciting developments in computer vision.

This reference guide has been completed through a team effort. We are most grateful for all the contributors and section editors who have made this project possible. Our special thanks go to Ms. Neha Thapa and other Springer colleagues for their assistance in the development and editing of this reference guide.

March 2014

Katsushi Ikeuchi, Editor in Chief
Yasuyuki Matsushita, Associate Editor in Chief
Rei Kawakami, Assistant Editor in Chief

Editor's Biography

Katsushi Ikeuchi is a professor at the University of Tokyo. He received his B.E. degree in mechanical engineering from Kyoto University in 1973 and Ph.D. in information engineering from the University of Tokyo in 1978. After working at the MIT Artificial Intelligence Laboratory as a postdoctoral fellow for 3 years, the MITI Electro-Technical Laboratory as a research fellow for 5 years, and the CMU Robotics Institute as a faculty member for 10 years, he joined the Institute of Industrial Science, University of Tokyo, as a professor in 1996.

His research activities span computer vision, computer graphics, robotics, and intelligent transportation systems.

In the computer vision area, he is considered as one of the founders of physics-based vision: modeling image forming processes, using physics and optics laws, and applying the inverse models in recovering shape and reflectance from observed brightness in a rigorous manner. He developed the "smoothness constraint," a constraint to force neighboring points to have similar surface orientations. The constraint optimization method based on the smoothness constraint, later referred to as the regularization method, has evolved into one of the fundamental paradigms, commonly employed in various low-level vision algorithms. In 1992, his paper with Prof. B.K.P. Horn, "Numerical Shape from Shading with Occluding Boundaries," was the original paper to describe the constraint-minimization algorithm with the smoothness constraint, [Ikeuchi K, Horn BKP (1981) Numerical shape from shading and occluding boundaries. Artif Intell 17(1–3):141–184] and was selected as one of the most influential papers to have appeared in the Artificial Intelligence Journal within the past 10 years.

Dr. Ikeuchi and his students developed a technique to automatically generate a virtual reality model by observing actual objects along the line of the physics-based vision paradigm [Sato Y, Wheeler MD, Ikeuchi K (1997) Object shape and reflectance modeling from observation. Computer graphics proceedings, SIGGRAPH97, Los Angeles, pp 379–387]. This early work, which appeared in SIGGRAPH1997, is considered one of the starting points of the area later referred to as "image-based modeling." After returning to Japan, he and his team began to apply the image-based modeling technique to model various cultural heritage sites. This project has become to be known as the e-Heritage project. They succeeded in modeling all of the three big Buddha statues in Japan as well as the complicated stone temple, Bayon, in Angkor Ruin, to name a few [Ikeuchi K, Miyazaki D (2008) Digitally archiving cultural objects. Springer, New York]. Through these efforts, Dr. Ikeuchi received the IEICE Distinguished Achievement Award in 2012.

Dr. Ikeuchi has also been working on robot vision. In this area, he has been concentrating research on how to reduce the cost of production by using robot vision technologies. This includes how to make efficient production lines and, more importantly, how to reduce the cost of making robot programs to be used in such production lines. In the early 1980s and even today, one of the obstacles to introduce robot technologies to production lines is the so-called bin-picking problem: how to pick up one part from a stack of similar parts. Using shape-from-shading techniques, he was successful in making a robot system that could pick up a mechanical part from a stack [Horn BKP, Ikeuchi K (1984) The mechanical manipulation of randomly oriented parts. Sci Am 251(2):100–111].

It was evident that the next obstacle was the cost of programming after completing the bin-picking system. In the early 1990s, he began a project to make a robot program which learns motions from observing human operators' movements [Ikeuchi K, Suehiro T (1994) Toward an assembly plan from observation, Part I: Task recognition with polyhedral objects. IEEE Trans Robot Autom 10(3):368–385]. He and his team demonstrated that this method, programming-by-demonstration, can be applied to handle not only assembling block-world objects, but also machine parts as well as flexible objects, such as rope knotting tasks [Takamatsu J et al (2006) Representation for knot-tying tasks. IEEE Trans Robot 22(1):65–78]. Along with his students, he further extended the method in the domain of whole-body motions by a humanoid robot [Nakaoka S et al (2007) Learning from observation paradigm: leg task models for enabling a biped humanoid robot to imitate human dance. Int J Robot Res 26(8):829–844]. They succeeded in making a dancing robot, which was capable of learning and mimicking Japanese folk dance from observation. He received several best paper awards in this line of work, including IEEE KS Fu memorial best transaction paper award and RSJ best transaction paper award.

Besides these research activities, he has also devoted his time to community service. He chaired a dozen major conferences, including 1995 IEEE-IROS (General), 1996 IEEE-CVPR (Program), 1999 IEEE-ITSC (General), 2001 IEEE-IV (General), 2003 IEEE-ICCV (Program), 2009 IEEE-ICRA (program), and 2010 IAPR-ICPR (program). His community service also includes IEEE RAS Adcom (98-04, 06-08), IEEE ITSS BOG, IEEE Fellow Committee (2010–2012), and 2nd VP of IAPR. He is an editor in chief of the *International Journal of Computer Vision*, and a fellow of IEEE, IEICE, IPSJ, and RSJ.

Through these research activities and community services, Dr. Ikeuchi received the IEEE PAMI-TC Distinguished Researcher Award (2011) and the Shiju Hou Sho (Medal of Honor with purple ribbon) from the Japanese Emperor (2012).

Yasuyuki Matsushita received his B.S., M.S., and Ph.D. in Electrical Engineering and Computer Science (EECS) from the University of Tokyo in 1998, 2000, and 2003, respectively. He joined Microsoft Research Asia in April 2003, where he is a senior researcher in the Visual Computing Group. His interests are in physics-based computer vision (photometric techniques, such as radiometric calibration, photometric stereo, shape-from-shading), computational photography, and general 3D reconstruction methodologies. He is on the editorial board of *International Journal of Computer Vision* (IJCV), *IEEE Transactions on Pattern Analysis and Machine Intelligence* (TPAMI), *The Visual Computer*, and associate editor in chief of the IPSJ Transactions on Computer Vision and Applications *Journal of Computer Vision and Applications* (CVA). He served/is serving as a program co-chair of Pacific-Rim Symposium on Image and Video Technology (PSIVT) 2010, The first joint 3DIM/3DPVT conference (3DIMPVT, now called 3DV) 2011, Asian Conference on Computer Vision (ACCV) 2012, International Conference on Computer Vision (ICCV) 2017, and a general co-chair of ACCV 2014. He has been serving as a guest associate professor at Osaka University (April 2010–), visiting associate professor at National Institute of Informatics (April 2011–) and Tohoku University (April 2012–), Japan.

Rei Kawakami is an assistant professor at The University of Tokyo, Tokyo, Japan. She received her B.S., M.S., and Ph.D. degrees in information science and technology from the University of Tokyo in 2003, 2005, and 2008, respectively. After working as a researcher at the Institute of Industrial Science, The University of Tokyo, for 3 years, at the University of California, Berkeley, for 2 years, and at Osaka University for a year, she joined the University of Tokyo as an assistant professor in 2014. Her research interests are in color constancy, spectral analysis, and physics-based computer vision. She has been serving as a program committee member and a reviewer for International Conference on Computer Vision (ICCV), International Conference on Computer Vision and Pattern Recognition (CVPR), European Conference on Computer Vision (ECCV), Asian Conference on Computer Vision (ACCV), and as a reviewer of the *International Journal of Computer Vision* (IJCV) and the *IEEE Transactions on Pattern Analysis and Machine Intelligence* (TPAMI).

Senior Editors

Peter N. Belhumeur Professor, Department of Computer Science, Columbia University, New York, New York, USA

Larry S. Davis Professor, Computer Vision Laboratory, Center for Automation Research, University of Maryland, College Park, MD, USA

Martial Hebert Professor, The Robotics Institute, School of Computer Science, Carnegie Mellon University, Pittsburgh, PA, USA

Jitendra Malik Arthur J. Chick Professor of EECS, University of California, Berkeley, CA, USA

Tomas Pajdla Assistant Professor, Center for Machine Perception, Department of Cybernetics, Faculty of Electrical Engineering, Czech Technical University, Prague, Czech Republic

James M. Rehg Professor, School of Interactive Computing, College of Computing, Georgia Institute of Technology, Atlanta, GA, USA

Zhengyou Zhang Research Manager/Principal Researcher, Microsoft Research, Redmond, WA, USA

Section Editors

Rama Chellappa Minta Martin Professor of Engineering, Chair, Department of Electrical Engineering and Computer Engineering and UMIACS, University of Maryland, College Park, MD, USA

Daniel Cremers Professor for Computer Science and Mathematics, Chair for Computer Vision and Pattern Recognition, Managing Director, Department of Computer Science, Technische Universität München, Garching, Germany

Koichiro Deguchi Professor Emeritus of Tohoku University, Katahira, Aoba-ku, Sendai, Japan

Hervé Delingette Research Director, Project Asclepios, INRIA, Sophia-Antipolis, France

Andrew Fitzgibbon Principal Researcher, Microsoft Research, Cambridge, England

Luc Van Gool Professor of Computer Vision, Computer Vision Laboratory, ETH, Zürich, Switzerland

Kenichi Kanatani Professor Emeritus of Okayama University, Okayama, Japan

Kyros Kutulakos Professor, Department of Computer Science, University of Toronto, Toronto, ON, Canada

In So Kweon Professor, Robotics and Computer Vision Laboratory, Korea Advanced Institute of Science and Technology (KAIST), Daejeon, Korea

Sang Wook Lee Professor, Department of Media Technology, Sogang University, Seoul, Korea

Atsuto Maki Associate Professor, Computer Vision and Active Perception Laboratory (CVAP), School of Computer Science and Communication, Kungliga Tekniska Högskolan (KTH), Stockholm, Sweden

Yasuyuki Matsushita Senior Researcher, Microsoft Research, Beijing, China

Gerard G. Medioni Professor, Institute for Robotics and Intelligence Systems, University of Southern California, Los Angeles, CA, USA

Ram Nevatia Director and Professor, Institute for Robotics and Intelligence Systems, University of Southern California, Los Angeles, CA, USA

Ko Nishino Associate Professor, Department of Computing, College of Computing and Informatics, Drexel University, Philadelphia, PA, USA

Nikolaos Papanikolopoulos Distinguished McKnight University Professor, Department of Computer Science and Engineering, University of Minnesota, Minneapolis, MN, USA

Shmuel Peleg Professor, School of Computer Science and Engineering, The Hebrew University of Jerusalem, Jerusalem, Israel

Marc Pollefeys Professor and the Head, Computer Vision and Geometry Lab (CVG) – Institute of Visual Computing, Department of Computer Science, ETH Zürich, Zürich, Switzerland

Jean Ponce Professor, Departement d'Informatique, Ecole Normale Supérieure [ENS], Paris, France

Long Quan Professor, The Department of Computer Science and Engineering, The Hong Kong University of Science and Technology, Kowloon, Hong Kong, China

Yoav Y. Schechner Associate Professor, Department of Electrical Engineering, Technion – Israel Institute of Technology, Haifa, Israel

Cordelia Schmid INRIA Research Director, Head of the LEAR Project, Montbonnot, France

Jun Takamatsu Associate Professor, Graduate School of Information Science, Nara Institute of Science and Technology (NAIST), Ikoma, Japan

Xiaoou Tang Professor, Department of Information Engineering, The Chinese University of Hong Kong, Hong Kong SAR

Song-Chun Zhu Professor, Statics Department and Computer Science Department, University of California, Los Angeles, CA, USA

Todd Zickler William and Ami Kuan Danoff Professor of Electrical Engineering and Computer Science, School of Engineering and Applied Sciences, Harvard University, Cambridge, MA, USA

Contributors

Hanno Ackermann Leibniz University Hannover, Hannover, Germany

J. K. Aggarwal Department of Electrical and Computer Engineering, The University of Texas at Austin, Austin, TX, USA

Amit Agrawal Mitsubishi Electric Research Laboratories, Cambridge, MA, USA

Daniel C. Alexander Centre for Medical Image Computing, Department of Computer Science, University College London, London, UK

Marina Alterman Department of Electrical Engineering, Technion – Israel Institute of Technology, Haifa, Israel

Yali Amit Department of Computer Science, University of Chicago, Chicago, IL, USA

Gary A. Atkinson Machine Vision Laboratory, University of the West of England, Bristol, UK

Ruzena Bajcsy Department of Electrical and Computer Sciences, College of Engineering University of California, Berkeley, CA, USA

Simon Baker Microsoft Research, Redmond, WA, USA

Dana H. Ballard Department of Computer Sciences, University of Texas at Austin, Austin, TX, USA

Richard G. Baraniuk Department of Electrical and Computer Engineering, Rice University 2028 Duncan Hal, Houston, TX, USA

Adrian Barbu Department of Statistics, Florida State University, Tallahassee, FL, USA

Nick Barnes Australian National University, Canberra, Australia

Ronen Basri Department of Computer Science And Applied Mathematics, Weizmann Institute of Science, Rehovot, Israel

Anup Basu Department of Computing Science, University of Alberta, Edmonton, AB, Canada

Rodrigo Benenson K.U. Leuven departement Elektrotechniek – ESAT, Centrum voor beeld- en spraakverwerking – PSI/VISICS, Heverlee, Belgium

Marcelo Bertalmío Universitat Pompeu Fabra, Barcelona, Spain

Jürgen Beyerer Fraunhofer Institute of Optronics, System Technologies and Image Exploitation IOSB, Karlsruhe, Germany

Irving Biederman Department of Computer Science, University of Toronto, Toronto, ON, Canada

Departments of Psychology, Computer Science, and the Neuroscience Program, University of Southern California, Los Angeles, CA, USA

Tom Bishop Image Algorithms Engineer, Apple, Cupertino, CA, USA

James Bonaiuto California Institute of Technology, Pasadena, CA, USA

Yuri Boykov Department of Computer Science, University of Western Ontario, London, ON, Canada

Christoph Bregler Courant Institute, New York University, New York, NY, USA

Michael H. Brill Datacolor, Lawrenceville, NJ, USA

Matthew Brown Dept of Computer Science, University of Bath, Bath, UK

Thomas Brox Department of Computer Science, University of Freiburg, Freiburg, Germany

Frank Michael Caimi IEEE OES, Vero Beach, FL, USA

Fabio Camilli Dipartimento SBAI, "Sapienza", Università di Roma, Rome, Italy

John N. Carter School of Electronics and Computer Science, University of Southampton, Southampton, Hampshire, UK

Vicent Caselles Universitat Pompeu Fabra, Barcelona, Spain

Shing Chow Chan Department of Electrical and Electronic Engineering, The University of Hong Kong, Hong Kong, China

Manmohan Chandraker NEC Labs America, Cupertino, CA, USA

Visesh Chari Institut National de Recherche en Informatique et en Automatique (INRIA), Le Chesnay Cedex, France

François Chaumette Inria, Rennes, France

Rama Chellappa Department of Electrical Engineering and Computer Engineering and UMIACS, University of Maryland, College Park, MD, USA

Liang Chen Department of Computer Science, University of Northern British Columbia, Prince George, Canada

Irene Cheng University of Alberta, Edmonton, AB, Canada

James J. Clark Department of Electrical and Computer Engineering, McGill University, Montreal, QC, Canada

Michael F. Cohen Microsoft Research, One Microsoft Way, Redmond, WA, USA

Kristin Dana Department of Electrical and Computer Engineering, Rutgers University, The State University of New Jersey, Piscataway, NJ, USA

Larry S. Davis Computer Vision Laboratory, Center for Automation Research, University of Maryland, College Park, MD, USA

Fatih Demirci Department of Computer Engineering, TOBB University of Economics and Technology, Sogutozu, Ankara, Turkey

Sven J. Dickinson Department of Computer Science, University of Toronto, Toronto, ON, Canada

Leo Dorst Intelligent Systems Laboratory Amsterdam, Informatics Institute, University of Amsterdam, Amsterdam, The Netherlands

Mark S. Drew School of Computing Science, Simon Fraser University, Vancouver, BC, Canada

Marc Ebner Institut für Mathematik und Informatik, Ernst Moritz Arndt Universität Greifswald, Greifswald, Germany

Jan-Olof Eklundh School of Computer Science and Communication, KTH - Royal Institute of Technology, Stockholm, Swedon

James H. Elde Centre for Vision Research, York University, Toronto, ON, Canada

Francisco J. Estrada Department of Computer and Mathematical Sciences, University of Toronto at Scarborough, Toronto, ON, Canada

Paolo Favaro Department of Computer Science and Applied Mathematics, Universität Bern, Switzerland

Pedro Felzenszwalb School of Engineering, Brown University, Providence, RI, USA

Robert B. Fisher School of Informatics, University of Edinburgh, Edinburgh, UK

Boris Flach Department of Cybernetics, Czech Technical University in Prague, Faculty of Electrical Engineering, Prague 6, Czech Republic

Gian Luca Foresti Department of Mathematics and Computer Science, University of Udine, Udine, Italy

Wolfgang Förstner ETH Zürich, Zürich, Switzerland

Universität Bonn, Bonn, Germany

Kazuhiro Fukui Department of Computer Science, Graduate School of Systems and Information Engineering, University of Tsukuba, Tsukuba, Japan

Yasutaka Furukawa Google Inc., Seattle, WA, USA

Juergen Gall Max Planck Institute for Intelligent Systems, Tübingen, Germany

David Gallup Google Inc., Seattle, WA, USA

Abhijeet Ghosh Institute for Creative Technologies, University of Southern California, Playa Vista, CA, USA

Michael Goesele GCC - Graphics, Capture and Massively Parallel Computing, TU Darmstadt, Darmstadt, Germany

Bastian Goldluecke Department of Computer Science, Technische Universität München, München, Germany

Gaopeng Gou Beihang University, Beijing, China

Mohit Gupta Department of Computer Science, Columbia University, New York, NY, USA

Bohyung Han Department of Computer Science and Engineering, Pohang University of Science and Technology (POSTECH), Pohang, South Korea

Richard Hartley Department of Engineering, Australian National University, ACT, Australia

Samuel W. Hasinoff Google, Inc., Mountain View, CA, USA

Nils Hasler Graphics, Vision & Video, MPI Informatik, Saarbrücken, Germany

Vaclav Hlavac Department of Cybernetics, Czech Technical University in Prague, Faculty of Electrical Engineering, Prague 6, Czech Republic

Andrew Hogue Faculty of Business and Information Technology, University of Ontario Institute of Technology, Oshawa, ON, Canada

Takahiko Horiuchi Graduate School of Advanced Integration Science, Chiba University, Inage-ku, Chiba, Japan

Berthold K. P. Horn Department of Electrical Engineering and Computer Science, MIT, Cambridge, MA, USA

Zhanyi Hu National Laboratory of Pattern Recognition, Institute of Automation, Chinese Academy of Sciences, Beijing, China

Ivo Ihrke MPI Informatik, Saarland University, Saarbrücken, Germany

Michael R. M. Jenkin Department of Computer Science and Engineering, York University, Toronto, ON, Canada

Jiaya Jia Department of Computer Science and Engineering, The Chinese University of Hong Kong, Shatin, N.T., Hong Kong, China

Zhuolin Jiang Noah's Ark Lab, Huawei Tech. Investment Co., LTD., Shatin, Hong Kong, China

Micah K. Johnson Computer Science and Artificial Intelligence Laboratory, Massachusetts Institute of Technology, Cambridge, MA, USA

Neel Joshi Microsoft Corporation, Redmond, WA, USA

Avinash C. Kak School of Electrical and Computer Engineering, Purdue University, West Lafayette, IN, USA

Kenichi Kanatani Professor Emeritus of Okayama University, Okayama, Japan

Sing Bing Kang Microsoft Research, Redmond, WA, USA

Peter Karasev Schools of Electrical and Computer and Biomedical Engineering, Georgia Institute of Technology, Atlanta, GA, USA

Jun-Sik Kim Korea Institute of Science and Technology, Seoul, Republic of Korea

Ron Kimmel Department of Computer Science, Technion – Israel Institute of Technology, Haifa, Israel

Eric Klassen Ohio State University, Columbus, OH, USA

Reinhard Koch Institut für Informatik Christian-Albrechts-Universität, Kiel, Germany

Jan J. Koenderink Faculty of EEMSC, Delft University of Technology, Delft, The Netherlands

The Flemish Academic Centre for Science and the Arts (VLAC), Brussels, Belgium

Laboratory of Experimental Psychology, University of Leuven (K.U. Leuven), Leuven, Belgium

Pushmeet Kohli Department of Computer Science And Applied Mathematics, Weizmann Institute of Science, Rehovot, Israel

Ivan Kolesov Schools of Electrical and Computer and Biomedical Engineering, Georgia Institute of Technology, Atlanta, GA, USA

Sanjeev J. Koppal Harvard University, Cambridge, MA, USA

Kevin Köser Institute for Visual Computing, ETH Zurich, Zürich, Switzerland

Sanjiv Kumar Google Research, New York, NY, USA

Takio Kurita Graduate School of Engineering, Hiroshima University, Higashi-Hiroshima, Japan

Sebastian Kurtek Ohio State University, Columbus, OH, USA

Annika Lang Seminar für Angewandte Mathematik, ETH Zürich, Zürich, Switzerland

Michael S. Langer School of Computer Science, McGill University, Montreal, QC, Canada

Fabian Langguth GCC - Graphics, Capture and Massively Parallel Computing, TU Darmstadt, Darmstadt, Germany

Longin Jan Latecki Department of Computer and Information Sciences, Temple University, Philadelphia, PA, USA

Denis Laurendeau Faculty of Science and Engineering, Department of Electrical and Computer Engineering, Laval University, QC, Canada

Jason Lawrence Department of Computer Science, School of Engineering and Applied Science University of Virginia, Charlottesville, VA, USA

Svetlana Lazebnik Department of Computer Science, University of Illinois at Urbana-Champaign, Urbana, IL, USA

Longzhuang Li Department of Computing Science, Texas A and M University at Corpus Christi, Corpus Christi, TX, USA

Wanqing Li University of Wollongong, Wollongong, NSW, Australia

Stephen Lin Microsoft Research Asia, Beijing Sigma Center, Beijing, China

Zhe Lin Advanced Technology Labs, Adobe Systems Incorporated, San Jose, CA, USA

Tony Lindeberg School of Computer Science and Communication, KTH Royal Institute of Technology, Stockholm, Sweden

Yanxi Liu Computer Science and Engineering, Penn State University, University Park, PA, USA

Yonghuai Liu Department of Computer Science, Aberystwyth University, Ceredigion, Wales, UK

Zhi-Yong Liu Institute of Automation, Chinese Academy of Sciences, Beijing, P. R. China

Zicheng Liu Microsoft Research, Microsoft Corporation, Redmond, WA, USA

Songde Ma Ministry of Science & Technology, Beijing, China

Eisaku Maeda NTT Communication Science Laboratories, Soraku-gun, Kyoto, Japan

Michael Maire California Institute of Technology, Pasadena, CA, USA

Pascal Mamassian Laboratoire Psychologie de la Perception, Université Paris Descartes, Paris, France

Ralph R. Martin School of Computer Science and Informatics Cardiff University, Cardiff, UK

Simon Masnou Institut Camille Jordan, Universitè Lyon 1, Villeurbanne, France

Darko S. Matovski School of Electronics and Computer Science, University of Southampton, Southampton, Hampshire, UK

Yasuyuki Matsushita Microsoft Research, Beijing, China

Larry Matthies Jet Propulsion Laboratory, California Institute of Technology, Pasadena, CA, USA

Stephen J. Maybank Department of Computer Science and Information Systems, Birkbeck College University of London, London, UK

Peter Meer Department of Electrical and Computer Engineering, Rutgers University, Piscataway, NJ, USA

Christian Micheloni Department of Mathematics and Computer Science, University of Udine, Udine, Italy

Sushil Mittal Department of Statistics, Columbia University, New York, NY, USA

Daisuke Miyazaki Graduate School of Information Sciences, Hiroshima City University, Asaminami-ku, Hiroshima, Japan

Vlad I. Morariu Computer Vision Laboratory, University of Maryland, College Park, MD, USA

Joseph L. Mundy Division of Engineering, Brown University Rensselaer Polytechnic Institute, Providence, RI, USA

Hiroshi Murase Faculty of Economics and Information, Gifu Shotoku Gakuen University, Gifu, Japan

Kazuo Murota Department of Mathematical Informatics, University of Tokyo, Bunkyo-ku, Tokyo, Japan

Bernd Neumann University of Hamburg, Hamburg, Germany

Mark S. Nixon School of Electronics and Computer Science, University of Southampton, Southampton, Hampshire, UK

Takayuki Okatani Graduate School of Information Sciences, Tohoku University, Sendai-shi, Japan

Vasu Parameswaran Microsoft Corporation, Sunnyvale, CA, USA

Johnny Park School of Electrical and Computer Engineering, Purdue University, West Lafayette, IN, USA

Samunda Perera Canberra Research Laboratory, NICTA, Canberra, Australia

Matti Pietikäinen Department of Computer Science and Engineering, University of Oulu, Oulu, Finland

Tomaso Poggio Department of Brain and Cognitive Sciences, McGovern Institute, Massachusetts Institute of Technology, Cambridge, MA, USA

Marc Pollefeys Computer Vision and Geometry Lab (CVG) – Institute of Visual Computing, Department of Computer Science, ETH Zürich, Zürich, Switzerland

S. C. Pont Industrial Design Engineering, Delft University of Technology, Delft, The Netherlands

Emmanuel Prados INRIA Rhône-Alpes, Montbonnot, France

Jerry L. Prince Electrical and Computer Engineering, Johns Hopkins University, Baltimore, MD, USA

Srikumar Ramalingam Mitsubishi Electric Research Laboratories, Cambridge, MA, USA

Rajeev Ramanath DLP® Products, Texas Instruments Incorporated, Plano, TX, USA

Dikpal Reddy Nvidia Research, Santa Clara, CA, USA

Xiaofeng Ren Intel Science and Technology Center for Pervasive Computing, Intel Labs, Seattle, WA, USA

Xuejun Ren School of Engineering Liverpool John Moores University, Liverpool, UK

Szymon Rusinkiewicz Department of Computer Science, Princeton University, Princeton, NJ, USA

Aswin C. Sankaranarayanan ECE Department, Rice University, Houston, TX, USA

Guillermo Sapiro Electrical and Computer Engineering, Computer Science, and Biomedical Engineering, Duke University, Durham, NC, USA

Silvio Savarese Department of Electrical and Computer Engineering, University of Michigan, Ann Arbor, MI, USA

Davide Scaramuzza Artificial Intelligence Lab - Robotics and Perception Group, Department of Informatics, University of Zurich, Zurich, Switzerland

Konrad Schindler ETH Zürich, Zürich, Switzerland

David C. Schneider Image Processing Department, Fraunhofer Heinrich Hertz Institute, Berlin, Germany

William Robson Schwartz Department of Computer Science, Universidade Federal de Minas Gerais, Belo Horizonte, MG, Brazil

Guna Seetharaman Air Force Research Lab RITB, Rome, NY, USA

Chunhua Shen School of Computer Science , The University of Adelaide, Adelaide, SA, Australia

Rui Shen University of Alberta, Edmonton, AB, Canada

Ali Shokoufandeh Department of Computer Science, Drexel University Philadelphia, PA, USA

Jamie Shotton Microsoft Research Ltd, Cambridge, UK

Kaleem Siddiqi School of Computer Science, McGill University, Montreal, PQ, Canada

Leonid Sigal Disney Research, Pittsburgh, PA, USA

Manish Singh Department of Psychology, Rutgers University, Piscataway, NJ, USA

Sudipta N. Sinha Microsoft Research, Redmond, WA, USA

Arnold W. M. Smeulders Centre for Mathematics and Computer Science (CWI), University of Amsterdam, Amsterdam, The Netherlands

Intelligent Systems Lab Amsterdam, Informatics Institute University of Amsterdam, Amsterdam, The Netherlands

Cees G. M. Snoek University of Amsterdam, Amsterdam, The Netherlands

Intelligent Systems Lab Amsterdam, Informatics Institute University of Amsterdam, Amsterdam, The Netherlands

Gunnar Sparr Centre for Mathematical Sciences, Lund University, Lund, Sweden

Anuj Srivastava Florida State University, Tallahassee, FL, USA

Gaurav Srivastava School of Electrical and Computer Engineering, Purdue University, West Lafayette, IN, USA

Peer Stelldinger Computer Science Department, University of Hamburg, Hamburg, Germany

Peter Sturm INRIA Grenoble Rhône-Alpes, St Ismier Cedex, France

Kokichi Sugihara Graduate School of Advanced Mathematical Sciences, Meiji University, Kawasaki, Kanagawa, Japan

Min Sun Department of Electrical and Computer Engineering, University of Michigan, Ann Arbor, MI, USA

Richard Szeliski Microsoft Research, One Microsoft Way, Redmond, WA, USA

Yu-Wing Tai Department of Computer Science, Korean Advanced Institute of Science and Technology (KAIST), Yuseong-gu, Daejeon, South Korea

Tomokazu Takahashi Faculty of Economics and Information, Gifu Shotoku Gakuen University, Gifu, Japan

Birgi Tamersoy Department of Electrical and Computer Engineering, The University of Texas at Austin, Austin, TX, USA

Ping Tan Department of Electrical and Computer Engineering, National University of Singapore, Singapore, Singapore

Robby T. Tan Department of Information and Computing Sciences, Utrecht University, Utrecht, CH, The Netherlands

Allen Tannenbaum Schools of Electrical and Computer and Biomedical Engineering, Georgia Institute of Technology, Atlanta, GA, USA

Marshall F. Tappen University of Central Florida, Orlando, FL, USA

Federico Tombari DEIS, University of Bologna, Bologna, Italy

Shoji Tominaga Graduate School of Advanced Integration Science, Chiba University, Inage-ku, Chiba, Japan

Lorenzo Torresani Computer Science Department, Dartmouth College, Hanover, NH, USA

Tali Treibitz Department of Computer Science and Engineering, University of California, San Diego, La Jolla, CA, USA

Yanghai Tsin Corporate R&D, Qualcomm Inc., San Diego, CA, USA

Pavan Turaga Department of Electrical and Computer Engineering, Center for Automation Research University of Maryland, College Park, MD, USA

Matthew Turk Computer Science Department and Media Arts and Technology Graduate Program, University of California, Santa Barbara, CA, USA

Tinne Tuytelaars KULEUVEN Leuven, ESAT-PSI, iMinds, Leuven, Belgium

Shimon Ullman Department of Brain and Cognitive Sciences, McGovern Institute, Massachusetts Institute of Technology, Cambridge, MA, USA

Anton van den Hengel School of Computer Science, The University of Adelaide, Adelaide, SA, Australia

Pramod K. Varshney Department of Electrical Engineering and Computer Science, Syracuse University, Syracuse, NY, USA

Andrea Vedaldi Oxford University, Oxford, UK

Ashok Veeraraghavan Department of Electrical and Computer Engineering, Rice University, Houston, TX, USA

David Vernon Informatics Research Centre, University of Skövde, Skövde, Sweden

Ramanarayanan Viswanathan Department of Electrical Engineering, University of Mississippi, MS, USA

Xiaogang Wang Department of Electronic Engineering, Chinese University of Hong Kong, Shatin, Hong Kong

Isaac Weiss Center for Automation Research, University of Maryland at College Park, College Park, MD, USA

Gregory F. Welch Institute for Simulation & Training, The University of Central Florida, Orlando, FL, USA

Michael Werman The Institute of Computer Science, The Hebrew University of Jerusalem, Jerusalem, Israel

Tien-Tsin Wong Department of Computer Science and Engineering, The Chinese University of Hong Kong, Hong Kong SAR, China

Robert J. Woodham Department of Computer Science, University of British Columbia, Vancouver, BC, Canada

John Wright Visual Computing Group, Microsoft Research Asia, Beijing, China

Ying Nian Wu Department of Statistics, UCLA, Los Angeles, CA, USA

Chenyang Xu Siemens Technology-To-Business Center, Berkeley, CA, USA

David Young School of Informatics, University of Sussex, Falmer, Brighton, UK

Guoshen Yu Electrical and Computer Engineering, University of Minnesota, Minneapolis, MN, USA

Christopher Zach Computer Vision and Geometry Group, ETH Zürich, Zürich, Switzerland

Alexander Zelinsky CSIRO, Information Sciences, Canberra, Australia

Zhengyou Zhang Microsoft Research, Redmond, WA, USA

Bo Zheng Computer Vision Laboratory, Institute of Industrial Science, The University of Tokyo, Meguro-ku, Tokyo, Japan

Zhigang Zhu Computer Science Department, The City College of New York, New York, NY, USA

Todd Zickler School of Engineering and Applied Science, Harvard University, Cambridge, MA, USA

A

Action Prototype Trees

►Prototype-Based Methods for Human Movement Modeling

Action Recognition

►Activity Recognition
►Affordances and Action Recognition

Active Calibration

Rui Shen[1], Gaopeng Gou[2], Irene Cheng[1] and Anup Basu[3]
[1]University of Alberta, Edmonton, AB, Canada
[2]Beihang University, Beijing, China
[3]Department of Computing Science, University of Alberta, Edmonton, AB, Canada

Synonyms

Active camera calibration; Pan-tilt camera calibration; Pan-tilt-zoom camera calibration; PTZ camera calibration

Related Concepts

►Camera Calibration

Definition

Active calibration is a process that determines the geometric parameters of a camera (or cameras) using the camera's controllable movements.

Background

Camera calibration aims to establish the best possible correspondence between the used camera model and the realized image acquisition with a given camera [12], i.e., accurately recover a camera's geometric parameters, such as focal length and image center/ principal point, from the captured images. The classical calibration techniques (e.g., [10, 16]) require predefined patterns and static cameras and often involve solving complicated equations. Taking advantage of a camera's controllable movements (e.g., pan, tilt, and roll), active calibration techniques can automatically calibrate the camera.

Theory

The pinhole camera model is one of the most commonly used models, as shown in Fig. 1. $\mathbf{p} = (x, y)^T$ is the 2D projection of the 3D point $\mathbf{P} = (X, Y, Z)^T$ on the image plane. Using homogeneous coordinates, $\tilde{\mathbf{p}}$ and $\tilde{\mathbf{P}}$ have the following relationship:

$$\lambda\tilde{\mathbf{p}} = \mathbf{K}(\mathbf{R} \mid \mathbf{t})\tilde{\mathbf{P}} \tag{1}$$

K. Ikeuchi (ed.), *Computer Vision*, DOI 10.1007/978-0-387-31439-6,

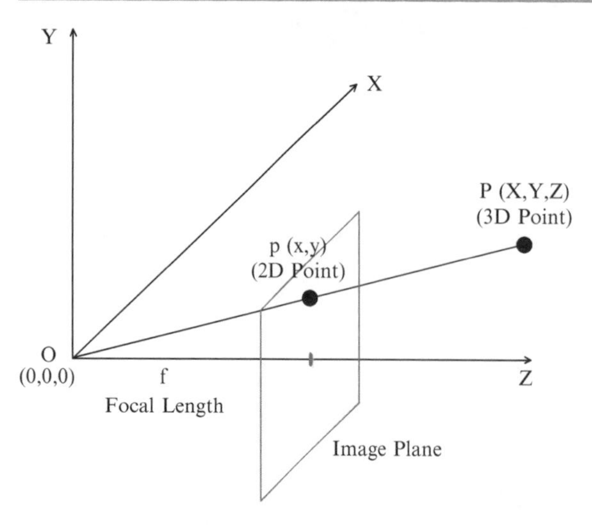

Active Calibration, Fig. 1 The pinhole camera model

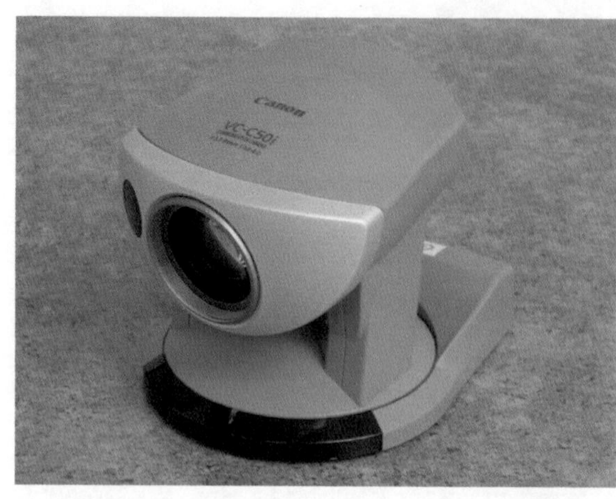

Active Calibration, Fig. 2 Canon VC-C50i PTZ camera

where $\mathbf{K} = \begin{pmatrix} f_x & s & x_0 \\ 0 & f_y & y_0 \\ 0 & 0 & 1 \end{pmatrix}$ is the camera calibration matrix; λ is the depth (when $\mathbf{R}$ is the identity matrix and $\mathbf{t}$ is a zero-vector, $\lambda = Z$); and $\mathbf{R}$ and $\mathbf{t}$ are the 3×3 rotation matrix and the 3×1 translation vector, respectively. f_x and f_y are the focal lengths in pixels in the x- and y-directions, respectively. They are proportional to the focal length f shown in Fig. 1, which is normally measured in millimeters. (x_0, y_0) are the coordinates of the image center/principal point on the image plane, i.e., the intersection of the lens' optical axis and the image plane. s is the skew. Normally, the sensor element is assumed rectangular; therefore, s becomes 0. $\gamma = f_x/f_y$ is the aspect ratio. If the sensor element is quadratic, γ becomes 1. The five parameters in $\mathbf{K}$ are called the camera intrinsic parameters; the three Euler angles in $\mathbf{R}$ and the three offsets in $\mathbf{t}$ are called the camera extrinsic parameters.

Figure 2 shows an active camera. When a camera rotates (no translation), the new image points can be obtained as transformations of the original image points [11]. The camera parameters can be related to the image points before and after rotation and to the angles of rotation. The equations describing these relations are simple and easy to solve. Thus, by considering different movements of the camera (pan, tilt, and roll with measurement of the angles), the intrinsic parameters of a camera can be estimated.

Basu [1] proposed an active calibration technique utilizing the edge information of a static scene. This technique was extended in [2–4] to give more robust estimation of the intrinsic parameters. No special patterns are required, but the observed scene should have strong and stable edges. Four strategies are introduced and validated through experiments [4]. Strategies A and B only utilize pan and tilt movements; strategy C utilizes pan, tilt, and roll movements; strategy D is a special case of strategy C when the roll angle is equal to 180°. The procedure of strategy C is outlined asfollows:

- Using the pan and tilt movement of the camera and a single image contour, obtain the values of f_x and f_y by solving Eqs. 2 and 3.
- Using the roll movement of the camera and a single image contour, obtain the values of δ_x and δ_y by Eqs. 4 and 5.

$$f_y = \frac{\overline{y_t} - \overline{y}(1+\theta_t^2)}{2\theta_t} + \frac{1}{2}\sqrt{\left(\frac{\overline{y}(1+\theta_t^2) - \overline{y_t^2}}{\theta_t}\right)^2 - 4\overline{y^2}} \qquad (2)$$

$$f_x = \frac{\overline{x_p} - \overline{x}(1+\theta_p^2)}{2\theta_p} + \frac{1}{2}\sqrt{\left(\frac{\overline{x}(1+\theta_p^2) - \overline{x_p^2}}{\theta_p}\right)^2 - 4\overline{x^2}} \qquad (3)$$

where (x_p, y_p) and (x_t, y_t) denote image coordinates after pan and tilt movements, respectively; θ_t and θ_p are the tilt angle and pan angle, respectively.

$$\delta_x(1-\cos(\theta_r)) + \delta_y\left(-\frac{f_x}{f_y}\sin(\theta_r)\right) = \overline{x_r} - \cos(\theta_r)\overline{x} - \frac{f_x}{f_y}\sin(\theta_r)\overline{y} \quad (4)$$

$$\delta_x\left(-\frac{f_x}{f_y}\sin(\theta_r)\right) + \delta_y(1-\cos(\theta_r)) = \overline{y_r} - \cos(\theta_r)\overline{y} + \frac{f_x}{f_y}\sin(\theta_r)\overline{x} \quad (5)$$

where (x_r, y_r) denotes image coordinates after roll movement; (δ_x, δ_y) denotes the error in the estimated principal point (e.g., taking the geometric center of the image plane as the estimated principal point); and θ_r is the roll angle.

Davis and Chen [9] introduced a new pan-tilt camera motion model, in which the pan and tilt axes are not necessarily orthogonal or aligned to the image plane. A tracked object is used to form a large virtual calibration object that covers the whole working volume. A set of pre-calibrated static cameras is needed to record the trajectory. The intrinsic parameters are recovered by minimizing the projection errors between the observed 2D data and the calculated 2D locations of the tracked object using the proposed camera model.

McLauchlan and Murray [13] applied the variable state-dimension filter to calibrating a single camera mounted on a robot by tracking the trajectories of an arbitrary number of tracked corner features and utilizing accurate knowledge of the camera rotation. The camera's intrinsic parameters are updated in real time.

Different zoom settings (focal lengths) can also be employed in active calibration. Seales and Eggert [14] calibrate a camera via a fully automated 4-stage global optimization process using a sequence of images of a known calibration target obtained at different mechanical zoom settings. Collins and Tsin [8] proposed a parametric camera model and calibration procedures for an outdoor active camera system with pan, tilt, and zoom control. Intrinsic parameters are recovered by fitting the camera model with the optic flow produced by the camera's movements. Extrinsic parameters are estimated as a pose estimation problem using sparsely deployed landmarks. Borghese et al. [5] proposed a technique to compute camera focal lengths by zooming a single point, assuming the principal point is in a fixed and known position. Sinha and Pollefeys [15] propose a camera model that incorporates the variation of radial distortion with camera zoom. The intrinsic parameters are first computed at the lowest zoom level from a captured panorama. Then, the intrinsic and radial distortion parameters are estimated at sequentially increased zoom levels, taking into account the influence of camera zoom.

Application

Active calibration has been receiving more and more attention with the increasing use of active systems in various applications, such as object tracking, surveillance, and video conference. More generally, camera calibration is important for any application that involves relating a 2D image to the 3D world. Such applications include pose estimation, 3D motion estimation, automated assembly, close-range photogrammetry, and so on.

Open Problems

Recent research is more focused on automatic active calibration of a multi-camera system without using a predefined calibration pattern/object. Chippendale and Tobia [7] presented an autocalibration system for the estimation of extrinsic parameters of active cameras in indoor environments. One constraint of the camera deployment is that each camera must be able to observe at least one other camera to form an observation chain. The extrinsic parameters are estimated using the circular shape of the camera lenses and a predetermined moving pattern of a particular camera. The accuracy of the algorithm is mainly affected by the distance between cameras. Brückner and Denzler [6] proposed a three-step multi-camera calibration algorithm. The extrinsic parameters of each camera are first roughly estimated using a probability distribution based on the captured images. Then, each camera pair rotates and zooms in a way that maximizes image similarity, and the extrinsic parameters are reestimated based on point correspondence. A final calibration is carried out using the probabilities and the reestimated

Active Calibration, Table 1 Results of image center estimation

Angle (strategy C)	Ground truth δ_x	δ_y	$\sigma = 0$ δ_x	δ_y	$\sigma = 5$ δ_x	δ_y	$\sigma = 10$ δ_x	δ_y
20°	10	20	11	22	12	23	12	23
40°	10	20	11	21	12	22	12	19
60°	10	20	11	21	12	21	13	21
80°	10	20	11	21	11	21	11	22
100°	10	20	10	21	10	21	11	21
120°	10	20	11	21	11	21	11	21
140°	10	20	11	21	11	21	11	21
160°	10	20	10	21	11	21	11	21
180°	10	20	10	20	10	21	10	21
Strategy D	10	20	10	20	11	20	11	20

Active Calibration, Table 2 Results of focal length estimation

Strategy	Ground truth f_x	f_y	$\sigma = 0$ f_x	f_y	$\sigma = 5$ f_x	f_y
Strategy C	400	600	403	602	396	603
Strategy D	400	600	401	601	403	599

extrinsic parameters. This method achieves relatively high accuracy and robustness, but one drawback is the high computational cost.

Experimental Results

Some experimental results using Strategies C and D from [4] are presented below.

Tables 1 and 2 summarize the results of computing the image center and focal lengths using simulated data, along with the ground truths. The simulated data contain an image contour consisting of 50 points. The pan and tilt angles were fixed at 3°. For the experiment on image center calculation, additive Gaussian noise with standard deviation σ of 0 (no noise), 5, and 10 pixels were added to test the robustness of the algorithms. It can be seen that strategy C performs reasonably in determining the image center even when σ is as large as 15. The results of strategy D are similar to those produced by strategy C when the roll angle is 180°.

For the experiment on focal length calculation, additive Gaussian noise with standard deviation σ of 0 (no noise) and 5 pixels were added. Strategy D is a little more accurate as the equations are obtained directly without using the estimates of f_x and f_y. The focal lengths obtained by strategy D are similar to those produced by strategy C. But when the noise is increased, strategy D produces more reliable results than strategy C.

Strategies C and D were also tested on a real camera in an indoor environment. The estimates for δ_x and δ_y obtained by strategy C (90° roll) were 3 and 29 pixels, while the values obtained by strategy D were 2 and 30 pixels, which demonstrates the stability of the active calibration algorithms in real situations. The estimated values of f_x and f_y were 908 and 1,126, respectively.

References

1. Basu A (1993) Active calibration. In: ICRA'93: proceedings of the 1993 IEEE international conference on robotics and automation, Atlanta, vol 2, pp 764–769
2. Basu A (1993) Active calibration: alternative strategy and analysis. In: CVPR'93: proceedings of the 1993 IEEE computer society conference on computer vision and pattern recognition (CVPR), New York, pp 495–500
3. Basu A (1995) Active calibration of cameras: theory and implementation. IEEE Trans Syst Man Cybern 25(2): 256–265
4. Basu A, Ravi K (1997) Active camera calibration using pan, tilt and roll. IEEE Trans Syst Man Cybern B 27(3):559–566
5. Borghese NA, Colombo FM, Alzati A (2006) Computing camera focal length by zooming a single point. Pattern Recognit 39(8):1522–1529
6. Brückner M, Denzler J (2010) Active self-calibration of multi-camera systems. In: Proceedings of the 32nd DAGM conference on pattern recognition, Darmstadt, pp 31–40
7. Chippendale P, Tobia F (2005) Collective calibration of active camera groups. In: AVSS'05: proceedings of the IEEE conference on advanced video and signal based surveillance, Como, pp 456–461
8. Collins RT, Tsin Y (1999) Calibration of an outdoor active camera system. In: CVPR'99: proceedings of the 1999 IEEE computer society conference on computer vision and pattern recognition (CVPR), Ft. Collins, pp 528–534
9. Davis J, Chen X (2003) Calibrating pan-tilt cameras in wide-area surveillance networks. In: ICCV'03: proceedings of the 9th IEEE international conference on computer vision, Nice, pp 144–149
10. Horaud R, Mohr R, Lorecki B (1992) Linear camera calibration. In: ICRA'92: proceedings of the IEEE international conference on robotics and automation, Nice, vol 2, pp 1539–1544
11. Kanatani K (1987) Camera rotation invariance of image characteristics. Comput Vis Graph Image Process 39(3):328–354
12. Klette R, Schlüns K, Koschan A (1998) Computer vision: three-dimensional data from images, 1st edn. Springer, New York/Singapore
13. McLauchlan PF, Murray DW (1996) Active camera calibration for a head-eye platform using the variable state-dimension filter. IEEE Trans Pattern Anal Mach Intell 18(1):15–22

14. Seales WB, Eggert DW (1995) Active-camera calibration using iterative image feature localization. In: CAIP'95: proceedings of the 6th international conference on computer analysis of images and patterns, Prague, pp 723–728
15. Sinha SN, Pollefeys M (2006) Pan-tilt-zoom camera calibration and high-resolution mosaic generation. Comput Vis Image Underst 103(3):170–183
16. Tsai R (1987) A versatile camera calibration technique for high-accuracy 3d machine vision metrology using off-the-shelf tv cameras and lenses. IEEE J Robot Autom 3(4): 323–344

Active Camera Calibration

▶Active Calibration

Active Contours

▶Numerical Methods in Curve Evolution Theory

Active Sensor (Eye) Movement Control

James J. Clark
Department of Electrical and Computer Engineering, McGill University, Montreal, QC, Canada

Synonyms

Gaze control

Related Concepts

▶Evolution of Robotic Heads; ▶Visual Servoing

Definition

Active sensors are those whose generalized viewpoint (such as sensor aperture, position, and orientation) is under computer control. Control is done so as to improve information gathering and processing.

Background

The *generalized viewpoint* [1] of a sensor is the vector of values of the parameters that are under the control of the observer and which affect the imaging process. Most often, these parameters will be the position and orientation of the image sensor, but may also include such parameters as the focal length, aperture width, and the nodal point to image plane distance, of the camera. The definition of generalized viewpoint can be extended to include illuminant degrees of freedom, such as the illuminant position, wavelength, intensity, spatial distribution (for structured light applications), and angular distribution (e.g., collimation) [2].

Changes in observer viewpoint are used in active vision systems for a number of purposes. Some of the more important uses are:

- Tracking a moving object to keep it in the field of view of the sensor
- Searching for a specific item in the observer's environment
- Inducing scene-dependent optical flow to aid in the extraction of 3D structure of objects and scenes
- Avoiding "accidental" or nongeneric viewpoints, which can result in sensor-saturating specularities or information-hiding occlusions
- Minimizing sensor noise and maximizing novel information content
- Increasing the dynamic range of the sensor, through adjustment of parameters such as sensor sensitivity, aperture, and focus
- Mapping the observer's environment

Theory

LowLevel Camera Motion Control Systems Most robotic active vision control systems act mainly to produce either smooth pursuit motions or rapid saccadic motions. Pursuit motions cause the camera to move so as to smoothly track a moving object, maintaining the image of the target object within a small region (usually in the center) of the image frame. Saccadic motions are rapid, usually large, jumps in the position of the camera, which center the camera field of view on different parts of the scene being imaged. This type of motion is used when scanning a scene, searching for objects or information, but can also be used to recover from a loss of tracking of an object during visual pursuit.

Much has been learned about the design of pursuit and saccadic motion control systems from the study of primate oculomotor systems. These systems have a rather complicated architecture distributed among many brain areas, the details of which are still subject to vigorous debate [3]. The high-level structure, however, is generally accepted to be that of a feedback system. A very influential model of the human oculomotor control system is that of Robinson [4], and many robotic vision control systems employ aspects of the Robinson model.

The control of an active camera system is both simple and difficult at the same time. Simplicity arises from the relatively unchanging characteristics of the load or "plant" being controlled. For most systems the moment of inertia of the camera changes only minimally over the range of motion, with slight variations arising when zoom lenses are used. The mass of the camera and associated linkages does not change. Inertial effects become more important for control of the "neck" degrees of freedom due to the changing orientation and position of the camera bodies relative to the neck. The specifications on the required velocities and control bandwidth for the neck motions are typically much less stringent than those for the camera motions, so that the inertial effects for the neck are usually neglected. The relatively simple nature of the oculomotor plant means that straightforward proportional-derivative (PD) or proportional-integral-derivative (PID) control systems are often sufficient for implementing tracking or pursuit motion. Some systems have employed more complex optimal control systems (e.g., [5]) which provide improved disturbance rejection and trajectory following accuracy compared to the simpler approaches.

There is a serious difficulty in controlling camera motion systems, however, caused by delays in the control loops. Such delays include the measurement delay due to the time needed to acquire and digitize the camera image and subsequent computations, such as feature extraction and target localization. There is also a delay or latency arising from the time needed to compute the controller output signal [6]. If these delays are not dealt with, a simple PD or PID controller can become unstable, leading to excessive vibration or shaking and loss of target tracking.

There are a number of approaches to dealing with delay. PID or PD systems can be made robust to delays simply by increasing system damping by reducing the proportional feedback gain to a sufficiently low value [7]. This results in a system that responds to changes in target position very slowly, however, and is unacceptable for most applications. For control of saccadic motion, a *sample/hold* can be used, where the position error is sampled at the time a saccade is triggered, and held in a first-order hold (integrator) [8]. In this way, the position error seen by the controller is held constant until the saccadic motion is completed. The controller is insensitive to any changes in the actual target position until the end of this *refractory* period. This stabilizes the controller, but has the drawback that if the target moves during the refractory period, the position error at the end of the refractory period can be large. In this case, another, corrective or secondary, saccadic motion may need to be triggered. For stabilization of pursuit control systems in the presence of delay, an internal positive feedback loop can be employed [4, 8]. This positive feedback compensates for delays in the negative feedback servo loop created by the time taken to acquire an image and compute the target velocity error. The positive feedback loop sends a delayed *efference copy* of the current commanded camera velocity (which is the output of the pursuit controller) back to the velocity error comparator where it is added to the measured velocity error. The positive feedback delay is set so that it arrives at the velocity error comparator at the same time as the measurement of the effect of the current control command, effectively canceling out the negative feedback and producing a new target velocity for the controller. Another delay handling technique is to use *predictive control*, such as the Smith Predictor, where the camera position and controller states are predicted for a time T in the future, where T is the controller delay, and control signals appropriate for those states are computed and applied immediately [6, 7]. Predictive methods make strong assumptions on changes in the external environment (e.g., that all objects in the scene are static or traversing known smooth trajectories). Such methods can perform poorly when these assumptions are violated.

The Next-Look Problem and Motion Planning The control of pursuit and saccadic motions are usually handled by different controllers. While pursuit or tracking behavior can be implemented using frequent small saccade-like motions, this can produce jumpy images which may degrade subsequent processing operations. With multiple controllers, there needs to be a way for the possibly conflicting commands from the controllers to be integrated and arbitrated. The simplest approach uses the output of the pursuit control system by default, with a switch over to the output of the saccade control system whenever the position error is greater than some threshold and switching back to pursuit control when the position error drops below another (lower) threshold.

Pursuit or tracking of visual targets is just one type of motor activity. Activities such as visual search may require large shifts of camera position to be executed based on a complex analysis of the visual input. The process of determining the active vision system controller set point is often referred to as *sensor planning* [1] or the *next-look problem* [9]. The next-look problem can be interpreted as determining sensor positions which increase or maximize the information content of subsequent measurements. In a visual search task, for example, the next-look may be specified to be a location which is expected to maximally discriminate between target and distractor. One principle that has been successfully employed in next-look processes is that of entropy minimization over viewpoints. In an object recognition or visual search task, this approach takes as the next viewpoint that which is maximally informative relative to the most probable hypotheses [10]. A common approach to the next-view problem in robotic systems is to employ an *attention mechanism* to provide the location of the next view. Based on models of mammalian vision systems, attention mechanisms determine *salient* regions in visual input, which compete or interact in winner-takes-all fashion to select a single location as the target for the subsequent motion [8].

Application

In the late 1980s and early 1990s, commercial camera motion platforms lacked the performance needed by robotics researchers and manufacturers. This led many universities to construct their own platforms and develop control systems for them. These were generally binocular camera systems with pan and tilt degrees of freedom for each camera. Often, to simplify the design, a common tilt action was employed for both cameras, and the pan actions were sometimes linked together to provided vergence and/or version motions only. Examples include the UPenn head [11], generally recognized as the first of its kind, the Harvard head [12], the KTH head [13], the TRISH head from the University of Toronto [14], the Rochester head [15], the SAGEM-GEC-Inria-Oxford head [16], the Surrey head [17], the LIFIA head [18], the LIA/AUC head [19], and the Technion head [5]. These early robotic heads generally used PD servo loops, some with delay compensation mechanisms as described above, and were capable of speeds up to 180 degrees per second. The pan axis maximum rotational velocities were usually higher than those of the tilt and vergence speeds. The axes were most often driven either by DC motors or by stepper motors.

A more recent example of a research system is the head of the iCub humanoid robot [20]. Unlike the early robotic heads, which were one-off systems limited to use in a single laboratory, this robot was developed by a consortium of European institutions and is used in many different research laboratories. It has independent pan and common tilt for two cameras as well as three neck degrees of freedom. The maximum pan speed is 180 degrees per second, and the maximum tilt speed is 160 degrees per second.

Currently, most robotic active vision systems are based on commercially available monocular pan-tilt platforms. The great majority of commercial platforms are designed for surveillance applications and are relatively slow. There are a few systems with specifications that are suitable for robotic active vision systems. Perhaps the most commonly used of these fast platforms are made by FLIR Motion Control Systems, Inc. (formerly Directed Perception). These are capable of speeds up to 120 degrees per second and can handle loads of up to 90 lbs. Commercial systems generally lack torsional motion and hence are not suitable for precise stereo vision applications.

The fastest current commercial pan/tilt units, as well as the early research platforms, only reach maximum speeds of around 200 degrees per second. This is sufficient to match the speeds of human pursuit eye movements, which top out around 100 degrees per second. However, if these speeds are compared to the maximum speed of 800 degrees per second for human saccadic motions, it can be seen that the performance of robotic active vision motion platforms still has room for improvement.

References

1. Tarabanis K, Tsai RY, Allen PK (1991) Automated sensor planning for robotic vision tasks. In: Proceedings of the 1991 IEEE conference on robotics and automation, Sacramento, pp 76–82
2. Yi S, Haralick RM, Shapiro LG (1990) Automatic sensor and light source positioning for machine vision. In: Proceedings of the computer vision and pattern recognition conference (CVPR), Atlantic City, June 1990, pp 55–59
3. Kato R, Grantyn A, Dalezios Y, Moschovakis AK (2006) The local loop of the saccadic system closes downstream of the superior colliculus. Neuroscience 143(1):319–337
4. Robinson DA (1968) The oculomotor control system: a review. Proc IEEE 56(6):1032–1049
5. Rivlin E, Rotstein H (2000) Control of a camera for active vision: foveal vision, smooth tracking and saccade. Int J Comput Vis 39(2):81–96
6. Brown C (1990) Gaze controls with interactions and delays. IEEE Trans Syst Man Cybern 20(1):518–527
7. Sharkey PM, Murray DW (1996) Delays versus performance of visually guided systems. IEE Proc Control Theory Appl 143(5):436–447
8. Clark JJ, Ferrier NJ (1992) Attentive visual servoing. In: Blake A, Yuille AL (eds) An introduction to active vision. MIT, Cambridge, pp 137–154
9. Swain MJ, Stricker MA (1993) Promising directions in active vision. Int J Comput Vis 11(2):109–126
10. Arbel T, Ferrie FP (1999) Viewpoint selection by navigation through entropy maps. In: Proceedings of the seventh IEEE international conference on computer vision, Kerkyra, pp 248–254
11. Krotkov E, Bajcsy R (1988) Active vision for reliable ranging: cooperating, focus, stereo, and vergence. Int J Comput Vis 11(2):187–203
12. Ferrier NJ, Clark JJ (1993) The Harvard binocular head. Int J Pattern Recognit Artif Intell 7(1):9–31
13. Pahlavan K, Eklundh J-O (1993) Heads, eyes and head-eye systems. Int J Pattern Recognit Artif Intell 7(1):33–49
14. Milios E, Jenkin M, Tsotsos J (1993) Design and performance of TRISH, a binocular robot head with torsional eye movements. Int J Pattern Recognit Artif Intell 7(1):51–68
15. Coombs DJ, Brown CM (1993) Real-time binocular smooth pursuit. Int J Comput Vis 11(2):147–164
16. Murray DW, Du F, McLauchlan PF, Reid ID, Sharkey PM, Brady M (1992) Design of stereo heads. In: Blake A, Yuille A (eds) Active vision. MIT, Cambridge, Massachusetts, USA, pp 155–172
17. Pretlove JRG, Parker GA (1993) The Surrey attentive robot vision system. Int J Pattern Recognit Artif Intell 7(1): 89–107
18. Crowley JL, Bobet P, Mesrabi M (1993) Layered control of a binocular camera head. Int J Pattern Recognit Artif Intell 7(1):109–122
19. Christensen HI (1993) A low-cost robot camera head. Int J Pattern Recognit Artif Intell 7(1):69–87
20. Beira R, Lopes M, Praga M, Santos-Victor J, Bernardino A, Metta G, Becchi F, Saltaren R (2006) Design of the robot-cub (iCub) head. In: Proceedings of the 2006 IEEE international conference on robotics and automation, Orlando, Florida, USA, pp 94–100

Active Stereo Vision

Andrew Hogue[1] and Michael R. M. Jenkin[2]
[1]Faculty of Business and Information Technology, University of Ontario Institute of Technology, Oshawa, ON, Canada
[2]Department of Computer Science and Engineering, York University, Toronto, ON, Canada

Related Concepts

►Camera Calibration

Definition

Active stereo vision utilizes multiple cameras for 3D reconstruction, gaze control, measurement, tracking, and surveillance. Active stereo vision is to be contrasted with passive or dynamic stereo vision in that passive systems treat stereo imagery as a series of independent static images while active and dynamic systems employ temporal constraints to integrate stereo measurements over time. Active systems utilize feedback from the image streams to manipulate camera parameters, illuminants, or robotic motion controllers in real time.

Background

Stereo vision uses two or more cameras with overlapping fields of view to estimate 3D scene structure from 2D projections. Binocular stereo vision – the most common implementation – uses exactly two cameras, yet one can utilize more than two at the expense of computational speed within the same algorithmic framework.

The "passive" stereo vision problem can be described as a system of at least two cameras attached rigidly to one another with constant intrinsic calibration parameters (assumed), and the stereo pairs are considered to be temporally independent. Thus no assumptions are made, nor propagated, about camera motion within the algorithmic framework. Passive vision systems are limited to the extraction of metric information from a single set of images taken from different locations in space (or at different times) and treat individual frames in stereo video sequences independently. Dynamic stereo vision systems are characterized by the extraction of metric information from sequences of imagery (i.e., video) and employ temporal constraints or consistency on the sequence (e.g., optical flow constraints). Thus, dynamic stereo systems place assumptions on the camera motion such as its smoothness (and small motion) between subsequent frames. Active stereo vision systems subsume both passive and dynamic stereo vision systems and are characterized by the use of robotic camera systems (e.g., stereo heads) or specially designed illuminant systems (e.g., structured light) coupled with a feedback system (see Fig. 1) for motor control. Although systems can be designed with more modest goals – object tracking, for example – the common computational goal is the construction of large-scale 3D models of extended environments.

Theory

Fundamentally, active stereo systems (see [1]) must solve three rather complex problems: (1) spatial correspondence, (2) temporal correspondence, and (3) motor/camera/illuminant control. Spatial correspondence is required in order to infer 3D depth information from the information available in camera images captured at one time instant, while temporal correspondence is necessary to integrate visual information over time. The spatial and temporal correspondences can either be treated as problems in isolation or integrated within a common framework. For example, stereo correspondence estimation can be seeded using an ongoing 3D representation using temporal coherence (e.g., [2, 3]) or considered in isolation using standard disparity estimation algorithms (see [4]).

Motor or camera control systems are necessary to move (rotate and translate) the cameras so they look in the appropriate direction (i.e., within a tracking or surveillance application), change their intrinsic camera parameters (e.g., focal length or zoom), or to tune the image processing algorithm to achieving higher accuracy for a specific purpose. Solving these three problems in an active stereo system enables one to develop "intelligent" algorithms that infer egomotion [5], autonomously control vehicles throughout the world [6], and/or reconstruct 3D models of the environment [7, 8]. Examples of the output of such a system is shown in Fig. 2, and [9] provides an example of an active system that interleaves the vergence and focus control of the cameras with surface estimation. The system uses an adaptive self-calibration method that integrates the estimation of camera parameters with surface estimation using prior knowledge of the calibration, the motor control, and the previous estimate of the 3D surface properties. The resulting system is able to automatically fixate on salient visual targets in order to extend the surface estimation volume.

Although vision is a powerful sensing modality, it can fail. This is a critical issue for active stereo vision where data is integrated over time. The use of complementary sensors – traditionally Inertial Measurement Units (see [10]) – augment the camera hardware system with the capability to estimate the system dynamics using real-world constraints. Accelerometers, gyroscopes, and compasses can provide timely and accurate information either to assist in temporal correspondences and ego-motion estimation or as a replacement when visual information is unreliable or absent (i.e., dead reckoning).

Relation to Robotics and Mapping

A wide range of different active and dynamic stereo systems have been built (e.g., [7, 8, 11, 12]). Active systems are often built on top of mobile systems (e.g., [7]) blurring the distinction between active and dynamic systems. In robotics, active stereo vision has been used for vehicle control in order to create 2D or 3D maps of the environment. Commonly the vision system is complemented by other sensors. For instance in [13], active stereo vision is combined with sonar sensors to create 2D and 3D models of the environment. Murray and Little [14] use a trinocular stereo system to create occupancy maps of the environment for in-the-loop path planning and robot navigation. Diebel et al. [15] employ active stereo vision for simultaneous estimation of the robot location and 3D map construction, and [7], describes a vision system used for in-the-loop mapping and navigational control for an aquatic robot. Davison, in [6], was one of the first to effectively demonstrate the use of active stereo vision technology as part of the navigation loop. The system used a stereo head to selectively fixate scene features that improve the quality of the estimated map and trajectory. This involved using knowledge of the

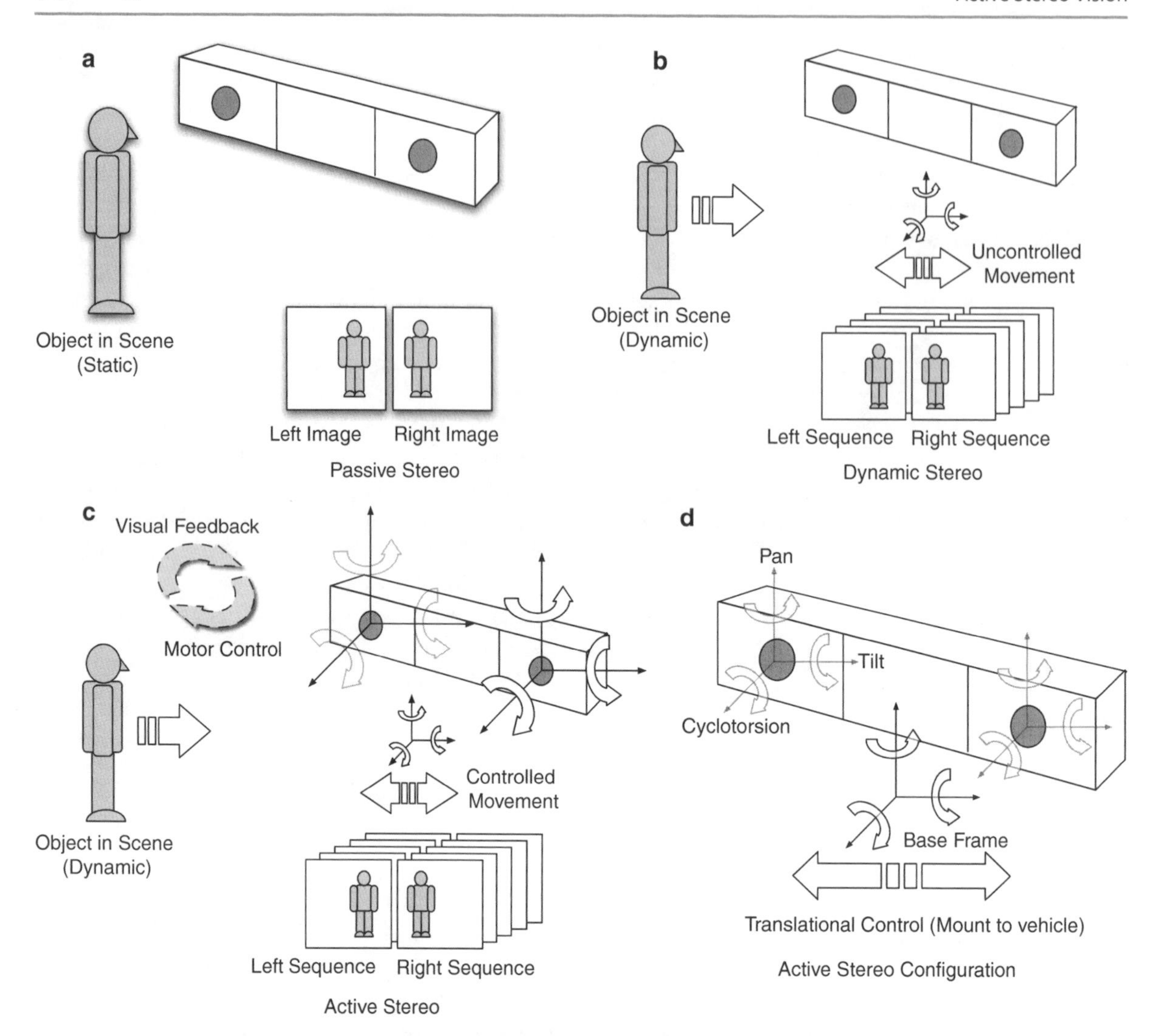

Active Stereo Vision, Fig. 1 Different types of stereo systems

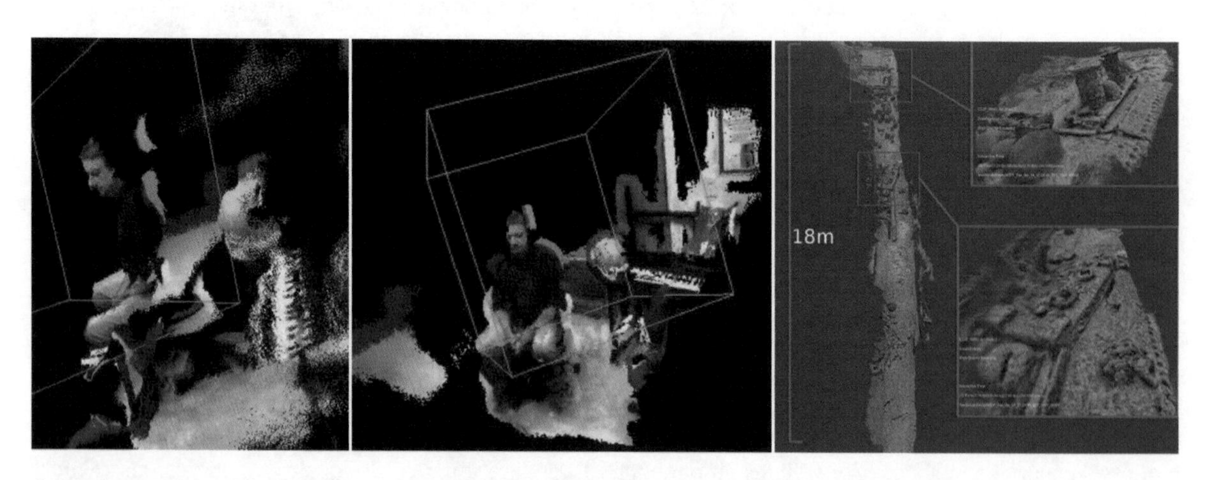

Active Stereo Vision, Fig. 2 Point cloud datasets obtained by the active stereo system described in [7]

current map of the environment to point the camera system in the direction where it should find salient features that it had seen before, move the robot to a location where the features are visible, and then searching visually to find image locations corresponding to these features.

Active Stereo Heads

The development of hardware platforms to mimic human biology has resulted in a variety of different designs and methods for controlling binocular sets of cameras. These result in what is known as "stereo heads" (see entry on *the evolution of stereo heads* in this volume). These hardware platforms all have a common set of constraints, i.e., the systems consist of two cameras (binocular) with camera intrinsics/extrinsics that may be controlled. In [16], an active stereo vision system is developed that mimics human biology that uses a bottom-up saliency map model coupled with a selective attention function to select salient image regions. If both left and right cameras estimate similar amounts of saliency in the same areas, the vergence of the cameras are controlled so that the cameras are focused on this particular landmark.

Autocalibration

A fundamental issue with active stereo vision is the need to establish and maintain calibration parameters online. Intrinsics and extrinsics are necessary to the 3D estimation process as they define the epipolar constraints which enable efficient disparity estimation algorithms [17, 18]. Each time the camera parameters are modified (e.g., vergence of the cameras, change of focus), the epipolar geometry must be re-estimated. Although kinematic modeling of motor systems provide good initial estimates of changes in camera pose, this is generally insufficiently accurate to be used by itself to update camera calibration. Thus, autocalibration becomes an important task within active stereo vision. Approaches to autocalibration are outlined in [17, 19]. In [17], the autocalibration algorithm operates on pairs of stereo images taken in sequence. A projective reconstruction for motion and structure of the scene is constructed. This is performed for each pair of stereo images individually for the same set of features (thus they must be matched in the stereo pairs as well as tracked temporally). The projective solutions can be upgraded to an affine solution (ambiguous up to a rigid rotation/translation/scale) by noting these features should match in 3D space as well as in 2D space. A transformation can be linearly estimated that constrains the projective solution to an affine reconstruction. Once the plane at infinity is known, the affine solution may be upgraded to a metric solution. In order to achieve the desired accuracy in the intrinsics, a nonlinear minimization scheme is employed to improve the solution. If one trusts the accuracy of the camera motion control system, the extrinsics can be seeded with this information in a nonlinear optimization scheme that minimizes the reprojection error of the image matching points and their 3D triangulated counterparts. This nonlinear optimization is known as bundle adjustment [20] and is used in a variety of forms in the structure-from-motion literature (see [17, 18]).

Relation to Other Types of Stereo Systems

Since active stereo systems are characterized by the use of visual feedback to inform motor control systems (or higher-level vehicular navigational systems), they are related to a wide range of research areas and hardware systems. Mounting a stereo system to a robotic vehicle is common in the robotics literature to inform the navigation system about the presence of obstacles [21] and to provide input to mapping algorithms [22]. The use of such active systems are applicable directly to autonomous systems as they provide a high amount of controllable accuracy and dense measurements at relatively low computational cost. One significant example is the use of active stereo in the Mars Rover autonomous vehicles [12].

Estimating 3D information from stereo views is problematic due to the lack of (or ambiguous) texture in man-made environments. This can be alleviated with the use of active illumination [23]. Projecting a known pattern, rather than uniform lighting, into the scene enables the estimation of a more dense disparity field using standard stereo disparity estimation algorithms due to the added texture in textureless regions (see [24]). The illumination may be controlled actively depending on perceived scene texture, the desired range, or the ambient light intensity of the environment. The illumination may be within the visible light spectrum or in the infrared spectrum as most camera sensors are sensitive to IR light. This has the added advantage that humans in the environment are not affected by the additional illumination.

Application

Active stereo vision is characterized by the use of visual feedback in multi-camera systems to control the intrinsics and extrinsics of the cameras (or vehicular platforms). Active stereo vision systems find a wide range of application in autonomous vehicle navigation, gaze tracking, and surveillance. A host of hardware systems exist and commonly utilize two cameras for binocular stereo and motors to control the gaze/orientation of the system. Visual attentive processes (e.g., [25, 26]) may be used to determine the next viewpoint for a particular task, and dense stereo algorithms can be used for estimating 3D structure of the scene. Fundamental computational issues include autocalibration of the sensor with changes in its configuration and the development of active stereo control and reconstruction algorithms.

References

1. Vieville T (1997) A few steps towards 3d active vision. Springer, New York/Secaucus
2. Leung C, Appleton B, Lovell B, Sun C (2004) An energy minimisation approach to stereo-temporal dense reconstruction. In: Proceedings of the 17th international conference on pattern recognition, vol 4, Cambridge, pp 72–75
3. Min D, Yea S, Vetro A (2010) Temporally consistent stereo matching using coherence function. In: 3DTV-conference: the true vision – capture, transmission and display of 3D video (3DTV-CON), 2010, Tampere, pp 1–4
4. Scharstein D, Szeliski R (2002) A taxonomy and evaluation of dense two-frame stereo correspondence algorithms. Int J Comput Vis 47(1–3):7–42
5. Olson CF, Matthies LH, Schoppers M, Maimone MW (2001) Stereo ego-motion improvements for robust rover navigation. In: Proceedings IEEE international conference on robotics and automation, vol 2, Seoul, pp 1099–1104
6. Davison AJ (1998) Mobile robot navigation using active vision. PhD thesis, University of Oxford
7. Hogue A, Jenkin M (2006) Development of an underwater vision sensor for 3d reef mapping. In: IEEE/RSJ international conference on intelligent robots and systems, Beijing, pp 5351–5356
8. Se S, Jasiobedzki P (2005) Instant scene modeler for crime scene reconstruction. In: 2005 IEEE computer society conference on computer vision and pattern recognition workshop on safety and security applications, vol III, San Diego, pp 123–123
9. Ahuja N, Abbott A (1993) Active stereo: integrating disparity, vergence, focus, aperture and calibration for surface estimation. IEEE Trans Pattern Anal Mach Intell 15(10): 1007–1029
10. Everett HR (1995) (ECCV) Sensors for mobile robots: theory and application. A. K. Peters, Ltd., Natick
11. Grosso E, Tistarelli M, Sandini G (1992) Active/dynamic stereo for navigation. In: Second European conference on computer vision, Santa Margherita Ligure, pp 516–525
12. Maimone MW, Leger PC, Biesiadecki JJ (2007) Overview of the mars exploration rovers autonomous mobility and vision capabilities. In: IEEE international conference on robotics and autonomsou (ICRA) space robotics workshop, Rome
13. Wallner F, Dillman R (1995) Real-time map refinement by use of sonar and active stereo-vision. Robot Auton Syst 16(1):47–56. Intelligent robotics systems SIRS'94
14. Murray D, Little JJ (2000) Using real-time stereo vision for mobile robot navigation. Auton Robot 8(2): 161–171
15. Diebel J, Reutersward K, Thrun S, Davis J, Gupta R (2004) Simultaneous localization and mapping with active stereo vision. In: IEEE/RSJ international conference on intelligent robots and systems, vol 4, Sendai, pp 3436–3443
16. Jung BS, Choi SB, Ban SW, Lee M (2004) A biologically inspired active stereo vision system using a bottom-up saliency map model. In: Rutkowski L, Siekmann J, Tadeusiewicz R, Zadeh LA (eds) Artificial intelligence and soft computing – ICAISC 2004. Volume 3070 of lecture notes in computer science. Springer, Berlin/Heidelberg, pp 730–735
17. Hartley RI, Zisserman A (2000) Multiple view geometry in computer vision. Cambridge University Press, Cambridge
18. Faugeras O, Luong QT, Papadopoulou T (2001) The geometry of multiple images: the laws that govern the formation of images of a scene and some of their applications. MIT, Cambridge
19. Horaud R, Csurka G (1998) Self-calibration and euclidean reconstruction using motions of a stereo rig. In: Sixth international conference on computer vision, Bombay, pp 96–103
20. Triggs B, McLauchlan P, Hartley R, Fitzgibbon A (2000) Bundle adjustment – a modern synthesis. In: Triggs B, Zisserman A, Szeliski R (eds) Vision algorithms: theory and practice. Volume 1883 of lecture notes in computer science. Springer, New York, pp 298–372
21. Williamson T, Thorpe C (1999) A trinocular stereo system for highway obstacle detection. In: IEEE international conference on robotics and automation, Detroit, pp 2267–2273
22. Schleicher D, Bergasa LM, Ocaña M, Barea R, López E (2010) Real-time hierarchical stereo visual slam in large-scale environments. Robot Auton Syst 58:991–1002
23. Se S, Jasiobedzki P, Wildes R (2007) Stereo-vision based 3d modeling of space structures. In: Proceedings of the SPIE conference on sensors and systems for space applications, vol 6555, Orlando
24. Rusinkiewicz S, Hall-Holt O, Levoy M (2002) Real-time 3d model acquisition. ACM Trans Graph 21(3):438–446
25. Frintrop S, Rome E, Christensen HI (2010) Computational visual attention systems and their cognitive foundations: a survey. ACM Trans Appl Percept 7(1):1–39
26. Tsotsos JK (2001) Motion understanding: task-directed attention and representations that link perception with action. Int J Comput Vis 45(3):265–280

Active Vision

▶Animat Vision

Activity Analysis

▶Multi-camera Human Action Recognition

Activity Recognition

Wanqing Li[1], Zicheng Liu[2] and Zhengyou Zhang[3]
[1]University of Wollongong, Wollongong, NSW, Australia
[2]Microsoft Research, Microsoft Corporation, Redmond, WA, USA
[3]Microsoft Research, Redmond, WA, USA

Synonyms

Action recognition

Related Concepts

▶Gesture Recognition

Definition

Activity recognition refers to the process of identifying the types of movement performed by humans over a certain period of time. It is also known as action recognition when the period of time is relatively short.

Background

The classic study on visual analysis of biological motion using moving light display (MLD) [1] has inspired tremendous interests among the computer vision researchers in the problem of recognizing human motion through visual information. The commonly used devices to capture human movement include human motion capture (MOCAP) with or without markers, multiple video camera systems, and single video camera systems. A MOCAP device usually works under controlled environment to capture the three-dimensional (3D) joint locations or angles of human bodies; multiple camera systems provide a way to reconstruct 3D body models from multiple viewpoint images. Both MOCAP and multiple camera systems have physical limitations on their use, and single camera systems are probably more practical for many applications. The latter, however, captures least visual information and, hence, is the most challenging setting for activity recognition. In the past decade, research in activity recognition has mainly focused on single camera systems. Recently, the release of commodity depth cameras, such as Microsoft Kinect Sensors, provides a new feasible and economic way to capture simultaneously two-dimensional color information and depth information of the human movement and, hence, could potentially advance the activity recognition significantly.

Regardless of which capturing device is used, a useful activity recognition system has to be independent of anthropometric differences among the individuals who perform the activities, independent of the speed at which the activities are performed, robust against varying acquisition settings and environmental conditions (for instance, different viewpoints and illuminations), scalable to a large number of activities, and capable of recognizing activities in a continuous manner. Since a human body is usually viewed as an articulated system of the rigid links or segments connected by joints, human motion can be considered as a continuous evolution of the spatial configuration of the segments or body posture, and effective representation of the body configuration and its dynamics over time has been the central to the research of human activity recognition.

Theory

Let $O = \{o_1, o_2, \cdots, o_n\}$ be a sequence of observations of the movement of a person over a period of time. The observations can be a sequence of joint angles, a sequence of color images or silhouettes, a sequence of depth maps, or a combination of them. The task of activity recognition is to label O into one of the L classes $C = \{c_1, c_2, \cdots, c_L\}$. Therefore, solutions to the problem of activity recognition are often based on machine learning and pattern recognition approaches, and an activity recognition system usually involves extracting features from the observation sequence O, learning a classifier from training samples, and classifying O using the trained classifier. However, the spatial and temporal complexity of human activities has led researchers to cast the problem from different perspectives. Specifically, the existing techniques for activity recognition can be divided into two categories based on whether the dynamics of the activities is implicitly or explicitly modeled.

In the first category [2–9], the problem of activity recognition is cast from a temporal classification problem to a static classification one by representing activities using descriptors. A descriptor is extracted from the observation sequence O, which intends to capture both spatial and temporal information of the activity and, hence, to model the dynamics of the activity implicitly. Activity recognition is achieved by a conventional classifier such as Support Vector Machines (SVM) or K-nearest neighborhood (KNN). There are three commonly used approaches to extract activity descriptors.

The first approach builds motion energy images (MEI) and motion history images (MHI), proposed by Bobick and Davis [2], by stacking a sequence of silhouettes to capture where and how the motion is performed. Activity descriptors are extracted from the MEI and MHI. For instance, seven Hu moments were extracted in [2] to serve as action descriptors and recognition was based on the Mahalanobis distance between the moment descriptors of the trained activities and the input activity.

The second approach considers a sequence of silhouettes as a spatiotemporal volume, and an activity descriptor is computed from the volume. Typical examples are the work by Yilmaz and Shah [3] which computes the differential geometric surface properties (i.e., Gaussian curvature and mean curvature); the work by Gorelick et al. [4] which extracts space-time saliency, action dynamics, and shape structure and orientation; and the work by Mokhber et al. [5] which calculates the 3D moments of the volume.

The third approach describes an activity using a set of spatiotemporal interest points (STIPs). The general concept is first to detect STIPs from the observations O which is usually a video sequence. Features are then extracted from a local volume around each STIP, and a descriptor can be formed by simply aggregating the local features together to become a bag of features or by classifying the STIPs into a set of vocabulary (i.e., a bag of visual words) and calculating the histogram of the occurrence of the vocabulary within the observation sequence O. There are two commonly used STIP extraction techniques. One extends Harris corner detection and automatic scale selection in 2D space to 3D space and time [6] and the other is based on a pair of one-dimensional (1D) Gabor filters applied temporally and spatially [7]. Recently, another STIP detector has been proposed by decomposing an image sequence into spatial components and motion components using nonnegative matrix factorization and detecting STIPs in 2D spatial and 1D motion space using difference of Gaussian (DoG) detectors [8]. In terms of the classifier for STIP-based descriptors, besides SVM and KNN, latent topic models such as the probabilistic latent semantic analysis (pLSA) model and latent Dirichlet allocation (LDA) were used in [9]. STIP-based descriptors have a few practical advantages including being applicable to image sequences in realistic conditions, not requiring foreground/background separation or human tracking, and having the potential to deal with partial occlusions [10]. In many realistic applications, an activity may occupy only a small portion of the entire space-time volume of a video sequence. In such situations, it does not make sense to classify the entire video. Instead, one needs to locate the activity in space and time. This is commonly known as the activity detection or action detection problem. Techniques have been developed for activity detection using interest points [11].

In the second category [12–17], the proposed methods usually follow the concept that an activity is a temporal evolution of the spatial configuration of the body parts and, hence, emphasize more on the dynamics of the activities than the methods in the first category. They usually extract a sequence of feature vectors, each feature vector being extracted from a frame, or a small neighborhood, of the observation sequence O. The two commonly used approaches are temporal templates and graphical models.

The temporal-template-based approach, typically, directly represents the dynamics through exemplar sequences and adopts dynamic time warping (DTW) to compare an input sequence with the exemplar sequences. For instance, Wang and Suter [18] employed locality preserving projection (LPP) to project a sequence of silhouettes into a low-dimensional space to characterize the spatiotemporal property of the activity and used DTW and temporal Hausdorff distance for similarity matching.

In the graphical model-based approach, both generative and discriminative models have been extensively studied for activity recognition. The most prominent generative model is the hidden Markov model (HMM), where sequences of observed features are grouped into similar configuration, i.e., states, and both the probability distribution of the observations at each state and the temporal transitional functions between these states are learned from training samples. The first work on action recognition using HMM is probably by Yamato et al. [12], where a discrete HMM is used to represent sequences over a set of vector-quantized silhouette features of tennis footage. HMM is a powerful tool to model a small number of short-term activities since a practical HMM is usually a fixed- and low-order Markov chain. Notable early extensions to overcome this drawback of the HMM are the variable-length Markov models (VLMM) and layered HMM. For details, the readers are referred to [13, 14], respectively. Recently, a more general generative graphical model, referred to as an action graph, has been established in [15], where nodes of the action graph represents salient postures that are used to characterize activities and shared by different activities, and weight between two nodes measures the transitional probability between the two postures represented by the two nodes. An activity is encoded by one or multiple paths in the action graph. Due to the sharing mechanism, the action graph can be trained and also easily expanded to new actions with a small number of training samples. In addition, the action graph does not need special nodes representing beginning and ending postures of the activities and, hence, allows continuous recognition.

The generative graphical models often rely on an assumption of statistical independence of observations to compute the joint probability of the states and the observations. This makes it hard to model the long-term contextual dependencies which is important to the recognition of activities over a long period of time. The discriminative models, such as conditional random fields (CRF), offer an effective way to model long-term dependency and compute the conditional probability that maps the observations to the motion class labels. The linear chain CRF was employed in [16] to recognize ten different human activities using features of combined shape-context and pair-wise edge features extracted at a variety of scales on the silhouettes and 3D joint angles. The results have shown that CRFs outperform the HMM and are also robust against the variability of the test sequences with respect to the training samples. More recently, Wang and Mori [17] modeled a human action by a flexible constellation of parts conditioned on image observations using hidden conditional random fields (HCRF) and achieved highly accurate frame-based action recognition.

Despite the extensive effort and progress in activity recognition research in the past decade, continuous recognition of activities under realistic conditions, such as with viewpoint invariance and large number of activities, remains challenging.

Application

Activity recognition has many potential applications. It is one of the key enabling technologies in security and surveillance for automatic monitoring of human activities in a public space and of activities of daily living of elderly people at home. Robust understanding and interpretation of human activities also allow a natural way for humans to interact with machines. A proper modeling of the spatial configuration and dynamics of human motion would enable realistic synthesis of human motion for gaming and movie industry and help train humanoid robots in a flexible and economic way. In sports, activity recognition technology has also been used in training and in the retrieval of video sequences.

References

1. Johansson G (1973) Visual perception of biological motion and a model for its analysis. Percept Psychophys 14(2): 201–211
2. Bobick A, Davis J (2001) The recognition of human movement using temporal templates. IEEE Trans Pattern Anal Mach Intell 23(3):257–267
3. Yilmaz A, Shah M (2008) A differential geometric approach to representing the human actions. Comput Vision Image Underst 109(3):335–351
4. Gorelick L, Blank M, Shechtman E, Irani M, Basri R (2007) Actions as space-time shapes. IEEE Trans Pattern Anal Mach Intell 29(12):2247–2253
5. Mokhber A, Achard C, Milgram M (2008) Recognition of human behavior by space-time silhouette characterization. Pattern Recogn 29(1):81–89

6. Laptev I, Lindeberg T (2003) Space-time interest points. In: International conference on computer vision, Nice, pp 432–439
7. Dollar P, Rabaud V, Cottrell G, Belongie S (2005) Behavior recognition via sparse-temporal features. In: 2nd joint IEEE international workshop on visual surveillance and performance evaluation of tracking and surveillance, Beijing, pp 65–72
8. Wong SF, Cipolla R (2007) Extracting spatiotemporal interest points using global information. In: International conference on computer vision, Rio de Janeiro, pp 1–8
9. Niebles JC, Wang H, Fei-Fei L (2008) Unsupervised learning of human action categories using spatial-temporal words. Int J Comput Vision 79(3):299–318
10. Liu J, Luo J, Shah M (2009) Recognizing realistic actions from videos "in the wild". In: International conference on computer vision and pattern recognition (CVPR), Miami, pp 1–8
11. Yu G, Goussies NA, Yuan J, Liu Z (2011) Fast action detection via discriminative random forest voting and top-K subvolume search. IEEE Trans Multimedia 13:507–517
12. Yamato J, Ohya J, Ishii K (1992) Recognizing human action in time-sequential images using hidden Markov model. In: International conference on computer vision and pattern recognition (CVPR), Champaign, pp 379–385
13. Galata A, Johnson N, Hogg D (2001) Learning variable-length Markov models of behaviour. Comput Vision Image Underst 81:398–413
14. Oliver N, Garg A, Horvits E (2004) Layered representations for learning and inferring office activity from multiple sensory channels. Comput Vision Image Underst 96:163–180
15. Li W, Zhang Z, Liu Z (2008) Expandable data-driven graphical modeling of human actions based on salient postures. IEEE Trans Circuits Syst Video Technol 18(11): 1499–1510
16. Sminchisescu C, Kanaujia A, Metaxas D (2006) Conditional models for contextual human motion recognition. Comput Vision Image Underst 104:210–220
17. Wang Y, Mori G (2011) Hidden part models for human action recognition: probabilistic versus max margin. IEEE Trans Pattern Anal Mach Intell 33(7):1310–1323
18. Wang L, Suter D (2007) Learning and matching of dynamic shape manifolds for human action recognition. IEEE Trans Image Process 16:1646–1661

AdaBoost

Paolo Favaro[1] and Andrea Vedaldi[2]
[1]Department of Computer Science and Applied Mathematics, Universität Bern, Switzerland
[2]Oxford University, Oxford, UK

Synonyms

Adaptive boosting; Discrete adaBoost

Definition

AdaBoost is an algorithm that builds a classifier by combining additively a set of weak classifiers. The weak classifiers are incorporated sequentially one at a time so that their combination reduces the empirical exponential loss.

Background

Boosting is a procedure to combine several classifiers with weak performance into one with arbitrarily high performance [1, 2] and was originally introduced by Robert Schapire in the machine learning community [3]. AdaBoost is a popular implementation of boosting for binary classification [4]. The enthusiasm generated by boosting, and in particular by AdaBoost, in machine learning can be highlighted via a quote of Breiman [1] saying that AdaBoost with trees is the "best off-the-shelf classifier in the world." In practice, much of the popularity of AdaBoost is due to both its performance being in the same league as support vector machines [5] and its algorithmic simplicity. In the computer vision community, AdaBoost has been made very popular by the work of Viola and Jones in face detection [6]. What attracted much of the attention was that, by using a cascade of AdaBoost-trained classifiers and the notion of integral image and Haar wavelets [6, 7], Viola and Jones were able to detect faces in real time.

The boosting framework is fairly general and several implementations have been proposed, among which are AdaBoost [4]; Real AdaBoost, LogitBoost, and GentleBoost [1]; Regularized AdaBoost [8]; or extensions to multiple classes such as AdaBoost.MH [9].

Theory

This section describes the AdaBoost algorithm as originally given by Freund and Schapire [4]. This particular variant, also known as Discrete AdaBoost [1], is summarized in Algorithm 1.

The purpose of AdaBoost is to learn a *binary classifier* that is a function $H(\mathbf{x}) = y$ that maps data $\mathbf{x} \in \mathcal{X}$ (e.g., a scrap of text, an image, or a sound wave) to its class label $y \in \{-1, +1\}$. In AdaBoost the classifier H is obtained as the sign of an additive combination of simple classifiers $h(\mathbf{x}) \in \{-1, +1\}$, called *weak hypotheses*:

$$H(\mathbf{x}) \doteq \text{sign}\left(\sum_{t=1}^{m} \alpha_t h_t(\mathbf{x})\right), \qquad (1)$$

where the coefficients $\alpha_t \in \mathbb{R}$. The input to AdaBoost is a set $\mathcal{H}$ of weak hypotheses and n data-label pairs $(\mathbf{x}_1, y_1), \ldots, (\mathbf{x}_n, y_n)$; the output is a combination H of m weak hypothesis in $\mathcal{H}$ and coefficients $\alpha_1, \ldots, \alpha_m$ that fit the data, i.e., $H(\mathbf{x}_i) = y_i$ for most $i = 1, \ldots, n$.

Let us denote with H_m the classifier H with m weak hypotheses shown in Eq. (1). AdaBoost operates sequentially by adding to H_{m-1} one new weak hypothesis (h_m, α_m). While any weak hypothesis with performance better than chance can be used, it is more common to select the weak hypothesis h_m in the set $\mathcal{H}$ that minimizes the weighted *empirical error* $\epsilon(h; \mathbf{w})$, i.e.,

$$h_m = \operatorname*{argmin}_{h \in \mathcal{H}} \epsilon(h; \mathbf{w}),$$

where

$$\epsilon(h; \mathbf{w}) \doteq \frac{\sum_{i=1}^{n} w_i [y_i \neq h(\mathbf{x}_i)]}{\sum_{i=1}^{n} w_i}$$

Algorithm 1 Discrete AdaBoost

1: Initialize $F_0(\mathbf{x}) = 0$ for all $\mathbf{x} \in \mathcal{X}$.
2: Initialize $w_i = 1$ for all $i = 1, 2, \ldots, N$.
3: **for** $t = 1$ to m **do**
4: Find the weak hypothesis $h_t \in \mathcal{H}$ that minimizes

$$\epsilon(h_t; \mathbf{w}) \propto \sum_{i=1}^{N} w_i [h_t(\mathbf{x}_i) \neq y_i].$$

5: Let

$$\alpha_t \leftarrow \frac{1}{2} \log \frac{1 - \epsilon(h_t; \mathbf{w})}{\epsilon(h_t; \mathbf{w})};$$

6: Update the weights

$$w_i \leftarrow w_i e^{-y_i \alpha_t h_t(\mathbf{x}_i)};$$

7: Update the function $F_t = F_{t-1} + \alpha_t h_t$.
8: **end for**
9: Return the classifier $H(\mathbf{x}) = \text{sign}\, F_m(\mathbf{x})$.

with positive data weights $\mathbf{w} = (w_1, \ldots, w_n)$. Here $[y_i \neq h(\mathbf{x}_i)]$ is equal to 1 if $y_i \neq h(\mathbf{x}_i)$ and 0 otherwise. Hence, the empirical error $\epsilon(h; \mathbf{w})$ is the average number of incorrect classifications of the weak hypothesis h on the weighted training data. The selected weak hypothesis h_m is then added to the current combination H_{m-1} with coefficient

$$\alpha_m = \frac{1}{2} \log \frac{1 - \epsilon(h_m; \mathbf{w})}{\epsilon(h_m; \mathbf{w})}. \qquad (2)$$

While AdaBoost minimizes the empirical error of the weak hypothesis h_m at each iteration, the weights $\mathbf{w} = (w_1, \ldots, w_n)$ are chosen so that the empirical error of H_m is reduced as well. AdaBoost starts with uniform weights $\mathbf{w} = (1, \ldots, 1)$ and updates them according to the rule

$$w_i \leftarrow w_i e^{-y_i \alpha_m h_m(x_i)}, \quad i = 1, \ldots, n. \qquad (3)$$

One intuitive interpretation of this rule is that it gives more importance to examples that are incorrectly classified. A formal justification is given in the next paragraph.

AdaBoost as Stagewise Minimization Denote by $F_m(\mathbf{x}) = \sum_{t=1}^{m} \alpha_t h_t(\mathbf{x})$ the additive combination of the weak hypotheses, so that the classifier $H_m(\mathbf{x})$ can be written as sign $F_m(\mathbf{x})$. AdaBoost performs a stagewise minimization of the cost

$$\frac{1}{n} \sum_{i=1}^{n} e^{-y_i F_m(\mathbf{x}_i)}.$$

This cost is known as the *empirical exponential loss* and is a convex upper bound to the empirical classification error of H_m:

$$\epsilon(H_m) \doteq \frac{1}{n} \sum_{i=1}^{n} [y_i \neq H_m(\mathbf{x}_i)] \leq \frac{1}{n} \sum_{i=1}^{n} e^{-y_i F_m(\mathbf{x}_i)}.$$

To understand the effect of the AdaBoost update on the empirical exponential loss, let $F_m(\mathbf{x}) = F_{m-1}(\mathbf{x}) + \alpha_m h_m(\mathbf{x})$ be the updatedadditive combination at iteration m. As the parameters of F_{m-1} are fixed, the empirical exponential loss is a function E of α_m and h_m:

$$E(\alpha_m, h_m) \doteq \frac{1}{n}\sum_{i=1}^{n} w_i e^{-y_i \alpha_m h_m(\mathbf{x}_i)},$$
$$\text{where } w_i = e^{-y_i F_{m-1}(\mathbf{x}_i)}. \tag{4}$$

By taking the derivative of E with respect to α_m and by setting it to zero, one obtains the optimality condition

$$0 = \sum_{i=1}^{n} w_i y_i h_m(\mathbf{x}_i) e^{-y_i h(\mathbf{x}_i)} e^{\alpha_m}$$
$$= \sum_{i:y_i \neq h_m(\mathbf{x}_i)} w_i e^{\alpha_m} - \sum_{i:y_i = h_m(\mathbf{x}_i)} w_i e^{-\alpha_m}$$

which results in the optimal coefficient given in Eq. (2):

$$\alpha_m(h_m) = \frac{1}{2}\log\frac{\sum_{i=1}^{n} w_i [y_i = h_m(\mathbf{x}_i)]}{\sum_{i=1}^{n} w_i [y_i \neq h_m(\mathbf{x}_i)]}$$
$$= \frac{1}{2}\log\frac{1-\epsilon(h_m;\mathbf{w})}{\epsilon(h_m;\mathbf{w})}.$$

By substituting this expression back in the cost (4), one obtains

$$E(\alpha_m(h_m), h_m) = 2\sqrt{\epsilon(h_m;\mathbf{w})(1-\epsilon(h_m;\mathbf{w}))}\sum_{i=1}^{n} w_i$$

which achieves its smallest value when the empirical classification error $\epsilon(h_m;\mathbf{w})$ approaches either 0, its minimum, or 1, its maximum. Notice that if the error $\epsilon(h_m;\mathbf{w}) > 1/2$, then the corresponding weight α_m is negative. In other words, when the weak hypothesis h_m makes more mistakes than correct classifications, AdaBoost automatically swaps the sign of the output label so that $\epsilon(-h_m;\mathbf{w}) < 1/2$. Finally, the weight update Eq. (3) follows from

$$w_i \leftarrow e^{-y_i F_m(\mathbf{x}_i)} = e^{-y F_{m-1}(\mathbf{x}_i)} e^{-y_i \alpha_m h_m(\mathbf{x}_i)}$$
$$= w_i e^{-y_i \alpha_m h_m(\mathbf{x}_i)}.$$

Application

One of the main uses of AdaBoost is for the recognition of patterns in data. Recognition can be formulated as a binary classification problem: Find whether data points match the pattern of interest or not. In computer vision, AdaBoost was popularized by its application to object detection, where the task is not only to recognize but also to localize within an image an object of interest (e.g., a face). Most of the ideas summarized in this section where first proposed by Viola and Jones [6].

A common technique for object detection is the *sliding window* detector. This method reduces the object detection problem to the task of classifying all possible image windows (i.e., patches) to find which ones are centered around the object of interest. In practice, windows may be sampled not only at all spatial locations but also at all scales and rotations. This results in a very large number of evaluations of the classifier function for each input image. Therefore, the computational efficiency of the classifier is of paramount importance.

Classifiers computed with AdaBoost can be made very computationally efficient by using weak hypotheses that are fast to compute and by letting AdaBoost select a small set of hypotheses most useful to the given problem. For example, in the Viola-Jones face detector, a weak hypothesis is computed by thresholding the output of a linear filter that computes averages over rectangular areas of the image. These filters are known as Haar wavelets and, because of their special structure, can be computed in constant time by using the *integral image* [6].

In order to further improve the speed of a sliding window detector, AdaBoost classifiers are often combined in a *cascade* [6]. A cascade exploits the fact that the vast majority of image windows are *not* centered around the object of interest and that, furthermore, most of these negative windows are easy to recognize as such. A cascade is built by appending one AdaBoost classifier after another. Classifiers are evaluated sequentially and an image window is rejected as soon as the response of a classifier is negative. All the classifiers are tuned to almost never reject a window that matches the object of interest (i.e., high recall). However, the first classifiers in the cascade are allowed to return several false positives (i.e., low precision) in exchange for a significantly reduced evaluation cost, obtained, for instance, by limiting the number of weak hypotheses in them. By using this scheme, the computationally costly and highly accurate classifiers are evaluated only on the most challenging cases: Windows that resemble the object of interest and that therefore either contain the object (i.e., a positive sample) or a visual structure that can be easily confused with it (i.e., a hard negative sample).

Finally, since each weak hypothesis is usually associated to an elementary feature, AdaBoost is also often used for *feature selection*. In some cases, feature selection improves the *interpretability* of the classifier. For instance, in the Viola-Jones face detector, the first few Haar wavelets selected by AdaBoost usually capture semantically meaningful anatomical structures such as the eyes and the nose.

References

1. Friedman JH, Hastie T, Tibshirani R (2000) Additive logistic regression: a statistical view of boosting. Ann Stat 28(2): 337–374
2. Hastie T, Tibshirani R, Friedman J (2001) The elements of statistical learning. Springer, New York
3. Schapire RE (1990) The strength of weak learnability. Mach Learn 5(2):197–227
4. Freund Y, Schapire RE (1996) Experiments with a new boosting algorithm. In: Proceedings of the 13th international conference on machine learning, Bari, 148–156
5. Vapnik V (1995) The nature of statistical learning theory. Springer, New York
6. Viola P, Jones M (2001) Robust real-time object detection. In: Proceedings of IEEE workshop on statistical and computational theories of vision, Vancouver, Canada
7. Papageorgiou CP, Oren M, Poggio T (1998) A general framework for object detection. In: International conference on computer vision, Bombay, pp 555–562
8. Sun Y, Li J, Hager W (2004) Two new regularized adaboost algorithms. In: Machine learning and applications, Louisville, pp 41–48
9. Schapire RE, Singer Y (1998) Improved boosting algorithms using confidence-rated predictions. In: Computational learning theory, Springer, New York, pp 80–91

Adaptive Boosting

▸AdaBoost

Adaptive Gains

▸von Kries Hypothesis

Affine Alignment

▸Affine Registration

Affine Camera

Zhengyou Zhang
Microsoft Research, Redmond, WA, USA

Synonyms

Affine projection

Related Concepts

▸Perspective Camera; ▸Perspective Transformation; ▸Weak Perspective Projection

Definition

An *affine camera* is a linear mathematical model to approximate the perspective projection followed by an ideal pinhole camera.

Background

Perspective projections give accurate models for a wide range of existing cameras, especially after calibration for a limited volume of workspace. However, the relationship from a 3D point to its 2D image point is nonlinear due to a scalar factor dependent of each individual point (see entry "▸Perspective Camera" for details). *Affine cameras* are introduced to make the projection model more mathematically tractable. An affine camera model is a first-order approximation obtained from Taylor expansion of the perspective camera model around a reference point. The reference point can be any point, but it may be set to the centroid of the 3D points, which results in a more accurate approximation. There are three important instances of an affine camera when the camera's intrinsic parameters are known: orthographic, weak perspective, and paraperspective projections. The reader is referred to entry "▶Weak Perspective Projection" for a detailed description of the affine camera model and its three instances.

Affine Invariants

Michael Werman
The Institute of Computer Science, The Hebrew University of Jerusalem, Jerusalem, Israel

Concept

Images of even the same object undergo various transformations depending on changes in the camera and its settings, the lighting, and the object itself. One of the ways to handle some of these changes, for tracking, search, and understanding images, is to use a description of the object that is oblivious to some of the above-mentioned transformations. Affine invariants are commonly used for this purpose.

There is a vast literature relating to affine invariants and only a small selection will be mentioned [4, 5, 11, 14, 17].

Affine Transformation

In order to have a property of an object that is invariant to an affine transformation, affine invariants can be used.

$$x \Rightarrow Lx + t$$

is an affine transformation where $x \in R^n$ is a vector, $L \in R^{n \times n}$ a matrix, and $t \in R^n$ a vector. L is a linear transformation, and t is a translation [2].

Affine transformations are used to describe different changes that images can undergo, such as an affine transformation of the (r, g, b) color values of an object under different lighting conditions or the transformation the shape of the image of an object undergoes when the camera and object are in different relative positions. The affine transformation in these cases does not necessarily model the exact physical distortion that is seen but often is a good approximation.

In order to have a property of an object that is invariant to an affine transformation, affine invariants can be used.

Affine Invariant

Let f be a function such that $f(a, b, \ldots) = f(T(a), T(b), \ldots)$ for any affine transformation T and $a, b, \ldots \in R^n$, f is an affine invariant.

First example Let p, q, r be real numbers then an affine transformation is of the form $x \Rightarrow \alpha x + \beta$; it is easy to check that for any three numbers p, q, r, and affine transformation parameters α, β,

$$\frac{p - q}{p - r} = \frac{(\alpha p + \beta) - (\alpha q + \beta)}{(\alpha p + \beta) - (\alpha r + \beta)}$$

[16, 18]

Second example We can generalize the first example to any dimension, d. The ratio of the volume of any two sets is an affine invariant (constant Jacobian, $|L|$) and is proportional to the $d + 1 \times d + 1$ determinant:

$$\begin{vmatrix} p & q & \ldots & r \\ 1 & 1 & \ldots & 1 \end{vmatrix}$$

Algebraic curves Not only points can be used to define affine invariance; other common examples are the parameters of curves, such as the equations of lines, conics, or other algebraic curves. For degree two curves in the plane, all ellipses are affinely equivalent.

Affine differential geometry There are affine invariant analogues of arc length and curvature. In general it is possible to find affine invariants involving points and their derivatives [3, 10, 13, 15].

Other parameters of the object Fourier coefficients and moments.

An affine transformation has $d^2 + d$ parameters, so in order to have a nontrivial affine invariant, one needs a function with more than that many arguments, where the result can be computed using algebraic elimination [9]. For example, two simplices in R^d have at least $d + 2$ different points which is $d^2 + 2d$ arguments.

In a certain sense, the number of independent invariants is the #(of parameters of a configuration) $-(d^2 + d)$.

There are other properties that are affine invariant, such as incidence, parallelism, centroids, barycentric coordinates, convexity, tangency, bi-tangency, Euler number, and connectivity.

If an objects area is A, then integrating by the area element divided by A is affine invariant, for example, $\frac{1}{A}\int_{Object} g(I(x,y))dxdy$ is affine invariant, g being any function of the pixel's color.

Affine Invariant Feature Detection

There are a number of affine invariant feature detectors that find affine invariant local features in an image [7, 17]. One of the successes in recent practice of computer vision was the SIFT, a similarity invariant feature, and its extension to an affine invariant feature, ASIFT [6, 8].

Normalization

There are normalizations that can be done to an object that remove all/some of the possible affine variation.

The center of gravity is a linear invariant, so it is possible to translate the object so that the object's center is at a fixed coordinate, thus removing the translation term, changing the problem from one of finding an affine invariant to a linear invariant.

The Whitening transform canonically transforms a set of points using an affine transformation so that the average is 0 and the covariance matrix is the identity, I.

Grassmannians

The set of labeled points modulo affine transformations are isomorphic to Grassmannians, using this one can define a geometry of affine invariant point sets and, for example, measure the distance between *affine invariant point sets* [1].

Correspondence

Another thing to notice is that usually there needs to be some correspondence of the objects in order to use these invariants, namely, the order of the arguments of the function f needs to be known, and the way to overcome this problem is summing over all permutations making this function invariant both to permutations. Another possibility is using permutation invariant features, for example, moments, which are built from summing over all the points.

Noise

Even though the group of affine transformation may be only a subgroup of the possible transformations, for example, projective transformations for an image of planar scene, being affinely invariant maybe an overkill as most of the radical transformations never really happen. Thus, noise can be explained by affine transformations making too many things close to each other.

References

1. Begelfor E, Werman M (2006) Affine invariance revisited. In: IEEE conference on computer vision pattern recognition (CVPR), New York
2. Gallier JH (2001) Geometric methods and applications for computer science and engineering. Springer, New York
3. Gotsman C, Werman M (1993) Recognition of affine transformed planar curves by extremal geometric properties. Int J Comput Geom Appl 3(2):183–202
4. Govindu VM, Werman M (2004) On using priors in affine matching. Image Vis Comput 22(14):1157–1164
5. Hartley RI, Zisserman A (2004) Multiple view geometry in computer vision, 2nd edn. Cambridge University Press, Cambridge/New York. ISBN:0521540518
6. Lowe DG (2004) Distinctive image features from scale-invariant keypoints. Int J Comput Vis 60:91–110
7. Mikolajczyk K, Tuytelaars T, Schmid C, Zisserman A, Matas J, Schaffalitzky F, Kadir T, Van Gool L (2005) A comparison of affine region detectors. Int J Comput Vis 65(1):43–72
8. Morel J-M, Yu G (2009) Asift: a new framework for fully affine invariant image comparison. SIAM J Imaging Sci 2(2):438–469
9. Olver PJ (2009) Equivalence, invariants and symmetry. Cambridge University Press, Cambridge
10. Olver PJ (2009) Moving frames and differential invariants in centroaffine geometry. Technical report
11. Petrou M, Kadyrov A (2004) Affine invariant features from the trace transform. IEEE Trans Pattern Anal Mach Intell 26:30–44
12. Rahtu E, Salo M, Heikkil J (2005) Affine invariant pattern recognition using multiscale autoconvolution. IEEE Trans Pattern Anal Mach Intell 27:908–918
13. Reid M, Szendroi B (2005) Geometry and topology. Cambridge University Press, Cambridge/New York
14. Sapiro G (2000) Geometric partial differential equations and image analysis. Cambridge University Press, Cambridge

15. Simon U (2000) Affine differential geometry [Chapter 9]. In: Handbook of differential geometry, vol 1. North-Holland, pp 905–961
16. Sprinzak J, Werman M (1994) Affine point matching. Pattern Recognit Lett 15(4):337–339
17. Tuytelaars T, Mikolajczyk K (2008) Local invariant feature detectors: a survey. Found Trends Comput Graph Vis 3: 177–280
18. Werman M, Weinshall D (1995) Similarity and affine invariant distances between 2D point sets. IEEE Trans Pattern Anal Mach Intell 17(8):810–814

Affine Projection

Zhengyou Zhang
Microsoft Research, Redmond, WA, USA

Synonyms

Affine camera

Related Concepts

▸Affine Camera; ▸Perspective Camera; ▸Perspective Transformation; ▸Weak Perspective Projection

Definition

An *affine projection* is a linear mathematical model to describe the projection performed by an affine camera. See entry "▸Affine Camera" for details.

Affine Registration

Kevin Köser
Institute for Visual Computing, ETH Zurich, Zürich, Switzerland

Synonyms

Affine alignment

Related Concepts

▸Affine Camera; ▸Rigid Registration

Definition

The goal of affine registration is to find the affine transformation that best maps one data set (e.g., image, set of points) onto another.

Background

In many situations data is acquired at different times, in different coordinate systems, or from different sensors. Such data can include sparse sets of points and images both in 2D and 3D, but the concepts generalize also to higher dimensions and other primitives. *Registration* means to bring these data sets into alignment, i.e., to find the "best" transformation that maps one set of data onto another, here using an *affine transformation*. For the sake of brevity, in this entry only the registration of two data sets is discussed, although approaches exist for finding consistent transformations that align more than two sets at once (e.g., [1]). While intuitively in 1D affine transformations compensate for scale and offset, in any dimension they can represent the first-order Taylor approximation (local linearization) for nonlinear functions.

For more complicated transformations, often a first affine registration is used as an initial solution that roughly aligns the data sets, followed by some (potentially local or nonlinear) search for more complicated parameters. In contrast to rigid registration, which allows only an offset and a rotation between the data sets, the affine model allows also for the full set of linear shape changes, including (nonisotopic) scale and shear. In particular, locally this approximates perspective effects or other nonlinear warps. On the other hand, affine registration uses still a single global transformation with a set of a few global parameters, which makes it mathematically easier to handle but also less powerful as compared to general nonrigid registration of deformable or articulated objects.

Theory

Two important cases can be distinguished for registration, the purely geometrical case, where two (finite) sets of points have to be registered, or the continuous/functional case, where the similarity of two functions must be maximized by an affine transformation of the functions' domain.

The Purely Geometrical Case

Let X and Y be two sets of points from $\mathbb{R}^m$ and $\mathbb{R}^n$, respectively, and without loss of generality, it is assumed that $m \geq n$. For now it is assumed that the sets are of the same size and that there exists an (unknown) affine transformation, such that each element in X is mapped to an element in Y. Then, the goal of affine registration is to find this transformation, i.e., the matrix $A \in \mathbb{R}^{n \times m}$ and the offset $t \in \mathbb{R}^n$, such as to minimize an energy:

$$E_d = \sum_{x \in X} \min_{y \in Y} d_g(Ax + t, y) \qquad (1)$$

Here $d_g(a, b)$ encodes the distance between a and b. Usually the points in X or Y are not available directly but only their noisy observations (e.g., when y are observed 2D projections of known 3D points and the parameters of the affine camera [2] are sought). In that case affine registration is based on these observations, and d_g should be chosen according to the noise distribution. If the sets are not of the same size or not each point in X corresponds to a point in Y, or in case there are gross errors in the data, still E_d can be minimized, e.g., with a robust cost function d_g to obtain some alignment between the sets [3]. Intuitively, in all these cases, A and t are sought which minimize the average distance for a transformed point of X to the closest point in Y.

In case bounds on A and t are known, a simple way to find an approximate solution is to sample the (continuous) space of all possible matrices A and offsets t, but this is usually computationally intractable. If it is known which points in X correspond to which others in Y, i.e., there is a set of pairs (x_i, y_i), the energy can be rewritten as

$$E_c = \sum_{x_i \in X} d_g(Ax_i + t, y_i) \qquad (2)$$

In case $d_g(a, b) = ||(a - b)||_2^2$, this can be solved explicitly. A contains mn unknowns, t contains n unknowns, and from each correspondence n observations can be obtained, so from counting it is clear that at least $m + 1$ correspondences are required to uniquely determine a solution (assuming they are in a general configuration and not, e.g., only on a line). In case the correspondences contain gross errors (mismatches, outliers), then the solution can be obtained using robust estimation techniques such as least median of squares [4] or RANSAC [5].

In case correspondences are not known, but that the data sets are already approximately aligned, it can be assumed that the closest neighbors are in correspondence. In this case the problem can be solved by using the iterative closest points (ICP) method [6–8]. Locally, for each point the offset to the closest point in the other data set is computed, and by collecting all the local offsets, a global affine transformation is estimated that aligns these sets (in a least squares sense according to Eq. 2). This transformation is applied to recalculate the nearest neighbor correspondences and the estimation step is repeated with these. In [9] efficient implementations and practical aspects of ICP have been studied. However, since ICP relies on local neighborhoods, it cannot cope well with situations where initially close points must be moved far away.

A technique proposed for this is based on normalization (as suggested, e.g., by [10]): Here, the means of the data sets are computed individually and t is defined as the difference of the means. Then, for each data set the (unbiased) covariance is computed and the matrices A_X,A_Y are searched that bring the respective point distributions to a unit covariance matrix (whitening of covariance). The two data sets now only differ by an orthogonal matrix that may be obtained by sampling or finding other characteristics in the data (cf. also [11]). However, this normalization approach assumes that the data sets overlap fully.

In general, similar concepts can be applied as for rigid registration, however, with a slightly different set of parameters.

The Continuous/Functional Case

Let I and J be functions (images) from $\mathbb{R}^n$ to $\mathbb{R}^d$, assigning some color value to each position, typically in the plane or in space. The function value will be referred to as the color hereafter, regardless of its physical meaning. In this case the affine registration can be stated as the minimization of an energy:

$$E = \int_x d_c(I(Ax + t), J(x))dx \qquad (3)$$

Here, d_c is a distance between two colors that should be chosen according to the expected measurement uncertainty of the colors. Very similar to the discrete case, a naive strategy to minimize this energy would be sampling; however, this is again computationally expensive and requires bounds on the parameters.

A possible solution is to compute local image features, such as corners, blobs, and so on (cf. to [12]), and – if possible – find correspondences among these features. There has been a huge body of work to particularly define features that can be detected reliably with affine changes of image coordinates. A comparison can be found in [13]. Given those the discrete methods of the previous section can be used to find an alignment A, t.

Afterwards, or in case A and t were approximately known from the beginning and if I and J are smooth, then the alignment can be performed by local linearization as proposed by Lucas and Kanade [14]. The assumption is that locally the image can be represented by its first-order Taylor approximation, i.e., the local color and the gradient:

$$I(x) \approx I(x_0) + \frac{\partial I}{\partial x}\underbrace{(x_0 - x)}_{\Delta x} \qquad (4)$$

Consequently, given two almost aligned images, at position x_0

$$\underbrace{J(x_0) - I(x_0)}_{\Delta I_0} \approx \left.\frac{\partial I}{\partial x}\right|_{x_0} \Delta x \qquad (5)$$

Stacking r of these equations on top of each other, an equation system is obtained:

$$\begin{pmatrix} \Delta I_0 \\ \Delta I_1 \\ \cdots \\ \Delta I_{r-1} \end{pmatrix} \approx \begin{pmatrix} \left.\frac{\partial I}{\partial x}\right|_{x_0} \\ \left.\frac{\partial I}{\partial x}\right|_{x_1} \\ \cdots \\ \left.\frac{\partial I}{\partial x}\right|_{x_{r-1}} \end{pmatrix} \Delta x \qquad (6)$$

This can be solved in a least squares sense to obtain Δx.

Similar to the local image gradient with respect to position, also the partial derivatives with respect to affine transformation parameters can be computed [14]. In this case, Δx (and also ∂x) contains $n(n+1)$ parameters (n^2 for the linear shape change and n for the offset). At least $n(n+1)$ equations are required to uniquely solve this. Often, to increase the basin of convergence, coarse-to-fine registration is applied. This can be implemented using image pyramids in case the uncertainty lies mostly in the offset parameters or by propagation of the affine parameter uncertainty to position uncertainty in the image and appropriate smoothing [15]. After having applied the estimated update Δx on the parameters, the steps can be applied repeatedly to register the two images. In [16] Baker and Matthews compare different formulations of such iterative, gradient-based image alignment, particularly the question of how to compose and parameterize the warps across multiple iterations.

On top of transformations on the domain of the images, often the two images differ in target (e.g., the colors of corresponding positions are related by some brightness offset), in which case also parameters for the change of color need to be estimated. In case the corresponding image colors are only statistically related but no explicit transformation model between colors is known, the concept of mutual information [17] might be used, where the entropy of the joint color histogram is minimized. An overview of image-based alignment can be found in [18, 19].

Application

Affine cameras [2] approximate a real camera by an affine mapping of 3D points to 2D image coordinates. For tracking local regions through videos, it has been shown [20] that keeping track of the affine deformations of local regions can help detecting tracking failures. Such affine warps, or those implied by correspondences of affine features between different images, represent the local linearization of a potentially nonlinear image warp (e.g., of perspective effects). If the structure of this nonlinear warp is known, the affine registration can allow inferring the global warp directly [21] or provide more constraints than just using the position of a region correspondence.

In general, affine registrations often provide a reasonable solution to align mean, linear shape, and orientation of data without making the transformation too problem specific, or the affine solution can serve as a basis for further more advanced alignment. Furthermore, multiple independent or coupled local affine registrations can help in registering articulated or deformable models.

References

1. Eggert DW, Fitzgibbon AW, Fisher RB (1998) Simultaneous registration of multiple range views for use in reverse engineering of CAD models. Comput Vis Image Underst 69(3): 253–272
2. Mundy JL, Zisserman A (eds) (1992) Geometric invariance in computer vision. MIT, Cambridge
3. Fitzgibbon AW (2003) Robust registration of 2d and 3d point sets. Image and Vis Computing 21:1145–1153
4. Rousseeuw PJ (1984) Least median of squares regression. J Am Stat Assoc 79(388):871–880
5. Fischler M, Bolles R (1981) RANdom SAmpling Consensus: a paradigm for model fitting with application to image analysis and automated cartography. Commun ACM 24(6):381–395
6. Besl PJ, McKay ND (1992) A method for registration of 3-d shapes. IEEE Trans Pattern Anal Mach Intell 14:239–256
7. Zhang Z (1994) Iterative point matching for registration of free-form curves and surfaces. Int J Comput Vis 13:119–152
8. Feldmar J, Ayache N (1994) Rigid and affine registration of smooth surfaces using differential properties. In: Eklundh JO (ed) European conference on computer vision (ECCV '94). Volume 801 of lecture notes in computer science. Springer, Berlin/Heidelberg, pp 396–406. doi:10.1007/BFb0028371
9. Rusinkiewicz S, Levoy M (2001) 3-D digital imaging and modeling. Proceedings Third International Conference on, Efficient variants of the ICP algorithm, 145–152. doi:10.1109/IM.2001.924423
10. Obdrzálek S, Matas J (2002) Local affine frames for image retrieval. Proceedings of the International Conference on Image and Video Retrieval, CIVR '02. Springer, London, pp 318–327
11. Ho J, Yang MH (2011) On affine registration of planar point sets using complex numbers. Comput Vis Image Underst 115(1):50–58
12. Tuytelaars T, Mikolajczyk K (2008) Local invariant feature detectors: A survey. Found Trends Comput Graph Vis 3(3):177–280
13. Mikolajczyk K, Tuytelaars T, Schmid C, Zisserman A, Matas J, Schaffalitzky F, Kadir T, van Gool L (2005) A comparison of affine region detectors. Int J Comput Vis 65(1–2):43–72
14. Lucas BD, Kanade T (1981) An iterative image registration technique with an application to stereo vision. Proceedings of the 7th international joint conference on Artificial intelligence, 2(6):674–679
15. Köser K, Koch R (2008) Exploiting uncertainty propagation in gradient-based image registration. Proc Br Mach Vis Conf 83–92
16. Baker S, Matthews I (2004) Lucas-kanade 20 years on: A unifying framework. Int J Comput Vis 56(1): 221–255
17. Viola PA III, WMW (1997) Alignment by maximization of mutual information. Int J Comput Vis 24(2):137–154
18. Zitov B, Flusser J (2003) Image registration methods: a survey. Image Vis Comput 21:977–1000
19. Szeliski R (2006) Image alignment and stitching: A tutorial. Found Trends Comput Graph Comput Vis 2
20. J Shi, Tomasi C (1994) Good features to track. IEEE Conf Comput Vis Pattern Recognit 593–600
21. Köser K, Koch R (2008) Differential spatial resection – pose estimation using a single local image feature. Eur Conf 278 Comput Vis (LNCS 5302Ű5305), 312–325

Affordances

►Affordances and Action Recognition

Affordances and Action Recognition

James Bonaiuto
California Institute of Technology, Pasadena, CA, USA

Synonyms

Action Recognition; Affordances

Related Concepts

►Gait Recognition; ►Gesture Recognition; ►Multi-camera Human Action Recognition

Definition

Affordances are opportunities for action that are directly perceivable in an organism's environment without higher-level cognitive functions. Action recognition is the result of mapping an observed action onto an internal motor or semantic representation.

Background

Affordances are defined by Gibson [1] as opportunities for action that are directly perceivable without the need for higher-level cognitive functions such as object recognition. The concept of affordances for action has generated significant interest in the computer vision and robotics community. More recently, links between this concept and that of action recognition have been explored, suggesting that the two may share common mechanisms.

Affordances. In robotics, early use of the term affordances dealt with the extraction of features from the visual environment that signal the possibility for action. However, neural models of vision and action have suggested a more active role for affordances in monitoring the ongoing control of an action and in predicting the effects of an action on a target object [2–4].

Because the affordances available to an agent depend on its embodiment, their representation needs to be grounded in its perceptual and motor repertoire. Most approaches therefore learn affordances through a stage of motor babbling with objects, where different actions are randomly attempted on an object while recording the initial perception of the object and any effects that the action has on it.

Typically, affordance formalisms involve storing entity, action, and effects relations, allowing the agent to associate objects with possible actions to perform on them and the effects of these actions. This formulation is therefore useful for planning by allowing an agent to search for an action that will produce a desired effect. More powerful formalisms store relations between entity, action, and effect equivalence classes [5]. These allow an agent to filter incoming sensory data to extract invariant affordance cues and to generalize learned affordances to novel objects.

Action Recognition. Action recognition can take place on several hierarchical scales of organization. At the lowest level, this consists of recognizing basic actions like reaching and grasping. Recognition of these actions can be used to recognize more complex actions and sequences of actions such as putting sugar in a cup. The highest level of action recognition involves inferring the goals, intentions, or the task of the observed agent (e.g., making coffee).

The two main approaches to low- and mid-level action recognition in computer vision are model-based and template-based or holistic. Model-based action recognition involves tracking body parts using a parametric model of the body kinematics. Body parts are typically recognized by low-level features, and then geometric constraints from the body model are imposed on them. Actions are typically represented in joint space. Template-based action recognition directly models actions using spatiotemporal features and represents actions as spatio-temporal shapes in 3 or 4d space. These features could be static features based on edges and limb shapes, dynamic features based on optical flows, or space-time volumes.

Neural models of low-level action recognition commonly use self-observation during action performance to associate the motor representation used to generate an action with the visual representation elicited by observation of it [6, 7]. An object-centered representation is hypothesized to exist for feedback-based control of actions, which causes self-observation during performance of an action and observation of another agent performing the same action to elicit similar visual representations [6, 8]. This link between affordance-activated motor representations and visual representations of observed actions is thought to ground understanding of the goals and intentions of other agents.

In order to recognize more complex actions and action sequences, the results of low-level action recognition are typically combined using some sequence parsing technique. Previous approaches have used hidden Markov models (HMMs) [9], probabilistic parsing using context-free grammars [10], and Bayesian inference on graphical models [11]. Models of action recognition at the highest level use lower level motor mechanisms in a simulation mode to infer the intention of the observed actor [12] or Bayesian inverse reinforcement learning to infer the reward function, or task, of a demonstration [13].

Theory

The close relationship between affordance and action recognition is made apparent by the related ideas that affordances play an active role in online control of an action and that the effects of an action on an object are included in the affordance representation. The same mechanisms used to monitor and control an agent's ongoing action can therefore also be leveraged to recognize the same action performed by another agent. Mechanisms for affordance perception and action recognition can interact through several ways. Recognition of an action performed by another agent can allow an agent to learn affordance cues or the effects of an affordance by observation. Observation of the effects of an affordance can aid in action recognition when the action itself appears ambiguous. Finally, extraction of affordance cues can similarly aid in action recognition by narrowing the space of possible actions.

Several systems use previously learned affordance information to recognize observed actions by searching for an action that is afforded by the observed object [14] or that would have produced the observed effect on the object [11, 13]. In both cases, the agent interprets the observed actions of other agents in terms of its own affordance representations. This can improve recognition accuracy when the details of the motion of the body parts during a particular is too difficult to disambiguate, but the effects of the action on the object are clearly perceivable.

Given a system for recognizing low-level body part movements, the affordances of objects can be approximately learned by observation [15]. This requires that the robot have a similar embodiment to that of the observed agent or a mechanism to map from observed body parts onto its own body. In this case, the features that indicate an affordance and the effects of the observed action can easily be recorded. If the observed action is novel and cannot be recognized at a high level, the recognized low-level subcomponents can be used to guide subsequent attempts at reproducing it by trying to replicate its effects on an object.

References

1. Gibson JJ (1966) The senses considered as perceptual systems. Houghton Mifflin, Boston
2. Fagg AH, Arbib MA (1998) Modeling parietal-premotor interactions in primate control of grasping. Neural Netw 11(7–8):1277–1303
3. Paletta L, Fritz G, Kintzler F, Irran J, Dorffner G (2007) Learning to perceive affordances in a framework of developmental embodied cognition. IEEE International Conference on Development and Learning, London, pp 110–115
4. Arbib MA (1997) From visual affordances in monkey parietal cortex to hippocampoparietal interactions underlying rat navigation. Phil Trans R Soc Lond B 352(1360): 1429–1436
5. Sahin E, Cakmak M, Dogar MR, Ugur E, Ucoluk G (2007) To afford or not to afford: a new formalization of affordances toward affordance-based robot control. Adapt Behav 15(4):447–472
6. Oztop E, Arbib MA (2002) Schema design and implementation of the grasp-related mirror neuron system. Biol Cybern 87(2):116–140
7. Metta G, Sandini G, Natale L, Craighero L, Fadiga L (2006) Understanding mirror neurons: a bio-robotic approach. Interact Stud 7(2):197–232
8. Bonaiuto J, Rosta E, Arbib MA (2007) Extending the mirror neuron system model, I: audible actions and invisible grasps. Biol Cybern 96:9–38
9. Yamato J, Ohya J, Ishii K (1992) Recognizing human action in time-sequential images using hidden Markov model. In: Proceedings of computer vision and pattern recognition (CVPR), Champaign, IL, pp 379–385
10. Bobick AF, Ivanov YA (1998) Action recognition using probabilistic parsing. In: Proceedings of computer vision and pattern recognition (CVPR), Santa Barbara, pp 196–202
11. Gupta A, Davis LS (2007) Objects in action: an approach for combining action understanding and object perception. In: Proceedings of computer vision and pattern recognition (CVPR), Minneapolis, pp 1–8
12. Oztop E, Wolpert D, Kawato M (2005) Mental state inference using visual control parameters. Cogn Brain Res 22:129–151
13. Lopes M, Melo FS, Montesano L (2007) Affordance-based imitation learning in robots. In: IEEE/RSJ international conference on intelligent robots and systems, San Diego, CA, pp 1015–1021
14. Moore DJ, Essa IA, Hayes MH (1999) Exploiting human actions and object context for recognition tasks. Proc Int Conf Comput Vis (ICCV) 1:80–86
15. Kjellstrom H, Romeroa J, Kragic D (2011) Visual object-action recognition: inferring object affordances from human demonstration. Comput Vis Image Underst 115(1): 81–90

Algebraic Curve

Bo Zheng
Computer Vision Laboratory, Institute of Industrial Science, The University of Tokyo, Meguro-ku, Tokyo, Japan

Synonyms

Implicit polynomial curve

Related Concepts

►Algebraic Surface

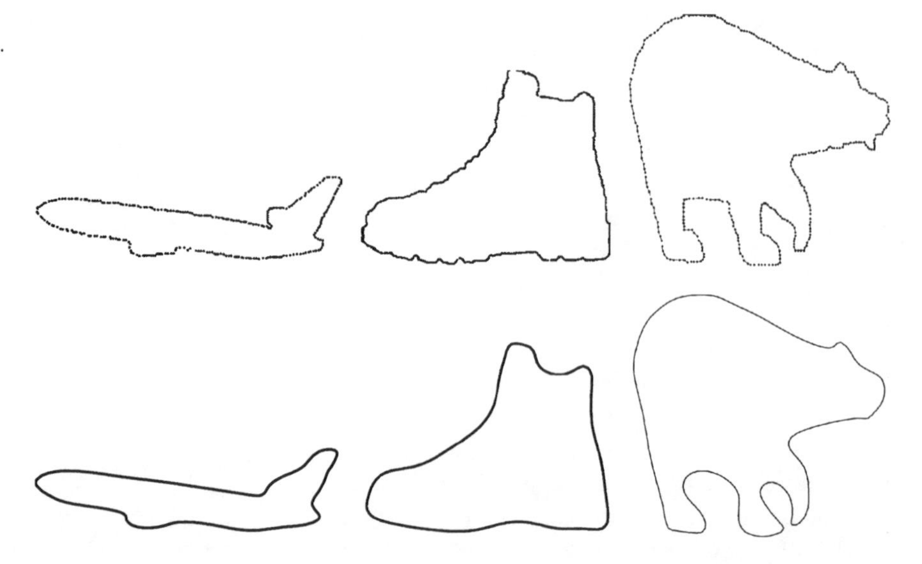

Algebraic Curve, Fig. 1 Examples of algebraic curves. *Top row*: original data sets; *bottom row*: represented algebraic curves

Definition

An *algebraic curve* is a curve determined by a 2-D *implicit polynomial* (IP) of degree n:

$$f_n(\mathbf{x}) = \sum_{0 \le i,j; i+j \le n} a_{ij} x^i y^j = \underbrace{(1\ x\ \ldots\ y^n)}_{\mathbf{m}^{\mathrm{T}}} \underbrace{(a_{00}\ a_{10}\ \ldots\ a_{0n})^{\mathrm{T}}}_{\mathbf{a}} = 0, \quad (1)$$

where $\mathbf{x} = (x, y)^{\mathrm{T}}$ is the coordinate of a point on a curve. That is, the curve is always represented by f_n's zero level set: $\{\mathbf{x} | f_n(\mathbf{x}) = 0\}$. The polynomial function is usually denoted by an inner product between two vectors: monomial vector $\mathbf{m}$ and coefficient vector $\mathbf{a}$. For the entries in these vectors, indices $\{i, j\}$ can be arranged in different orders, such as *lexicographical order* or *inverse lexicographical order*. In addition, the *homogeneous binary polynomial* of degree r in x and y, $\sum_{i+j=r} a_{ij} x^i y^j$, is called the r-th degree form of the IP. The form of degree n is called the *leading form*. The degree of an algebraic curve is the degree of the polynomial (e.g., n). An algebraic curve of degree 2 is called a conic, degree 3 a cubic, degree 4 a quartic, and so on.

Background

In computer vision, representing 2-D data sets with algebraic curves has been studied extensively for the past three decades. It is attractive for vision applications due to its applicability to object recognition, pose estimation, and registration. In contrast to the curves represented by other functions such as splines, Fourier, rational Gaussian, and radial basis function, algebraic curve is superior in such areas as fast fitting, few parameters, algebraic/geometric invariants, and robustness against noise and occlusion. Algebraic curve is also capable of modeling non-star shapes, open curves, curves that contain gaps, and unordered curve data. However, algebraic curve representation still suffers from some major issues such as the lack of local accuracy and global stability when representing a complex 2-D shape (see [7]). Figure 1 shows some example algebraic curves successfully used to represent closed 2-D curves.

Application and Theory

Algebraic curve representation is mainly attractive for vision applications such as fast shape registration or pose estimation [3, 6, 8, 9, 12] and recognition [2, 4–6, 9–11]. To achieve these purposes, many efforts have been made in three topics: curve fitting, algebraic/geometric invariants, and curve registration. The first is about solving the problem of accurately and stably fitting an algebraic curve to a complex shape, the second is on extracting algebraic or geometric invariants from the algebraic curve representing a shape, and the third concerns estimating the Euclidean transformation(s) between two or more algebraic curves representing different instances of the same shape.

Curve Fitting. There have been great improvements concerning algebraic curve fitting with its increased use during the late 1980s and early 1990s [8]. Recently, new robust and consistent fitting methods like 3L fitting [1], gradient-one fitting with Rigid regression [7], and degree-adaptive fitting [13] that are suitable for vision applications have been introduced.

Algebraic/Geometric Invariants. The main advantage of the use of algebraic curves for recognition is the existence of algebraic/geometric invariants, which are functions of the polynomial coefficients that do not change after a coordinate transformation. To some major contributions, the global Euclidean invariants are found by Taubin and Cooper [9], Teral and Cooper [6], and Keren [2], which can be expressed as simple explicit functions of the IP coefficients. Wolovich et al. [11] also introduced a set of invariants from covariant conic decompositions of implicit polynomials.

Curve Registration. In prior literatures [6, 9], global shape registration is performed through a single (non-iterative) computation using the central and oriented information extracted from the polynomial coefficients of two algebraic curves. Recently, an iterative method for aligning partially matched curves that uses the distance measurement of the polynomial gradient field together with a fast polynomial transformation has been introduced [12].

References

1. Blane M, Lei ZB, Cooper DB (2000) The 3L algorithm for fitting implicit polynomial curves and surfaces to Data. IEEE Trans Pattern Anal Mach Intell 22(3):298–313
2. Keren D (1994) Using symbolic computation to find algebraic invariants. IEEE Trans Pattern Anal Mach Intell 16(11):1143–1149
3. Keren D, Cooper D, Subrahmonia J (1994) Describing complicated objects by implicit polynomials. IEEE Trans Pattern Anal Mach Intell 16(1):38–53
4. Oden C, Ercil A, Buke B (2003) Combining implicit polynomials and geometric features for hand recognition. Pattern Recognit Lett 24(13):2145–2152
5. Subrahmonia J, Cooper DB, Keren D (1996) Practical reliable bayesian recognition of 2D and 3D objects using implicit polynomials and algebraic invariants. IEEE Trans Pattern Anal Mach Intell 18(5):505–519
6. Tarel J, Cooper DB (2000) The complex representation of algebraic curves and its simple exploitation for pose estimation and invariant recognition. IEEE Trans Pattern Anal Mach Intell 22(7):663–674
7. Tasdizen T, Tarel J-P, Cooper DB (2000) Improving the stability of algebraic curves for applications. IEEE Trans Imag Proc 9(3):405–416
8. Taubin G (1991) Estimation of planar curves, surfaces and nonplanar space curves defined by implicit equations with applications to edge and range image segmentation. IEEE Trans Pattern Anal Mach Intell 13(11):1115–1138
9. Taubin G, Cooper DB (1992) Symbolic and numerical computation for artificial intelligence, chapter 6, Computational mathematics and applications. Academic, London
10. Unel M, Wolovich WA (2000) On the construction of complete sets of geometric invariants for algebraic curves. Adv Appl Math 24:65–87
11. Wolovich WA, Unel M (1998) The determination of implicit polynomial canonical curves. IEEE Trans Pattern Anal Mach Intell 20(10):1080–1090
12. Zheng B, Ishikawa R, Oishi T, Takamatsu J, Ikeuchi K (2009) A fast registration method using IP and its application to ultrasound image registration. IPSJ Trans Comput Vis Appl 1:209–219
13. Zheng B, Takamatsu J, Ikeuchi K (2010) An adaptive and stable method for fitting implicit polynomial curves and surfaces. IEEE Trans Pattern Anal Mach Intell 32(3): 561–568

Algebraic Surface

Bo Zheng
Computer Vision Laboratory, Institute of Industrial Science, The University of Tokyo, Meguro-ku, Tokyo, Japan

Synonyms

Implicit polynomial surface

Related Concepts

▸Algebraic Curve

Definition

Similar to an algebraic curve, an *algebraic surface* is determined by a 3-D *implicit polynomial* (IP) of degree n:

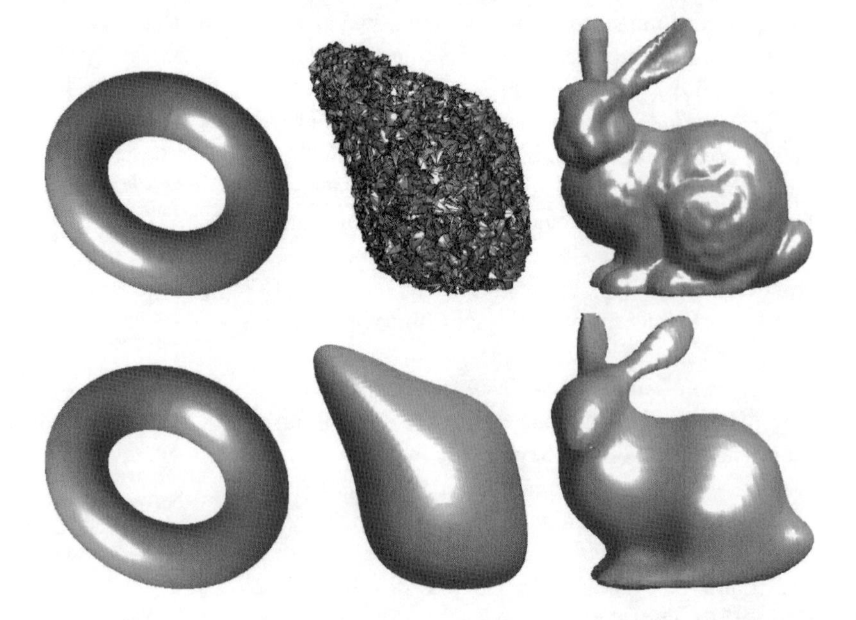

Algebraic Surface, Fig. 1 Examples of algebraic surfaces. *Top row*: original 3-D data sets of torus, simple shape with noise, and bunny; *bottom row*: resulting algebraic surface fits of degree 4, 4, and 8, respectively

$$\begin{aligned} f_n(\mathbf{x}) &= \sum_{0 \le i,j,k; i+j+k \le n} a_{ijk} x^i y^j z^k \\ &= \underbrace{(1\ x\ y\ z \ldots\ z^n)}_{\mathbf{m}^{\mathrm{T}}} \underbrace{(a_{000}\ a_{100}\ a_{010}\ a_{001}\ \ldots\ a_{00n})}_{\mathbf{a}}^{\mathrm{T}} \\ &= 0, \end{aligned} \tag{1}$$

where $\mathbf{x} = (x, y, z)$ is a 3-D point on a surface, that is, the surface is always represented by f_n's zero level set: $\{\mathbf{x} | f_n(\mathbf{x}) = 0\}$. The polynomial function can be denoted by an inner product of two vectors: monomial vector **m** and coefficient vector **a**. For the entries in these vectors, indices $\{i, j, k\}$ can be arranged in different orders, such as *lexicographical order* or *inverse lexicographical order*. In addition, the *homogeneous binary polynomial* of degree r in x, y, and z, $\sum_{i+j+k=r} a_{ijk} x^i y^j z^k$, is called the r*-th degree form* of the IP. The n-th form is also called *leading form*. The degree of algebraic surface is the degree of polynomial: n. An algebraic surface of degree 2 is called a quadric surface, degree 3 a cubic surface, degree 4 a quartic surface, and so on.

Background

In computer vision, representing 3-D surface data sets with algebraic surfaces has been also well studied. It is attractive for vision applications such as 3-D object recognition, pose estimation, and registration. In contrast to other surfaces represented by the functions such as Splines, Fourier, Rational Gaussian, and radial basis function, algebraic surface is superior in such areas as fitting efficiency, few parameters, convenience for calculating algebraic/geometric invariants, and robustness against noise and occlusion. Algebraic surface is also capable of modeling nonstar shapes, open curves, curves that contain gaps, and unordered curve data. However, algebraic surface representation still suffers from some major issues such as the lack of accuracy and stability when representing a complex 3-D shape. Figure 1 shows some examples of algebraic surfaces representing for 3-D surface data sets.

Application and Theory

Algebraic surface representation is mainly attractive for vision applications such as 3-D object registration or pose estimation [2, 4, 8, 10, 11, 13] and recognition [3, 6, 8, 11]. To achieve those purposes, many efforts have been made in three topics: surface fitting, algebraic/geometric invariants, and 3-D object registration. The first topic faces the problem of how to fit an algebraic surface to a complex 3-D shape accurately and stably; the second topic focuses on the problem of how to extract algebraic or geometric invariants from a 3-D shape-representing polynomial; and the third topic concentrates on the task of how to estimate the rigid transformation relationship between two algebraic surfaces representing the same object in different positions.

Surface Fitting

There have been great improvements concerning algebraic surface fitting with its increased use during the late 1980s and early 1990s [10]. Recently, new robust and consistent fitting methods such as 3L fitting [1], gradient-one fitting with Rigid regression [5, 9], and degree-adaptive fitting [14] have been proposed to make the algebraic surface representation more feasible for vision applications.

Algebraic/Geometric Invariants

The main advantage of algebraic surfaces for recognition is the existence of algebraic/geometric invariants, which are functions of the polynomial coefficients that do not change after a coordinate transformation. The algebraic/geometric invariants that are found by Taubin and Cooper [11], Teral and Cooper [8], and Keren [3] are global invariants and are expressed as simple explicit functions of the coefficients. Another set of invariants that have been mentioned by Wolovich et al. is derived from the covariant conic decompositions of implicit polynomials [12].

3-D Object Registration

In prior literatures such as [7, 11], the global shape registration is performed through single (non-iterative) computation after obtaining the central and oriented information extracted from polynomial coefficients. An iterative method in [13] is proposed by using the distance metric generated from polynomial gradient field and fast polynomial coefficient transformation.

References

1. Blane M, Lei ZB, Cooper DB (2000) The 3L algorithm for fitting implicit polynomial curves and surfaces to data. IEEE Trans Pattern Anal Mach Intell 22(3):298–313
2. Forsyth D, Mundy JL, Zisserman A, Coelho C, Heller A, Rothwell C (1991) Invariant descriptors for 3D object recognition and pose. IEEE Trans Pattern Anal Mach Intell 13(10):971–992
3. Keren D (1994) Using symbolic computation to find algebraic invariants. IEEE Trans Pattern Anal Mach Intell 16(11):1143–1149
4. Keren D, Cooper D, Subrahmonia J (1994) Describing complicated objects by implicit polynomials. IEEE Trans Pattern Anal Mach Intell 16(1):38–53
5. Sahin T, Unel M (2005) Fitting globally stabilized algebraic surfaces to range data. Proc IEEE Conf Int Conf Comp Visi 2:1083–1088
6. Subrahmonia J, Cooper DB, Keren D (1996) Practical reliable bayesian recognition of 2D and 3D objects using implicit polynomials and algebraic invariants. IEEE Trans Pattern Anal Mach Intell 18(5):505–519
7. Tarel J, Cooper DB (2000) The complex representation of algebraic curves and its simple exploitation for pose estimation and invariant recognition. IEEE Trans Pattern Anal Mach Intell 22(7):663–674
8. Tarel J-P, Civi H, Cooper DB (1998) Pose estimation of free-form 3D objects without point matching using algebraic surface models. In: Proceedings of IEEE Workshop Model Based 3D Image Analysis, Mumbai, pp 13–21
9. Tasdizen T, Tarel J-P, Cooper DB (2000) Improving the stability of algebraic curves for applications. IEEE Trans Imag Process 9(3):405–416
10. Taubin G (1991) Estimation of planar curves, surfaces and nonplanar space curves defined by implicit equations with applications to edge and range image segmentation. IEEE Trans Pattern Anal Mach Intell 13(11):1115–1138
11. Taubin G, Cooper DB (1992) Symbolic and numerical computation for artificial intelligence, chapter 6. In: Donald BR, Kapur D, Mundy JL (eds) Computational Mathematics and Applications. Academic, London
12. Wolovich WA, Unel M (1998) The determination of implicit polynomial canonical curves. IEEE Trans Pattern Anal Mach Intell 20(10):1080–1090
13. Zheng B, Ishikawa R, Oishi T, Takamatsu J, Ikeuchi K (2009) A fast registration method using IP and its application to ultrasound image registration. IPSJ Trans Comput Vision Appl 1:209–219
14. Zheng B, Takamatsu J, Ikeuchi K (2010) An adaptive and stable method for fitting implicit polynomial curves and surfaces. IEEE Trans Pattern Anal Mach Intell 32(3): 561–568

Analytic Reflectance Functions

►Reflectance Models

Animat Vision

Dana H. Ballard
Department of Computer Sciences, University of Texas at Austin, Austin, TX, USA

Synonyms

Active vision; Purposive vision

Definition

Animat vision is the computational study of the visual systems used by animals, with special attention to the binocular systems used by humans. For human vision, the goal is to show how the characteristics of the human eye movement system can be used to make the computation of needed information more efficient.

Background

The field of computer vision was given an enormous impetus by the publication of Rosenfeld's seminal book *Picture Processing by Computer* [28] in 1969, but in the 1970s the research focus shifted to human vision with the exciting new formulations of *early vision* that recognized that the human visual system devoted enormous resources to extracting cues such as binocular disparity, color, and motion from their composite representation in the initial image. Two groups were especially influential: the group at MIT headed by David Marr and Tomas Poggio [21] and the group at SRI headed by Harry Barrow and Martin Tanenbaum [8]. David Marr in particular had an enormous effect on the field and his book, *Vision*, is a classic [20].

While the early vision paradigm had a wonderful impact of defining computation in vision, by the early 1980s it was apparent that the computations defined on static images were mathematically delicate and could only be tamed with exceptional ingenuity. Thus the idea evolved that perhaps a moving camera, with known movement parameters, would help. An early effort was undertaken at MIT, but the first complete working system was built at the University of Pennsylvania by Ruzena Bajcsy who coined the term *active perception* to describe it [5]. That system was unveiled at a computer vision conference in northern Michigan run by Avi Kak and had an instantaneous acceptance among the researchers present.

Very shortly afterwards, Christopher Brown and the author built a similar system that had a significant advantage. Brown was tracking video processing pipeline computers and realized that this evolving computer architecture, when combined with a servo-driven binocular camera system, would allow the new computations to be realized in real time. The complete system is shown in Fig. 1. Subsequently the appearance of video-rate graphics cards capable of doing real-time image operations served to spur progress. Originally driven by the needs of the computer games industry, researchers quickly realized that now a large amount of the expensive visual calculations could be done in real time. The net result is that animat/active vision moved to lower-cost mobile robotic platforms with the result that robots using a moving cameras on mobile platforms are now commonplace.

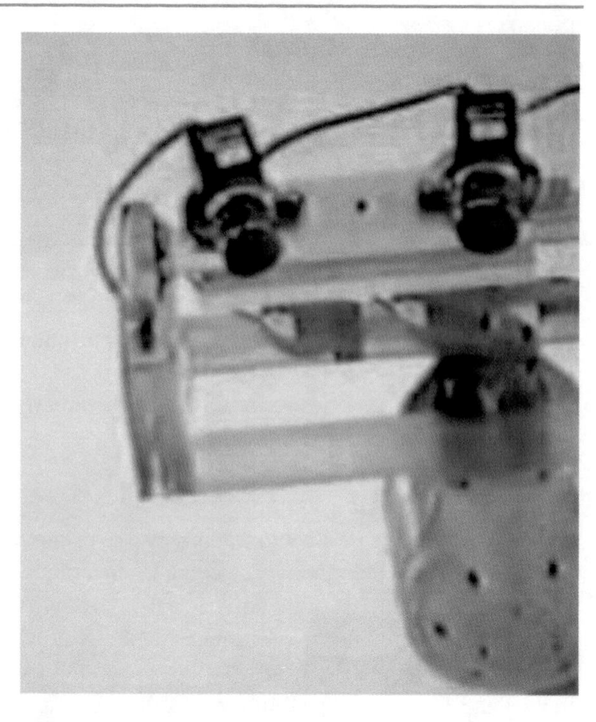

Animat Vision, Fig. 1 The University of Rochester real-time servo-driven robotic "head" mounted on its large PUMA "body." Similar systems were built at KTH in Stockholm, Carnegie Mellon University, MIT, the University of Pennsylvania, as well as many other places

Along the mainstream path of robotic animats, Rodney Brooks at MIT, perhaps inspired by Shimon Ullman's concept of *visual routines*, realized that the jobs that vision had to do now became preeminent and built MIT's humanoid robot *Cog* to focus on task-based vision. The particular architecture advocated may not have caught on, but the point was made and the system has been enormously influential. Two diverse communities – robotics and psychology – have been working on cognitive architectures for managing complex tasks that take a more integrated approach to vision and action, and both have recognized that the ultimate model architecture will have a hierarchical structure, e.g., [3, 11, 12, 16, 23]. Robotics researchers have gravitated to a three-tiered structure that models strategic, tactical, and detailed levels in complex behavior [10].

Theory

Animat vision, like its larger cousin active vision, is a paradigm with a huge number of important papers outstanding, with the consequence that it is only possible to provide the barest of outlines here. The interested reader is referred to some of the early papers [2, 6, 32]. Here we will demonstrate the impact on the calculation of early vision and introduce some more recent developments.

Consequences for Early Vision

Consider the problem of computing just one of the useful early vision representations, that of *optic flow*. Three-dimensional motion due to a moving observer induces the projection of two-dimensional motion on the retina. If the time-varying image function $f(x, y, t)$ represents only this effect, then the differential equations that represent the relationship between optic flows $(u(x, y, t), v(x, y, t))$ can be related to changes in photometric intensities $\frac{\partial f}{\partial x}, \frac{\partial f}{\partial y}, \frac{\partial f}{\partial t}$ with the equation, captured at the sensor to the image intensity M by

$$\frac{\partial f}{\partial x}u + \frac{\partial f}{\partial y}v + \frac{\partial f}{\partial t} = 0 \quad (1)$$

The conundrum of early vision can be easily apparent: At each point in the image x, y at time t, there is a single equation in two unknowns u, v. Poggio famously characterized this as an "ill-posed problem." A plethora of solution methods were tried, but they almost all involved integrating information across the optic array, a very delicate process. In contrast, moving the camera view point immediately solves this problem. If the motion is under the control of the human observer, then, say for the case of horizontal motion, to a first approximation one can assume a reliable estimate for u, and of course the system reduces to a well-behaved two equations in two unknowns. This kind of simplification surfaces again and again in early vision, and many novel instances of this kind of constraint still remain to be discovered.

The Importance of Eye Fixations

Yarbus's original work in gaze recordings [37] in the 1950s and 1960s revealed the enormous importance of gaze in revealing the underlying structure of human cognition. From this perspective, it is somewhat surprising that the first significant computational theory of vision [20] postponed the study of gaze as well as any influence of cognition of the extraction of information from the retinal array. In his "principle of least commitment," Marr argued the case for the role of the cortex in building elaborate goal-independent descriptions of the physical world. Perhaps as a consequence, when researchers took on the task of defining the mechanisms for directing gaze deployment, these turned out to be predominantly image based [18, 19, 36]. These theories have been compelling, but have many drawbacks. They usually cannot predict the exact landing points and typically leave more than 30 % of the fixations unaccounted for.

Recent experiments show that fixations are extracting very specific information needed by the human subject's ongoing *task* [14, 34]. The task context introduces enormous economies into this process that are very obvious: If a subject needs to pick up a red object, the search for that object can be limited to just red portions of the image; vast amounts of extraneous detail can be neglected. The visual information-gathering specificity of almost every portion of every task will introduce similar economies. Knowledge of task also has the promise of interpreting a substantial literature devoted to "change blindness." Subjects fail to notice large changes between successive images or movie frames. While the exact reason for this has been the subject of controversy [22], the problem may be resolvable if one has access to the viewer's cognitive agenda. On agenda changes are noticed and off agenda changes are not.

Theoretical Breakthroughs in Task Modeling

What experiments testing the information extracted during a fixation have lacked is a theory that accounts for the role of the cognitive processes that are controlling the subject's behaviors. What form should such a theory take? There have been several enormous theoretical advances, mostly from the fast emerging field of machine learning, that promise to have an enormous impact on quantitatively testable theories of cognition. The requirements of animat vision suggest that such a cognitive theory will have three important elements: (1) probabilistic representations, (2) the use of reward in learning, and (3) embodied cognitive architectures.

1. *Probabilistic methods.* There is rapidly increasing recognition that the brain is a probabilistic device and maintains a variety of mechanisms for calculating the statistical model of the world around it and its actions upon that world. To handle this a major new representational formalism has been developed that goes under the name *graphical models*. Originally developed by Pearl [26], such models have seen refinement as a general way to factor complex statistical interdependencies into to locally calculable quantities. The result is that the maintenance of elaborate statistical dependencies has become practical. Furthermore, Bayesian models of these interdependencies have proven their worth in characterizing many different observations in visual perceptions [35, 38].
2. *Reinforcement Learning.* A second breakthrough has been the development of model-making algorithms that are programmed by reward. It has long been appreciated that the brain must have mechanisms to learn complex behaviors and that these mechanisms must be steered by some scalar affinity signal. For the dominant effects the neurotransmitter dopamine has been implicated as the major signaling mechanism. Schultz, Glimcher, and others have made the connection between dopamine and reinforcement learning algorithms [17, 27, 33], the latter which by themselves have seen rapid development [1, 9]. Reinforcement learning algorithms are in their infancy but hold the promise of being and integral part of a comprehensive theory of animat vision learning.
3. *Embodied Cognition.* As emphasized by a number of researchers, the brain cannot be understood in isolation as so much of its structure is dictated by the body it finds itself in and the world that the body has to survive in [7, 13, 24, 25, 29]. This has special important implications, particularly for the cognitive architectures, because the brain can be dramatically simpler that it could ever be without these encasing milieus. The reason is that the brain does not have to replicate the natural structure of the world or the special ways of interacting with it taken by the body but instead can have an internal structure that implicitly and explicitly anticipates these commitments. The brain just has to have an interface that allows successful interactions with the world, but does not have to explicitly model all the detailed consequences of the actions taken. This realization opens up a way to address the challenge of making the leap from the apparent simplicity of the observed behaviors to the complexity of the brain-body-world system that produces them and that is to see the behaving body itself as a laboratory instrument. From this vantage point, the momentary disposition of the eyes, head, and hands during the course of behavior reveals essential details about the underlying cognitive program that is making those choices.

Open problems

Although the importance of body in cognition has been stressed at least since Merleau-Ponty, until the middle of 1980s, it was only practical to study very controlled circumstances such as those made by an experimental subject seated in front of a small display.

The research program at Rochester pioneered the study of embodied, visually driven behaviors by the development of innovative laboratory equipment and techniques. With Pelz at the Rochester Institute of Technology [4], they were the first laboratory that were able to track the eyes inside a head-mounted display. This capability allowed the exploitation of another recent development: Virtual Reality(VR). It is now straightforward to render scenes in real time from a moving observer's vantage point that are extraordinarily close to the real thing. Thus a person can have the compelling illusion of being in a fictional world that at the same time is under experimental control. This capability, in turn, has allowed researchers to address many new experimental questions for the first time. For example, one can study a person's disposition of visual resources in these virtual worlds by using the eye trackers inside the head-mounted display to manipulate the information that is available at each fixation [14, 30, 34].

Now flexible portable instrumentation can be attached to the body during the course of extended natural behaviors. Eye tracking capability that started out requiring subjects to be restrained in a bite bar has evolved to the point where portable trackers can be worn during a squash match. Head, hand, and body movements, even those of the facial muscles during expressions, can be reliably captured at high data rates during tea making, athletics, and everyday conversation. The new instrumentation opens up the possibility of acquiring large amounts of such data at millisecond time scales during these natural behaviors and thus provides access to the essential choices made in directing behavior under natural circumstances.

Obtaining such data from behavior and modeling it has led to another new question: How does one become confident that the models one builds are accurate? Answering this question has led to another new development and that is simulated human modeling. It is now possible to create models of humans that have the degrees of freedom of the skeletal system and also the capabilities of the binocular vision system [15, 31]. Thus one can build a human avatar that acts out the cognitive models obtained by fitting human data. A bonus is that one can test the models in completely new situations that were not part of the original human data gathering and observe their performance. This in turn can lead to an iterative refinement of the models and new experiments. However the most important aspect of this animat vision research avenue is the testing of the embodied cognition hypothesis with a suitably rich model. Our everyday experience and introspection as to the nature of the execution of everyday tasks has proven very misleading as to the brain's underlying representations owing to the artfulness of conscious experience.

References

1. Abeel P, Quigley M, Ng AY (2006) Using inaccurate models in reinforcement learning. In: International conference on machine learning, Pittsburgh
2. Aloimonos J, Bandopadhay A, Weiss I (1988) Active vision. Int J Comput Vis 1(4):333–356
3. Arkin R (1998) Behavior based robotics. MIT, Cambridge
4. Babcock JS, Pelz JB, Peak J (2003) The wearable eye-tracker: a tool for the study of high-level visual tasks. In: Proceedings of the MSS-CCD2003
5. Bajcsy R (1988) Active perception. Proc IEEE 76:966–1005
6. Ballard DH (1991) Animate vision. Artif Intell 48(1): 57–86
7. Ballard D, Hayhoe M, Pook P (1997) Deictic codes for the embodiment of cognition. Behav Brain Sci 20:723–767
8. Barrow HG, Tanenbaum JM (1978) Computer vision systems. Academic, New York, pp 3–26
9. Barto AG, Mahadevan S (2004) Recent advances in hierarchical reinforcement learning. Discret Event Dyn Syst 13:341–379
10. Bonasso RP, Firby RJ, Gat E, Kortenkamp D, Miller DP, Slack MG (1997) Experiences with an architecture for intelligent reactive agents. J Exp Theor Artif Intell 9:237–256
11. Brooks RA (1986) A robust layered control system for a mobile robot. IEEE J Robot Autom RA-2(1):14–23
12. Bryson JJ, Stein LA (2001) Modularity and design in reactive intelligence. In: International joint conference on artificial intelligence, Seattle
13. Clark A (1999) An embodied model of cognitive science? Trends Cogn Sci 3:345–351
14. Droll JA, Hayhoe MM, Triesch J, Sullivan BT (2005) Task demands control acquisition and storage of visual information. J Exp Psychol Hum Percept Perform 31: 1415–1438
15. Faloutsos P, van de Panne M, Terzopoulos D (2001) The virtual stuntman: dynamic characters with a repertoire of motor skills. Comput Graph 25:933–953
16. Firby RJ, Kahn RE, Prokopowicz PN, Swain MJ (1995) An architecture for vision and action. In: Fourteenth international joint conference on artificial intelligence, Montréal, pp 72–79
17. Hayden BY, Platt ML (2007) Temporal discounting predicts risk sensitivity in rhesus macaques. Curr Biol 17:49–53
18. Itti L, Baldi P (2005) A principled approach to detecting surprising events in video. In: IEEE international conference on computer vision and pattern recognition (CVPR), San Diego, vol 1, pp 631–637
19. Koch C, Ullman S (1985) Shifts in selective visual attention: towards the underlying neural circuitry. Hum Neurobiol 4:219–227
20. Marr D (1982) Vision: a computational investigation into the human representation and processing of visual information. Freeman, San Francisco
21. Marr D, Poggio T (1979) A computational theory of human stereo vision. Proc R Soc Lond B 204:301–328
22. Most SB, Scholl BJ, Clifford ER, Simons DJ (2005) What you see is what you set: sustained inattentional blindness and the capture of awareness. Psychol Rev 112:217–242
23. Newell A (1990) Unified theories of cognition. Harvard University Press, Cambridge
24. Noe A (2005) Action in perception. MIT, Cambridge/London
25. O'Regan JK, Noe A (2001) A sensorimotor approach to vision and visual consciousness. Behav Brain Sci 24: 939–973
26. Pearl J (1988) Probabilistic reasoning in intelligent systems: networks of plausible inference. Morgan Kaufmann, San Mateo

27. Platt ML, Glimcher PW (1999) Neural correlates of decision variables in parietal cortex. Nature 400:233–238
28. Rosenfeld A (1969) Digital picture processing. Academic, Orland
29. Roy DK, Pentland AP (2002) Learning words from sights and sounds: a computational model. Cogn Sci 26:113–146
30. Shinoda H, Hayhoe MM, Shrivastava A (2001) What controls attention in natural environments? Vis Res 41: 3535–3546
31. Sprague N, Ballard D (2003) Multiple-goal reinforcement learning with modular sarsa(0). Technical report 798, Computer Science Department, University of Rochester
32. Terzopoulos D, Rabie TF (1997) Animat vision: active vision in artificial animals. Videre J Comput Vis Res 1(1): 2–19
33. Tobler PN, Fiorillo CD, Schultz W (2005) Adaptive coding of reward value by dopamine neurons. Science 307: 1642–1645
34. Triesch J, Ballard D, Hayhoe M, Sullivan B (2003) What you see is what you need. J Vis 3:86–94
35. Weiss Y, Simoncelli EP, Adelson EH (2002) Motion illusions as optimal percepts. Nat Neurosci 5:598–604
36. Wolfe J (1994) Guided search 2.0. a revised model of visual search. Psychon Bull 1:202–238
37. Yarbus AL (1967) Eye movements and vision. Plenum Press, New York
38. Yuille A, Kersten D (2006) Vision as bayesian inference: analysis by synthesis? Trends Cogn Sci 10:301–308

Anisotropic Diffusion

►Diffusion Filtering

Aperture Ghosting

►Lens Flare and Lens Glare

Appearance Scanning

►Recovery of Reflectance Properties

Appearance-Based Human Detection

William Robson Schwartz
Department of Computer Science, Universidade Federal de Minas Gerais, Belo Horizonte, MG, Brazil

Synonyms

Appearance-based pedestrian detection

Related Concepts

►Object Detection

Definition

Human detection may be seen as a classification problem with two classes: human and nonhumans, in which the latter class is composed of background samples containing anything but humans. When the appearance-based human detection is employed, a large number of examples of human and nonhumans are considered to capture different poses, backgrounds, and occlusion situations through the extraction of feature descriptors so that a machine learning method can be used to classify samples as belonging to either one of the classes.

Background

Due to the large number of applications that require information regarding people's location, such as autonomous vehicles, surveillance, and robotics, finding people in images or videos presents large interest of the community. Even though widely studied in recent years [1], the human detection problem is still a challenge due to the wide variety of poses, clothing, background, and partial occlusions, which generate a large number of person's appearances.

Two main approaches have been explored in the human detection literature. The first class of methods consists of a generative process that combines detected parts of the human body according to a model. The second class considers statistical analysis through the use of machine learning techniques to classify a feature vector composed of low-level feature descriptors extracted from a detection window. This approach, also referred to as appearance-based, captures the appearance information and focuses on the discrimination between human and nonhuman samples.

Theory

Appearance-based human detection presents two important aspects: feature extraction and classification. Once both aspects have been considered, the training and test (detection) steps can be performed.

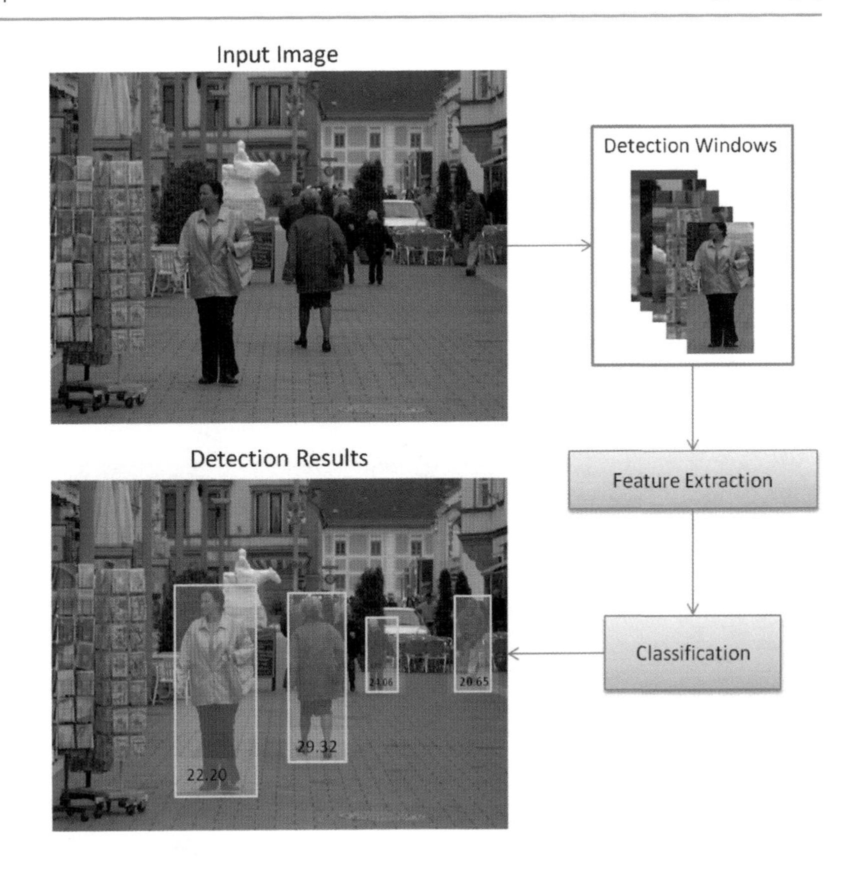

Appearance-Based Human Detection, Fig. 1 Example of the human detection process. Image sample extracted from the INRIA Person dataset [2]

The feature extraction is responsible for capturing the visual information from the scene, such as the presence of strong vertical edges, homogeneous textured clothing, or color constancy in the face. Such characteristics, useful for human detection, will be extracted using low-level feature descriptors. It has been shown that the combination of these characteristics improves detection results [3]. Among the most used feature descriptors are the histograms of oriented gradients (HOG) [2], local binary patterns (LBP) [4], and Haar wavelet-based features [5].

The second relevant aspect is the choice of a machine learning method capable of classifying between humans and non-humans by giving higher importance to those descriptors that best distinguish between the two classes. Among the most employed methods are the linear discriminant analysis (LDA), neural networks (NN), support vector machines (SVM), and partial least squares (PLS).

The training step is responsible for learning parameters of the machine learning methods such that the differences between the two classes can be properly captured. For that, features are extracted from multiple samples from both classes, and the descriptors are stored in feature vectors. It is important to emphasize that a good training set is important to assure that variations of the human appearances are captured. Each classification method presents a different way of learning the differences. For example, while SVM finds support vectors that maximize the margins between both classes, PLS will give more weight to those dimensions of the feature vector that best discriminate between the classes. In addition, it is important to note that a good training set is also important to assure that variations of the human appearances are captured.

Once the training has been performed, a sliding window is passed in the image at multiple scales to locate humans at different locations and scales. For each location, features are extracted and stored in a feature vector, which is then presented to the classifier. The output for each detection window is a value that reflects the probability or confidence in which a human is located inside the detection window. Figure 1 illustrates the detection process of a typical appearance-based human detection method.

Application

In general, the human detection is of interest in any application that falls inside the *Looking at People* [6] (domain which focuses primarily in analyzing images and videos containing humans). For example, a human detector can be used to provide the location of each agent in a scene so that tasks such as tracking, re-identification, action, and activity recognition can be executed by a surveillance system. In addition, a human detector can be executed in the domain on autonomous navigation, where the location of the pedestrians will be used as information for path planning. Furthermore, the use of human detection systems embedded in vehicles may be very useful to assure pedestrian safety [7].

References

1. Enzweiler M, Gavrila DM (2009) Monocular pedestrian detection: survey and experiments. IEEE Trans Pattern Anal Mach Intell 31(12):2179–2195
2. Dalal N, Triggs B (2005) Histograms of oriented gradients for human detection. In: IEEE conference Computer Vision and Pattern Recognition (CVPR), San Diego, pp 886–893
3. Schwartz WR, Kembhavi A, Harwood D, Davis LS (2009) Human detection using partial least squares analysis. In: IEEE International Conference on Computer Vision, Kyoto, pp 24–31
4. Wang X, Han TX, Yan S (2009) An HOG-LBP human detector with partial occlusion handling. In: IEEE International Conference on Computer Vision, Kyoto, pp 32–39
5. Viola P, Jones MJ, Snow D (2005) Detecting pedestrians using patterns of motion and appearance. Int J Comput Vis 63(2):153–161
6. Gavrila DM (1999) The visual analysis of human movement: a survey. Comput Vis Image Underst 73(1):82–98
7. Gandhi T, Trivedi M (2007) Pedestrian protection systems: issues, survey, and challenges. IEEE Trans Intell Transp Syst 8(3):413–430

Appearance-Based Human Tracking

Bohyung Han
Department of Computer Science and Engineering, Pohang University of Science and Technology (POSTECH), Pohang, South Korea

Synonyms

Human appearance modeling and tracking

Related Concepts

►Human Pose Estimation

Definition

Appearance-based human tracking is a human tracking algorithm, where the measurement is based on the appearance of human such as color, texture, shape, and their combination.

Background

Various human tracking algorithms have been proposed so far, but the focus of each algorithm is different. Appearance-based human tracking is an algorithm to track human based on the similarity between the existing appearance model and the observation in image; the control and search algorithm in tracking can be arbitrary. Several different features have been used for human tracking including color, edge, gradient, texture, and shape, and multiple features are often integrated together for more robust observations. The appearance of human based on the features is represented by density function, histogram, template, and other descriptors. For the representation of human body, the spatial layout of the features is typically employed to model the appearance of human. The appearance models can be constructed for the entire human body, a few subregions of the human body, or the individual articulated human body parts separately, depending on the target state spaces and tracking algorithms.

Generic tracking algorithms can be employed to track a blob-based human with a simple appearance model in a low-dimensional state space. However, in case of articulated human body tracking, the dimensionality of target state space is typically very high, and the inference procedure of the complex human body configuration is generally complex; more efficient and specialized tracking algorithms are required.

Theory

Methodologies of Appearance Models

Two aspects – integrated features and mathematical representation techniques – are considered to characterize the appearance-based human tracking.

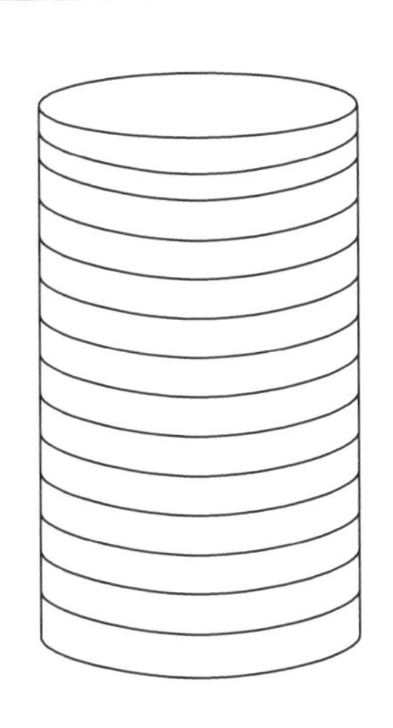

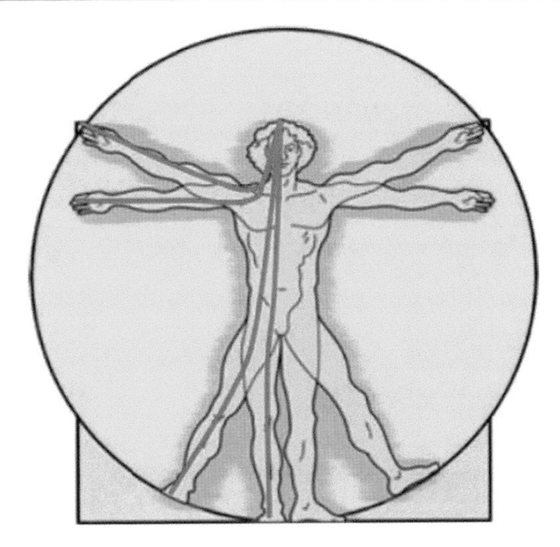

Appearance-Based Human Tracking, Fig. 1 Some examples to model spatial layout of a person in human appearance modeling. (*Left*) The appearance model is developed for each height slice of a person. (Fig. 2 in [1]) (*Right*) The path length is the distance from the *top* of the head to a given point on the path. (Fig. 2 in [9])

Features Integrated

The appearance for human tracking is modeled with color [1–3], silhouette [4, 5], shape [6, 7], edge [4, 8], or texture.

Representation Method

Some appearance modeling techniques assume that the appearance of human body is consistent horizontally. With the assumption, human body is represented with multiple histograms based on a cylinder model as in the left subfigure in Fig. 1 [1] or density function [3]. Another method is a path-length model [9], where the spatial variations are modeled by the distance from head along the shape of the person as shown in the right subfigure in Fig. 1 and spatial-feature distribution is constructed for appearance modeling. Template is also frequently used [7, 10], and probabilistic template is integrated in [11].

Acquisition and Maintenance of Appearance Models

The appearances may be fixed throughout the sequence or adaptive to the variations of human body appearance. The initialization of the appearance can be performed based on (manual or automatic) human detection. In [12], the appearance of each body part of human is learned in an online manner based on simple features obtained in off-line process.

Tracking Control and Search

Human tracking can be classified into two types based on the description method of human body; one is blob-based tracking, and the other is articulated human body tracking. In case of blob-based tracking, tracking algorithm is simple, and the state space of the target is typically low dimensional. The algorithms in this type have no big difference from generic tracking algorithms for other objects; a major difference is that human tracking algorithms often divide target into a few sub-regions based on appearance consistency to improve measurement accuracy. However, the articulated human body tracking involves very high-dimensional state space (typically more than 20 dimension) and complicated probabilistic inference procedures; efficient tracking control and search algorithms are required to handle the challenges such as annealing [4, 11] and covariance sampling [5] in particle filter framework.

Application

Appearance-based human tracking has a lot of potential applications such as event detection and video understanding, pedestrian detection and tracking for intelligent vehicles and vision-based user interface in computer games.

References

1. Mittal A, Davis L (2003) M2tracker: a multi-view approach to segmenting and tracking people in a cluttered scene. Int J Comput Vis 51(3):189–203
2. Ramanan D, Forsyth D, Zisserman A (2005) Strike a pose: tracking people by finding stylized poses. In: IEEE conference on computer vision and pattern recognition (CVPR), vol 1. IEEE Computer Society, Los Alamitos, pp 271–278
3. Kim K, Davis LS (2006) Multi-camera tracking and segmentation of occluded people on ground plane using search-guided particle filtering. In: European conference on computer vision (ECCV), Graz, vol 3, pp 98–109
4. Deutscher J, Reid I (2005) Articulated body motion capture by stochastic search. Int J Comput Vis 61:185–205
5. Sminchisescu C, Triggs B (2003) Estimating articulated human motion with covariance scaled sampling. Int J Robot Res 22(6):371–392
6. Haritaoglu I, Harwood D, David LS (2000) W4: real-time surveillance of people and their activities. IEEE Trans Pattern Anal Mach Intell 22:809–830
7. Lim H, Camps O, Sznaier M, Morariu V (2006) Dynamic appearance modeling for human tracking. In: IEEE conference on computer vision and pattern recognition (CVPR), New York, vol 1, pp 751–757
8. Poon E, Fleet D (2002) Hybrid Monte Carlo filtering: edge-based people tracking. In: IEEE workshop on motion and video computing, Orlando, pp 151–158
9. Yoon K, Harwood D, Davis LS (2006) Appearance-based person recognition using color/path-length profile. J Vis Commun Image Represent 17(3):605–622
10. Cham TJ, Rehg J (1999) A multiple hypothesis approach to figure tracking. In: IEEE conference on computer vision and pattern recognition (CVPR), Fort Collins, vol 1, pp 239–245
11. Balan A, Black M (2006) An adaptive appearance model approach for model-based articulated object tracking. In: IEEE conference on computer vision and pattern recognition (CVPR), New York, vol 1, pp 758–765
12. Ramanan D, Forsyth D, Zisserman A (2007) Tracking people by learning their appearance. IEEE Trans Pattern Anal Mach Intell 29(1):65–81

Appearance-Based Pedestrian Detection

►Appearance-Based Human Detection

Articulated Pose Estimation

►Human Pose Estimation

Asperity Scattering

S. C. Pont
Industrial Design Engineering, Delft University of Technology, Delft, The Netherlands

Synonyms

►Surface scattering; ►Velvety reflectance

Related Concepts

►Retroreflection; ►Lambertian Reflectance; ►Surface Roughness

Definition

The "asperities" can be of various nature, hairtips, dust, "fluff", local high curvature spots or ridges (the term derives from scattering by powdered materials where the "asperities" are sharp edges like on broken glass).

Background

The reflectance of natural, opaque, and rough surfaces [1, 2] can be described by the Bidirectional Reflectance Distribution Function (BRDF) [6]. BRDFs that are common and well-known are those of Lambertian, perfectly diffusely scattering surfaces and of specular surfaces. Such surfaces scatter light in all directions (diffuse scattering) or primarily in the mirror direction (specular reflection). However, natural surfaces can scatter light in many other ways. Asperity scattering adds a "surface lobe" to the usual diffuse, backscatter, and specular lobes of rough surfaces. It is an important effect in many materials that are covered with a thin layer of sparse scatterers such as dust or hairs. In the common case that single scattering predominates, asperity scattering adds important contributions to the structure of the occluding contour and the edge of the body shadow. This is the case because the BRDF is inversely proportional to the cosines of both illumination and viewing angles. The BRDF is generally low (and typically negligible), except when either the illuminating rays or visual directions graze the surface.

Because asperity scattering selectively influences the edges in the image of an object, it has a disproportionally (as judged by photometric magnitudes) large effect on (human) visual appreciation. It is a neglected but often decisive visual cue in the rendering of human skin. Its effect is to make smooth cheeks to look "velvety" or "peachy" (the appearances of both velvet and "peachy" skin are dominated by asperity scattering), that is to say, soft. This is a most important aesthetic and emotional factor that is lacking from Lambertian (looks merely dullish, paperlike), "skin type" BRDF (looks like glossy plastic), or even translucent (looks "hard", vitreous) types of rendering.

Theory

Asperity scattering is due to scattering by a sparse "cloud cover" of the surface with essentially point scatterers. In sparse distributions of scatterers, one may assume that single scattering predominates. Then parameters of interest are the geometry of the cloud and the nature of the single scatterers. For this case, a physical, geometrical optical model was derived [3] and experimental data gathered [5].

It is also possible to fit asperity scattering characteristics in a convenient, simplified formula (note that basic physical constraints should hold, e.g., non-negativity, energy conservation, and Helmholtz reciprocity) [4]. For instance, for a surface element with unit (outward) normal n , irradiated from the direction (unit vector) i and viewed from the direction (unit vector) j, the following BRDF model

$$V(\mathbf{i},\mathbf{j},\mathbf{n},a) = \frac{1}{\pi}\frac{a}{a+(\mathbf{i}\cdot\mathbf{n})(\mathbf{j}\cdot\mathbf{n})}, \qquad (1)$$

describes a "surface lobe" such as one observes in black velvet cloth or peach skin. The parameter a determines the width of the lobe. (A similar behavior results if one substitutes $(\mathbf{i}\cdot\mathbf{n})+(\mathbf{j}\cdot\mathbf{n})$ for $(\mathbf{i}\cdot\mathbf{n})(\mathbf{j}\cdot\mathbf{n})$.) The albedo is found to be

$$A_V(\mathbf{i},\mathbf{n},a) = \frac{2a}{(\mathbf{i}\cdot\mathbf{n})^2}\left(\mathbf{i}\cdot\mathbf{n} + a\log\frac{a}{a+\mathbf{i}\cdot\mathbf{n}}\right), \qquad (2)$$

which has a lowest value

$$2a\left(1+\log\left(\frac{a}{1+a}\right)^a\right) \approx 2a + 2\log a a^2 + \ldots \qquad (3)$$

at normal incidence and rises monotonically to unity at grazing incidence. (For black velvet $a \ll 1$.) Other possibilities for simplified formulations may be found in graphics as so-called velvet shaders. However, care should be taken that many of these rendering applications do not fulfill the above mentioned basic physical constraints.

Open Problems

BRDFs of natural surfaces can probably be categorized into about a dozen different modes. Currently, only the forward, backward, diffuse, and surface scattering modes have been described by formal optical models.

Reflectance estimation from images suffers from image ambiguities. Prior knowledge on the reflectance statistics of natural materials plus formal descriptive models for the common modes of natural BRDFs can constrain this problem.

References

1. CUReT: Columbia–Utrecht reflectance and texture database. http://www.cs.columbia.edu/CAVE/curet
2. Dana KJ, van Ginneken B, Nayar SK, Koenderink, JJ (1999) Reflectance and texture of real-world surfaces, ACM Trans on Graphics, 18(1):1–34
3. Koenderink JJ, Pont SC (2003) The secret of velvety skin. Mach Vis Appl 14:260–268
4. Koenderink JJ, Pont SC (2008) Material properties for surface rendering. Int J Comput Vis Biomech 1(1):43–53

5. Lu R, Koenderink JJ, Kappers AML (1998) bidirectional reflection distribution functions) of velvet. Appl Opt 37(25):5974–5984
6. Nicodemus FE, Richmond JC, Hsia JJ (1977) Geometrical considerations and nomenclature for reflectance. National bureau of standard US monograph, 160

Automatic Gait Recognition

►Gait Recognition

Automatic Scale Selection

►Scale Selection

Automatic White Balance (AWB)

►White Balance

B

Backscattering

►Retroreflection

Bas-Relief Ambiguity

Manmohan Chandraker
NEC Labs America, Cupertino, CA, USA

Synonyms

Generalized bas-relief (GBR) transformation

Related Concepts

►Illumination Estimation, Illuminant Estimation; ►Lambertian Reflectance; ►Photometric Stereo; ►Shape from Shadows

Definition

Members of the equivalence class of convex Lambertian surfaces that produce the same set of orthographic images under arbitrary combinations of distant point light sources are related by elements of a three-parameter subgroup of $GL(3)$, called generalized bas-relief (GBR) transformations. This inherent ambiguity in determining the three-dimensional shape of an object from shading and shadow information is called the bas-relief ambiguity.

Background

For a surface $f(x, y)$, the GBR-transformed surface is given by $\bar{f}(x, y) = \mu x + \nu y + \lambda f(x, y)$, where $\mu, \nu \in \mathbb{R}$ and $\lambda \in \mathbb{R}_{++}$. The orthographic image of an object with Lambertian reflectance, illuminated by an arbitrary set of distant point light sources, remains unchanged when the object shape is transformed by a GBR, with an inverse transformation applied on the set of light sources and a corresponding pointwise transformation on the albedos. Further, any continuous transformation that preserves the shading and shadowing configuration for a convex surface must belong to the GBR group [1].

Thus, for a Lambertian surface, any reconstruction or recognition algorithm based on shading and shadow information alone can at best enunciate the shape, albedo, or lighting up to a "bas-relief ambiguity." The ambiguity derives its name from the corresponding low-relief sculpture technique (Italian: *basso rilievo*), which can be understood as a special case of the GBR, where $\lambda < 1$.

Theory

Under orthographic projection, each point on a surface may be represented as $[x, y, f(x, y)]^\top$, where $(x, y) \in \mathbb{R}^2$ is a point on the image plane and f is a piecewise differentiable function. The unit surface normal is given by

$$\hat{\mathbf{n}} = \frac{[-f_x, -f_y, 1]^\top}{\sqrt{f_x^2 + f_y^2 + 1}}. \quad (1)$$

K. Ikeuchi (ed.), *Computer Vision*, DOI 10.1007/978-0-387-31439-6,

A GBR transformation that maps a surface $f(x, y)$ to $\bar{f}(x, y) = \mu x + \nu y + \lambda f(x, y)$ and the corresponding inverse GBR transformation may be represented as 3×3 linear transformations:

$$\mathbf{G} = \begin{bmatrix} 1 & 0 & 0 \\ 0 & 1 & 0 \\ \mu & \nu & \lambda \end{bmatrix}, \quad \mathbf{G}^{-1} = \frac{1}{\lambda}\begin{bmatrix} \lambda & 0 & 0 \\ 0 & \lambda & 0 \\ -\mu & -\nu & 1 \end{bmatrix}. \tag{2}$$

Under the matrix product operation, the set of GBR transformations forms a subgroup of $GL(3)$, the group of 3×3 invertible linear transformations. The unit surface normals of the GBR-transformed surface $\bar{\mathbf{f}}$ are $\frac{\mathbf{G}^{-\top}\hat{\mathbf{n}}}{\|\mathbf{G}^{-\top}\hat{\mathbf{n}}\|}$.

Shadows and Shading

The image formation equation at a point $\mathbf{p} = [x, y, f(x, y)]^\top$ on a Lambertian surface is given by

$$I(x, y) = \mathbf{n}^\top \mathbf{s} \tag{3}$$

where I is the intensity, $\mathbf{n}$ is the product of albedo a and unit surface normal $\hat{\mathbf{n}}$, while $\mathbf{s}$ is the light source direction, scaled by its strength. The point $\mathbf{p}$ lies in an attached shadow if $\mathbf{s}^\top \hat{\mathbf{n}} < 0$, while it lies on a cast shadow boundary if there exists a point $\mathbf{p}'$ on the surface, with unit normal $\hat{\mathbf{n}}'$, such that

$$\mathbf{s}^\top \hat{\mathbf{n}}' = 0, \quad \mathbf{p} - \mathbf{p}' = k\mathbf{s}, \text{ for some } k \in \mathbb{R}_{++}. \tag{4}$$

A point $\mathbf{p} = [x, y, f(x, y)]^\top$ lies in an attached shadow or on a cast shadow boundary in an image produced by the light source $\mathbf{s}$ if and only if the point $\bar{\mathbf{p}} = \mathbf{Gp}$ does so in an image produced by the light source $\bar{\mathbf{s}} = \mathbf{Gs}$, where $\mathbf{G}$ is a GBR transformation given by Eq. (2). Further, the image of a surface $f(x, y)$ with albedo $a(x, y)$, when illuminated by a light source $\mathbf{s}$, is equivalent to the image under the light source $\bar{\mathbf{s}} = \mathbf{Gs}$ of the GBR-transformed surface $\bar{f}$, with a pointwise albedo transformation given by

$$\bar{a} = \frac{a}{\lambda}\left(\frac{(\lambda f_x + \mu)^2 + (\lambda f_y + \mu)^2 + 1}{f_x^2 + f_y^2 + 1}\right)^{\frac{1}{2}}. \tag{5}$$

It follows that the set of images of a Lambertian surface-albedo pair $\{f, a\}$, under all possible combinations of distant light sources, is identical to that of any GBR-transformed surface-albedo pair $\{\bar{f}, \bar{a}\}$ [1] (see Fig. 1). Thus, the illumination cones of surfaces related by a GBR transformation are identical [2].

Existence and Uniqueness

It is shown in [1] that any two convex, smooth surfaces with visible occluding contours that produce the same set of attached shadow boundaries must be related by a GBR transformation. Thus, the GBR transformation is the only one that preserves the set of all images of an object.

While the existence result for the bas-relief ambiguity does not explicitly require convexity of the surface, in practice, the image formation model for concave regions must account for interreflections. It has been shown that modeling diffuse interreflections uniquely determines the shape and lighting [3].

Integrability

In traditional photometric stereo, given images of a point $\mathbf{p}$ under three or more known light sources, one may recover its surface normal $\hat{\mathbf{n}}$ using Eq. (3). However, in uncalibrated photometric stereo where the light sources are unknown, the surface normal and the light sources can be recovered only up to an arbitrary, invertible 3×3 linear transformation, since $\mathbf{n}^* = \mathbf{A}^\top \mathbf{n}$ and $\mathbf{s}^* = \mathbf{A}^{-1}\mathbf{s}$ satisfy Eq. (3) for any $\mathbf{A} \in GL(3)$.

For the recovered normal field to correspond to a surface, it must satisfy the integrability constraint [4]:

$$\frac{\partial}{\partial y}\left(\frac{n_1^*}{n_3^*}\right) = \frac{\partial}{\partial x}\left(\frac{n_2^*}{n_3^*}\right). \tag{6}$$

It is shown in [1] that requiring the recovered normal field to be integrable restricts $\mathbf{A}$ to lie in the group of GBR transformations.

Generalizations

Under perspective projection, the shadows produced by an object under distant or proximal point light sources are the same as those produced by a surface transformed by a generalized perspective bas-relief (GPBR) transformation, with an inverse transformation applied on the light sources [5]. The GPBR is a three-dimensional elation [6] and, in the limiting case

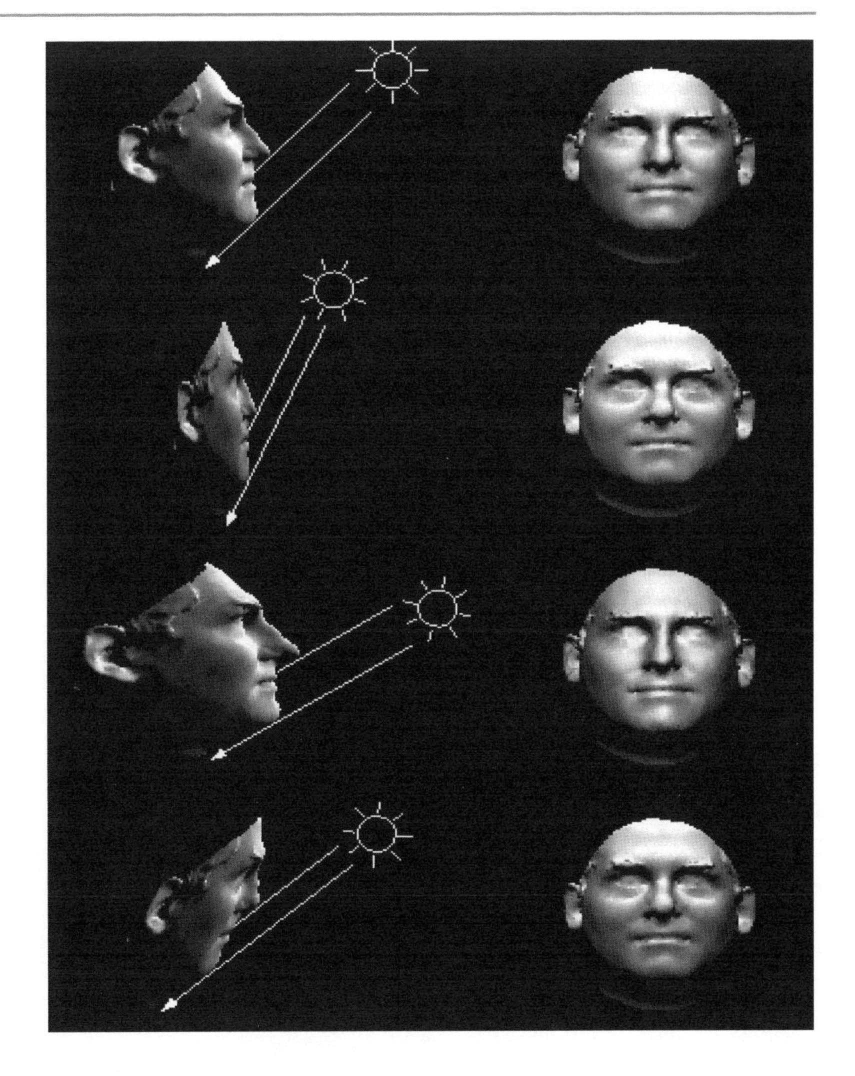

Bas-Relief Ambiguity, Fig. 1 The *left column* shows various GBR transformations applied to a surface, with the corresponding inverse transformation applied to the light source direction. The *top row* is the true shape. The *right column* shows that the shading and shadows produced in an orthographic image of the surface are identical for any GBR transformation (Figure reproduced in part from [1], courtesy of the authors)

of orthographic projection, reduces to the definition of GBR in Eq. (2).

Under orthographic projection from an unknown viewpoint, there exists an ambiguity that corresponds to the group of three-dimensional affine transformations, called the Klein generalized bas-relief (KGBR) ambiguity, such that the set of images of an object is preserved under the action of a KGBR transformation on the shape, lighting, albedos, and viewpoint [7]. In the limiting case of a fixed viewpoint, the KGBR ambiguity reduces to the bas-relief ambiguity.

Application

The bas-relief ambiguity in computer vision explains psychophysical observations of similar unresolved ambiguities in human visual perception [8]. An important consequence of the existence of the bas-relief ambiguity is that any image-based computer vision algorithm, relying on inference based solely on shading and shadow information, can only describe the object up to an arbitrary GBR transformation. Further, it has been established that for any infinitesimal motion of a surface f, there exists a motion for the GBR-transformed surface $\bar{f}$ that produces the same motion field [1]. Thus, an infinitesimal motion does not provide additional cues for disambiguation.

Surface reconstruction up to a GBR transformation can be performed by imposing integrability in uncalibrated photometric stereo [9]. The bas-relief ambiguity may be resolved in practice by incorporating additional information, for instance, priors on albedo distribution [10, 11]. Alternatively, the presence of non-Lambertian effects – such as specular highlights [12], a Torrance-Sparrow reflectance [13], or a spatially unvarying, isotropic, additive non-Lambertian reflectance component [14] – eliminates the GBR ambiguity.

References

1. Belhumeur P, Kriegman D, Yuille A (1999) The bas-relief ambiguity. Int J Comput Vis 35(1):33–44
2. Belhumeur P, Kriegman D (1998) What is the set of images of an object under all possible illumination conditions? Int J Comput Vis 28(3):245–260
3. Chandraker M, Kahl F, Kriegman D (2005) Reflections on the generalized bas-relief ambiguity. In: Proceedings of IEEE conference on computer vision and pattern recognition (CVPR). IEEE Computer Society, San Diego, CA, pp 788–795
4. Horn B, Brooks M (1986) The variational approach to shape from shading. Comput Vis Graph Image Process 33: 174–208
5. Kriegman D, Belhumeur P (2001) What shadows reveal about object structure. J Opt Soc Am A 18(8):1804–1813
6. Coxeter H (1969) Introduction to geometry, 2nd edn. Wiley, New York
7. Yuille A, Coughlan J, Konishi S (2003) The KGBR viewpoint-lighting ambiguity. J Opt Soc Am A 20(1):24–31
8. Koenderink J, van Doorn A, Christon C, Lappin J (1996) Shape constancy in pictorial relief. Perception 25(2):151–164
9. Yuille A, Snow D (1997) Shape and albedo from multiple images using integrability. In: Proceedings of IEEE conference on computer vision and pattern recognition (CVPR), San Juan, pp 158–164
10. Alldrin N, Mallick S, Kriegman D (2007) Resolving the generalized bas-relief ambiguity by entropy minimization. In: Proceedings of IEEE conference on computer vision and pattern recognition (CVPR), Minneapolis, pp 1–7
11. Hayakawa H (1994) Photometric stereo under a light source with arbitrary motion. J Opt Soc Am A 11(11):3079–3089
12. Drbohlav O, Sara R (2002) Specularities reduce ambiguity of uncalibrated photometric stereo. In: Proceedings of European conference on computer vision (ECCV), Copenhagen, pp 46–60
13. Georghiades A (2003) Incorporating the Torrance-Sparrow model in uncalibrated photometric stereo. In: Proceedings of IEEE conference on computer vision (ICCV), Nice, pp 816–823
14. Tan P, Mallick S, Quan L, Kriegman D, Zickler T (2007) Isotropy, reciprocity and the generalized bas-relief ambiguity. In: Proceedings of IEEE conference on computer vision and pattern recognition (CVPR), Minneapolis, pp 1–8

Behavior Understanding

►Multi-camera Human Action Recognition

Bidirectional Texture Function and 3D Texture

Kristin Dana
Department of Electrical and Computer Engineering, Rutgers University, The State University of New Jersey, Piscataway, NJ, USA

Synonyms

Mesostructure; Microgeometry; Relief texture; Solid texture; Surface roughness; Volumetric texture

Related Concepts

►Bidirectional Texture Function and 3D Texture; ►Bidirectional Feature Histogram; ►Light Field; ►Texture Recognition

Definition

In the context of computer vision, texture often refers to a variation of image intensity or color, where the variation exhibits some type of repetition. The terms *2D texture* and *3D texture* provide a more precise definition of texture. A *2D texture* may be a color or shade variation such as a paisley print or zebra stripes. A textured surface can also exhibit geometric variations on the surface such as gravel, grass, or any rough surface. This type of texture is termed *3D texture* [1, 2]. Algorithms developed for 2D texture are generally not useful for 3D texture because appearance varies as a function of viewing and illumination direction.

The difference between 2D and 3D texture is readily apparent when considering photometric effects due to illumination direction and geometric effects due to viewing direction. Consider Fig. 1 with four images of the same surface under different surface tilt angles. The surface geometry does not change, but the illumination and viewing direction is different in each of the images. With a 2D texture model, these changes could be misinterpreted as changes in the texture class. As shown in Fig. 2, the appearance of the texture also changes significantly over the surface of a 3D

Bidirectional Texture Function and 3D Texture, Fig. 1 Four images of the same 3D-textured surface. As the surface tilt and illumination direction varies, surface appearance changes

object. The photometry of 3D texture causes shading and shadowing that vary with illumination direction. The geometry of 3D texture causes a variation in foreshortening and occlusions along the imaged surface. Consider Fig. 3 which illustrates oblique viewing of a 3D-textured surface patch. A similar oblique view of a 2D-textured surface patch gives a uniformly compressed or downsampled version of the frontal view. However, for an obliquely viewed 3D-textured surface patch, there is a non uniform resampling of the frontal view. Consequently, some texture features are compressed in the oblique view, while others expand. Computer graphics algorithms for texture mapping traditionally characterize the texture with a single image. To synthesize oblique views, these texture-mapping algorithms apply a uniform resampling which clearly cannot account for the spatially varying foreshortening and occlusions.

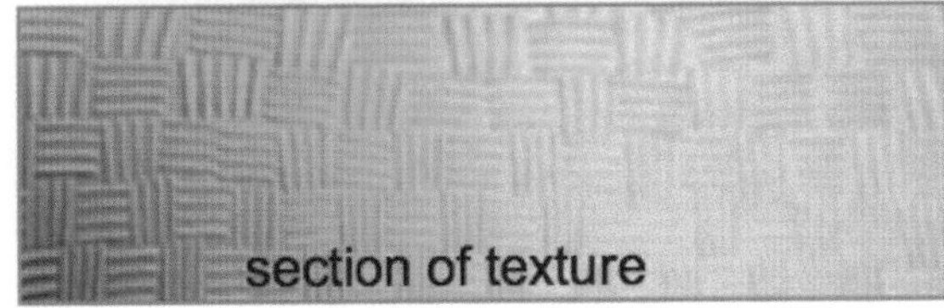

Bidirectional Texture Function and 3D Texture, Fig. 2 The local appearance of a 3D-textured object. Notice that local foreshortening, shadowing, and occlusions change across the texture because of the differences in global surface orientation and illumination direction

Background

Measurement of 3D texture with a bidirectional texture function (BTF) was introduced in [1, 2]. This work created a database of 3D texture called the CUReT database (Columbia-Utrecht Reflectance and Texture database). This publicly available collection of measurements from real-world surfaces served as a starting point for subsequent work in 3D texture.

Theory

Histogram Models for 3D Texture

Numerous texture models for 2D texture have been developed since the early 1970s and are used in areas like texture mapping, texture synthesis, shape-from-texture, and texture classification and segmentation. These representations include co-occurrence matrices, histograms, power spectra, Gabor filter outputs, textons, wavelets, and Markov random fields. In the late 1990s, the emphasis of texture research expanded to include 3D textures. Analytical models of intensity histograms of 3D-textured surface are developed in [3] and [4]. Intensity histograms are a very basic tool to represent texture but are too simple for most computer vision tasks. A standard framework for texture representations consists of a primitive and a statistical distribution (histogram) of this primitive over space. Intensity is the simplest primitive; however, image features are better primitives to characterize the spatial

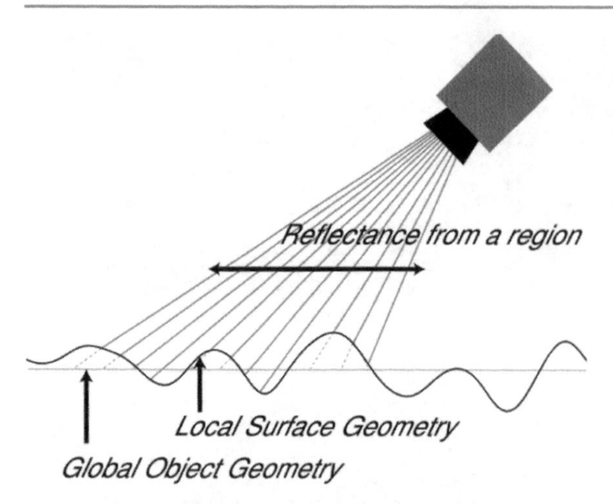

Bidirectional Texture Function and 3D Texture, Fig. 3 Geometry of 3D texture. For oblique views of a 3D-textured surface, the sampling rate of the surface depends on the local geometry

relationship of pixels. In order to account for changes with imaging parameters (view/illumination direction), either the primitive or the statistical distribution should be a function of the imaging parameters. Two methods to represent 3D texture are the bidirectional feature histogram (BFH) [5] and the 3D texton method [6]. In the 3D texton method, the primitive is a function of imaging parameters, while in the BFH method, the histogram is a function of imaging parameters. The advantage of the BFH approach is that there is no need for aligning images obtained under different imaging parameters.

Appearance-Based Representations

In computer vision, the precise surface geometry that comprises a 3D texture is often unknown. Instead, images of the 3D texture, i.e., appearance, are used to represent the textured surface. The term bidirectional texture function (BTF) is the appearance of texture as a function of viewing and illumination direction. BTF is typically captured by imaging the surface at a set of possible viewing and illumination directions. Therefore, measurements of BTF are collections of images. The term BTF was first introduced in [2, 7], and similar terms have since been introduced including BSSRDF [8] and SBRDF (spatial BRDF) [9]. SBRDF has a very similar definition to BTF, i.e., BTF is also a spatially varying BRDF.

BTF measurements can be very large because the measurements typically consist of a high-resolution image for every possible viewing and illumination direction. Dense sampling of the illumination and viewing space results in extremely large datasets to represent the surface. For example, if a 3 Mb image is captured for each of the 100 sampled viewing directions and 100 illumination directions, the resulting dataset is 30 Gb. Compact representations of the BTF are clearly important for efficient storage, rendering, and recognition. Methods for compact representations and compression of the BTF are presented in [10–13].

Geometry-Based Representations from Computer Graphics

In computer vision, image-based representations are standard because the surface geometry is typically unknown. However, in computer graphics, the precise geometry of the 3D-textured surface may be known. Rendering 3D textures using a volumetric representation of surface geometry is a common approach. Many of the rendering packages such as OpenGL and Blender use the term 3D texture to refer to volumetric texture. In this definition, the 3D texture is defined by opacity in a volume, instead of the definition here which refers to a surface height variation. In recent work [14], a volumetric representation is used for texture rendering, where the volume is a stack of semitransparent layers obtained using measured BTF data. Historically, volumetric texture methods are also referred to as solid texturing [15–18].

Application

The main applications for 3D texture representations are in recognition, synthesis, and rendering. While many applications use image-based representations, or assume a known geometric model, other applications need to capture the local geometry of 3D texture. Digital archiving of art is an example of such an application where fine-scale surface detail can enhance geometric models. Often the 3D texture is not easily captured with standard laser scanning devices. For example, in capturing the geometry of sculptures, researchers devised ways to capture high-resolution 3D texture such as tool marks [19–21]. An additional method for capturing high-resolution 3D texture geometry uses a specialized texture camera based on the optical properties of curved mirrors [22]. For texture recognition, 3D texture methods have been used in material recognition [6, 23–25]. One of the popular recognition tasks is the recognition of materials from the CUReT database. Another real-world recognition task is the measurement and recognition of skin texture [26–28]. For texture synthesis and rendering, the main problem is to synthesize the appearance of 3D texture. Several authors have developed methods to synthesize and render 3D-textured surfaces using the BTF representation [29–35]. Other authors used an image-based approach to capture and render complex surfaces by direct photography of the full object under various illumination and viewing directions, simultaneously capturing object shape and surface light fields [36–38]. Synthesizing 3D textures via texture morphing enables creating and rendering novel 3D texture [39]. In computer graphics, 3D-textured surfaces were traditionally rendered by bump mapping [40], albeit with limited realism. In more recent years, several new methods have been developed that can efficiently render 3D texture detail. These include view-dependent displacement mapping [41], relief texture mapping [42], the polynomial texture map [43], and a Blender-based rendering method [44].

References

1. Dana KJ, van Ginneken B, Nayar SK, Koenderink JJ (1997) Reflectance and texture of real world surfaces. In: Proceedings of the IEEE conference on computer vision and pattern recognition (CVPR), San Juan, Puerto Rico, pp 151–157
2. Dana KJ, van Ginneken B, Nayar SK, Koenderink JJ (1999) Reflectance and texture of real world surfaces. ACM Trans Graph 18(1):1–34
3. van Ginneken B, Koenderink JJ, Dana KJ (1999) Texture histograms as a function of irradiation and viewing direction. Int J Comput Vis 31(2–3):169–184
4. Dana KJ, Nayar SK (1998) Histogram model for 3d textures. In: Proceedings of the IEEE conference on computer vision and pattern recognition (CVPR), Santa Barbara, California, pp 618–624
5. Cula OG, Dana KJ (2004) 3D texture recognition using bidirectional feature histograms. Int J Comput Vis 59(1): 33–60
6. Leung T, Malik J (2001) Representing and recognizing the visual appearance of materials using three-dimensional textons. Int J Comput Vis 43(1):29–44
7. Dana KJ, van Ginneken B, Nayar SK, Koenderink JJ (1996) Reflectance and texture of real-world surfaces. Columbia University Technical Report CUCS-048-96
8. Jensen HW, Marschner SR, Levoy M, Hanrahan P (2001) A practical model for subsurface light transport. In: ACM SIGGRAPH, Los Angeles, California, pp 511–518
9. McAllister DK, Lastra AA, Cloward BP, Heidrich W (2002) The spatial bi-directional reflectance distribution function. In: ACM SIGGRAPH 2002 conference abstracts and applications, SIGGRAPH'02. ACM, New York, pp 265–265
10. Cula OG, Dana KJ (2001) Compact representation of bidirectional texture functions. In: Proceedings of the IEEE conference on computer vision and pattern recognition (CVPR), Kauai, Hawaii, vol 1. pp 1041–1067
11. Haindl M, Filip J (2007) Extreme compression and modeling of bidirectional texture function. IEEE Trans Pattern Anal Mach Intell 29(10):1859–1865
12. Ruiters R, Klein R (2009) Btf compression via sparse tensor decomposition. Proc EGSR Comput Graph Forum 28(4):1181–1188
13. Wu H, Dorsey J, Rushmeier H (2011) A sparse parametric mixture model for btf compression, editing and rendering. Proc Eurograp Comput Graph Forum 30(2):465–473
14. Magda S, Kriegman D (2006) Reconstruction of volumetric surface textures for real-time rendering. In: Proceedings of the Eurographics symposium on rendering, Nicosia, Cyprus, pp 19–29
15. Perlin K (1985) An image synthesizer. ACM SIGGRAPH 19(3):287–296
16. Kajiya JT, Kay TL (1989) Rendering fur with 3d textures. ACM SIGGRAPH 23(3):271–80
17. Jagnow R, Dorsey J, Rushmeier H (2004) Stereological techniques for solid textures. In: ACM SIGGRAPH 2004 Papers, SIGGRAPH '04. ACM, New York, pp 329–335
18. Kopf J, Fu CW, Cohen-Or D, Deussen O, Lischinski D, Wong TT (2007) Solid texture synthesis from 2d exemplars. In: ACM SIGGRAPH 2007 papers, SIGGRAPH '07. ACM, New York
19. Bernardini F, Martin IM, Rushmeier H (2001) High-quality texture reconstruction from multiple scans. IEEE Trans Vis Comput Graph 7(4):318–332
20. Bernardini F, Martin I, Mittleman J, Rushmeier H, Taubin G (2002) Building a digital model of michelangelo's florentine pieta. IEEE Comput Graph Appl 1(22):59–67
21. Levoy M, Pulli K, Curless B, Rusinkiewicz S, Koller D, Pereira L, Ginzton M, Anderson S, Davis J, Ginsberg J, Shade J, Fulk D (2000) The digital michelangelo project: 3d scanning of large statues. In: Proceedings of the 27th annual conference on Computer graphics and interactive techniques. SIGGRAPH '00. ACM/Addison-Wesley, New York, pp 131–144
22. Wang J, Dana KJ (2006) Relief texture from specularities. IEEE Trans Pattern Anal Mach Intell 28(3):446–457
23. Varma M, Zisserman A (2002) Classifying images of materials. In: Proceedings of the European Conference on Computer Vision (ECCV), Copenhagen, Denmark, pp 255–271
24. Cula OG, Dana KJ (2001) Recognition methods for 3d textured surfaces. Proc SPIE Conf Human Vis Electron Imaging VI 4299:209–220
25. Chantler M, Petrou M, Penirsche A, Schmidt M, McGunnigle G (2005) Classifying surface texture while

simultaneously estimating illumination direction. Int J Comput Vis 62(1–2):83–96
26. Cula O, Dana K, Murphy F, Rao B (2004) Bidirectional imaging and modeling of skin texture. IEEE Trans Biomed Eng 51(12):2148–2159
27. Cula OG, Dana KJ, Murphy FP, Rao BK (2005) Skin texture modeling. Int J Comput Vis 62(1/2):97–119
28. Weyrich T, Matusik W, Pfister H, Bickel B, Donner C, Tu C, McAndless J, Lee J, Ngan A, Jensen HW, Gross M (2006) Analysis of human faces using a measurement-based skin reflectance model. ACM Trans Graph 25:1013–1024
29. Liu X, Yu Y, Shum H (2001) Synthesizing bidirectional texture functions for real world surfaces. In: ACM SIGGRAPH, Los Angeles, pp 97–106
30. Tong X, Zhang J, Liu L, Wang X, Guo B, Shum H (2002) Synthesis of bidirectional texture functions on arbitrary surfaces. ACM Trans Graph 21(3):665–672
31. Koudelka ML, Magda S, Belhumeur PN, Kriegman DJ (2003) Acquisition, compression and synthesis of bidirectional texture functions. In: 3rd international workshop on texture analysis and synthesis (Texture 2003), Nice, France, pp 59–64
32. Meseth J, Muller G, Klein R (2003) Preserving realism in real- time rendering of bidirectional texture functions. In: Proceedings of the openSG symposium, Darmstadt, Germany, pp 89–96
33. Vasilescu MAO, Terzopoulos D (2004) Tensortextures: multilinear image-based rendering. In: ACM SIGGRAPH 2004 Papers, SIGGRAPH '04. ACM, New York, pp 336–342
34. Wang J, Tong X, Chen Y, Guo B, Shum H, Snyder J (2005) Capturing and rendering geometry details for btf-mapped surfaces. Vis Comput 21:559–568
35. Kautz J, Boulos S, Durand F (2007) Interactive editing and modeling of bidirectional texture functions. In: ACM SIGGRAPH 2007 papers, SIGGRAPH'07. ACM, New York
36. Matusik W, Pfister H, Ngan A, Beardsley P, Ziegler R, McMillan L (2002) Image-based 3d photography using opacity hulls. In: Proceedings of the 29th annual conference on Computer graphics and interactive techniques. SIGGRAPH'02. ACM, New York, pp 427–437
37. Debevec P, Hawkins T, Tchou C, Duiker H, Sarokin W, Sagar M (2002) Acquiring the reflectance field of a human face. In: ACM SIGGRAPH, San Antonio, pp 145–156
38. Nishino K, Sato Y, Ikeuchi K (2001) Eigentexture method: Appearance compression and synthesis based on a 3d model. IEEE Trans Pattern Anal Mach Intell 23(11): 1257–1265
39. Matusik W, Zwicker M, Durand F (2005) Texture design using a simplicial complex of morphable textures. ACM Trans Graph 25(3):784–794
40. Blinn JF (1978) Simulation of wrinkled surfaces. ACM SIGGRAPH 12(3):286–292
41. Wang L, Wang X, Tong X, Lin S, Hu S, Guo B, Shum H (2003) View-dependent displacement mapping. ACM Trans Graph 22(3):334–339
42. Oliveira MM, Bishop G, McAllister D (2000) Relief texture mapping. In: Proceedings of the 27th annual conference on Computer graphics and interactive techniques, SIGGRAPH'00. ACM/Addison-Wesley, New York, pp 359–368
43. Malzbender T, Gelb D, Wolters H (2001) Polynomial texture maps. In: Proceedings of the 28th annual conference on Computer graphics and interactive techniques, SIGGRAPH'01. ACM, New York, pp 519–528
44. Hatka M, Haindl M (2011) Btf rendering in blender. In: Proceedings of the 10th International Conference on Virtual Reality Continuum and Its Applications in Industry, VRCAI'11. ACM, New York, pp 265–272

Blackbody Radiation

Rajeev Ramanath[1] and Mark S. Drew[2]
[1]DLP® Products, Texas Instruments Incorporated, Plano, TX, USA
[2]School of Computing Science, Simon Fraser University, Vancouver, BC, Canada

Synonyms

Thermal Radiator

Related Concepts

▸Planckian Locus

Definition

A blackbody is an idealized object that absorbs all electromagnetic radiation incident on it. The absorbed energy is incandescently emitted with a spectral power distribution that is a function of its temperature. It is called a blackbody partly because it appears black to the human observer when it is cold, as it emits mostly infrared radiation.

Background

Blackbody radiation, in general, stood as a major challenge to the scientists in the nineteenth century as they were pushing the limits of classical physics. Several physicists studied blackbody radiators, including Lord Rayleigh, James Jeans, Josef Stefan, Gustav Kirchhoff, Ludwig Boltzmann, Wilhelm Wien, and finally, Max Planck, who arguably broke the way for quantum physics.

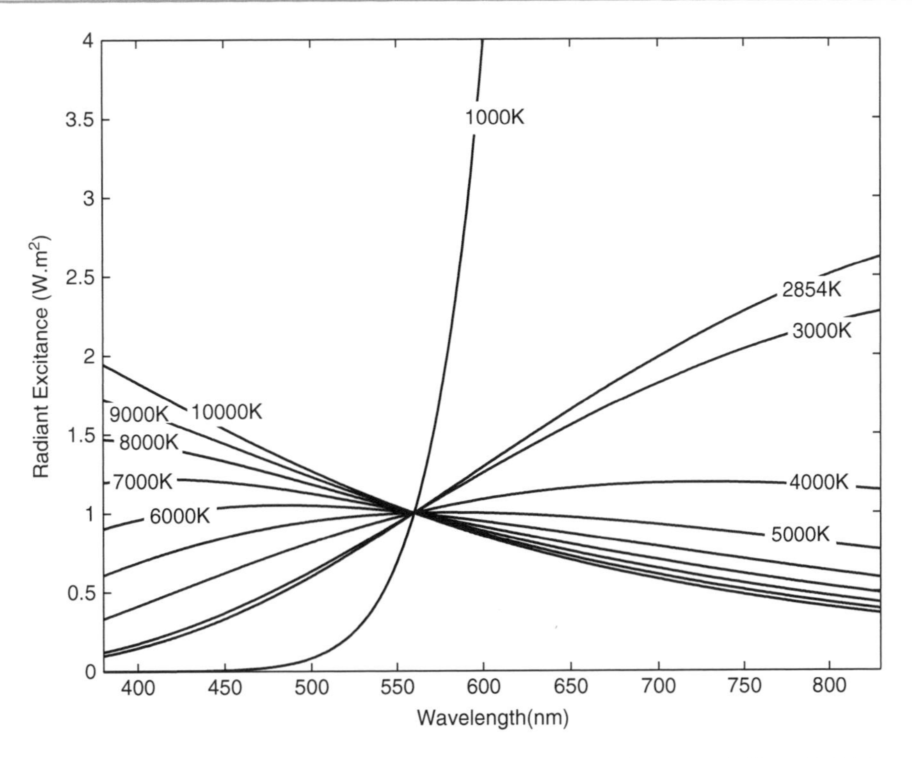

Blackbody Radiation, Fig. 1 Spectral power distributions of various blackbody radiators from 1,000 to 10,000 K, with all spectral distributions normalized to unity at 560 nm

Wien's approximation is used to approximately describe the spectral content of radiation from a blackbody. This approximation was first derived by Wilhelm Wien in 1893 [8]. This law was found to be accurate only for short wavelengths of emission spectra of blackbody radiators and is used today only as an approximation. Wien stated that the radiant excitance of a blackbody may be given by

$$M_e(\lambda, T) = \frac{2\pi h \nu^3}{c^2} \exp -\frac{h\nu}{kT} \tag{1}$$

$$= \frac{2\pi h c^2}{\lambda^5} \exp -\frac{hc}{kT\lambda} \tag{2}$$

The above equation holds in the specific case when $\exp -\frac{hc}{kT\lambda} \gg 1$, which typically occurs when the wavelength is short. The work by Wien was soon replaced by the findings of Max Planck, in 1901.

Another important finding regarding blackbody radiators was made by Josef Stefan in 1879 and later formalized by Ludwig Boltzmann. Known as the *Stefan-Boltzmann law*, it states that the total radiant excitance of a blackbody radiator is proportional to the fourth power of its absolute temperature. In other words,

$$\int_\lambda M_e(\lambda, T) = \sigma T^4, \tag{3}$$

where $\sigma = 5.674 \times 10^{-8} W \cdot \text{m}^{-1} \cdot \text{K}^{-4}$ is known as the Stefan-Boltzmann constant.

Max Planck, in 1901, stated a more general law, known as *Planck's radiation law*. This describes the spectral distribution of radiant excitance M_e as a function of wavelength (These equations are often given in terms of frequencies instead of wavelength, as shown in this article. This is easily converted back and forth using $\nu \cdot \lambda = c$, where ν denotes frequency, λ the wavelength, and c the speed of light in vacuum.) λ and temperature T and is given by

$$M_e(\lambda, T) = \frac{c_1}{\lambda^5 \left[\exp\left(\frac{c_2}{\lambda T}\right) - 1 \right]} \tag{4}$$

where $c_1 = 2\pi h c^2 = 3.74183 \times 10^{-16} W \cdot \text{m}^{-2}$, $c_2 = h \cdot c/k = 1.4388 \times 10^{-2} \text{m} \cdot \text{K}$ (c is the speed of light in vacuum: $2.99792458 \times 10^8 \text{m} \cdot \text{s}^{-1}$, h is Planck's constant: $6.62606896 \times 10^{-34} \text{J} \cdot \text{s}$, k is Boltzmann's constant: $1.38065 \times 10^{-23} \text{J} \cdot \text{K}^{-1}$) [4], and the excitance is defined in units of $W \cdot \text{m}^{-3}$ [3, 5, 6, 9].

Figure 1 shows the spectral power distribution of various blackbody radiators from 1,000 to 10,000 K, with all spectral power distributions normalized to unity at 560 nm. As the blackbody gets hotter (T increases), one can see that the red content in the spectrum reduces and the blue content increases – an indication of the color as would be seen by the human observer. Both Wien's approximation and the Stefan-Boltzmann law can be derived from Planck's law.

Wien's displacement law may be derived by differentiating the equation for Planck's law Eq. (4) with respect to wavelength λ and equating the result to zero to find the maximum. It states that the spectral distribution of the spectral excitance of a blackbody reaches a maximum at a wavelength λ_m and that the product of the this maximum wavelength and the temperature of a blackbody is a constant, given by:

$$\lambda_m \cdot T = \frac{h \cdot c}{4.965114\, k} = 2.8977685 \times 10^{-3} \qquad (5)$$

where h denotes the Planck constant, c denotes the speed of light in vacuum, and k denotes the Boltzmann constant. The corresponding value of the spectral excitance is given by:

$$M_{e\lambda_m} = T^5\, 1.286673 \times 10^{-5} W \text{m}^{-3} \qquad (6)$$

Blackbody radiators are the select few sources of illumination that match "standard illuminant" spectral power distributions – this may be seen as valid in the case of the equivalence of standard illuminant "A" and a blackbody with a temperature 2,856 K.

References

1. 1998c CIE (1998) CIE standard illuminants for colorimetry. CIE, Vienna. Also published as ISO 10526/CIE/S006/E1999
2. CIE 15:2004 (2004) Colorimetry. CIE, Vienna
3. Longair MS (1984) Theoretical concepts in physics: an alternative view of theoretical reasoning in physics. Cambridge University Press, Cambridge
4. Mohr PJ, Taylor BN, Newell DB (2007) CODATA recommended values of the fundamental physical constants:2006. National Institute of Standards and Technology, Gaithersburg
5. Planck M (1901) On the law of distribution of energy in the normal spectrum. Annalen der Physik 4:553
6. Rybicki G, Lightman AP (1985) Radiative processes in astrophysics. Wiley-Interscience, New York
7. Wien W (1893) Situngsber. d. Berliner Akad, pp 55–62
8. Wien W (1896) Über die energievertheilung im emissionsspectrum eines schwaarzen körpers. Annalen der Physik 58:662–669
9. Wyszecki G, Stiles WS (1982) Color science: concepts and methods, quantitative data and formulas, 2nd edn. Wiley, New York

Blackbody Radiator

►Planckian Locus

Blind Deconvolution

Tom Bishop[1] and Paolo Favaro[2]
[1]Image Algorithms Engineer, Apple, Cupertino, CA, USA
[2]Department of Computer Science and Applied Mathematics, Universität Bern, Switzerland

Synonyms

Deblurring; Deconvolution; Kernel estimation; Motion deblurring; PSF estimation

Related Concepts

►Denoising; ►Image-Based Modeling; ►Inpainting

Definition

Blind image deconvolution is the problem of recovering a sharp image (such as that captured by an ideal pinhole camera) from a blurred and noisy one, without exact knowledge of how the image was blurred. The unknown blurring operation may result from camera motion, scene motion, defocus, or other optical aberrations.

Background

A correct photographic exposure requires a trade-off in exposure time and aperture setting. When illumination is poor, the photographer can choose to use a long exposure time or a large aperture. The first setting results in motion blur when the camera moves relative to objects in the scene during the exposure. The second setting results in out-of-focus blur for objects at depths away from the focal plane. Furthermore, these effects may be exacerbated by the user due to camera shake, incorrect focus settings, or other distortions such as atmospheric turbulence.

Under local approximations, these processes can be modeled as convolution operations between an ideal "sharp" image and a *kernel* or point-spread function (PSF). This (PSF) represents how a point of light in the scene would be imaged onto the camera's sensor. In general the (PSF) may be space-varying or depth-varying, such that each point in 3D space has its own response in the kernel function. In general, the effect of the kernel is to blur and remove information from the image (see Fig. 1).

When the (PSF) is known, deconvolution algorithms can be used to remove the effect of these degradations. Deconvolution may be performed using direct (e.g., Fourier-based) or iterative (e.g., gradient descent or conjugate gradient based) algorithms. Essentially, a large linear system must be inverted to recover the sharp image, and depending on the conditioning of the matrix representing the blurring, the solution may be obtained to a greater or lesser accuracy. Observation noise also hinders exact invertibility, and for these reasons *regularization* of the solution is required. Such regularization typically imparts prior knowledge about the expected statistics of the sharp image, such as smoothness, sparseness of its gradients, or compressibility in some domain, and is typically key to obtaining well-behaved solutions.

In practice, the PSF is rarely known from calibration, and in a practical scenario it must be estimated from the blurred image itself. While many algorithms have been proposed to tackle the blind deconvolution problem with some success, a universal solution is not yet available and it is still an active area of research. The difficulty owes in part to the dimensionality of the unknowns in the problem, and its extreme ill-posedness, with the potential for multiple local solutions arising from non-convex optimization problems. Progress has been made in both priors to describe images and blurs and better constrain the solution and estimation methods to better approximate the intractable inference problems.

Related methods also exist to recover the sharp image from multiple observations, for example, a blurred image and a sharp but noisy image, or multiple blurred images, in which case the problem is much more well posed and the solution is more readily obtained; this is also closely related to the problem of super-resolution where the sharp image is estimated at a higher resolution than the input images.

Theory

Imaging Model

The following linear spatially varying (LSV) observation model represents the formation of a general blurred image on the camera sensor:

$$g(\boldsymbol{x}) = \sum_{\substack{\boldsymbol{s}\in\mathcal{S}_f,\\ (\boldsymbol{x},\boldsymbol{s})\in\mathcal{S}_H}} h(\boldsymbol{x},\boldsymbol{s}) f(\boldsymbol{s}) + w(\boldsymbol{x}), \qquad \boldsymbol{x}\in\mathcal{S}_g, \tag{1}$$

where $f(\boldsymbol{s})$ is the *true* or *sharp* image, $g(\boldsymbol{x})$ the observed blurred image, $w(\boldsymbol{x})$ is additive noise, and $h(\boldsymbol{x},\boldsymbol{s})$ is the kernel function describing the blur. The position $\boldsymbol{x}$ lies in the blurred image support $\mathcal{S}_g \subset \mathbb{R}^2$, while the position $\boldsymbol{s}$ lies in the true image support $\mathcal{S}_f \subset \mathbb{R}^2$. Notice that $\mathcal{S}_H \subset \mathbb{R}^4$ denotes the support of the kernel h. In the case the blur does not vary with position $\boldsymbol{s}$ in the true image and the depth is constant, the kernel may be reduced to a stationary PSF, $h(\boldsymbol{x},\boldsymbol{s}) = h(\boldsymbol{x}-\boldsymbol{s})$. Then the image model becomes a convolution:

$$g(\boldsymbol{x}) = [h \star f](\boldsymbol{x}) + w(\boldsymbol{x}) \tag{2}$$

With discretization and appropriate *lexicographic ordering*, or raster scanning of the images into vectors, either (1) or (2) may also be expressed in matrix-vector form:

$$\boldsymbol{g} = \boldsymbol{H}\boldsymbol{f} + \boldsymbol{w}, \qquad \text{or equivalently,} \tag{3}$$

$$\boldsymbol{g} = \boldsymbol{F}\boldsymbol{h} + \boldsymbol{w}. \tag{4}$$

a
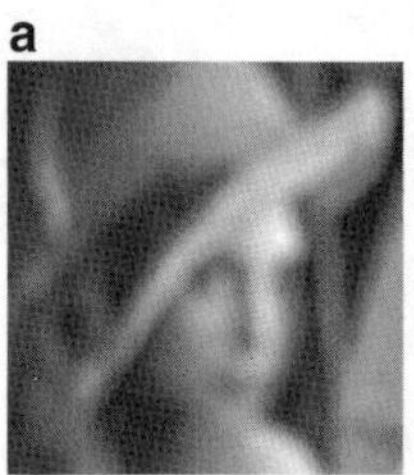
Photographed out-of-focus image (on a plane)

b
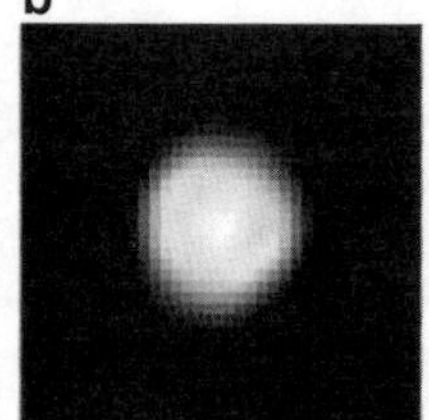
Estimated PSF

c
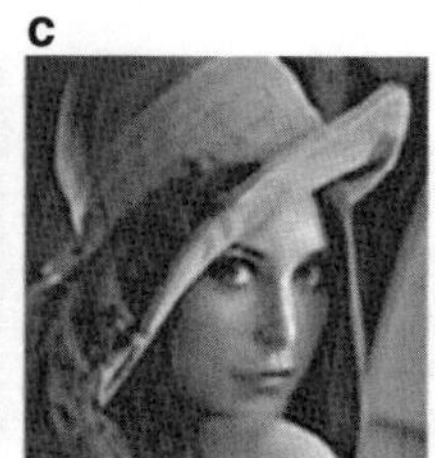
Restored image

d

Actual in-focus image for comparison

Blind Deconvolution, Fig. 1 Example of blind deconvolution using the method in [1] where a single image on a plane has been captured by a defocused camera. (**a**) Photographed out-of-focus image (on a plane). (**b**) Estimated PSF. (**c**) Restored image. (**d**) Actual in-focus image for comparison

With the spatially invariant degradation model, the matrices $\boldsymbol{F}$ and $\boldsymbol{H}$ acquire a special structured block form, termed block Toeplitz with Toeplitz blocks (BTTB), with constant block entries on each block diagonal and constant diagonals within each block. Sometimes these matrices are approximated as circulant, implying circular convolution of the sharp image, and then they may be diagonalized using the discrete Fourier transform (DFT), enabling fast calculations to be performed.

Probabilistic Formulation

With uncertainty in the observation model (3), it is natural to estimate the most likely solution for the sharp image using a probabilistic approach. The Bayesian framework provides a unifying way to tackle such ill-posed inverse problems. Here a *likelihood* $\mathrm{p}\,(\boldsymbol{g}\,|\,\cdot)$ is specified from the imaging model and combined with a *prior* $\mathrm{p}\,(\boldsymbol{f}, \boldsymbol{h}\,|\,\cdot)$ on the image and blur to be estimated, ensuring that only reasonable solutions are obtained. The resulting posterior distribution

$$\mathrm{p}\,(\boldsymbol{f}, \boldsymbol{h}\,|\,\boldsymbol{g}, \boldsymbol{\Omega}\,) = \frac{\mathrm{p}\,(\boldsymbol{g}\,|\,\boldsymbol{f}, \boldsymbol{h}, \boldsymbol{\Omega}\,)\,\mathrm{p}\,(\boldsymbol{f}\,|\,\boldsymbol{\Omega}\,)\,\mathrm{p}\,(\boldsymbol{h}\,|\,\boldsymbol{\Omega}\,)}{\mathrm{p}\,(\boldsymbol{g})} \tag{5}$$

is used for inference of the unknowns, where $\boldsymbol{\Omega}$ denotes *hyperparameters* of the model, such as noise variances or regularization parameters; these are usually considered known, but correctly estimating them is often critical for accurate blind deconvolution. The additive noise $\boldsymbol{w}$ is commonly assumed to be Gaussian or sometimes Poisson distributed. In the independent white Gaussian noise (WGN) case with variance σ_w, the distribution is $\mathrm{p}_W\,(\boldsymbol{w}) = \mathcal{N}\,(\boldsymbol{w}\,|\,0, \sigma_w \boldsymbol{I}\,)$. Thus, the likelihood of $\boldsymbol{g}$ conditioned on $\boldsymbol{h}, \boldsymbol{f}$ is given by

$$\begin{aligned}\mathrm{p}_G\,(\boldsymbol{g}\,|\,\boldsymbol{f}, \boldsymbol{h}, \sigma_w) &= \mathrm{p}_W\,(\boldsymbol{g} - \boldsymbol{H}\boldsymbol{f}) \\ &= (2\pi\sigma_w)^{-\frac{L_g}{2}} \exp\left[-\frac{1}{2\sigma_w^2}\|\boldsymbol{g} - \boldsymbol{H}\boldsymbol{f}\|^2\right],\end{aligned} \tag{6}$$

where L_g is the size of the vector $\boldsymbol{g}$.

Bayesian Inference Methods

Due to the complexity of the chosen prior models, it may not be possible to obtain an exact analytic solution of Eq. (5). A common approximation is to compute a point estimate of the unknowns $\boldsymbol{f}$ and $\boldsymbol{h}$ via an optimization procedure. However, these estimates work well only with highly peaked distributions. With more uncertainty in the parameters, it is better to estimate the whole parameters *distribution*. Unfortunately, strategies that do so in the Bayesian framework are typically more computationally demanding [2–4]. Finally, as the estimated posterior distribution of the parameters must have a finite representation, further approximations or simulations must be introduced.

Maximum A Posteriori and Maximum Likelihood

The maximum a posteriori (MAP) solution is one common point estimate where it is possible to prescribe our prior knowledge about the unknowns. It is defined as the values $\hat{\boldsymbol{f}}, \hat{\boldsymbol{h}}$, and $\hat{\boldsymbol{\Omega}}$ that maximize the posterior probability density:

$$\begin{aligned}&\{\hat{\boldsymbol{f}}, \hat{\boldsymbol{h}}, \hat{\boldsymbol{\Omega}}\}_{\mathrm{MAP}} \\ &\quad = \underset{\boldsymbol{f}, \boldsymbol{h}, \boldsymbol{\Omega}}{\operatorname{argmax}}\, \mathrm{p}\,(\boldsymbol{g}\,|\,\boldsymbol{f}, \boldsymbol{h}, \boldsymbol{\Omega}\,)\,\mathrm{p}\,(\boldsymbol{f}\,|\,\boldsymbol{\Omega}\,)\,\mathrm{p}\,(\boldsymbol{h}\,|\,\boldsymbol{\Omega}\,)\,\mathrm{p}\,(\boldsymbol{\Omega}).\end{aligned} \tag{7}$$

A very related method is *maximum likelihood (ML)*, where one looks for

$$\{\hat{f}, \hat{\boldsymbol{h}}, \hat{\boldsymbol{\Omega}}\}_{\mathrm{ML}} = \underset{f, \boldsymbol{h}, \boldsymbol{\Omega}}{\operatorname{argmax}}\ \mathrm{p}\,(\boldsymbol{g} \mid f, \boldsymbol{h}, \boldsymbol{\Omega})\,. \tag{8}$$

Notice that the *maximum likelihood (ML)* method is essentially the *maximum a posteriori (MAP)* method where the prior distributions are uniform (uninformative). Despite this equivalence, the *maximum likelihood (ML)* is usually referred to as a non-Bayesian method. The advantage of using the *maximum a posteriori (MAP)* approach is that we can also encode the case where parameters are entirely or partly known by using degenerate distributions (Dirac deltas), that is, $\mathrm{p}\,(\boldsymbol{\Omega}) = \delta\,(\boldsymbol{\Omega} - \boldsymbol{\Omega}_0)$. Then, the *MAP* and *ML* formulations become, respectively,

$$\{\hat{f}, \hat{\boldsymbol{h}}\}_{\mathrm{MAP}} = \underset{f, h}{\operatorname{argmax}}\ \mathrm{p}\,(\boldsymbol{g} \mid f, \boldsymbol{h}, \boldsymbol{\Omega}_0)\,\mathrm{p}\,(\boldsymbol{f} \mid \boldsymbol{\Omega}_0)\,\mathrm{p}\,(\boldsymbol{h} \mid \boldsymbol{\Omega}_0) \tag{9}$$

$$\{\hat{f}, \hat{\boldsymbol{h}}\}_{\mathrm{ML}} = \underset{f, h}{\operatorname{argmax}}\ \mathrm{p}\,(\boldsymbol{g} \mid f, \boldsymbol{h}, \boldsymbol{\Omega}_0)\,. \tag{10}$$

The Bayesian framework can be used to describe several deconvolution methods and to emphasize their differences in terms of choice of likelihood, priors on the image, blur, and hyperparameters. Further differences can be found in how the maximization problem is solved.

A typical example of methods that can be formulated in the Bayesian framework is regularized approaches based on the L_2 norm. One such method is Tikhonov regularization, where a linear system is solved in least-squares sense by introducing an additional L_2 constraint on the unknowns. More in general, the blind deconvolution task is formulated as a constrained minimization where several regularization constraint terms are added.

A common choice is to always use a term in the form of $\|\boldsymbol{g} - \boldsymbol{H} \boldsymbol{f}\|^2$, called *data fidelity term*. The additional regularization terms encode the constraints on the unknowns. For example, one may want to impose smoothness of the image and the blur. To do so, a term that penalizes small variations of the image and the blur can be used. As the solution will be a trade-off between the data fidelity term and the regularization terms, regularization parameters are used to adjust their weight.

An important example that illustrates this procedure is [5]. In that work the classical regularized image deconvolution formulation [6, 7] was extended to the blind image deconvolution (BID) case by adding regularization on the blur parameters. The problem is formulated as

$$\hat{f}, \hat{\boldsymbol{h}} = \underset{f, h}{\operatorname{argmin}} \Big[\|\boldsymbol{g} - \boldsymbol{H} \boldsymbol{f}\|^2_{Q_w^{-1}} + \lambda_1 \|\boldsymbol{L}_f \boldsymbol{f}\|^2 + \lambda_2 \|\boldsymbol{L}_h \boldsymbol{h}\|^2\Big], \tag{11}$$

where λ_1 and λ_2 are the Lagrange multipliers for each constraint, and $\boldsymbol{L}_f$ and $\boldsymbol{L}_h$ are the regularization operators. To avoid oversmoothing the edges each $\boldsymbol{L}$ operator is the Laplacian multiplied by space-varying weights. These weights are obtained from the local image variance as in [7–10].

Alternating Minimization or Iterated Conditional Modes

One of the main difficulties in MAP is to simultaneously recovering both $\boldsymbol{f}$ and $\boldsymbol{h}$. The problem can be already observed in the image model Eq. (1), which is bilinear in both $\boldsymbol{f}$ and $\boldsymbol{h}$. One common way to address this challenge is alternating minimization (AM). When applied to Eq. (11), it performs the minimization by working on one variable at a time while the others are fixed. As a result, the minimization of Eq. (11) becomes a sequence of linear problems. A related method is iterated conditional modes (ICM) proposed by Besag [11].

Minimum Mean-Squared Error

As mentioned earlier on, the MAP estimate is only a point estimate of the whole posterior probability density function (PDF). While this is not a problem when the posterior is highly peaked about the maximum, in the case of high observation noise or a broad (heavy-tailed) posterior, this estimate is likely to be unreliable (as chances of obtaining values that are different from the maximum are very likely). Indeed, in a high-dimensional Gaussian distribution, most of the probability *mass* lies away from the probability *density* peak [12].

One way to correct for this shortcoming is to use the minimum mean-squared error (MMSE) estimate.

The rationale is to find the optimal parameter values as those that minimize the expected mean-squared error between the estimates and the true values. This requires to compute the mean value of $\mathrm{p}(\boldsymbol{f}, \boldsymbol{h}, \boldsymbol{\Omega}|\boldsymbol{g})$. However, computing MMSE estimates analytically is generally difficult. A more practical solution is to use sampling-based methods (see next paragraph).

Markov Chain Monte Carlo Sampling

A general technique to perform inference is to simulate the posterior distribution in Eq. (5), by drawing samples. Provided that we have obtained enough independent samples, this strategy allows us to deal with arbitrarily complex models in high-dimensional spaces, where no analytic solution is available. Markov chain Monte Carlo (MCMC) methods approximate the posterior distribution by the statistics of samples generated from a Markov chain. Widely used Markov chain Monte Carlo (MCMC) algorithms are the Metropolis-Hastings or Gibbs Samplers (see, e.g., [3, 13–15]).

The samples can then be used in Monte Carlo integration to obtain point estimates or other distribution statistics. For instance, in the BID problem the MMSE estimate of the $\boldsymbol{f}$ can be readily obtained by taking the mean of the samples, $\frac{1}{n}\sum_{t=1}^{n}\boldsymbol{f}^{(t)}$.

Markov chain Monte Carlo (MCMC) can provide better solutions than AM or any other method. However, there are some limitations. First, they are very computationally intensive in comparison to the point estimate methods. Second, convergence to the posterior can be theoretically guaranteed, but in practice it can be hard to tell when this has occurred, and it may require a long time to explore the parameter space.

Marginalizing Hidden Variables

In the discussion so far the aim was to recover all the unknowns. However, in most cases one is interested in recovering only the sharp image $\boldsymbol{f}$. This leads to another approach to the BID problem where undesired unknowns are marginalized and inference is performed on the remaining variables (i.e., a subset of $\boldsymbol{f}, \boldsymbol{h}$, and $\boldsymbol{\Omega}$). Therefore, one can approach the BID inference problem in two steps. First, one can calculate

$$\hat{\boldsymbol{h}}, \hat{\boldsymbol{\Omega}} = \underset{\boldsymbol{h},\boldsymbol{\Omega}}{\operatorname{argmax}} \int_{\boldsymbol{f}} \mathrm{p}(\boldsymbol{g}|\boldsymbol{\Omega}, \boldsymbol{f}, \boldsymbol{h})\mathrm{p}(\boldsymbol{f}, \boldsymbol{h}|\boldsymbol{\Omega})\mathrm{p}(\boldsymbol{\Omega})\mathrm{d}\boldsymbol{f} \tag{12}$$

and, second, one selects the sharp image

$$\hat{\boldsymbol{f}}\Big|_{\hat{\boldsymbol{h}},\hat{\boldsymbol{\Omega}}} = \underset{\boldsymbol{f}}{\operatorname{argmax}}\, \mathrm{p}(\boldsymbol{g}|\hat{\boldsymbol{\Omega}}, \boldsymbol{f}, \hat{\boldsymbol{h}})\mathrm{p}(\boldsymbol{f}|\hat{\boldsymbol{\Omega}}). \tag{13}$$

Alternatively, one can also marginalize $\boldsymbol{h}$ and $\boldsymbol{\Omega}$ to obtain

$$\hat{\boldsymbol{f}} = \underset{\boldsymbol{f}}{\operatorname{argmax}} \int_{\boldsymbol{h},\boldsymbol{\Omega}} \mathrm{p}(\boldsymbol{g}|\boldsymbol{\Omega}, \boldsymbol{f}, \boldsymbol{h})\mathrm{p}(\boldsymbol{f}, \boldsymbol{h}|\boldsymbol{\Omega})\mathrm{p}(\boldsymbol{\Omega})\mathrm{d}\boldsymbol{h}\cdot \mathrm{d}\boldsymbol{\Omega}. \tag{14}$$

Deconvolution Under a Gaussian Prior (MAP)

If the point-spread function (PSF) is known and assuming a Gaussian prior $\mathrm{p}\left(\boldsymbol{f} \mid \boldsymbol{\Sigma}_f\right) = \mathcal{N}\left(\boldsymbol{f} \mid 0, \boldsymbol{\Sigma}_f\right)$ for $\boldsymbol{f}$, with a given covariance matrix $\boldsymbol{\Sigma}_f$, the posterior for $\boldsymbol{f}$ is found as

$$\mathrm{p}(\boldsymbol{f} \mid \boldsymbol{g}) \propto \mathrm{p}(\boldsymbol{g} \mid \boldsymbol{f})\mathrm{p}(\boldsymbol{f} \mid \boldsymbol{\Sigma}_f) \tag{15}$$

$$\propto \exp\Big(-\frac{1}{2}\big[\boldsymbol{f}^T(\sigma_w^{-2}\boldsymbol{H}^T\boldsymbol{H} + \boldsymbol{\Sigma}_f^{-1})\, \boldsymbol{f} - 2\boldsymbol{f}^T(\sigma_w^{-2}\boldsymbol{H}^T\boldsymbol{g}) + \sigma_w^{-2}\boldsymbol{g}^T\boldsymbol{g}\big]\Big) \tag{16}$$

which is a Gaussian

$$\mathrm{p}(\boldsymbol{f} \mid \boldsymbol{g}) \propto \mathcal{N}\left(\boldsymbol{f} \,\middle|\, \boldsymbol{\mu}_{\hat{f}}, \boldsymbol{\Sigma}_{\hat{f}}\right) \tag{17}$$

$$\propto \exp\Big(-\frac{1}{2}\big[\boldsymbol{f}^T\boldsymbol{\Sigma}_{\hat{f}}^{-1}\boldsymbol{f} - 2\boldsymbol{f}^T\boldsymbol{\Sigma}_{\hat{f}}^{-1}\boldsymbol{\mu}_{\hat{f}} + \boldsymbol{\mu}_{\hat{f}}^T\boldsymbol{\Sigma}_{\hat{f}}^{-1}\boldsymbol{\mu}_{\hat{f}}\big]\Big). \tag{18}$$

By comparison of (16) and (18), the parameters are given as

$$\boldsymbol{\Sigma}_{\hat{f}}^{-1} = \sigma_w^{-2}\boldsymbol{H}^T\boldsymbol{H} + \boldsymbol{\Sigma}_f^{-1} \qquad \boldsymbol{\mu}_{\hat{f}} = \boldsymbol{\Sigma}_{\hat{f}}\left(\sigma_w^{-2}\boldsymbol{H}^T\boldsymbol{g}\right). \tag{19}$$

The mean of this distribution, which is also the maximum, is just

$$\hat{\boldsymbol{f}} = \boldsymbol{\mu}_{\hat{f}} = \left(\boldsymbol{H}^T\boldsymbol{H} + \sigma_w^2\boldsymbol{\Sigma}_f^{-1}\right)^{-1}\boldsymbol{H}^T\boldsymbol{g}. \tag{20}$$

In practice, as solving the above equation involves inverting a large linear system, one employs iterative methods.

Application

Blind deconvolution methods are commonly used to restore images that have been distorted by motion blur, out-of-focus blur, and turbulence. As these methods provide an estimate of blur, other uses include digital refocusing, that is, digitally changing the focus setting of a camera after the snapshot, changing the camera bokeh, and obtaining a 3D model of the scene (from the out-of-focus blur).

Acronyms

PSF point-spread function
MC Markov chain
MCMC Markov chain Monte Carlo
ICM iterated conditional modes
PDF probability density function
ML maximum likelihood
MAP maximum a posteriori
AM alternating minimization
BID blind image deconvolution
MMSE minimum mean-squared error

References

1. Bishop TE, Molina R, Hopgood JR (2008) Blind restoration of blurred photographs via AR modelling and MCMC. In: IEEE international conference on image processing (ICIP), San Diego
2. Gelman A, Carlin JB, Stern HS, Rubin DB (2004) Bayesian data analysis, 2nd edn. Chapman & Hall, London
3. Neal RM (1993) Probabilistic inference using Markov chain Monte Carlo methods. Technical report CRG-TR-93-1, Department of Computer Science, University of Toronto, University of Toronto available online at http://www.cs.toronto.edu/~radford/res-mcmc.html
4. Jordan MI, Ghahramani Z, Jaakola TS, Saul LK (1998) An introduction to variational methods for graphical models. Machine Learning, Kluwer Academic Publishers Hingham, MA, USA, Bari, Italy, 37(2):183–233
5. You YL, Kaveh M (1996) A regularization approach to joint blur identification and image restoration. IEEE Trans Image Process 5(3):416–428
6. Lagendijk RL, Biemond J, Boekee DE (1988) Regularized iterative image restoration with ringing reduction. IEEE Trans Acoust Speech Signal Process 36(12):1874–1887
7. Katsaggelos AK (1985) Iterative image restoration algorithms. PhD thesis, Georgia Institute of Technology, School of Electrical Engineering, Bombai, India
8. Katsaggelos AK, Biemond J, Schafer RW, Mersereau RM (1991) A regularized iterative image restoration algorithm. IEEE Trans Signal Process, Louisville, Kentucky, USA, 39(4):914–929
9. Efstratiadis SN, Katsaggelos AK (1999) Adaptive iterative image restoration with reduced computational load. Machine Learning - The Eleventh Annual Conference on computational Learning Theory archive, Kluwer Academic Publishers Hingham, MA, USA, 37(3):297–336
10. Kang MG, Katsaggelos AK (1995) General choice of the regularization functional in regularized image restoration. IEEE Trans Image Process 4(5):594–602
11. Besag J (1986) On the statistical analysis of dirty pictures. J R Stat Soc B 48(3):259–302
12. Molina R, Katsaggelos AK, Mateos J (1999) Bayesian and regularization methods for hyperparameter estimation in image restoration. IEEE Trans Image Process 8(2):231–246
13. Andrieu C, de Freitras N, Doucet A, Jordan M (2003) An introduction to MCMC for machine learning. Mach Learn 50:5–43
14. Gilks W, Richardson S, Spiegelhalter D (eds) (1995) Markov chain Monte Carlo in practice: interdisciplinary statistics. Machine Learning, Kluwer Academic Publishers Hingham, MA, USA, 37(2):183–233
15. Ó Ruanaidh JJ, Fitzgerald W (1996) Numerical Bayesian methods applied to signal processing, 1st edn. Springer series in statistics and computing. Springer, New York. ISBN:0-387-94629-2

Blur Estimation

Yu-Wing Tai
Department of Computer Science, Korean Advanced Institute of Science and Technology (KAIST), Yuseong-gu, Daejeon, South Korea

Synonyms

Blur Kernel estimation; Point spread function estimation

Related Concepts

▸Defocus Blur; ▸Image Enhancement and Restoration; ▸Motion Blur

Definition

Blur estimation is a process to estimate the point spread function (a.k.a. blur kernel) from an image which suffered from either the motion blur or the defocus blur effects.

Background

When taking a photo with long exposure time, or with wrong focal length, the captured image will look blurry. This is because during the exposure period, the lights captured for a pixel are mixed with the lights captured for the other pixels within a local neighborhood. Such effect is modeled by the point spread function which describes how the lights are mixed during the exposure period.

In motion blur, the point spread function describes the relative motions between the camera and the scene. In defocus blur, the point spread function is related to the distance of a scene point from the focal plane of the camera. Recovering the point spread function is an important step in image deblurring in which the goal is to recover the sharp and clear image from the input blurry image. Also, the estimated point spread function can be used as a feature to evaluate the quality of a photo or to segment the in-focus regions from an image with large depth of field effects.

Theory

Representation

Mathematically, the effects of blurriness can be described by the following equation:

$$B(x, y) = \sum_{(m,n)\in\mathcal{N}(x,y)} I(x-m, y-n)k_{(x,y)}(m,n) \tag{1}$$

where B is the blurry image, I is the sharp image, $k_{(x,y)}$ is the point spread function at (x, y), and $\mathcal{N}$ is the local neighborhood of a pixel, respectively. In general, the point spread function $k_{(x,y)}$ is spatially varying. This means that every pixel can carry different point spread function. This happens when the scene has large depth disparity (for defocus blur) or when the scene contains a moving object (for motion blur) or when the camera exhibits rotational motion (for motion blur) during exposure period.

While the point spread function is spatially varying, the variation of the point spread function is spatially smooth. To simplify the problem, in case of the camera motion, many previous works have assumed the point spread function is spatially invariant, which reduces Eq. (1) into a convolution equation:

$$B = I \otimes k \tag{2}$$

where $\otimes$ is the convolution operator. Recently, [1, 2] generalized Eq. (2) and proposed the projective motion blur model which uses a sequence of homographies to model the camera motion:

$$B(x, y) = \frac{1}{N}\sum_{i=1}^{N} I(H_i[x \;\; y \;\; 1]^t) \tag{3}$$

where H_i is the homography, $[x\ y\ 1]^t$ is the set of homogeneous coordinates of a pixel at (x, y), and N is the number of homographies used to approximate the camera motion in discrete domain. Note that when there is image noise, an additional noise variable will be appended in Eqs. (1)–(3) under the assumption that noise is additive and is independent to the blur process.

Methodology

In order to estimate the point spread function, there are two main methods. The first method relies on hardware modification or calibration. The second method is purely based on software which typically requires assumptions on image prior or regularization to achieve reliable estimation.

If the effect of blur is caused by the camera internal setting, such as lens aberration, the point spread function can be calibrated. To calibrate the point spread function, the simplest method is to capture an image of a spotlight in a dark room. When there is no blur, the image of the captured spotlight (ideally) should occupy only one pixel. When there is blur, the image of the captured spotlight will occupy more than one pixel and the shape of the recorded spotlight is the point spread function. Similarly, for defocus blur, the focal length of the camera can be adjusted in order to obtain a set of defocus point spread function [3]. In the case of the motion blur, Ben-Ezra and Nayer [4] and Tai et al. [5] proposed a hybrid camera system which estimates the point spread function through integration of optical flows from the auxiliary high-speed camera. Yuan et al. [6] use a noisy/blurry image pair to estimate the point spread function for deblurring. Joshi et al. [7] use motion inertia sensor to measure the camera motion in 3D world.

There are many previous works targeting blur estimation using software approaches. In defocus blur, if the point spread function is a Gaussian kernel and the edges are step edges, a defocus map can be obtained through the analysis of edge orientation and edge sharpness [8–10]. The defocus map indicates the scale of the defocus point spread function. In motion deblurring, blur estimation is usually coupled with the motion deblurring process. Fergus et al. [11] and Whyte et al. [2] proposed a multiscale variational Bayesian framework to estimate the point spread function. Jia [12] and Dai et al. [13] analyzed the edge alpha matte to obtain the marginal probability of the point spread function. Shan et al. [14] and Cho and Lee [15] used alternating optimization to iteratively refine the estimated point spread function and the deblurred image. Xu and Jia [16] proposed an edge selection method which improves the performance of blur estimation algorithms. A study on the blind motion deblurring algorithm with analysis on the point spread function can be found in [17].

Application

Blur estimation continuously receives a lot of attention in research area due to its application on deblurring. Since motion blur and defocus blur are common artifacts in imaging system, its applications range from astronomy telescope to satellite imaging, to medical imaging, and to common consumer-level camera. Besides deblurring, point spread function can also be used to evaluate image quality and to identify moving objects from a scene.

References

1. Tai YW, Tan P, Brown M (2011) Richardson-lucy deblurring for scenes under a projective motion path. IEEE Trans PAMI 33(8):1603–1618
2. Whyte O, Sivic J, Zisserman A, Ponce J (2010) Non-uniform deblurring for shaken images. In: IEEE conference on computer vision pattern recognition (CVPR), San Francisco
3. Levin A, Fergus R, Durand F, Freeman WT (2007) Image and depth from a conventional camera with a coded aperture. ACM Trans Graph 26(3):70
4. Ben-Ezra M, Nayar S (2003) Motion deblurring using hybrid imaging. In: IEEE conference on computer vision pattern recognition (CVPR), Madison, vol I, pp 657–664
5. Tai YW, Du H, Brown M, Lin S (2008) Image/video deblurring using a hybrid camera. In: IEEE conference on computer vision pattern recognition (CVPR), Anchorage
6. Yuan L, Sun J, Quan L, Shum H (2007) Image deblurring with blurred/noisy image pairs. 26(3):1
7. Joshi N, Kang S, Zitnick L, Szeliski R (2010) Image deblurring with inertial measurement sensors. ACM Trans Graph 29(3):30
8. Bae S, Durand F (2007) Defocus magnification. Computer Graphics Forum 26(3):571–579 (Proc. of Eurographics)
9. Sun J, Sun J, Xu Z, Shum HY (2008) Image super-resolution using gradient profile prior. In: IEEE conference on computer vision pattern recognition (CVPR), Anchorage
10. Joshi N, Szeliski R, Kriegman D (2008) Psf estimation using sharp edge prediction. In: IEEE conference on computer vision pattern recognition (CVPR), Anchorage
11. Fergus R, Singh B, Hertzmann A, Roweis ST, Freeman WT (2006) Removing camera shake from a single photograph. ACM Trans Graph 25(3):787–794
12. Jia J (2007) Single image motion deblurring using transparency. In: IEEE conference on computer vision pattern recognition (CVPR), Minneapolis
13. Dai S, Wu Y (2008) Motion from blur. In: IEEE conference on computer vision pattern recognition (CVPR), Anchorage
14. Shan Q, Jia J, Agarwala A (2008) High-quality motion deblurring from a single image. ACM Trans Graph 27(3):73
15. Cho S, Lee S (2009) Fast motion deblurring. ACM SIGGRAPH ASIA 28(5):145
16. Xu L, Jia J (2010) Two-phase kernel estimation for robust motion deblurring. In: European conference on computer vision (ECCV), Heraklion
17. Levin A, Weiss Y, Durand F, Freeman W (2009) Understanding and evaluating blind deconvolution algorithms. In: IEEE conference on computer vision pattern recognition (CVPR), Miami

Blur Kernel Estimation

►Blur Estimation

Body Configuration Recovery

►Human Pose Estimation

Boundary Detection

Xiaofeng Ren
Intel Science and Technology Center for Pervasive Computing, Intel Labs, Seattle, WA, USA

Synonyms

Boundary extraction; Contour detection

Boundary Detection, Fig. 1 Examples of scenes in the Berkeley Segmentation Dataset [1]. Each photo is labeled by multiple human subjects, and the boundaries are shown as being stacked together. There are variations across human subjects, but the marked boundaries are largely consistent especially for the salient ones

Related Concepts

▸Edge Detection

Definition

Boundary detection is the process of detecting and localizing salient boundaries between objects in a scene.

Background

Boundary detection is closely related to, but not identical with, edge detection. Edge detection is a classical problem in computer vision which aims at finding brightness discontinuities. Edge detection is usually viewed as a low-level process of feature extraction that works under the assumption of ideal edge models (such as step and ridge edges).

In comparison, boundary detection is usually viewed as a mid-level process of finding boundaries of (and between) objects in scenes, thus having close ties with both grouping/segmentation and object shape. A large-scale dataset of natural images with human-marked groundtruth boundaries, the *Berkeley Segmentation Dataset* (BSDS) [1, 2], was established in 2001 and quickly became the standard benchmark for both boundary detection and segmentation (see examples in Fig. 1). The Berkeley Segmentation Dataset helped defining the problem of boundary detection and clarifying several fundamental issues:

1. It directly addressed the complexities of real-world scenes by using a variety of photos from the Corel database.
2. It defined boundary detection as a perceptual problem by using human-marked boundaries as the groundtruth.
3. It showed that boundary detection is well defined by demonstrating that boundaries marked by human subjects are consistent.
4. It illustrated many challenges of boundary detection, including those of real-world texture, complex object appearance, and low-contrast boundaries.

By clarifying the task and establishing quantitative evaluation metrics, the Berkeley benchmark has witnessed and motivated large progresses in boundary detection in the recent years.

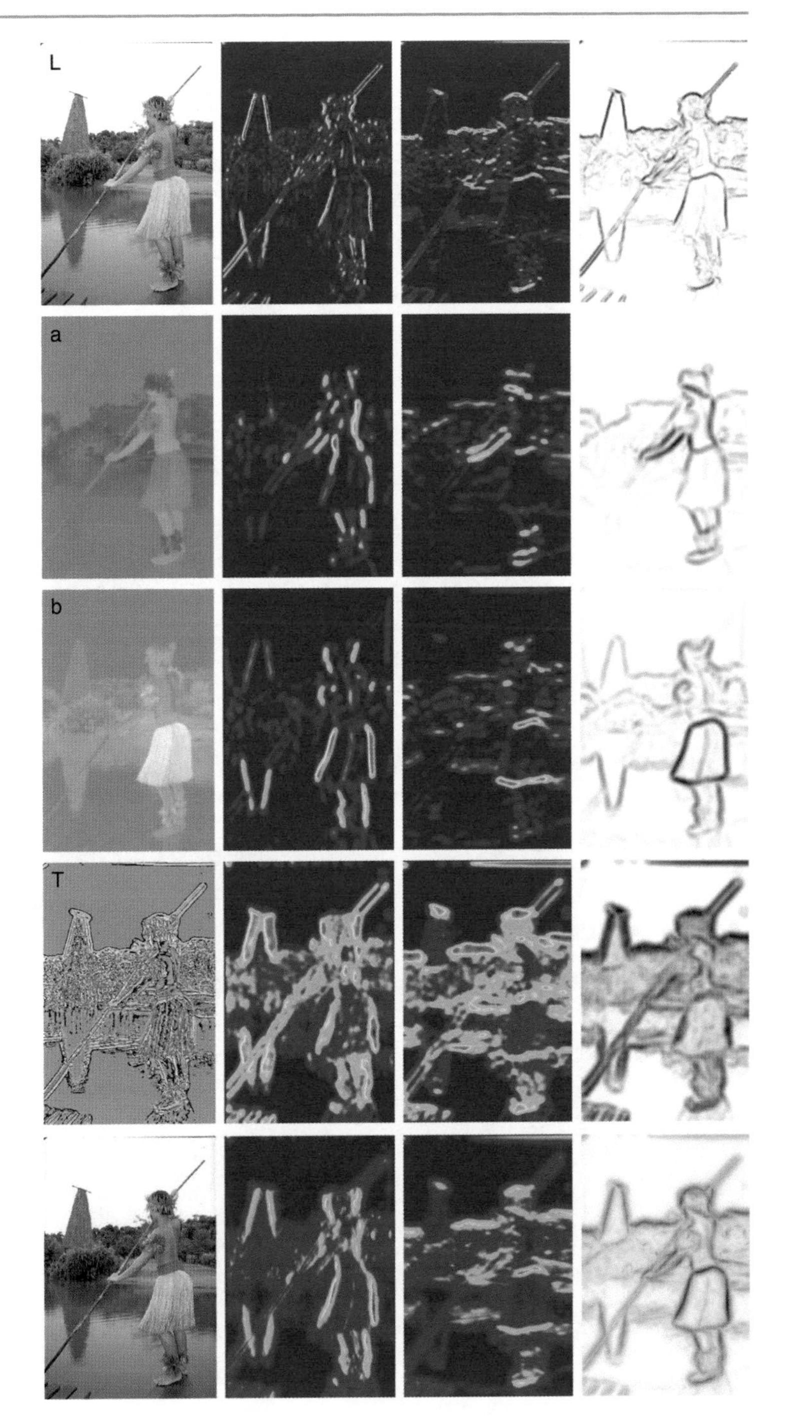

Boundary Detection, Fig. 2 Boundary detection combines multiple types of contrast (Courtesy of [9], see details there). Top-performing boundary detectors integrate together local contrast measurements from multiple channels (row 1, brightness; 2 and 3, color; and 4, texture, through textons) and multiple orientations (column 2, vertical; and 3, horizontal). *Red* color means high probability being a boundary and *blue* otherwise. The combined boundary contrast (last row, last column) is much better than any individual channel

Local Boundary Detection

Early approaches to edge detection used local derivative filters such as the Roberts, Sobel, or Prewitt filters [3]. More advanced solutions included that of the zero crossing of Marr and Hildreth [4], the optimal filter design and non-maximum suppression in the Canny detector [5], and the use of quadrature filter pairs in oriented energy [6]. Scale (of the filter) is an important issue in edge detection and

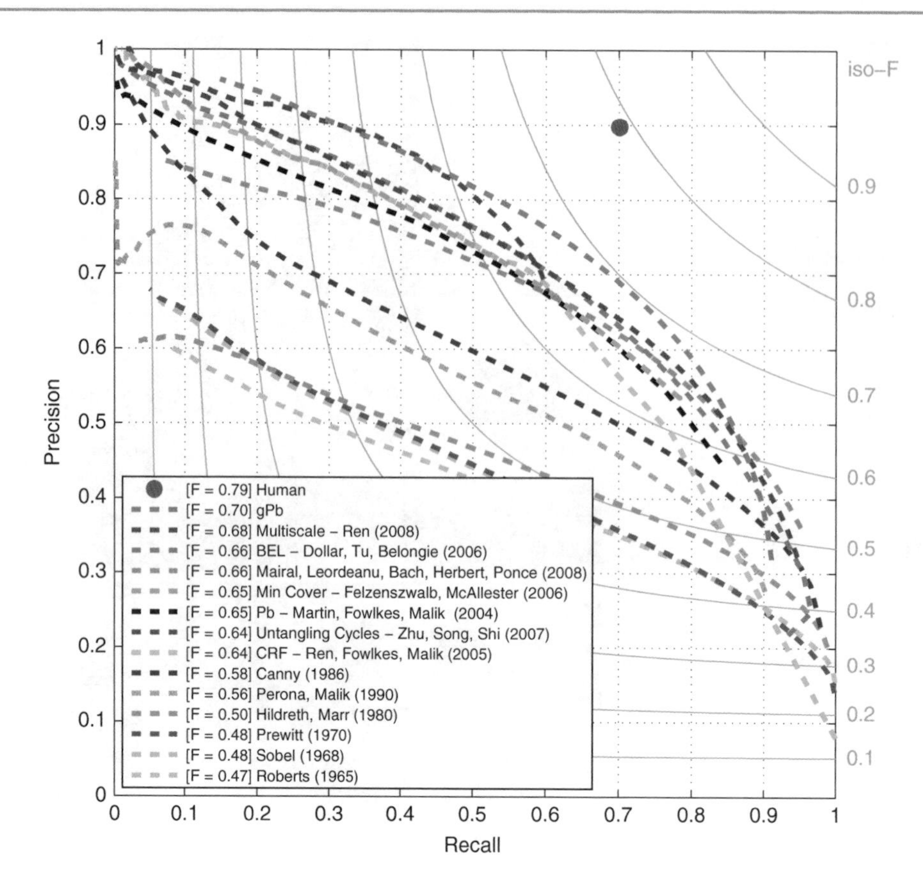

Boundary Detection, Fig. 3 Precision-recall curves and F-measures of classical and modern boundary detection algorithms on the Berkeley benchmark (Courtesy of [9], also see [1]). A variety of approaches have been proposed and evaluated on the benchmark. One can typically observe qualitative improvements in the boundary detection accuracy as the F-measure increases. State-of-the-art boundary detectors perform much better than, for instance, the Canny detector

Lindeberg proposed a mechanism for automatic scale selection [7].

The key concept in boundary detection is that of *contrast*: regions on two sides of a boundary tend to have different appearances; consequently, there tends to be a high contrast at a boundary location. To a large extent, this contrast can be captured and measured locally in an image neighborhood (e.g., a disk with a fixed radius). Local contrast can be measured in a number of ways, such as using linear filters or computing distances between histograms. To handle real-world scenes, modern boundary detectors utilize contrast information from multiple channels (including brightness, color, and texture), multiple orientations, and multiple scales (see examples in Fig. 2). Good examples of these contrast operators can be found in the *Pb* work (probability-of-boundary) of Martin et al. [8] and the *gPb* work (global probability-of-boundary) of Arbelaez et al. [9].

Given the complexities of contrast cues and the availability of labeled images, local boundary detection is often formulated as learning a binary classifier of boundary vs. non-boundary, which will produce a soft boundary "likelihood" at each pixel. Such a dense boundary map can be used directly or converted to a sparse boundary map through non-maximum suppression. A number of supervised machine learning techniques were used and tested in [8] to combine a small set of handcrafted contrast cues. Others have taken a more direct learning approach, such as using boosting trees to combine thousands of simple features over patches [10].

Global Boundary Detection

Boundaries are not local phenomena that occur independently at pixels. In fact, boundaries are defined at the object level, and boundary pixels tend to form long, smooth contours, as evident from the examples in Fig. 1. Considerable efforts have been devoted to extracting boundaries globally, closely related to the classical problem of contour completion in perceptual organization. Algorithms for global boundary detection can be quantitatively evaluated based on their precision recall on the boundary detection task, same as for local boundary detection.

A variety of very different formulations have been proposed for global boundary detection and contour extraction, including classical works such as the Mumford-Shah functional [11]. Several recent approaches have successfully demonstrated, through benchmarking, that globalization greatly improves boundary detection accuracy over local detectors. Ren et al. [12] applied constrained Delaunay triangulation (CDT) to decompose locally detected contours into pieces and used conditional random fields (CRF) and belief propagation to integrate local contrast cues through interactions at junctions. Zhu et al. [13] computed complex eigenvectors of a normalized random walk matrix, using circular embedding, to detect topologically closed cycles. In the *gPb* work of Arbelaez et al. [9], eigenvectors of the affinity matrix were first computed, as in Normalized Cuts [14], and then the gradients of these eigenvectors were added to the local contrast cues to produce a single contrast map.

[Optional]: One related but different form of global boundary detection can be found in the case of *top-down object segmentation* [15], where the algorithm has access to the knowledge (such as shape or texture) of the objects that are in the scene. It is beyond the scope of the discussion here, as boundary detection typically refers to the *bottom-up* case where no high-level object knowledge is needed.

Application

Boundary detection is fundamentally connected to both image segmentation and object shape, and there should be no surprise that advances in boundary detection have led to many interesting applications in segmentation and object recognition.

For image segmentation, the use of *intervening contour* [16] allows one to convert any boundary map to pairwise affinities for use in the Normalized Cuts framework, and many systems (e.g., [17]) have been using modern boundary detectors such as the *Pb* operator [8]. *Pb* is also used in [18], combined with the Watershed algorithm, to produce superpixels. Arbelaez et al. [9] proposed a hierarchical segmentation algorithm that, using the *gPb* boundary operator, produces compelling segmentation results and at the same time further improves the boundary detection accuracy.

For object recognition, boundary detectors such as *Pb* and *gPb* are often used to produce a boundary map, which is in turn used to compute shape descriptors. For instance, the work of Berg et al. [19] used *Pb* boundary maps with Geometric Blur for object and face recognition. The work of Ferrari et al. [20] used *Pb* to produce contour segments as a basis for shape matching. There are many segmentation-based approaches to recognition that also heavily rely on the quality of boundary detection (e.g., [21]).

State-of-the-art boundary detectors are sophisticated and require fairly intensive computation, which limits their applicability. There have been studies and efforts to speed up boundary detectors. In particular, the GPU-based detector of Catanzaro et al. [22] achieved a two-orders-of-magnitude improvement of speed over *gPb* without suffering any loss in boundary quality.

References

1. Martin D, Fowlkes C, Malik J (2002) Berkeley segmentation dataset. http://www.cs.berkeley.edu/projects/vision/bsds
2. Martin D, Fowlkes C, Tal D, Malik J (2001) A database of human segmented natural images and its application to evaluating segmentation algorithms and measuring ecological statistics. In: Proceedings of international conference on computer vision (ICCV), Vancouver, vol 2, pp 416–423
3. Faugeras O (1993) Three-dimensional computer vision: a geometric viewpoint. MIT, Cambridge
4. Marr D, Hildreth E (1980) Theory of edge detection. Proc R Soc Lond B 207(1167):187–217
5. Canny J (1986) A computational approach to edge detection. IEEE Trans Pattern Anal Mach Intell 8:679–698
6. Morrone M, Owens R (1987) Feature detection from local energy. Pattern Recognit Lett 6:303–13
7. Lindeberg T (1998) Edge detection and ridge detection with automatic scale selection. Int J Comput Vis 30: 117–156
8. Martin D, Fowlkes C, Malik, J (2004) Learning to detect natural image boundaries using local brightness, color and texture cues. IEEE Trans Pattern Anal Mach Intell 26(5): 530–549

9. Arbelaez P, Maire M, Fowlkes C, Malik J (2010) Contour detection and hierarchical image segmentation. IEEE Trans Pattern Anal Mach Intell 33(5):898–916
10. Dollar P, Tu Z, Belongie S (2006) Supervised learning of edges and object boundaries. In: Proceedings of computer vision and pattern recognition (CVPR), New York, vol 2, pp 1964–1971
11. Mumford D, Shah J (1989) Optimal approximation by piecewise smooth functions and associated variational problems. Commun Pure Appl Math 42:577685
12. Ren X, Fowlkes C, Malik J (2008) Learning probabilistic models for contour completion in natural images. Int J Comput Vis 77(1–3):47–64
13. Zhu Q, Song G, Shi J (2007) Untangling cycles for contour grouping. In: Proceedings of international conference on computer vision (ICCV), Vancouver, pp 1–8
14. Malik J, Belongie S, Leung T, Shi J (2001) Contour and texture analysis for image segmentation. Int J Comput Vis 43(1):7–27
15. Borenstein E, Ullman S (2002) Class-specific, top-down segmentation. In: Proceedings of european conference on computer vision (ECCV), Copenhagen, vol 2, pp 109–124
16. Leung T, Malik J (1998) Contour continuity in region-based image segmentation. In: Proceedings of european conference on computer vision (ECCV), Freiburg, Germany. vol 1, pp 544–559
17. Mori G, Ren X, Efros A, Malik J (2004) Recovering human body configurations: combining segmentation and recognition. In: Proceedings of computer vision and pattern recognition (CVPR), Washington, vol 2, pp 326–333
18. Hoiem D, Efros A, Hebert M (2007) Recovering occlusion boundaries from a single image. In: Proceedings of international conference on computer vision (ICCV), Rio de Janeiro
19. Berg A, Berg T, Malik J (2005) Shape matching and object recognition using low distortioncorrespondence. In: Proceedings of computer vision and pattern recognition (CVPR), San Diego, vol 1, pp 26–33
20. Ferrari V, Tuytelaars T, Gool LV (2006) Object detection by contour segment networks. In: Proceedings of european conference on computer vision (ECCV), Graz, pp 14–28
21. Gu C, Lim J, Arbeláez P, Malik J (2009) Recognition using regions. In: Proceedings of computer vision and pattern recognition (CVPR), Miami
22. Catanzaro B, Su BY, Sundaram N, Lee Y, Murphy M, Keutzer K (2009) Efficient, high-quality image contour detection. In: Proceedings of international conference on computer vision (ICCV), Kyoto, pp 2381–2388
23. Maire M, Arbelaez P, Fowlkes C, Malik J (2008) Using contours to detect and localize junctions in natural images. Proceedings of computer vision and pattern recognition (CVPR), Anchorage, pp 1–8

Boundary Extraction

▸Boundary Detection

BRDF Measurement

▸Recovery of Reflectance Properties

BRDF Models

▸Reflectance Models

C

Calculus of Variations

▶Variational Analysis

Calibration

Zhengyou Zhang
Microsoft Research, Redmond, WA, USA

Related Concepts

▶Camera Calibration; ▶Hand-Eye Calibration

Definition

According to McGraw-Hill Encyclopedia of Science and Technology [1], calibration is the process of determining the performance parameters of an artifact, instrument, or system by comparing it with measurement standards. Adjustment may be a part of a calibration, but not necessarily. A calibration assures that a device or system will produce results which meet or exceed some defined criteria with a specified degree of confidence.

Background

In computer vision, there are multiple calibration problems. The most fundamental one is the camera calibration, which determines the intrinsic and extrinsic parameters of a camera. It is the first step toward 3D computer vision. Other problems include hand-eye calibration, color calibration, and photometric calibration.

As stated in [1], two important measurement concepts related to calibration are precision and accuracy. Precision refers to the minimum discernible change in the parameter being measured, while accuracy refers to the actual amount of error that exists in a calibration. All measurement processes used for calibration are subject to various sources of error. It is common practice to classify them as random or systematic errors. When a measurement is repeated many times, the results will exhibit random statistical fluctuations which may or may not be significant. Systematic errors are offsets from the true value of a parameter and, if they are known, corrections are generally applied, eliminating their effect on the calibration. If they are not known, they can have an adverse effect on the accuracy of the calibration. High-accuracy calibrations are usually accompanied by an analysis of the sources of error and a statement of the uncertainty of the calibration. Uncertainty indicates how much the accuracy of a calibration could be degraded as a result of the combined errors.

References

1. Parker SP (ed) (1982) McGraw-Hill encyclopedia of science and technology, 5th edn. McGraw-Hill, New York. http://www.answers.com/topic/calibration

K. Ikeuchi (ed.), *Computer Vision*, DOI 10.1007/978-0-387-31439-6,

Calibration of a Non-single Viewpoint System

Peter Sturm
INRIA Grenoble Rhône-Alpes, St Ismier Cedex, France

Synonyms

Non-central camera calibration

Related Concepts

▶Camera Calibration; ▶Center of Projection

Definition

A non-single viewpoint system refers to a camera for which the light rays that enter the camera and contribute to the image produced by the camera, do not pass through a single point. The analogous definition holds for *models* for non-single viewpoint systems. Hence, a non-single viewpoint camera or model does not possess a single center of projection. Nevertheless, a non-single viewpoint model (NSVM), like any other camera model such as the pinhole model, enables to project points and other geometric primitives, into the image and to back-project image points or other image primitives, to 3D. Calibration of a non-single viewpoint model consists of a process that allows to compute the parameters of the model.

Background

There exist a large variety of camera technologies ("regular" cameras, catadioptric cameras, fish-eye cameras, etc.) and camera models designed for these technologies. Often, technologies are developed in order to accommodate a desired model, such as for example, to provide a uniform spatial resolution.

Most cameras used in computer vision and other areas, can be well modeled by so-called single viewpoint, or central, camera models. These usually model the 3D-to-2D mapping carried out by a camera, via lines of sight (or, camera rays) that all pass through a single point (the center of projection or optical center), and a mapping from these lines of sight to the image points where they hit the image plane.

Some camera types, especially some cameras having a wide field of view, but not only these, cannot be modeled very well using a single viewpoint model. This may be the case because a camera was designed to possess lines of sight that do not pass through a single point. This is, for example, the case for catadioptric cameras where the mirror surface is of conical shape: even if the camera looking at the mirror is positioned on the mirror's axis, the lines of sight of the system do not converge to a single point, rather there exists a *viewpoint locus*. Another example is a single-lens stereo system consisting of a pinhole camera and two planar mirrors, such that the obtained images represent two perspective images acquired from two different effective viewpoints.

A camera may also be unintentionally of the non-single viewpoint type, for example, catadioptric cameras that were designed to have a single viewpoint but that due to a bad alignment of the camera and the mirror of the system, lose the single viewpoint property. Another example are fish-eye cameras; fish-eye optics are complex, and in principle, one can probably consider them as non-single viewpoint systems. However, in this and the previous example, it is not clear without further investigation of the actual system under consideration, if a single viewpoint model or an NSVM is better suited. This indeed depends on "how much" the system deviates from having a single viewpoint, how close the scene is in a typical application, how much image resolution is available, and so forth. This issue is further discussed in [1].

In the following, it is supposed that calibration is performed by acquiring one or several images of a calibration object, whose geometry is known and whose characteristic features (for simplicity, points shall be considered) can be extracted and identified in images.

Theory

There are different types of NSVM's. One usually distinguishes parametric from non-parametric such models. For example, for non-single viewpoint catadioptric systems, if the shape of the system's mirror is known or is known to belong to a parametric family of shapes, then the entire system can be described by few parameters: intrinsic parameters of the camera looking at the mirror, relative pose of mirror and camera, and possibly, shape parameters for the mirror. If such a parametric model is considered, calibration is, conceptually speaking, analogous to that of pinhole cameras. The main difference to calibration of pinhole cameras usually concerns the initialization process that allows to compute initial estimates of the camera parameters. Other than that, one may, in general, formulate calibration by a bundle adjustment type optimization of camera parameters, by minimizing, for example, the reprojection error, that is, a measure related to the distance between the predicted projections of points of the calibration object and those extracted in the images. Examples of parametric NSVM's are the so-called two-plane and GLC models [2–5], where lines of sight are parameterized by linear or higher-order transformations applied to points in two basis planes, other similar models where lines of sight are parameterized by linear transformations [6–9], models for pushbroom and X-slit cameras [10–13], and others [14].

A different concept consists in using non-parametric models to calibrate cameras. An example is the raxel model introduced by Grossberg and Nayar [15]. It essentially associates, to each pixel, a ray in 3D supporting the line of sight, and possibly properties such as its own radiometric response function. Importantly, one may use such a model without making any assumption about a parametric relationship between the position of pixels and the position and orientation of the associated lines of sight. Rather, one may store the coordinates of the lines of sight of all pixels, in a look-up table. Simplified versions of this model (without considering optical properties for individual pixels) have been used in several similar calibration approaches, for example, [15–18].

The principle of these approaches is thus to compute, for every camera pixel, a line of sight in 3D. To do so, at least two images of a calibration object are required. The simplest scenario considers the case where the calibration object is displaced by some known motion, between the image acquisitions. For each image, one has to estimate correspondences between camera pixels and points on the calibration object. One way of achieving such dense correspondences is to use structured light principles, for instance, to use as calibration object a computer screen and to display a series of black and white patterns on it that encodes each pixel of the screen by a unique sequence of black and white intensities (e.g., Grey codes). Once correspondences of camera pixels and points of the calibration object (pixels of the computer screen in the above example) are known, the lines of sight can be computed by simply fitting straight 3D lines to the matched points on the calibration object. To do so, the latter must be expressed in the same 3D coordinate system, which is possible since it was assumed above that the motion of the calibration object between different acquisitions, is known. This approach was proposed independently by different researchers [15–17].

The above approach requires a minimum of two images, for different positions of the calibration object, and knowledge of the object's displacements. An extension to the case of unknown displacements was proposed in [18]. That approach requires at least three images; from matches of camera pixels and points on the calibration object, it first recovers the displacements of the object using an analysis of this scenario's multi-view geometry, and then computes lines of sight as above. Other approaches following this line are [19, 20].

The above approaches compute, for each camera pixel, an individual line of sight. If one assumes that the relation between pixels and lines of sight is particular, for instance, radially symmetric about an optical axis, then alternative solutions become possible. Such a possibility is to use a non-parametric representation of the distortion or undistortion function of a camera, that is, a function that maps viewing angles (angles between lines of sight and the optical axis) to distances in the image, between image points and the principal point or a distortion center. This can be done for both, single viewpoint and non-single viewpoint models. In the former case, it is assumed that all lines of sight pass through a single center of projection, whereas in the latter case, the model usually includes a mapping from viewing angles to the position of the intersection between lines of sight and the optical axis. Approaches of the latter type include [21, 22]. Besides making and using the assumption that the camera is radially symmetric, these calibration approaches resemble those explained above.

Application

All approaches described above, be they parametric or non-parametric, allow to perform 3D-to-2D projection and/or 2D-to-3D back projection, the latter meaning the mapping from an image point to the associated line of sight. By definition, the parametric models give analytical expressions to perform these operations. As for non-parametric ones, projection and back projection usually imply some interpolation and, possibly, a search. For instance, if a non-parametric model consists of a look-up table that gives, for each pixel, its line of sight, then back-projection of an image point with non-integer coordinates requires interpolation, whereas projection of a 3D point requires the search of the closest line(s) of sight in the look-up table and again an interpolation stage.

Other than these particular aspects, NSVM's can be used for many structure-from-motion computations completely analogously to other camera models, in particular, the pinhole model. Among the essential building blocks of structure from motion, there are pose estimation, motion estimation, and 3D point triangulation for calibrated cameras. As for pose and motion estimation (and other tasks), one usually requires two types of methods in an application: so-called minimal methods, which perform the estimation task from the minimum required number of point matches and which can be efficiently embedded in robust estimation schemes such as RANSAC, and non-linear optimization methods that refine initial estimates obtained from minimal methods. Minimal methods for pose [23–25] and motion estimation [26] with NSVM's are formulated perfectly analogously to those for the pinhole model, although their algebraic complexity is higher. All that is required by these methods from the NSVM is to compute lines of sight of interest points that are extracted and matched to another image (for motion estimation) or to a reference object (for pose estimation). As for the non-linear optimization stage, the minimization of the reprojection errors requires 3D-to-2D projections to be carried out, which, as explained above, may require search and interpolation, in which case the computation of the cost function's derivatives may have to rely on numerical differentiation. Other than that, there is no major conceptual difference compared to pose/motion estimation with pinhole cameras.

Another essential structure-from-motion task is 3D point triangulation. Here again, suboptimal methods work with lines of sight computed by the camera model for interest points in the images, and optimal methods perform the non-linear optimization of reprojection errors, where the same considerations hold as above for pose and motion estimation.

References

1. Sturm P, Ramalingam S, Tardif JP, Gasparini S, Barreto J (2011) Camera models and fundamental concepts used in geometric computer vision. Found Trends Comput Graph Vis 6(1–2):1–183
2. Chen NY (1979) Visually estimating workpiece pose in a robot hand using the feature points method. PhD thesis, University of Rhode Island, Kingston
3. Chen NY, Birk J, Kelley R (1980) Estimating workpiece pose using the feature points method. IEEE Trans Autom Cont 25(6):1027–1041
4. Martins H, Birk J, Kelley R (1981) Camera models based on data from two calibration planes. Comput Graph Image Process 17:173–180
5. Yu J, McMillan L (2004) General linear cameras. Proceedings of the 8th European conference on computer vision (ECCV), Prague, Czech Republic. pp 14–27
6. Pajdla T (2002) Stereo with oblique cameras. Int J Comput Vis 47(1–3):161–170
7. Ponce J (2009) What is a camera? Proceedings of the IEEE conference on computer vision and pattern recognition (CVPR), Miami, USA
8. Seitz S, Kim J (2002) The space of all stereo images. Int J Comput Vis 48(1):21–38
9. Batog G, Goaoc X, Ponce J (2010) Admissible linear map models of linear cameras. Proceedings of the IEEE conference on computer vision and pattern recognition (CVPR), San Francisco, USA
10. Gupta R, Hartley R (1997) Linear pushbroom cameras. IEEE Trans Pattern Anal Mach Intell 19(9):963–975
11. Pajdla T (2002) Geometry of two-slit camera. Technical Report CTU-CMP-2002-02, Center for Machine Perception, Czech Technical University, Prague
12. Zomet A, Feldman D, Peleg S, Weinshall D (2003) Mosaicing new views: the crossed-slit projection. IEEE Trans Pattern Anal Mach Intell 25(6):741–754
13. Feldman D, Pajdla T, Weinshall D (2003) On the epipolar geometry of the crossed-slits projection. Proceedings of the 9th IEEE international conference on computer vision, Nice, France. pp 988–995
14. Gennery D (2006) Generalized camera calibration including fish-eye lenses. Int J Comput Vis 68(3):239–266
15. Grossberg M, Nayar S (2005) The raxel imaging model and ray-based calibration. Int J Comput Vis 61(2):119–137
16. Gremban K, Thorpe C, Kanade T (1988) Geometric camera calibration using systems of linear equations. Proceedings of the IEEE international conference on robotics and automation, Philadelphia, Pennsylvania, USA. pp 562–567

17. Champleboux G, Lavallée S, Sautot P, Cinquin P (1992) Accurate calibration of cameras and range imaging sensors: the NPBS method. Proceedings of the IEEE international conference on robotics and automation, Nice, France. pp 1552–1558
18. Sturm P, Ramalingam S (2004) A generic concept for camera calibration. Proceedings of the 8th European conference on computer vision (ECCV), Prague, Czech Republic. pp 1–13
19. Ramalingam S, Sturm P, Lodha S (2005) Towards complete generic camera calibration. Proceedings of the IEEE conference on computer vision and pattern recognition (CVPR), San Diego, USA, vol 1. pp 1093–1098
20. Dunne A, Mallon J, Whelan P (2010) Efficient generic calibration method for general cameras with single centre of projection. Comput Vis Image Underst 114(2):220–233
21. Tardif JP, Sturm P, Trudeau M, Roy S (2009) Calibration of cameras with radially symmetric distortion. IEEE Trans Pattern Anal Mach Intell 31(9):1552–1566
22. Ying X, Hu Z (2004) Distortion correction of fisheye lenses using a non-parametric imaging model. Proceedings of the Asian conference on computer vision, Jeju Island, Korea. pp 527–532
23. Chen CS, Chang WY (2004) On pose recovery for generalized visual sensors. IEEE Trans Pattern Anal Mach Intell 26(7):848–861
24. Ramalingam S, Lodha S, Sturm P (2004) A generic structure-from-motion algorithm for cross-camera scenarios. Proceedings of the 5th workshop on omnidirectional vision, camera networks and non-classical cameras, Prague, Czech Republic. pp 175–186
25. Nistér D, Stewénius H (2007) A minimal solution to the generalised 3-point pose problem. J Math Imaging Vis 27(1):67–79
26. Stewénius H, Nistér D, Oskarsson M, Åström K (2005) Solutions to minimal generalized relative pose problems. Proceedings of the 6th workshop on omnidirectional vision, camera networks and non-classical cameras, Beijing, China

Calibration of Multi-camera Setups

Jun-Sik Kim
Korea Institute of Science and Technology, Seoul, Republic of Korea

Synonyms

Multi-camera calibration

Related Concepts

▶Camera Calibration; ▶Camera Parameters (Intrinsic, Extrinsic)

Definition

Calibration of multi-camera setups is a process to estimate parameters of cameras which are fixed in a setup. It usually refers to the process to find relative poses of the cameras in a single coordinate system under the assumption of known intrinsic camera parameters.

Background

Many computer vision methods including 3D reconstruction from stereo cameras utilize the multiple cameras in a system, assuming that the relative poses of cameras in a single coordinate system is already known. While the camera calibration using a planar pattern [1] simplifies calibration process for intrinsic and extrinsic parameters of each camera, estimating camera poses in *a fixed global coordinate system* is still required.

The term "multi-camera setup" includes many different camera configurations such as a stereo, inward-looking cameras, outward-looking cameras, or camera sensor networks. Because each camera setup has different viewpoint and field of view (FOV) configuration, one single calibration method is not able to deal with all the multi-camera setups. Depending on the camera configuration, different calibration approach should be considered.

Theory

A projection matrix $\mathbf{P}_i$ of a camera i in a multi-camera setup is given as

$$\mathbf{P}_i = \mathbf{K}_i \, [\mathbf{R}_i \; \mathbf{T}_i] \,. \qquad (1)$$

The matrices $\mathbf{R}_i$ and $\mathbf{T}_i$ represent the pose of the camera i in a *predetermined* fixed coordinate system. More specifically, the matrices express a transformation between the fixed global coordinate system and the local camera coordinate system. The goal of the multi-camera calibration is to estimate the matrices $\mathbf{R}_i$ and $\mathbf{T}_i$ for all i in the system. The camera matrix $\mathbf{K}_i$ represents intrinsic parameters of the camera i, which can be assumed to be known by calibrating the intrinsic parameters of each camera independently in advance.

Multi-camera systems can be categorized into three configurations: inward-looking cameras, outward-looking cameras and large camera networks.

Inward-looking cameras The case that all the cameras in the system have a FOV shared. A stereo camera is considered *inward-looking* because the two cameras should see the same scene.

Outward-looking cameras The case that all the cameras in the system do not share their FOV. Because no FOVs are overlapped, it is usually called *non-overlapping cameras*.

Camera networks The case that some cameras share their FOVs but there are no common FOV for all cameras. Distributed cameras are usually in this category. Most likely, nearby cameras have a common FOV, farther cameras can not see it.

Each camera configuration has different constraints used in the multi-camera calibration, and the resulting calibration method becomes different to each other.

Inward-Looking Cameras

When all the cameras in the multi-camera setup have a common FOV, the multi-camera calibration is relatively simple. A calibration object is placed in the common FOV as shown in Fig. 1 so that each camera can see it, and the pose of the cameras with respect to the object coordinate system is estimated by using conventional pose estimation methods [2–4]. In this case, the common coordinate system of the multi-camera configuration is set to be the coordinate system of the calibration object.

For pose estimation of each camera, a planar pattern is preferable because it provides better visibility for all cameras. Note that, however, it is not limited to a planar pattern when the object visibility from every camera is ensured.

Sinha et al. [5] present an automatic calibration method using object silhouettes. In this method, epipolar geometry between cameras is estimated from dynamic silhouettes and projective structure is recovered. Following self-calibration completes the Euclidean reconstruction. This aims especially for shape-from-silhouette or visual hull reconstruction.

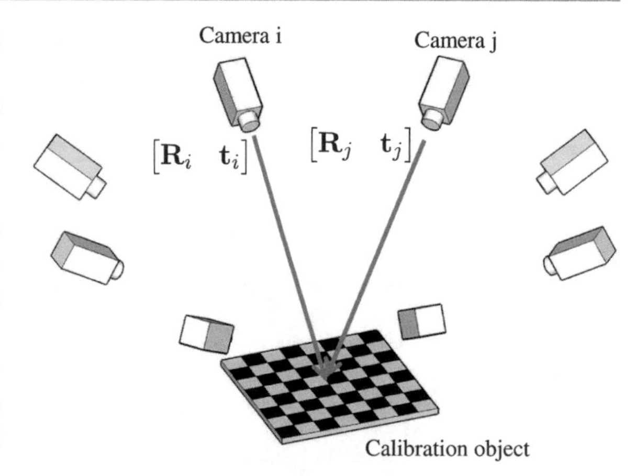

Calibration of Multi-camera Setups, Fig. 1 Calibration of inward-looking cameras

Outward-Looking Cameras

When the FOVs of cameras in the system are not overlapped, it is impossible to place a calibration object which is observable from multiple cameras. The pose between cameras can be estimated by utilizing the fact that the transformation between cameras are fixed in motions called a rigidity constraint [6].

Assume that the coordinate systems of cameras i and j are transformed by a transformation $\mathbf{R}_{ij}$ and $\mathbf{t}_{ij}$. When the camera i moves with a transformation $\Delta\mathbf{R}_i$ and $\Delta\mathbf{t}_i$, the motion of the camera j with a rotation $\Delta\mathbf{R}_j$ and a translation $\Delta\mathbf{t}_j$ is given as

$$\begin{bmatrix} \Delta\mathbf{R}_j & \Delta\mathbf{t}_j \\ \mathbf{0}^\top & 1 \end{bmatrix} = \begin{bmatrix} \mathbf{R}_{ij} & \mathbf{t}_{ij} \\ \mathbf{0}^\top & 1 \end{bmatrix} \begin{bmatrix} \Delta\mathbf{R}_i & \Delta\mathbf{t}_i \\ \mathbf{0}^\top & 1 \end{bmatrix} \begin{bmatrix} \mathbf{R}_{ij} & \mathbf{t}_{ij} \\ \mathbf{0}^\top & 1 \end{bmatrix}^{-1}, \tag{2}$$

and this equation can be rewritten in a $\mathbf{AX} = \mathbf{XB}$ form on the Euclidean group as

$$\begin{bmatrix} \Delta\mathbf{R}_j & \Delta\mathbf{t}_j \\ \mathbf{0}^\top & 1 \end{bmatrix} \begin{bmatrix} \mathbf{R}_{ij} & \mathbf{t}_{ij} \\ \mathbf{0}^\top & 1 \end{bmatrix} = \begin{bmatrix} \mathbf{R}_{ij} & \mathbf{t}_{ij} \\ \mathbf{0}^\top & 1 \end{bmatrix} \begin{bmatrix} \Delta\mathbf{R}_i & \Delta\mathbf{t}_i \\ \mathbf{0}^\top & 1 \end{bmatrix}. \tag{3}$$

Solving the unknowns $\mathbf{R}_{ij}$ and $\mathbf{t}_{ij}$ on the Euclidean group is known as a *hand-eye calibration* in the robotics community [7–9]. Two or more motions of the camera rig provide enough number of constraints.

One practical problem is to estimate *motions* $\Delta\mathbf{R}$ and $\Delta\mathbf{t}$ of each camera. One stable way is to use calibration objects for each camera. First place a calibration object for each camera and take pictures while moving the camera rig. At each time frame,

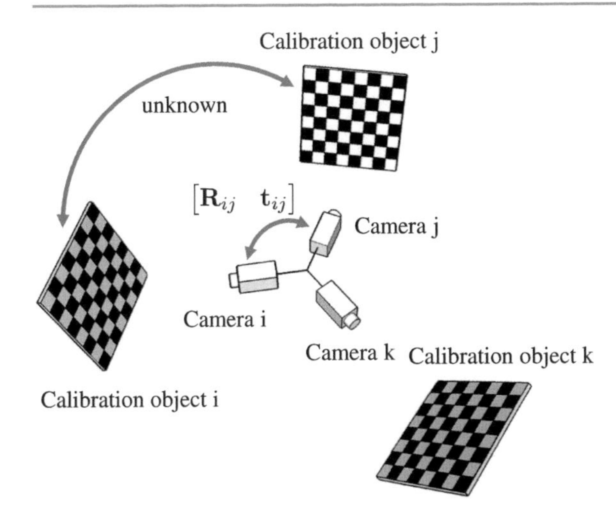

Calibration of Multi-camera Setups, Fig. 2 Calibration of outward-looking cameras

the poses of each camera can be estimated by using a conventional pose estimation method. The motions of cameras are computed by calculating the difference of the poses at different time frames. Note that the geometric relation between calibration objects is not required in calculating each camera motions. The only requirement is that the calibration objects are fixed in motions. Figure 2 shows the calibration objects for three non-overlapping cameras.

By solving the $\mathbf{AX} = \mathbf{XB}$ equation on the Euclidean group, the transformation between two cameras is obtained. If the multi-camera system has more than two cameras, every camera can be registered in the fixed global coordinate system by chaining the transformations pairwisely. However, the pairwise chaining of transformation does not guarantee the globally consistent registration. In addition, the transformation between cameras may be inconsistent depending on the data provided to the $\mathbf{AX} = \mathbf{XB}$ solver. Dai et al. [10] represent a rotation averaging strategy to improve the consistency of the estimated transformation, and use a global bundle adjustment [11] for final polishing. Lebraly et al. [12] more focus on an sparse implementation of the global bundle adjustment to ensure the consistency in the fixed global coordinate system.

Kumar et al. [13] present completely different approach to use a mirror so that the cameras can observe the mirrored calibration pattern, and show successful calibration result for the ladybug camera.

Camera Networks

One general configuration is a camera network, which usually has many fixed cameras seeing in different directions. When every camera shares its FOV with any other camera in the network, relative transformations between the cameras can be estimated using calibration objects, and they are registered by chaining the transformation. In fact, this calibration process is cast to a conventional structure from motion problem. Once all the relative transformations are obtained, globally consistent localization of the cameras in the fixed coordinate system is achieved by using bundle adjustment [11]. Figure 3 shows a possible placement of calibration objects in calibrating a camera network. Devarajan et al. [14] introduce a *vision graph* to describe the feature visibility between cameras, and try to optimize the graph network by belief propagation.

Baker and Aloimonos [15] present a method based on the multi-frame structure from motion algorithm, and use a rod with two LEDs at each end as a calibration object. The LEDs provides accurate and easily detectable correspondences for the precisely synchronized cameras by waiving the rod.

Svoboda et al. [16] propose a convenient and complete self-calibration method using a laser pointer including intrinsic parameter estimation. The method is based on the stratification of the transformations; at first projective reconstruction is achieved by factorization and later upgraded to Euclidean space by imposing geometric constraints such as a square pixel assumption. Their source codes are available for public use.

If there is no camera sharing its FOV with others, it is challenging to establish the common global coordinate system. One idea is to use a mobile robot carrying a calibration object [17]. The location of the object is estimated by the SLAM of the mobile robot. However, this is not stable enough for practical use yet.

Application

Multi-camera calibration is essential in constructing a system using multiple cameras depending on the

Calibration of Multi-camera Setups, Fig. 3 Calibration of a camera network

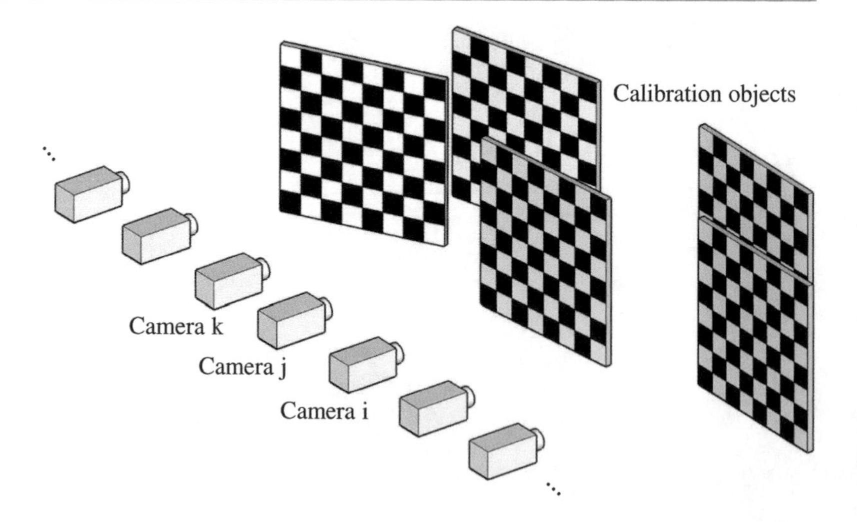

applications and sensor configurations. The *inward-looking configuration* is generally used in many 3D reconstruction tasks such as stereo and visual hull reconstruction. The *outward-looking configuration* is useful to enlarge the effective FOV of the system, and especially for structure from motion applications. The most general *camera network* has diverse applications such as 3D reconstruction, surveillance, environmental monitoring and so on. Note that the camera network includes the inward-looking configuration of cameras.

References

1. Zhang Z (2000) A flexible new technique for camera calibration. IEEE Trans Pattern Anal Mach Intell 22(11): 1330–1334
2. Dementhon DF, Davis LS (1995) Model-based object pose in 25 lines of code. Int J Comput Vis 15:123–141
3. Quan L, Lan Z (1999) Linear n-point camera pose determination. IEEE Trans Pattern Anal Mach Intell 21(8): 774–780
4. Ababsa F, Mallem M (2004) Robust camera pose estimation using 2d fiducials tracking for real-time augmented reality systems. Proceedings of the 2004 ACM SIGGRAPH international conference on virtual reality continuum and its applications in industry. ACM, New York, pp 431–435
5. Sinha S, Pollefeys M, McMillan L (2004) Camera network calibration from dynamic silhouettes. Proceedings of the IEEE computer society conference on computer vision and pattern recognition (CVPR), vol 1. IEEE, Silver Spring, pp I–195
6. Esquivel S, Woelk F, Koch R (2007) Calibration of a multi-camera rig from non-overlapping views. Proceedings of the 29th DAGM conference on pattern recognition. Springer, Berlin/New York, pp 82–91
7. Andreff N, Horaud R, Espiau B (1999) On-line hand-eye calibration. Proceedings of the second international conference on 3-D digital imaging and modeling, 1999. Ottawa, Canada, IEEE, pp 430–436
8. Dornaika F, Horaud R (1998) Simultaneous robot-world and hand-eye calibration. IEEE Trans Robot Autom 14(4): 617–622
9. Park F, Martin B (1994) Robot sensor calibration: solving ax= xb on the euclidean group. IEEE Trans Robot Autom 10(5):717–721
10. Dai Y, Trumpf J, Li H, Barnes N, Hartley R (2010) Rotation averaging with application to camera-rig calibration. Asian conference on computer vision. Springer, Berlin/New York, pp 335–346
11. Triggs B, McLauchlan P, Hartley R, Fitzgibbon A (2000) Bundle adjustment – modern synthesis. Vision algorithms: theory and practice. Springer-Verlag, London, UK, pp 298–372
12. Lebraly P, Royer E, Ait-Aider O, Deymier C, Dhome M (2011) Fast calibration of embedded non-overlapping cameras. IEEE international conference on robotics and automation (ICRA). Shanghai, China, IEEE, pp 221–227
13. Kumar R, Ilie A, Frahm J, Pollefeys M (2008) Simple calibration of non-overlapping cameras with a mirror. IEEE conference on computer vision and pattern recognition (CVPR). Anchorage, AK, USA, IEEE, pp 1–7
14. Devarajan D, Cheng Z, Radke R (2008) Calibrating distributed camera networks. Proc IEEE 96(10): 1625–1639
15. Baker P, Aloimonos Y (2000) Complete calibration of a multi-camera network. Proceedings of the IEEE workshop on omnidirectional vision, 2000. Hilton Head Island, SC, USA, IEEE, pp 134–141
16. Svoboda T, Martinec D, Pajdla T (2005) A convenient multicamera self-calibration for virtual environments. Presence 14(4):407–422
17. Rekleitis I, Dudek G (2005) Automated calibration of a camera sensor network. IEEE/RSJ international conference on intelligent robots and systems. Edmonton, Canada, IEEE, pp 3384–3389

Calibration of Projective Cameras

Zhengyou Zhang
Microsoft Research, Redmond, WA, USA

Synonyms

Lens distortion correction

Related Concepts

▶Calibration; ▶Camera Calibration; ▶Geometric Calibration; ▶Perspective Camera

Definition

Calibration of a projective camera is the process of determining an adjustment on the camera so that, after adjustment, it follows the pinhole or perspective projection model.

Background

A projective camera follows pinhole or perspective projection, which is also known as rectilinear projection because straight lines in a scene remain straight in an image. A real camera usually uses lenses with finite aperture, especially for low-end cameras (such as WebCams) or wide-angle cameras. Lens distortion also arises from imperfect lens design and manufacturing, as well as camera assembly. A line in a scene is not seen as a line in the image. A point in 3D space, its corresponding point in image, and the camera's optical center are not collinear. The linear projective equation is sometimes not sufficient, and lens distortion has to be considered or corrected beforehand.

Theory

According to [1], there are four steps in camera projection including lens distortion:

Step 1: *Rigid transformation* from world coordinate system (X_w, Y_w, Z_w) to camera one (X, Y, Z):

$$[X \quad Y \quad Z]^T = \mathbf{R}\,[X_w \quad Y_w \quad Z_w]^T + \boldsymbol{t}$$

Step 2: *Perspective projection* from 3D camera coordinates (X, Y, Z) to *ideal* image coordinates (x, y) under pinhole camera model:

$$x = f\frac{X}{Z}, \qquad y = f\frac{Y}{Z}$$

where f is the effective focal length.

Step 3: *Lens distortion*:

$$\breve{x} = x + \delta_x\,, \qquad \breve{y} = y + \delta_y$$

where $(\breve{x}, \breve{y})$ are the *distorted* or *true* image coordinates and (δ_x, δ_y) are distortions applied to (x, y). Note that the lens distortion described here is different from Tsai's treatment. Here, we go from ideal to real image coordinates, similar to [2].

Step 4: *Affine transformation* from real image coordinates $(\breve{x}, \breve{y})$ to *frame buffer* (pixel) image coordinates $(\breve{u}, \breve{v})$:

$$\breve{u} = d_x^{-1}\breve{x} + u_0\,, \qquad \breve{v} = d_y^{-1}\breve{y} + v_0\,,$$

where (u_0, v_0) are coordinates of the principal point and d_x and d_y are distances between adjacent pixels in the horizontal and vertical directions, respectively.

There are two types of distortions:

Radial distortion: It is symmetric; ideal image points are distorted along radial directions from the distortion center. This is caused by imperfect lens shape.

Decentering distortion: This is usually caused by improper lens assembly; ideal image points are distorted in both radial and tangential directions.

The reader is referred to [3–6] for more details.

The distortion can be expressed as power series in radial distance $r = \sqrt{x^2 + y^2}$:

$$\delta_x = x(k_1 r^2 + k_2 r^4 + k_3 r^6 + \cdots) + \left[p_1(r^2 + 2x^2) + 2p_2 xy\right](1 + p_3 r^2 + \cdots),$$

$$\delta_y = y(k_1 r^2 + k_2 r^4 + k_3 r^6 + \cdots) + \left[2p_1 xy + p_2(r^2 + 2y^2)\right](1 + p_3 r^2 + \cdots),$$

where k_is are coefficients of radial distortion and p_js are coefficients of decentering distortion.

Based on the reports in the literature [1, 2, 4], it is likely that the distortion function is totally dominated by the radial components and especially dominated by the first term. It has also been found that any more elaborated modeling not only would not help (negligible when compared with sensor quantization) but also would cause numerical instability [1, 2].

Denote the ideal pixel image coordinates by $u = x/d_x$ and $v = y/d_y$. By combining Steps 3 and 4 and if only using the first two radial distortion terms, we obtain the following relationship between $(\breve{u}, \breve{v})$ and (u, v):

$$\breve{u} = u + (u - u_0)\left[k_1(x^2 + y^2) + k_2(x^2 + y^2)^2\right] \quad (1)$$

$$\breve{v} = v + (v - v_0)\left[k_1(x^2 + y^2) + k_2(x^2 + y^2)^2\right]. \quad (2)$$

Lens distortion parameters can be determined as an integrated part of geometric calibration [7]. This can be done by observing a known 3D target [1, 2, 6], by observing a 2D planar pattern [8], by observing a linear point pattern [9], or by moving the camera through a rigid scene [10]. The nonlinearity of the integrated projection and lens distortion model does not allow for a direct calculation of all the parameters of the camera model. Camera calibration including lens distortion can be performed by minimizing the distances between the image points and their predicted positions, i.e.,

$$\min_{\mathbf{A},\mathbf{R},\boldsymbol{t},k_1,k_2} \sum_i \|\boldsymbol{m}_i - \breve{\boldsymbol{m}}(\mathbf{A}, \mathbf{R}, \boldsymbol{t}, k_1, k_2, \boldsymbol{M}_i)\|^2 \quad (3)$$

where $\breve{\boldsymbol{m}}(\mathbf{A}, \mathbf{R}, \boldsymbol{t}, k_1, k_2, \boldsymbol{M}_i)$ is the projection of $\boldsymbol{M}_i$ onto the image according to the pinhole model, followed by distortion according to Eq. (1) and Eq. (2). The minimization is performed through an iterative approach such as using the Levenberg-Marquardt method.

An alternative approach is to perform lens distortion correction as a separate process. Invariance properties under projective transformation are exploited. One is the "plumb line" constraint [4], which is based on the fact that a line in a scene remains a line in an image. Another is the cross-ratio constraint [2], which states that, for four collinear points with known distances between each other in 3D, their corresponding image points are collinear, and their cross-ratio remains the same. Due to lens distortion, projective invariants are not preserved, and we can use the variance to compute the distortion.

References

1. Tsai RY (1987) A versatile camera calibration technique for high-accuracy 3D machine vision metrology using off-the-shelf tv cameras and lenses. IEEE J Robot Autom 3(4):323–344
2. Wei G, Ma S (1994) Implicit and explicit camera calibration: theory and experiments. IEEE Trans Pattern Anal Mach Intell 16(5):469–480
3. Slama CC (ed) (1980) Manual of photogrammetry, 4th edn. American Society of Photogrammetry, Falls Church
4. Brown DC (1971) Close-range camera calibration. Photogramm Eng 37(8):855–866
5. Faig W (1975) Calibration of close-range photogrammetry systems: mathematical formulation. Photogramm Eng Remote Sens 41(12):1479–1486
6. Weng J, Cohen P, Herniou M (1992) Camera calibration with distortion models and accuracy evaluation. IEEE Trans Pattern Anal Mach Intell 14(10):965–980
7. Zhang Z (2004) Camera calibration. In: Medioni G, Kang S (eds) Emerging topics in computer vision. Prentice Hall Professional Technical Reference, Upper Saddle River, pp 4–43
8. Zhang Z (2000) A flexible new technique for camera calibration. IEEE Trans Pattern Anal Mach Intell 22(11): 1330–1334
9. Zhang Z (2004) Camera calibration with one-dimensional objects. IEEE Trans Pattern Anal Mach Intell 26(7): 892–899
10. Zhang Z (1996) On the epipolar geometry between two images with lens distortion. In: International conference on pattern recognition, Vienna, Austria, vol I, pp 407–411

Calibration of Radiometric Falloff (Vignetting)

Stephen Lin
Microsoft Research Asia, Beijing Sigma Center, Beijing, China

Synonyms

Vignetting estimation

Related Concepts

▶Irradiance; ▶Radiometric Calibration; ▶Radiance; ▶Vignetting

Definition

Calibration of radiometric falloff is the measurement of brightness attenuation away from the image center for a given camera, lens, and camera settings.

Background

Several mechanisms may be responsible for radiometric falloff. One is the optics of the camera, which may have a smaller effective lens opening for light incident at greater off-axis angles (i.e., irradiance toward the edges of an image). Radiometric falloff also occurs naturally due to foreshortening of the lens when viewed at increasing angles from the optical axis. A third cause is mechanical in nature, where light arriving at oblique angles is partially obstructed by camera components such as the field stop or lens rim. Digital sensors may also contribute to this falloff because of angle-dependent sensitivity to light. The profile of the radiometric falloff field varies with respect to camera, lens, and camera settings such as focal length and aperture.

Many computer vision algorithms assume that the image irradiance measured at the camera sensor is equal to the scene radiance that arrives at the camera. However, this assumption often does not hold because of radiometric falloff. It is therefore important to measure or estimate the radiometric falloff, and remove its effects from images.

Theory

Radiometric falloff, or vignetting, may be modeled as a function f that represents the proportion of image brightness I at an image position (x, y) relative to that at the image center (x_0, y_0):

$$f(x, y) = \frac{I(x, y)}{I(x_0, y_0)}. \tag{1}$$

Because of approximate radial symmetry in the optical systems of most cameras, the radiometric falloff function may alternatively be expressed in terms of image distance r from the image center:

$$f(r) = \frac{I(r)}{I(0)}, \tag{2}$$

where $r = \sqrt{x^2 + y^2}$. The purpose of radiometric falloff calibration is to recover f, so that its inverse function f^{-1} can be applied to an image i, recorded by the same camera and camera settings, to obtain an image $\tilde{i}$ without radiometric falloff:

$$\tilde{i}(x, y) = f^{-1}(i(x, y)). \tag{3}$$

The effect of radiometric falloff calibration is illustrated in Fig. 1.

Methods

A basic method for calibration of radiometric falloff is to capture a reference image consisting of a uniform radiance field [1–4]. Since the scene itself contains no brightness variation, intensity differences in the image can be attributed solely to radiometric falloff. It must be noted that in these and other methods of falloff calibration, it is assumed that the camera response function is known.

Another approach examines image sequences with overlapping views of an arbitrary static scene [5–9]. In overlapping image regions, corresponding points are assumed to have the same scene radiance. Differences in their intensities are therefore a result of different radiometric falloff at their respective image positions. From the positions and relative intensities among each set of corresponding points, the radiometric falloff field can be recovered without knowledge of scene content. Most of these methods are designed to recover both the radiometric falloff field and the camera response function in a joint manner [6–9].

The radiometric falloff field may alternatively be estimated from a single arbitrary input image. To infer the falloff field in this case, the intensity variation caused by falloff needs to be distinguished from that due to scene content. This has been done using a segmentation-based approach that identifies image regions with reliable data for falloff estimation [10], and by examining the effect of falloff on radial gradient distributions in the image [11].

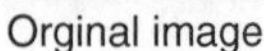

Calibration of Radiometric Falloff (Vignetting), Fig. 1 Image calibrated for radiometric falloff

Application

Calibration of radiometric falloff is of importance to algorithms such as shape-from-shading and photometric stereo that infer scene properties from image irradiance values. It is also essential in applications such as image mosaicing and segmentation that require photometric consistency of the same scene point appearing in different images, or different scene points within the same image.

A measured radiometric falloff field may be used to locate the optical center of the image, since radiometric falloff generally exhibits radial symmetry. The spatial variation of light transmission may also be exploited in sensing, such as to capture high dynamic range intensity values of scene points viewed from a moving camera [12].

References

1. Sawchuk AA (1977) Real-time correction of intensity nonlinearities in imaging systems. IEEE Trans Comput 26(1):34–39
2. Asada N, Amano A, Baba M (1996) Photometric calibration of zoom lens systems. In: IEEE international conference on pattern recognition, Washington DC, pp 186–190
3. Kang SB, Weiss R (2000) Can we calibrate a camera using an image of a flat textureless lambertian surface? In: European conference on computer vision 2000 (ECCV), vol II. Springer, London, pp 640–653
4. Yu W (2004) Practical anti-vignetting methods for digital cameras. IEEE Trans Consum Electron 50:975–983
5. Jia J, Tang CK (2005) Tensor voting for image correction by global and local intensity alignment. IEEE Trans Pattern Anal Mach Intell 27(1):36–50
6. Litvinov A, Schechner YY (2005) Addressing radiometric nonidealities: a unified framework. In: IEEE computer vision and pattern recognition (CVPR), IEEE Computer Society, Silver Spring, pp 52–59
7. Litvinov A, Schechner YY (2005) A radiometric framework for image mosaicing. J Opt Soc Am A 22:839–848
8. Goldman DB, Chen JH (2005) Vignette and exposure calibration and compensation. In: IEEE international conference on computer vision, Beijing, pp 899–906
9. Kim SJ, Pollefeys M (2008) Robust radiometric calibration and vignetting correction. IEEE Trans Pattern Anal Mach Intell 30(4):562–576
10. Zheng Y, Lin S, Kambhamettu C, Yu J, Kang SB (2009) Single-image vignetting correction. IEEE Trans Pattern Anal Mach Intell 31(12):2243–2256
11. Zheng Y, Yu J, Kang SB, Lin S, Kambhamettu C (2008) Single-image vignetting correction using radial gradient symmetry. IEEE Computer Vision and Pattern Recognition (CVPR), Los Alamitos
12. Schechner YY, Nayar SK (2003) Generalized mosaicing: high dynamic range in a wide field of view. Int J Comput Vis 53(3):245–267

Camera Calibration

Zhengyou Zhang
Microsoft Research, Redmond, WA, USA

Related Concepts

▸Calibration; ▸Calibration of Projective Cameras; ▸Geometric Calibration

Definition

Camera calibration is the process of determining certain parameters of a camera in order to fulfill desired tasks with specified performance measures. The reader is referred to entry ▸Calibration for a general discussion on calibration.

Background

There are multiple camera calibration problems. The most common one, which almost becomes the synonym of camera calibration, is geometric calibration (see entry ▸Geometric Calibration). It consists in determining the intrinsic and extrinsic parameters of a camera.

Other camera calibration problems include:

Stereo calibration. A stereo (or stereovision) system consists of multiple cameras. Stereo calibration determines the relative geometry (rotation and translation) between cameras. The intrinsic parameters of each camera can be determined separately as in camera calibration or jointly with the relative geometry.

Photometric calibration. Photometry concerns the measurement of quantities associated with light. Photometric calibration of a camera is a process of determining a function which converts the pixel values to photometric quantities such as SI (*Système International* in French) light units. A test chart of patches with known relative luminances is usually used for photometric calibration.

Color calibration. The pixel values of a color camera depend not only on the surface reflectance but also on the illuminating source. White balance is a common color calibration task, which uses a standard test target with known reflectance to remove the influence of lighting on the scene. Another common task is to calibrate multiple seemingly identical cameras which are not due to tolerance in fabrication.

Camera Extrinsic Parameters

Zhengyou Zhang
Microsoft Research, Redmond, WA, USA

Synonyms

Camera pose; Extrinsic parameters

Related Concepts

▸Camera Parameters (Intrinsic, Extrinsic); ▸Intrinsics

Definition

extrinsic, short for *extrinsic parameters*, refer to the parameters not forming the essential part of a thing, which is usually a camera in computer vision. The extrinisc parameters of a camera include its pose (rotation and translation) with respect to a reference coordinate system.

See entry "▸Camera Parameters (Intrinsic, Extrinsic)" for more details.

Camera Model

Zhengyou Zhang
Microsoft Research, Redmond, WA, USA

Related Concepts

▸Camera Calibration; ▸Camera Parameters (Intrinsic, Extrinsic); ▸Intrinsics; ▸Perspective Camera

Definition

A *camera* is a device that records lights coming from the scene and saves them in images. These images may be still photographs or moving images (videos). The *camera model* describes the mathematical relationship between the 3D coordinates of a point in the scene from which the light comes from and the 2D coordinates of its projection onto the image plane. The ideal camera model is known as the pinhole camera model or perspective camera model, but other camera models exist such as thin and thick cameras.

Background

The term *camera* comes from the *camera obscura* (in Latin for "dark chamber") [1]. A camera obscura is a dark room, consisting of a darkened chamber or box, into which light is admitted through a pinhole (later a convex lens), forming an image of external objects on a surface of wall, paper, or glass. A modern camera generally consists of an enclosed hollow with an opening (aperture) at one end for light to enter and a recording or viewing surface (such as a CCD or CMOS sensor) for capturing the light on the other end. A majority of cameras have a lens positioned in front of the camera's opening to gather the incoming light and focus all or part of the image on the recording surface.

A camera may work with the light of the visible spectrum. If it records each of the red, blue, and green primary colors at each pixel, then the camera is called a color camera; if it only records the shades of black and white (the grey levels of the light intensity), the camera is called a black and white camera.

A camera may also work with the light outside of the visible spectrum, e.g., with the infrared (IR) light, and the camera is called the IR camera.

Pinhole Camera Model

A pinhole camera can be ideally modeled as the *perspective projection*. This is by far the most popularly used model in the computer vision community. The reader is referred to the entry ▸Perspective Camera for details.

Thin and Thick Lens Camera Models

Although pinhole cameras model quite well most of the cameras we use in the computer vision community, they cannot be used physically in a real imaging system. This is for two reasons:

- An ideal pinhole, having an infinitesimal aperture, does not allow to gather enough amount of light to produce measurable image brightness (called *image irradiance*).
- Because of the wave nature of light, diffraction occurs at the edge of the pinhole and the light spread over the image [2]. As the pinhole is made smaller and smaller, a larger and larger fraction of the incoming light is deflected far from the direction of the incoming ray.

To avoid these problems, a real imaging system usually uses lenses with finite aperture. This appendix aims at having the reader know that there are other camera models available. One should choose an appropriate model for a particular imaging device [3, Sect. 2.A.1].

For an ideal lens, which is known as the *thin lens*, all optical rays parallel to the optical axis converge to a point on the optical axis on the other side of the lens at a distance equal to the so-called *focal length* f (see Fig. 1).

The light ray through the center of the lens is undeflected; thus a thin lens produces the same projection as the pinhole. However, it gathers also a finite amount of light reflected from (or emitted by) the object (see Fig. 2). By the familiar *thin lens law*, rays from points at a distance Z are focused by the lens at a distance $-F$, and Z and $-F$ satisfy

$$\frac{1}{Z} + \frac{1}{-F} = \frac{1}{-f}, \tag{1}$$

where f is the focal length.

If we put an image plane at the distance $-F$, then points at other distances than Z are imaged as small blur circles. This can be seen by considering the cone of light rays passing through the lens with apex at the point where they are correctly focused [2]. The size of the blur circle can be determined as follows. A point at distance $\hat{Z}$ is focused if it is imaged at a point $-\hat{F}$ from the lens (see Fig. 3), where

$$\frac{1}{\hat{Z}} + \frac{1}{-\hat{F}} = \frac{1}{-f}.$$

It gives rise to a blur circle on the image plane located at distance $-F$. The diameter of the blur circle, e, can be computed by triangle similarity

$$\frac{e}{d} = \frac{|F - \hat{F}|}{\hat{F}},$$

which gives

$$e = \frac{d}{\hat{F}} |F - \hat{F}| = \frac{fd}{\hat{Z}} \frac{|Z - \hat{Z}|}{Z - f},$$

where d is the diameter of the lens. If the diameter of blur circles, e, is less than the resolution of the image, then the object is well focused and its image is clean. The range of distances over which objects are focused "sufficiently well" is called the *depth of field*. It is clear

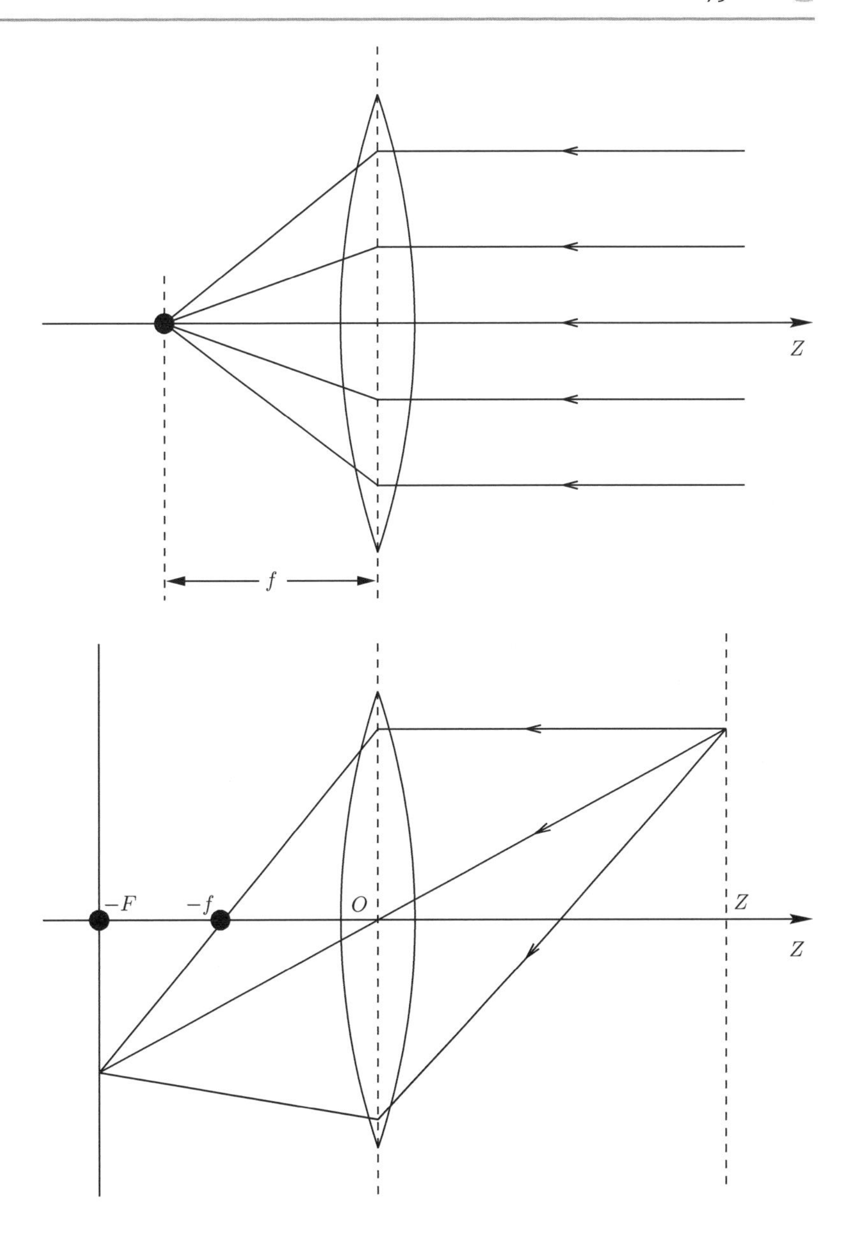

Camera Model, Fig. 1 Cross-sectional view of a thin lens sliced by a plane containing the optical axis. All light rays parallel to the optical axis converge to a point at a distance equal to the focal length

Camera Model, Fig. 2 A thin lens gathers light from a finite area and produces a well-focused image at a particular distance

that the larger the lens aperture d, the less the depth of field.

From (1), it is seen that for objects relatively distant from the lens (*i.e.*, $Z \gg f$), we have $F = f$. If the image plane is located at distance f from the lens, then the camera can be modeled reasonably well by the pinhole.

It is difficult to manufacture a perfect lens. In practice, several simple lenses are carefully assembled to make a compound lens with better properties. In an imaging device with mechanism of focus and zoom, the lenses are allowed to move. It appears difficult to model such a device by a pinhole or thin lens. Another model, called the *thick lens*, is used by more and more researchers [4, 5].

An ideal thick lens is illustrated in Fig. 4. It is composed of two lenses, each having two opposite surfaces, one spherical and the other plane. These two planes p_1 and p_2, called the *principal planes*, are perpendicular to the optical axis and are separated by a distance t, called the *thickness of the lens*. The principal planes intersect the optical axis at two points, called the *nodal points*. The thick lens produces the same perspective projection as the ideal thin lens,

Camera Model, Fig. 3 Focus and blur circles

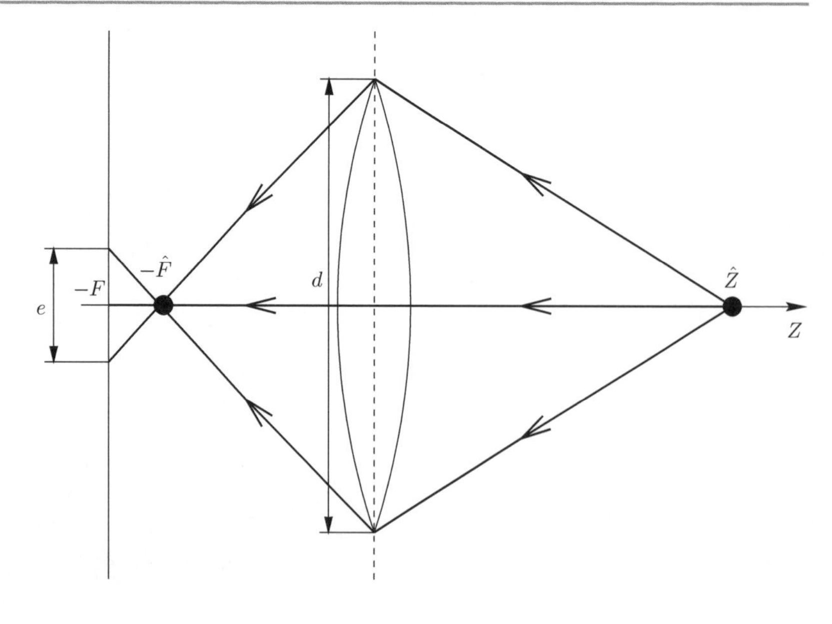

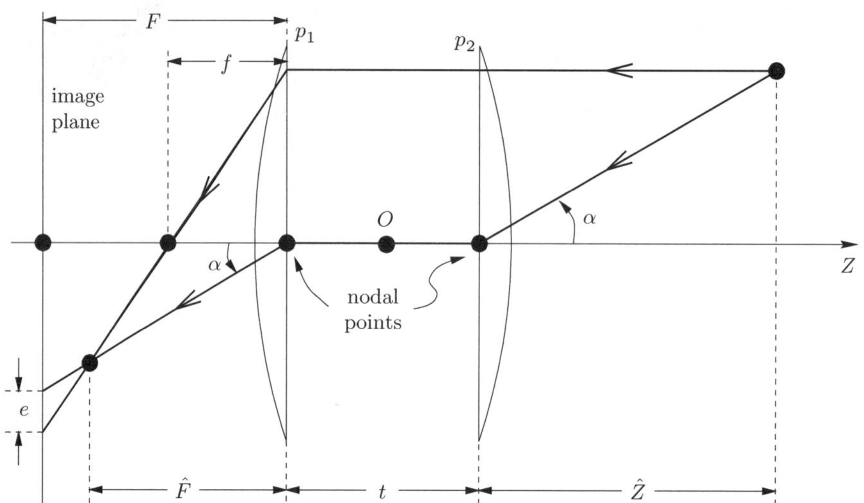

Camera Model, Fig. 4 Cross-sectional view of a thick lens sliced by a plane containing the optical axis

except for *an additional offset* equal to the lens thickness t along the optical axis. A light ray arriving at the first nodal point leaves the rear nodal point without changing direction. A thin lens can then be considered as a thick lens with $t = 0$.

It is thus clear that a thick lens can be considered as a thin lens if the object is relatively distant to the camera compared to the lens thickness (*i.e.*, $\hat{Z} \gg t$). It can be further approximated by a pinhole only when the object is well focused (*i.e.*, $F \approx \hat{F}$), and this is valid only locally.

References

1. Wikipedia (2011) History of the camera. http://en.wikipedia.org/wiki/History_of_the_camera
2. Horn BKP (1986) Robot vision. MIT, Cambridge
3. Xu G, Zhang Z (1996) Epipolar geometry in stereo, motion and object recognition. Kluwer Academic, Dordrecht
4. Pahlavan K, Uhlin T, Ekhlund JO (1993) Dynamic fixation. In: Proceedings of the 4th international conference on computer vision, Berlin, Germany. IEEE Computer Society, Los Alamitos, pp 412–419
5. Lavest J (1992) Stéréovision axiale par zoom pour la robotique. Ph.D. thesis, Université Blaise Pascal de Clermont-Ferrand, France

Camera Parameters (Internal, External)

▶Camera Parameters (Intrinsic, Extrinsic)

Camera Parameters (Intrinsic, Extrinsic)

Zhengyou Zhang
Microsoft Research, Redmond, WA, USA

Synonyms

Camera model; Camera parameters (internal, external)

Related Concepts

▶Camera Calibration; ▶Calibration of Projective Cameras; ▶Camera Parameters (Intrinsic, Extrinsic); ▶Depth Distortion; ▶Perspective Transformation; ▶Perspective Transformation

Definition

Camera parameters are the parameters used in a camera model to describe the mathematical relationship between the 3D coordinates of a point in the scene from which the light comes from and the 2D coordinates of its projection onto the image plane. The *intrinsic parameters*, also known as *internal parameters*, are the parameters intrinsic to the camera itself, such as the focal length and lens distortion. The *extrinsic parameters*, also known as *external parameters* or *camera pose*, are the parameters used to describe the transformation between the camera and its external world.

Background

In computer vision, in order to understand the environment surrounding us with a camera, we have to know first the camera parameters. Depending on the accuracy we need to achieve and on the quality of the camera, some parameters can be neglected. For example, with a high-quality camera, the lens distortion can usually be ignored in most of the applications.

Theory

In the entry ▶Perspective Camera, we describe the mathematical model of a perspective camera with only a single parameter, the focal length f. The relationship between a 3D point and its image projection is described by

$$s\widetilde{\mathbf{m}} = \mathbf{P}\widetilde{\mathbf{M}}\,, \tag{1}$$

where $s = S$ is an arbitrary nonzero scalar and $\mathbf{P}$ is a *projective projection matrix* given by

$$\mathbf{P} = \begin{bmatrix} f & 0 & 0 & 0 \\ 0 & f & 0 & 0 \\ 0 & 0 & 1 & 0 \end{bmatrix}.$$

Before proceeding, the reader needs to review the entry ▶Perspective Camera.

Extrinsic Parameters

In the projective projection matrix described in the entry ▶Perspective Camera, recapitulated above, we assumed that 3D points are expressed in the camera coordinate system. In practice, they can be expressed in any 3D coordinate system, which is sometimes referred as the *world coordinate system*. As shown in Fig. 1, we go from the old coordinate system centered at the optical center C (camera coordinate system) to the new coordinate system centered at point O (world coordinate system) by a rotation $\mathbf{R}$ followed by a translation $\mathbf{t} = CO$. Then, for a single point, its coordinates expressed in the camera coordinate system, M_c, and those expressed in the world coordinate system, M_w, are related by

$$\mathbf{M}_c = \mathbf{R}\mathbf{M}_w + \mathbf{t}\,,$$

or more compactly

$$\widetilde{\mathbf{M}}_c = \mathbf{D}\widetilde{\mathbf{M}}_w\,, \tag{2}$$

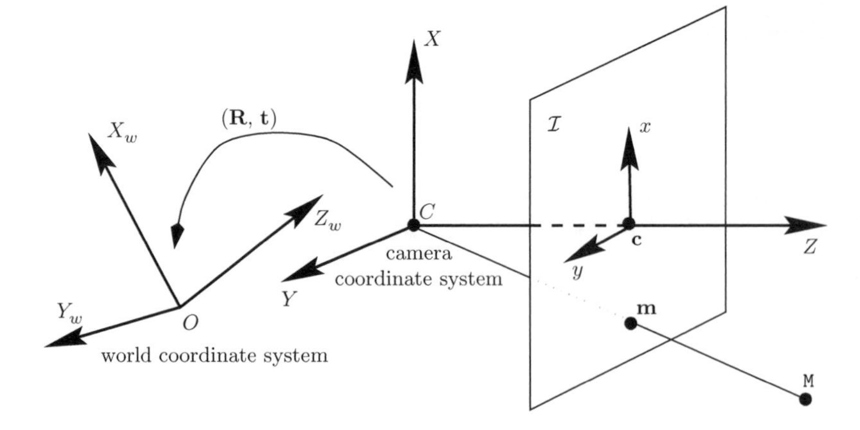

Camera Parameters (Intrinsic, Extrinsic), Fig. 1 World coordinate system and camera extrinsic parameters

where $\mathbf{D}$ is a Euclidean transformation of the three-dimensional space

$$\mathbf{D} = \begin{bmatrix} \mathbf{R} & \mathbf{t} \\ \mathbf{0}_3^T & 1 \end{bmatrix} \quad \text{with } \mathbf{0}_3 = [0,0,0]^T \,. \tag{3}$$

The matrix $\mathbf{R}$ and the vector $\mathbf{t}$ describe the orientation and position of the camera with respect to the new world coordinate system. They are called the *extrinsic parameters* of the camera.

From (1) and (2), we have

$$\widetilde{\mathbf{m}} = \mathbf{P}\widetilde{\mathbf{M}}_c = \mathbf{P}\mathbf{D}\widetilde{\mathbf{M}}_w \,.$$

Therefore the new perspective projection matrix is given by

$$\mathbf{P}_{\text{new}} = \mathbf{P}\mathbf{D} \,. \tag{4}$$

This tells us how the perspective projection matrix $\mathbf{P}$ changes when we change coordinate systems in the three-dimensional space: We simply multiply it on the right by the corresponding Euclidean transformation.

Intrinsic Parameters and Normalized Camera

This section considers the transformation in image coordinate systems. It is very important in practical applications because:

- We do not know the origin of the image plane in advance. It generally does not coincide with the intersection of the optical axis and the image plane.
- The units of the image coordinate axes are not necessarily equal, and they are determined by the sampling rates of the imaging devices.

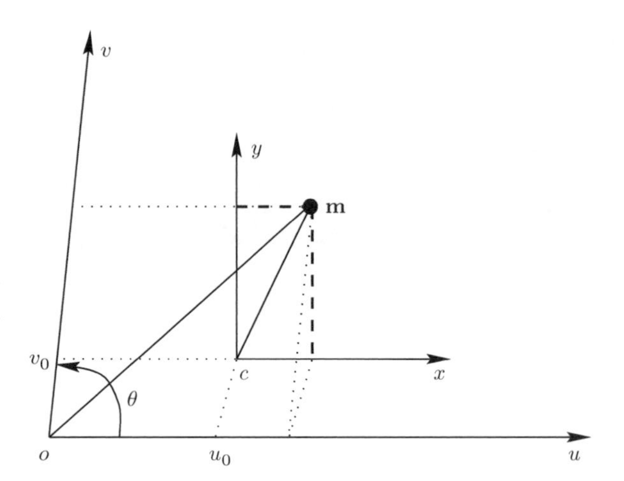

Camera Parameters (Intrinsic, Extrinsic), Fig. 2 Camera intrinsic parameters

- The two axes of a real image may not form a right angle.

To handle these effects, we introduce an affine transformation.

Consider Fig. 2. The original image coordinate system (c, x, y) is centered at the principal point c and has the same units on both x- and y-axes. The coordinate system (o, u, v) is the coordinate system in which we address the pixels in an image. It is usually centered at the upper left corner of the image, which is usually not the principal point c. Due to the electronics of acquisition, the pixels are usually not square. Without loss of generality, the u-axis is assumed to be parallel to the x-axis. The units along the u- and v-axes are assumed to be k_u and k_v with respect to the unit used in (c, x, y). The u- and v-axes may not be exactly orthogonal, and we denote their angle by θ. Let the coordinates of the principal point c in (o, u, v) be $[u_0, v_0]^T$. These five parameters do not depend on the position and orientation of the cameras and are thus called the camera *intrinsic parameters*.

For a given point, let $\mathbf{m}_{\text{old}} = [x, y]^T$ be the coordinates in the original coordinate system; let $\mathbf{m}_{\text{new}} = [u, v]^T$ be the pixel coordinates in the new coordinate system. It is easy to see that

$$\widetilde{\mathbf{m}}_{\text{new}} = \mathbf{H}\widetilde{\mathbf{m}}_{\text{old}} ,$$

where

$$\mathbf{H} = \begin{bmatrix} k_u & k_u \cot\theta & u_0 \\ 0 & k_v / \sin\theta & v_0 \\ 0 & 0 & 1 \end{bmatrix} .$$

Since, according to (1), we have

$$s\widetilde{\mathbf{m}}_{\text{old}} = \mathbf{P}_{\text{old}}\widetilde{\mathbf{M}} ,$$

we conclude that

$$s\widetilde{\mathbf{m}}_{\text{new}} = \mathbf{H}\mathbf{P}_{\text{old}}\widetilde{\mathbf{M}} ,$$

and thus

$$\mathbf{P}_{\text{new}} = \mathbf{H}\mathbf{P}_{\text{old}} = \begin{bmatrix} fk_u & fk_u \cot\theta & u_0 & 0 \\ 0 & fk_v / \sin\theta & v_0 & 0 \\ 0 & 0 & 1 & 0 \end{bmatrix} . \tag{5}$$

Note that it depends on the products fk_u and fk_v, which means that a change in the focal length and a change in the pixel units are indistinguishable. We thus introduce two parameters $\alpha_u = fk_u$ and $\alpha_v = fk_v$.

We will now define a special coordinate system that allows us to normalize the image coordinates [1]. This coordinate system is called the *normalized coordinate system* of the camera. In this "normalized" camera, the image plane is located at a unit distance from the optical center (*i.e.* $f = 1$). The perspective projection matrix of the normalized camera is given by

$$\mathbf{P}_N = \begin{bmatrix} 1 & 0 & 0 & 0 \\ 0 & 1 & 0 & 0 \\ 0 & 0 & 1 & 0 \end{bmatrix} . \tag{6}$$

For a world point $[X, Y, Z]^T$ expressed in the camera coordinate system, its normalized coordinates are

$$\hat{x} = \frac{X}{Z}, \quad \hat{y} = \frac{Y}{Z} . \tag{7}$$

A matrix $\mathbf{P}$ defined by (5) can be decomposed into the product of two matrices:

$$\mathbf{P}_{\text{new}} = \mathbf{A}\mathbf{P}_N , \tag{8}$$

where

$$\mathbf{A} = \begin{bmatrix} \alpha_u & \alpha_u \cot\theta & u_0 \\ 0 & \alpha_v / \sin\theta & v_0 \\ 0 & 0 & 1 \end{bmatrix} . \tag{9}$$

The matrix $\mathbf{A}$ contains only the intrinsic parameters and is called *camera intrinsic matrix*. It is thus clear that the normalized image coordinates are given by

$$\begin{bmatrix} \hat{x} \\ \hat{y} \\ 1 \end{bmatrix} = \mathbf{A}^{-1} \begin{bmatrix} u \\ v \\ 1 \end{bmatrix} . \tag{10}$$

Through this transformation from the available pixel image coordinates, $[u, v]^T$, to the imaginary normalized image coordinates, $[\hat{x}, \hat{y}]^T$, the projection from the space onto the normalized image does not depend on the specific cameras. This frees us from thinking about characteristics of the specific cameras and allows us to think in terms of ideal systems in stereo, motion, and object recognitions.

The General Form of Perspective Projection Matrix

The camera can be considered as a system that depends upon the intrinsic and the extrinsic parameters. There are five intrinsic parameters: the scale factors α_u and α_v, the coordinates u_0 and v_0 of the principal point, and the angle θ between the two image axes. There are six extrinsic parameters, three for the rotation and three for the translation, which define the transformation from the world coordinate system to the standard coordinate system of the camera.

Combining (4) and (8) yields the general form of the perspective projection matrix of the camera:

$$\mathbf{P} = \mathbf{A}\mathbf{P}_N\mathbf{D} = \mathbf{A}\,[\mathbf{R}\ \mathbf{t}] . \tag{11}$$

The projection of 3D world coordinates $\mathbf{M} = [X, Y, Z]^T$ to 2D pixel coordinates $\mathbf{m} = [u, v]^T$ is then described by

$$s\widetilde{\mathbf{m}} = \mathbf{P}\widetilde{\mathbf{M}}\,, \tag{12}$$

where s is an arbitrary scale factor. Matrix $\mathbf{P}$ has $3{\times}4 = 12$ elements but has only 11 degrees of freedom because it is defined up to a scale factor.

Let p_{ij} be the (i, j) entry of matrix $\mathbf{P}$. Eliminating the scalar s in (12) yields two nonlinear equations:

$$u = \frac{p_{11}X + p_{12}Y + p_{13}Z + p_{14}}{p_{31}X + p_{32}Y + p_{33}Z + p_{34}} \tag{13}$$

$$v = \frac{p_{21}X + p_{22}Y + p_{23}Z + p_{24}}{p_{31}X + p_{32}Y + p_{33}Z + p_{34}}\,. \tag{14}$$

Camera *calibration* is the process of estimating the intrinsic and extrinsic parameters of a camera or the process of first estimating the matrix **P** and then deducing the camera parameters from **P**. A wealth of work has been carried out in this domain by researchers either in photogrammetry [2, 3] or in computer vision and robotics [4–9] (see [10] for a review). The usual method of calibration is to compute camera parameters from one or more images of an object of *known size and shape*, for example, a flat plate with a regular pattern marked on it. From (12) or (13) and (14), we have two nonlinear equations relating 2D to 3D coordinates. This implies that each pair of an identified image point and its corresponding point on the calibration object provides two constraints on the intrinsic and extrinsic parameters of the camera. The number of unknowns is 11. It can be shown that, given N points ($N \geq 6$) in general position, the camera can be calibrated. The presentation of calibration techniques is beyond the scope of this book. The interested reader is referred to the above-mentioned references.

Once the perspective projection matrix **P** is given, we can compute the coordinates of the optical center C of the camera in the world coordinate system. We first decompose the 3×4 matrix **P** as the concatenation of a 3×3 submatrix **B** and a 3-vector **b**, that is, $\mathbf{P} = [\mathbf{B}\,\mathbf{b}]$. Assume that the rank of **B** is 3. In the entry ▶Perspective Camera, we explained that, under the pinhole model, the optical center projects to $[0, 0, 0]^T$ (*i.e.* $s = 0$). Therefore, the optical center can be obtained by solving

$$\mathbf{P}\widetilde{C} = \mathbf{0}, \qquad \text{that is, } [\mathbf{B}\,\mathbf{b}]\begin{bmatrix} C \\ 1 \end{bmatrix} = \mathbf{0}\,.$$

The solution is

$$C = -\mathbf{B}^{-1}\mathbf{b}\,. \tag{15}$$

Given matrix **P** and an image point **m**, we can obtain the equation of the 3-D semi-line defined by the optical center C and point **m**. This line is called the *optical ray* defined by **m**. Any point on it projects to the single point **m**. We already know that C is on the optical ray. To define it, we need another point. Without loss of generality, we can choose the point D such that the scale factor $s = 1$, that is,

$$\widetilde{\mathbf{m}} = [\mathbf{B}\,\mathbf{b}]\begin{bmatrix} D \\ 1 \end{bmatrix}.$$

This gives $D = \mathbf{B}^{-1}(-\mathbf{b} + \widetilde{\mathbf{m}})$. A point on the optical ray is thus given by

$$\mathbf{M} = C + \lambda(D - C) = \mathbf{B}^{-1}(-\mathbf{b} + \lambda\widetilde{\mathbf{m}})\,,$$

where λ varies from 0 to ∞.

References

1. Faugeras O (1993) Three-dimensional computer vision: a geometric viewpoint. MIT, Cambridge
2. Brown DC (1971) Close-range camera calibration. Photogramm Eng 37(8):855–866
3. Faig W (1975) Calibration of close-range photogrammetry systems: mathematical formulation. Photogramm Eng Remote Sens 41(12):1479–1486
4. Tsai R (1986) Multiframe image point matching and 3d surface reconstruction. IEEE Trans Pattern Anal Mach Intell 5:159–174
5. Faugeras O, Toscani G (1986) The calibration problem for stereo. In: Proceedings of the IEEE conference on computer vision and pattern recognition, Miami Beach, FL. IEEE, Los Alamitos, pp 15–20
6. Lenz R, Tsai R (1987) Techniques for calibrating of the scale factor and image center for high accuracy 3D machine vision metrology. In: International conference on robotics and automation, Raleigh, NC, pp 68–75
7. Toscani G (1987) Système de Calibration optique et perception du mouvement en vision artificielle. Ph.D. thesis, Paris-Orsay
8. Wei G, Ma S (1991) Two plane camera calibration: a unified model. In: Proceedings of the IEEE conference on

computer vision and pattern recognition (CVPR), Hawaii, pp 133–138
9. Weng J, Cohen P, Rebibo N (1992) Motion and structure estimation from stereo image sequences. IEEE Trans RA 8(3):362–382
10. Tsai R (1989) Synopsis of recent progress on camera calibration for 3D machine vision. In: Khatib O, Craig JJ, Lozano-Pérez T (eds) The robotics review. MIT, Cambridge, pp 147–159

Camera Pose

Zhengyou Zhang
Microsoft Research, Redmond, WA, USA

Synonyms

Camera extrinsic parameters

Related Concepts

▸Camera Calibration; ▸Camera Parameters (Intrinsic, Extrinsic); ▸Intrinsics; ▸Perspective Camera

Definition

Camera pose is referred to the position and orientation of a camera with respect to a reference coordinate system, which is usually known as the world coordinate system.

Background

Determining the camera pose is usually a first step toward perceiving the surrounding environment. In structure from motion where a camera is moving through the environment, one needs to determine the successive camera poses at different instants in order to reconstruct the surrounding environment in 3D. In a multi-camera (two or more) system, one needs to determine the relative camera pose, i.e., how one camera is related to other cameras.

The reader is referred to entry "▸Camera Parameters (Intrinsic, Extrinsic)" for details.

Camera Response Function

▸Radiometric Response Function

Camera-Shake Blur

▸Motion Blur

Catadioptric Camera

Srikumar Ramalingam
Mitsubishi Electric Research Laboratories, Cambridge, MA, USA

Synonyms

▸Catoptrics; ▸Dioptrics

Related Concepts

▸Center of Projection; ▸Field of View; ▸Omnidirectional Camera

Definition

A catadioptric system is a camera configuration where both lenses and mirrors are jointly used to achieve specialized optical properties. These configurations are referred to as *catadioptric*, where "cata" comes from mirrors (reflective) and "dioptric" comes from lenses (refractive).

Background

In 1637, René Descartes observed that the refractive and reflective "ovals" (conical lenses and mirrors) have the ability to focus light into one single point on illumination from a chosen point [1]. It was reported that the same results were derived by Feynman et al. [2] and

Drucker and Locke [3]. In computer vision community, Baker and Nayar presented the complete class of single viewpoint catadioptric configurations with detailed solutions and degenerate cases [4]. Some of these results have been independently derived by Bruckstein and Richardson [5]. Survey of various catadioptric cameras, a review and details of their calibration, and 3D reconstruction algorithms can also be found in [6, 7].

Theory and Classification

The combination of mirrors and lenses provides a wide range of design possibilities leading to interesting applications in computer vision. Most catadioptric configurations have larger field of view compared to conventional pinhole cameras. Other important design goals are compactness of a sensor, a single effective viewpoint, image quality, focusing properties, or a desired projection function. The catadioptric cameras may be classified in many ways. In [6], Sturm et al. classify the catadioptric cameras in to five different types:

- Single-mirror central systems, having a single effective viewpoint
- Central systems using multiple mirrors
- Noncentral systems
- Single-lens stereo systems
- Programmable devices

In what follows, catadioptric cameras are classified into central and noncentral systems. Most of these catadioptric configurations were proposed by researchers along with specified calibration and 3D reconstruction algorithms.

Central Catadioptric Configurations

It requires a very careful choice of the shape of the mirrors and their positioning to obtain a single effective viewpoint in catadioptric imaging. The single viewpoint design goal is important because it allows the generation of pure perspective images from the catadioptric images. Furthermore, it allows one to solve motion estimation and 3D reconstruction algorithms in the same way as perspective cameras:

- *Planar mirror*: In [4, 8], it can be observed that by using planar mirrors along with a perspective camera, one can obtain a single viewpoint configuration. Since planar mirrors do not increase the field of view of the system, they are not very interesting for building omnidirectional cameras. Using four planar mirrors in a pyramidal configuration along with four perspective cameras, Nalwa [9] produced an omnidirectional sensor of field of view of $360° \times 50°$. The optical centers of the four cameras and the angles made by the four planar faces are adjusted to obtain a single effective viewpoint for the system.
- *Conical mirrors*: By positioning the optical center of a perspective camera at the apex of a cone, one can obtain a single center configuration. Nevertheless, the only light rays reaching the camera after a reflection in the mirror are those grazing the cone. This case is thus not useful to enhance the field of view while conserving a single center of projection. However, in the work [10], it was proved that conical mirrors can be used to construct a nondegenerate single viewpoint omnidirectional cameras. The outer surface of the conical mirror forms a virtual image corresponding to the real scene behind the conical mirror. On placing the optical center of the pinhole camera at the vertex of the cone, the camera sees the world through the reflection on the outer surface of the mirror. In other words, the cone is not blocking the view. On the other hand, the cone is the view.
- *Spherical mirror*: If the optical center of a perspective camera is fixed at the center of a spherical mirror, one can obtain a single viewpoint configuration. Unfortunately, all that the perspective camera sees is its own reflection. As a result the spherical mirror produces a degenerate configuration without any advantage. Remember that by positioning the perspective camera outside the sphere, one can obtain a useful *noncentral* catadioptric camera.
- *Parabolic mirror*: Figure 1 shows a single viewpoint catadioptric system with a parabolic mirror and an orthographic camera. It is easier to study a catadioptric configuration by considering the back projection rather than the forward projection. Consider the back projection of an image point **p**. The back-projected ray from the image pixel **p**, starting from the optical center at infinity, is parallel to the axis of the parabolic mirror. This ray intersects and reflects from the surface of the mirror. The reflection is in accordance with the laws of reflection. This

Catadioptric Camera, Fig. 1 Parabolic mirror + orthographic camera [8]. **P** refers to the 3D scene point. **F**, the focus of the parabolic mirror, is the effective viewpoint

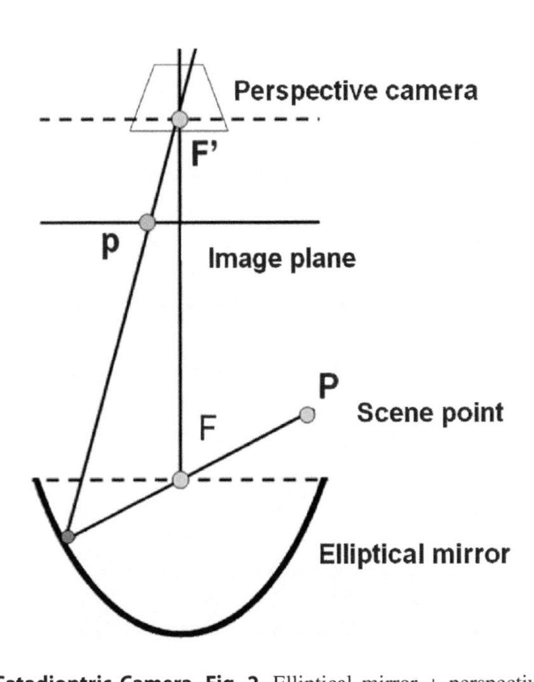

Catadioptric Camera, Fig. 2 Elliptical mirror + perspective camera [4]. **P** refers to the 3D scene point. **F** and **F′** refer to the two foci of the mirror and **p** refers to the image point. **F** is the effective viewpoint

reflected light ray is nothing but the incoming light ray from a scene point **P** in forward projection. The incoming ray passes through the focus **F** if extended on the inside of the mirror. This point where all the incoming light rays intersect (virtually) is called the effective viewpoint.

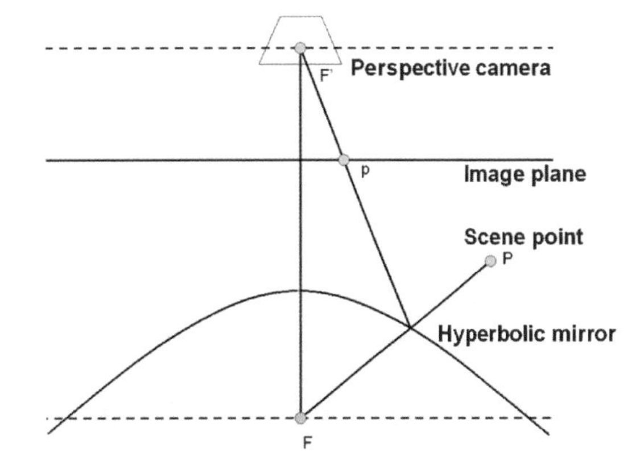

Catadioptric Camera, Fig. 3 Hyperbolic mirror + perspective camera [4]. **P** refers to the 3D scene point. **F** and **F′** refer to the two foci of the mirror and **p** refers to the image point. **F** is the effective viewpoint

– *Elliptical mirror*: Figure 2 shows a central catadioptric system with an elliptical mirror and a perspective camera. The optical center of the perspective camera is placed at the upper focus of the elliptical mirror. By back-projecting an image point **p**, one can observe the following. The back-projected ray, starting from the optical center at the upper focus of the elliptical mirror, intersects and reflects from the surface of the elliptical mirror. The reflected back-projected ray, or the incoming light ray, virtually passes through the lower focus of the mirror. Thus, the lower focus (**F**) is the effective viewpoint of the system.
– *Hyperbolic mirror*: In Fig. 3, a catadioptric system is shown with a hyperbolic mirror and a perspective camera. The optical center of the perspective camera is placed at the external focus of the mirror **F′**. The back-projected ray of the image point **p** starts from the optical center, which is the external focus **F′** of the mirror, of the perspective camera. Using the same argument as above, one can observe that the lower focus **F** is the effective viewpoint. The first known work to use a hyperbolic mirror along with a perspective camera at the external focus of the mirror to obtain a single effective viewpoint configuration is [11]. Later in 1995, a similar implementation was proposed in [12].

Noncentral Catadioptric Cameras

Single viewpoint configurations are extremely delicate to construct, handle, and maintain. By relaxing this single viewpoint constraint, one can obtain greater flexibility in designing novel systems. In fact most real catadioptric cameras are geometrically noncentral, and even the few restricted central catadioptric configurations are usually noncentral in practice [13]. For example, in the case of para-catadioptric cameras, the telecentric lens is never truly orthographic and it is difficult to precisely align the mirror axis and the axis of the camera. In hyperbolic or elliptic configurations, precise positioning of the optical center of the perspective camera in one of the focal points of the hyperbolic or elliptic mirror is practically infeasible. In [14], Ramalingam et al. show that most of the practically used catadioptric configurations fall under an axial camera model where all the projection rays pass through a single line rather than a single point in space. A few noncentral catadioptric configurations are mentioned below. Analogous to the single viewpoint in central cameras, there is a *viewpoint locus* in non-central cameras. It can be defined as follows: a curve or other set of points such that all projection rays cut at least one of the points in the viewpoint locus. Usually, one tries to find the "simplest" such set of points:

- *Conical mirror*: On using a conical mirror in front of a perspective camera, one can obtain an omnidirectional sensor [15, 16]. Nevertheless this configuration does not obey the single viewpoint restriction (besides in the degenerate case of the perspective optical center being located at the cone's vertex). If the optical center lies on the mirror axis, then the viewpoint locus is a circle in 3D, centered in the mirror axis (it can be pictured as a halo over the mirror). An alternative choice of viewpoint locus is the mirror axis. Otherwise, the viewpoint locus is more general.
- *Spherical mirror*: On using a spherical mirror along with a perspective camera, one can enhance the field of view of the imaging system [16–18]. Again this configuration does not obey the single viewpoint restriction (besides in the degenerate case of the perspective optical center being located at the sphere center).
- *Digital micro-mirror array*: Another interesting camera is the recently introduced programmable imaging device using a digital micro-mirror array [19]. A perspective camera is made to observe a scene through a programmable array of micro-mirrors. By controlling the orientations and positions of these mirrors, one can obtain an imaging system with complete control (both in terms of geometric and radiometric properties) over the incoming light ray for every pixel. However, there are several practical issues which make it difficult to realize the full potential of such an imaging system. First, current hardware constraints prohibit the usage of more than two possible orientations for each micro-mirror. Second, arbitrary orientations of the micro-mirrors would produce a discontinuous image which is unusable for many image processing operations.
- *Oblique cameras*: An ideal example for a noncentral camera is an oblique camera. No two rays intersect in an oblique camera [20]. In addition to developing multi-view geometry for oblique cameras, Pajdla also proposed a physically realizable system which obeys oblique geometry. The practical system consists of a rotating catadioptric camera that uses a conical mirror and a telecentric optics. The viewpoint locus is equivalent to a two-dimensional surface or a set of points, where each of the projection rays passes through at least one of the points. Different catadioptric configurations come with different calibration and 3D reconstruction algorithms. Recently, there has been a lot of interest in unifying different camera models and developing generic calibration and 3D reconstruction algorithms [21–24].

Application

Due to enhanced field of view, catadioptric cameras are mainly used in surveillance, car navigation, image-based localization, and augmented reality applications.

References

1. Descartes R, Smith D (1637) The geometry of René Descartes. Dover, New York. Originally published in Discours de la Méthode
2. Feynman R, Leighton R, Sands M (1963) The feynman lectures on physics. Mainly mechanics, radiation, and heat, Addison-Wesley, Reading, vol 1
3. Drucker D, Locke P (1996) A natural classification of curves and surfaces with reflection properties. Math Mag 69(4):249–256
4. Baker S, Nayar S (1998) A theory of catadioptric image formation. In: International conference on computer vision (ICCV), Bombay, pp 35–42

5. Bruckstein A, Richardson T (2000) Omniview cameras with curved surface mirrors. In: IEEE workshop on omnidirectional vision, Hilton Head Island, pp 79–84
6. Sturm P, Ramalingam S, Tardif JP, Gasparini S, Barreto J (2011) Camera models and fundamental concepts used in geometric computer vision. Found Trends Comput Graph Vis 6(1–2):1–183
7. Ramalingam S (2006) Generic imaging models: calibration and 3d reconstruction algorithms. PhD thesis, INRIA Rhone Alpes
8. Nayar S (1997) Catadioptric omnidirectional camera. In: Conference on computer vision and pattern recognition (CVPR), San Juan, pp 482–488
9. Nalwa V (1996) A true omnidirectional viewer. Technical report, Bell Laboratories, Holmdel, NJ, USA
10. Lin S, Bajcky R (2001) True single view cone mirror omnidirectional catadioptric system. In: International conference on computer vision (ICCV), Vancouver, vol 2, pp 102–107
11. Rees DW (1970) Panoramic television viewing system. US Patent 3,505,465
12. Yamazawa K, Yagi Y, Yachida M (1995) Obstacle avoidance with omnidirectional image sensor hyperomni vision. In: International conference on robotics and automation, Nagoya, pp 1062–1067
13. Micusik B, Pajdla T (2004) Autocalibration and 3d reconstruction with non-central catadioptric cameras. In: Computer vision and pattern recognition (CVPR), San Diego, pp 748–753
14. Ramalingam S, Sturm P, Lodha S (2006) Theory and calibration algorithms for axial cameras. In: ACCV, Hyderabad
15. Yagi Y, Kawato S (1990) Panoramic scene analysis with conic projection. In: International conference on robots and systems (IROS), Cincinnati
16. Bogner S (1995) Introduction to panoramic imaging. In: IEEE SMC, New York, vol 54, pp 3100–3106
17. Hong J (1991) Image based homing. In: International conference on robotics and automation, Sacramento
18. Murphy JR (1995) Application of panoramic imaging to a teleoperated lunar rover. In: IEEE SMC conference, Vancouver, pp 3117–3121
19. Nayar S, Branzoi V, Boult T (2004) Programmable imaging using a digital micromirror array. In: International conference on computer vision and pattern recognition (CVPR), Washington, DC, pp 436–443
20. Pajdla T (2002) Stereo with oblique cameras. Int J Comput Vis 47(1):161-170
21. Geyer C, Daniilidis K (2000) A unifying theory of central panoramic systems and practical implications. In: European conference on computer vision (ECCV), Dublin, pp 159–179
22. Grossberg M, Nayar S (2001) A general imaging model and a method for finding its parameters. In: International conference on computer vision (ICCV), Vancouver, vol 2, pp 108–115
23. Sturm P, Ramalingam S (2004) A generic concept for camera calibration. In: European conference on computer vision (ECCV), Prague, vol 2, pp 1–13
24. Ramalingam S, Sturm P, Lodha S (2005) Towards complete generic camera calibration. In: Conference on computer vision and pattern recognition (CVPR), San Diego

Catoptrics

▶Catadioptric Camera

Center of Projection

Srikumar Ramalingam
Mitsubishi Electric Research Laboratories, Cambridge, MA, USA

Synonyms

Optical center; Single viewpoint

Related Concepts

▶Field of View

Definition

Center of projection is a single 3D point in space where all the light rays sampled by a conventional pinhole camera intersect.

Background

Albrecht Dürer, a German artist, published a treatise on measurement using a series of illustrations of drawing frames and perspective machines. On the left side of Fig. 1, an apparatus for drawing a lute is shown. One end of a thread is attached to a pointer and the other end to a pulley on a wall. The thread also passes through a frame in between the lute and the pulley. When the pointer is fixed at different points on the lute, the vertical and horizontal coordinates of the thread, as it passes through the frame, are marked. By meticulously marking the coordinates for each point on the lute, the perspective image of the lute is created. It is obvious to see the intuition behind this setup, i.e., its similarity to a pinhole camera. The pulley is equivalent to the single viewpoint or the center of projection, the frame replaces the image plane, and finally, the thread is nothing but the light ray emerging from the scene.

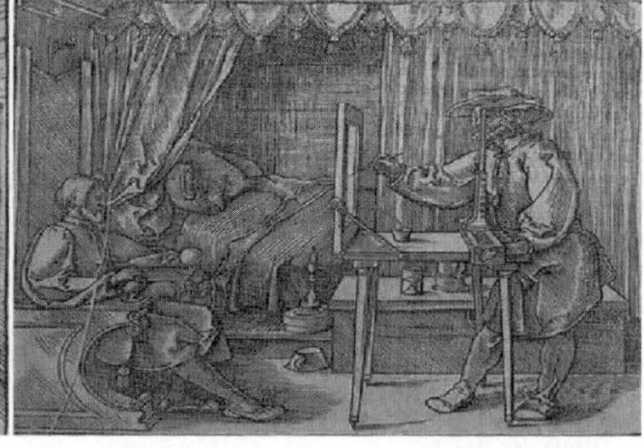

Center of Projection, Fig. 1 *Left*: One of Albrecht Dürer's perspective machines, which was used to draw a lute in the year 1525 [1]. *Right*: An artist uses the perspective principle to accurately draw a man sitting on a chair, by looking at him through a peep hole with one eye, and tracing his features on a glass plate

Though the principle is correct, the procedure is quite complicated. On the right side of Fig. 1, another perspective machine is shown. One can observe an artist squinting through a peep hole with one eye to keep a single viewpoint and tracing his sitter's features onto a glass panel. The idea is to trace the important features first and then transfer the drawing for further painting.

Theory

Pinhole Camera Model

Consider the perspective model that is shown in Fig. 2. Every 3D scene point $\mathbf{P}(X, Y, Z)$ gets projected onto the image plane to a point $\mathbf{p}(x, y)$ through the optical center $\mathbf{C}$. The optical axis is the perpendicular line to the image plane passing through the optical center. The center of radial symmetry in the image or principal point, i.e., the point of intersection of the optical axis and the image plane, is given by $\mathbf{O}$. The distance between $\mathbf{C}$ (optical center) and the image plane is the focal length f. The optical center of the camera is the origin of the coordinate system. The image plane is parallel to the XY plane, held at a distance of f from the origin. Using the basic laws of trigonometry, one can observe the following:

$$x = \frac{fX}{Z}, \; y = \frac{fY}{Z}$$

Once expressed in homogeneous coordinates, the above relations transform to the following:

$$\begin{bmatrix} x \\ y \\ 1 \end{bmatrix} \sim \begin{bmatrix} f & 0 & 0 & 0 \\ 0 & f & 0 & 0 \\ 0 & 0 & 1 & 0 \end{bmatrix} \begin{bmatrix} X \\ Y \\ Z \\ 1 \end{bmatrix}$$

where the relationship $\sim$ stands for "equal up to a scale."

Practically available CCD cameras deviate from the perspective model. First, the principal point (u_0, v_0) does not necessarily lie on the geometrical center of the image. Second, the horizontal and vertical axes (u and v) of the image are not always perfect perpendicular. Let the angle between the two axes be θ. Finally, each pixel is not a perfect square and consequently, f_u and f_v are the two focal lengths that are measured in terms of the unit lengths along u and v directions. By incorporating these deviations in the camera model, one can obtain the following scene (X, Y, Z) to image (u, v) transformation:

$$\begin{bmatrix} u \\ v \\ 1 \end{bmatrix} \sim \begin{bmatrix} f_x & f_v \cot\theta & u_0 & 0 \\ 0 & \frac{f_v}{\sin\theta} & v_0 & 0 \\ 0 & 0 & 1 & 0 \end{bmatrix} \begin{bmatrix} X \\ Y \\ Z \\ 1 \end{bmatrix}$$

In practice, the 3D point is available in some world coordinate system that is different from the camera

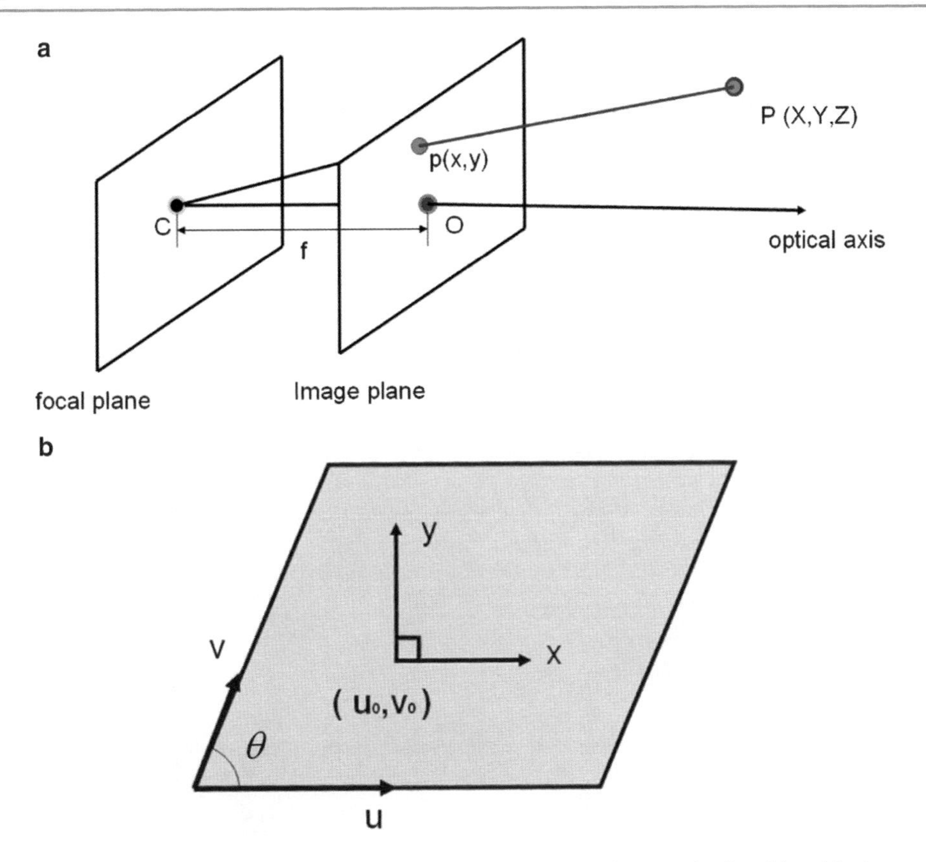

Center of Projection, Fig. 2 (**a**) Perspective camera model. (**b**) The relationship between (u, v) and (x, y) is shown

coordinate system. The motion between these coordinate systems is given by $(\mathsf{R}, \mathbf{t})$:

$$\begin{bmatrix} u \\ v \\ 1 \end{bmatrix} \sim \begin{bmatrix} f_x & f_v \cot\theta & u_0 \\ 0 & \frac{f_v}{\sin\theta} & v_0 \\ 0 & 0 & 1 \end{bmatrix} \begin{bmatrix} \mathsf{R} & -\mathsf{R}\mathbf{t} \end{bmatrix} \begin{bmatrix} X \\ Y \\ Z \\ 1 \end{bmatrix} \quad (1)$$

$$\mathsf{M} = \begin{bmatrix} f_x & f_v \cot\theta & u_0 \\ 0 & \frac{f_v}{\sin\theta} & v_0 \\ 0 & 0 & 1 \end{bmatrix} \begin{bmatrix} \mathsf{R} & -\mathsf{R}\mathbf{t} \end{bmatrix}$$

$$\mathsf{K} = \begin{bmatrix} f_x & f_v \cot\theta & u_0 \\ 0 & \frac{f_v}{\sin\theta} & v_0 \\ 0 & 0 & 1 \end{bmatrix}$$

The 3×4 matrix M that projects a 3D scene point $\mathbf{P}$ to the corresponding image point $\mathbf{p}$ is called the projection matrix. The 3×3 matrix K that contains the internal parameters $(u_0, v_0, \theta, f_x, f_y)$ is generally referred to as the *intrinsic* matrix of a camera.

In back projection, given an image point $\mathbf{p}$, the goal is to find the set of 3D points that project to it. The back projection of an image point is a ray in space. One can compute this ray by identifying two points on this ray. The first point can be the optical center $\mathbf{C}$, since it lies on this ray. Since $\mathsf{M}\mathbf{C} = \mathbf{0}$, $\mathbf{C}$ is nothing but the right nullspace of M. Second, the point M^+p, where M^+ is the pseudoinverse of M, lies on the back-projected ray because it projects to point $\mathbf{p}$ on the image. Thus, the back projection of $\mathbf{p}$ can be computed as follows:

$$\mathsf{P}(\lambda) = \mathsf{M}^+\mathbf{p} + \lambda\mathbf{C}$$

The parameter λ allows to get different points on the back-projected ray.

Caustics

In a single viewpoint imaging system, the geometry of the projection rays is given by the effective viewpoint and the direction of the projection rays. In a noncentral imaging system, *caustic*, a well-known terminology in the optics community, can be utilized for representing the geometry of projection rays [2]. A caustic refers to the loci of viewpoints in 3D space to represent a noncentral imaging system. Concretely, the envelope of all incoming light rays that are eventually imaged is defined as the caustic. A caustic is referred to as diacaustic for dioptric (lens-based systems) and catacaustic (mirror-based systems) for catadioptric systems. A complete study of conic catadioptric systems has been done [3]. Once the caustic is determined, each point on the caustic represents a light ray by providing its position and the direction. Position is given by the point on the caustic, and orientation is related to the concept of tangent. Figure 3 shows the caustic for several noncentral imaging systems. For a single viewpoint imaging system, the caustic is a degenerate one being a single point. Simple methods exist for the computation of the caustic from the incoming light rays such as local conic approximations [4] and the so-called Jacobian method [5], A few examples for caustics are shown in Fig. 3.

Generalized and Multi-perspective Imaging Models

Many novel camera models have multiple centers of projection and they cannot be explained by a simple parametric pinhole model. In computer vision, there has been significant interest in generalizing the camera models to reuse the existing calibration and 3D reconstruction algorithms for novel cameras. In order to do this, first renounce on parametric models and adopt the following very general model: a camera acquires images consisting of pixels; each pixel captures light that travels along a ray in 3D. The camera is fully described by:

- The coordinates of these rays (given in some local coordinate system).
- The mapping between pixels and rays; this is basically a simple indexing.

The generic imaging model is shown in Fig. 4. This allows to describe all above models and virtually any camera that captures light rays traveling along straight lines. The above imaging model has already been used, in more or less explicit form, in various works [3, 6–16], and is best described in [6]. There are conceptual links to other works: acquiring an image with a camera of the general model may be seen as sampling the plenoptic function [17], and a light field [18] or lumigraph [19] may be interpreted as a single image, acquired by a camera of an appropriate design. More details of generic imaging model, their calibration, and 3D reconstruction algorithms can also be found in [20].

Taxonomy of Generic Imaging Models

Central Model: All the projection rays go through a single point, the optical center. Examples are mentioned below:

- The conventional perspective camera forms the classical example for a central camera.
- Perspective+radial or decentering distortion.
- Central catadioptric configurations using parabolic, hyperbolic, or elliptical mirrors.
- Fish-eye cameras can be considered as approximate central cameras.

Axial Model [21]: All the projection rays go through a single line in space, the *camera axis*. Examples of cameras falling into this class are:

- Stereo systems consisting of 2, 3, or more central cameras with collinear optical centers.
- Noncentral catadioptric cameras of the following type: the mirror is any surface of revolution and the optical center of the central camera looking at it (can be any central camera, not only perspective) lies on its axis of revolution. It is easy to verify that in this case, all the projection rays cut the mirror's axis of revolution, i.e., the camera is an axial camera, with the mirror's axis of revolution as camera axis. Note that catadioptric cameras with a spherical mirror and a central camera looking at it are always axial ones.
- X-slit cameras [22] (also called two-slit or crossed-slit cameras), and their special case of linear push-broom cameras [23].

Noncentral Cameras: A noncentral camera may have completely arbitrary projection rays. Common examples are given below:

- Multi-camera system consisting of 3 or more cameras, all of whose optical centers are not collinear.
- Oblique camera: This is an ideal example for a noncentral camera. No two rays intersect in an oblique camera [8].
- Imaging system using a micro-mirror array [24]. A perspective camera is made to observe a scene through a programmable array of micro-mirrors.

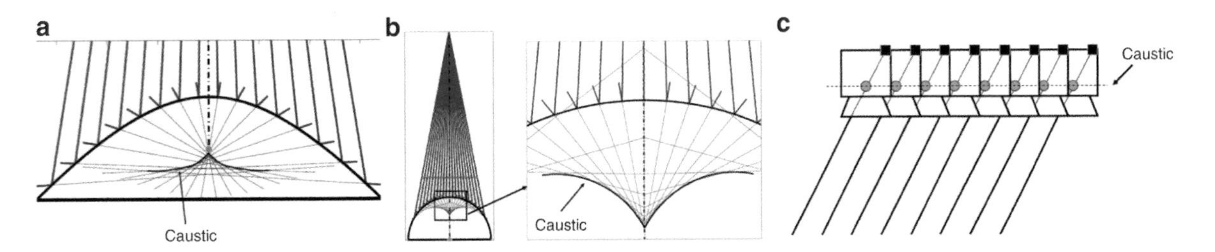

Center of Projection, Fig. 3 Caustics for several imaging systems **(a)** Hyperbolic catadioptric system **(b)** Spherical catadioptric system **(c)** Pushbroom camera

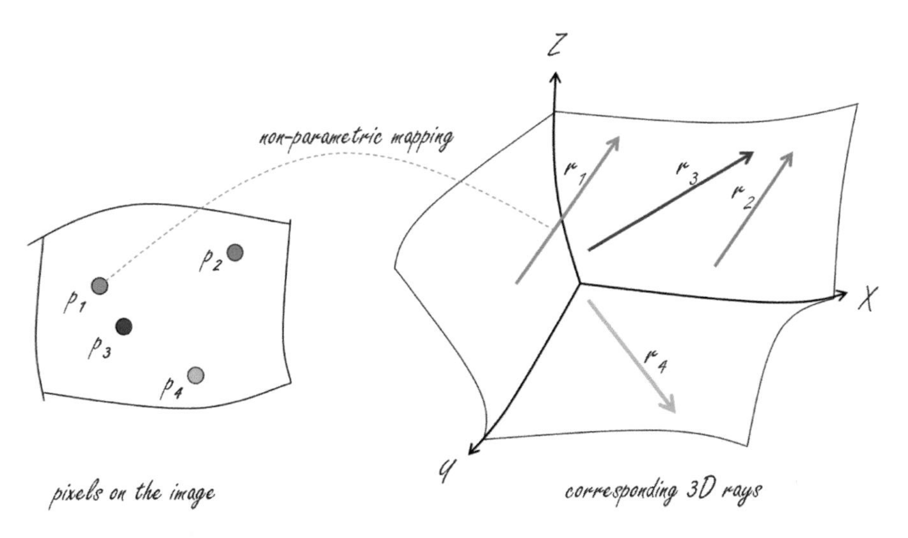

Center of Projection, Fig. 4 The main idea behind the generic imaging model: The relation between the image pixels ($\mathbf{p}_1, \mathbf{p}_2, \mathbf{p}_3, \mathbf{p}_n$) and their corresponding projection rays ($\mathbf{r}_1, \mathbf{r}_2, \mathbf{r}_3, \mathbf{r}_n$) is *non-parametric*

By controlling the orientations and positions of these mirrors, one can obtain an imaging system with complete control (both in terms of geometric and radiometric properties) over the incoming light ray for every pixel.

- Noncentral mosaic: An image sequence is captured by moving the optical center of a perspective camera in a circular fashion [3]. The center columns of the captured images are concatenated to create a noncentral mosaic image.
- Center strip mosaic: The optical center of the camera is moved [3]. The center columns of the captured images are concatenated to form a center strip mosaic. The resulting mosaic corresponds to a noncentral camera.

These three classes of camera models may also be defined as existence of a linear space of d dimensions that has an intersection with all the projection rays: $d = 0$ defines central, $d = 1$ axial, and $d = 2$ general noncentral cameras.

A detailed survey of various camera models, calibration, and 3D reconstruction algorithms is given in [25].

References

1. Dürer A (1525) Underweysung der Messung (Instruction in measurement), Book with more than 150 woodcuts
2. Born M, Wolf E (1965) Principles of optics. Permagon, Oxford
3. Swaminathan R, Grossberg M, Nayar S (2003) A perspective on distortions. In: Conference on computer vision and pattern recognition (CVPR), Madison, vol 2, p 594
4. Bruce JW, Giblin PJ, Gibson CG (1981) On caustics of plane curves. Am Math Mon 88:651–667
5. Burkhard D, Shealy D (1973) Flux density for ray propagation in geometrical optics. J Opt Soc Am 63(2):299–304
6. Grossberg M, Nayar S (2001) A general imaging model and a method for finding its parameters. In: International conference on computer vision (ICCV), Vancouver, vol 2, pp 108–115

7. Neumann J, Fermüller C, Aloimonos Y (2003) Polydioptric camera design and 3d motion estimation. In: Conference on computer vision and pattern recognition (CVPR), Madison, vol 2, pp 294–301
8. Pajdla T (2002) Stereo with oblique cameras. Int J Comput Vis 47(1):161–170
9. Peleg S, Ben-Ezra M, Pritch Y (2001) Omnistereo: panoramic stereo imaging. IEEE Trans Pattern Anal Mach Intell 23:279–290
10. Pless R (2003) Using many cameras as one. In: Conference on computer vision and pattern recognition (CVPR), Madison, pp 587–594
11. Seitz S (2001) The space of all stereo images. In: International conference on computer vision (ICCV), Vancouver, vol 1, pp 26–33
12. Shum H, Kalai A, Seitz S (1999) Omnivergent stereo. In: International conference on computer vision (ICCV), Corfu, pp 22–29
13. Sturm P, Ramalingam S (2004) A generic concept for camera calibration. In: European conference on computer vision (ECCV), Prague, vol 2, pp 1–13
14. Ramalingam S, Sturm P, Lodha S (2005) Towards complete generic camera calibration. In: Conference on computer vision and pattern recognition (CVPR), San Diego
15. Wexler Y, Fitzgibbon A, Zisserman A (2003) Learning epipolar geometry from image sequences. In: Conference on computer vision and pattern recognition (CVPR), Madison, vol 2, pp 209–216
16. Wood D, Finkelstein A, Hughes J, Thayer C, Salesin D (1997) Multiperspective panoramas for cell animation. In: SIGGRAPH, Los Angeles, pp 243–250
17. Adelson E, Bergen J (1991) The plenoptic function and the elements of early vision. In: Computational models of visual processing, edited by Michael S. Landy and J. Anthony Movshon, MIT Press, Cambridge, MA, pp 3–20
18. Levoy M, Hanrahan P (1996) Light field rendering. In: SIGGRAPH, New Orleans, pp 31–42
19. Gortler S, Grzeszczuk R, Szeliski R, Cohen M (1996) The Lumigraph. In: SIGGRAPH, New Orleans, pp 43–54
20. Ramalingam S (2006) Generic imaging models: calibration and 3d reconstruction algorithms. PhD Thesis, INRIA Rhone Alpes
21. Ramalingam S, Sturm P, Lodha S (2006) Theory and calibration algorithms for axial cameras. In: ACCV, Hyderabad
22. Feldman D, Pajdla T, Weinshall D (2003) On the epipolar geometry of the crossed-slits projection. In: International conference on computer vision (ICCV), Nice
23. Gupta R, Hartley R (1997) Linear pushbroom cameras. In: IEEE transactions on pattern analysis and machine intelligence (PAMI) 19(9):963–975
24. Nayar S, Branzoi V, Boult T (2004) Programmable imaging using a digital micromirror array. In: International conference on computer vision and pattern recognition (CVPR), Washington, pp 436–443
25. Sturm P, Ramalingam S, Tardif JP, Gasparini S, Barreto J (2011) Camera models and fundamental concepts used in geometric computer vision. Found Trends Comp Gr Vis 6(1–2):1–183

Change Detection

Joseph L. Mundy
Division of Engineering, Brown University
Rensselaer Polytechnic Institute, Providence, RI, USA

Synonyms

Normalcy modeling

Definition

A determination that there are significant differences between visual scenes

Background

Change detection is a key task for computer vision algorithms. The goal is to compare two or more visual scenes and report any significant differences between the scenes. As with many vision tasks, the meaning of *significant* is application dependent. The change detection task can be rendered somewhat more concrete by considering the types of changes that are not typically of interest. Examples of changes that are usually irrelevant are:

- Camera viewpoint
- Varying illumination
- Wind-based motion
- Weather, e.g., snow and rain

The implementation of algorithms that can detect interesting changes while ignoring trivial changes such as these is a very difficult problem, and only quite limited change detection capabilities have achieved to date. It is also the case that the change detection task, when viewed broadly, overlaps the scope of many other vision tasks such as visual inspection and moving object detection.

Early attempts at change detection were based on simple strategies such as thresholding the magnitude of intensity differences between a reference image and an image that manifests change. This simple approach is only practical if the scene and imaging conditions are closely controlled so that the only scene changes are due to events of interest. Such highly controlled conditions can be found in industrial applications where the lighting and camera pose is accurately maintained. For example, missing components in a circuit board can be detected by comparing the image intensity of a high-quality *master* board with images of boards with potentially missing components or other flaws. In this application, images are accurately registered to the master image, so that intensity differences correspond to physical differences in the boards. The detected changes correspond to missing components or additional material such as excess solder.

A more advanced change detection principle is to classify elements of a scene into categories of interest. Then two scenes can be compared as to the presence or absence of various category instances. Change detection based on this principle can be successful if the classification process is insensitive to the types of irrelevant scene variations mentioned above. This classification approach has been applied extensively to the detection of changes in multispectral aerial and satellite imagery [1]. Multispectral images typically have four or more color bands, so that each pixel is a feature vector that can be subjected to standard classifiers. Image regions are classified into types such as roads, vegetation, forest, and water. Images can then be compared to determine the change in area of each category. This approach requires pixel-level image-to-image registration, which in turn is only effective if the viewpoint change between the images is small and the spatial resolution of the images is low.

Given the complexity of scene appearance and contextual variations, the great majority of change detection algorithms have adopted a learning approach based on the observation of a number of images. In this learning paradigm, it is assumed that changes are rare so that a statistical model for *normal* image appearance can be formed without removing regions of the image that represent change.

A good example of this principle is learning the normal appearance of a road surface in a set of images that depict moving vehicles on the road. In any given image, some of the road surface is visible and contributes correct image appearance information to the model. In other images different parts of the surface may be visible. The assumption is that over a large set of training images, the frequency of vehicle appearance at any given road surface point will be small compared to the frequency of the road appearance. This approach can also be denoted as *background modeling* because a model is being constructed for the scene background rather than the moving or changing objects, i.e., *foreground*. The approach can also be considered as developing a model for *normalcy* where changes are infrequent, i.e., abnormal.

This statistical modeling principle has received widespread acceptance with the advent of cheap computing power and the significance of video data, which provides large training samples. A classical reference is the work of Stauffer and Grimson, who implemented a moving object detection system in online video streams [2]. An important aspect of change detection in video is that illumination and atmospheric properties vary relatively slowly compared to the video frame rate. Under this condition, statistical appearance models can gradually adapt to such variations and not manifest them as change.

The effects of viewpoint and illumination can be overcome through the use of active 3-d range sensors such as LIDAR (LIght Detection And Ranging) or laser triangulation. Scene illumination is provided by the sensor itself and therefore has known direction and spatial extent. The result is a 3-d point cloud of samples from scene surfaces and associated surface reflectance values. In this approach, change is detected by measuring the distance between two 3-d scene point sets after they have been accurately registered together. Missing points or points that are significantly far from the reference point set are considered to be change. Such change points can be grouped into a connected region to further characterize the change. An example of this strategy is the work of Girardeau-Montaut et al. [3].

Theory

The applications of change detection are so broad that a complete theoretical background is beyond the scope of an encyclopedia entry. Moreover, the success of change detection is critically dependent on accurate registration of the reference and current scenes so that change is due solely to the actual scene differences. However, the registration of scene images or 3-d data is a topic in its own right and will not be considered

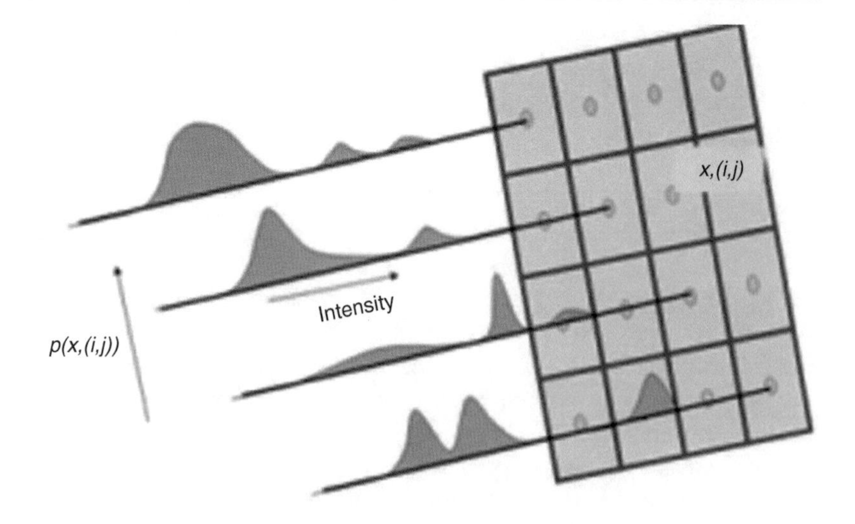

Change Detection, Fig. 1 The appearance of each pixel is modeled by a Gaussian mixture distribution

here. Instead, two important statistical methods will be described that have achieved considerable success in image-based change detection: the joint probability method by Carlotto [4] and the background modeling method by Stauffer and Grimson [2].

Joint Probability Method

This approach can accommodate scenes with significant differences in illumination, and even between images taken with different sensor modalities such as visible and IR wave lengths. The method typically is applied to two images, $x_1(i, j)$ and $x_2(i, j)$, where the pixel intensity values, x, are considered to be random variables. It is assumed that the images are registered and the joint histogram, $p(x_1, x_2)$, of the intensity values at each pixel is accumulated. The expected intensity of a pixel in image 2, given the image value in image 1, is defined by

$$\tilde{x}_2(x_1) = \int x_2\, p(x_2|x_1)\, dx_2, \tag{1}$$

where

$$p(x_2|x_1) = \frac{p(x_1, x_2)}{p(x_1)}.$$

Change is then defined as values of

$$c_{21}(i, j) = \|x_2(i, j) - \tilde{x}_2(x_1(i, j))\| \tag{2}$$

that exceed a decision threshold. The same analysis can be applied in reverse to compute

$$c_{12}(i, j) = \|x_1(i, j) - \tilde{x}_1(x_2(i, j))\|. \tag{3}$$

Gaussian Mixture Method

A second approach to modeling normal scene appearance is to associate a probability distribution with each image pixel location. This distribution accounts for normal variations in image intensity for each pixel across a set of images. It is assumed that the images are spatially registered so that a pixel in each image corresponds to the same surface element in the scene. An example of such registration is provided by a fixed video camera viewing a dynamic scene with moving objects and a stationary background.

The intensity of a given pixel will sometimes be due to background and sometimes foreground moving objects. Thus, an appropriate probability distribution for the overall appearance variation is a Gaussian mixture as shown in Fig. 1. The mixture distribution is defined by

$$p(x) = \sum_i \frac{w_i}{\sqrt{2\pi}\sigma_i} e^{-\frac{1}{2}\left(\frac{(x-\mu_i)}{\sigma_i}\right)^2}. \tag{4}$$

The distribution parameters, $\{w_i, \mu_i, \sigma_i\}$, are learned using a continuous online update algorithm where a new intensity sample, $x_{N+1}(i, j)$, is associated with an existing mixture component if it is within a few standard deviations of the component mean or a new mixture is initiated. The update procedure for a scalar pixel intensity sample after N observations is defined by the following equations:

$$w_{N+1} = \frac{N w_N}{N+1} + \frac{1}{N+1}$$
$$\mu_{N+1} = \frac{N \mu_N}{N+1} + \frac{x_N}{N+1}$$
$$\sigma^2_{N+1} = \frac{N \sigma^2_N}{N+1} + \frac{(x_N - \mu_n)^2}{N+1}. \quad (5)$$

The update is applied if the sample, x_{N+1}, is with a specified number of standard deviations from one of the mixture component means. If a new sample is not within the capture range of one of the mixture components, then a new mixture component is added with user-specified default weight, w_i, and standard deviation. When the limit on the number of mixture components is reached, the component with the smallest value of $\frac{w_i}{\sigma_i}$ is discarded. This component is the least informative (greatest entropy) of the mixture.

The update can be applied continuously over a video sequence, and the mixture parameters will adapt to the normal appearance of the scene. Change is detected as pixel intensities having low probability density according to the mixture distribution. It is the case that such change objects will introduce new modes into the mixture. However, these modes will typically have low weight compared to appearance mixture components corresponding to the stable scene background.

The probability of change can be computed as a Bayesian posterior based on the scene background mixture. That is,

$$P(\text{change}|x) = \frac{P(\text{change})}{P(\text{change}) + p(x|\text{background})(1 - P(\text{change}))}. \quad (6)$$

Here x is the observed intensity and $p(x|\text{background})$ is the Gaussian mixture learned from the sequence of images. The probability $P(\text{change})$ is the prior belief that the pixel exhibits change, and $P(\text{change}|x)$ is the posterior belief after observing the current image intensity value. This result is based on the assumption that image intensities are normalized to the range $[0, 1]$ and the probability density for foreground (change) intensities is uniform, i.e., $p(x|\text{change}) = 1$. Change is detected for pixels having $P(\text{change}|x)$ greater than a threshold.

Open Problems

In spite of considerable research on change detection, the current state-of-the-art algorithms only perform well in scenes where appearance is highly consistent and viewpoint variations can be eliminated by using a fixed camera or by direct comparison of 3-d data. One possible way forward is the work of Pollard [5] where Gaussian appearance models are maintained in a full 3-d volumetric grid. This representation can account for variable viewpoint but is still limited to fixed or slowly varying illumination. Even if the viewpoint and illumination issues are overcome, accounting for the vast diversity of types of change that may or may not be of interest remains a significant challenge.

Experimental Results

An example of detected changes using the joint probability method is shown in Fig. 2. Note that significantly different appearance of the orchard on the right and different overall image contrast does not result in change since the differences are accumulated into the joint probability distribution. The method does not completely account for specular reflections of some of the building roofs (e.g., center and lower left) since these occur relatively infrequently. However, the cars in the roadway at the center are detected with reasonable accuracy.

An example of change detection using the Gaussian mixture model is shown in Fig. 3. In this example a sequence of aerial video frames are registered to a common ground plane and provide updates to the Gaussian mixture model at each pixel. Some false change probability can be seen at occluding boundaries and for metal building roofs that cause large variation in image intensity as the viewpoint changes. The actual changes in the scene are the moving vehicles on the roadways, which exhibit highly salient change probability values.

Change Detection, Fig. 2 Two satellite images taken at different times. (**a**) May 2006. (**b**) November 2006. (**c**) The resulting change map from the joint probability algorithm. Bright intensities indicate high values of $c(i, j)$. The result is by Pollard [5] (Images copyright Digital Globe)

Change Detection, Fig. 3 An example of change probability computed using a Gaussian background mixture model. (**a**) A typical video frame. (**b**) The value of $P(\text{change}|x)$ displayed with white = 1

References

1. Congalton R, Green K (2009) Assessing the accuracy of remotely sensed data: principles and practices, 2nd edn. CRC Press, Boca Raton, FL
2. Stauffer C, Grimson W (1999) Adaptive background mixture models for real-time tracking. In: Proceedings of the international conference on computer vision and pattern recognition (CVPR), Fort Collins, Colorad, New York, vol 2, pp 246–252
3. Girardeau-Montau D, Roux M, Marc R, Thibault G (2005) Change detection on points cloud data acquired with a ground laser scanner. Remote sensing and spatial information sciences 36 (part3/W19), Enschede, The Netherland, pp 30–35
4. Carlotto MJ (2005) Detection and analysis of change in remotely sensed imagery with application to wide area surveillance. IEEE trans on image process, New York, 2(3), pp 189–202
5. Pollard T (2009) Comprehensive 3-d change detection using volumetric appearance modeling. Brown University, Providence, RI

Chromaticity

Michael H. Brill
Datacolor, Lawrenceville, NJ, USA

Related Concepts

▸Trichromatic Theory

Definition

Chromaticity is a representation of the tristimulus values of a light (see entry on ▸Trichromatic Theory) by only two numbers, computed so as to suppress via ratios the absolute intensity of the light. The two numbers (called chromaticity coordinates) define a space (called chromaticity space) in which any additive mixture of two lights lies on a straight line between those two lights.

Theory

Most often, chromaticity is derived from a standardized tristimulus coordinate system representing human vision (e.g., a CIE XYZ system [1, 2]). If a light has tristimulus values X, Y, and Z, its chromaticity coordinates are conventionally defined as the ratios

$x = X/(X+Y+Z)$ and $y = Y/(X+Y+Z)$. More generally, the chromaticity coordinates could be defined as the ratios $q_1 = (aX+bY+cZ)/(pX + qY + rZ)$ and $q_2 = (dX+eY+fZ)/(pX + qY + rZ)$, where a, b, c, d, e, f, p, q, r are constants that do not depend on the light. The above mapping from (X,Y,Z) to (q_1, q_2) is called a homogeneous central projection. It has the property of showing straight lines in (X,Y,Z) as straight lines in (q_1, q_2). Hence it is obvious from looking at the chromaticity coordinates of three lights whether one of the lights could have been additively mixed from the other two. This property is useful for visualizing the color gamut (set of producible colors) of a self-luminous trichromatic display such as a cathode-ray tube (CRT) or a liquid-crystal display (LCD). The primaries (say, red, green, and blue) for such displays define three points in chromaticity space, and the triangle they generate spans the chromaticities producible by the display.

To illustrate the geometry of chromaticity, the figure below shows the 1931 CIE (x,y) diagram (Fig. 1). The horseshoe-shaped curve (from 380 nm at the left-hand end to 720 nm on the right-hand end) is the set of all monochromatic lights within the space, and is called the spectrum locus. The straight line connecting the ends of the curve is the line of purples. (There are no monochromatic purples.) Finally, the vertices of the triangle inside the spectrum locus represent the R, G, and B display primaries prescribed by ITU Recommendation 709. This set of primaries is only one example of many prescribed by standards bodies [3], and of even more sets manifested in real displays.

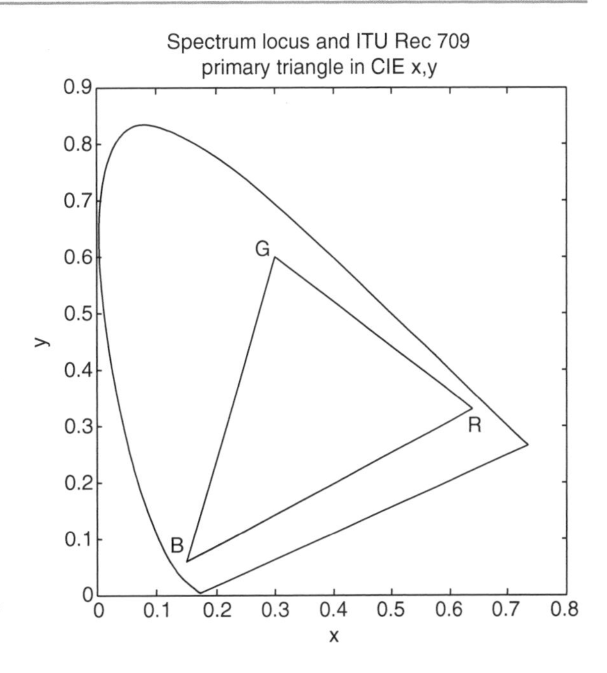

Chromaticity, Fig. 1 CIE (x, y) chromaticity diagram with spectrum locus and RGB display primaries

Various useful constructions are possible in chromaticity space. For example, the dominant wavelength of a light is defined by the point on the spectrum locus intercepted by a ray from the agreed-upon white chromaticity through the chromaticity of the light in question. Another example is the set of lights defined by the blackbody radiators (parameterized by their temperature). These lights collectively define the blackbody locus, which of course lies within the spectrum locus. The set of conventional daylights also forms a curve that is close to the blackbody locus. As another example, it is possible to transform from one tristimulus primary set to another such set using only information in the chromaticity domain [4].

Because a digital camera is a trichromatic device, one can also think of a camera's response to light as having a chromaticity. Such chromaticities are useful because they suppress spatial variations of light intensity in an image (e.g., at shadow boundaries) and thereby facilitate image segmentation by object color (see entry on Band Ratios). Of course, band-ratio pairs are more general than chromaticities because the ratios can be separately defined without the mixture-on-a-straight-line constraint required for a chromaticity.

It should be noted that camera chromaticity is not at all the same as the human-vision chromaticity and must not be confused with it. In fact, two lights that have the same human-vision chromaticity can have different camera chromaticities, and vice versa. This effect is related to metameric matching as noted in the entry on Trichromatic Theory. It should also be noted that, for the same reason, camera-derived values I, H, S used by computer-vision and image-processing applications are also not transformable to human perceptual attributes (e.g., intensity, hue, and saturation).

For both cameras and humans, chromaticity generalizes to a non-trichromatic system (i.e., one that has a number of different sensor types that is different from three). A sensor system with N sensor types delivers N-stimulus values for each light, the chromaticity space has N-1 dimensions (so as to suppress the light intensity), and the chromaticity coordinates comprise a homogeneous central projection out of the space of N-stimulus values.

References

1. Wyszecki G, Stiles WS (1982) Color science: concepts and methods, quantitative data and formulae, Chaps. 3 and 5, 2nd edn. Wiley, New York
2. Fairman HS, Brill MH, Hemmendinger H (1997) How the CIE 1931 color-matching functions were derived from Wright-Guild data. Color Res Appl 22:11–23
3. Poynton CA (1996) A technical introduction to digital video, Chap. 7. Wiley, New York
4. Brill MH (2008) Transformation of primaries using four chromaticity points and their matches. Color Res Appl 33:506–508
5. Fairchild M (2005) Color appearance models, Chap. 3, 2nd edn. Wiley, Chichester
6. MacAdam DL (1985) Color measurement: theme and variations. Springer, Berlin/New York
7. Koenderink JJ (2010) Color for the sciences, Chap. 4. MIT, Cambridge
8. Berns RS (2000) Billmeyer and Saltzman's principles of color technology, 3rd edn. Wiley, New York
9. Hunt RWG (1998) The measurement of colour, 3rd edn. Fountain Press, Kingston-Upon-Thames
10. Hunt RWG (2004) The reproduction of colour, 6th edn. Wiley, Chichester
11. Wandell BA (1995) Foundations of vision. Sinauer, Sunderland

CIE Standard Illuminant

►Standard Illuminants

Clifford Algebra

►Geometric Algebra

Clipping

►Saturation (Imaging)

Cluster Sampling

►Swendsen-Wang Algorithm

Coefficient Rule

►von Kries Hypothesis

Cognitive Agent

►Cognitive System

Cognitive System

David Vernon
Informatics Research Centre, University of Skövde, Skövde, Sweden

Synonyms

Cognitive agent

Related Concepts

►Cognitive Vision

Definition

A cognitive system is an autonomous system that can perceive its environment, learn from experience, anticipate the outcome of events, act to pursue goals, and adapt to changing circumstances.

Background

There are several scientific perspectives on the nature of cognition and on how it should be modeled. All fall under the general umbrella of cognitive science which embraces the disciplines of neuroscience, artificial intelligence, cognitive psychology, linguistics, and epistemology. Among these differing perspectives, however, there are two broad classes: the *cognitivist* approach based on symbolic information processing representational systems, and the *emergent systems* approach, encompassing connectionist systems, dynamical systems, and enactive systems, all based to a lesser or greater extent on principles of self-organization [1–4]. A third class – *hybrid systems* – attempts to combine something from each of the cognitivist and emergent paradigms. All three approaches have their origins in cybernetics [5] which in the decade from 1943 to 1953 made the first efforts to formalize what had up to that point been purely psychological and philosophical treatments of cognition. The intention of the early cyberneticians was to create a science of mind, based on logic. Examples of the application of cybernetics to cognition include the seminal paper by McCulloch and Pitts "A logical calculus immanent in nervous activity" [6] and Ashby's "Design for a Brain" [7].

Theory

The initial attempt in cybernetics to create a science of cognition was followed by the development of an approach referred to as *cognitivism*. The birth of the cognitivist paradigm, and its sister discipline of Artificial Intelligence, dates from a conference held at Dartmouth College, New Hampshire, in July and August 1956 and attended by people such as John McCarthy, Marvin Minsky, Allen Newell, Herbert Simon, and Claude Shannon. Cognitivism holds that cognition is achieved by computation performed on internal symbolic knowledge representations in a process whereby information about the world is abstracted by perception, and represented using some appropriate symbolic data-structure, reasoned about, and then used to plan and act in the world. The approach has also been labeled by many as the information processing or symbol manipulation approach to cognition [1, 8–10]. In most cognitivist approaches concerned with the creation of artificial cognitive systems, the symbolic representations are the descriptive product of a human designer. This is significant because it means that they can be directly accessed and interpreted by humans and that semantic knowledge can be embedded directly into and extracted directly from the system. In cognitivism, the goal of cognition is to reason symbolically about these representations in order to effect the required adaptive, anticipatory, goal-directed behavior. Typically, this approach to cognition will deploy machine learning and probabilistic modeling in an attempt to deal with the inherently uncertain, time-varying, and incomplete nature of the sensory data that is used to drive this representational framework. Significantly, in the cognitivist paradigm, the instantiation of the computational model of cognition is inconsequential: any physical platform that supports the performance of the required symbolic computations will suffice [8]. This principled separation of operation from instantiation is referred to as functionalism.

In the emergent paradigm, cognition is the process whereby an autonomous system becomes viable and effective in its environment. It does so through a process of self-organization by which the system continually maintains its operational identity through the moderation of mutual system-environment interaction. In other words, the ultimate goal of an emergent cognitive system is to maintain its own autonomy. In achieving this, the cognitive process determines what is real and meaningful for the system: the system constructs its reality – its world and the meaning of its perceptions and actions – as a result of its operation in that world. Consequently, the system's understanding of its world is inherently specific to the form of the system's embodiment and is dependent on the system's history of interactions, i.e., its experiences. This mutual-specification of the system's reality by the system and its environment is referred to as co-determination [11] and is related to the concept of radical constructivism [12]. This process of making sense of its environmental interactions is one of the foundations of the enactive approach to cognition [13]. Cognition is also the means by which the system compensates for the immediate nature of perception, allowing it to anticipate environmental interaction that occurs over longer time scales, i.e., cognition is intrinsically linked with the ability of an agent to act prospectively: to deal with what might be, not just with what is. Many emergent approaches adhere to the principle that the primary model for cognitive learning is anticipative skill construction rather than knowledge acquisition. Thus, processes which guide action and improve the capacity to guide action form the root capacity of all intelligent systems [14].

As noted already, the emergent paradigm embraces connectionist systems, dynamical systems, and enactive systems. Connectionist systems rely on parallel processing of non-symbolic distributed activation patterns using statistical properties, rather than logical rules, to process information and achieve effective behavior [15]. In this sense, the neural network instantiations of the connectionist model are dynamical systems that capture the statistical regularities in training data [16]. Dynamical systems theory has been used to complement classical approaches in artificial intelligence [17] and it has also been deployed to model natural and artificial cognitive systems [10, 18, 19]. Although dynamical systems theory approaches often differ from connectionist systems on several fronts, it is better perhaps to consider them complementary ways of describing cognitive systems, dynamical systems addressing macroscopic behavior at an emergent level, and connectionist systems addressing microscopic behavior at a mechanistic level [20]. Enactive systems take the emergent paradigm even further. Enaction [13, 21–23] asserts that cognition is a process whereby the issues that are important for the continued existence of a cognitive entity are brought out or enacted: co-determined by the entity and the environment in which it is embedded. Thus, enaction entails that a cognitive system operates autonomously, that it generates its own models of how the world works, and that the purpose of these models is to preserve the system's autonomy.

Considerable effort has gone into developing hybrid approaches which combine aspects of cognitivist and emergent systems. Typically, hybrid systems exploit symbolic knowledge to represent the agent's world and logical rule-based systems to reason about this knowledge in order to achieve goals and select actions, while at the same time using emergent models of perception and action to explore the world and construct this knowledge. Thus, hybrid systems still use cognitivist representations and representational invariances but they are constructed by the system itself as it interacts with and explores the world rather than through a priori specification or programming. Consequently, as with emergent systems, the agent's ability to understand the external world is dependent on its ability to interact flexibly with it, and interaction is the organizing mechanism that establishes the association between perception and action.

Cognitivism and artificial intelligence research are strongly related. In particular, Newell and Simon's "Physical Symbol System" approach to artificial intelligence [8] has been extremely influential in shaping how we think about intelligence, natural as well as computational. In their 1976 paper, two hypotheses are presented: the *Physical Symbol System Hypothesis* and the *Heuristic Search Hypothesis*. The first hypothesis is that a physical symbol system has the necessary and sufficient means for general intelligent action. This implies that any system that exhibits general intelligence is a physical symbol system *and* any physical symbol system of sufficient size can be configured to exhibit general intelligence. The second hypothesis states that the solutions to problems are represented as symbol structures and that a physical-symbol system exercises its intelligence in problem-solving by search, i.e., by generating and progressively modifying symbol structures in an effective and efficient manner until it produces a solution structure. This amounts to an assertion that symbol systems solve problems by heuristic search, i.e., the successive generation of potential solution structures. The task of intelligence, then, is to avert the ever-present threat of the exponential explosion of search. Subsequently, Newell defined intelligence as the degree to which a system approximates the ideal of a knowledge-level system [24]. A knowledge-level system is one which can bring to bear *all* its knowledge onto *every* problem it attempts to solve (or, equivalently, every goal it attempts to achieve). Perfect intelligence implies complete utilization of knowledge. It brings this knowledge to bear according to the *principle of maximum rationality* which was proposed by Newell in 1982 [25] as follows: "If an agent has knowledge that one of its actions will lead to one of its goals, then the agent will select that action." Anderson [26] later offered a slightly different principle, the *principle of rationality*, sometimes referred to as rational analysis, stated as follows: "the cognitive system optimizes the adaptation of the behavior of the organism." Note that Anderson's principle considers optimality to be necessary for rationality, something that Newell's principle does not.

Cognitivist and emergent approaches are normally contrasted on the basis of the symbolic or non-symbolic nature of their computational operation and representational framework. Cognitivist systems typically use production systems to effect rule-based manipulation of symbol tokens whereas emergent systems exploit dynamical processes of self-organization in which representations are encoded in global system states. However, the distinction between cognitivist and emergent is not restricted to the issue of symbolic representation and they can be contrasted on the basis of several other characteristics such as semantic grounding, temporal constraints, inter-agent epistemology, embodiment, perception, action, anticipation, adaptation, motivation, autonomy, among others [27].

The differences between the cognitivist and the emergent paradigm can be traced to their underlying distinct philosophies [28]. Broadly speaking, cognitivism is dualist, functionalist, and positivist. It is dualist in the sense that there is a fundamental distinction between the mind (the computational processes) and the body (the computational infrastructure and, if required, the physical structure that instantiates any physical interaction). It is functionalist in the sense that the actual instantiation and computational infrastructure is inconsequential: any instantiation that supports the symbolic processing is sufficient. It is positivist in the sense that they assert a unique and absolute empirically-accessible external reality that is apprehended by the senses and reasoned about by the cognitive processes. In contrast, emergent systems are neither dualist nor functionalist, since the system's embodiment is an intrinsic component of the cognitive process, nor positivist, since the form and meaning of the system's world is dependent in part on the system itself. The emergent paradigm, and especially the enactive approach, can trace its roots to the philosophy of phenomenology [28, 29].

A criticism often leveled at cognitivist systems is that they are relatively poor at functioning effectively outside well-defined problem domains because they tend to depend on in-built assumptions and embedded knowledge arising from design decisions. Emergent systems should in theory be much less brittle because they develop through mutual specification and co-determination with the environment. However, the ability to build artificial cognitive systems based on emergent principles is very limited at present, and cognitivist and hybrid systems currently have more advanced capabilities within a narrower application domain.

Any cognitive system is inevitably going to be complex. Nonetheless, it is also the case that it will exhibit some degree of structure. This structure is often encapsulated in what is known as a cognitive architecture [30]. Although used freely by proponents of the cognitivist, emergent, and hybrid approaches to cognitive systems, the term "cognitive architecture" originated with the seminal cognitivist work of Newell et al. [25]. Consequently, the term has a very specific meaning in this paradigm where cognitive architectures represent attempts to create unified theories of cognition [24, 31], i.e., theories that cover a broad range of cognitive issues, such as attention, memory, problem-solving, decision-making, learning, from several aspects including psychology, neuroscience, and computer science. In the cognitivist paradigm, the focus of a cognitive architecture is on the aspects of cognition that are constant over time and that are independent of the task. Since cognitive architectures represent the fixed part of cognition, they cannot accomplish anything in their own right and need to be provided with or acquire knowledge to perform any given task. For emergent approaches to cognition, which focus on development from a primitive state to a fully cognitive state over the lifetime of the system, the architecture of the system is equivalent to its phylogenetic configuration: the initial state from which it subsequently develops through ontogenesis.

Open Problems

The study of cognitive systems is a maturing discipline with contrasting approaches. Consequently, there are several open problems. These include the role of physical embodiment, the need for development, the system's cognitive architecture, the degree of autonomy required, the issue of symbol grounding, the problem of goal specification, the ability to explain the rationale for selection actions, the problem of generating generalized concepts and transferring knowledge from one context to another, and the interdependence of perception and action. The nature of any resolution of these problems is inextricably linked to the choice of paradigm: cognitivist, emergent, or hybrid.

The role of physical embodiment in a cognitive system [32–34] depends strongly on the chosen paradigm. Due to their functionalist characteristics, cognitivist systems do not depend on physical embodiment to operate successfully but there is nothing to prevent them from being embodied if that is what the task in hand requires. Emergent systems, by definition, require embodiment since the body plays a key role in the way a cognitive system comes to understand – make sense of – its environment. If a body is required, the form of embodiment must still be specified [35]. This is significant because, in the emergent paradigm at least, the ability of two cognitive agents to communicate effectively requires them to have similar embodiments so that they have a shared history of interaction and a common epistemology.

The extent to which a cognitive system requires a capacity for development and, if so, the mechanisms by which development can take place are both open problems. In natural systems, growth is normally associated with development. However, growth in artificial systems remains a distant goal, although one whose achievement would open up many avenues of fruitful enquiry in cognitive systems. For current state-of-the-art cognitive systems, one can define development as the process by which a system discovers for itself the models that characterize its interactions with its environment. This contrasts with learning as the process whereby the parameters of an existing model are estimated or improved. Development, then, requires a capacity for self-modification [36] and in embodied emergent systems leads to an increased repertoire of effective actions and a greater ability to anticipate the need for and outcome of future actions [27].

The capacity to develop introduces another open issue: the minimal phylogenetic configuration – the perceptual, cognitive, and motoric capabilities with which a system is endowed at "birth" – that is required to facilitate subsequent ontogenesis – development and learning through exploration and social interaction [27]. This issue is related to the specification of the system's cognitive architecture and the necessary and sufficient conditions that must be satisfied for cognitive behavior to occur in a system. In addressing these issues, there is a trade-off between the initial phylogeny and the potential for subsequent development. This trade-off is reflected by the existence of two types of species in nature: precocial and altricial. Precocial species are those that are born with well-developed behaviors, skills, and abilities which are the direct result of their genetic make-up (i.e., their phylogenic configuration). As a result, precocial species tend to be quite independent at birth. Altricial species, on the other hand, are born with poor or undeveloped behaviors and skills, and are highly dependent for support. However, in contrast to precocial species, they proceed to learn complex cognitive skills over their lifetime (i.e., through ontogenetic development). The precocial and the altricial effectively define a spectrum of possible configurations of phylogenetic configuration and ontogenetic potential [37]. The problem is to identify a feasible point in this spectrum that will yield a cognitive system capable of developing the skills we require of it.

Autonomy is a crucial issue for cognitive systems [38] but the degree of autonomy required is unclear. To an extent, it depends on how autonomy is defined and which paradigm of cognition is being considered. Definitions range from self-regulation and homeostasis to the ability of a system to contribute to its own persistence [39]. In the former case, self-regulation is often cast as a form of self-control so that the systems can operate without interference from some outside agent, such as a human user. In the latter case, autonomy is the self-maintaining organizational feature of living creatures that enables them to use their own capacities to manage their interactions with the world in order to remain viable [14]. Cognitivist systems tend to adopt the former definition, emergent systems, the latter.

Broadly speaking, cognitivist systems exploit symbolic representations while emergent systems exploit sub-symbolic state-based representations, with hybrid systems using both. The manner in which cognitivist and hybrids systems ground their symbolic representations in experience is still an open issue [40], with some arguing for a bottom-up approach [41] and others for a process of learned association, where meaning is attached rather than grounded [37].

The opening definition of a cognitive system states that it can act to achieve goals. The specification of these goals poses a significant challenge due to the autonomous nature of cognitive systems. It is more easily resolved for cognitivist systems since the goals can be hard-wired into the cognitive architecture. It is less clear how goals can be specified in an emergent system since the over-arching goal here is the maintenance of the system's autonomy. The goals of such a system reflect its intrinsic motivations and its associated value system [42]. The problem is to understand how to engineer this value system to ensure that the system is motivated to act in a way that satisfies goals which are external to the system and to decide how these goals can be communicated to the system.

Ideally, in addition to the characteristics of a cognitive system listed in the opening definition – autonomy, perception, learning, anticipation, goal-directed action, and adaptation – a cognitive computer system should also be able to say what it is doing and why it is doing it, i.e., it should be able to explain the reasons for an action [43]. This would enable the system to identify potential problems which might appear when carrying out a task and it would know when it needed new information in order to complete that task. Consequently, a cognitive system would be able to view a problem in several different ways and to look at different alternative ways of tackling it. In a sense, this is something similar to the issue discussed above about cognition involving an ability to anticipate the need for actions and their outcome. The difference in this case is that the cognitive system is considering not just one but many possible sets of needs and outcomes. In a sense, it is adapting *before* things do not go according to plan. From this point of view, cognition also involves a sense of self-reflection.

Cognitive systems also learn from experience and adapt to changing circumstances. To do this, the system must have some capacity for generalization so that concepts can be formed from specific instances and so that knowledge and know-how can be transferred from one context to another. This capacity would allow the system to adapt to new application scenarios and to explore the hypothetical situations that arise from the self-reflection mentioned above. It is unclear at present how such generalized conceptual knowledge and know-how should be generated, represented, and incorporated into the system dynamics.

Perception and action have been demonstrated to be co-dependent in biological systems. Perceptual development depends on what actions an infant is capable of and what use objects and events afford in the light of these capabilities. This idea of the action-dependent perceptual interpretation of an object is referred to as its affordance [44]. In neuroscience, the tight relationship between action and perception is exemplified by the presence of mirror neurons, neurons that become active when an action is performed and when the action or a similar action is observed being performed by another agent. It is significant that these neurons are specific to the goal of the action and not the mechanics of carrying it out. The related Ideomotor Theory [45] asserts the existence of such a common or co-joint representational framework for perception and action. Such a framework would facilitate the inference of intention and the anticipation of an outcome of an event due to the goal-oriented nature of the action. The realization of and effective co-joint perception-action framework remains an important challenge for cognitivist and emergent approaches alike.

Although clearly there are some fundamental differences between the cognitivist and the emergent paradigms, the gap between the two shows some signs of narrowing. This is mainly due to (1) a recent movement on the part of proponents of the cognitivist paradigm to assert the fundamentally important role played by action and perception in the realization of a cognitive system [32]; (2) the move away from the view that internal symbolic representations are the only valid form of representation [2]; and (3) the weakening of the dependence on embedded a priori knowledge and the attendant-increased reliance on machine learning and statistical frameworks both for tuning system parameters and the acquisition of new knowledge. This suggests that hybrid approaches may be the way forward, especially if a principled synthesis of cognitivist and emergent approaches is possible, such as "dynamic computationalism" [2] or "computational mechanics" [46]. Hybrid approaches appear to many to offer the best of both worlds – the adaptability of emergent systems and the advanced starting point of cognitivist systems – since the representational invariances and representational frameworks need not be learned but can be designed in and since the system populates these representational frameworks through learning and experience. However, it is uncertain that one can successfully combine what are ultimately highly incompatible underlying philosophies. Opinion is divided, with arguments both for (e.g., [2, 40, 46]) and against (e.g., [47]).

References

1. Varela FJ (1992) Whence perceptual meaning? A cartography of current ideas. In: Varela FJ, Dupuy JP (eds) Understanding origins – contemporary views on the origin of life, mind and society. Boston studies in the philosophy of science. Kluwer Academic Publishers, Dordrecht, pp 235–263
2. Clark A (2001) Mindware – an introduction to the philosophy of cognitive science. Oxford University Press, New York
3. Freeman WJ, Núñez R (1999) Restoring to cognition the forgotten primacy of action, intention and emotion. J Conscious Stud 6(11–12):ix–xix
4. Winograd T, Flores F (1986) Understanding computers and cognition – a new foundation for design. Addison-Wesley Publishing Company, Inc., Reading
5. Ross Ashby W (1957) An introduction to cybernetics. Chapman and Hall, London
6. McCulloch WS, Pitts W (1943) A logical calculus of ideas immanent in nervous activity. Bull Math Biophys 5:115–133
7. Ross Ashby W (1954) Design for a brain. Chapman and Hall, London
8. Newell A, Simon HA (1976) Computer science as empirical inquiry: symbols and search. Commun Assoc Comput Mach 19:113–126. Tenth Turing award lecture, ACM, 1975
9. Haugland J (ed) (1997) Mind design II: philosophy, psychology, artificial intelligence. MIT, Cambridge, MA
10. Kelso JAS (1995) Dynamic patterns – the self-organization of brain and behaviour, 3rd edn. MIT, Cambridge, MA
11. Maturana H, Varela F (1987) The tree of knowledge – the biological roots of human understanding. New Science Library, Boston & London

12. von Glaserfeld E (1995) Radical constructivism. RouteledgeFalmer, London
13. Stewart J, Gapenne O, Di Paolo EA (2011) Enaction: toward a new paradigm for cognitive science. MIT, Cambridge, MA
14. Christensen WD, Hooker CA (2000) An interactivist-constructivist approach to intelligence: self-directed anticipative learning. Philos Psychol 13(1):5–45
15. Medler DA (1998) A brief history of connectionism. Neural Comput Surv 1:61–101
16. Smolensky P (1996) Computational, dynamical, and statistical perspectives on the processing and learning problems in neural network theory. In: Smolensky P, Mozer MC, Rumelhart DE (eds) Mathematical perspectives on neural networks. Erlbaum, Mahwah, NJ, pp 1–15
17. Reiter R (2001) Knowledge in action: logical foundations for specifying and implementing dynamical systems. MIT, Cambridge, MA
18. Thelen E, Smith LB (1994) A dynamic systems approach to the development of cognition and action. MIT press/Bradford books series in cognitive psychology. MIT, Cambridge, MA
19. Port RF, van Gelder T (1995) Mind as motion – explorations in the dynamics of cognition. Bradford Books, MIT, Cambridge, MA
20. McClelland JL, Vallabha G (2006) Connectionist models of development: mechanistic dynamical models with emergent dynamical properties. In: Spencer JP, Thomas MSC, McClelland JL (eds) Toward a new grand theory of development? Connectionism and dynamic systems theory re-considered. Oxford University Press, New York
21. Maturana HR, Varela FJ (1980) Autopoiesis and cognition – the realization of the living. Boston studies on the philosophy of science. D. Reidel Publishing Company, Dordrecht
22. Varela F, Thompson E, Rosch E (1991) The embodied mind. MIT, Cambridge, MA
23. Vernon D (2010) Enaction as a conceptual framework for development in cognitive robotics. Paladyn J Behav Robot 1(2):89–98
24. Newell A (1990) Unified theories of cognition. Harvard University Press, Cambridge, MA
25. Newell A (1982) The knowledge level. Artif Intell 18(1): 87–127
26. Anderson J (1999) Cognitive architectures in rational analysis. In: van Lehn K (ed) Architectures for intelligence. Lawrence Erlbaum Associates, Hillsdale, pp 1–24
27. Vernon D, von Hofsten C, Fadiga L (2010) A roadmap for cognitive development in humanoid Robots. Volume 11 of cognitive systems monographs (COSMOS). Springer, Berlin
28. Vernon D, Furlong D (2007) Philosophical foundations of enactive AI. In: Lungarella M, Iida F, Bongard JC, Pfeifer R (eds) 50 years of AI. Volume LNAI 4850. Springer, Heidelberg, pp 53–62
29. Froese T, Ziemke T (2009) Enactive artificial intelligence: investigating the systemic organization of life and mind. Artif Intell 173:466–500
30. Langley P, Laird JE, Rogers S (2009) Cognitive architectures: research issues and challenges. Cogn Syst Res 10(2):141–160
31. Anderson JR, Bothell D, Byrne MD, Douglass S, Lebiere C, Qin Y (2004) An integrated theory of the mind. Psychol Rev 111(4):1036–1060
32. Anderson ML (2003) Embodied cognition: a field guide. Artif Intell 149(1):91–130
33. Steels L (2007) Fifty years of AI: from symbols to embodiment – and back. In: Lungarella M, Iida F, Bongard JC, Pfeifer R (eds) 50 years of AI. Volume LNAI 4850. Springer, Heidelberg, pp 18–28
34. Vernon D (2008) Cognitive vision: the case for embodied perception. Image Vis Comput 26(1):127–141
35. Ziemke T (2003) What's that thing called embodiment? In: Alterman R, Kirsh D (eds) Proceedings of the 25th annual conference of the cognitive science society. Lund university cognitive studies. Lawrence Erlbaum, Mahwah, pp 1134–1139
36. Weng J (2004) Developmental robotics: theory and experiments. Int J Humanoid Robot 1(2):199–236
37. Sloman A, Chappell J (2005) The altricial-precocial spectrum for robots. In: IJCAI '05 – 19th international joint conference on artificial intelligence, Edinburgh
38. Varela F (1979) Principles of biological autonomy. Elsevier North Holland, New York
39. Bickhard MH (2000) Autonomy, function, and representation. Artif Intell Spec Issue Commun Cogn 17(3–4): 111–131
40. Barsalou LW (2010) Grounded cognition: past, present, and future. Top Cogn Sci 2:716–724
41. Harnad S (1990) The symbol grounding problem. Phys D 42:335–346
42. Edelman GM (2006) Second nature: brain science and human knowledge. Yale University Press, New Haven and London
43. Brachman RJ (2002) Systems that know what they're doing. IEEE Intell Syst 17(6):67–71
44. Gibson JJ (1979) The ecological approach to visual perception. Houghton Mifflin, Boston
45. Stock A, Stock C (2004) A short history of ideo-motor action. Psychol Res 68(2–3):176–188
46. Crutchfield JP (1998) Dynamical embodiment of computation in cognitive processes. Behav Brain Sci 21(5):635–637
47. Christensen WD, Hooker CA (2004) Representation and the meaning of life. In: Clapin H, Staines P, Slezak P (eds) Representation in mind: new approaches to mental representation. Elsevier, Oxford, pp 41–70

Cognitive Vision

David Vernon
Informatics Research Centre, University of Skövde, Skövde, Sweden

Related Concepts

►Cognitive System; ►Visual Cognition

Definition

Cognitive vision refers to computer vision systems that can pursue goals, adapt to unexpected changes of the visual environment, and anticipate the occurrence of objects or events.

Background

The field of cognitive vision grew from the broader area of computer vision in response to a need for vision systems that are more widely applicable, that are able adapt to novel scenes and tasks, that are robust to unexpected variations in operating conditions, and that are fast enough to deal with the timing requirements of these tasks [1]. Adaptability entails the ability to acquire knowledge about the application domain, thereby removing the need to embed all the required knowledge in the system when it is designed. Robustness allows the system to be tolerant to changes in environmental conditions so that system performance is not negatively impacted by them when carrying out a given task. Speed and the ability to pay attention to critical events are essential when providing feedback to users and devices in situations which change unexpectedly [2, 18]

While computer vision systems routinely address signal processing of sensory data and reconstruction of 3D scene geometry, cognitive vision goes beyond this by providing a capability for conceptual characterization of the scene structure and dynamics using qualitative representations. Having knowledge about the scene available in conceptual form allows the incorporation of consistency checks through the use of, e.g., logic inference engines. These checks can be applied both to the knowledge that is embedded in the system at the outset and the knowledge that the system learns for itself. Consistency checking applies across several scales of space and time, requiring cognitive vision to have an ability to operate with past, present, *and* future events. These consistency checks are one way in which the robustness associated with cognitive vision can be achieved [3]. Furthermore, the conceptual knowledge generated by cognitive vision can, if required, be communicated to a human user in natural language [4]. This linguistic communication is one manifestation of an autonomous system demonstrating its understanding of the visual events in its environment [3].

Theory

Cognitive vision entails abilities to anticipate future events and to interpret a visual scene in the absence of complete information. To achieve this, a cognitive system must have the capacity to acquire new knowledge and to use it to fill in gaps that are present in what is being made immediately available by the visual sensors: to extrapolate in time and space to achieve a more robust and effective understanding of the underlying behavior of the sensed world. In the process, the system learns, anticipates, and adapts. These three characteristics of learning, anticipation, and adaptivity are the hallmarks of cognition, in general, and cognitive vision, in particular [6, 7].

A key property of cognitive vision is its capacity to exhibit a robust performance even in scenarios that were not anticipated when it was designed. The degree to which a system can deal with unexpected circumstances will vary. Systems that can adapt autonomously to arbitrary situations are unrealistic at present but it is plausible that they should be able to deal with new variants of visual form, function, and behavior, and also incremental changes in context. Ideally, a cognitive vision system should be able to recognize and adapt to novel variations in the current visual environment, generalize to new contexts and application domains, interpret and predict the behavior of agents detected in the system's environment, and communicate an understanding of the environment to other systems, including humans.

A cognitive vision system is a visually enabled cognitive system, defined in this encyclopedia as "an autonomous system that can perceive its environment, learn from experience, anticipate the outcome of events, act to pursue goals, and adapt to changing circumstances." Since cognitive vision is principally a mode of perception, physical action – with the possible exception of the camera movements associated with active vision – usually falls outside its scope. However, speech acts may be involved when communicating an interpretation of the scene in conceptual terms through language [5]. Since cognitive vision is a particular type of cognitive system, all the issues identified in the cognitive system entry in this encyclopedia apply equally to cognitive vision. They will not be revisited here apart from noting that there are several scientific perspectives on the nature of cognition and on how it should be modeled. Among these differing perspectives, there are two broad classes: the *cognitivist* approach based on symbolic information processing representational systems, and the *emergent systems* approach, encompassing connectionist systems, dynamical systems, and enactive systems, all based to a lesser or greater extent on principles of self-organization. A third class – *hybrid systems* – attempts to combine something from each of the cognitivist and emergent paradigms. The vast majority of cognitive vision systems adopt either a cognitivist or a hybrid approach, with matching cognitive architectures [8].

The term *visual cognition* is strikingly similar to the term *cognitive vision*. However, they are not equivalent. Visual cognition is a branch of cognitive psychology concerned with research on visual perception and cognition in humans [9, 10, 19]. It addresses several distinct areas such as object recognition, face recognition, scene understanding, visual attention (including visual search, change blindness, repetition blindness, and the control of eye movements), short-term and long-term visual memory, and visual imagery. It is also concerned with the representation and recognition of visual information currently being perceived by the senses and with reasoning about memorized visual imagery. Thus, visual cognition addresses many visual mechanisms that are relevant to cognitive vision but without necessarily treating the entire cognitive system or the realization of these mechanisms in artificial systems.

Application

Several applications of cognitive vision may be found in [11–13]. Examples include natural language description of traffic behavior [5], autonomous control of cars [14], and observation and interpretation of human activity [15, 16].

Open Problems

All of the open problems associated with cognitive systems apply equally to cognitive vision. Three which are particularly relevant are highlighted here.

The first concerns embodiment [17]. There is no universal agreement on whether or not a cognitive vision system must be embodied. Even if it is, several forms of embodiment are possible. One form is a physical robot capable of moving in space, manipulating the environment, and experiencing the physical forces associated with that manipulation. Other forms of embodiment do not involve physical contact and simply require the system to be able to change the state of its visual environment, for example, a surveillance system which can control ambient lighting. These alternative forms of embodiment are consistent with the cognitivist and hybrid paradigms of cognition but do not satisfy the requirements of the emergent approach.

Learning in cognitive vision presents several significant challenges. Since cognitive vision systems do not have all the knowledge required to carry out their tasks, they need to learn. More specifically, they need to be able to learn in an incremental, continuous, open-ended, and robust manner, with learning and recognition being interleaved, and with both improving over time. Since the learning process will normally be effected autonomously without supervision, the learning technique needs to be able to distinguish between good and bad data, otherwise bad data may corrupt the representation and cause errors to become embedded and to propagate. Furthermore, the use of learning in several domains is required, including perceptual (spatiotemporal) and conceptual (symbolic) domains, as well as in mapping between them. The mapping from perceptual to conceptual facilitates communication, categorization, and reasoning, whereas the mapping from conceptual to perceptual facilitates contextualization and embodied action. Learning may be interpreted in a restricted sense as the estimation of the parameter values that govern the behavior of models that have been designed into the system, or in a more general sense as the autonomous generation of mappings that represent completely new models.

The identification and achievement of goals in cognitive vision presents a further challenge. With cognitivist approaches, goals are specified explicitly by designers or users in terms of the required outcome of cognitive behavior. With emergent approaches, goals are more difficult to specify since cognitive behavior is an emergent consequence of the system dynamics. Consequently, they have to be specified in terms of constraints or boundary conditions on the system configuration, either through phylogenetic configuration or ontogenetic development, or both. It is a significant challenge to understand how implicit goals can be specified and incorporated, and how externally communicated goals can be introduced to the system from its environment or from those interacting with it, e.g., through some form of conditioning, training, or communication.

References

1. Christensen HI, Nagel HH (2006) Introductory remarks. In: Christensen HI, Nagel HH (eds) Cognitive vision systems: sampling the spectrum of approaches. Volume 3948 of LNCS. Springer, Heidelberg, pp 1–4
2. Tsotsos JK (2006) Cognitive vision need attention to link sensing with recognition. In: Christensen HI, Nagel HH (eds) Cognitive vision systems: sampling the spectrum of approaches. Volume 3948 of LNCS. Springer, Heidelberg, pp 25–35
3. Nagel HH (2003) Reflections on cognitive vision systems. In: Crowley J, Piater J, Vincze M, Paletta L (eds) Proceedings of the third international conference on computer vision systems, ICVS 2003. Volume LNCS 2626. Springer, Berlin, Heidelberg, pp 34–43
4. Arens M, Ottlik A, Nagel HH (2002) Natural language texts for a cognitive vision system. In: Harmelen FV (ed) Proceedings of the 15th European conference on artificial intelligence (ECAI-2002). IOS Press, Amsterdam, pp 455–459
5. Nagel HH (2004) Steps toward a cognitive vision system. AI Magazine 25(2):31–50
6. Auer P et al (2005) A research roadmap of cognitive vision. ECVision: European network for research in cognitive vision systems. http://www.ecvision.org
7. Vernon D (2006) The space of cognitive vision. In: Christensen HI, Nagel HH (eds) Cognitive vision systems: sampling the spectrum of approaches. Volume 3948 of LNCS. Springer, Heidelberg, pp 7–24
8. Granlund G (2006) Organization of architectures for cognitive vision systems. In: Christensen HI, Nagel HH (eds) Cognitive vision systems: sampling the spectrum of approaches. Volume 3948 of LNCS. Springer, Heidelberg, pp 37–55
9. Pinker S (1984) Visual cognition: an introduction. Cognition 18:1–63
10. Coltheart V (ed) (2010) Tutorials in visual cognition. Macquarie monographs in cognitive science. Psychology Press, London
11. Neumann B (ed) (2005) Künstliche intelligenz, special issue on cognitive computer vision, vol 2. Böttcher IT Verlag, Bremen
12. Christensen HI, Nagel HH (2006) Cognitive vision systems: sampling the spectrum of approaches. Volume 3948 of LNCS. Springer, Heidelberg
13. Buxton H (ed) (2008) Image vision comput, special issue on cognitive vision 26(1)
14. Dickmanns ED (2004) Dynamic vision-based intelligence. AI Magazine 25(2):10–29
15. Sage K, Howell J, Buxton H (2005) Recognition of action, activity and behaviour in the actIPret project. Künstliche Intell Spec Issue Cogn Comput Vis 19(2):30–34
16. Crowley JL (2006) Things that see: context-aware multimodal interaction. In: Christensen HI, Nagel HH (eds) Cognitive vision systems: sampling the spectrum of approaches. Volume 3948 of LNCS. Springer, Heidelberg, pp 183–198
17. Vernon D (2008) Cognitive vision: the case for embodied perception. Image Vis Comput 26(1):127–141
18. Tsotsos JK (2011) A computational perspective on visual attention. MIT Press, Cambridge
19. Cavanagh P (2011) Visual cognition. Vision Res 51(13): 1538–1551

Color Adaptation

▸von Kries Hypothesis

Color Appearance Models

▸Color Spaces

Color Balance

▸White Balance

Color Constancy

Marc Ebner
Institut für Mathematik und Informatik, Ernst Moritz Arndt Universität Greifswald, Greifswald, Germany

Definition

Color constancy is the ability to perceive colors as approximately constant even though the light entering the eye varies with the illuminant. Color constancy also names the field of research investigating the extent of this ability, that is, the conditions under which a color is actually perceived as constant and which factors influence color constancy. Computer scientists working in the field of color constancy try to mimic this ability in order to produce images which are independent of the illuminant, that is, color constant. Simple color-constancy algorithms, also known under the name automatic white balance, are used in digital cameras to compute a color-corrected output image. The input of a color constancy algorithm is often one image taken under an arbitrary illuminant and the output of a color constancy algorithm is frequently the image as it would appear had it been taken with a canonical illuminant such as CIE Standard Illuminant D65 or a spectrally uniform illuminant.

Background

An introduction to computational color constancy is given by Ebner [1]. Maloney [2] also reviews different algorithms for surface color perception. Color is a product of the brain [3]. When an observer perceives an object, processing starts with the retinal sensors. The sensors in the retina measure the light entering the eye. However, the light entering the eye is dependent on both the spectral characteristics of the illuminant and the reflectance properties of the object. Therefore, without any additional processing, the measured light varies with the illuminant.

In the eye, two different types of retinal sensors exist: rods and cones. The rods are used for viewing when little light is available. The cones are mostly used in bright light conditions for color vision. Three types of cones can be distinguished which absorb light primarily in the red, green, and blue parts of the spectrum. Similarly, a digital sensor often measures the incident light in three different parts of the spectrum and uses red, green, and blue sub-sensors. However, cameras with four sub-sensors, for example, red, green, blue, and cyan also exist.

Suppose that an observer views a diffusely reflecting surface which uniformly reflects the incident light. Now assume that the surface is illuminated by a candle. A candle emits more light toward the red part of the spectrum. The candle light will reach the surface where part of the light will be absorbed and the remainder will be reflected. Part of the reflected light enters the eye where it is measured. The sensitivity function of the sensors in combination with the amount of light entering the eye will determine how strongly the sensors respond. Now consider another illuminant with a higher color temperature, for example, daylight or an electronic flash. Such an illuminant will emit more light toward the blue spectrum compared to the candle. If the same surface is viewed with an illuminant that has a high color temperature, then the sensors shift their response toward the blue part of the spectrum. Assuming a normalized response of all three sensors, then the reflected light will have a red color cast when the surface is illuminated by a candle. The measured light will appear white during daylight and it will have a bluish color cast for an illuminant with a high color temperature. The observer, however, will be able to call out the correct color of the surface independent of the illuminant. Color constancy algorithms try to mimic this ability by computing an image which is independent of the illuminant.

Theory

Let $I(\lambda, x, y)$ be the irradiance captured by either a digital sensor or by the eye at position (x, y) for wavelength λ. Let $S_i(\lambda)$ be the response function of sensor i. Then the response of the sensor $c_i(x, y)$ at position (x, y) is given by:

$$c_i(x, y) = \int S_i(\lambda) I(\lambda, x, y) d\lambda.$$

The integration is done over all wavelengths to which the sensor responds. Assuming three receptors with sensitivity in the red, green, and blue parts of the spectrum, then $i \in \{r, g, b\}$. In this case, the measurement of the sensor is a three component vector $\mathbf{c} = [c_r, c_g, c_b]$.

The irradiance I falling onto the sensor is a result of the light reflected from an object patch. Let $L(\lambda, x, y)$ be the irradiance falling onto a diffusely reflecting object patch which is imaged at position (x, y) of the sensor arrangement. Let $R(\lambda, x, y)$ be the reflectance of the imaged object patch. Thus,

$$I(\lambda, x, y) = G(x, y) R(\lambda, x, y) L(\lambda, x, y)$$

where $G(x, y)$ is a geometry factor which takes the orientation between the surface and the light source into account. For a diffusely reflecting surface, $G(x, y) = \cos\alpha$ where α is the angle between the unit vector which points from the surface into the direction of the light source and the normal vector at the corresponding surface position. Thus, the sensor response can be written as:

$$c_i(x, y) = G(x, y) \int S_i(\lambda) R(\lambda, x, y) L(\lambda, x, y) d\lambda.$$

From this equation, it is apparent that the sensor response depends on the orientation of the patch relative to the light source (because of $G(x, y)$); it depends on the sensitivity S_i of the sensor i, the reflectance of the object patch $R(\lambda, x, y)$, and on the illuminant $L(\lambda, x, y)$. Some color constancy algorithms are based on a set of basis functions to model illuminants and reflectances. See Maloney [2] for an introduction.

Color constancy algorithms frequently assume that the sensitivity of the sensors is very narrow band. Assuming that they have the shape of a delta function, $S_i = \delta(\lambda - \lambda_i)$, it holds that:

$$c_i(x, y) = G(x, y) \int \delta(\lambda - \lambda_i) R(\lambda, x, y) L(\lambda, x, y) d\lambda,$$
$$c_i(x, y) = G(x, y) R(\lambda_i, x, y) L(\lambda_i, x, y).$$

This equation is often written in the form:

$$c_i(x, y) = G(x, y) R_i(x, y) L_i(x, y)$$

where the only difference to the previous equation is that the index i is used instead of the parameter λ_i. In this treatment, sensor response, reflectance, and the illuminant is considered only for three distinct wavelengths, that is, color bands, with $i \in \{r, g, b\}$.

A color constancy algorithm tries to discount the illuminant by computing a color constant descriptor $d(\mathbf{c})$ which is independent of the illuminant $\mathbf{L}(x, y) = [L_r(x, y), L_g(x, y), L_b(x, y)]$. Geusebroek et al. [4] have derived several different descriptors which can be computed from $\mathbf{c}$ and are invariant to some imaging conditions such as viewing direction, surface orientation, highlights, illumination direction, illumination intensity, or illumination color. Finlayson and Hordley [5] have shown how a color constant descriptor can be computed provided that the illuminant can be approximated by a black body radiator. Apart from computing a color constant descriptor, a color constancy algorithm usually tries to output an image of the scene which would either correspond to the perception of a human photographer observing the scene or it would correspond to the image that would have resulted if a spectrally uniform illuminant or illuminant D65 had been used.

An ideal solution to the problem of color constancy would be to compute $\mathbf{R}(x, y) = [R_r(x, y), R_g(x, y), R_b(x, y)]$ from the sensor responses. It is of course clear that this problem cannot be solved without making additional assumptions, because for each position on the sensor array, one only has three measurements but there are seven unknowns (shading, reflectance, and illumination components). Note that the above model for image generation is already a simple model assuming narrow band sensor responses.

A frequently made assumption is that the illuminant varies slowly over the image while reflectance is able to change abruptly between sensor responses. Since color constancy algorithms are based on certain assumptions, they will not work correctly if the assumptions are violated. In many cases, it is possible to find images where the color constancy algorithm does not perform as intended. The goal is to develop algorithms which perform well on most everyday scenes.

Simple Algorithms

If a single illuminant illuminates the scene uniformly, that is, $L_i(x, y) = L_i$, then it suffices to estimate a three component vector $\tilde{\mathbf{L}}$ from all the measured sensor responses with $\tilde{\mathbf{L}} \approx \mathbf{L}$. Given an estimate $\tilde{\mathbf{L}}(x, y)$, a color constant descriptor can be computed by dividing the sensor response by the estimate of the illuminant.

$$\begin{aligned} \frac{c_i(x, y)}{\tilde{L}_i(x, y)} &= \frac{G(x, y) R_i(x, y) L_i(x, y)}{\tilde{L}_i(x, y)} \\ &\approx \frac{G(x, y) R_i(x, y) L_i(x, y)}{L_i(x, y)} \\ &= G(x, y) R_i(x, y) \end{aligned} \quad (1)$$

Such an output image will be a shaded reflectance image. In other words, a diagonal color transform suffices if the sensor response is very narrow band. Some color constancy algorithms, however, also use a general 3×3 matrix transform to compute a color-corrected output image.

It is also possible to transform a given image taken under one illuminant $\mathbf{L}$ to another image taken under a different illuminant $\mathbf{L}'$. This can be done by multiplying each sensor response vector by a diagonal matrix whose elements are set to $k_i = \frac{L'_i}{L_i}$. The coefficients k_i are called von Kries coefficients [6]. Necessary and sufficient conditions on whether von Kries chromatic adaptation gives color constancy have been derived by West and Brill [7].

A simple algorithm to estimate the color of the illuminant is the white patch Retinex algorithm. It is a simplified version of the parallel Retinex algorithm [8]. In order to understand how this algorithm works, suppose that a white patch, that is, a uniformly reflecting patch, is contained in the imaged scene which is uniformly illuminated. Assuming a normalized sensor response, then the response of the sensors on the white patch will be an estimate of the illuminant. For the white patch, it holds that $R_i = 1$ which leads to $c_i(\text{white patch}) = GL_i$. The white patch algorithm treats each color band separately and searches for the maximum response which is assumed to be an estimate of the illuminant.

$$\tilde{L}_i = \max_{x,y} c_i(x, y)$$

Instead of locating the maximum response, one can also compute a histogram of the sensor responses and then set the estimate of the illuminant at some percentage from above. This will lead to a more robust estimate of the illuminant. Fig. 1b shows the output of the white patch Retinex algorithm for a sample image shown in Fig. 1a. The illuminant was estimated at 5% from above using a histogram approach for each color band.

Another simple algorithm is based on the gray world assumption which is due to Buchsbaum [9]. According to Buchsbaum, the world is gray on average. Let a_i be the global average of all sensor responses for color channel i where n is the number of sensors.

$$a_i = \frac{1}{n}\sum_{x,y} c_i(x, y) = \frac{1}{n}\sum_{x,y} G(x, y)R_i(x, y)L_i(x, y)$$

Assuming a uniform illuminant $L_i(x, y) = L_i$ and an independence between shading and reflectance, then:

Color Constancy, Fig. 1 (**a**) sample input image (**b**) results for the white patch Retinex algorithm using a histogram (**c**) results for the gray world assumption

$$\begin{aligned} a_i &= L_i \frac{1}{n}\sum_{x,y} G(x, y)R_i(x, y) \\ &= L_i E[G(x, y)R_i(x, y)] = L_i E[G(x, y)][R_i(x, y)] \end{aligned} \tag{2}$$

where $E[x]$ denotes the expected value of x. Suppose that a large number of differently colored objects are contained in the scene. Thus, a uniform distribution

Color Constancy, Fig. 2 (**a**) image of a leaf from a banana plant (**b**) results for the gray world assumption

of colors is assumed. This results in $E[R_i(x, y)] = \frac{1}{n}\sum_{x,y} R_i(x, y) = \frac{1}{2}$ assuming a range of [0, 1] for reflectances and $E[G(x, y)] = c$ where c is a constant. The result is:

$$a_i = \frac{c}{2}L_i.$$

Hence, the illuminant is proportional to global space average color $L_i \propto a_i$ and an estimate of the illuminant can be obtained by setting:

$$\tilde{L}_i = 2a_i.$$

Using $c = 1$ assumes that all patches are frontally oriented or alternatively, that the geometry factor is subsumed into a combined reflectance and geometry factor. Fig. 1c shows the results for the gray world assumption on a sample image. Fig. 2 shows the results for another image where the assumption, that a large number of different colored objects are contained in the scene, is not fulfilled. In this case, the gray world assumption will not work correctly.

Additional Color Constancy Algorithms

Other important algorithms include gamut constraint algorithms originally developed by Forsyth [10]. A gamut constraint algorithm looks at the gamut of colors contained in a sample image and then transforms this color gamut to a color gamut of a canonical image. Forsyth's algorithm operates on three color channels. Finlayson [11] has developed a variant which operates with a projected color gamut (2D gamut constraint algorithm). Van de Weijer et al. [12] have introduced the gray-edge hypothesis. While the gray-world assumption suggests that the world is gray on average, the gray-edge hypothesis suggests that image derivatives are gray on average. Brainard and Freeman [13] have addressed the problem of color constancy using Bayesian decision theory.

Uniform Versus Nonuniform Illumination

Many color constancy algorithms assume that the scene is uniformly illuminated by a single illuminant. If multiple light sources are distributed over the scene, then it is assumed that these light sources can be combined into a single average illuminant. This is possible provided that the light sources are sufficiently distant from the scene. If the illumination is uniform, then only a three component descriptor has to be estimated from the input image. However, in practice, many scenes are illuminated nonuniformly. Very often, one has several different illuminants. For instance, daylight may be falling through a window while artificial lights have already been turned on inside a building. Nonuniform illumination may also be present outside during a sunny day. Consider a family sitting in the garden under a red umbrella and a photographer taking a photograph of the family members. The family would be illuminated by light reflected from the red umbrella while the surrounding would be illuminated by direct sunlight. A digital camera usually corrects for a single illuminant. Thus, either the family members would have a red color cast to them or the background colors would not look right in the resulting image.

Algorithms have also been developed which can cope with a spatially varying illuminant. Land and McCann's Retinex algorithm [8] is a parallel algorithm for color constancy which allows for a nonuniform illumination. They only considered one-dimensional paths over the sensor array. Horn [14] extended Land's Retinex algorithm to two dimensions.

Blake [15] provided additional improvements. Barnard et al. [16] extended the 2D gamut constraint algorithm to scenes with varying illumination.

Ebner [1] showed how a grid of processing elements is able to estimate the color of the illuminant locally using the gray world assumption. Each processing element receives the measurement c_i from the sensor and computes local space average color. The processing elements are laterally connected to neighboring processing elements. Let $a_i(x, y)$ be the current estimate of local space average color for a processing element located at position (x, y). Each processing element receives the estimate of local space average color from neighboring elements and averages the neighboring estimates to update its own estimate. A small component from the sensor measurement is then added to this estimate. Let p be a small percentage and let $N(x, y)$ be the neighborhood of the processing element at position (x, y), then the computation of a processing element consists of the following two updates.

$$a_i'(x, y) = \frac{1}{|N(x, y)|} \sum_{(x',y')\in N(x,y)} a_i(x', y')$$

$$a_i(x, y) = (1 - p)a_i'(x, y) + pc_i(x, y)$$

The two updates are carried out iteratively. This process converges to local space average color which is an estimate of the illuminant, $\tilde{L}_i(x, y) = 2a_i(x, y)$. Figure 3 shows local space average color after 0, 200, and 2,000 iterations using $p = 0.0001$ given the input image shown in Fig. 4b. The extent of the averaging is determined by the parameter p. Figure 4 shows the output when local space average color is used to estimate the illuminant locally. Local average color was computed for a downscaled image (25% in each direction), the original image had 768×512 pixels. Ebner [1] suggested that a similar algorithm may be used by the brain for color perception.

0

200

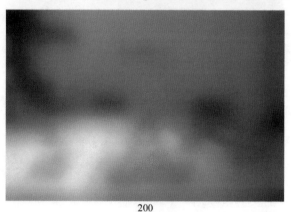

2000

Color Constancy, Fig. 3 Local space average color after 0, 200 and 2,000 iterations

Advanced Reflectance Models

The theoretical model of color image formation, that has been given above, assumed that objects diffusely reflect the incident light. This is not the case for all surfaces, for example, brushed metal or plastics. Especially for plastic objects or objects covered with gloss varnish, a more elaborate reflectance model is more appropriate. The dichromatic reflection model [17, 18] assumes that reflectance is composed of interface reflectance, which occurs at the boundary between the object's surface and air, and body reflectance which is due to the scattering of light below the object's surface. In other words, the reflection of light from the object's surface is assumed to be partially specular

Color Constancy, Fig. 4 (**a** and **b**) input images (**c** and **d**) results using local space average color

and partially diffuse. Color constancy algorithms have also been developed using the dichromatic reflection model, for example, by Risson [19].

Application

Color constancy algorithms are ideal for color correction in digital photography. In digital photography, the goal is to obtain a color-corrected image that corresponds nicely to human perception. A printed photograph or a photograph viewed on a computer display should appear in exactly the same way that the human observer (the photographer) perceived the scene. Besides digital photography, color constancy algorithms can be applied in the context of most computer vision tasks. For many tasks, one should try to estimate object reflectances. For instance, image segmentation would not be as difficult, if object reflectance could be correctly determined. Similarly, color-based object recognition is easier if performed on reflectance information. Thus, color constancy algorithms should often be applied as a preprocessing step. This holds especially for autonomous mobile robots equipped with a vision system because autonomous mobile robots need to operate in different environments under different illuminants.

Experimental Results and Datasets

A comparison of computational color constancy algorithms is given by Barnard et al. [20] for synthetic as well as real image data. Another detailed comparison of color constancy algorithms along with pseudocode for many algorithms is given by Ebner [1]. Data for computer vision and computational color science can be found at the Simon Frasier University, Canada (www.cs.sfu.ca/~colour/data/). A repository has also been created by the color group at the University of East Anglia, UK (www.colour-research.com). A database for spectral color science is available from the University of Eastern Finland, Finland (spectral.joensuu.fi/).

References

1. Ebner M (2007) Color constancy. Wiley, England
2. Maloney LT (1999) Physics-based approaches to modeling surface color perception. In: Gegenfurtner KR, Sharpe LT (eds) Color vision: from genes to perception. Cambridge University Press, Cambridge, pp 387–422
3. Zeki S (1993) A vision of the brain. Blackwell, Oxford
4. Geusebroek JM, van den Boomgaard R, Smeulders AWM, Geerts H (2001) Color invariance. IEEE Trans Pattern Anal Mach Intell 23(12):1338–1350
5. Finlayson GD, Hordley SD (2001) Color constancy at a pixel. J Opt Soc Am A 18(2):253–264
6. Richards W, Parks EA (1971) Model for color conversion. J Opt Soc Am 61(7):971–976
7. West G, Brill MH (1982) Necessary and sufficient conditions for von Kries chromatic adaptation to give color constancy. J Math Biol 15:249–258
8. Land EH, McCann JJ (1971) Lightness and retinex theory. J Opt Soc Am 61(1):1–11
9. Buchsbaum G (1980) A spatial processor model for object colour perception. J Frankl Inst 310(1):337–350
10. Forsyth DA (1990) A novel algorithm for color constancy. Int J Comput Vis 5(1):5–36
11. Finlayson GD (1996) Color in perspective. IEEE Trans Pattern Anal Mach Intell 18(10):1034–1038
12. van de Weijer J, Gevers T, Gijsenij A (2007) Edge-based color constancy. IEEE Trans Image Process 16(9): 2207–2214
13. Brainard DH, Freeman WT (1997) Bayesian color constancy. J Opt Soc Am A 14(7):1393–1411
14. Horn BKP (1974) Determining lightness from an image. Comput Graph Image Process 3:277–299
15. Blake A (1985) Boundary conditions for lightness computation in mondrian world. Comput Vis Graph Image Process 32:314–327
16. Barnard K, Finlayson G, Funt B (1997) Color constancy for scenes with varying illumination. Comput Vis Image Underst 65(2):311–321
17. Klinker GJ, Shafer SA, Kanade T (1988) The measurement of highlights in color images. Int J Comput Vis 2:7–32
18. Tominaga S (1991) Surface identification using the dichromatic reflection model. IEEE Trans Pattern Anal Mach Intell 13(7):658–670
19. Risson VJ (2003) Determination of an illuminant of digital color image by segmentation and filtering. United States Patent Application, Pub. No. US 2003/0095704 A1
20. Barnard K, Cardei V, Funt B (2002) A comparison of computational color constancy algorithms – Part I and II. IEEE Trans Image Process 11(9):972–996

Color Difference

►Color Similarity

Color Discrepancy

►Color Similarity

Color Dissimilarity

►Color Similarity

Color Management

►Gamut Mapping

Color Model

Shoji Tominaga
Graduate School of Advanced Integration Science, Chiba University, Inage-ku, Chiba, Japan

Synonyms

Color spaces; Color specification systems

Definition

A color model is a mathematical model describing the way colors are specified by three-dimensional coordinates in such terms as three numerical values, three perceptual attributes, or three primary colors.

Background

The human retina contains three types of cones L, M, and S as photoreceptors, which respond best to light of long, medium, and short wavelengths, respectively, in the visible range. Since the perception of color depends on the response of these cones, it follows in principle that visible color can be mapped into a three-dimensional space in terms of three numbers.

Trichromatic color vision theory is closely related to such three types of color sensors. The trichromatic theory by Young and Helmholtz suggests that any colors and spectra in the visible range can be visually matched using mixtures of three primary colors. These primary colors are usually red, green, and blue. Most image output devices such as television screens, computer and video displays, and image projectors create visible colors by using additive mixtures of the three primary colors.

Digital cameras also capture color by using the same principle of trichromatic color vision. They usually use three different types of sensors which primarily respond to the red, green, and blue parts of the visible spectrum. Therefore, a color model mathematically describing the color coordinate system is crucial for analysis, evaluation, and the rendering of color images.

Theory

Color Model for RGB Color Signals

Most color models used in computer vision and image processing are based on the RGB primaries. This is because color images captured by digital cameras are all represented by the RGB primaries, and image output devices such as displays and projectors are based on an additive mixture of the RGB primaries. Therefore, the simplest color model (called RGB color space) uses a Cartesian coordinate system defined by (R, G, B) triples. This system, however, is not available for colorimetry, because the spectral-sensitivity curves of cameras are generally not coincident with the color-matching functions.

The RGB space is not a perceptually uniform color space. In computer graphics and image processing applications, approximately uniform spaces derived from RGB are defined in terms of the three attributes: hue (H), saturation (S), and value (V) (representing lightness). For example, the HSV model by Smith [9] was defined as a nonlinear transform of RGB given in 8 bits:

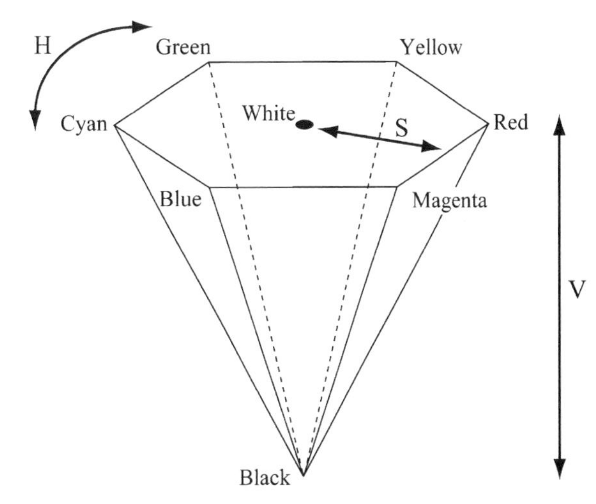

Color Model, Fig. 1 Single hexagonal color space

$$\begin{aligned} H &= \tan^{-1}\left\{\frac{\sqrt{3}(g-b)}{(r-g)+(r-b)}\right\} \\ S &= 1 - Min/V \\ V &= Max \end{aligned} \quad (1)$$

where

$$r = R/255, \quad g = G/255, \quad b = B/255,$$
$$Max = \max(r, g, b), \qquad Min = \min(r, g, b).$$

This model represents a hexagonal color space as seen in Fig. 1, where the saturation decreases monotonically as the value decreases. Figure 2 represents the improved HSV model to a double hexagonal space, which is defined as:

$$\begin{aligned} H &= \tan^{-1}\left\{\frac{\sqrt{3}(g-b)}{(r-g)+(r-b)}\right\} \\ S &= Max - Min \\ V &= (R+g+b)/3 \end{aligned} \quad . \quad (2)$$

Color Model for Colorimetry

The CIE-XYZ color system for colorimetry was created by the International Commission on Illumination (CIE). This color system was derived from a series of visual experiments of color matching [1]. The tristimulus values of a color were used for representing

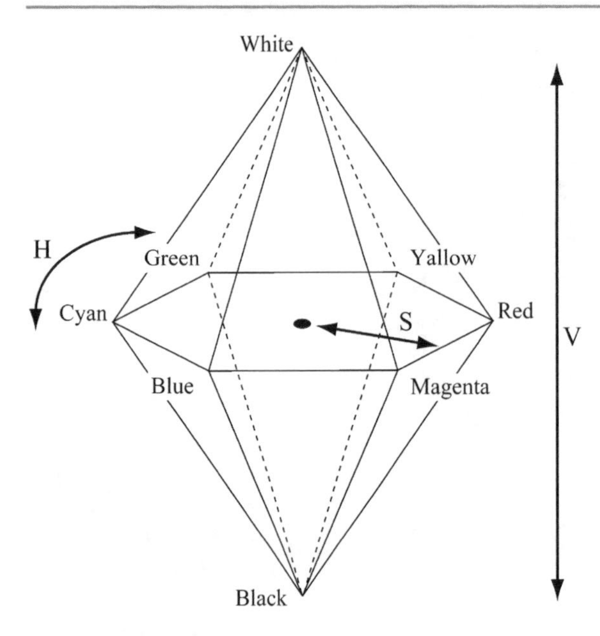

Color Model, Fig. 2 Double hexagonal color space

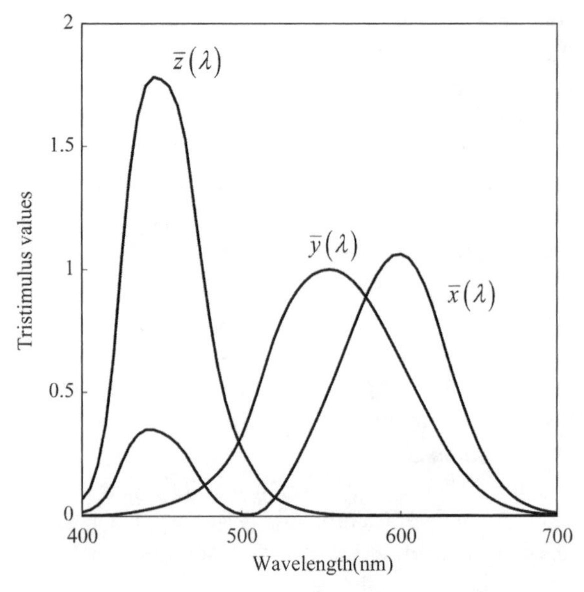

Color Model, Fig. 3 CIE 1931 2° Standard Observer color-matching functions

the amounts of three primary colors needed to match the test color. The tristimulus values depend on the observer's field of view. Therefore, the CIE defined the standard observer and a set of three color-matching functions, called $\bar{x}(\lambda)$, $\bar{y}(\lambda)$, and $\bar{z}(\lambda)$. The chromatic response of the standard observer is characterized by the three color-matching functions. Figure 3 shows the CIE 1931 2° standard observer color-matching functions.

The tristimulus values for a color signal with a spectral-power distribution $L(\lambda)$ are then given by:

$$
\begin{aligned}
X &= k \int_{\lambda} L(\lambda)\bar{x}(\lambda)d\lambda \\
Y &= k \int_{\lambda} L(\lambda)\bar{y}(\lambda)d\lambda \\
Z &= k \int_{\lambda} L(\lambda)\bar{z}(\lambda)d\lambda
\end{aligned} \tag{3}
$$

where the integration on the wavelength λ is normally calculated over the range of visible spectrum 400–700 nm, and the coefficient k is a normalizing constant. In application to photometry, the Y tristimulus value becomes the luminance of the color signal. In application to object color, the tristimulus values become:

$$
\begin{aligned}
X &= k \int_{\lambda} E(\lambda)S(\lambda)\bar{x}(\lambda)d\lambda \\
Y &= k \int_{\lambda} E(\lambda)S(\lambda)\bar{y}(\lambda)d\lambda \\
Z &= k \int_{\lambda} E(\lambda)S(\lambda)\bar{z}(\lambda)d\lambda
\end{aligned} \tag{4}
$$

where $S(\lambda)$ is the surface-spectral reflectance of the illuminated object, and $E(\lambda)$ is the spectral-power distribution of the light source illuminating the object. In this case, the Y tristimulus value becomes the luminance factor of the object-color stimulus. The constant factor is usually chosen as:

$$
k = 100 \bigg/ \int_{\lambda} S(\lambda)\bar{x}(\lambda)d\lambda. \tag{5}
$$

The chromaticity coordinates (x, y) of a given color are defined as:

$$
x = \frac{X}{X+Y+Z}, \quad y = \frac{Y}{X+Y+Z}. \tag{6}
$$

Figure 4 depicts the chromaticity diagram, where all visible chromaticities are located within the horseshoe-shaped region. The outer curved boundary is the spectral locus, corresponding to monochrome light with the

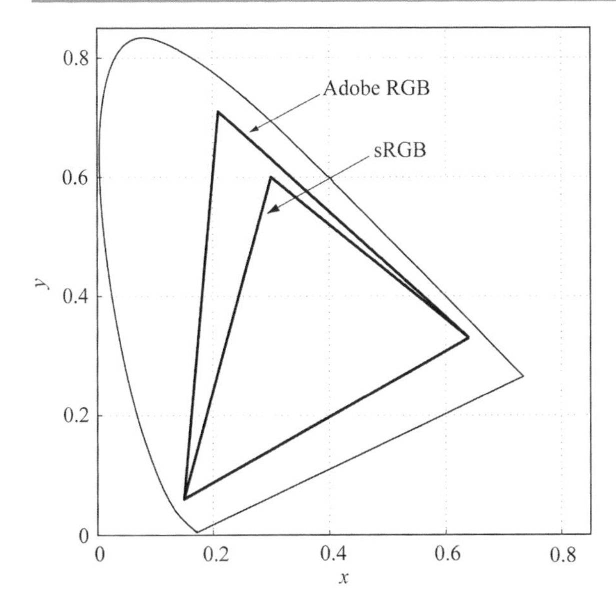

Color Model, Fig. 4 CIE 1931 chromaticity diagram

most saturated color. Less saturated colors are located in the inside of the region with white at the center. The triangle of sRGB represents the gamut of a standard RGB color space proposed by HP and Microsoft in 1996 [10]. Recently, usual monitors, printers, and the Internet are based on this standard space. The triangle of Adobe RGB represents the gamut of an RGB color space developed by Adobe Systems in 1998 [11]. This color space, improving the gamut of sRGB primarily in cyan-green, is used for desktop publishing.

It should be noted that the CIE XYZ tristimulus values were not defined for color differences. Two colors with a constant difference in the tristimulus values may look very different, depending on where the two colors are located in the (x, y) chromaticity diagram.

Color Model for Uniform Color Space

Many people believe that a uniform color space is most useful for applications where perceptual errors are evaluated. The CIE 1976 L*a*b* color system (also called CIE LAB color system) was designed to approximate a perceptually uniform color space in terms of the tristimulus values XYZ. It is defined as follows:

$$\begin{aligned} L^* &= 116(Y/Y_n)^{1/3} - 16 \\ a^* &= 500[(X/X_n)^{1/3} - (Y/Y_n)^{1/3}] \\ b^* &= 200[(Y/Y_n)^{1/3} - (Z/Z_n)^{1/3}] \end{aligned} \quad (7)$$

where (X_n, Y_n, Z_n) are the tristimulus values of the reference white point. This system provides an opponent-color space with dimension L^* for lightness, and a^* and b^* for opponent-color dimensions. $L^*=0$ and $L^*=100$ indicate black and white, respectively. The axes of a* and b* take the position between red and green, and the position between yellow and blue, respectively.

The two-dimensional chromaticity diagram is expressed in the orthogonal coordinate system of (a*, b*). The chromaticity components can also be expressed in a cylindrical system of chroma and hue, where the chroma and hue-angle are defined respectively by:

$$\begin{aligned} C^*_{ab} &= (a^{*2} + b^{*2})^{1/2} \\ h_{ab} &= \arctan^{-1}(b^*/a^*) \end{aligned} . \quad (8)$$

The color difference between two color stimuli 1 and 2 is defined as follows:

$$\Delta E^*_{ab} = [(\Delta L^*)^2 + (\Delta a^*)^2 + (\Delta b^*)^2]^{1/2}, \quad (9)$$

where $\Delta L^* = L^*_1 - L^*_2$, $\Delta a^* = a^*_1 - a^*_2$, and $\Delta b^* = b^*_1 - b^*_2$. Figure 5 depicts the Munsell color system with value 5 on the (a*, b*) chromaticity coordinates under the reference condition of illuminant D_{65}. The color system approximates roughly the Munsell uniform color space.

Color Model for Color Order System

Color order systems were created to order colors in terms of intuitive perceptual attributes. Most colored samples, used as a reference in many design and engineering applications, are made in equal steps of perceptual attributes.

The Munsell color order system is one of the most widely used systems in the world. This system is based on the three perceptual attributes: hue, value (representing lightness), and chroma (representing saturation), and is organized in a cylindrical coordinate system as shown in Fig. 6. Hue is arranged as a circle,

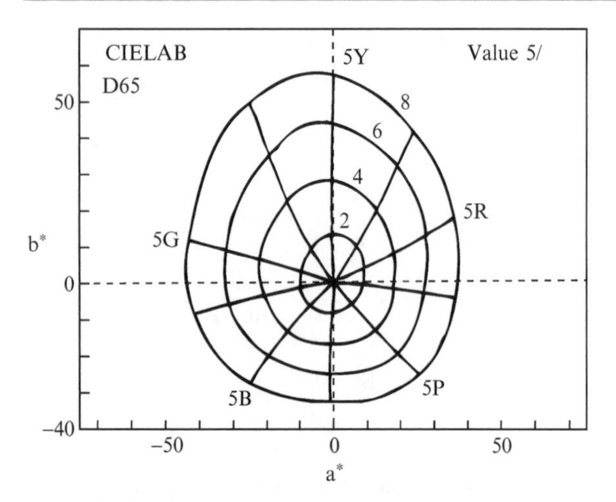

Color Model, Fig. 5 Munsell color system with Value 5 on the (a*, b*) chromaticity coordinates

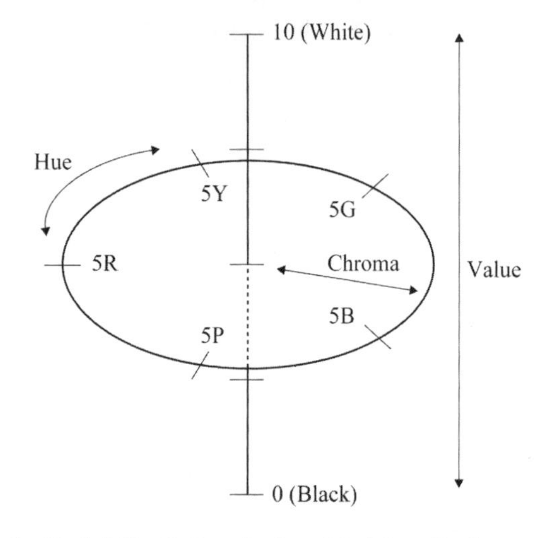

Color Model, Fig. 6 Organization of the Munsell color system

value as the axis perpendicular to the hue circle, and chroma as the distance from the center.

The Munsell color samples are arranged in equal steps of each attribute, where each sample is described using a three-part syntax of hue chroma/value. For example, 5YR 8/4 indicates a color sample with the hue of 5YR (yellow-red), the chroma of level 4, and the value of level 8. A conversion table to reproduce the Munsell color samples in terms of the CIE tristimulus values was given by Newhall et al. [12]. The table contains the luminance factor Y and chromaticity coordinates (x, y) equivalent to the Munsell renotation system (H, V, C), under the condition of the CIE standard illuminant C. It should be noted that there is no formula for a color notation conversion between the Munsell and CIE system, except for the table. Therefore, the conversion was carried out by interpolating the table data.

Obviously, this method is not efficient from the point of view of mass data processing by computers. Tominaga [13] developed a method for color notation conversion, between the Munsell and CIE systems by means of neural networks. In the neural network method, it is not necessary to use the special database of color samples. This is because the knowledge of conversion between the two color spaces is stored in a small set of the weighting parameters in a multilayer feedforward network.

References

1. Wyszecki G, Stiles WS (1982) Color science: concepts and methods, quantitative data and formulae. Wiley, New York
2. Hall R (1989) Illumination and color in computer-generated imagery. Springer, New York
3. Foley JD, van Dam A, Feiner SK, Hughes JF (1990) Computer graphics: principles and practice, 2nd edn. Addison-Wesley, Reading
4. Wandell BA (1995) Foundations of vision. Sinauer Associates, Sunderland
5. Shevell SK (2003) The science of color, 2nd edn. Elsevier, Oxford
6. Lee H-C (2005) Introduction to color imaging science. Cambridge University Press, Cambridge
7. Fairchild MD (2005) Color appearance models. Wiley, Chichester
8. Shanda J (2007) Colorimetry. Wiley, Hoboken
9. Smith AR (1978) Color gamut transform pairs. Comput Graph 12(3):12–19
10. Stokes M, Anderson M, Chandrasekar S, Motta R (1996) A standard default color space for the Internet: sRGB. Version 1.10, International Color Consortium
11. Adobe Systems (2005) Adobe RGB(1998)b color image encoding. Version 2005-5, Adobe Systems
12. Newhall SM, Nickerson D, Judd DB (1943) Final report of the O.S.A. subcommittee on the spacing of the munsell colors. J Opt Soc Am 33(7):385–411
13. Tominaga S (1992) Color classification of natural color images. Color Res Appl 17(4):230–239
14. Brainard DH (1989) Calibration of a computer controlled color monitor. Color Res Appl 14:23–34
15. Pratt WK (1991) Digital image processing, 2nd edn. Wiley, New York
16. Ohta N, Robertson AR (2005) Colorimetry. Wiley, Chichester
17. Hunt RWG (2004) The reproduction of colour, 6th edn. Wiley, Chichester
18. Green P (2010) Color management. Wiley, Chichester
19. Watt A (2000) 3D computer graphics, 3rd edn. Addison Wesley, Reading

Color Similarity

Takahiko Horiuchi and Shoji Tominaga
Graduate School of Advanced Integration Science,
Chiba University, Inage-ku, Chiba, Japan

Synonyms

Color difference; Color discrepancy; Color dissimilarity

Definition

Color similarity is a measure that reflects the strength of relationship between two colors.

Background

A measure of similarity between colors is needed when one color region is matched to another region among color images, for example, a purple region is more similar to a blue region than a green region. Therefore, the similarity between two colors is an important metric in not only color science but also imaging science and technology (including computer vision). If color similarity is measured, it allows people to determine the strength of relationship in terms of numerical values; otherwise, it will be determined in nonnumerical ways.

Thus, the benefit of color similarity is its application to various techniques of image analysis, such as noise removal filters, edge detectors, object classification, and image retrieval.

Theory

Color similarity is usually represented in the range of either interval [−1, 1] or interval [0, 1]. Let $\mathbf{x}_i = [x_{i1}, x_{i2}, x_{i3}]$ be a three-dimensional vector specifying a color feature at location i in a color image. The similarity measure $s(\mathbf{x}_i, \mathbf{x}_j)$ between two color features is a symmetric function whose value is larger when color features $\mathbf{x}_i$ and $\mathbf{x}_j$ are closer. The color feature is specified by three coordinates in color space (or color model). Color space is defined in many different ways in color and image science, where color spaces such as RGB, L*a*b*, L*u*v*, YIQ, and HSI are often used.

On the other hand, color dissimilarity is also used for measuring the discrepancy between two colors. This quantity is usually represented in a positive range of [0, 1] because the dissimilarity is measured by a "normalized distance" between two colors. A dissimilarity measure $d(\mathbf{x}_i, \mathbf{x}_j)$ is a symmetric function whose value is larger when color features $\mathbf{x}_i$ and $\mathbf{x}_j$ are more dissimilar. Therefore, a relationship between two measures of color similarity and color dissimilarity is given by:

$$s(\mathbf{x}_i, \mathbf{x}_j) = 1 - d(\mathbf{x}_i, \mathbf{x}_j), \quad (1)$$

where the color similarity $s(\mathbf{x}_i, \mathbf{x}_j)$ is bounded by [0, 1]. Note that when the color similarity is one (i.e., exactly same), the color dissimilarity is zero, and when color similarity is zero (i.e., quite different), the color dissimilarity is one. If the color similarity is defined in the range [−1, 1], then:

$$s(\mathbf{x}_i, \mathbf{x}_j) = 1 - 2d(\mathbf{x}_i, \mathbf{x}_j). \quad (2)$$

Note in this case that the color dissimilarity of 1 (i.e., quite different) corresponds to the color similarity of −1, and when the color dissimilarity of 0 (i.e., exactly same) corresponds to the color similarity of 1. In many cases, measuring dissimilarity is easier than measuring similarity. Once dissimilarity is measured, it can easily be normalized and converted to the similarity measure. Therefore, color dissimilarity measures are first described in the following.

Color Dissimilarity Measures

The most commonly used dissimilarity measure to quantify the distance between two color vectors $\mathbf{x}_i$ and $\mathbf{x}_j$ is the weighted Minkowski metric:

$$d(\mathbf{x}_i, \mathbf{x}_j) = c\left(\sum_{k=1}^{3} \xi_k \left|x_{ik} - x_{jk}\right|^L\right)^{1/L}, \quad (3)$$

where c is the nonnegative scaling parameter for $d(\mathbf{x}_i, \mathbf{x}_j) \in [0, 1]$ and the exponent L defines the nature of the distance metric. The parameter ξ_k, for $\sum_k \xi_k = 1$, measures the weight assigned to the color channel k. Usually it is determined $\xi_k = 1/3, \forall k$ (e.g., [1, 2]). In the case $L = 1$ in (Eq. 3), the measure is known as Manhattan distance, which is also called city-block distance, absolute value distance, and taxicab distance [3].

$$d(\mathbf{x}_i, \mathbf{x}_j) = c \sum_{k=1}^{3} \left| x_{ik} - x_{jk} \right|. \quad (4)$$

The dissimilarity measure represents distance between points in a city road grid. It examines the absolute differences between a pair of colors. In the case $L = 2$ in (Eq. 3), the measure is known as Euclidean distance which is the most common use of distance.

$$d(\mathbf{x}_i, \mathbf{x}_j) = c \left(\sum_{k=1}^{3} \left(x_{ik} - x_{jk} \right)^2 \right)^{1/2}. \quad (5)$$

In the case $L \to \infty$ in (Eq. 3), the measure is known as Chebyshev distance which is also called Kolmogorov-Smirnov Statistic [4], chess-board distance, and maximum value distance. It examines the absolute magnitude of the differences between coordinates of a pair of objects.

$$d(\mathbf{x}_i, \mathbf{x}_j) = c \max_k \left| x_{ik} - x_{jk} \right|. \quad (6)$$

As a well-known dissimilarity metric based on the Minkowski metric, [6] color difference ΔE^*_{ab} is useful [5, 7]. ΔE^*_{ab} is the Euclidean distance between two colors in CIEL*a*b* color space as follows:

$$\Delta E^*_{ab}(\mathbf{x}_i, \mathbf{x}_j) = \left(\left(L^*_i - L^*_j \right)^2 + \left(a^*_i - a^*_j \right)^2 + \left(b^*_i - b^*_j \right)^2 \right)^{1/2}, \quad (7)$$

where $\mathbf{x}_i = \left[L^*_i, a^*_i, b^*_i \right]$ and $\mathbf{x}_j = \left[L^*_j, a^*_j, b^*_j \right]$. Please note that ΔE^*_{ab} is used as a "distance" with $c = 1$ in (Eq. 5). When $\Delta E^*_{ab} \approx 2.3$, it corresponds to a just noticeable difference (JND) of surface colors [8, 9]. Perceptual nonuniformities in the underlying CIEL*a*b* color space, however, have led to the CIE refining their 1976 definition over the years. The refined color difference ΔE^*_{94} is defined in the L*C*h color space [10]. Given a reference color $\mathbf{x}_i = \left[L^*_i, C^*_i, h^*_i \right]$ and another color $\mathbf{x}_j = \left[L^*_j, C^*_j, h^*_j \right]$, the difference is:

$$\Delta E^*_{94}(\mathbf{x}_i, \mathbf{x}_j) = \left(\left(\frac{L^*_i - L^*_j}{K_L} \right)^2 + \left(\frac{C^*_i - C^*_j}{1 + K_1 C^*} \right)^2 + \left(\frac{h^*_i - h^*_j}{1 + K_2 C^*} \right)^2 \right)^{1/2} \quad (8)$$

where C* represents the geometrical average of chroma, and K_L, K_1, and K_2 represent the weighting factors which depend on the application (in original definition, $K_L = 1$, $K_1 = 0.045$, and $K_2 = 0.015$ for graphic arts). Since the 1994 definition did not adequately resolve the perceptual uniformity issue, the CIE refined their definition, adding five corrections:

1. A hue rotation term (RT), to deal with the problematic blue region
2. Compensation for neutral colors
3. Compensation for lightness
4. Compensation for chroma
5. Compensation for hue

The definition of the CIEDE2000 color difference is explained in Ref. [11].

Color Similarity Measure

The simplest color similarity measure is an angular separation similarity. It represents a cosine angle between two color vectors, which is often called as coefficient of correlation. In this measure, similarity in orientation is expressed through the normalized inner product as:

$$s(\mathbf{x}_i, \mathbf{x}_j) = \mathbf{x}_i \cdot \mathbf{x}_j / |\mathbf{x}_i| \left| \mathbf{x}_j \right| \quad (9)$$

which corresponds to the cosine of the angle between the two color vectors $\mathbf{x}_i$ and $\mathbf{x}_j$. Since similar colors have almost parallel orientations, (*while significantly different colors point in different directions in a 3-D color space*) the normalized inner product can be used to quantify orientation similarity between the two color vectors.

The correlation coefficient is a standardized angular separation by centering the coordinates to their mean value. Let $\bar{x}_i = \frac{1}{p} \sum_{k=1}^{p} x_{ik}$ and $\bar{x}_j = \frac{1}{p} \sum_{k=1}^{p} x_{jk}$; the correlation coefficient between these two vectors $\mathbf{x}_i$ and $\mathbf{x}_j$ is given as follows:

$$s(\mathbf{x}_i, \mathbf{x}_j) = \frac{\sum_{k=1}^{p} |x_{ik} - \bar{x}_i| \, |x_{jk} - \bar{x}_j|}{\left(\sum_{k=1}^{p} (x_{ik} - \bar{x}_i)^2\right)^{1/2} \left(\sum_{k=1}^{p} (x_{jk} - \bar{x}_j)^2\right)^{1/2}}. \quad (10)$$

In (Eq. 10) the color similarity between two points is shown as $p = 3$ shows. Then the correlation coefficient can be applied to color region similarity, such as textured regions by setting $p = 3 \times n$ (where n is the number of pixels).

There are many other methods to measure the similarity between two color vectors. So depending on the nature and the objective of the problem at hand, it is possible that one method is more appropriate than the other.

Application

The formulation of similarity between two color vectors is of paramount importance for the development of the vector processing techniques. These include noise removal filters, edge detectors, image zoomers, and image retrievals.

References

1. Kruskal JB (1964) Multidimensional scaling by optimizing goodness of fit to a nonmetric hypothesis. Psychometrika 29(1):1–27. doi:10.1007/BF02289565
2. Duda RO, Hart PE (1973) Pattern classification and scene analysis. Wiley, New York
3. Krause EF (1987) Taxicab geometry: adventure in non-euclidean geometry. Dover, New York. ISBN 0–486–25202–7
4. Geman D, Geman S, Graffigne C, Dong P (1990) Boundary detection by constrained optimization. IEEE Trans Pattern Anal Mach Intell 12(7):609–628
5. Pauli H (1976) Proposed extension of the CIE recommendation on uniform color spaces, color difference equations, and metric color terms. J Opt Soc Am 66(8):866–867. doi:10.1364/JOSA.66.000866
6. CIE (2004) Colorimetry. Vienna, CIE Pb. 15:2004, ISBN 978-3-901906-33-6
7. CIE Publication (1978) Recommendations on uniform color spaces, color difference equations, psychometric color terms. 15(Suppl 2), Paris, (E.-1.3.1) 1971/(TC-1.3)
8. Sharma G (2003) Digital color imaging handbook, 1.7.2 edn. CRC, Florida. ISBN 084930900X
9. Mahy M, Eycken LV, Oosterlinck A (1994) Evaluation of uniform color spaces developed after the adoption of CIELAB and CIELUV. Color Res Appl 19(2): 105–121
10. CIE Publication (1995) Industrial colour-difference evaluation. Vienna, CIE 116–1995, ISBN 978-3-900734-60-2
11. CIE Publication (2001) Improvement to industrial colour-difference evaluation. Vienna, CIE 142-2001, ISBN 978-3-901906-08-4

Color Spaces

Rajeev Ramanath[1] and Mark S. Drew[2]
[1]DLP® Products, Texas Instruments Incorporated, Plano, TX, USA
[2]School of Computing Science, Simon Fraser University, Vancouver, BC, Canada

Synonyms

Color appearance models; Color model

Definition

A color space describes the range of colors – the *gamut* – that an imaging device or software has to work with. Consequently, the design of these color spaces allows a user to modify images in a predefined manner based on the specific needs of the application.

Theory

Color spaces may be generally separated into those that are defined for analysis of color by color scientists (*colorimetric color spaces*) and those that are used for image/color editing.

Color Spaces for Colorimetric Analysis

Color spaces for colorimetric analysis are typically based on the human observer. Central to such color spaces is the CIEXYZ *color matching functions*, which are based on CIERGB color matching functions – based in turn on the LMS cone fundamentals. CIE is an abbreviation for the International Commission on Illumination (Commission International de L'Eclairage) that is the body that is responsible for standardization of data in this field. However, the ability to directly measure the cone functions of an observer is only a recent development. Researchers had originally inferred the cone functions based on psychophysical experiments measuring the RGB color matching functions. Consequently, the CIE had standardized the color matching functions long before the cone functions could be directly measured. Details of how the color matching functions were developed are well documented in the book by Wyszecki and Stiles that is arguably a cornerstone of colorimetry [27].

The RGB color matching functions, denoted $\{\bar{r}(\lambda), \bar{g}(\lambda), \bar{b}(\lambda)\}$, are shown in Fig. 1a.

Note that these color matching functions have negative excursions, due to the fact that there are some colors that reside outside the triangle formed by these three primaries – a negative excursion in the primaries' weights is the only way to represent a color outside this triangle. In order to address this problem, the CIE also published a set of color matching functions with nonnegative values, denoted $\bar{x}(\lambda)$, $\bar{y}(\lambda)$, $\bar{z}(\lambda)$ and shown in Fig. 1b. These are known generally as *the* CIEXYZ color matching functions. Note that these are now fictitious, nonphysical primaries. The equations for computing these special "tristimulus" values, in XYZ color space, are:

$$\begin{aligned} X &= k \int_{\lambda} \bar{x}(\lambda) i_r(\lambda) d\lambda \\ Y &= k \int_{\lambda} \bar{y}(\lambda) i_r(\lambda) d\lambda \\ Z &= k \int_{\lambda} \bar{z}(\lambda) i_r(\lambda) d\lambda. \end{aligned} \tag{1}$$

where $i_r(\lambda)$ denotes the spectral power distribution of the light energy incident on the retina, and k denotes a normalization factor that is set to 683 lumens/Watt in the case of absolute colorimetry and to $100/\int_{\lambda} \bar{y}(\lambda) i(\lambda) d\lambda$ for relative colorimetry. In the case of relative colorimetry, this means a value of $Y = 100$ denotes the brightest color – the illuminant reflecting from a perfect reflecting diffuser.

The first set of color matching functions published by the CIE were originally empirically determined for a 2° field – the bipartite field used for matching a subtended 2° angle on the observers' retina. Following the 1931 publication, W. S. Stiles and J. M. Burch conducted experiments [25] to measure color matching functions for larger fields of view. This was combined with the findings of Speranskaya [24] into the publication by the CIE of a 10° observer in 1964 [5]. The difference between these two standard observers is significant enough to warrant a clear specification of which observer color matching functions are used in experimental work. More specifically, the 10° observer has noticeable shifts of the color matching functions in the blue direction due to the fact that the subtense of the stimulus encompasses a larger portion of the retina and hence more S cones and also increased macular pigment absorption.

In the CIE colorimetric system, an XYZ tristimulus value uniquely specifies a color. However, a convenient 2D representation of the tristimulus values led to the projection of the tristimulus values by normalizing by the sum of the three values. These "chromaticity" values are given by:

$$\begin{aligned} x &= \frac{X}{X + Y + Z} \\ y &= \frac{Y}{X + Y + Z}. \end{aligned} \tag{2}$$

A color may be specified uniquely by its (x, y) chromaticity coordinates and its luminance Y, and is often used to describe a color since the tristimulus values are straightforward to derive from the (x, y, Y) values. The biggest advantage of the (x, y) chromaticity coordinates is that they specify a magnitude-independent hue and purity of a color. A chromaticity diagram (see Fig. 2) is typically used to specify a color using its chromaticity coordinates. In Fig. 2, the horseshoe-shaped locus denotes the locus of monochromatic stimuli visible to the standard 2° observer (the gamut of visible colors). Shorter wavelength stimuli (starting at 380 nm, eliciting a relatively strong blue response) reside in the lower left of this horseshoe shape while the longer wavelengths (ending at 830 nm, eliciting a relatively strong red response) reside on the lower right, with the top of the horseshoe curve around 520 nm (eliciting a strong green response). The line connecting the blue and red corners is referred to as the *line of purples*. Colors on this line, although on the border of the gamut, have no counterpart in monochromatic sources of light and hence have no wavelengths associated with them.

The (x, y) chromaticity diagram is perceptually nonuniform: Unit vectors in the chromaticity space do not correspond to a unit change in perception even if the luminance is kept constant. In an attempt to improve the uniformity of the chromaticity diagram, in 1976, the CIE published a uniform chromaticity scale (UCS) diagram that scaled and normalized the XYZ

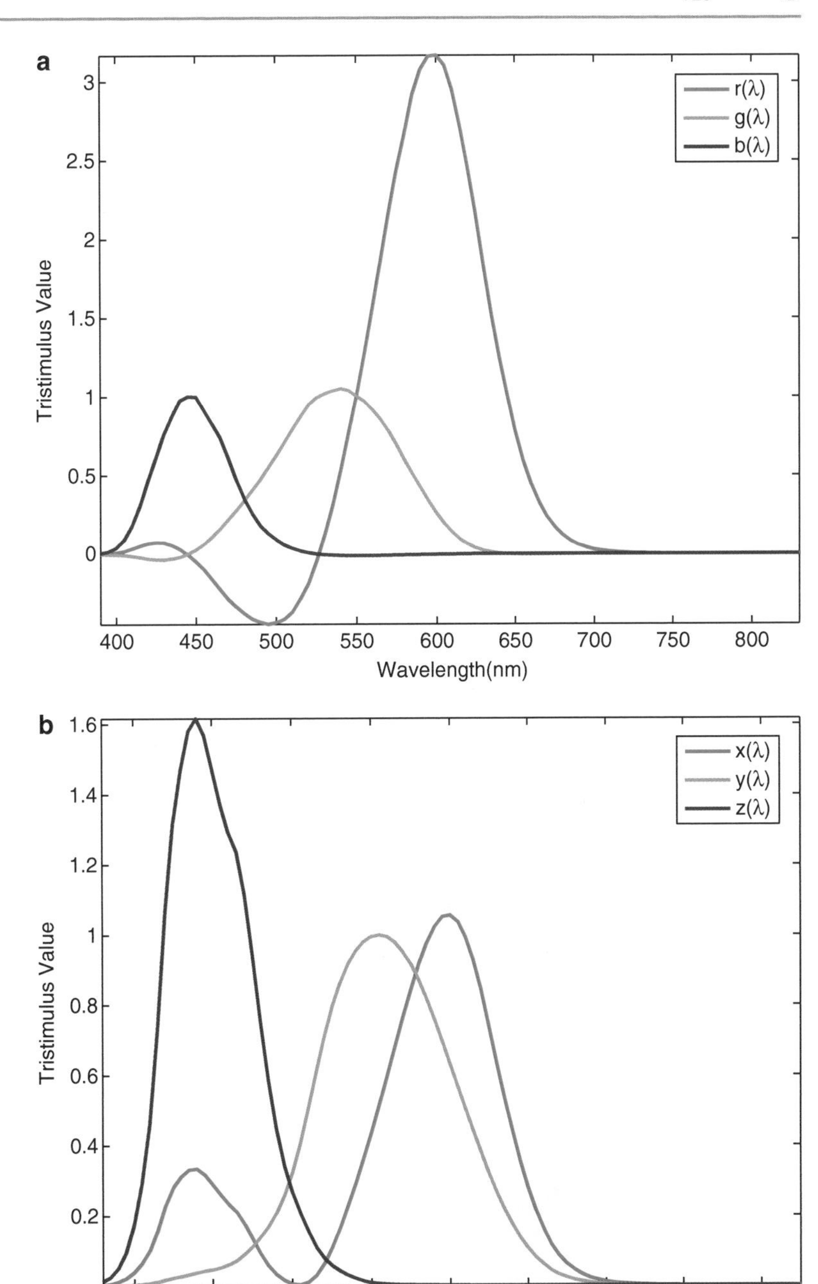

Color Spaces, Fig. 1 (**a**) RGB color matching functions (**b**) XYZ color matching functions including Judd-Vos modifications

tristimulus values [6]. This chromaticity diagram is denoted by u', v' axes, which are related to the XYZ tristimulus values by the following equations:

$$u' = \frac{4X}{X + 15Y + 3Z}$$
$$v' = \frac{9Y}{X + 15Y + 3Z}. \quad (3)$$

Figure 3 shows the UCS. Just as in the case of Fig. 2, in this figure, the horseshoe-shaped locus represents the gamut of visible colors.

The CIEXYZ color space does not have perceptual correlates that would make it useful for common use. In an attempt to add perceptual behavior to color spaces, based on earlier works of many researchers, the CIE proposed a lightness scale along with two

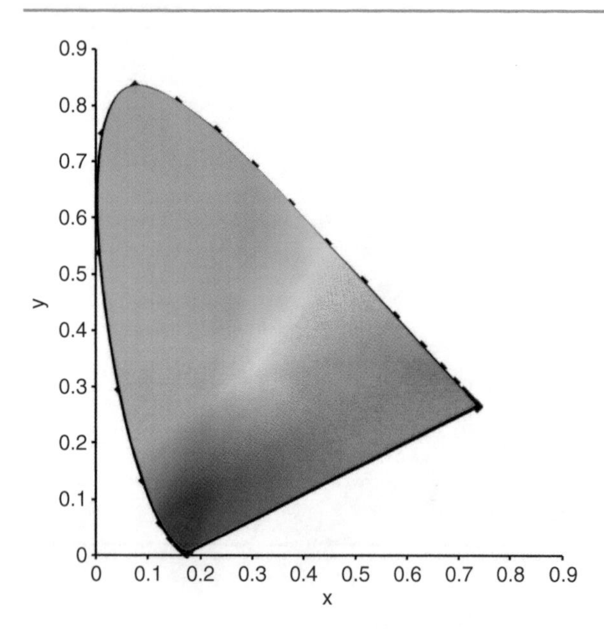

Color Spaces, Fig. 2 CIE xy chromaticity diagram for a 2° observer

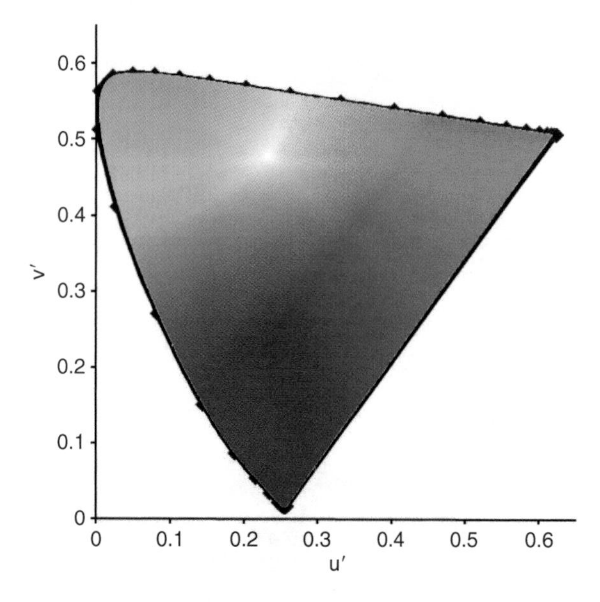

Color Spaces, Fig. 3 CIE UCS u′v′ chromaticity diagram for a 2° observer

chromatic scales. The CIELAB color space is one such space, where the axes are denoted by L* (Lightness), a* (redness–greenness), and b* (yellowness–blueness). For a stimulus given by a tristimulus value X, Y, and Z, the CIELAB coordinates are given by:

$$
\begin{aligned}
L^* &= 116\, f(Y/Y_n) - 16 \\
a^* &= 500\, [f(X/X_n) - f(Y/Y_n)] \\
b^* &= 200\, [f(Y/Y_n) - f(Z/Z_n)], \quad \text{where} \\
f(t) &= t^{1/3}, \quad \text{for} \quad t > 0.008856 \\
f(t) &= 7.787\, t + 16/116 \quad \text{otherwise.}
\end{aligned} \tag{4}
$$

In the above equations, the subscript n denotes the tristimulus values corresponding to the reference white chosen, which makes the CIELAB color space a relative color space. Given the CIELAB coordinates in a three-dimensional space, correlates of chroma and hue may be derived as follows:

$$C^*_{ab} = \left(a^{*2} + b^{*2}\right)^{1/2} \tag{5}$$

$$h^*_{ab} = \tan^{-1}\left(b^*/a^*\right). \tag{6}$$

Under highly controlled viewing conditions, a CIELAB ΔE difference of 1 correlates with a single just noticeable difference in color. It is to be noted though that the CIELAB color difference measure was designed for color differences between uniform color patches in isolation.

In a similar construct, the CIE also recommended a CIELUV color space, based on the uniform chromaticity scale (UCS). This uses a subtractive shift from the reference white instead of the normalization based on division that is used in the CIELAB space. The equations to transform a tristimulus value from u^*,v^* coordinates to CIELUV are given by:

$$
\begin{aligned}
L^* &= 116\, f(Y/Y_n) - 16 \\
u^* &= 13\, L^*(u' - u'_n) \\
v^* &= 13\, L^*(v' - v'_n), \quad \text{where} \\
f(t) &= t^{1/3}, \quad \text{for} \quad t > 0.008856 \\
f(t) &= 7.787\, t + 16/116 \quad \text{otherwise.}
\end{aligned} \tag{7}
$$

The u',v' coordinates for a tristimulus value are computed using (Eq. 3). As in the CIELAB definitions, the subscript n denotes the u',v' coordinates of the reference white being used. The descriptions of the u* and v* axes are similar to those in CIELAB: approximating redness–greenness and yellowness–blueness directions.

Based on these correlates, the CIE recommends that color difference measures in the two uniform-perception spaces CIELAB and CIELUV be given by the Euclidean difference between the coordinates of two color samples:

$$\Delta E^*_{ab} = \left[(\Delta L^*)^2 + (\Delta a^*)^2 + (\Delta b^*)^2\right]^{1/2}$$
$$\Delta E^*_{uv} = \left[(\Delta L^*)^2 + (\Delta u^*)^2 + (\Delta v^*)^2\right]^{1/2} \tag{8}$$

where the differences are given between the corresponding color coordinates in the CIELAB and CIELUV spaces between the standard and test samples.

Many improvements to this basic color difference measure have been proposed and adopted over the years involving scaling the lightness, chroma, and hue differences appropriately based on the application and the dataset of samples to which the color difference measure has been adapted or improved [7]. Typically, color difference thresholds are dependent on application and thresholds for *perceptibility* judgments are significantly lower than thresholds for *acceptability* judgments. These color spaces were designed for threshold color differences and their application to supra-threshold (larger than about 5 units of ΔE) color differences is to be handled with care [17]. Many other color difference measures have been proposed and more recently, the CIE DE2000 has been adopted as a measure of color difference, again for uniform color patches under highly controlled viewing conditions, and is slowly gaining acceptance [7, 15].

These color spaces provide a powerful tool to model and quantify color stimuli and are used both in color difference modeling for color patches and, as well, have more recently been used to describe color appearance (see [19]). Models based on describing colors based on lightness, chroma, and hue are powerful in their abilities to enable communication of color stimuli, as well.

Color spaces for colorimetric analysis have the advantage that the need for communicating these colors to others can be done without much ambiguity, if the viewing conditions (reference white) are specified and well controlled. Arguably, the biggest disadvantage is that it is not straightforward to understand what a certain color coordinate in a certain color space would mean to an observer without having to use mathematical equations.

Color Spaces for Editing

Color spaces for editing are specifically designed with the following key requirements:

- Colors have perceptual correlates that are easily understood, such as hue, saturation, brightness, purity, etc.
- Colors described in these spaces are reproducible – within reason – across different media: monitors, printers, etc.
- Colors in these spaces are defined based on primaries – the axes:
  - On the display side of the world of color applications, these spaces are additive, defined by red, green, and blue primaries: Equal amounts of red and green will give yellow.
  - On the printing side, primaries are defined roughly by cyan, magenta, yellow, and black, and these spaces are subtractive: Knowing the print surface, equal densities of cyan and yellow will give green.

A Brief Note on "Gamma"

If a color system is linear, then for additive combinations of colors, a unit input of a color corresponds to a globally constant unit of the output signal; whereas for nonlinear combinations, a transfer function would determine the input–output relationship. This nonlinearity is often used in digital systems with limited available bits and in signal compression systems to take advantage of the available output signal bit depth by stretching small input codes over a larger range of output codes and compressing the larger input codes into a smaller output dynamic range. This is referred to as *gamma encoding*.

From an encoding perspective, in its simplest form, the input–output relationship is typically given by a gain-offset-gamma model, given by:

$$y = \text{round}\left[\left(2^N - 1\right)\left(\alpha x + \beta\right)^\gamma\right] \tag{9}$$

where α denotes a scalar gain, N denotes the number of bits in a system, β is a scalar offset, γ denotes a power law with values larger than 1 (typically around 2.2), and x and y denote the normalized input and output signals respectively [1]. In encoding systems, the three channels typically have the same parameters α, β, γ, N. Display systems based on cathode ray tubes (CRTs) have an inherent response that follows the inverse relationship – large steps in input signal at the low end of the input signal cause a small change in output whereas at the upper end of the signal range, small steps caused large output changes. It so happens that gamma encoding (using a power law of γ on the linear luminance input) the input prior to transmitting the data to a CRT display causes the display luminance to follow similar steps resulting in a net unity transfer function. This is also a useful means of encoding data to maximize bit-depth usage while reducing visibly apparent contouring on the output data and display [10, 18]. In the case of a quantized color space, for reasons of perceptual uniformity, it is preferable to establish a nonlinear relationship between color values and intensity or luminance.

RGB Color Model

The specification of a color in the RGB color space implies a linear (after gamma is removed) or nonlinear (gamma applied) combination of the red, green, and blue primaries in varying strengths. In RGB color spaces, the manner in which colors are reproduced varies from device to device. For example, a color specified as an RGB triplet is more than likely going to look different from one display device to another when the exact same RGB triplet is provided as input, due to differences in the "color" of the primaries and also differences in gamma curves. This makes the RGB space a device-dependent color space. Specifying colors in device-dependent color spaces, although not preferred from a color-reproduction perspective, is often resorted to due to its ease in comprehension. An RGB color model can represent any color within an RGB color cube, as shown in Fig. 4. This color model is most commonly used in display applications where data is additive in nature.

For example, a full-strength yellow color is specified by (1.0, 1.0, 0.0), denoting the use of the red and green primaries at full strength and the blue primary completely turned off. In an 8-bit system, this will correspond to a code value of (255,255,0). A three-primary display with three independent color primaries (typically denoted by their CIE x,y chromaticity values along with that of white) is specified by:

$$\begin{bmatrix} x_R & x_G & x_B & x_W \\ y_R & y_G & y_B & y_W \end{bmatrix}. \tag{10}$$

An introduction to how specifications such as those in (Eq. 10) may be used to generate a transformation matrix from RGB to XYZ for linear RGB is given in the book [18].

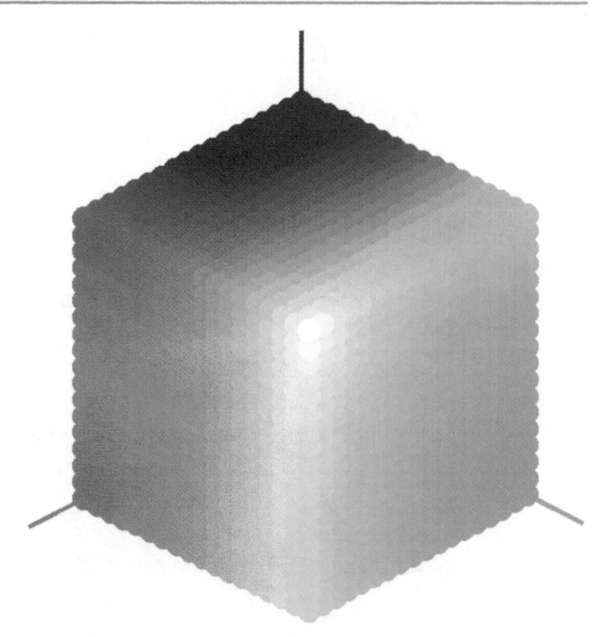

Color Spaces, Fig. 4 RGB color cube commonly used in display applications

Different RGB color spaces differ in their gamma values and the chromaticity coordinates in (Eq. 10). A useful compilation of various RGB color spaces may be found in a Web site hosted by Bruce Lindbloom [11].

A color mixing matrix for additive colors mixing is shown in Table 1, stating, for example, that a cyan color would be created using maximum intensity of green and blue primaries and none of the red primary.

CMY/CMYK Color Model

Printers, on the other hand, create colors using inks that are deposited on paper, in which case the manner in which they create color is called subtractive color mixing. The inks selectively absorb wavelengths of incident light and reflect the remainder. As a beam of light passes through an absorbing medium, the amount of light absorbed is proportional to the intensity of the incident light times the coefficient of absorption (at a given wavelength). This is often referred to as *Beer-Lambert-Bouguer law* and is given by:

$$A(\lambda) = \log_{10} \varepsilon(\lambda) c(\lambda) l(\lambda) \tag{11}$$

Color Spaces, Table 1 Color mixing matrix for additive primaries

	Color displayed							
Primary used	Red	Green	Blue	Cyan	Yellow	Magenta	White	Black
Red	1	0	0	0	1	1	1	0
Green	0	1	0	1	1	0	1	0
Blue	0	0	1	1	0	1	1	0

where $\varepsilon(\lambda)$ is denotes absorptivity, $c(\lambda)$ denotes the concentration, and $l(\lambda)$ denotes the path length for the beam of light. Stated differently, the higher the concentration or thickness or absorptivity of a certain absorptive material, the higher is absorption – the intensity of reflected or transmitted beam of light will be reduced [20]. The simplest model for printer inks is called the *block-dye model*, an idealized system where, for example, cyan ink absorbs all the red but none of the green or blue, with rectangular shapes for absorptivity as a function of wavelength.

In a subtractive-color setup, different thicknesses of the three primary inks may be deposited on top of each other to result in a final color to the observer. The colorant amounts required to print a stimulus designated by RGB emissions are given by $Y = 1 - X$, where $Y \in \{C, M, Y\}$ and $X \in \{R, G, B\}$, all normalized to unity. Real primary inks, however, do not correspond to these ideal functions and, hence, more sophisticated models need to include not just the spectral absorptions/reflectances of the inks, but the density (or area) of the inks and the characteristics of the media (paper) involved. The Kubelka-Munk equations describe the absorption and scattering of light as it passes through layers of ink and the substrate, for example, paper. Various extensions are used in practice that account for the shortcomings of the basic Kubelka-Munk analysis, considering issues such as nonlinearities in ink deposition, interactions between inks, etc., [8, 20]. In subtractive color mixing, the primaries are typically cyan (C), yellow (Y), and magenta (M). The color mixing matrix for subtractive color is shown in Table 2.

Much like RGB color systems, where the reference white made a difference in the appearance of a certain color, depending upon the kind of paper and inks used for printing, the rendered color can be significantly different from one printer to another.

Most printers use a "K" channel, denoting black ink, primarily because a black generated by mixing cyan, yellow, and magenta is not black enough in appearance. Additionally, to complicate matters, in order to print black, a printer would need to lay cyan, magenta, and yellow inks on top of each other, making ink drying a cause for concern and additionally the limits of ink absorption by the substrate, for example, paper. Additionally, using 1 unit of black ink instead of 1 unit each of cyan, yellow, and magenta inks can lead to significant cost savings.

HSL/HSV Color Model

In order to make the representation of colors intuitive, colors may be ordered along three independent dimensions corresponding to the perceptual correlates of lightness, hue, and chroma. In device-dependent color spaces, there are many commonly used variants of these perceptual correlates: HSV is by far the most common. H stands for the perceptual correlate of hue; S stands for the saturation of a color, defined by the chroma of a color divided by its luminance (the more desaturated the color the closer it is to gray); and V stands for value (a perceptual correlate of lightness). This color model is commonly used in image processing and editing software. However, the HSV color model has two visualization representations, one of which is a cylinder with black at the bottom and pure full-intensity colors on the top, and the other is a representation by a cone, with black at the apex and white on the base. The equations used to convert RGB data into the HSV color space are given by:

$$V = \max \tag{12}$$

$$S = \begin{cases} 0 & \text{if } V = 0 \\ \left(V - \min\right)/V & \text{if } V > 0 \end{cases} \tag{13}$$

$$H = \begin{cases} 0 & \text{if } S = 0 \\ 60(G - B)/(\max - \min) & \text{if } (\max = R \text{ and } G \geq B \\ 60(G - B)/(\max - \min) + 360 & \text{if } (\max = R \text{ and } G < B \\ 60(B - R)/(\max - \min) + 120 & \text{if } \max = G \\ 60(R - G)/(\max - \min) + 240 & \text{if } \max = B \end{cases} \tag{14}$$

where max and min denote the maximum and minimum of the (R, G, B) triplet. These two representations are shown in Fig. 5. From the figure, it is apparent that saturation is not dependent on the

Color Spaces, Table 2 Color mixing matrix for subtractive primaries

	Color displayed							
Primary used	Red	Green	Blue	Cyan	Yellow	Magenta	White	Black
Cyan	0	1	1	1	0	0	0	1
Yellow	1	1	0	0	1	0	0	1
Magenta	1	0	1	0	0	1	0	1

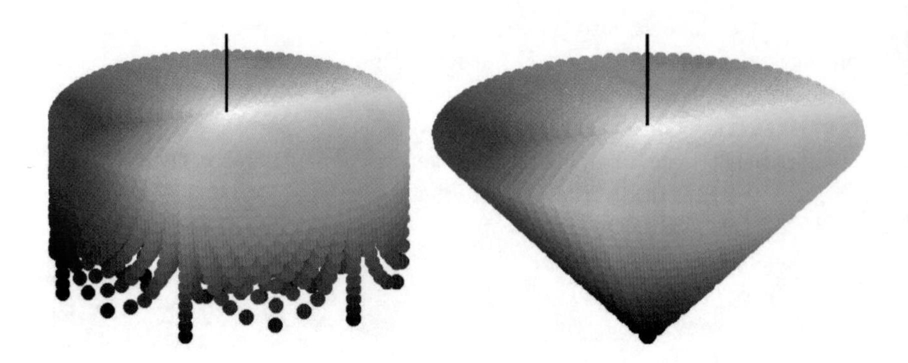

Color Spaces, Fig. 5 The HSV color model represented as a cylinder and as a cone

intensity of the signal. It is, however, often useful in image processing applications to have an indicator of saturation given by a function of the intensity of the signal, resulting in a conical-shaped HSV color space (Fig. 5). When the conical representation is preferred, S is given by $(\max - \min)/(2^N - 1)$ where $2^N - 1$ denotes the largest possible value for R, G, or B. Other variants of the HSV color space also exist and are used as an intuitive link to RGB color spaces (HSB, or HLS denoting various correlates of hue, saturation, and brightness/lightness).

Other Color Spaces

Color spaces designed for editing and communication needs are typically formulated such that colors are encoded/transmitted in the color space of a reference device. Colors spaces fitting such a description include the sRGB color space, the YCC, YUV, YIQ color transmission spaces, the SWOP CMYK color space, Adobe RGB, and ProPhoto RGB, to list a few.

A popular mechanism to standardize colors across electronic devices such as printers, monitors, and the Internet is the use of the sRGB color space. Originally, this was proposed by Hewlett-Packard and Microsoft, and was later standardized by the International Electrotechnical Commission under IEC 61966-2-1 [12]. There are two primary parts to the sRGB standard: the viewing conditions and the necessary colorimetric definitions and transformations. The sRGB reference viewing environment corresponds to conditions typical of monitor display viewing conditions and thus may not be as well suited for print material, due to the various proprietary gamut mapping algorithms in most printers that take advantage of each printer's color gamut. The colorimetric definitions provide the transforms necessary to convert between the sRGB color space and the CIEXYZ tristimulus color space as defined for a standard 2° observer. More specifically, the standard is written for a standard reference monitor that has Rec. 709 primaries and a D65 white point. An overview of the technical advantages and challenges of the sRGB color space may be found in Refs. [21, 26]. As was mentioned earlier, color spaces for video directly make use of the gamma-corrected signals, denoted R′, G′, B′, from camcorders, without any attempt to correlate to the linear signals used in color science, such as those in (Eq. 1). For still imaging as well as video, this problem can be mitigated by the use of the transform built into the sRGB standard, which includes a function for transforming from nonlinear signals I' to linear ones. On a scale of 0.0–1.0, for each of $I = R, G, B$, the following function is applied:

$$\begin{cases} I = I'/12.92, & \text{if } I' < 0.04045; \\ I = ((I' + 0.055)/1.055)^{2.4} & \text{otherwise.} \end{cases} \tag{15}$$

In the video industry, a common mode of communication is the YCbCr color space (YPbPr in the analog domain) that converts RGB signal information into an opponent luma-chroma color space. A nonlinear transfer function is applied to linear-light R, G, B values and a weighted sum of the resulting R', G', B' values is used in the Y, Cb, and Cr signals. In the television domain, these signals have dynamic ranges (on an 8-bit scale) of 16–235 for the luma signal and 16–240 in the Cb and Cr signals. This is to allow for signal noise and potential signal processing noise, giving some head- and foot-room. The weights are different depending upon the color space that the data is being created for. For example, encoding R', G', B' signals with a 16–235 dynamic range into a color space defined by the NTSC primaries (often referred to as ITU-R BT.601), is given by:

$$\begin{bmatrix} Y \\ Cb \\ Cr \end{bmatrix} = \begin{bmatrix} 0.299 & 0.587 & 0.114 \\ -0.169 & -0.331 & 0.500 \\ 0.500 & -0.419 & -0.081 \end{bmatrix} \begin{bmatrix} R' \\ G' \\ B' \end{bmatrix} + \begin{bmatrix} 0 \\ 128 \\ 128 \end{bmatrix} \tag{16}$$

whereas when using HDTV (referred to as ITU-R BT.709) primaries, is given by:

$$\begin{bmatrix} Y \\ Cb \\ Cr \end{bmatrix} = \begin{bmatrix} 0.213 & 0.715 & 0.072 \\ -0.117 & -0.394 & 0.511 \\ 0.511 & -0.464 & -0.047 \end{bmatrix} \begin{bmatrix} R' \\ G' \\ B' \end{bmatrix} + \begin{bmatrix} 0 \\ 128 \\ 128 \end{bmatrix}. \tag{17}$$

A useful reference for further details on this topic is a book by Keith Jack [14]. The Y channel typically contains most of the information in the image, as defined by spatial frequencies, and is hence sampled at much higher rates than the chroma signals. This greatly helps in the ability of the transmission system to compress luma and chroma data with low overheads when compared to luma-only systems. To aid compression formats, color images in the JPEG and JPEG2000 file formats also convert the R', G', B' information into the YCbCr color space prior to compression.

In the printing industry, a commonly specified color space is the SWOP (Specifications for Web Offset Publications) CMYK color space. The SWOP CMYK [2] is a proofing specification that has a well-established relationship between the CMYK input to a standard printer and its CIELAB values (an approximation of the perceptual coordinates of a color) and for a standardized dataset. Specifying images in the SWOP CMYK color space allows the printing house and the content creator to preview images on a common baseline prior to printing. Most image editing software available nowadays allows the user to preview images in the SWOP CMYK color space.

Open Problems

Depending upon the application, color spaces have their individual, optimized uses. Device-independent color models like the CIEXYZ, CIELAB, CIELUV, and their derivatives are used most often to communicate color either between devices or between different color processing teams across the world. The International Color Consortium (ICC) has been extremely successful in standardizing device-independent color spaces between displays, printers, and capture devices [9, 13]. The color profiles that are stored and communicated in ICC profiles use an intermediate profile connection space (PCS) such as CIEXYZ or CIELAB. ICC profiles also store color transformation profiles to and from different color devices (say, from an input device such as a scanner to CIEXYZ, and from CIEXYZ to an output device such as a printer). For example, an sRGB ICC profile incorporates the color space transform from sRGB to the PCS and a SWOP CMYK ICC profile would incorporate the color space transform from the PCS to the CMYK output color space for a printer. Furthermore, depending upon the rendering intent (how the colors need to be represented on the output device), different transformations may be specified in the ICC profile. Interested readers are referred to more detailed discussions of this subject such as the comprehensive books by Hunt [10], BillMeyer and Saltzman [1], Kuehni [16], Fairchild [19], Sharma [23], and Green and MacDonald [8].

References

1. Berns RS (2000) Billmeyer and Saltzman's principles of color technology, 3rd edn. Wiley, New York
2. CGATS TR 001 (1995) Graphic technology – color characterization data for type 1 printing. American National Standards Institute, Washington, DC
3. Commission Internationale de l'Eclairage (1926) The basis of physical photometry, CIE proceedings 1924. Cambridge University Press, Cambridge
4. Commission Internationale de l'Eclairage (1931) Proceedings international congress on illumination. Cambridge University Press, Cambridge

5. Commission Internationale de l'Eclairage (1964) CIE proceedings (1964) Vienna session, committee report E-1.4.1. Bureau Central de la CIE, Paris
6. Commission Internationale de l'Eclairage (1986) CIE publication 15.2, colorimetry. Central Bureau CIE, Vienna
7. Commission Internationale de l'Eclairage (2004) Colorimetry, publication CIE 15:2004, 3rd edn. CIE, Vienna
8. Green P, MacDonald L (eds) (2002) Colour engineering: achieving device independent colour. Wiley, Chichester
9. http://www.color.org. International Color Consortium
10. Hunt RWG (2004) The reproduction of color, 6th edn. Wiley, Chichester/Hoboken
11. Information About RGB Working Spaces. http://www.brucelindbloom.com/WorkingSpaceInfo.html
12. International Electrotechnical Commission (1999) IEC 61966-2-1: multimedia systems and equipment – colour measurement and management – Part 2–1: colour management – default RGB colour space – sRGB. IEC, Geneva
13. International Organization for Standardization (2005) ISO 15076–1, image technology colour management – Architecture, profile format and data structure – Part 1:Based on ICC.1:2004–10, ISO, Geneva
14. Jack K (2001) Video demystified – a handbook for the digital engineer, 3rd edn. LLH Technology Publishing, Eagle Rock
15. Kuehni RG (2002) CIEDE2000, milestone or a final answer? Color Res Appl 27(2):126–127
16. Kuehni RG (2003) Color space and its divisions: color order from antiquity to the present. Wiley, Hoboken
17. Kuehni RG (2004) Color: an introduction to practice and principles, 2nd edn. Wiley, Hoboken
18. Li Z-N, Drew MS (2004) Fundamentals of multimedia. Prentice-Hall, Upper Saddle River
19. Fairchild MD (2005) Color appearance models, 3rd edn. http:// www.wiley.com/WileyCDA/WileyTitle/product Cd-1119967031.html
20. McDonald R (1997) Colour physics for industry, 2nd edn. Society of Dyers and Colourists, Bradford
21. Microsoft Corporation (2001) Colorspace interchange using srgb. http://www.microsoft.com/whdc/device/display/color/sRGB.mspx
22. Nassau K (1983) The physics and chemistry of color: the fifteen causes of color. Wiley, New York
23. Sharma G (ed) (2003) Digital color imaging handbook. CRC, Boca Raton
24. Speranskaya NI (1959) Determination of spectrum color coordinates for twenty-seven normal observers. Opt Spectrosc 7:424–428
25. Stiles WS, Burch JM (1959) Npl colour-matching investigation: final report (1958). Opt Acta 6:1–26
26. Stokes M, Anderson M, Chandrasekar S, Motta R (1996) A standard default color space for the internet: sRGB. http://www.color.org/sRGB.html
27. Wyszecki G, Stiles WS (2000) Color science: concepts and methods, quantitative data and formulae, 2nd edn. Wiley, New York

Color Specification Systems

►Color Model

Compressed Sensing

►Compressive Sensing

Compressive Sensing

Aswin C. Sankaranarayanan[1] and Richard G. Baraniuk[2]
[1]ECE Department, Rice University, Houston, TX, USA
[2]Department of Electrical and Computer Engineering, Rice University 2028 Duncan Hal, Houston, TX, USA

Synonyms

Compressed sensing

Related Concepts

►Dimensionality Reduction

Definition

Compressed sensing refers to parsimonious sensing, recovery, and processing of signals under a sparse prior.

Background

The design of conventional sensors is based heavily on the Shannon-Nyquist sampling theorem which states that a signal $\mathbf{x}$ band limited to W Hz is determined completely by its discrete time samples provided the sampling rate is greater than $2W$ samples per second. This theorem is at the heart of modern signal processing as it enables signal processing in the discrete time or digital domain without any loss of information. However, for many applications, the Nyquist sampling rate is high as well as redundant and unnecessary. As a motivating example, in modern cameras, the high resolution of the CCD sensor reflects the large amount of data sensed to capture an image. A 10 megapixel camera, in effect, takes 10 million linear measurements of the scene. Yet, almost immediately after capture, redundancies in the image are exploited to compress the image significantly, often by compression ratios of 100:1 for visualization and even higher for detection and classification tasks. This suggests immense wastage in the overall design of the conventional camera.

Compressive sensing (CS) refers to a sampling paradigm where additional structure on the signal is exploited to enable sub-Nyquist sampling rates. The structure most commonly associated with CS is that of signal sparsity in a transform basis. As an example, the basis behind most image compression algorithms is that images are sparse (or close to sparse) in transform bases such as wavelets and DCT. In such a scenario, a CS camera takes *under-sampled* linear measurements of the scene. Given these measurements, the image of the scene is recovered by searching for the image that is sparsest in the transform basis (wavelets or DCT) while simultaneously satisfying the measurements. This search procedure can be shown to be convex. Much of CS literature revolves around the design of linear measurement matrices, characterizing the number of measurements required and the design of image/signal recovery algorithms.

Theory

Compressive sensing [1–3] enables reconstruction of sparse signals from under-sampled linear measurements. A vectors is termed K sparse if it has at most K nonzero components, or equivalently, if $\|\mathbf{s}\|_0 \leq K$, where $\|\cdot\|_0$ is the ℓ_0 norm or the number of nonzero components. Consider a signal (e.g., an image or a video) $\mathbf{x} \in \mathbb{R}^N$, which is sparse in a basis Ψ, that is, $\mathbf{s} \in \mathbb{R}^N$, defined as $\mathbf{s} = \Psi^T\mathbf{x}$, is sparse. Examples of the sparsifying basis Ψ for images includes DCT and wavelets. The main problem of interest is that of sensing the signal $\mathbf{x}$ from linear measurements. With no additional knowledge about $\mathbf{x}$, N linear measurements of $\mathbf{x}$ are required to form an invertible linear system. In a conventional digital camera, an identity sensing matrix is used so that each pixel is sensed directly. For sensing reflectance fields, optimal linear sensing matrices have been designed [4].

The theory of compressive sensing shows that it is possible to reconstruct $\mathbf{x}$ from M linear measurements even when $M \ll N$ by exploiting the sparsity of $\mathbf{s} = \Psi^T\mathbf{x}$. Consider a measurement vector $\mathbf{y} \in \mathbb{R}^M$ obtained using an $M \times N$ measurement matrix Φ, such that

$$\mathbf{y} = \Phi\mathbf{x} + e = \Phi\Psi\mathbf{s} + e = \Theta\mathbf{s} + e, \qquad (1)$$

where e is the measurement noise (see Fig. 1) and $\Theta = \Phi\Psi$. The components of the measurement vector $\mathbf{y}$ are called the *compressive measurements* or compressive samples. For $M < N$, estimating $\mathbf{x}$ from the linear measurements is an ill-conditioned problem. However, when $\mathbf{x}$ is K sparse in the basis Ψ, then CS enables recovery of $\mathbf{s}$ (or alternatively, $\mathbf{x}$) from $M = O(K\log(N/K))$ measurements, for certain classes of matrices Θ. The guarantees on the recovery of signals extend to the case when $\mathbf{s}$ is not exactly sparse but compressible. A signal is termed compressible if its sorted transform coefficients delay according to power law, that is, the sorted coefficient of $\mathbf{s}$ decay rapidly in magnitude [5].

Restricted Isometry Property (RIP)

The condition for stable recovery for both sparse and compressible signals is that the matrix Θ satisfies the following property. Given S-sparse vector $\mathbf{s}$, $\exists\delta_S$, $0 < \delta_S < 1$ such that

$$(1-\delta_S)\|\mathbf{s}\|_2 \leq \|\Theta\mathbf{s}\|_2 \leq (1+\delta)\|\mathbf{s}\|_2. \qquad (2)$$

Stable recovery is guaranteed when (Eq. 2) is satisfied for $S = 2K$. This is referred to as the *Restricted Isometry Property* (RIP) [2, 7]. In particular, when Ψ is a fixed basis, it can be shown that using a randomly generated sub-Gaussian measurement matrix Φ, ensures that Θ satisfies RIP with a high probability provided $M = O(K\log(N/K))$. Typical choices for Φ (or equivalently, Θ) are matrices whose entries are independently generated using the Radamacher or the sub-Gaussian distribution.

Signal Recovery

Estimating K-sparse vectors that satisfy the measurement equation of (Eq. 1) can be formulated as the following ℓ_0 optimization problem:

$$(P0): \ \min \|\mathbf{s}\|_0 \text{ s.t. } \|\mathbf{y} - \Phi\Psi\mathbf{s}\|_2 \leq \epsilon, \qquad (3)$$

with ϵ being a bound for the measurement noise e in (Eq. 1). While this is a NP-hard problem in general, the equivalence between ℓ_0 and ℓ_1 norm for such

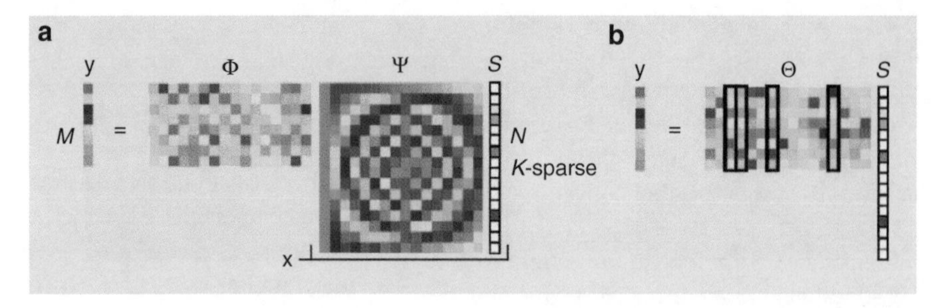

Compressive Sensing, Fig. 1 (**a**) Compressive sensing measurement process with a random Gaussian measurement matrix Φ and discrete cosine transform (DCT) matrix Ψ. The vector of coefficients **s** is sparse with $K = 4$. (**b**) Measurement process with $\mathbf{y} = \Phi\mathbf{x}$. There are four columns that correspond to nonzero $\mathbf{s}_i$ coefficients; the measurement vector **y** is a linear combination of these columns (Figure from [6])

systems [8] allows us to reformulate the problem as one of ℓ_1 norm minimization.

$$(P1) : \widehat{\mathbf{s}} = \arg\min \|\mathbf{s}\|_1 \text{ s.t. } \|\mathbf{y} - \Phi\Psi\mathbf{s}\| \leq \epsilon. \quad (4)$$

It can be shown that, when $M = O(K \log(N/K))$, the solution to the $(P1) - \widehat{\mathbf{s}} -$ is, with overwhelming probability, the K-sparse solution to $(P0)$. In particular, the estimation error can be bounded as follows:

$$\|\mathbf{s} - \widehat{\mathbf{s}}\|_2 \leq C_0 \|\mathbf{s} - \mathbf{s}_K\| / \sqrt{K} + c_1 \epsilon, \quad (5)$$

where $\mathbf{s}_K$ is the best K-sparse approximation of **s**.

There exist a wide range of algorithms that solve $(P1)$ to various approximations or reformulations [9, 10]. One class of algorithms model $(P1)$ as a convex problem and recast it as a linear program (LP) or a second order cone program (SOCP) for which there exist efficient numerical techniques. Another class of algorithms employ greedy methods [11] which can potentially incorporate other problem-specific properties such as structured supports [12].

Hardware Implementations

Compressive sensing has been successfully applied to sense various visual signals such as images [13, 14], videos [15] and light transport matrices [16]. The single-pixel camera (SPC) [13] for CS of images is designed as follows. The SPC consists of a lens that focuses the scene onto a digital micro-mirror device (DMD), which takes the place of the CCD array in a conventional camera. Each micro-mirror can be oriented in one of two possible angles. Micro-mirrors that are in the same angle/direction are focused onto a photodetector (hence, the single pixel) using a second lens (see Fig. 2). Each measurement that is obtained corresponds to an inner product of the scene with a set of 1s and 0s corresponding to the state of the micro-mirrors. By flipping the micro-mirrors randomly according to a Bernoulli distribution, multiple compressive measurements are obtained.

The SPC is most useful when sensing is costly. One such example is in sensing at wavelengths beyond the visible spectrum, where exotic and costly materials are needed to detect light. For such applications, the cost of building a focal plane array of even VGA resolution is prohibitive. However, the SPC needs only a single photodiode tuned to the spectrum of interest, with otherwise minimal changes to the underlying architecture.

Implications for Computer Vision

Novel Sensors

Compressive sensing enables a new sensing design that measures random linear projection of a scene and subsequently reconstructs the image of the scene using these measurements. The reconstructed image is no different from one captured from a suitably placed camera, barring reconstruction artifacts. However, compressive cameras enable parsimony in sensing – which can be extremely useful in problems where sensing is costly. This could be due to storage,

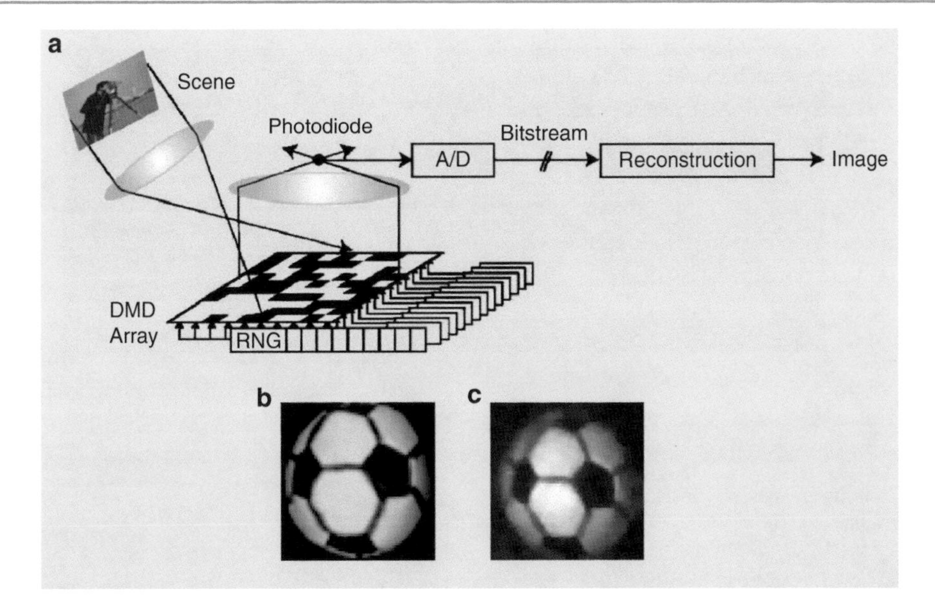

Compressive Sensing, Fig. 2 (**a**) Single-pixel, compressive sensing camera. (**b**) Conventional digital camera image of a soccer ball. (**c**) 64×64 *black*-and-*white* image of the same ball ($N = 4,096$ pixels) recovered from $M = 1,600$ random measurements taken by the camera in (**a**). The images in (**b**) and (**c**) are not meant to be aligned (Figure from [6])

as in *the data deluge* problem in large sensor networks. Compressive sensing is extremely valuable in time sensitive applications such as medical imaging, sensing and modeling fluid dispersions, time varying reflectance fields, and in general, high speed imaging. Finally, sensing could be costly due to adverse affects caused by the process of sensing. In electron microscopy, there are fundamental limitations on the number of images of a thin layer of tissue that comes as the tissue is progressively destroyed in the process of imaging. In all such applications, where sensing is costly in some manner, CS offers better trade-offs over traditional linear sensing.

Tailoring Vision Algorithms for Compressive Data

In many cases, it is beneficial to work on the compressive measurements directly without reconstructing the images. As an example, in background subtraction, the silhouettes are spatially sparse and in many applications far sparser than the original images (in a transform basis). Given a static background image $\mathbf{x}_s$, silhouettes are recovered by identifying $(\mathbf{y} - \Phi\mathbf{x}_s)$ as compressive measurements of the silhouettes [17]. This can lead to silhouette recovery at extremely high compression ratios. An other example is in video CS for dynamic textures, where the linear dynamical system parameters of the scene can be recovered by a suitably designed acquisition device [15].

Beyond Sparse Models

While the theory of CS relies on assumptions of sparsity or compressibility of the signal in a transform basis, it can be extended to signals that exhibit additional structures. One such idea is that of model-based CS [12] where in the signal exhibits sparsity in a space of models which encompasses models such as block sparsity, union of subspaces, and wavelet tree models. Another idea is in the use of non-sparse models such as Markov Random fields [18] for CS – with the eventual goal of using the Ising model for sensing images and background subtracted silhouettes. Such progress hints at interesting avenues for future research governed by the use of rich models in existing vision literature for the task of sensing.

Machine Learning and Signal Processing on Compressive Data

Restricted Isometry Property (see (Eq. 2)) for random matrices implies that, for K-sparse signals, distances are approximately preserved under random projections. A host of machine learning and signal processing algorithms depend on pairwise distances of points as opposed to their exact location. For such algorithms, almost identical results can be obtained by applying them to randomly projected data as opposed to the data in the original space [19, 20]. This has tremendous advantages as the random projected data lies on a much lower dimensional space, and can be directly obtained from a compressive imager. Such ideas have been applied for detection and classified of signals in compressed domain [21].

References

1. Candès E, Romberg J, Tao T (2006) Robust uncertainty principles: exact signal reconstruction from highly incomplete frequency information. IEEE Trans Inf Theory 52(2): 489–509
2. Candès E, Tao T (2006) Near optimal signal recovery from random projections: universal encoding strategies? IEEE Trans Inf Theory 52(12):5406–5425
3. Donoho D (2006) Compressed sensing. IEEE Trans Inf Theory 52(4):1289–1306
4. Schechner Y, Nayar S, Belhumeur P (2003) A theory of multiplexed illumination. International conference on computer vision, Nice, France, pp 808–815
5. Haupt J, Nowak R (2006) Signal reconstruction from noisy random projections. IEEE Trans Inf Theory 52(9):4036–4048
6. Baraniuk R (2007) Compressive sensing. IEEE Signal Process Mag. 24(4):118–120, 124
7. Davenport M, Laska J, Boufouons P, Baraniuk R (2009) A simple proof that random matrices are democratic. Technical report TREE 0906, Rice University, ECE Department
8. Donoho D (2006) For most large underdetermined systems of linear equations, the minimal ℓ_1-norm solution is also the sparsest solution. Comm Pure Appl Math 59(6):797–829
9. Candès E, Tao T (2005) Decoding by linear programming. IEEE Trans Inf Theory 51(12):4203–4215
10. Tibshirani R (1996) Regression shrinkage and selection via the lasso. J R Stat Soc B 58(1):267–288
11. Needell D, Tropp J (2009) CoSaMP: iterative signal recovery from incomplete and inaccurate samples. Appl Comput Harmon Anal 26(3):301–321
12. Baraniuk R, Cevher V, Duarte M, Hegde C (2010) Model-based compressive sensing. IEEE Trans Inf Theory 56(4):1982–2001
13. Duarte M, Davenport M, Takhar D, Laska J, Sun T, Kelly K, Baraniuk R (2008) Single-pixel imaging via compressive sampling. IEEE Signal Process Mag 25(2):83–91
14. Wagadarikar A, John R, Willett R, Brady D (2008) Single disperser design for coded aperture snapshot spectral imaging. Appl Opt 47(10):B44–B51
15. Sankaranarayanan AC, Turaga P, Baraniuk R, Chellappa R (2010) Compressive acquisition of dynamic scenes. In: ECCV, Heraklion, Greece, pp 129–142
16. Peers P, Mahajan D, Lamond B, Ghosh A, Matusik W, Ramamoorthi R, Debevec P (2009) Compressive light transport sensing. ACM Trans Graph 28(1):1–3
17. Cevher V, Sankaranarayanan A, Duarte M, Reddy D, Baraniuk R, Chellappa R (2008) Compressive sensing for background subtraction. In: Proceedings of the European conference on computer Vision (ECCV), Marseille, France
18. Cevher V, Duarte M, Hegde C, Baraniuk R (2008) Sparse signal recovery using markov random fields. Neural Info Proc Sys, Vancouver, BC, Canada, pp 257–264
19. Davenport M, Boufounos P, Wakin M, Baraniuk R (2010) Signal processing with compressive measurements. IEEE J Sel Top Signal Process 4(2):445–460
20. Hegde C, Wakin M, Baraniuk R (2007) Random projections for manifold learning, Vancouver, BC
21. Davenport M, Duarte M, Wakin M, Laska J, Takhar D, Kelly K, Baraniuk R (2007) The smashed filter for compressive classification and target recognition, San Jose, CA

Computational Symmetry

Yanxi Liu
Computer Science and Engineering, Penn State University, University Park, PA, USA

Synonyms

Symmetry-based X; Symmetry detection

Definition

Computational symmetry is a branch of research using computers to model, analyze, synthesize, and manipulate symmetries in digital forms, imagery, or otherwise [1].

Background

Symmetry is a pervasive phenomenon presenting itself in all forms and scales, from galaxies to microscopic biological structures, in nature and man-made environments. Much of one's understanding of the world is based on the perception and recognition of recurring patterns that are generalized by the mathematical concept of symmetries [2–4]. Humans and animals have an innate ability to perceive and take advantage of symmetry in everyday life [5–8], while harnessing this powerful insight for machine intelligence remains a challenging task for computer scientists.

Interested readers can find several influential symmetry-related papers below to gain a historic perspective: the wonderful exposition on the role of symmetry in "Biological Shape and Visual Science" by Blum in 1973 [9]; in 1977, the "Description and Recognition of Curved Objects" reported by Nevatia and Binford, where bilateral symmetry of an object about different axes is examined [10]; the method of detecting angle/side regularities of closed curves and plateaus in one-dimensional patterns by Davis, Blumenthal, and Rosenfeld [11, 12]; the introduction of the term *skewed symmetry* by Takeo Kanade in 1981 [13]; the exposition on "Smoothed Local Symmetries and Their Implementation" by Brady and Asada [14]; the theory of *recognition-by-components* (RBC) proposed by Biederman in 1985 [15]; "Perceptual Grouping and the Natural Representation of Natural Form" using superquadrics as restricted *generalized cylinders* (GC) by Pentland in 1986 [16]; "Perceptual Organization and Visual Recognition" by Lowe [17], where the non-coincidental appearance of symmetry in the real world was noted; and the "Symmetry-Seeking Models for 3D Object Reconstruction" (1987) illustrated by Terzopoulos, Witkin, and Kass [18].

A computational model for symmetry is especially pertinent to computer vision, computer graphics, and machine intelligence in general because symmetry is:

- *Ubiquitous*: Both the physical and digital worlds are filled with various forms of symmetry, near-symmetry, and distorted symmetry patterns.
- *Essential*: Intelligent beings perceive and interact with the chaotic real world in the most efficient and effective manner by capturing its essential structures and substructures – the generators of symmetry, near-symmetry, distorted symmetry, and/or recurring patterns.
- *Compact*: The recognition of symmetries is the first step towards minimizing redundancy, often leading to drastic reductions in computation.
- *Aestho-physiological*: From a butterfly to an elephant, from a tea cup to a building, symmetry or deviation from it has been a time-honored principle for design (by nature or by human) that can guide machine perception, detection, recognition, and synthesis of the real world.

Figure 1 from [19] shows the statistics of published papers during the period of 1974–2009 (36 years) in several major computer vision/graphics conferences and journals. An increasing level of interests in computational symmetry, and a dominant role reflection symmetry played in the past several decades, can be observed in both computer vision and computer graphics literature.

Theory

From the spirit of Felix Klein's Erlangen program [20] that described geometry as the study of a space that is invariant under a given transformation group to the Gestalt principles of perception [21], symmetries and group theory play an important role in describing the geometry and appearance of an object. Informally, one may think of symmetry as expressing the notion that a figure or an object is made from multiple copies of smaller unit that are interchangeable. Mathematically, this notion is formalized by examining the effect of transformations on the object in a certain space such that its subparts map to each other.

More formally, in a metric space M, a *symmetry* $g \in G$ of a set $S \subseteq M$ is an isometry (a distance preserving transformation) that maps S to itself (an automorphism), $g(S) = S$. The transformation g keeps S invariant as a whole while permuting its parts. Symmetries G of S form a mathematical *group* $\{G, *\}$, closed under transformation composition $*$, called the *symmetry group* of S [3].

Basic Concepts

The basic definitions of symmetry and group theory can be found in most general mathematic textbooks on modern algebra. In particular the following books are recommended: *Geometry* by Coxeter [22], *Generators and relations for discrete groups* by Coxeter and Moser [3], *Symmetry* by Weyl [2], and *Tilings and Patterns* by Grünbaum and Shephard [23].

Some key concepts and computationally relevant definitions [19] are provided below:

Definition 1 *Let S be a subset of R^n. Then an isometry (a distance preserving mapping) g is a symmetry of S if and only if $g(S) = S$.*

Definition 2 *A symmetry g for a set $S \in R^n$ is a primitive symmetry if and only if (1) $g \neq e$ where e is an identity mapping; and (2) if $g = g_1 g_2, g_1 \neq e, g_2 \neq e$, then neither g_1 nor g_2 is a symmetry of S.*

For example, in 2D Euclidean space R^2, there are four types [2–4] of *primitive symmetries* $g(S) = S$.

a

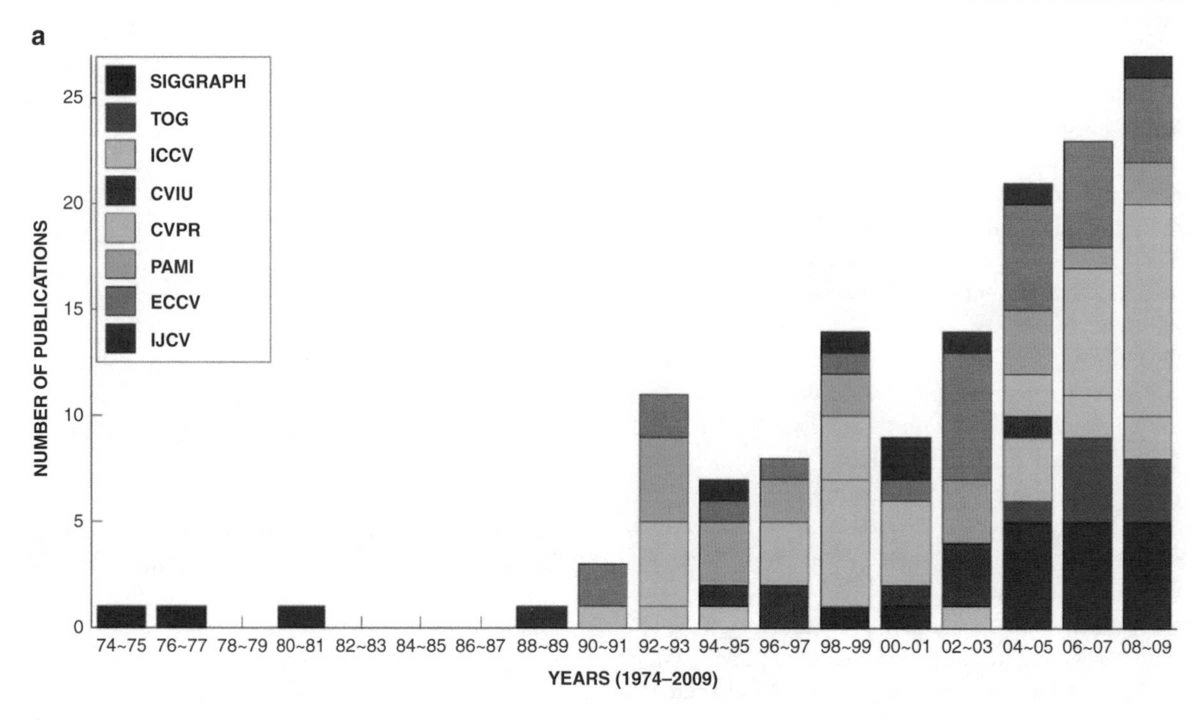

b

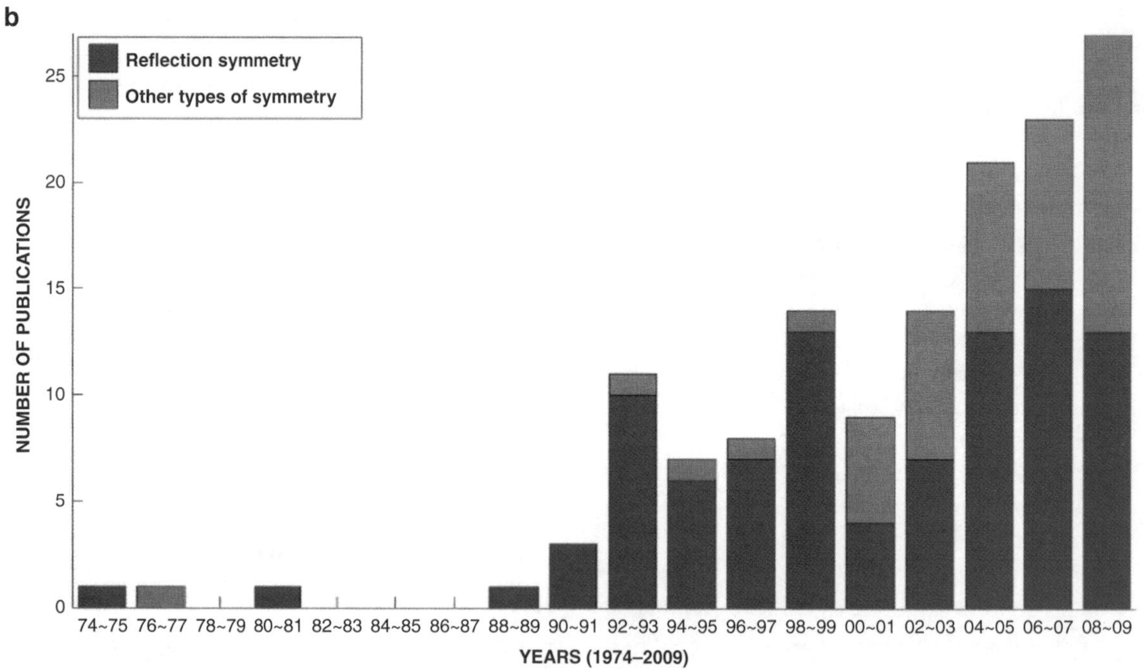

Computational Symmetry, Fig. 1 From the publication statistics (data collected in [19]), it is obvious that (1) there is an increasing trend of interests in computational symmetry and (2) research on reflection symmetry has been dominating the field in the past several decades, though there is a recent growing awareness of the whole symmetry spectrum. A similar reflection-symmetry-dominating trend has also been observed in the psychology literature for human perception of symmetries [8] (**a**) Publications on symmetry detection and applications in major computer vision and computer graphics conferences/journals (**b**) Dividing papers into reflection symmetry alone versus other types of symmetries (rotation, translation, and glide reflection)

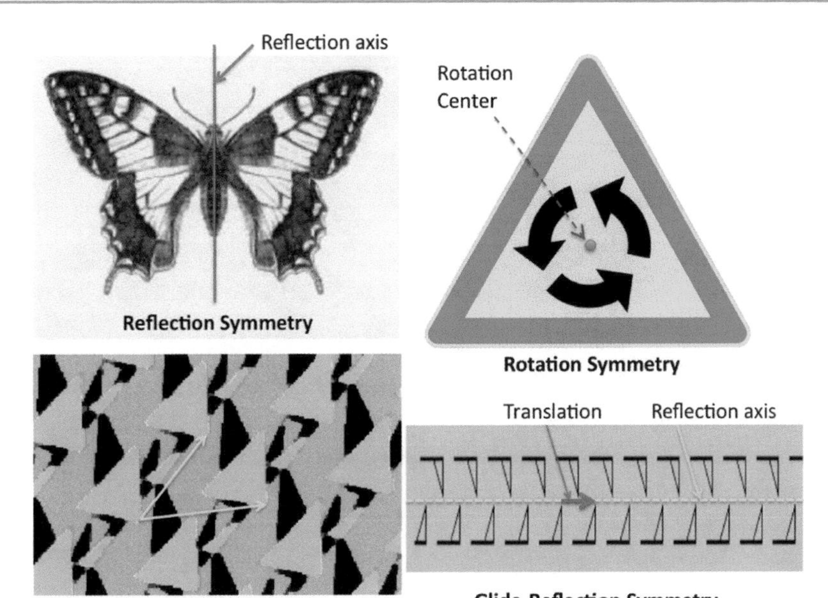

Computational Symmetry, Fig. 2 An illustration of patterns with the four different primitive symmetries, respectively, in 2D Euclidean space. Reflection symmetry has the reflection axis as its point-wise invariance. Rotation symmetry has the *center* of rotation as its invariant point. An n-fold rotational symmetry with respect to a particular point (in 2D) or axis (in 3D) means that, setwise, a rotation by an angle of 360/n does not change the object. Glide reflection is composed of a translation that is $1/2$ of the smallest translation symmetry t and a reflection r with respect to a reflection axis along the direction of the translation. There are no invariant points under translation and glide-reflection symmetries

They are, without loss of generality, for the four images $f(x, y)$ shown in Fig. 2:

1. *Reflection*: $f(x, y) = f(-x, y)$; its reflection axis (plane) remains invariant under the reflection.
2. *Rotation*: $f(x, y) = f(r \cos(2\pi/n), r \sin(2\pi/n))$, $r = \sqrt{(x^2 + y^2)}$, n is an integer ($n = 3$ in the top right of Fig. 2), and its rotation center point (axis) remains invariant under the rotation.
3. *Translation*: $f(x, y) = f(x + \Delta x, y + \Delta y)$, for some $\Delta x, \Delta y \in R$, no invariant points exist.
4. *Glide Reflection*: $f(x, y) = f(x + \Delta x, -y)$, for some $\Delta x \in R$, no invariant points exist. A glide reflection g can be expressed as $g = tr$, where t is a translation and r is a reflection whose axis of reflection is along the direction of the translation. Note: neither t nor r alone is a symmetry of S; thus, g is a primitive symmetry of S.

Definition 3 *Let G be a nonempty set with a well-defined binary operation $*$ such that for each ordered pair $g_1, g_2 \in G$, $g_1 * g_2$ is also in G. $(G, *)$ is a* **group** *if and only if:*

1. There exists an *identity* element $e \in G$ such that $e * g = g = g * e$ for all $g \in G$.
2. Any element g in G has an *inverse* $g^{-1} \in G$ such that $g * g^{-1} = g^{-1} * g = e$.
3. The binary operation $*$ is associative: $a * (b * c) = (a * b) * c$ for all $a, c, b \in G$.

Using the composition of transformations on R^n as the binary operation $*$, one can prove that *symmetries of a subset $S \subset R^n$ form a group*, which is called the *symmetry group* of S. Figure 3 and Table 1 illustrate the four distinct types of symmetry groups in R^2.

Definition 4 *All the symmetries of R^n form the* **Euclidean group** $\mathcal{E}$.

Proposition 5 *Symmetries of a subset $S \subset R^n$ form a symmetry group G_S of S where $G_S \subset \mathcal{E}$ (see a proof in [24]).*

Definition 6 *A* **point group** *G is a symmetry group that leaves a single point fixed. In other words, $G = G_x$, here G_x is the stabilizer subgroup on x.*

Definition 7 *A* **space group** *is a group G of operations which leaves the infinitely extended, regularly repeating lattice pattern $L \subset R^n$ invariant while for all nontrivial $g \in G$, i.e., $g \neq e$, and $x \in R^n$, $g(x) \neq x$.*

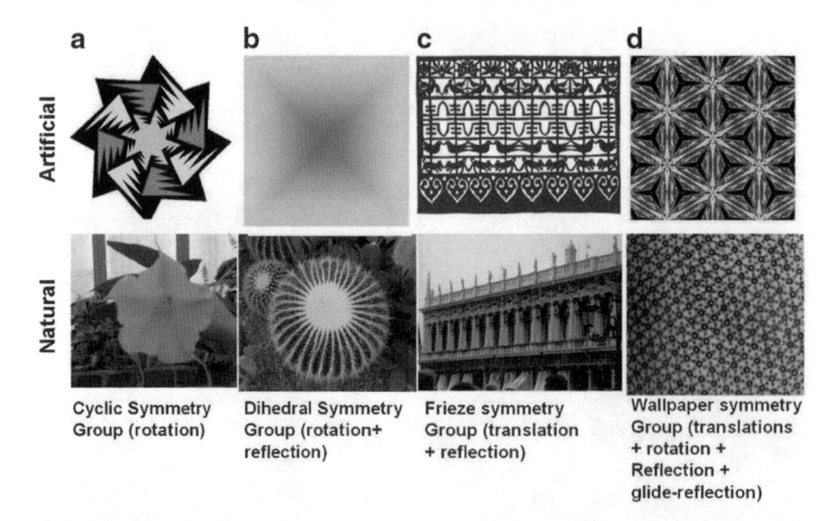

Computational Symmetry, Fig. 3 Sample images of symmetry categorized by their respective ideal symmetry groups: column (**a**) cyclic group containing rotation symmetries only, (**b**) dihedral group (reflection and rotation), (**c**) frieze group (translation plus reflection and rotation), and (**d**) wallpaper group (translation, rotation, reflection, and glide-reflection). The *top row* contains synthetic patterns, while the *bottom row* shows photos of real-world scenes (*bottom right* is an image of a transverse slice of skeletal muscle magnified with a transmitter microscope 800,000 times)

Computational Symmetry, Table 1 Discrete symmetry groups in R^2 (See Fig. 3 for sample images)

Name	Group type	Symbol	Order	Primitive symmetry	Example
Cyclic	Point	C_n	n	Rotation	Fig. 3a
Dihedral	Point	D_n	$2n$	Rotation and reflection	Fig. 3b
Frieze	Space	G_{frieze}	∞	All four 2D-primitive symmetry types	Fig. 3c
Wallpaper	Space	$G_{\text{wallpaper}}$	∞	All four 2D-primitive symmetry types	Fig. 3d

Crystallographic Groups

An n-dimensional periodic pattern is formed by repeating a unit pattern in equal intervals along $k \leq n$ directions. A mature mathematical theory for symmetries of periodic patterns in n-dimensional Euclidean space has been known for over a century [25–28], namely, the *crystallographic groups* [2, 22, 23]. An important mathematical discovery is the answer to the first part of the Hilbert's 18th problem [29]: regardless of dimension n and despite an infinite number of possible instantiations of periodic patterns, the number of distinct symmetry groups for periodic patterns in any Euclidean space R^n is always *finite*. For 2D monochrome patterns, there are *7 frieze-symmetry groups* translating along one direction (strip patterns) [22] (Fig. 3c) and *17 wallpaper groups* covering the whole plane [23, 25] (Fig. 3d). In 3D, there are 230 different *space groups* [30] generated by three linearly independent translations (regular crystal patterns).

Frieze Symmetry Group

A frieze pattern is a 2D strip in the plane that is periodic along one dimension. Any frieze pattern P is associated with one of the seven unique symmetry groups (Fig. 4). These seven symmetry groups, denoted by crystallographers as $l1$, lg, ml, $l2$, mg, lm, and mm [22], are called the *frieze groups*. Without loss of generality, assuming the direction of translation symmetry of a frieze pattern is horizontal, the frieze pattern can then exhibit five different types of symmetries (Fig. 4a):

1. Horizontal translation
2. Two-fold rotation
3. Horizontal reflection (reflection axis is placed horizontally)
4. Vertical reflection
5. Horizontal glide reflection composed of a half-unit translation followed by a horizontal reflection

The primitive symmetries in each group (the inner structure of a frieze group) and the relationship

Computational Symmetry, Fig. 4 The five types of primitive symmetries (left table of (**a**)) and the inner (right of (**a**)) and inter-structures (**b**) of the seven frieze groups

a

Symmetry Group	translation	2-fold rotation	Horizontal reflection	Vertical reflection	Glide reflection	Seven Frieze Patterns
F1 11	yes	no	no	no	no	
F2 1g	yes	no	no	no	yes	
F3 m1	yes	no	no	yes	no	
F4 12	yes	yes	no	no	no	
F5 mg	yes	yes	no	yes	yes	
F6 1m	yes	no	yes	no	no	
F7 mm	yes	yes	yes	yes	no	

b

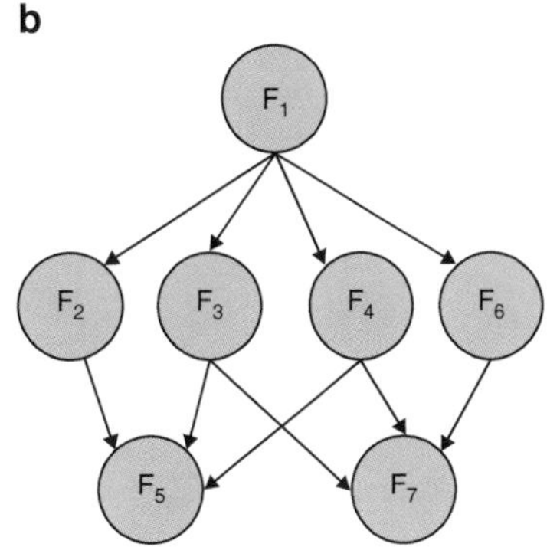

among the seven frieze groups (inter-structure of frieze groups) are depicted in Fig. 4a, b, respectively. Each frieze pattern is associated with one of the seven possible frieze groups, subject to the combination of these five primitive symmetries in the pattern (Fig. 4a).

Wallpaper Symmetry Group

A wallpaper pattern is a 2D periodic pattern extending along two linearly independent directions [22, 31] (Fig. 5). Any wallpaper pattern is associated with one of the 17 wallpaper groups. Wallpaper group theory [23] states that all translationally symmetric patterns P_r can be generated by a pair of linearly independent, shortest (among all possible) vectors t_1, t_2 applied to a minimum tile. The *orbits* of this pair of translation symmetry generator vectors form a 2D *quadrilateral lattice*, which simultaneously defines all 2D tiles (partitions the space into its smallest generating regions by its translation subgroup), and a topological lattice structure relating all tiles (Fig. 5a).

Motifs of Wallpaper Patterns

When the translational subgroup of a periodic pattern is determined, it fixes the size, shape, and orientation of the unit lattice, but leaves open the question of where the unit lattice should be anchored in the pattern. Any parallelogram of the same size and shape carves out an equally good tile that can be used to tile the plane. However, from a perception point of view, some parallelograms produce tiles that are better descriptors of the underlying symmetry of the overall pattern than others. For example, if the whole pattern has some rotation symmetries, a tile located on the rotation centers can represent the global symmetry property of the wallpaper pattern instantly (Fig. 5a). Such *motifs*, as representative tiles of a periodic pattern, can be defined mathematically (and discovered computationally [32, 33]):

Definition 8 *A* motif *of a wallpaper pattern is a tile that is cut out by a lattice whose lattice unit is centered on the fixed point of the largest stabilizer group.*

Application

Automatic detection of 2D primitive symmetries (reflection, rotation, translation, glide reflection) from 2D images has been a standing topic in computer vision. The earliest attempt at an algorithmic treatment of bilateral reflection symmetry detection predates computer vision itself [34]. Interested readers can find a long list of published symmetry detection

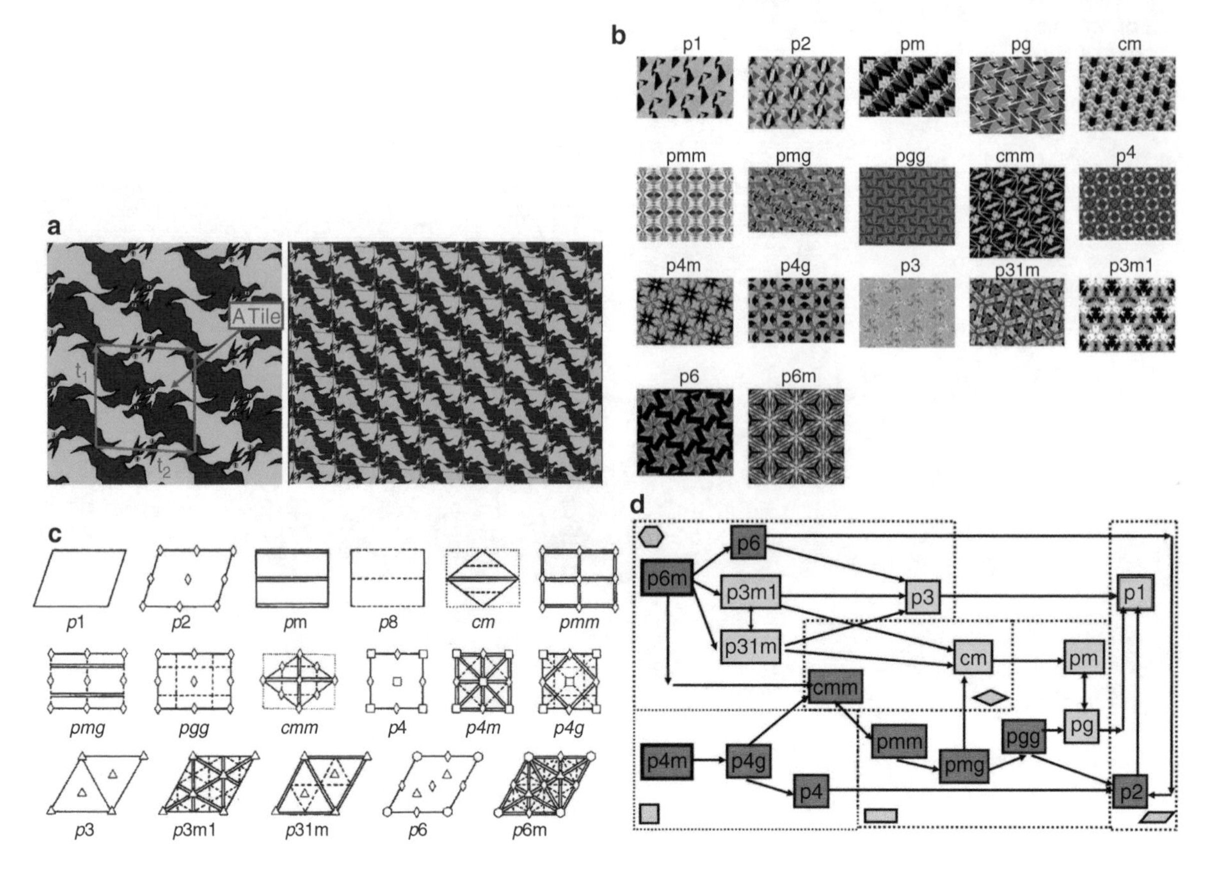

Computational Symmetry, Fig. 5 (**a**) A *tile* and a 2D lattice (*red*) determined simultaneously by the generators t_1, t_2 of the translation subgroup. (**b**) Sample wallpaper patterns associated with the 17 distinct wallpaper groups. The inner (**c**) and inter-structures (**d**) of the 17 wallpaper groups are shown. (**c**) The symbols *p1, p2, pm, pg, cm, pmm, pmg, pgg, cmm, p4, p4m, p4g, p3, p31m, p3m1, p6*, and *p6m* are crystallographers' representations for the wallpaper groups; the diagram is courtesy of [31]. The *diamond, triangle, square*, and *hexagon shapes* correspond to two-, three-, four-, and six-fold rotation centers. *Solid single line* and *dotted single line* and *double parallel lines* denote unit translation, glide-reflection, and reflection symmetries, respectively. (**d**) The subgroup hierarchy, where $A \rightarrow B$ means that B is a subgroup of A (information extracted from [22])

algorithms in [19]. The first quantitative benchmark on reflection, rotation, and translation symmetry detection algorithms can be found in [35, 36].

The first general-purpose, automatic frieze and wallpaper group classification algorithm on real images is published in [32, 33], followed by the classification of *skewed symmetry groups*, in particular wallpaper groups and rotation symmetry groups from affinely and projectively distorted images [37, 38]. More recently, aiming at capturing the translation subgroup of a wallpaper group, several algorithms [39–42] are developed for lattice detection (Fig. 5a) from various locally and/or globally distorted, unsegmented real images.

The applications of computational symmetry are diverse and cross-disciplinary [19]. Some sample applications in computer vision/graphics include texture replacement in real photos [43], human gait analysis [33, 44], static and dynamic near-regular texture analysis, synthesis, replacement/superimposition and tracking [45–50], image "de-fencing" [51, 52], automatic geotagging [53], fabric categorization [54], architectural facade in-painting [55], multi-view 3D geometry [56], and multi-target tracking [50, 57].

New applications of computational symmetry and group theory attract attentions in urban scene and architectural image analysis, synthesis and 3D-reconstruction [53, 55, 58, 59], multi-target tracking of topology-varying spatial patterns [60], image segmentation and grouping [58, 61–63], transformation estimation [64–66], facial asymmetry as a biometric [67, 68], and an expression and gender cue [69–71].

For biomedical applications, quantified symmetry has also been used to measure symmetry of moving objects such as molecules [72–83], pathological and degenerative volumetric neuroradiology images [84–89], and the firing fields of the grid cells of rats in the context of computational neurosciences [90].

Open Problems

1. Mathematical Symmetries versus Digitized Real-World Data: Well-defined and exhaustively categorized symmetries and symmetry groups become computationally brittle on real-world, noisy, and sometimes distorted digital data.
2. Robust Symmetry Detectors: In spite of years of effort, a robust, widely applicable "symmetry detector" for various types of real-world symmetries remains elusive.
3. Benchmarking: Though an initial effort has been made to quantify and compare the outcome of multiple symmetry detection algorithms [35, 36], the field is yet to see a large-scale, systematic, quantitative evaluation and a publicly available test-image database to gauge the progress in this important, widely applicable research direction.

References

1. Liu Y (2002) Computational symmetry. In: Hargittai I, Laurent T (eds) Symmetry 2000. Wenner-Gren international series, vol 80. Portland, London, pp 231–245. ISBN:I 85578 149 2
2. Weyl H (1952) Symmetry. Princeton University Press, Princeton
3. Coxeter H, Moser W (1980) Generators and relations for discrete groups, 4th edn. Springer, New York
4. Conway J, Burgiel H, Goodman-Strauss C (2008) The symmetries of things. AK Peters, Wellesley
5. Thompson DW (1961) On growth and form. Cambridge University Press, Cambridge
6. Giurfa M, Eichmann B, Menzel R (1996) Symmetry perception in an insect. Nature 382(6590):458–461
7. Rodríguez I, Gumbert A, Hempel de Ibarra N, Kunze J, Giurfa M (2004) Symmetry is in the eye of the 'beeholder': innate preference for bilateral symmetry in flower-naïve bumblebees. Naturwissenschaften 91(8):374–377
8. Tyler C (ed) (1996) Human symmetry perception and its computational analysis. VSP, Utrecht
9. Blum H (1973) Biological shape and visual science (part i). J Theor Biol 38:205–287
10. Nevatia R, Binford T (1977) Description and recognition of curved objects. Artif Intell 8(1):77–98
11. Davis L (1977) Understanding shape: angles and sides. IEEE Trans Comput 26(3):236–242
12. Blumenthal AF, Davis L, Rosenfeld A (1977) Detecting natural "plateaus" in one-dimensional patterns. IEEE Trans Comput 26(2):178–179
13. Kanade T (1981) Recovery of the 3-dimensional shape of an object from a single view. Artif Intell 17:75–116
14. Brady M, Asada H (1984) Smoothed local symmetries and their implementation. Int J Robot Res 3(3):36–61
15. Biederman I (1985) Human image understanding: recent research and a theory. Comput Vis Graph Image Process 32:29–73
16. Pentland A (1986) Perceptual organization and the representation of natural form. Artif Intell 28:293–331
17. Lowe D (1985) Perceptual organization and visual recognition. Kluwer Academic, Boston
18. Terzopoulos D, Witkin A, Kass M (1987) Symmetry-seeking models and 3D object reconstruction. Int J Comput Vis 1:211–221
19. Liu Y, Hel-Or H, Kaplan C, Van Gool L (2010) Computational symmetry in computer vision and computer graphics: a survey. Found Trends Comput Graph Vis 5(1/2):1–165
20. Greenberg MJ (1993) Euclidean and Non-Euclidean geometries: development and history, 3rd edn. WH Freeman, New York
21. Arnheim R (2004) Art and visual perception: a psychology of the creative eye. University of California Press, Berkeley/London
22. Coxeter H (1980) Introduction to geometry, 2nd edn. Wiley, New York
23. Grünbaum B, Shephard G (1987) Tilings and patterns. WH Freeman, New York
24. Liu Y (1990) Symmetry groups in robotic assembly planning. PhD thesis, University of Massachusetts, Amherst
25. Fedorov E (1885) The elements of the study of figures. (Russian) (2) 21. Zapiski Imperatorskogo S. Peterburgskogo Mineralogichesgo Obshchestva (Proc. S. Peterb. Mineral. Soc.). 1–289
26. Fedorov E (1891) Symmetry of finite figures. (Russian) (2) 28. Zapiski Imperatorskogo S. Peterburgskogo Mineralogichesgo Obshchestva (Proc. S. Peterb. Mineral. Soc.). 1–146
27. Fedorov E (1891) Symmetry in the plane. (Russian) (2) 28. Zapiski Imperatorskogo S. Peterburgskogo Mineralogichesgo Obshchestva (Proc. S. Peterb. Mineral. Soc.). 345–390
28. Bieberbach L (1910) Über die Bewegungsgruppen der n-dimensional en Euklidischen Räume mit einem endlichen Fundamental bereich. Göttinger Nachrichten 75–84
29. Milnor J (1976) Hilbert's problem 18. In: Proceedings of symposia in pure mathematics, vol 28. American Mathematical Society, Providence. ISBN:0-8218-1428-1 (Browder FE, Mathematical developments arising from Hilbert problems)
30. Henry N, Lonsdale K (eds) (1969) International tables for X-ray crystallography. Symmetry groups, vol 1. The Kynoch, England/The International Union of Crystallography, Birmingham
31. Schattschneider D (1978) The plane symmetry groups: their recognition and notation. Am Math Mon 85:439–450
32. Liu Y, Collins RT (2000) A computational model for repeated pattern perception using frieze and wallpaper

groups. In: Computer vision and pattern recognition conference (CVPR'00), Hilton Head, SC. IEEE Computer Society, pp 537–544. http://www.ri.cmu.edu/pubs/pub_3302.html
33. Liu Y, Collins R, Tsin Y (2004) A computational model for periodic pattern perception based on frieze and wallpaper groups. IEEE Trans Pattern Anal Mach Intell 26(3):354–371
34. Birkoff G (1932) Aesthetic measure. Harvard University Press, Cambridge
35. Chen P, Hays J, Lee S, Park M, Liu Y (2007) A quantitative evaluation of symmetry detection algorithms. Technical report PSU-CSE-07011 (also listed as technical report CMU-RI-TR-07-36), The Pennsylvania State University, State College, PA
36. Park M, Lee S, Chen P, Kashyap S, Butt A, Liu Y (2008) Performance evaluation of state-of-the-art discrete symmetry detection algorithms. In: IEEE conference on computer vision and pattern recognition (CVPR 2008), Anchorage, pp 1–8
37. Liu Y, Collins RT (2001) Skewed symmetry groups. In: Proceedings of IEEE computer society conference on computer vision and pattern recognition (CVPR'01), Kauai, HI. IEEE Computer Society, pp 872–879. http://www.ri.cmu.edu/pubs/pub_3815.html
38. Lee S, Liu Y (2010) Skewed rotation symmetry group detection. IEEE Trans Pattern Anal Mach Intell 32(9):1659–1672
39. Tuytelaars T, Turina A, Van Gool L (2003) Noncombinatorial detection of regular repetitions under perspective skew. IEEE Trans Pattern Anal Mach Intell 25(4):418–432
40. Hays J, Leordeanu M, Efros A, Liu Y (2006) Discovering texture regularity as a higher-order correspondence problem. In: European conference on computer vision (ECCV'06), Graz, Austria
41. Park M, Liu Y, Collins R (2008) Deformed lattice detection via mean-shift belief propagation. In: Proceedings of the 10th European conference on computer vision (ECCV'08), Marseille, France
42. Park M, Brocklehurst K, Collins R, Liu Y (2009) Deformed lattice detection in real-world images using mean-shift belief propagation. IEEE Trans Pattern Anal Mach Intell 31(10):1804–1816
43. Tsin Y, Liu Y, Ramesh V (2001) Texture replacement in real images. In: Proceedings of IEEE computer society conference on computer vision and pattern recognition (CVPR'01), Kauai. IEEE Computer Society, Los Alamitos, pp 539–544
44. Liu Y, Collins R, Tsin Y (2002) Gait sequence analysis using frieze patterns. In: Proceedings of the 7th European conference on computer vision (ECCV'02), a longer version can be found as CMU RI tech report 01–38 (2001), Copenhagen, Denmark
45. Liu Y, Tsin Y (2002) The promise and perils of near-regular texture. In: Texture 2002, Copenhagen, Denmark, in conjunction with European conference on computer vision (ECCV'02), pp 657–671
46. Liu Y, Lin W (2003) Deformable texture: the irregular-regular-irregular cycle. In: Texture 2003, the 3rd international workshop on texture analysis and synthesis, Nice, France, pp 65–70
47. Liu Y, Lin W, Hays J (2004) Near-regular texture analysis and manipulation. ACM Trans Graph 23(3):368–376
48. Liu Y, Tsin Y, Lin W (2005) The promise and perils of near-regular texture. Int J Comput Vis 62(1-2):145–159
49. Lin W, Liu Y (2006) Tracking dynamic near-regular textures under occlusion and rapid movements. In: Proceedings of the 9th European conference on computer vision (ECCV'06), Graz, Austria, vol 2, pp 44–55
50. Lin W, Liu Y (2007) A lattice-based mrf model for dynamic near-regular texture tracking. IEEE Trans Pattern Anal Mach Intell 29(5):777–792
51. Liu Y, Belkina T, Hays H, Lublinerman R (2008) Image defencing. In: IEEE computer vision and pattern recognition (CVPR 2008), Anchorage, pp 1–8
52. Park M, Brocklehurst K, Collins R, Liu Y (2010) Image defencing revisited. In: Asian conference on computer vision (ACCV'10), Queenstown. IEEE Computer Society, pp 1–13
53. Schindler G, Krishnamurthy P, Lublinerman R, Liu Y, Dellaert F (2008) Detecting and matching repeated patterns for automatic geo-tagging in urban environments. In: IEEE computer vision and pattern recognition (CVPR 2008), Anchorage, pp 1–8
54. Han J, McKenna S, Wang R (2008) Regular texture analysis as statistical model selection. In: ECCV08, Marseille
55. Korah T, Rasmussen D (2008) Analysis of building textures for reconstructing partially occluded facades. In: European conference on computer vision (ECCV08), Marseille, pp 359–372
56. Hong W, Yang AY, Ma Y (2004) On symmetry and multiple view geometry: structure, pose and calibration from a single image. Int J Comput Vis 60(3):241–265
57. Park M, Liu Y, Collins R (2008) Efficient mean shift belief propagation for vision tracking. In: Proceedings of computer vision and pattern recognition conference (CVPR'08), Anchorage. IEEE Computer Society
58. Park M, Brocklehurst K, Collins R, Liu Y (2010) Translation-symmetry-based perceptual grouping with applications to urban scenes. In: Asian conference on computer vision (ACCV'10), Queenstown. IEEE Computer Society, pp 1–14
59. Wu C, Frahm JM, Pollefeys M (2010) Detecting large repetitive structures with salient boundaries. In: Daniilidis K, Maragos P, Paragios N (eds) European conference on computer vision (ECCV 2010). Lecture notes in computer science, vol 6312. Springer, Berlin/Heidelberg, pp 142–155. doi:10.1007/978-3-642-15552-9_11
60. Liu J, Liu Y (2010) Multi-target tracking of time-varying spatial patterns. In: Proceedings of IEEE computer society conference on computer vision and pattern recognition (CVPR'10), San Francisco. IEEE Computer Society, pp 1–8
61. Sun Y, Bhanu B (2009) Symmetry integrated region-based image segmentation. In: Proceedings of IEEE computer society conference on computer vision and pattern recognition (CVPR'08). IEEE Computer Society, Anchorage, Alaska, pp 826–831
62. Yang A, Rao S, Huang K, Hong W, Ma Y (2003) Geometric segmentation of perspective images based on symmetry groups. In: Proceedings of the 10th IEEE international conference on computer vision (ICCV'03), Nice, vol 2, p 1251
63. Levinshtein A, Sminchisescu C, Dickinson S (2009) Multiscale symmetric part detection and grouping. In: ICCV, Kyoto

64. Makadia A, Daniilidis K (2006) Rotation recovery from spherical images without correspondences. IEEE Trans Pattern Anal Mach Intell 28:1170–1175
65. Tuzel O, Subbarao R, Meer P (2005) Simultaneous multiple 3D motion estimation via mode finding on lie groups. In: Proceedings of the 10th IEEE international conference on computer vision (ICCV'05), Beijing, vol I, pp 18–25
66. Begelfor E, Werman M (2005) How to put probabilities on homographies. IEEE Trans Pattern Anal Mach Intell 27:1666–1670
67. Liu Y, Schmidt K, Cohn J, Weaver R (2002) Facial asymmetry quantification for expression invariant human identification. In: International conference on automatic face and gesture recognition (FG'02), Washington, DC
68. Liu Y, Schmidt K, Cohn J, Mitra S (2003) Facial asymmetry quantification for expression invariant human identification. Comput Vis Image Underst J 91(1/2):138–159
69. Liu Y, Palmer J (2003) A quantified study of facial asymmetry in 3D faces. In: IEEE international workshop on analysis and modeling of faces and gestures, IEEE, Nice, pp 222–229
70. Mitra S, Liu Y (2004) Local facial asymmetry for expression classification. In: Proceedings of IEEE computer society conference on computer vision and pattern recognition (CVPR'04). IEEE Computer Society, Washington, DC, pp 889–894. http://www.ri.cmu.edu/pubs/pub_4640.html
71. Mitra S, Lazar N, Liu Y (2007) Understanding the role of facial asymmetry in human face identification. Stat Comput 17:57–70
72. Zabrodsky H, Avnir D (1993) Measuring symmetry in structural chemistry. In: Hargittai I (ed) Advanced molecular structure research, vol 1. JAI, Greenwich
73. Zabrodsky H, Avnir D (1995) Continuous symmetry measures, iv: chirality. J Am Chem Soc 117:462–473
74. Avnir D, Katzenelson O, Keinan S, Pinsky M, Pinto Y, Salomon Y, Hel-Or H (1997) The measurement of symmetry and chirality: conceptual aspects. In: Rouvray DH (ed) Concepts in chemistry. Research Studies, Somerset, pp 283–324
75. Kanis DR, Wong JS, Marks TJ, Ratner M, Zabrodsky H, Keinan S, Avnir D (1995) Continuous symmetry analysis of hyperpolarizabilities. characterization of second order nonlinear optical response of distorted benzene. J Phys Chem 99:11061–11066
76. Yogev-Einot D, Avnir D (2006) The temperature-dependent optical activity of quartz: from le châtelier to chirality measures. Tetrahedron Asymmetry 17:2723–2725
77. Yogev-Einot D, Avnir D (2004) Pressure and temperature effects on the degree of symmetry and chirality of the molecular building blocks of low quartz. Acta Crystallogr B60:163–173
78. Keinan S, Avnir D (2000) Quantitative symmetry in structure-activity correlations: the near c2 symmetry of inhibitor/hiv-protease complexes. J Am Chem Soc 122:4378–4384
79. Alvarez S, Alemany P, Casanova D, Cirera J, Llunell M, Avnir D (2005) Shape maps and polyhedral interconversion paths in transition metal chemistry. Coord Chem Rev 249:1693–1708
80. Pinsky M, Avnir D (1998) Continuous symmetry measures, v: the classical polyhedra. Inorg Chem 37:5575–5582
81. Steinberg A, Karni M, Avnir D (2006) Continuous symmetry analysis of NMR chemical shielding anisotropy. Chem Eur J 12:8534–8538
82. Pinto Y, Fowler P, Mitchell D, Avnir D (1998) Continuous chirality analysis of model stone-wales rearrangements in fullerenes. J Phys Chem 102:5776–5784
83. Keinan S, Avnir D (2001) Continuous symmetry analysis of tetrahedral/planar distortions: copper chlorides and other AB4 species. Inorg Chem 40:318–323
84. Liu Y, Dellaert F (1998) A classification-based similarity metric for 3D image retrieval. In: Proceedings of computer vision and pattern recognition conference (CVPR'98), Santa Barbara. IEEE Computer Society, pp 800–807
85. Liu Y, Collins R, Rothfus W (2001) Robust midsagittal plane extraction from normal and pathological 3D neuroradiology images. IEEE Trans Med Imaging 20(3):175–192
86. Liu Y, Dellaert F, Rothfus W, Moore A, Schneider J, Kanade T (2001) Classification-driven pathological neuroimage retrieval using statistical asymmetry measures. In: International conference on medical imaging computing and computer assisted intervention (MICCAI 2001), Utrecht. Springer, pp 655–665
87. Liu Y, Teverovskiy L, Carmichael O, Kikinis R, Shenton M, Carter C, Stenger V, Davis S, Aizenstein H, Becker J, Lopez O, Meltzer C (2004) Discriminative MR image feature analysis for automatic schizophrenia and alzheimer's disease classification. In: 7th international conference on medical imaging computing and computer assisted intervention (MICCAI 2004), Saint-Malo. Springer, pp 378–385
88. Liu Y, Teverovskiy L, Lopez O, Aizenstein H, Becker J, Meltzer C (2007) Discovery of "biomarkers" for Alzheimer's disease prediction from structural MR images. In: 2002 IEEE international symposium on biomedical imaging: macro to nano, Arlington, pp 1344–1347
89. Teverovskiy L, Becker J, Lopez O, Liu Y (2008) Quantified brain asymmetry for age estimation of normal and AD/MCI subjects. In: 2008 IEEE international symposium on biomedical imaging: nano to macro, Paris, pp 1509–1512
90. Chastain E, Liu Y (2007) Quantified symmetry for entorhinal spatial maps. Neurocomputing 70(10–12):1723–1727

Computer Vision

▶Rationale for Computational Vision

Computing Architectures for Machine Perception

▶High-Performance Computing in Computer Vision

Concept Languages

▶Description Logics

Concurrent Mapping and Localization (CML)

▶Exploration: Simultaneous Localization and Mapping (SLAM)

Conditional Random Fields

▶Discriminative Random Fields

Contour Detection

▶Boundary Detection

Convex Minimization

▶Semidefinite Programming

Cook-Torrance BRDF

▶Cook-Torrance Model

Cook-Torrance Model

Abhijeet Ghosh
Institute for Creative Technologies, University of Southern California, Playa Vista, CA, USA

Synonyms

Cook-Torrance BRDF

Definition

Cook-Torrance model is an analytic BRDF model that describes the wavelength dependent reflectance property of a surface based on the principles of microfacet theory.

Background

Accurate descriptions of how light reflects off a surface are a fundamental prerequisite for computer vision and graphics applications. Real world materials exhibit characteristic surface reflectance, such as glossy or specular highlights, anisotropy, or off-specular reflection, which need to be modeled for such applications. The surface reflectance of a material is formalized by the notion of the Bidirectional Reflectance Distribution Function (BRDF) [1], which describes the reflected response of a surface in a certain exitant direction to illumination from a certain incident direction over a hemisphere of directions.

Analytical reflection models attempt to describe certain classes of BRDFs using a mathematical representation involving a small number of parameters. The Cook-Torrance model [2] is an analytic isotropic BRDF model that falls under the category of a physics-based model and is based on the microfacet theory of inter-reflection of light at rough surfaces. It extends the Torrance-Sparrow reflectance model, originally developed in the field of applied optics [3], for modeling wavelength dependent effects of reflection. The model predicts both the directional distribution as well as spectral composition of reflected light.

Theory

Given a light source, a surface, and an observer, the Cook-Torrance model describes the intensity and spectral composition of the reflected light reaching the observer. The geometry of reflection is shown in Fig. 1. An observer is looking at a point P on a surface. V is the unit vector in the direction of the viewer, N is the unit normal to the surface, and L is the unit vector in the direction of a specific light source. H is the normalized half-vector between V and L given as

$$H = \frac{L + V}{\|L + V\|}, \tag{1}$$

and is the unit normal of a microfacet that would reflect light specularly from L to V. α is the angle between H and N, and θ is the angle btween H and V, so that $\cos\theta = V \cdot H = L \cdot H$.

The main components of the reflectance model are the directionally dependent diffuse and specular

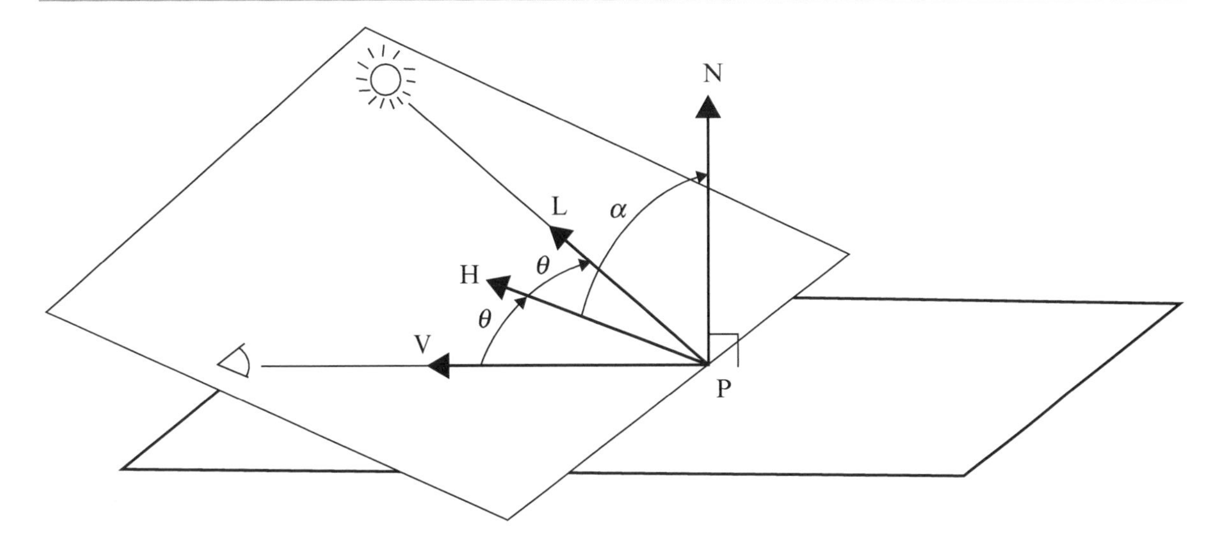

Cook-Torrance Model, Fig. 1 The geometry of reflection [2]

reflection terms, and a directionally independent ambient term. The specular component R_s represents light that is reflected from the surface of the material. The diffuse component R_d originates from internal scattering (in which the incident light penetrates beneath the surface of the material) or from multiple surface reflections (which occur if the surface is sufficiently rough). The directional reflectance is thus give as $R = sR_s + dR_d$, where $s + d = 1$. The model also includes a directionally independent ambient term R_a to approximate the effects of global illumination.

The total intensity of the light reaching the observer is the sum of the reflected intensities from all light sources plus the reflected intensity from any ambient illumination. The basic reflectance model then becomes

$$I_r = I_{ia} R_a + \Sigma_l (sR_s + dR_d) I_{il} (N \cdot L_l) d\omega_{il}. \quad (2)$$

This formulation accounts for the effect of light sources with different intensities and different projected areas which may illuminate a scene. The next two sub-sections consider the directional and wavelength dependence of the reflectance model.

Directional Distribution

The ambient R_a and diffuse R_d components of the model reflect light independent of the viewing direction. However, the specular component R_s does depend on the viewing direction. The angular spread of the specular component can be described by assuming that the surface consists of microfacets, each of which reflects specularly [3].

Only facets whose normal is in the direction H contribute to the specular component of reflection from L to V. The specular component is given as

$$R_s = \frac{F}{\pi} \frac{D \cdot G}{(N \cdot L)(N \cdot V)}, \quad (3)$$

where D is the facet slope distribution function, G is the geometric masking and shadowing attenuation term, and F accounts for Fresnel reflectance. The Fresnel term F describes the fraction of light that is reflected from each smooth microfacet. It is a function of incidence angle and wavelength of incident light and is discussed in the next section.

Microfacet Distribution

The facet slope distribution function D represents the fraction of the facets that are oriented in the direction H. Various facet slope distribution functions have been considered by Blinn [4] including the Gaussian model:

$$D = ce^{-(\alpha/m)^2}, \quad (4)$$

where c is the normalization constant.

Beckmann described a model that originated from the study of scattering of radar waves from rough surfaces [5], and is applicable to a wide range of surface conditions ranging from smooth to very rough.

For rough surfaces, the Beckmann distribution is

$$D = \frac{1}{m^2 \cos^4 \alpha} e^{-[(\tan \alpha)/m]^2} \quad (5)$$

In both the facet slope distribution functions, the spread of the specular component depends on the root mean square (rms) slope m. Small values of m signify gentle facet slopes and give a distribution that is highly directional around the specular direction, while large values of m imply steep facet slopes and give a distribution that is spread out with off-specular peak modeled by the Beckmann distribution (see Fig. 2).

For general surfaces with two or more scales of surface roughness, the slope m can be modeled by using a convex weighted combination of two or more distribution functions [6]:

$$D = \Sigma_j w_j D(m_j), \quad (6)$$

where m_j is the rms slope and w_j the weight of the jth distribution respectively.

Geometric Attenuation

The geometrical attenuation factor G accounts for the shadowing and masking of one facet by another and is discussed in detail in [3, 4]. The following expression is derived for G for microfacets in the shape of v-shaped grooves (see Fig. 3):

$$G = min \left\{ 1, \frac{2(N \cdot H)(N \cdot V)}{(V \cdot H)}, \frac{2(N \cdot H)(N \cdot L)}{(V \cdot H)} \right\}. \quad (7)$$

Spectral Composition

The ambient, diffuse, and specular reflectances all depend on wavelength. R_a, R_d, and the F term of R_s may be obtained from the appropriate reflectance spectra for the material. Reflectance spectra of thousands of materials can be found in the literature [7–10]. The reflectance data are usually for illumination at normal incidence. These values are normally measured for polished surfaces and must be multiplied by $1/\pi$ to obtain the bidirectional reflectance for a rough surface. Most materials are measured at only a few wavelengths in the visible range (typically around 10–15), so that values for intermediate wavelengths must be interpolated.

The spectral energy distribution of the reflected light is found by multiplying the spectral energy distribution of the incident light by the reflectance spectrum of the surface. An example of this is shown in Fig. 4. The spectral energy distributions of the sun and a number of CIE standard illuminants are available in [11].

Fresnel Reflectance

The reflectance F may be obtained theoretically from the Fresnel equation [12]. This equation expresses the reflectance of a perfectly smooth, mirrorlike surface in terms of the index of refraction n (for both metals and dielectrics) and the extinction coefficient k (for metals only) of the surface and the angle of incidence of illumination θ. The Fresnel equation for unpolarized incident light and a dielectric material ($k = 0$) is

$$F = \frac{1}{2} \frac{(g-c)^2}{(g+c)^2} \left\{ 1 + \frac{[c(g+c)-1]^2}{[c(g-c)+1]^2} \right\}, \quad (8)$$

where $c = cos\theta = V \cdot H$ and $g^2 = n^2 + c^2 - 1$. The dependence of the reflectance on wavelength and the angle of incidence implies that the color of the reflected light changes with the incidence angle, from color of the material at normal incidence to color of the illuminant at grazing incidence.

In general, both n and k vary with wavelength, but their values are frequently not known. On the other hand, experimentally measured values of the reflectance at normal incidence are frequently known. To obtain the spectral and angular variation of F, the following practical approach is adopted: If n and k are known, the Fresnel equation is used. If not, but the normal reflectance is known, the Fresnel equation is fit to the measured normal reflectance for a polished surface. For nonmetals or dieletrics, for which $k = 0$, this immediately gives an estimate of the index of refraction n. For metals, for which k is generally not 0, k is set to 0 to get an effective value for n from the normal reflectance. The angular dependence of F is then available from the Fresnel equation. The above procedure yields the correct value of F for normal incidence and a good estimate of its angular dependence, which is only weakly dependent on the extinction coefficient k.

To illustrate this procedure, consider a dielectric material ($k = 0$) at normal incidence. $\theta = 0$, so $c = 1$,

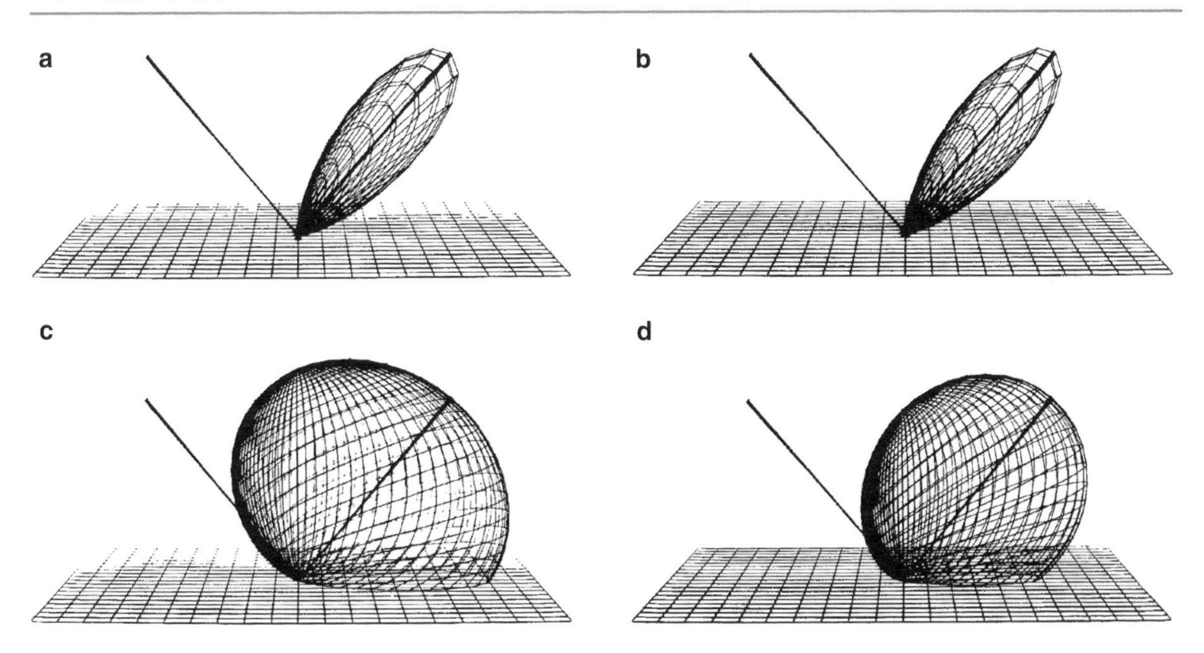

Cook-Torrance Model, Fig. 2 (**a**) Beckmann distribution for $m = 0.2$, (**b**) Gaussian distribution for $m = 0.2$, (**c**) Beckmann distribution for $m = 0.6$, (**d**) Gaussian distribution for $m = 0.6$ [2]

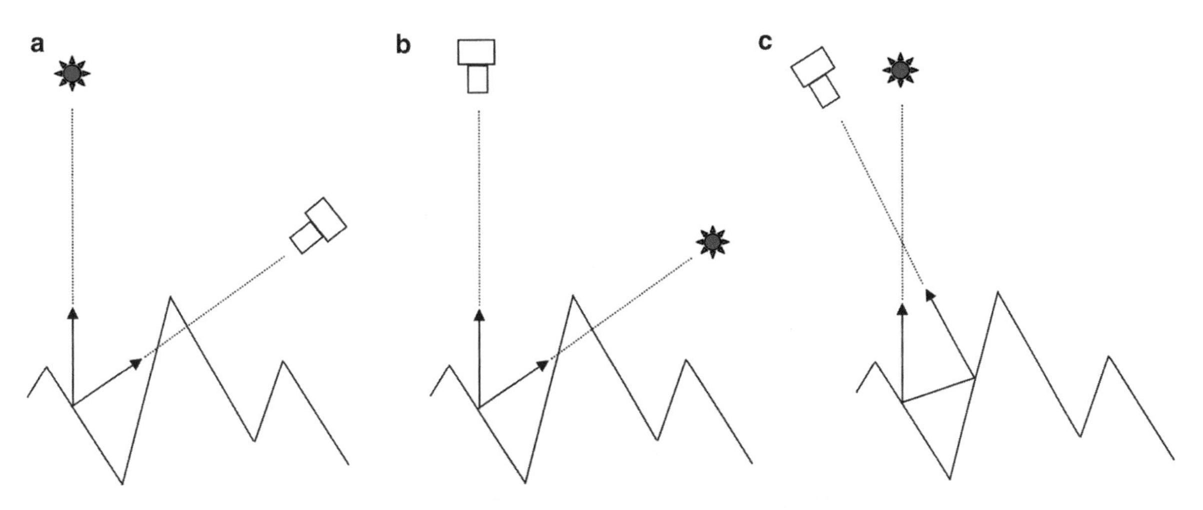

Cook-Torrance Model, Fig. 3 Geometric attenutation due to microfacets. (**a**) Masking. (**b**) Shadowing. (**c**) Inter-reflection

$g = n$ and Eq. 8 reduces to

$$F_0 = \frac{(n-1)^2}{(n+1)^2}. \tag{9}$$

Solving for n gives the equation

$$n = \frac{1+\sqrt{F_0}}{1-\sqrt{F_0}}. \tag{10}$$

Values of n determined in this way can then be substituted into the original Fresnel equation to obtain the reflectance F at other angles of incidence. The procedure may then be repeated at other wavelengths to obtain the spectral and directional dependence of the reflectance.

RGB Values

The laws of trichromatic color reproduction are finally used to convert the spectral energy distribution of the

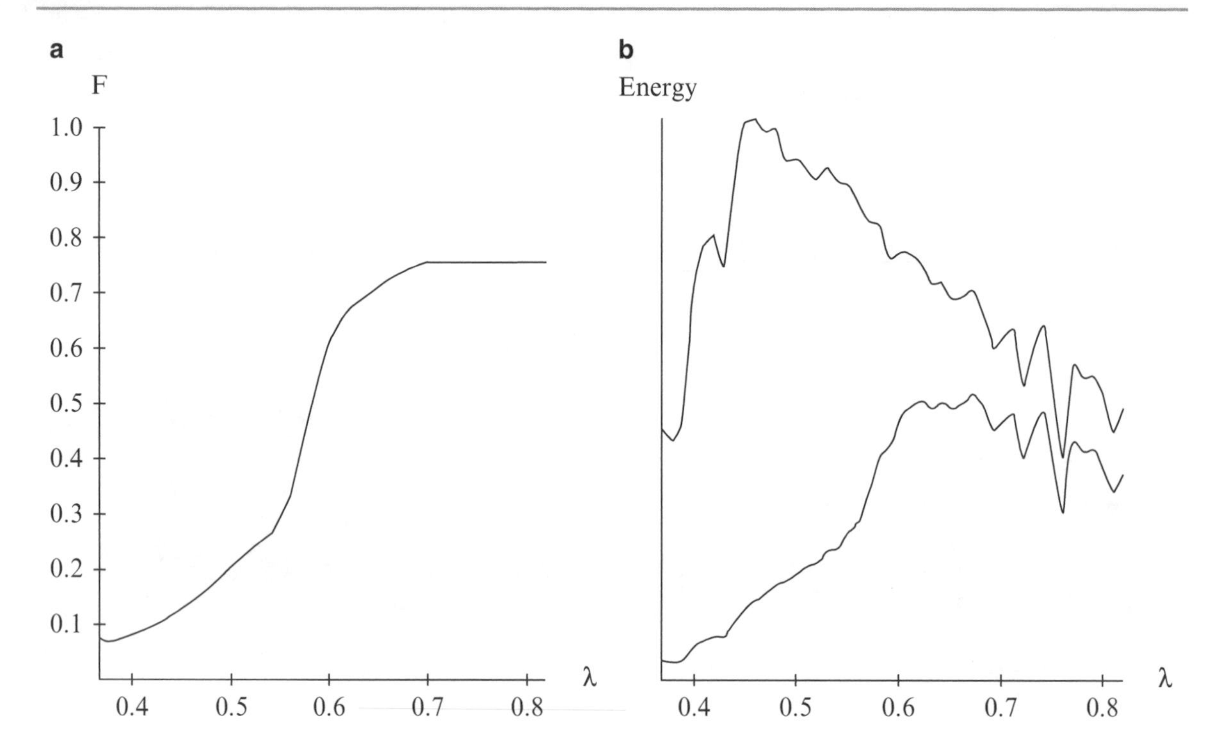

Cook-Torrance Model, Fig. 4 (**a**) Reflectance of a copper mirror for normal incidence. Wavelength is in micrometers (**b**) *Top curve*: Spectral energy distribution of CIE standard illuminant D6500. *Bottom curve*: Spectral energy distribution of light reflected from a copper mirror illuminated by D6500 [2]

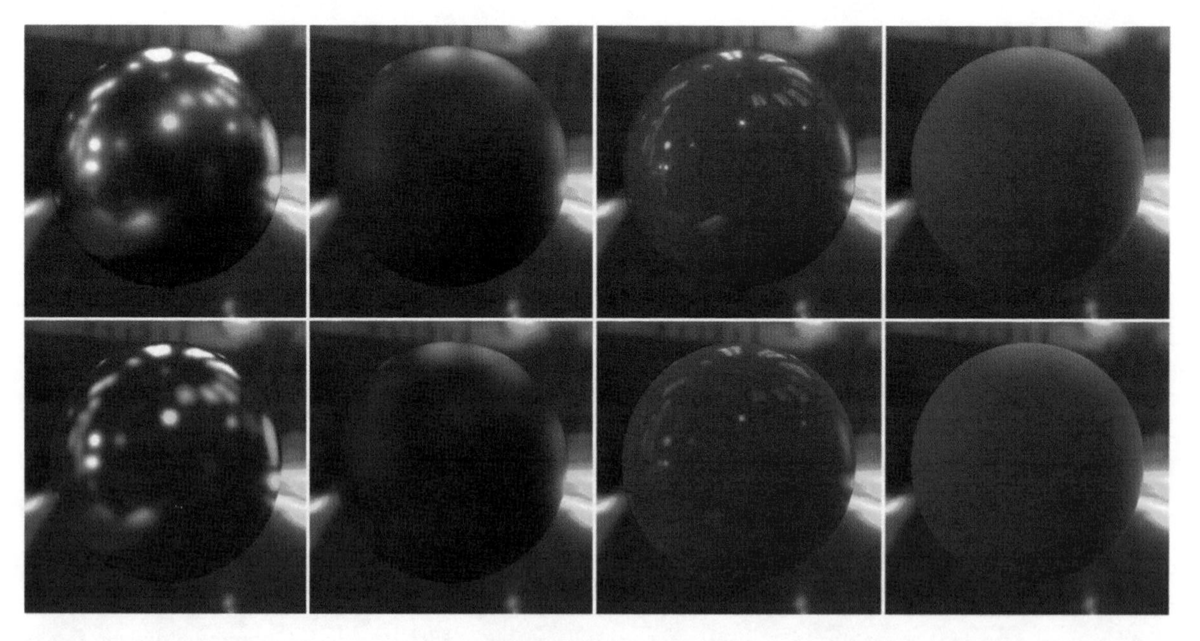

Cook-Torrance Model, Fig. 5 Measured isotropic BRDFs [14] (*top row*) and their Cook-Torrance fits [15] (*bottom-row*). Left to right: Nickel, Oxidized steel, Red plastic (specular), Dark red paint

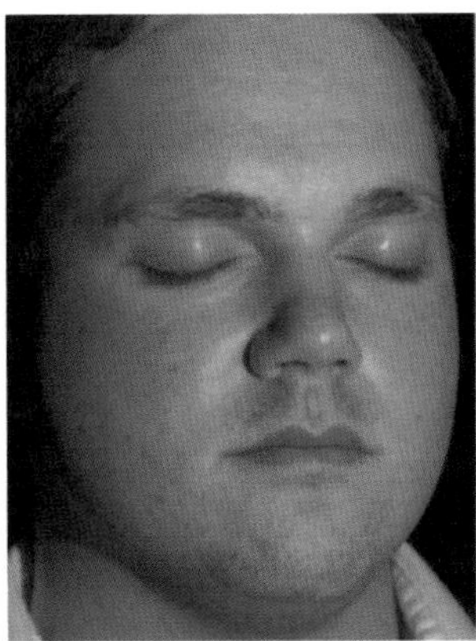
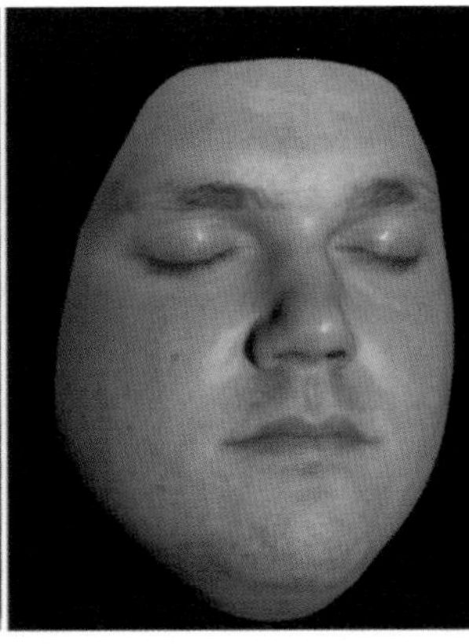
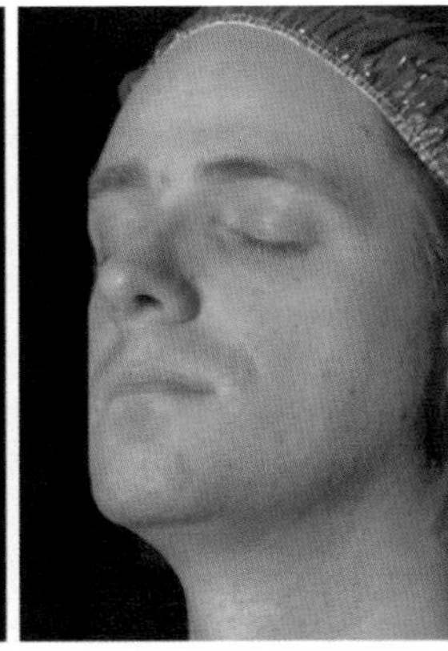
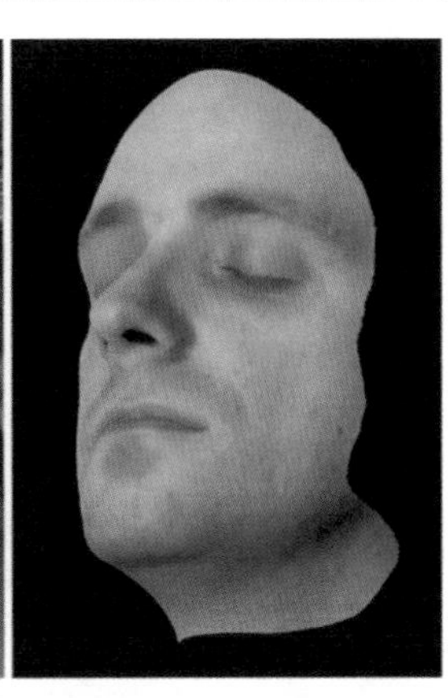

Cook-Torrance Model, Fig. 6 Photograph-rendering pairs of Torrance-Sparrow BRDF fits for modeling spatially varying specular reflectance on faces. *Left-pair*: [16]. *Right-pair*: [17]

reflected light to the appropriate RGB values for a particular display device. Every color sensation can be uniquely described by its location in a three dimensional color space. One such color space is called the XYZ space. A point in this space is specified by three coordinates, the color's XYZ tristimulus values. Each spectral energy distribution is associated with a point in the XYZ color space and thus with tristimulus values. If two spectral energy distributions are associated with the same tristimulus values, they produce the same color sensation and are called metamers. The goal, then, is to find the proportions of RGB that produce a spectral energy distribution that is a metamer of the spectral energy distribution of the reflected light.

These proportions are determined by calculating the XYZ tristimulus values that are associated with the spectral energy distribution of the reflected light [11], and then calculating the RGB values (with a linear transform from XYZ followed by a non-lienar transform based on display gamma curve) that produce a spectral energy distribution with these tristimulus values [13].

Application

Being a physics-based reflectance model, the Cook-Torrance model has been widely used in computer vision and graphics to model the appearance of real world materials. It has been shown to well approximate the reflectance of many measured isotropic BRDFs in the MERL database [14] ranging from metals, plastics, rubber and fabrics [15] (see Fig. 5). It has also been successfully applied in computer vision to the problem of uncalibrated photometric stereo [18], and in computer graphics to model the measured surface reflectance of human skin [16, 17] (see Fig. 6).

References

1. Nicodemus FE, Richmond JC, Hsia JJ, Ginsberg IW, Limperis T (1977) Geometric considerations and nomenclature for reflectance. National Bureau of Standards: NBS Monograph, vol 160
2. Cook R, Torrance KE (1982) A reflection model for computer graphics. ACM Trans Graph 1(1):7–24
3. Torrance KE, Sparrow EM (1967) Theory of off-specular reflection from roughened surfaces. J Opt Soc Am 57: 1104–1114
4. Blinn JF (1977) Models of light reflection for computer synthesized pictures. Comput Graph 11(2):192–198
5. Beckmann P, Spizzichino A (1963) The scattering of electromagnetic waves for rough surfaces. MacMillan, New York
6. Porteus JO (1963) Relation between the height distribution of a rough surface and the reflectance at normal incidence. J Opt Soc Am 53(12):1394–1402
7. Gubareff GG, Janssen JE, Torborg R (1960) Thermal radiation properties survey: a review of the literature. Honeywell Research Center, Minneapolis
8. Purdue University (1970) Thermophysical properties of matter. Vol. 7: thermal radiative properties of metals. Plenum, New York
9. Purdue University (1970) Thermophysical properties of matter. Vol. 8: thermal radiative properties of nonmetallic solids. Plenum, New York
10. Purdue University (1970) Thermophysical properties of matter. Vol. 9: thermal radiative properties of coatings. Plenum, New York

11. CIE International Commission on Illumination (1970) Official recommendations of the international commission on illumination, Colorimetry (E-1.3.1), CIE 15. Bureau Central de la CIE, Paris
12. Sparrow EM, Cess RD (1978) Radiation heat transfer. McGraw-Hill, New York
13. Meyer G, Greenberg D (1980) Perceptual color spaces for computer graphics. Computer Graph 14(3):254–261
14. Matusik W, Pfister H, Brand M, McMillan L (2003) A data-driven reflectance model. ACM Trans Graph 22(3): 759–769
15. Ngan A, Durand F, Matusik W (2005) Experimental analysis of BRDF models. In: Proceeding the Sixteenth Eurographics Conf on Rendering Techniques, Aire-la-Ville, Switzerland, pp 117–126
16. Weyrich T, Matusik W, Pfister H, Bickel B, Donner C, Tu C, McAndless J, Lee J, Ngan A, Jensen HW, Gross M (2006) Analysis of human faces using a measurement-based skin reflectance model. ACM Trans Graph 25(3): 1013–1024
17. Ghosh A, Hawkins T, Peers P, Frederiksen S, Debevec PE (2008) Practical modeling and acquisition of layered facial reflectance. ACM Trans Graph 27(5)
18. Georghiades A, (2003) Incorporating the Torrance and Sparrow model of reflectance in uncalibrated photometric stereo. In: Proceedings of the IEEE international conference on computer vision, Nice, pp 816–823

Coplanarity Constraint

▸Epipolar Constraint

Corner Detection

Michael Maire
California Institute of Technology, Pasadena, CA, USA

Related Concepts

▸Edge Detection

Definition

Corner detection is the process of locating points in an image whose surrounding local neighborhoods contain edges of different orientations that intersect at those points.

Background

A corner can be viewed as a special type of interest point. Interest points [1] are distinct local image regions with well-defined positions that are robust to various image deformations, such as changes in viewpoint or lighting. Many corner detection algorithms relax the strict requirement of edge intersection and, for example, instead locate centroids of windows containing high edge energy in multiple directions.

Theory

Most corner detection algorithms operate by scoring local image patches [2–6]. Nonmaximum suppression and thresholding steps can then be applied to localize corners by selecting the peak responses and retaining only those deemed sufficiently salient.

An early approach to corner detection scores patches based on their similarity to neighboring patches [7]. Comparing a patch centered on a corner to patches offset by several pixels in any direction should produce a low similarity score as the edges incident at the corner do not align once shifted. In contrast, patches in uniform regions are identical to their neighboring patches. Comparing patches using the sum of squared differences (SSD) yields score:

$$S_{u,v}(x, y) = \sum_{x_i} \sum_{y_i} [I(x_i + u, y_i + v) - I(x_i, y_i)]^2 \quad (1)$$

for comparing the patch in image I centered at pixel (x, y) to the one offset by u in the x-direction and v in the y-direction. The summation is over the pixels (x_i, y_i) belonging to the patch. An optional weighting factor can be used to decrease the importance of the outer patch region with respect to the center. Since patches centered on edge pixels which are not corners exhibit a large SSD when displaced orthogonal to the edge, but no difference when displaced along it, a robust measure of corner strength C takes the minimum over all possible displacements:

$$C(x, y) = \min_{u,v} S_{u,v}(x, y) \quad (2)$$

In the limit of small displacement, the difference in (Eq. 1) can be approximated by image derivatives:

$$I(x_i + \Delta x, y_i + \Delta y) - I(x_i, y_i) = I_x(x_i, y_i)\Delta x + I_y(x_i, y_i)\Delta y \quad (3)$$

yielding:

$$S(x,y) = \begin{bmatrix}\Delta x & \Delta y\end{bmatrix}\begin{bmatrix}\sum_{x_i,y_i}(I_x(x_i,y_i))^2 & \sum_{x_i,y_i} I_x(x_i,y_i)I_y(x_i,y_i) \\ \sum_{x_i,y_i} I_x(x_i,y_i)I_y(x_i,y_i) & \sum_{x_i,y_i}(I_y(x_i,y_i))^2\end{bmatrix}\begin{bmatrix}\Delta x \\ \Delta y\end{bmatrix} \quad (4)$$

Define:

$$A(x,y) = [\nabla I \nabla I^T]|_{(x,y)} \quad (5)$$

The eigenvalues λ_1 and λ_2 of $A(x,y)$ describe the behavior of the local neighborhood of point (x,y). This region is uniform if λ_1 and λ_2 are both small, an edge if exactly one of λ_1, λ_2 is large, and a corner if both are large.

The Harris corner detector [4] translates this observation into a measure of corner strength given by:

$$C(x,y) = \lambda_1\lambda_2 - k(\lambda_1 + \lambda_2)^2 \quad (6)$$

where k is a parameter. Shi and Tomasi [8] instead use:

$$C(x,y) = \min(\lambda_1, \lambda_2) \quad (7)$$

Lindeberg [5] extends the Harris detector to operate across multiple scales and adds automatic scale selection.

Wang and Brady [9] define corners in terms of curvature using the score:

$$C(x,y) = \nabla^2 I - k|\nabla I|^2 \quad (8)$$

where k is again a constant user-defined parameter (Fig. 1).

Forstner and Gulch [3] take a different approach to corner localization by finding points p whose distance to edges in their local neighborhood is minimal. Specifically,

$$p = \underset{(x',y')}{\operatorname{argmin}} \sum_{(x_i,y_i)} D((x_i,y_i),(x',y')) \quad (9)$$

where D is the distance from point (x', y') to the line passing through (x_i, y_i) with orientation orthogonal to the gradient at (x_i, y_i). The distance is further weighted by the gradient magnitude. A similar technique can be used to locate junctions with respect to discrete contour fragments [10].

Application

Interest points, including corners, have found use in a variety of computer vision applications such as image matching, object detection and tracking, and 3D reconstruction. The ability to localize interest points in different views of the same object makes matching and tracking feasible. By design, corner and other interest point detectors respond at locations with rich structure in the surrounding image neighborhood. Thus, these positions are natural choices at which to compute informative feature vectors that describe local image content. The ability to associate these descriptors with interest points further facilitates matching and detection algorithms. In general, the choice of feature descriptor to compute at corners can be application specific and may or may not be coupled with the choice of corner detection algorithm.

While a goal of corner and interest point detectors is to localize the same physical sites across different views, in practice this is only accomplished in a statistical sense. There is usually a significant chance of both missed and spurious detections. Consequently, algorithms built on these components must be robust to errors in detecting individual corners, and instead rely on average detector performance. A typical approach is to utilize a large number of corners or interest points per image for added redundancy.

It is not strictly necessary to use corners or interest points in the process of extracting feature descriptors from images. Alternatives include computing features at sampled edge points or on regions from a segmentation of the image. The sliding

Corner Detection, Fig. 1 Corners detected using the Harris operator followed by nonmaximum suppression and thresholding

window detection paradigm exhaustively scans the image, computing features over all possible image windows.

References

1. Mikolajczyk K, Schmid C (2004) Scale and affine invariant interest point detectors. In: Int J Comput Vis 60(1):63–86
2. Deriche R, Giraudon G (1993) A computational approach for corner and vertex detection. In: Int J Comput Vis 10(2):101–124
3. Forstner W, Gulch E (1987) A fast operator for detection and precise localization of distinct corners. In: ISPPR intercommission conference on fast processing of photogrammetric data, Interlaken, Switzerland pp 281–305
4. Harris C, Stephens M (1988) A combined corner and edge detector. In: Proceedings of the 4th alvey vision conference, University of Manchester, Manchester, UK
5. Lindeberg T (1998) Feature detection with automatic scale selection. In: Int J Comput Vis 30(2):79–116
6. Ruzon MA, Tomasi C (2001) Edge, junction, and corner detection using color distributions. In: IEEE Trans Pattern Analysis and Machine Intelligence, 23(11):1281–1295
7. Moravec H (1980) Obstacle avoidance and navigation in the real world by a seeing robot rover. Technical Report CMU-RI-TR-3, Carnegie-Mellon University
8. Shi J, Tomasi C (1994) Good features to track. In: International conference on computer vision pattern recognition (CVPR), Seattle, WA, pp 593–600
9. Wang H, Brady M (1995) Real-time corner detection algorithm for motion estimation. Image Vis Comput 13:695–703
10. Maire M, Arbeláez P, Fowlkes C, Malik J (2008) Using contours to detect and localize junctions in natural images. In: IEEE conference on computer vision pattern recognition, Anchorage, AK

Cross Entropy

Ying Nian Wu
Department of Statistics, UCLA, Los Angeles, CA, USA

Definition

Cross entropy is a concept in information theory to measure the independence of two probability distributions.

Theory

For two distributions $p(x)$ and $q(x)$ defined on the same space, the cross entropy is defined as

$$\begin{aligned} H(p,q) &= \mathrm{E}_p[-\log q(X)] \\ &= \mathrm{E}_p[-\log p(X)] + \mathrm{E}_p[\log(p(X)/q(X))] \\ &= H(p) + KL(p,q), \end{aligned}$$

where $H(p) = \mathrm{E}_p[-\log p(X)]$ is the entropy of p and $KL(p,q) = \mathrm{E}_p[\log(p(X)/q(X))]$ is the Kullback-Leibler divergence from p to q. The Kullback-Leibler divergence is also called relative entropy.

The cross entropy method is a Monte Carlo method for rare event simulation and stochastic optimization [2].

References

1. Cover TM, Thomas JA (1991) Elements of information theory. Wiley, New York
2. Rubinstein RY (1997) Optimization of computer simulation models with rare events. Eur J Oper Res 99:89–112

Curvature

Takayuki Okatani
Graduate School of Information Sciences, Tohoku University, Sendai-shi, Japan

Synonyms

Curvedness

Related Concepts

►Geodesics, Distance Maps, and Curve Evolution

Definition

Curvature is a fundamental concept of differential geometry that represents local "curvedness" of some object such as a curve, a surface, and a Riemannian space.

Background

Dealing with the shape of an object is a fundamental issue of computer vision. It is necessary, for example, to represent the two-dimensional or three-dimensional shape of an object, to extract the object shape from various types of images, and to measure similarity between two object shapes. The application of differential geometry to these problems is getting more and more common in modern computer vision. Curvature is one of the most fundamental concept of differential geometry, and its use can be seen throughout all sorts of related problems. This entry explains basic definitions of the curvature of a plane curve and a surface in Euclidean space and summarizes their applications to a few major applications. See [8] for the definition of curvatures of a Riemannian space.

Theory and Application

A plane curve C can be specified by the coordinates $(x(s), y(s))$ of each point, where s is a monotonic function of the arc length. Algebraically, the curvature $\kappa(s)$ of C at a point $(x(s), y(s))$ is defined as

$$\kappa(s) = \frac{\dot{x}\ddot{y} - \dot{y}\ddot{x}}{(\dot{x}^2 + \dot{y}^2)^{3/2}}, \tag{1}$$

where $\dot{x} = dx/ds$, $\ddot{x} = d^2x/ds^2$, etc. Note that the denominator becomes 1 when s is the arc length itself. Let $\phi(s)$ be the orientation of the tangent to C at $(x(s), y(s))$, that is, $\tan\phi(s) = \dot{y}/\dot{x}$. If s is the arc length, the curvature can be represented as $\kappa = \dot{\phi}(s)$.

Geometrically, the absolute value of κ is equal to the reciprocal of the radius r of the circle osculating the curve C at the point, that is, $|\kappa| = 1/r$, as shown in Fig. 1a; r is called the radius of curvature. The curve C can also be locally approximated by a second-order polynomial curve. Consider the local coordinates XY defined by the tangent and the normal vectors to C, as shown in Fig. 2b. The approximating quadratic curve will be given as $Y = \kappa X^2/2$. Note that there is freedom in the choice of the sign of the curvature; in its definition of Eq. (1), the sign depends on the parametrization of s.

The curvature of a plane curve is effectively used to analyze and/or represent two-dimensional shapes. In [2], the curvature primal sketch is proposed, in which significant changes in curvature along a curve are detected and used for shape representation. In [6], the concept of curvature scale space (CSS) is presented, where the zero-crossings of curvature (i.e., the inflection points of the curve) are used to represent the structure of the curve at varying levels of detail. This technique has been applied to various computer vision problems, such as feature extraction, shape retrieval, and object recognition [5].

The CSS technique is summarized as follows. Let $X(s;\sigma)$ and $Y(s;\sigma)$ be the convolutions of $x(s)$ and $y(s)$ with a Gaussian kernel $g(s,\sigma)$, respectively, that is, $X(s;\sigma) = x(s) * g(s,\sigma)$ and $Y(s;\sigma) = y(s) * g(s,\sigma)$. As shown in Fig. 2a, the curve given by $(X(s;\sigma), Y(s;\sigma))$ gradually becomes smoother, as σ increases. Figure 2b shows a $s - \sigma$ plot of the trajectories of the inflection points. The horizontal axis indicates the normalized arc length s that is scaled to the range [0–1].

The curvature of a plane curve serves as a basic measure of its smoothness. A smoother curve has smaller curvature at each point. Curve evolution like the one in CSS is used in many applications besides shape analysis/representation, where this property with curvature plays a central role.

The active contour model (ACM) [9] was developed to detect the contour of an object in an image by moving an elastic curve from a given initial position to nearby the object contour. This is performed by minimizing the sum of two energy terms, an external term modeling the similarity/dissimilarity to the image edges and an internal term modeling the "elasticity" of the curve. This latter term usually includes the arc length as well as the curvature of the curve to obtain a smoother curve that is more desirable in practice. ACM was later reformulated as a problem of finding local geodesics in a space with Riemannian metric computed from the image, which is known as the geodesic active contour model [4]. The same smoothing property associated with (Euclidean) curvature also plays a key role there.

Curvature, Fig. 1 (**a**) A plane curve (solid line) and an osculating circle (broken line). The curvature κ at P is the reciprocal of the radius r of the circle. (**b**) The approximating quadratic curve at P (broken line)

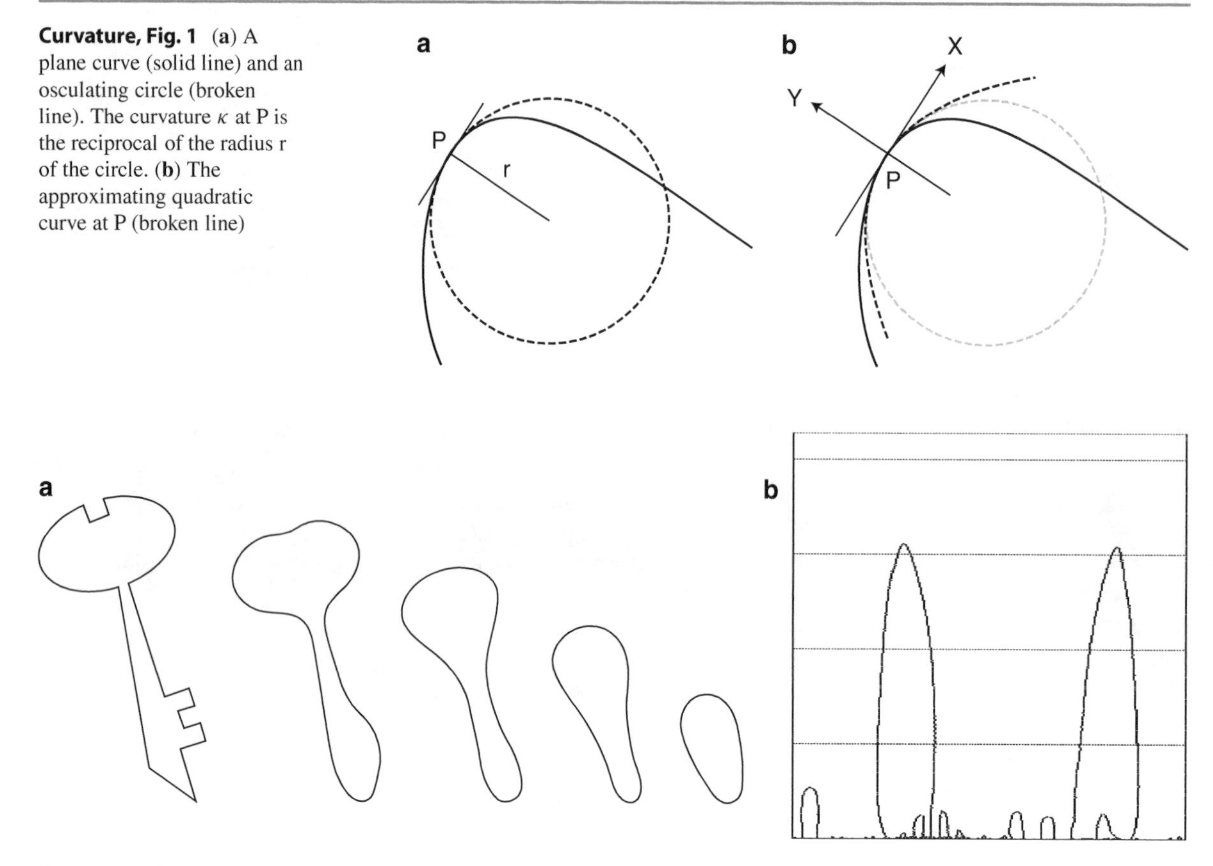

Curvature, Fig. 2 (**a**) The curvature scale space of a shape. (**b**) s—σ plot of the trajectories of the inflection points of the shape

The evolution of a curve can be fully represented by specifying the normal speed of the curve, that is, the evolution speed measured along the normal vector at each point. Consider a closed curve evolving with speed equal to its curvature, where the sign of the curvature is chosen so that a circle would shrink inward. Its generalization to higher-dimensional space is known as the mean curvature flow. This curve evolution has the properties of "smoothing"; the curve evolves so that its high curvature parts are smoothed out in a finite time. It is shown that any closed curve will become a convex curve and then shrink to a point (Grayson's theorem [7]).

The level set method (LSM) [17] is a numerical framework for computing such curve evolutions (as well as surface/manifold evolution), which has many advantages to previous methods, such as being able to handle topological changes of the curves. In LSM, a curve is represented as a zero-level curve of an auxiliary function ϕ as $\phi(x, y) = 0$. Then, the curve evolution with normal speed F is represented as $\partial\phi/\partial t = -F|\nabla\phi|$, a time-dependent evolution equation of ϕ. The curvature κ of each point of the level curve is computed using ϕ as

$$\kappa = \nabla \cdot \frac{\nabla\phi}{|\nabla\phi|} \tag{2}$$

where ∇ is the gradient operator $\nabla = (\partial/\partial x, \partial/\partial y)$. Thus, the mean curvature flow is represented as $\partial\phi/\partial t = -\kappa|\nabla\phi|$.

In some problems, the evolution equations of an image $I(x, y)$, which are similar to those of $\phi(x, y)$ above, are considered. An example is the diffusion equation [15] represented as $\partial I/\partial t = \nabla \cdot (c(x, y)\nabla I)$, which has many applications, for example, image denoising/restoration [16] and image inpainting [3]. Although the choice of $c(x, y)$ depends on each application, it often has a term of the curvature of the level curves of $I(x, y)$, which can be also computed by Eq. (2) (When $c(x, y)$ is constant, the resulting image evolution coincides with the Gaussian blurring).

The curvature of a smooth surface S in a three-dimensional space is defined as follows. Consider the

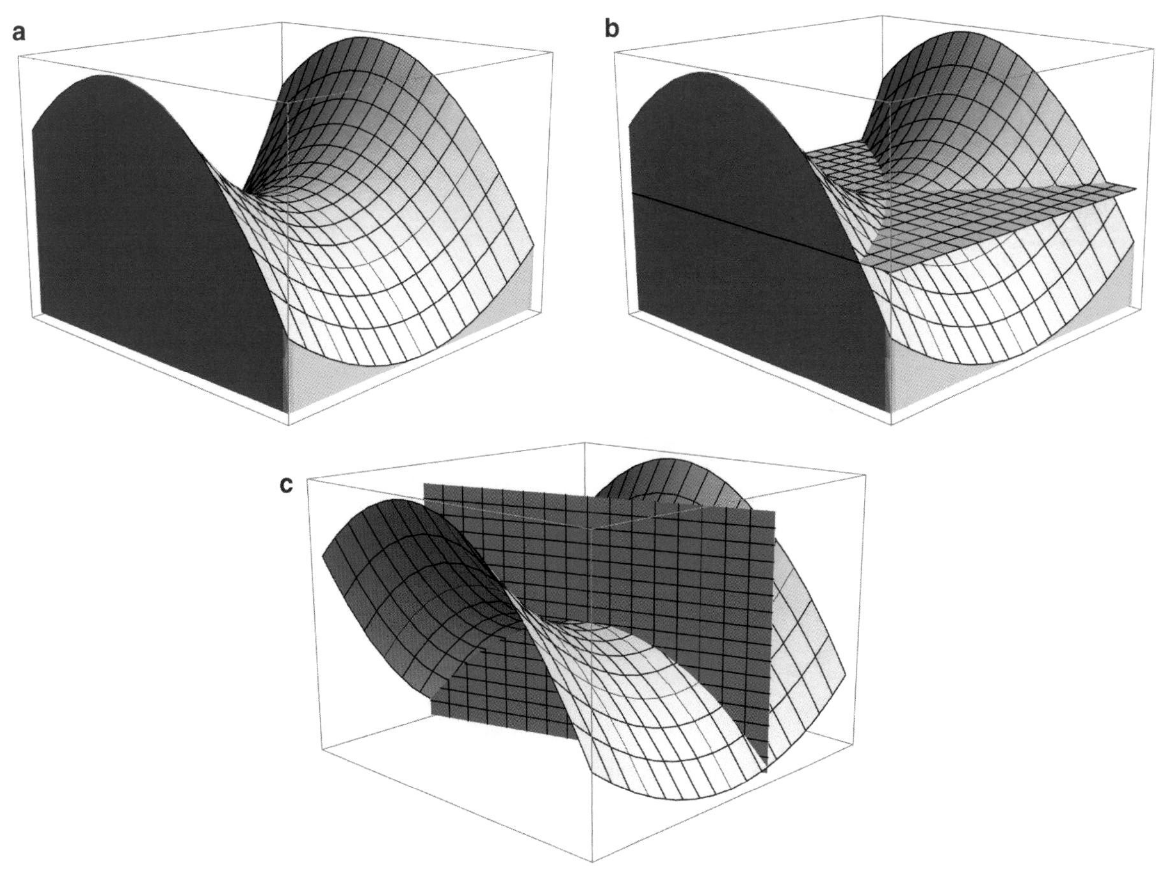

Curvature, Fig. 3 (**a**) A surface S. (**b**) The tangent plane to S at a point P in the center. (**c**) A plane containing the normal vector of S at P. The intersecting curve on the plane gives the normal curvature of S at P

tangent plane to S at a point P of S (Fig. 3b). The normal vector of S at P should be perpendicular to the tangent plane. Then consider a plane containing this normal vector (Fig. 3c). The intersection of the plane with S yields a plane curve on it. The curvature of this plane curve at P, defined in the same manner as above, is called the normal curvature of S at P. The plane has a one-dimensional rotational freedom around the normal vector, and the normal curvature is defined for each of such planes.

Let the maximum and minimum values of the normal curvature at P be k_1 and k_2. They are called the principal curvatures of S at P. Consider a local coordinate frame XYZ whose origin is located at P and Z axis coincides with the normal vector. The surface is locally approximated by a second-order polynomial surface

$$Z = \frac{1}{2}\begin{bmatrix} X & Y \end{bmatrix} \mathbf{H} \begin{bmatrix} X \\ Y \end{bmatrix}, \tag{3}$$

where $\mathbf{H}$ is a 2×2 symmetric matrix. Specifying a direction in the XY plane by a two-dimensional normalized vector $\mathbf{v}$ ($\|\mathbf{v}\|^2 = 1$), the normal curvature for the direction $\mathbf{v}$ is given by $\mathbf{v}^\top \mathbf{H}\mathbf{v}$. The eigenvalues of $\mathbf{H}$ are the same as the principal curvatures k_1 and k_2, and their associated eigenvectors the corresponding directions $\mathbf{v}$'s. The Gaussian curvature of S at P is defined as a product of principal curvatures k_1 and k_2,

$$\kappa = k_1 k_2, \tag{4}$$

and the mean curvature of S at P is defined as their mean,

$$h = \frac{k_1 + k_2}{2}. \tag{5}$$

Local shapes of a surface are classified by the signs of the principal curvatures k_1 and k_2, as shown in Fig. 4a [10]. A point at which k_1 and k_2 have the same sign, that is, the Gaussian curvature $\kappa > 0$, is called

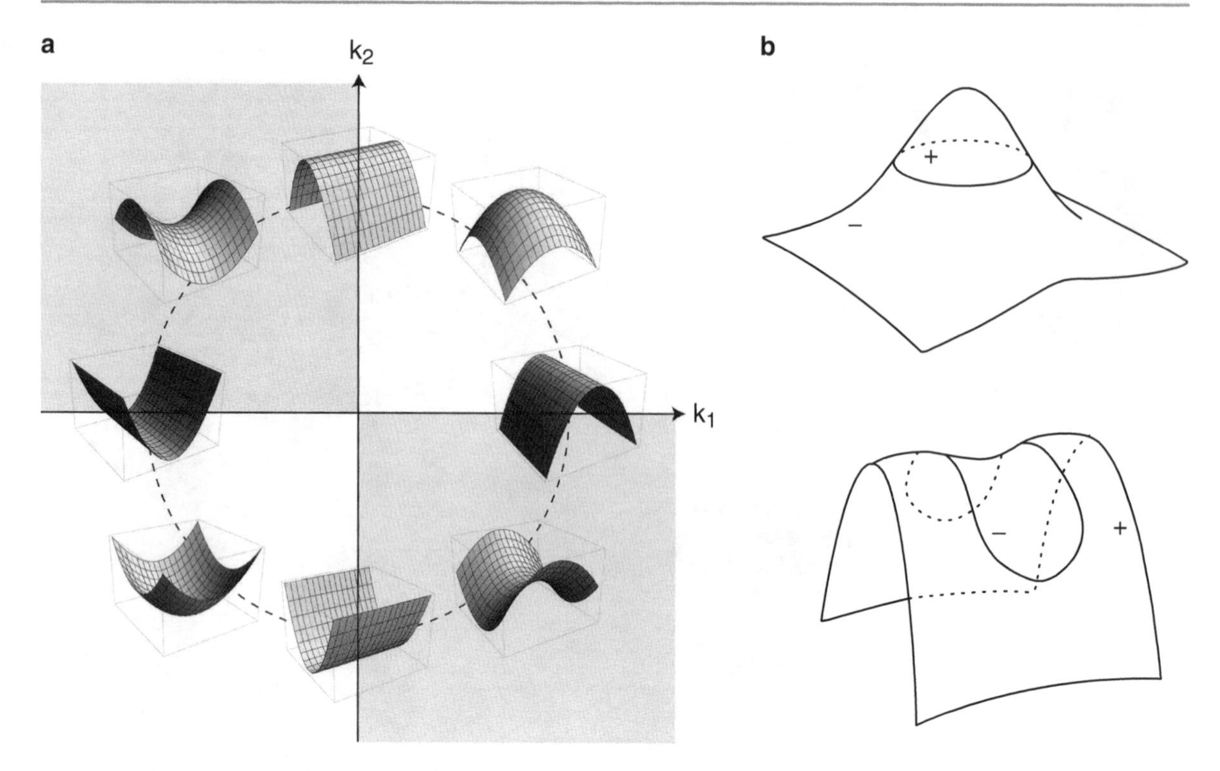

Curvature, Fig. 4 (**a**) Classification of local shapes according to principal curvatures k1 and k2. The first and third quadrants are where the surface point is elliptical, and the second and fourth quadrants are where it is hyperbolic. (**b**) Examples of the surface segmentation based on the sign of the Gaussian curvature

an elliptical point. If k_1 and k_2 have different signs, that is, $\kappa < 0$, then the point is called a hyperbolic point. If either k_1 or k_2 vanishes, that is, $\kappa = 0$, then the point is called a parabolic point. Using this classification, a smooth surface may be segmented into finite regions depending on the sign of the curvatures; Fig. 4b shows examples of the surface segmentation based on the Gaussian curvature signs.

This classification method is used in all sorts of applications. For example, it is used in the detection of features such as corners and edges in images, where the images or their variants are regarded as the surface whose local shape is classified; see, for example, [11]. In [18], the shape of an object obtained as range data is represented based on its curvature. It is shown in [1, 13] that similar curvature-based shape representation can be computed from multiple images taken under different illumination directions but without detailed knowledge on the process of the image formation.

More advanced forms of curvatures, such as the curvature of higher-dimensional submanifolds and the Riemannian curvature tensors, are used in recent studies. In [12], several Riemannian metrics on the space of two-dimensional shapes are studied. In [14], biases of the maximum likelihood estimates derived for the problems of estimating some geometric structure from images (e.g., the epipolar geometry) are related to the curvature of the hypersurfaces given by the geometric structure.

References

1. Angelopoulou E, Wolff LB (1998) Sign of gaussian curvature from curve orientation in photometric space. IEEE Trans Pattern Anal Mach Intell 20(10):1056–1066
2. Asada H, Brady M (1986) The curvature primal sketch. IEEE Trans Pattern Anal Mach Intell 8(1):2–14
3. Bertalmio M, Sapiro G, Caselles V, Ballester C (2000) Image inpainting. In: Proceedings of SIGGRAPH 2000, New Orleans, pp 417–424
4. Caselles V, Kimmel R, Sapiro G (1997) Geodesic active contours. Int J Comput Vis 22(1):61–79
5. Mokhtarian F, Bober M (2003) Curvature scale space representation: theory, Applications, and MPEG-7 Standardization. Kluwer Academic, Dordrecht
6. Mokhtarian F, Mackworth A (1992) A theory of multi-scale, curvature-based shape representation for planar curves. IEEE Trans Pattern Anal Mach Intell 14(8):789–805
7. Grayson MA (1987) The heat equation shrinks embedded plane curves to round points. J Differ Geom 26(2):285–314

8. Hazewinkel M (ed) (2002) Encyclopaedia of mathematics. Springer. http://eom.springer.de/C/c027320.htm
9. Kass M, Witkin A, Terzopoulos D (1988) Snakes: active contour models. Int J Comput Vis 1:321–331
10. Koenderink JJ (1990) Solid shape. MIT, Cambridge
11. Lowe DG (2004) Distinctive image features from scale-invariant keypoints. Int J Comput Vis 60(2):91–110
12. Michor PW, Mumford D (2004) Riemannian geometries on spaces of plane curves. J Eur Math Soc 8:1–48
13. Okatani T, Deugchi K (1999) Computation of the sign of the gaussian curvature of a surface from multiple unknown illumination images without knowledge of the reflectance property. Comput Vis Image Underst 76(2):125–134
14. Okatani T, Deugchi K (2009) On bias correction for geometric parameter estimation in computer vision. In: Proceedings of the IEEE computer society conference on computer vision and pattern recognition (CVPR), Miami, pp 959–966
15. Perona P, Malik J (1990) Scale-space and edge detection using anisotropic diffusion. IEEE Trans Pattern Anal Mach Intell 12(7):629–639
16. Rudin LI , Osher S, Fatemi E (1992) Nonlinear total variation based noise removal algorithms. Phys D 60:259–268
17. Sethian JA (1999) Level set methods and fast marching methods. Cambridge University Press, Cambridge
18. Vemuri BC, Mitiche A, Aggarwal JK (1986) Curvature-based representation of objects from range data. Image Vis Comput 4(2):107–114

Curvedness

►Curvature

Curves

►Curves in Euclidean Three-Space

Curves in Euclidean Three-Space

Jan J. Koenderink
Faculty of EEMSC, Delft University of Technology, Delft, The Netherlands
The Flemish Academic Centre for Science and the Arts (VLAC), Brussels, Belgium
Laboratory of Experimental Psychology, University of Leuven (K.U. Leuven), Leuven, Belgium

Synonyms

Curves; Space curves

Related Concepts

►Curvature; ►Differential Invariants; ►Euclidean Geometry; ►Osculating Paraboloids

Definition

Space curves are one-parameter manifolds immersed in Euclidean 3D space $\mathbf{r}(s) \subset \mathbb{E}^3$, where $s \in \mathbb{R}$. One requires differentiability to whatever order is necessary, and $\|\frac{\partial \mathbf{r}(s)}{\partial s}\|^2 \neq 0$. It is convenient to require $\|\frac{\partial \mathbf{r}(s)}{\partial s}\|^2 = 1$, though this can always be achieved through a reparameterization. Such curves are known as "rectified" or "parameterized by arc-length," and one writes $\dot{\mathbf{r}}(s)$ for the partial derivative with respect to arc-length ($\mathbf{r}'(t)$ will be used if the parameter t is not arc-length). As discussed below, in addition one requires $\ddot{\mathbf{r}}(s) \neq 0$ for a generic space curve. The very notion of "rectifiable" is of course Euclidean. Curves in non-Euclidean spaces (affine, projective) have to be handled in appropriate ways.

Background

The classical theory of curves starts with Newton and Leibniz; it was brought in the form presented here in the course of the eighteenth and nineteenth century.

Theory

In differential geometry the "shape" of a curve is defined locally as a set of differential invariants that are algebraic combinations of derivatives $\{\dot{\mathbf{r}}, \ddot{\mathbf{r}}, \dddot{\mathbf{r}}, \ldots\}$, and that are invariant with respect to Euclidean motions (notice that "congruencies" would assign the same shape to helices of opposite chirality). A complete set of such differential invariants, specified as a function of the parameter s (arc-length), allows one to construct the curve on the basis of this, up to arbitrary motions. Such a specification of the curve is known as its "natural equations."

In performing coordinate-wise operations one has to refer to a fiducial frame, most conveniently an orthonormal basis. All equations will take on their simplest form in a frame that is especially fit to the curve. The classical Frenet-Serret frame is one way to achieve this. As one moves along the curve the frame will rotate in various ways. The simplest description expresses the instantaneous motion of the frame in terms of the frame itself. The differential invariants have geometrically intuitive interpretations in such a system. This insight is due to Elie Cartan, though already implicit in the classical formulation.

These are the basic insights of the classical theory. A short derivation with appropriate geometrical interpretations follows.

The first derivative of position $\mathbf{r}(s)$ with respect to arc-length s is (by construction) a unit vector known as the *tangent* to the curve, denoted $\mathbf{t}(s) = \dot{\mathbf{r}}(s)$. Geometrically the tangent is the direction of the curve, it is the limit of the difference of two points $\mathbf{r}(s_1) - \mathbf{r}(s_2)$ (with $s_1 > s_2$) of the curve divided by the chord-length, that is, $\|\mathbf{r}(s_1) - \mathbf{r}(s_2)\|$, as the points approach each other infinitesimally. Thus the expression $\mathbf{r}(s_0) + (s - s_0)\mathbf{t}(s_0)$ is a first order approximation, the tangent line, to the curve.

Because the tangent is a unit vector, one has $\dot{\mathbf{t}} \cdot \mathbf{t} = 0$. (Note the Euclidean nature of this!) Thus the second-order derivative $\ddot{\mathbf{r}}$ is orthogonal to the tangent. We write $\ddot{\mathbf{r}} = \kappa\mathbf{n}$, where the unit vector $\mathbf{n}(s)$ is the "normal" to the curve, and $\kappa(s)$ the "curvature." Notice that the normal would be undefined in the case the tangent did not change direction. For a generic space curve we have to require $\kappa > 0$ throughout (though of course the choice of sign is arbitrary). This is different from planar curves for which a signed curvature makes sense. Thus planar curves may have points of inflection, whereas this notion makes no sense for space curves (Fig. 1).

The normal and the tangent define a plane, the so-called osculating plane of the curve. It is the limit of the plane spanned by three points of the curve as the points approach each other infinitesimally. One might say that, at least locally, the curve lies in its osculating plane (remember that "to osculate" means literally "to kiss"). The three points also define a circle (lying in the osculating plane), whose radius can be shown to be $1/\kappa$. (An easy way to show this is to write down the second order Taylor development of the curve.) Thus, locally, the curve is like a circular arc of radius $1/\kappa$ in the osculating plane, moving in the tangent direction. The curvature measures the degree to which the curve veers away from the tangent direction, into the direction of the normal. Notice that $\ddot{\mathbf{r}} \cdot \mathbf{n} = \kappa$, and that the curvature is a scalar that does not depend upon the coordinate frame. The curvature is the first example of a differential invariant.

In Euclidean space $\mathbb{E}^3$ the tangent and the normal imply a third vector orthonormal to them both. It is $\mathbf{b} = \mathbf{t} \times \mathbf{n}$, known as the "binormal." This is again a very Euclidean construction, the vector product being a Euclidean 3D concept. The orthonormal frame $\{\mathbf{t}, \mathbf{n}, \mathbf{b}\}$ is the Frenet-Serret frame. It is tightly connected to the curve and a complete basis of $\mathbb{E}^3$. Thus it is perfectly suited to describe the third order derivative $\dddot{\mathbf{r}}$. The obvious move is to express $\dot{\mathbf{n}}$ in terms of the tangent and the binormal (given the fact that the normal is a unit vector this should be possible). Thus one writes $\dot{\mathbf{n}} = -\kappa\mathbf{t} + \tau\mathbf{b}$, where τ is another differential invariant known as the torsion of the curve. The reason for the term $-\kappa\mathbf{t}$ is that $\dot{\mathbf{t}} = \kappa\mathbf{n}$: the frame turns about the binormal with angular speed κ (Fig. 2). The third derivative itself then is $\dddot{\mathbf{r}} = -\kappa^2\mathbf{t} + \dot{\kappa}\mathbf{n} + \tau\kappa\mathbf{b}$.

The torsion (sometimes called "second curvature") has a simple geometrical interpretation. It is the angular rate of change of the attitude of the osculating plane.

Notice that the metrical structure of $\mathbb{E}^3$ is used in an essential manner in all constructions thus far. The classical theory of curves cannot be used in spaces with different structures, even in homogeneous spaces such as affine or projective spaces. Of course, a theory of curves can be developed for such spaces too, but the differential invariants, although often denoted "curvature" and "torsion," will have meanings that are completely distinct from the curvature or torsion of curves in Euclidean-space. The reader should be keenly aware of this, as non-Euclidean spaces occur frequently and naturally in computer vision and image processing applications, the best known being the affine and projective 3D spaces, as well as "graph space."

The structure found thus far can be appreciated from a straightforward Taylor expansion:

$$\mathbf{r}(s) = \mathbf{r}(0) + \dot{\mathbf{r}}s + \ddot{\mathbf{r}}\frac{s^2}{2!} + \dddot{\mathbf{r}}\frac{s^3}{3!} + \mathrm{O}[s]^4, \tag{1}$$

which in terms of the Frenet-Serret frame is (the "canonical representation"):

$$\mathbf{r}(0) + \left(s - \kappa^2\frac{s^3}{3!} + \ldots\right)\mathbf{t} + \left(\kappa\frac{s^2}{2!} + \dot{\kappa}\frac{s^3}{3!} + \ldots\right)\mathbf{n} + \left(\kappa\tau\frac{s^3}{3!} + \ldots\right)\mathbf{b}, \tag{2}$$

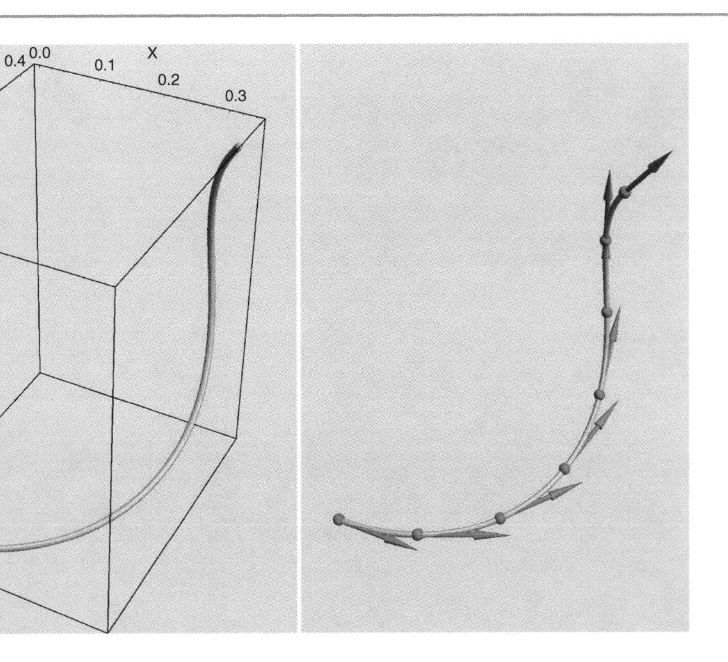

Curves in Euclidean Three-Space, Fig. 1 At *left* generic space curve. It is curved throughout and "winds" in a single sense. This curve was defined via its natural equations (see below), $\kappa(s) = 1 + 3s$, $\tau(s) = 1 + 5s$, $0 < s < 1$. At *right* the field of tangents along the curve

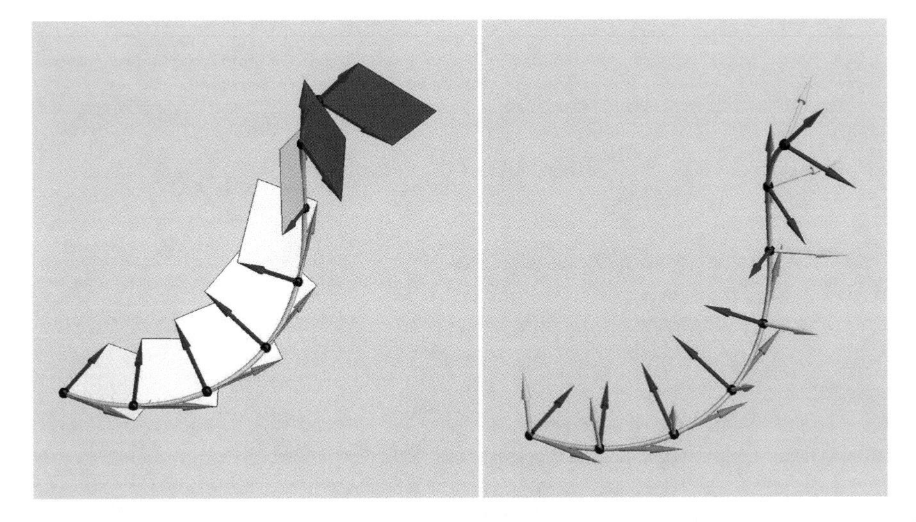

Curves in Euclidean Three-Space, Fig. 2 At *left* the field of osculating planes along the space curve. Notice how it rotates, revealing the curve to be a "twisted" one. At *right* the field of Frenet frames along the curve. Again, notice how it rotates as it moves along the curve

from which the habitus of the curve is readily gleaned: The projection on the osculating plane is (approximately) a parabolic arc, on the normal ($\mathbf{n} \times \mathbf{b}$) plane a cusp, and on the tangential ($\mathbf{b} \times \mathbf{t}$) plane an inflection (Figs. 3 and 4).

Notice that the third order includes a term in the rate of change of curvature, not merely the torsion. The meaning of this becomes evident from another geometrical construction. Four distinct points define a sphere, and in the limit one obtains the osculating sphereat a

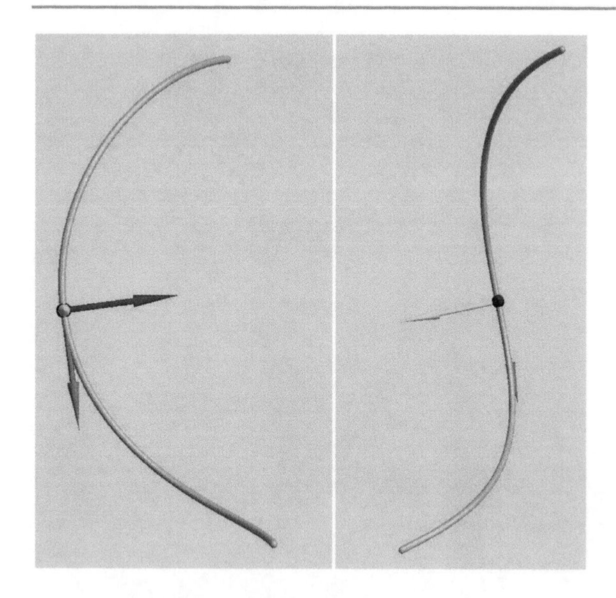

Curves in Euclidean Three-Space, Fig. 3 At *left* a view from the binormal, at *right* a view from the normal direction. From the binormal direction the curve shows its curvature, from the normal direction one sees an inflection

point of the curve. When the curve is twisted at the point ($\tau \neq 0$), the center of the osculating sphere is given by:

$$\mathbf{c}_{osc} = \mathbf{r}(s) + \frac{1}{\kappa(s)}\mathbf{n}(s) - \frac{\dot{\kappa}(s)}{\kappa(s)^2\tau(s)}\mathbf{b}(s), \quad (3)$$

and its radius of curvature $\varrho_{\text{osc}}(s)$ by:

$$\varrho_{osc} = \sqrt{\varrho^2(s) + \left(\frac{\dot{\varrho}(s)}{\tau(s)}\right)^2}, \quad (4)$$

where ϱ is the radius of the osculating circle. Thus only at "vertices" of the curve ($\dot{\kappa} = 0$) is the osculating circle a great circle of the osculating sphere. The osculating sphere always cuts the osculating plane in the osculating circle though.

The geometrical structure is formulated rather elegantly by the Frenet-Serret formulas (notice the antisymmetry of the matrix):

$$\begin{pmatrix} \dot{\mathbf{t}} \\ \dot{\mathbf{n}} \\ \dot{\mathbf{b}} \end{pmatrix} = \begin{pmatrix} 0 & +\kappa & 0 \\ -\kappa & 0 & +\tau \\ 0 & -\tau & 0 \end{pmatrix} \begin{pmatrix} \mathbf{t} \\ \mathbf{n} \\ \mathbf{b} \end{pmatrix}. \quad (5)$$

The "natural equations" simply specify $\kappa(s)$ and $\tau(s)$. Using the Frenet-Serret equations one constructs the curve by integration, specifying an arbitrary initial Frenet-Serret frame. Thus the curvature and torsion specify the curve up to arbitrary Euclidean motions.

A useful formalism that extends this is due to Darboux (Fig. 5). The "Darboux vector" is defined as $\mathbf{d} = \tau\mathbf{t} + \kappa\mathbf{b}$. Now one has $\dot{\mathbf{t}} = \mathbf{d} \times \mathbf{t}$, $\dot{\mathbf{n}} = \mathbf{d} \times \mathbf{n}$, and $\dot{\mathbf{b}} = \mathbf{d} \times \mathbf{b}$, thus the Darboux vector is the angular velocity of the "moving trihedron" $\{\mathbf{t}, \mathbf{n}, \mathbf{b}\}$. One immediately concludes that the curvature is the rate of turning about the binormal and the torsion the rate of turning about the tangent. This nicely "explains" the geometrical meaning of the differential invariants κ (the curvature) and τ (the torsion).

The Darboux formalism is by far the simplest to commit to memory. It completely sums up the structure of generic space curves in Euclidean three-space. The Darboux vector also lets one handle degenerate cases easily, for instance that of planar curves (binormal constant), or straight curves (or rather, lines; tangent constant). Finally, the Frenet-Serret equations are written in an easily remembered (because of the cyclic **t**-**n**-**b**–structure) form:

$$\dot{\mathbf{t}} = \mathbf{d} \times (\mathbf{n} \times \mathbf{b}), \quad (6)$$

$$\dot{\mathbf{n}} = \mathbf{d} \times (\mathbf{b} \times \mathbf{t}), \quad (7)$$

$$\dot{\mathbf{b}} = \mathbf{d} \times (\mathbf{t} \times \mathbf{n}). \quad (8)$$

There are a number of geometrical structures related to a curve that are of occasional use. A few of these are discussed below.

So-called spherical images are spherical curves – that are curves on the surface of the unit sphere – related to a curve (Fig. 6). One naturally considers the tangent, normal, and binormal spherical images that are the curves $\mathbf{t}(s)$, $\mathbf{n}(s)$, and $\mathbf{b}(s)$ (notice that these curves are not rectified!). Notice that for straight lines the tangent spherical images degenerate to a point, whereas the other spherical images are undefined. For planar curves the tangent and normal spherical images are degenerated to arcs of great circles, whereas the binormal spherical image degenerates to a point. Special points of the spherical images, like inflections or cusps, relate to special points of the original curve. For some problems of a physical nature it is the spherical image, rather than the curve itself that is of primary

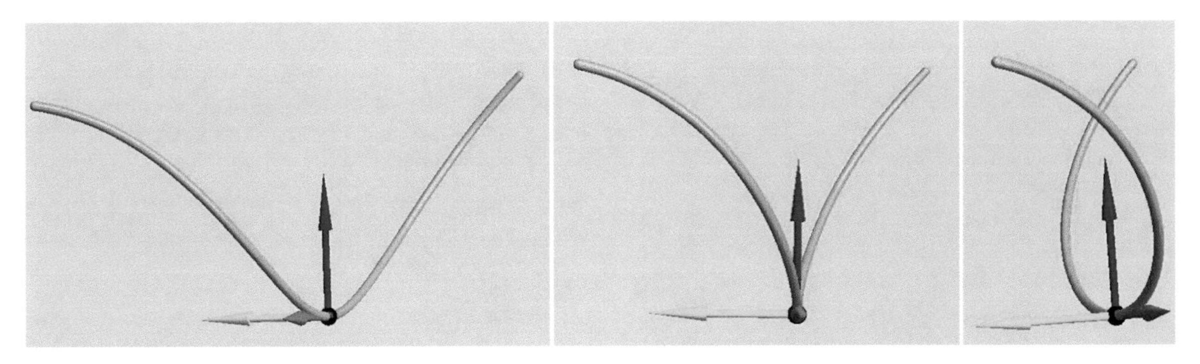

Curves in Euclidean Three-Space, Fig. 4 At *center* a view from the tangent direction. In this view the curve appears as a cusp. At *left* and *right* views from nearby directions

Curves in Euclidean Three-Space, Fig. 5 At *left* osculating circles along the curve. At *right* the Darboux vector field along the curve

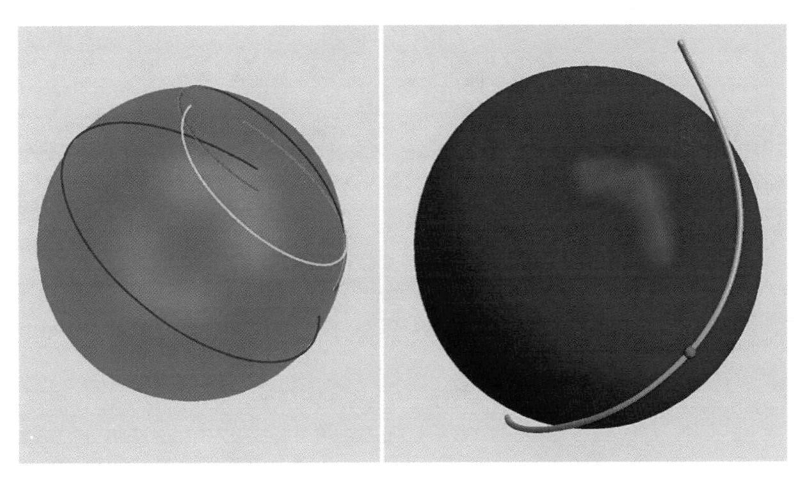

Curves in Euclidean Three-Space, Fig. 6 At *left* the spherical images associated with the curve (*red*: tangent image; *blue*: normal image; *yellow*: binormal image). At *right* an osculating sphere at the curve

interest. An example is that of specularities on tubular surfaces like human hairs.

The surfaces described by the lines defined by the tangent, normal, binormal, and Darboux vector are also of occasional interest. These are – by construction – ruled surfaces, though not necessarily developable ones. Their singular features, like the edge of regression in the case of developable surfaces or the line of striction in the case of skew surfaces, have useful geometrical relations to the curve. The best known

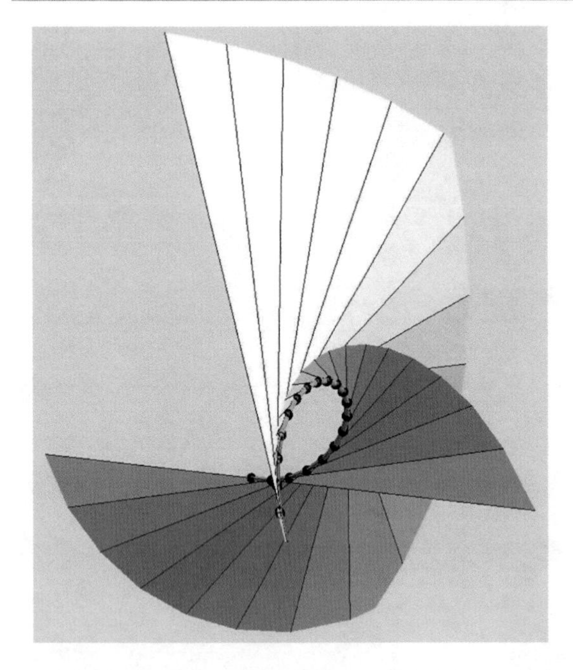

Curves in Euclidean Three-Space, Fig. 7 The surface described by the tangents to the curve is a developable surface, the curve being its edge of regression

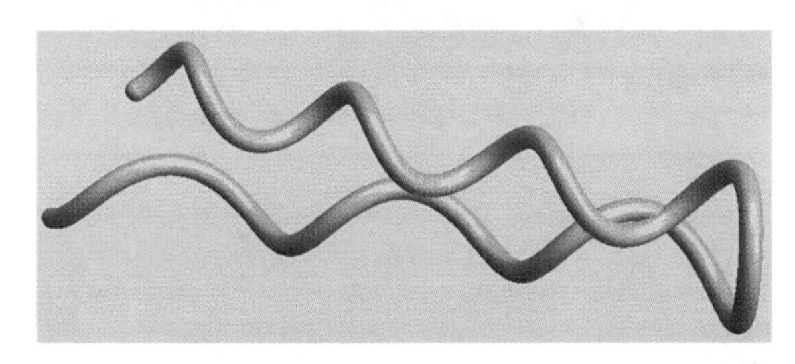

Curves in Euclidean Three-Space, Fig. 8 Example of a torsion zero. Notice the opposite chirality of the parts of the curves at either side of the torsion zero (which is at the *right* in this picture)

example is the surface of tangent lines, which happens to be developable, with the curve itself as the edge of regression, and the surface of normal lines, which is commonly used to describe surface strips (Fig. 7).

Special points on the curve may also be studied by direct means of course. Perhaps the most obvious instance is that of a torsion zero (Fig. 8). At a torsion zero the chirality of the curve changes. Whereas the curve generically osculates, but also pierces its osculating plane, the curve merely osculates, but fails to pierce the osculating plane, it being "deflected" by it. Such special points are often introduced by design in telephone cords, and many vines also have frequent torsion zeroes in their tendrils

Additional Problems

This entry describes the Euclidean differential geometry of space curves. In many problems the context is different from Euclidean though. Because the differential invariants introduced here are specific for the Euclidean transformation group, one needs to develop the differential geometry from scratch. Examples frequently occur in computer vision, for instance, and one often works in spaces with mere affine, or even projective structure. Spaces with even less structures are common. A common case involves "isotropic differential geometry" in "graph spaces."

References

1. Bruce JW, Giblin PJ (1992) Curves and singularities. Cambridge University Press, Cambridge, MA
2. do Carmo MP (1976) Differential geometry of curves and surfaces. Prentice–Hall, Englewood Cliffs
3. Eisenhart LP (2004) A treatise on the differential geometry of curves and surfaces. Dover, New York
4. Gray A (1997) Modern Differential Geometry of Curves and Surfaces with Mathematica, 2nd edn. CRC, Boca Raton
5. Guggenheimer H (1977) Differential geometry. Dover, New York
6. Kreyszig E (1991) Differential geometry, Chapter II. Dover Publications, New York
7. Lockwood EH (1961) Book of curves. Cambridge University Press, Cambridge, MA
8. Porteous I (2001) Geometric differentiation. Cambridge University Press, Cambridge, MA
9. Spivak M (1999) A Comprehensive introduction to differential geometry, vol 2. Publish or Perish, Houston

D

Data Augmentation

Ying Nian Wu
Department of Statistics, UCLA, Los Angeles, CA, USA

Definition

Data augmentation is a Markov chain Monte Carlo algorithm for sampling from a Bayesian posterior distribution

Background

Data augmentation was originally developed by Tanner and Wong [10] as a stochastic counterpart of the EM algorithm [1], and it is closely related to the Gibbs sampler [2]. Thus, the basic setup of data augmentation is similar to the EM algorithm.

Theory

Let y be the observed data and z be the missing data or latent variable. Let $p(y, z|\theta)$ be the probability distribution of the complete data (y, z), with θ being the unknown parameter. The marginal distribution of the observed data y is $p(y|\theta) = \int p(y, z|\theta)dz$. Let $p(\theta)$ be the prior distribution of θ. The goal is to draw Monte Carlo samples from the posterior distribution $p(\theta|y) \propto p(\theta)p(y|\theta)$.

The data augmentation algorithm is an iterative algorithm. It starts from an initial value θ_0. Let (θ_t, z_t) be the values of θ and z sampled in the t-th iteration. Then in the $(t + 1)$-st iteration, it goes through the following two steps:

Imputation step: Sample $z_{t+1} \sim p(z|y, \theta_t)$.
Posterior step: Sample $\theta_{t+1} \sim p(\theta|y, z_{t+1})$.

$p(z|y, \theta)$ is the predictive distribution for imputing the missing data z given y and θ. $p(\theta|y, z)$ is the posterior distribution of the complete-data model. The data augmentation algorithm capitalizes on the fact that both the $p(z|y, \theta)$ in the imputation step and $p(\theta|y, z)$ in the posterior step are often easy to sample from.

In correspondence to the EM algorithm, the imputation step corresponds to the E-step, and the posterior step corresponds to the M-step. The data augmentation algorithm can also be viewed as a two-component Gibbs sampler for sampling from $p(z, \theta|y)$, but the emphasis of the data augmentation algorithm is that it augments the missing data or latent variable z to simplify the computation. In that sense, it is related to the auxiliary variable algorithm [3], the most prominent example being the Swendsen-Wang algorithm [9].

The rate of convergence of the data augmentation algorithm is determined by a quantity called Bayesian fraction of missing information [5]. It is the Bayesian version of the fraction of missing information that determines the rate of convergence of the EM algorithm [1].

Meng and van Dyk [7] observed that for the same marginal model $p(y|\theta)$, it is possible to construct a class of complete-data models $p_a(y, z|\theta)$ indexed by a working parameter a so that $\int p_a(y, z|\theta)dz = p(y|\theta)$ for all a. One may then optimize the fraction of missing information over the working parameter a in order to

K. Ikeuchi (ed.), *Computer Vision*, DOI 10.1007/978-0-387-31439-6,

devise the EM or data augmentation algorithm with the optimal rate of convergence.

Inspired by Meng and van Dyk [7], Liu, Rubin and Wu observed that when augmenting the data from y to (y, z), it is possible to expand the parameter θ to θ, α so that the complete model becomes $p(y, z|\theta, \alpha)$, where both θ and α are identifiable in the complete-data model, but $\int p(y, z|\theta, \alpha)dz = p(y|\theta)$ so that the parameter α disappears in the observed-data model [6]. Based on this observation, they proposed a parameter-expanded EM (PX-EM) algorithm which has faster convergence rate than the original EM algorithm. Liu and Wu (2000) [4] proposed a parameter-expanded data augmentation (PX-DA) algorithm which has a faster convergence rate than the original data augmentation algorithm. Independently, Meng and van Dyk [8] proposed a similar algorithm called marginal augmentation.

Application

The data augmentation algorithm and its extensions can be used for sampling of posterior distributions from a wide range of Baysian models. Practically for any EM algorithm for maximum likelihood estimation, there is a corresponding data augmentation algorithm for posterior sampling. In fact, the imputation step of the data augmentation algorithm can be easier to implement than the E-step of the EM algorithm because it is often easier to sample from a distribution than calculating the expectation in closed form.

References

1. Dempster AP, Laird NM, Rubin DB (1977) Maximum likelihood from incomplete data via the EM algorithm. J R Stat Soc B 39:1–38
2. Geman S, Geman D (1984) Stochastic relaxation, Gibbs distributions, and the Bayesian restoration of images. IEEE Trans Pattern Anal Mach Intell 6:721–741
3. Higdon DM (1998) Auxiliary variable methods for Markov chain Monte Carlo with applications. J Am Stat Assoc 93:585–595
4. Liu JS, Wu YN (1999) Parameter expansion for data augmentation. J Am Stat Assoc 94(448):1264–1274
5. Liu JS, Wong WH, Kong A (1994) Covariance structure of the Gibbs sampler with applications to the comparisons of estimators and augmentation schemes. Biometrika 81:27–40
6. Liu C, Rubin DB, Wu YN (1998) Parameter expansion to accelerate EM: the PX-EM algorithm. Biometrika 85(4): 755–770
7. Meng XL, van Dyk D (1997) The EM algorithm – an old folk-song sung to a fast new tune. J R Stat Soc B 59:511–567
8. Meng XL, van Dyk D (1999) Seeking efficient data augmentation schemes via conditional and marginal augmentation. Biometrika 86:301–320
9. Swendsen RH, Wang J (1987) Nonuniversal critical dynamics in Monte Carlo simulations. Phys Rev Lett 58:86–88
10. Tanner MA, Wong WH (1987) The calculation of posterior distributions by data augmentation. J Am Stat Assoc 82: 528–540

Data Fusion

Ramanarayanan Viswanathan
Department of Electrical Engineering, University of Mississippi, MS, USA

Synonyms

Information fusion

Definition

Data fusion refers to combining data from multiple sources for achieving better understanding of a phenomenon of interest. Applications abound in engineering and applied sciences, including wireless sensor networks, computer vision, and biometrics.

Background

In several fields, combining different sets of information have taken place, although a more systematic study for the fusion of data is emerging since a decade [1]. The human brain is an example of a complex system which integrates data or signals from different sensory preceptors in the body. Building a machine-based system that can meaningfully integrate data from different sources for better understanding of a phenomenon of interest is the challenge faced in many fields. Since data emerges from different sensors with varying accuracy and coverage factor, benefits of data fusion include improved system reliability and/or redundancy, extended coverage, and possible shorter response time. Applications in human-machine interface area include robots in industrial automation, surveillance in military and commercial fields, battlefield management, medical diagnostics, biometrics, satellite imaging, image understanding, computer vision, target detection and tracking, wireless sensor networks, and wireless communications.

Application

In many vision-related applications, a single sensor will not provide complete information with respect to the scene that is being sensed. Also, sensors may have different fields of views and different sensors, such as IR and optical, may have different resolution capabilities. Better results are obtained when data from sensors are combined in an appropriate manner. For imaging applications, fusion of images can be done at various levels, viz., pixel, feature, and decision. In general, the latter two entail some loss of information as the fusion is performed after extraction of information from the original images. Pixel-level fusion methods include Laplacian pyramid, discrete wavelet transforms, and support vector machines [2]. In decision/classification systems, combining features or decisions may be appropriate when sensors are geographically dispersed, thereby requiring distributed processing at sensor sites. In video surveillance applications involving tracking of objects or persons of interest, fusion of data from multiple images would be needed [3]. Similar situations arise in military applications involving tracking of maneuvering enemy aircrafts. Track-to-track fusion of target position estimates derived from individual sensor data is a possible approach [4, 5].

Detection and estimation of targets using a set of geographically dispersed multiple sensors necessitate distributed signal processing at sensor sites [6, 7]. Although individual sensors process sensed information and possibly make inference regarding the presence or absence of a target (or a phenomenon of interest), a final determination that is based on the collective information is usually made at a central site called the fusion center. It is conceivable that, in some applications, one of the sensors could be the fusion center making the final decision. Thus, the traditional signal detection (estimation) paradigm is naturally extended to the situation of distributed processing. The terminology "distributed detection," "decentralized detection," or "distributed decision fusion" refers to such situations. Depending on the flow of information from the sensors to the fusion center and between the sensors, several configurations of sensor suites are possible: serial or tandem, and parallel, tree [8]. Formulation of Neyman-Pearson and Bayes optimization criteria leads to fundamental theoretical solutions in this area. Asymptotic solutions involving a very large number of sensors as well as computational approaches for obtaining optimal solutions are available [6–9]. Recently, there is interest in the estimation of parameters using processed data from multiple sensors [10].

In earlier works on distributed detection problems, the communications channels between the sensors and the fusion center were assumed ideal or error-free. However, with the pervasiveness of wireless sensor networks, the assumption of error-free links is not quite true. The inclusion of error-prone links in the distributed signal processing systems has further broadened the literature to provide practically achievable as well as theoretically possible solutions [11, 12]. Channel-aware solutions paradigm leads to several suboptimal solutions depending on the availability of knowledge of channel information. Although some similarity to traditional diversity reception in signal communication exists, the decentralized detection in wireless sensor network is distinctly different in the sense that the decisions made by the sensors regarding a phenomenon of interest need not all be identical unlike the case of diversity reception. Thus, the maximal ratio combining, which is optimal for diversity reception, is generally inferior in wireless sensor network case. Depending on the noisiness of the channel, combining based on individual sensor decisions arrived at the fusion center can outperform other suboptimal combiners based on received sensor data at the fusion center.

Cognitive radio has evolved over the last two decades. Cognitive radio now could be a simple receiver which senses and adapts to different signal modulations to a more complex modem where the radio is able to sense the non-presence of a user in a spectral frequency range, thereby adapting its operating frequency to be in that range for achieving signal connectivity [13]. The problem faced by a secondary user is the detection of the presence or absence of a primary licensed-user through spectrum sensing. The spectrum sensing is done cooperatively by several secondary users in order to opportunistically access the spectrum, when it becomes available. Presence of distributed radios, inability of some radios not being able to sense the primary user due to shadowing or "hidden terminal" problem, limitations on communication capacity between the secondary radios and the radio acting as the fusion center, and availability of signal processing capability within the radios have set the stage for the application of decentralized detection concepts to cognitive radios [14, 15].

Biometrics is used for authentication purposes in civilian and military applications. Biometrics for person identification could be fingerprints, facial image, voice, or iris. Use of several biometrics, when appropriately combined, can lead to better results than that can be obtained through any one biometrics. Also, multiple samples of any one biometrics can be combined for achieving better results. Combining data from multimodal biometrics is a challenging problem because of the need for proper normalization of biometrics scores before combining [16, 17]. The determination of optimal combining method needs to take into account possible correlation among different biometrics. Challenges exist due to stringent requirements in surveillance applications where false accept rates and false reject rates need to be kept small. Moreover, biometrics identification systems need to be robust to the extent possible due to possible fake biometrics posed by latex fingers, face masks, etc. The fusion of biometric can be at various levels: data, features, and classifiers (decisions).

References

1. Varshney PK (1997) Mutisensor data fusion. Electron Commun Eng J 9(12):245–253
2. Zheng S, Shi W-Z, Liu J, Zhu G-X, Tian J-W (2007) Multisource image fusion method using support value transform. IEEE Trans Image Process 16(7):1831–1839
3. Snidaro L, Niu R, Foresti GL, Varshney PK (2007) Quality-based fusion of multiple video sensors for video surveillance. IEEE Trans Syst Man Cybern Part B Cybern 37(4):1044–1051
4. Hall DL, Llinas J (1997) An introduction to multisensor data fusion. Proc IEEE 85(1):6–23
5. Chen H, Kirubarajan T, Bar-shalom Y (2003) Performance limits of track-to-track fusion versus centralized estimation: theory and application. IEEE Trans Aerosp Electron Syst 39(2):386–399
6. Viswanathan R, Varshney PK (1997) Distributed detection with multiple sensors: part I-fundamentals (invited paper). Proc IEEE 85(1):54–63
7. Blum RS, Kassam SA, Poor HV (1997) Distributed detection with multiple sensors: part II-advanced topics (invited paper). Proc IEEE 85(1):64–79
8. Dasarathy BV (1994) Decision fusion. IEEE Computer Society Press, Los Alamitos
9. Tay PW, Tsitsiklis JN, Win MZ (2008) On the subexponential decay of detection error probabilities in long tandems. IEEE Trans Inf Theory 54(10):4767–4771
10. Ribeiro A, Giannakis GB (2006) Bandwidth-constrained distributed estimation for wireless sensor networks- Part I: Gaussian case. IEEE Trans Signal Process 54(3):1131–1143
11. Chamberland J-F, Veeravalli VV (2007) Wireless sensors in distributed detection applications. IEEE Signal Process Mag 24(3):16–25
12. Chen B, Jiang R, Kasetkesam T, Varshney PK (2004) Channel aware decision fusion in wireless sensor networks. IEEE Trans Signal Process 52(12):3454–3458
13. Gandetto M, Regazzoni C (2007) Spectrum sensing: a distributed approach for cognitive terminals. IEEE J Sel Areas Commun 25(3):546–557
14. Unnikrishnan J, Veeravalli VV (2008) Cooperative sensing for primary detection in cognitive radio. IEEE J Sel Top Signal Process 2(1):18–27
15. Letaief KB, Zhang W (2009) Cooperative communications for cognitive radio networks. Proc IEEE 97(5):878–893
16. Jain AK, Chellappa R, Draper SC, Memon N, Phillips PJ, Vetro A (2007) Signal processing for biometric systems (DSP forum). IEEE Signal Process Mag 24(6):146–152
17. Basak J, Kate K, Tyagi V, Ratha N (2010) QPLC: a novel multimodal biometric score fusion method. In: Computer Vision and Pattern recognition Workshops (CVPRW), San Francisco, 2010. IEEE Computer Society Conference, pp 46–52

Deblurring

Yu-Wing Tai
Department of Computer Science, Korea Advanced Institute of Science and Technology (KAIST), Yuseong-gu, Daejeon, South Korea

Synonyms

Deconvolution

Related Concepts

►Blind Deconvolution; ►Blur Estimation; ►Image Enhancement and Restoration; ►Motion Blur

Definition

Deblurring is a process to recover a sharp and clear image from a blurry image which suffered from either the motion blur or the defocus blur effects.

Background

When taking a photo with long exposure time, or with wrong focal length, the captured image will look blurry. This is because during the exposure period, the lights captured for a pixel are mixed with the lights captured for the other pixels within a local neighborhood. Such effect is modeled by the point spread function (a.k.a. blur kernel) which describes how the lights are mixed during the exposure period. The aim of deblurring is to reverse the blur process to recover a sharp and clear image of the scene from the captured blurring image. Deblurring, however, is a severely ill-posed problem because the number of unknowns exceeds the number of equations that can be derived from the observed data.

The problem of deblurring can be further categorized into non-blind deblurring and blind deblurring. In non-blind deblurring, the point spread function is given and is used to recover the latent image from a blurry image. In blind deblurring, the point spread function is unknown and additional process is needed to estimate point spread function. However, the process to estimate point spread function usually requires information of the latent image, which is also unknown. A typical approach in blind deblurring consists of two interdependent steps: point spread function estimation and latent image restoration. These two steps are solved alternatingly and iteratively, and the solution of blind deblurring is only guaranteed to converge to a local minimum.

Theory

Representation

Mathematically, the effects of blurriness can be described by the following equation:

$$B(x, y) = \sum_{(m,n)\in\mathcal{N}(x,y)} I(x-m, y-n)k_{(x,y)}(m,n) \tag{1}$$

where B is the blurry image, I is the sharp image, $k_{(x,y)}$ is the point spread function (PSF) at (x, y), and $\mathcal{N}$ is the local neighborhood of a pixel respectively. In general, the point spread function $k_{(x,y)}$ is spatially varying. This means that every pixel can carry different point spread function. This happens when the scene has large depth disparity (for defocus blur) or when the scene contains a moving object (for motion blur) or when the camera exhibits rotational motion (for motion blur) during exposure period.

While the point spread function is spatially varying, the variation of the point spread function is spatially smooth. To simplify the problem, in case of the camera motion, many previous works have assumed that the point spread function is spatially invariant, which reduce (1) into a convolution equation:

$$B = I \otimes k \tag{2}$$

where $\otimes$ is the convolution operator. Recently, [1, 2] generalized (2) and proposed the projective motion blur model which uses a sequence of homographies to model the camera motion:

$$B(x, y) = \frac{1}{N}\sum_{i=1}^{N} I(H_i[x \;\; y \;\; 1]^t) \tag{3}$$

where H_i is the homography, $[x \;\; y \;\; 1]$ is the homogeneous coordinate of a pixel at (x, y), and N is the number of homography used to approximate the camera motion in discrete domain. Note that when there is image noise, an additional noise variable will be appended in (1)–(3) under the assumption that noise is addition and is independent to the blur process.

Methodology

Image deblurring is a long-standing problem. Many existing works targeting image blur due to camera shake motion has assumed a globally uniform point spread function which can be described by (2). Traditional non-blind deblurring algorithms include the well-known Richardson-Lucy algorithm [3, 4] and Wiener filter [5] (Refers to the entry on Richardson-Lucy Deconvolution and Wiener Filter for details.).

Due to poor blur kernel estimation or unrecoverable frequency loss from convolution, undesirable artifacts such as ringing and amplification of image noise can be introduced in deblurred results. The state-of-the-art approaches deal with these artifacts by including different image priors to regularize the point spread function estimation process, the latent image restoration process, or both. Representative techniques include [6–14]. For example, Chan and Wong [6] utilized total variation regularization to help ameliorate ringing and noise artifacts. Fergus et al. [7] demonstrated how to use a variational Bayesian approach combined with gradient-domain statistics to estimate a more accurate PSF. Raskar et al. [15, 16] coded the exposure to make the PSF more suitable for deconvolution. Jia [17] demonstrated how to use an object's alpha matte to better compute the PSF. Levin et al. [8] introduced a gradient sparsity prior to regularize results for images exhibiting defocus blur. This prior is also applicable to motion blurred images. Yuan et al. [11] proposed a multiscale approach to progressively recover blurred details while Shan et al. [12] introduced regularization based on higher-order partial derivatives to reduce image artifacts. Cho and Lee [13] proposed edge prediction step with GPU implementation to achieve almost real-time deblurring algorithm. Xu and Jia [14] further studied the scale of edges and proposed the edge selection step with the TV-l_1 deconvolution algorithm. A study on the blind motion deblurring algorithms for uniform point spread function can be found in [18].

As mentioned in [18], camera ego motion causes a spatially varying motion blur that cannot be accurately modeled with a uniform PSF. Prior work has recognized the need to handle nonuniform motion blur for camera ego motion, moving objects, and defocus blur. For example, early work by Sawchuk [19] addressed motion blur from a moving camera by first using a log-polar transform to transform the image such that the blur could be expressed as a spatially invariant PSF. The range of motion that could be addressed was limited to rotation and translation. Similarly, Shan et al. [20] handled spatially varying blur by restricting the relative motion between the camera and the scene to be a global in-plane rotation. Dai and Wu [21] used alpha matte to estimate spatially varying motion PSF.

When addressing moving objects, the input image can be segmented into multiple regions each with a constant PSF as demonstrated by Levin [22], Bardsley et al. [23], Cho et al. [24], and Li et al. [25]. Such segmented regions, however, should be small to make the constant PSF assumption valid for the spatially varying motion blur in camera shake motion. For example, Tai et al. [26] extended the hybrid camera framework used by Ben-Ezra and Nayar [27] to estimate a PSF per pixel using an auxiliary video camera.

Some recent techniques have tried to estimate the motion point spread function using the representation in (3) for deblurring. Tai et al. [28] use coded exposure with user scribbles to estimate the point spread function through homography alignment; Li et al. [29] use video sequences and constraint the transformation from one frame to another frame following a global homography transformation and Whyte et al. [2] limited the space of parameters with only rotational motions and use the variational Bayesian framework proposed by Fergus [7] to estimate the point spread function. Joshi et al. [30] directly measure the point spread function of the camera motion using motion inertia sensor. Detailed deblurring algorithm for the representation in (3) can be found in [1].

While there have been a lot of works targeting deblurring, this problem is still an open problem. As mentioned previously, solving (1–3) is mathematically ill-posed. In addition, many common photographic effects, such as saturation, noise, nonlinear camera response function and compression artifacts, were not probably modeled in (1–3) which limits the practicality of existing deblurring algorithms. There are also studies, such as [31], on the trade-off between deblurring and denoising in high dynamic range imaging.

Application

The main application of deblurring is on image restoration and enhancement. Since motion blur and defocus blur are common artifacts in imaging system, its applications range from astronomy telescope, to satellite imaging, to medical imaging, and to common customer level camera. A method to remove motion blur/defocus blur from a captured photograph is valuable for digital photography. Not only the blurred image information can be revealed, but the photos which were taken at important moment can also be recovered.

References

1. Tai YW, Tan P, Brown M (2011) Richardson-lucy deblurring for scenes under a projective motion path. IEEE Trans Pattern Anal Mach Intell 33(8):1603–1618
2. Whyte O, Sivic J, Zisserman A, Ponce J (2010) Non-uniform deblurring for shaken images. IEEE conference on computer vision pattern recognition (CVPR), San Francisco, California
3. Richardson W (1972) Bayesian-based iterative method of image restoration. J Opt Soc Am 62(1):55–59
4. Lucy L (1974) An iterative technique for the rectification of observed distributions. Astron J 79(6):745–754

5. Wiener N (1949) Extrapolation, interpolation, and smoothing of stationary time series. Wiley, New York
6. Chan TF, Wong CK (1998) Total variation blind deconvolution. IEEE Trans Image Process 7(3):370–375
7. Fergus R, Singh B, Hertzmann A, Roweis ST, Freeman WT (2006) Removing camera shake from a single photograph. ACM Trans Graph 25(3):787–794
8. Levin A, Fergus R, Durand F, Freeman WT (2007) Image and depth from a conventional camera with a coded aperture. ACM Trans Graph 26(3):70
9. Yuan L, Sun J, Quan L, Shum H (2007) Image deblurring with blurred/noisy image pairs. ACM Trans Graph 26(3):1
10. Chen J, Tang CK (2008) Robust dual motion deblurring. IEEE conference on computer vision pattern recognition (CVPR), Anchorage, Alaska
11. Yuan L, Sun J, Quan L, Shum HY (2008) Progressive inter-scale and intra-scale non-blind image deconvolution. ACM Trans Graph 27(3):74
12. Shan Q, Jia J, Agarwala A (2008) High-quality motion deblurring from a single image. ACM Trans Graph 27(3):73
13. Cho S, Lee S (2009) Fast motion deblurring. ACM SIGGRAPH Asia 28(5):145
14. Xu L, Jia J (2010) Two-phase kernel estimation for robust motion deblurring. European conference on computer vision (ECCV), Crete, Greece
15. Raskar R, Agrawal A, Tumblin J (2006) Coded exposure photography: motion deblurring using fluttered shutter. ACM Trans Graph **25**(3):795–804
16. Agrawal A, Raskar R (2007) Resolving objects at higher resolution from a single motion-blurred image. IEEE conference on computer vision pattern recognition (CVPR), Minneapolis, Minnesota
17. Jia J (2007) Single image motion deblurring using transparency. IEEE conference on computer vision pattern recognition (CVPR), Minneapolis, Minnesota
18. Levin A, Weiss Y, Durand F, Freeman W (2009) Understanding and evaluating blind deconvolution algorithms. IEEE conference on computer vision pattern recognition (CVPR), Miami, Florida
19. Sawchuk AA (1974) Space-variant image restoration by coordinate transformations. J Opt Soc Am 64(2):138–144
20. Shan Q, Xiong W, Jia J (2007) Rotational motion deblurring of a rigid object from a single image. ICCV, Rio de Janeiro, Brazil
21. Dai S, Wu Y (2008) Motion from blur. IEEE conference on computer vision pattern recognition (CVPR), Anchorage, Alaska
22. Levin A (2006) Blind motion deblurring using image statistics. NIPS. pp 841–848
23. Bardsley J, Jefferies S, Nagy J, Plemmons R (2006) Blind iterative restoration of images with spatially-varying blur. Opt Express 1767–1782
24. Cho S, Matsushita Y, Lee S (2007) Removing non-uniform motion blur from images. ICCV, Rio de Janeiro, Brazil
25. Li F, Yu J, Chai J (2008) A hybrid camera for motion deblurring and depth map super-resolution. IEEE conference on computer vision pattern recognition (CVPR), Anchorage, Alaska
26. Tai YW, Du H, Brown MS, Lin S (2010) Correction of spatially varying image and video motion blur using a hybrid camera. IEEE Trans Pattern Anal Mach Intell 32(6): 1012–1028
27. Ben-Ezra M, Nayar S (2004) Motion-based motion Deblurring. IEEE Trans PAMI 26(6):689–698
28. Tai Y, Kong N, Lin S, Shin S (2010) Coded exposure imaging for projective motion deblurring. IEEE conference on computer vision pattern recognition (CVPR), San Francisco, California
29. Li Y, Kang SB, Joshi N, Seitz S, Huttenlocher D (2010) Generating sharp panoramas from motion-blurred videos. IEEE conference on computer vision pattern recognition (CVPR), San Francisco, California
30. Joshi N, Kang S, Zitnick L, Szeliski R (2010) Image deblurring with inertial measurement sensors. ACM Trans Graph 29(3):30
31. Zhang L, Deshpande A, Chen X (2010) Denoising vs. deblurring: Hdr imaging techniques using moving cameras. IEEE conference on computer vision pattern recognition (CVPR), San Francisco, California

Deconvolution

▸Blind Deconvolution
▸Deblurring

Defocus Blur

Neel Joshi
Microsoft Corporation, Redmond, WA, USA

Synonyms

Out of focus blur

Related Concepts

▸Blur Estimation; ▸Motion Blur

Definition

Defocus blur is a loss of sharpness that occurs due to integrating light over an aperture with a nonzero area, where the light source is off of the image focal plane. The amount of blur that is visible in an image is a function of the lens aperture, the object and focal depth, and the camera pixel (or grain) size.

Background

Image blur can be described by a point spread function (PSF). A PSF models how an imaging system captures a single point in the world – it literally describes how a point spreads across an image. An entire image is then made up of a sum of the individual images of every scene point, where each point's image is affected by the PSF associated with that point. For an image to be "in focus" means that one ideally does not want any image blur at a particular depth of the scene. Thus, the PSF should be minimal, i.e., a delta function, where each scene point should correspond only to one image point. In practice, PSFs can take on a range of shapes and sizes depending on the properties of an imaging system. When this PSF is large relative to image resolution, an image with defocus blur is captured.

The fundamental cause of defocus blur is that a camera does not point-sample a scene, but instead captures images by integrating the light over fixed areas. Defocus blur is a function of aperture size and the relative position between camera and objects in the imaged scene. For very small apertures or "pinhole"-sized apertures, the blur or PSF can be insignificant relative to the resolution or pixel size of the camera. However, in light-limited situations or when a particular photographic affect is desired, such as shallow depth of field in a portrait photo, larger apertures are used. With a large aperture, the depth of field is shallow, and the PSF for objects off the focal plane is larger. The amount of blur is depth dependent: it depends on the focal length of the lens and the focal depth, and it grows with distance from the focal plane, as illustrated in Fig. 1.

Theory

Image blur is described by a point spread function (PSF). The PSF models how an imaging system captures a single point in the world.

The most commonly used model for blur is the a linear model, where the blurred image b is represented as a convolution of a kernel k, plus noise:

$$b = i \otimes k + n, \tag{1}$$

where $n \sim \mathcal{N}(0, \sigma^2)$, which represents and additive Gaussian noise model. In this model, the blur is assumed to be constant over the entire image, i.e., spatially invarient; however, that is often not true in practice [1, 2]. If there is depth variation in the scene, the blur will change with that depth. Similarly, in many lenses, the shape of the blur changes across the image plane, as illustrated in Fig. 1. In both of these cases, one can think of the blur kernel as being a function of image position, i.e., $k(x, y)$.

To model spatially varying blur, the spatially invariant kernel and convolution in (Eq. 1) can be replaced by a sparse re-sampling matrix that models the spatially variant blur, and the convolution process is now a matrix-vector product:

$$\boldsymbol{b} = A\boldsymbol{i} + \boldsymbol{i}. \tag{2}$$

Each column of A is the unraveled kernel for the pixel represented by that column. Thus, the blurred response at a pixel in the observed image is computed as a weighted sum, as governed by A, of the latent sharp image i formed into a column vector.

Representation

There are several common assumptions made on the form of a defocus blur kernel:

- The PSF is positive, i.e., all values in the kernel are nonnegative.
- The PSF is energy conserving, i.e., $\sum_i k_i = 1$.
- The PSF is symmetric – radially or along some cartesian axis.
- The PSF has a known parametric form.

The assumptions are listed in order from least to most restrictive. Positivity is a strong constraint, and the least restrictive in that does not eliminate any truly valid kernels, i.e., no true blur kernel can have negative values as blurring in is a purely additive process. Another way of thinking of this is that there is no "negative" light. Similarly the second constraint is equally not restrictive in that blurring does not remove light; thus, all true blur kernels should be energy conserving. Thus, the assumptions of positivity and energy conservation are ones that can be used by virtually all PSF models. In practice, whether a particular model uses them depends on the nature of the other assumptions.

The second two assumptions are much more restrictive. Symmetry is typically used when one wants to generalize a 2D PSF from some 1D cross section [3]. Assuming a parametric form is also very restrictive as

Defocus Blur, Fig. 1 With defocus blur, the amount of blur is depth dependent; it depends on the focal length of the lens and the focal depth, and it grows with distance from the focal plane. An example of defocus blur is shown in the *middle* (From [12]), and spatially varying defocus kernels (From [1]) are shown on the *right*

it assumes the entire shape of the blur kernel can be modeled by a low parameter mathematical model.

Defocus blur has two commonly used parametric models. A circular disk or "pillbox" function [4]:

$$k(x, y) = \begin{cases} 0 & \sqrt{x^2 + y^2} > r \\ \frac{1}{\pi r^2} & \sqrt{x^2 + y^2} \leq r \end{cases} \tag{3}$$

and a circularly symmetric 2D Gaussian [5]:

$$K(x, y) = C exp\left(-\frac{x^2 + y^2}{2\sigma^2}\right), \tag{4}$$

where C is a normalization constant. In both cases, a single parameter determines the PSF – r for the pillbox and σ for the 2D Gaussian.

Application

Estimation of defocus blur constitutes an extensively researched area [4, 6–9]. Estimated blur kernels are typically used for improving image quality by reducing blur using image deblurring and deconvolution methods [1, 10–12]. Models of defocus blur are also used in depth estimation methods such as depth from defocus [11, 13, 14].

References

1. Joshi N, Szeliski R, Kriegman DJ (2008) Psf estimation using sharp edge prediction. In: Computer vision and pattern recognition 2008. IEEE conference on computer vision pattern recognition (CVPR 2008). IEEE conference on, Anchorage, pp 1–8
2. Levin A, Weiss Y, Durand F, Freeman W (2009) Understanding and evaluating blind deconvolution algorithms. In: Computer vision and pattern recognition 2009. IEEE conference on computer vision pattern recognition (CVPR 2009). IEEE conference on, Miami (Beach), IEEE Computer Society, pp 1964–1971
3. Yoon J, Shin J, Paik JK (2001) Enhancement of out-of-focus images using fusion-based psf estimation and restoration. In: VCIP, San Jose, pp 819–829
4. Cannon M (1976) Blind deconvolution of spatially invariant image blurs with phase. IEEE Trans Acoust Speech Signal Process 24(1):58–63 (see also IEEE Trans Signal Process)
5. Banham MR, Katsaggelos AK (1997) Digital image restoration. IEEE Signal Process Mag 14(2):24–41
6. Gennery DB (1973) Determination of optical transfer function by inspection of frequency-domain plot. J Opt Soc Am 63(12):1571
7. Chang MM, Tekalp AEA (1991) Blur identification using the bispectrum. IEEE Trans Signal Process 39(10):2323–2325 (see also Acoust Speech Signal Process IEEE Trans)
8. Savakis A, Trussell H (1993) Blur identification by residual spectral matching. IEEE Trans Image Process 2(2):141–151
9. Rooms F, Philips WR, Portilla J (2004) Parametric PSF estimation via sparseness maximization in the wavelet domain. In: Truchetet F, Laligant O (eds) Wavelet applications in industrial processing II. Proceedings of the SPIE, Vol 5607, SPIE, Bellingham, pp 26–33. Volume 5607 of presented at the society of photo-optical instrumentation engineers (SPIE) conference, pp 26–33
10. Richardson WH (1972) Bayesian-based iterative method of image restoration. J Opt Soc Am (1917–1983) 62:55–59
11. Levin A, Fergus R, Durand F, Freeman WT (2007) Image and depth from a conventional camera with a coded aperture. In: SIGGRAPH '07: ACM SIGGRAPH 2007 papers, ACM, New York, p 70
12. Joshi N, Matusik W, Adelson EH, Kriegman DJ (2010) Personal photo enhancement using example images. ACM Trans Graph 29:12:1–12:15
13. Nayar S, Nakagawa Y (1990) Shape from focus: an effective approach for rough surfaces. In: International conference on robotics and automation, vol 1, Cincinnati, pp 218–225
14. Nayar S, Nakagawa Y (1994) Shape from focus. IEEE Trans Pattern Anal Mach Intell 16(8):824–831

Dehazing

►Descattering

Dehazing and Defogging

Robby T. Tan
Department of Information and Computing Sciences, Utrecht University, Utrecht, CH, The Netherlands

Synonyms

Visibility enhancement in bad weather

Related Concepts

►Descattering

Definition

Dehazing is a process to visually improve degraded visibility caused by atmospheric conditions where the horizontal visibility on the ground level is greater than 1 km. Defogging is a similar process yet focusing on fog, which, unlike haze, is a cloud of water droplets near ground level reducing the horizontal visibility to less than 1 km. Both dehazing and defogging are part of algorithms to enhance visibility in bad weather due to light being scattered and absorbed by atmospheric particles.

Background

Poor visibility in outdoor scenes generates significant problems for many applications of computer vision. Most automatic systems for surveillance, intelligent vehicles, object recognition, etc., assume the input images have clear visibility. Unfortunately, this assumption does not always hold, particularly in bad weather. Therefore, improving the degraded visibility is practically important.

In outdoor scenes, poor visibility is caused by the substantial presence of various atmospheric particles that have significant density in the participating medium (creating haze, fog, mist, smoke, dust, etc.). Light from the atmosphere and light reflected from objects are absorbed and scattered by those particles, causing the visibility of a scene to be degraded. Physically, fog is a cloud of water droplets near ground level that reduces the horizontal visibility to less than 1 km [1]. According to [2], the particle radius of the water droplets is between 1 and 10 μm with particle concentration between 100 and 10 cm^{-3}. Unlike fog, haze is not water droplets. It is composed of dry particles (of dust or salt) so small that they cannot be felt or seen individually with the naked eye, but the aggregate reduces horizontal visibility and gives the atmosphere an opalescent appearance (a bluish or yellowish veil depending on whether the background is dark or light, respectively) [1]. In hazy scenes, the horizontal visibility on the ground level is greater than 1 km but typically less than 5 km [3]. The particle radius is between 10^{-2} and 1 μm with particle concentration between 10^3 and 10 cm^{-3} [2].

Aside from fog and haze, there is also mist. In terms of visibility, mist is similar to haze (the horizontal visibility on the ground level is greater than 1 km but less than 5 km); however, the particle type is similar to that of fog. With respect to the opalescent appearance, mist can be discriminated from haze, since it gives a grayish cast to the sky [1].

Theory

In the field of optics and atmospheric sciences, there are a number of scattering models depending on the size of the particles and the complexity of the formulas. Most of them are derived from radiative transfer, i.e., the physical phenomenon of energy transfer in the form of electromagnetic radiation, which is affected by absorption, emission, and scattering processes [4]. Most of these models are rather complex. Hence, an approximated optical model for both haze and fog is more commonly used in computer vision (e.g., [5–7]), which is based on Lambert-Beer Law [8] and Koschmieder's equation [9]:

$$\mathbf{I}(x) = \mathbf{L}_1(x)\boldsymbol{\rho}(x)e^{-\beta d(x)} + \mathbf{L}_2(x)(1 - e^{-\beta d(x)}). \tag{1}$$

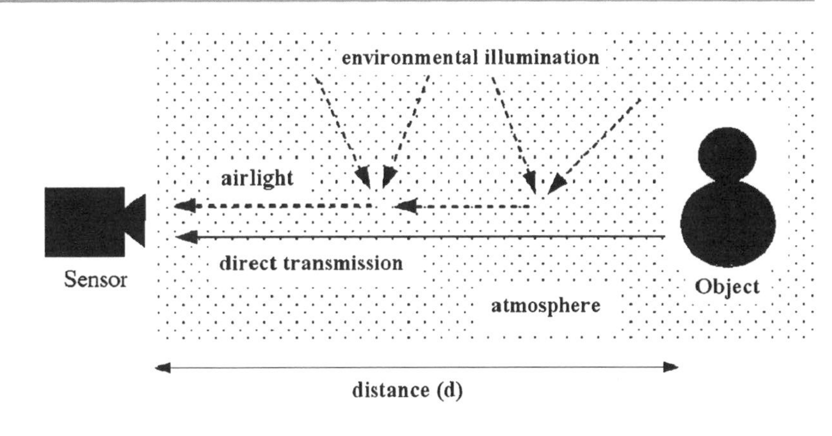

Dehazing and Defogging, Fig. 1 The pictorial description of the optical model in Eq. (1)

Dehazing and Defogging, Fig. 2 *Left*: an image plagued by atmospheric particles. *Right*: the result of enhancing visibility using the method introduced in Tan [16]

The first term is the direct attenuation (**D**), and the second term is the airlight (**A**):

$$\mathbf{D}(x) = \mathbf{L}_1(x)\rho(x)e^{-\beta d(x)} \quad (2)$$

$$\mathbf{A}(x) = \mathbf{L}_2(x)(1 - e^{-\beta d(x)}), \quad (3)$$

where **I** is the image intensity, x is the 2D spatial location, and ρ is the reflectance of an object that appears in the image. Here $\mathbf{L}_1$ is the irradiance on the object, while $\mathbf{L}_2$ is the radiance scattered by the medium to the camera without reaching the objects. A further assumption sometimes made is that $\mathbf{L}_1 = \mathbf{L}_2$ and both of them represent the intensity of the infinitely distant light sources and thus are assumed to be globally constant. The parameter β is the atmospheric attenuation coefficient, while d is the distance between an object in the image and the camera. Note that $\mathbf{I}$, $\mathbf{L}_1$, $\mathbf{L}_2$, $\boldsymbol{\rho}$ in the equation are color dependent.

Equation (2) is in principle the Lambert-Beer law [8] for transparent objects. Equation (1) is basically Koschmieder's equation [9] that takes into account only single scattering towards the camera. Consequently, the model fails to capture the effects of highly scattering media. In terms of degree of polarization (DOP), the airlight $\mathbf{A}$ is often partially polarized. Hence, two polarization components of $\mathbf{A}$ can be obtained using a polarizing filter attached to the camera. These components are parallel and perpendicular to the plane of incidence [6, 10, 11]. Figure 1 illustrates the model.

Problem Definition

Based on Eq. (1), the problem can be generally described as follows: Given an image or a sequence of images whose intensity is represented by $\mathbf{I}(x)$, estimate $\mathbf{R}(x)$ for every pixel, where $\mathbf{R}(x) = \mathbf{L}_1\rho(x)$ is the scene radiance had there not been particles along the line of sight.

Possible Assumptions and Solutions

For a single image, the problem described above is ill-posed. In the literature, a few approaches trying to tackle the problem have been proposed. The first approach is based on polarizing filters (e.g., [6, 10, 11]). The main idea of this approach is to exploit two or more images of the same scene obtained by rotating a polarizing filter attached to the camera. Having at least two different intensity values of each pixel of the scene, [11] transforms the problem into a form solvable using ICA. The main assumption is that the scene is static while the filter rotates.

Another approach is based on multiple images of a scene taken in different weather conditions, i.e., with different properties of the participating medium (e.g., [5, 7]). By deriving two equations from Eq. (1), which represent the two input images, [7] can obtain the absolute values of airlight by assuming the presence of a totally dark object in the scene.

Different approaches proposed by Tan and Oakley [12], Narasimhan and Nayar [13], Hautière et al. [14], and Kopf et al. [15] are based on a single image, yet they require geometric information about the input scene. This requirement might be impractical, and thus methods in [16–20] use a single image and rely on various statistical image priors. Many methods [5, 7, 16–19] assume that β in Eq. (1) is constant for across the light spectrum. This assumption is reasonable in fog (but away from fogbows and rainbows) [2]. Figure 2 shows the result of using the method of [16].

Open Problems

Despite the development in the recent years, one problem still remains, i.e., the inaccuracy of the model in Eq. (1) with regard to multiple scattering. The model is a single scattering model and thus cannot capture significant scattering effects (e.g., blur) that can occur particularly in dense foggy scenes.

References

1. Encyclopedia-britannica. http://www.britannica.com
2. McCartney EJ (1976) Optics of the atmosphere: scattering by molecules and particles. Wiley, New York
3. Kumar B, De Remer D, Marshall DM (2005) An illustrated dictionary of aviation. McGraw-Hill, New York
4. Chandrasekhar S (1960) Radiative transfer. Dover, New York
5. Nayar SK, Narasimhan SG (1999) Vision in bad weather. In: The proceedings of the seventh IEEE international conference on computer vision, Kerkyra, vol 2. IEEE, pp 820–827
6. Schechner YY, Narasimhan SG, Nayar SK (2001) Instant dehazing of images using polarization. In: Proceedings of the IEEE computer society conference on computer vision and pattern recognition (CVPR), Kauai, vol 1. IEEE, pp 325–332
7. Narasimhan SG, Nayar SK (2003) Contrast restoration of weather degraded images. IEEE Trans Pattern Anal Mach Intell 25(6):713–724
8. Beer A (1852) Bestimmung der absorption des rothen lichts in farbigen flussigkeiten. Ann Phys Chem 86(2):78–90
9. Koschmieder H (1924) Theorie der horizontalen sichtweite. eitr. Phys. Freien Atm 12(2):33–53
10. Shwartz S, Namer E, Schechner YY (2006) Blind haze separation. In: IEEE computer society conference on computer vision and pattern recognition (CVPR), New York, vol 2. IEEE, pp 1984–1991
11. Treibitz T, Schechner YY (2009) Polarization: beneficial for visibility enhancement? In: IEEE conference on computer vision and pattern recognition (CVPR), Miami. IEEE, pp 525–532
12. Tan K, Oakley JP (2001) Physics-based approach to color image enhancement in poor visibility conditions. JOSA A 18(10):2460–2467
13. Narasimhan SG, Nayar SK (2003) Interactive (de) weathering of an image using physical models. In: IEEE workshop on color and photometric methods in computer vision, Nice
14. Hautière N, Tarel JP, Aubert D (2007) Towards fog-free in-vehicle vision systems through contrast restoration. In: IEEE conference on computer vision and pattern recognition (CVPR), Minneapolis. IEEE, pp 1–8
15. Kopf J, Neubert B, Chen B, Cohen M, Cohen-Or D, Deussen O, Uyttendaele M, Lischinski D (2008) Deep photo: model-based photograph enhancement and viewing. ACM Trans Graph (TOG) 27:116
16. Tan RT (2008) Visibility in bad weather from a single image. IEEE conference on computer vision and pattern recognition (CVPR), Anchorage. IEEE, pp 1–8

17. Fattal R (2008) Single image dehazing. ACM Trans Graph (TOG) 27:72
18. He K, Sun J, Tang X (2009) Single image haze removal using dark channel prior. In: IEEE conference on computer vision and pattern recognition (CVPR), Miami. IEEE, pp 1956–1963
19. Tarel JP, Hautiere N (2009) Fast visibility restoration from a single color or gray level image. In: IEEE 12th international conference on computer vision, Kyoto. IEEE, pp 2201–2208
20. Kratz L, Nishino K (2009) Factorizing scene albedo and depth from a single foggy image. In: IEEE 12th international conference on computer vision, Kyoto. IEEE, pp 1701–1708

Denoising

Francisco J. Estrada
Department of Computer and Mathematical Sciences, University of Toronto at Scarborough, Toronto, ON, Canada

Synonyms

Noise removal

Related Concepts

▶Image Enhancement and Restoration

Definition

Denoising is the process of recovering a reference signal that has been corrupted by noise. In computer vision, the reference signal is typically assumed to be the undistorted image of an object or scene, and noise is introduced as a result of the imaging process. The amount and type of noise changes from application to application. An example of a typical noisy image and the result of performing image denoising on it are shown in Fig. 1.

Background

Noise is an unavoidable consequence of the imaging process. Sources of noise include measurement and quantization errors introduced during signal acquisition and processing, thermal noise from the sensor and electronics in digital imaging systems, photographic grain in the case of film, and the physical nature of light itself.

The importance of denoising stems from the fact that noise in the input image can adversely affect the result of all subsequent visual processing. The value of image-dependent quantities such as brightness gradients, the accuracy in the localization of image features, the presence or absence of object boundaries, and even the subjective perceptual properties of the image are all affected to some degree by noise. Applications such as medical imaging, astronomy, low-light or high-speed photography, and synthetic aperture radar (SAR) imaging are typically characterized by larger amounts of noise relative to the reference signal.

Depending on the application, preprocessing the input with a denoising algorithm can improve the results obtained from further stages of visual processing. However, the denoising method should be selected with care since the denoising process is imperfect and will invariably destroy part of the information contained in the reference signal.

Theory

For a grayscale image, the perceived image brightness can be approximated by [6]

$$I = f(L + n_s + n_c) + n_q, \qquad (1)$$

where I is the observed brightness value, $f(\cdot)$ is the sensor response function, L is the image irradiance, n_s represents the noise contribution from brightness-dependent sources, n_c represents a constant noise factor, and n_q is the quantization noise.

Estimating all the quantities in this model is too complex a problem if only I is known, so in practice the simplified relationship $I = I_r + N$ is used instead where I is the observed image brightness, I_r is reference image that the denoising process must estimate, and N is the noise component. Since there is only one known value (the observed brightness) and two unknowns, the problem is under-constrained.

Additional external constraints must be placed on the properties of I_r and N so that a solution can be computed. Typically N is assumed to be i.i.d. noise, and most denoising methods assume a zero mean

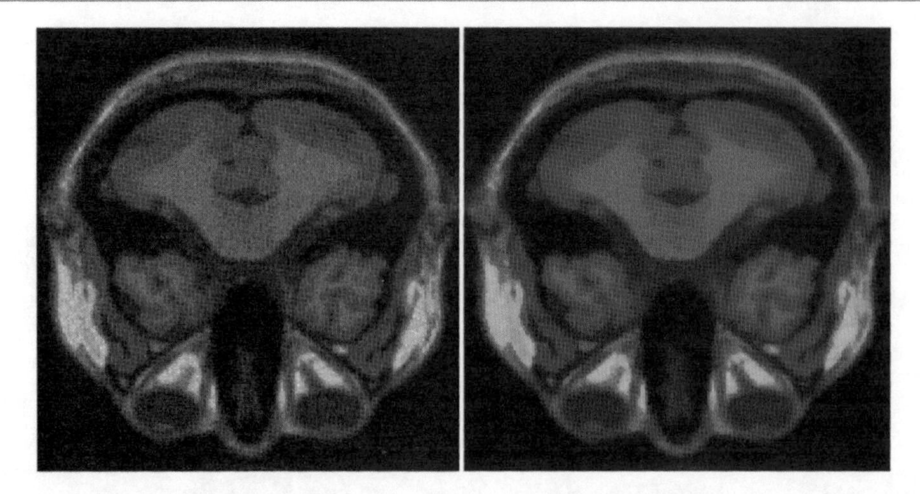

Denoising, Fig. 1 *Left* original, noisy MRI scan. *Right* result after denoising

Gaussian distribution with fixed standard deviation. Depending on the constraints placed on the properties of I_r, we can group denoising methods into a handful of major classes.

Classes of Denoising Algorithms

The first class of image denoising methods is based on averaging the value of pixels within small image neighborhoods under the assumption that the brightness away from image edges should be uniform. These algorithms perform anisotropic smoothing, so called because it has strong tendency to smooth uniform-looking regions in the image while preserving strong brightness discontinuities [13]. The original anisotropic smoothing algorithm [7], the bilateral filter [10], methods based on minimizing total variation [2], and the stochastic denoising algorithm [4] are all examples of denoising methods based on maximizing brightness uniformity across homogeneous regions.

A second class of algorithms is based on the analysis of image statistics. The underlying principle is that the statistical properties of natural images can be modelled and that given the model, denoising can be carried out by examining the statistics of a noisy image and transforming the image so that its statistics match those of learned model [11]. Common models include distributions of filter responses and pooled statistics for collections of image patches. Algorithms such as Gaussian scale mixtures [8], fields of experts [9], the nonlocal means method [1], and the block matching algorithm [3] rely on exploiting the statistical regularities of images or small image patches.

A third class of image denoising methods specifically designed to remove salt and pepper noise is based on outlier detection. The process involves estimating a distribution of brightness values around image neighborhoods and using this distribution to identify and remove outliers. The median filter [5] is an example of this class of denoising algorithms.

Regardless of the method, the goal of denoising is to remove as much of the noise as possible while preserving the information contained in the reference signal. For this reason, denoising algorithms are often evaluated using peak signal to noise ratio (pSNR) or the structured image similarity index (SSIM) [12] on images for which the noise-free reference is known.

Application

Denoising can be a useful preprocessing step for images from domains such as medical imaging, astronomy, or synthetic aperture radar. It is also applicable to digital photography under low illumination conditions and to the restoration of archival footage and photographs. Image editing and image processing programs typically include denoising modules.

Experimental Results

The quality of the results produced by different algorithms changes from image to image and across different domains. But recent benchmarks [14] indicate that the block matching algorithm achieves better performance overall on natural images with different amounts of Gaussian noise.

References

1. Buades A, Coll B, Morel J (2008) Nonlocal image and movie denoising. Int J Comput Vision 76:123–139
2. Chambolle A (2004) An algorithm for total variation minimization and applications. J Math Imaging Vis 20:89–97
3. Dabov K, Foi R, Katkovnik V, Egiazarian K (2006) Image denoising with block matching and 3D filtering. SPIE Electronic Imaging 6064A–30
4. Estrada F, Fleet D, Jepson A (2009) Stochastic Image Denoising. British Machine Vision Conference (no printed proceedings)
5. Juhola M, Katajainen J, Raita T (1991) Comparison of Algorithms for Standard Median Filtering. IEEE T Signal Proces (TSP) 39:204–208
6. Liu C, Szeliski R, Kang SB, Zitnick L, Freeman WT (2008) Automatic estimation and removal of noise from a single image. IEEE Trans Pattern Anal Mach Intell (PAMI) 30(2):299–314
7. Perona P, Malik J (1990) Scale-space and edge detection using anisotropic diffusion. IEEE Trans Pattern Anal Mach Intell (PAMI) 12(7):629–639
8. Portilla J, Strela V, Wainwright M, Simoncelli E (2003) Image denoising using scale mixtures of Gaussians in the wavelet domain. IEEE T Image Process (TIP) 12(11): 1338–1351
9. Roth S, Black M (2009) Fields of Experts. Int J Comput Vision (IJCV) 82(2):205–229
10. Tomasi C, Manduchi R (1998) Bilateral filtering for gray and color images. Int Conf Comput Vision 839–846
11. Torralba A, Oliva A (2003) Statistics of Natural Image Categories. Network-Comp Neural 391–412
12. Wang Z, Bovik A, Sheikh H, Simoncelli E (2004) Image quality assessment: From error visibility to structural similarity. IEEE T Image Process (TIP) 13(4):600–612
13. Weickert J (1998) Anisotropic Diffusion in Image Processing. Teubner-Verlag
14. Image Denoising Benchmark (2010) http://www.cs.utoronto.ca/~strider/Denoise/Benchmark

Dense 3D Modeling

▶Dense Reconstruction

Dense Reconstruction

Christopher Zach
Computer Vision and Geometry Group, ETH Zürich, Zürich, Switzerland

Synonyms

Dense 3D modeling; Multiview stereo

Related Concepts

▶Structure-from-Motion (SfM)

Definition

Dense reconstruction aims on determining the complete 3D geometry of a static environment solely from a set of provided images.

Background

Correspondences between images depicting a static scene cannot only be established for a small set of visually salient regions but can be also extended to the entire image domain (i.e., to all pixels). Under the assumption of a static environment (or equivalently, a rigidly moving scene captured by a static camera), corresponding points in images together with known camera calibration induce a 3D point. Thus, a dense set of correspondences implies a densely sampled surface in 3D. Establishing per-pixel correspondences between only two (or any small number of) narrow-baseline images is usually referred as computational stereo. *Dense reconstruction* addresses the more general problem of obtaining the full 3D geometry from a larger collection of images. In contrast to computational stereo, the returned surfaces are not 2.5D depth maps (as in stereo) but can possess arbitrary topology.

Theory

Determining the dense 3D geometry of a static scene from images can be achieved via direct or via indirect approaches:

- Direct approaches estimate the dense 3D geometry by using all available images simultaneously.
- Indirect methods estimate partial 3D models from a subset of (usually narrow-baseline) images (e.g., via computational stereo) and subsequently merge the partial models into a globally consistent geometry.

Direct Methods

Direct methods estimate the likelihood of the 3D surface being at a particular position via computation of a *photo-consistency score* incorporating all given images. Typically the 3D region of interest $\Omega \subseteq \mathbb{R}^3$ is discretized as a voxel space, and photo-consistency is computed with respect to the voxels. Since accurate visibility is not yet known, the photo-consistency score must be robust against potential occlusions. Approximate visibility of a surface patch in the images can be determined e.g., by view frustum and back-face culling, and when a coarse estimate of the 3D geometry (e.g., the corresponding visual hull) is available. By traversing the voxel space in a conservative order with respect to visibility and computing photo-consistency by only using certainly unoccluded images, one obtains the *space carving* method [1]. Sufficiently photo-inconsistent voxels are declared as empty and influence the visibility of not-yet-visited voxels in the images. Under certain assumptions, the set of occupied voxels (which is called the photo hull) contains the true 3D geometry. For objects with little texture, the space carving results are not satisfactory, and spatial smoothness assumptions need to be added. The volumetric graph cut approach [2] can be interpreted as regularized space carving. Since the state of a voxel (empty or occupied) is determined by a global optimization method, a robust photo-consistency score for each voxel is computed using static and approximate visibility information.

The knowledge of sparsely sampled surfaces, e.g., by triangulating correspondences between interest points, can be used to discretize Ω more adaptively by a general tetrahedral mesh. In such a representation, it is possible to employ a higher-resolution discretization where the true surface is expected [3, 4].

The task of determining the 3D geometry consistent with all given images can be formulated as mesh evolution approach, which is inspired, e.g., by deformable contour approaches used in image segmentation. The evolution of the mesh is guided by external forces based on the photo-consistency and optionally on silhouette data and internal forces regularizing the mesh [5, 6]. One drawback of such methods is that the obtained result may strongly depend on the initial mesh and only local minima of the underlying energy are usually reached.

Indirect Approaches

This set of methods first computes a collection of depth maps from a suitable narrow-baseline subset of images and fuses these depth images into a consistent 3D model. Since knowing the exact visibility of 3D locations in the images is not required for photo-consistency computation in a small baseline setting, these methods do not need a coarse geometry to estimate potential occlusions. Furthermore, sophisticated computational stereo methods are available to generate high-quality depth maps ultimately yielding photo-realistic reconstructions. On the other hand, indirect methods do not utilize the available image data to the full extent, since the original images are usually not considered in the depth map fusion step.

Merging multiple depth maps can be identified as a particular instance of surface fitting from unorganized point data (e.g., [7, 8]), but it is clearly beneficial to exploit the specific structure in the input data. The mathematical formulations employed for depth map fusion can be tracked back to the problem of merging partial range scans obtained by an active sensor device, e.g., laser data or from structured light. Early approaches use an explicit polygonal representation [9], but more current methods are usually based on implicit volumetric surface representations capable of handling surfaces with arbitrary genus. It is a common aspect of all volumetric approaches that they return watertight meshes, i.e., surfaces may be hallucinated in occluded regions. In general, depth map fusion using a volumetric representation can be formalized as determining a minimizer $u : \Omega \to \{-1, 1\}$ of a functional E,

$$E(u) = \phi(u; \{d_i\}) + \psi(u),$$

where $\Omega \subset \mathbb{R}^3$ is the region of interest (usually a 3D bounding box), $\phi(u; \{d_i\})$ is a compatibility (data fidelity) function measuring the agreement of the solution u with the set of input depth maps $\{d_i\}$, and $\psi(\cdot)$ denotes the spatial regularization of u. The interpretation of u is that $u(x) = 1$ if the 3D location x is occupied (solid) and $u(x) = -1$ if x corresponds to empty space. Often u is allowed to attain fraction values, e.g., $u : \Omega \to [-1, 1]$. In the following, let $D_i : \Omega \to \mathbb{R}$ be the signed distance transform induced by the depth map d_i, and $F : \Omega \to \mathbb{R}^3$ is a 3D vector field with $F(x)$ corresponding to a (smoothed) surface normal induced by the sampled depth maps. $F(x) = \mathbf{0}$ if x is distant to any of the depth map surfaces. By appropriate choice of ϕ and ψ, one obtains

- *Volumetric range image integration* [10] (with $\phi(u; \{d_i\}) = \int_\Omega \sum_i (u(x) - D_i(x))^2 \, dx$ and $\psi \equiv 0$)
- *Poisson surface reconstruction* [11] (with $\phi(u) = \int \left\| F - \nabla u \right\|^2 dx$ and $\psi \equiv 0$)
- *Global shape fitting* [12] (with $\phi(u) = \int F^T \nabla u \, dx$ and $\psi(u) = \int \|\nabla u\| \, dx \doteq TV(u)$)
- *TV-L^1 depth map fusion* [13] (with $\phi(u; \{d_i\}) = \int_\Omega \sum_i |u(x) - D_i(x)| \, dx$ and $\psi(u) = TV(u)$)

The above mentioned methods use a uniformly sampled voxel space to discretize the domain Ω. Determining a minimizer u for these energies is relatively simple due to their intrinsic convexity and amounts, e.g., to per-voxel averaging [10] or solving a sparse linear system of equations [11]. The final triangular mesh can be extracted from the implicit volumetric representation as the zero-level set, e.g., via the marching cubes method [14].

Application

Dense reconstruction plays a fundamental role in fully automatic 3D content creation from image data. Its application ranges from 3D city modeling and cultural heritage preservation to obstacle avoidance in autonomous systems navigation and 3D face reconstruction for character animation. Dense reconstruction can achieve an accuracy comparable to actively measured geometry (using laser range scanners or structured light) by relying only on passive imaging sensors [15, 16]. Accurate dense scene geometry can be augmented with texture images or more general appearance properties enabling photo-realistic rendering of virtual 3D representations.

References

1. Kutulakos K, Seitz S (2000) A theory of shape by space carving. Int J Computer Vision, 38(3):198–218
2. Vogiatzis G, Hernandez C, Torr P, Cipolla R (2007) Multiview stereo via volumetric graph-cuts and occlusion robust photo-consistency. IEEE trans pattern analysis and machine intelligence, 29(12):2241–2246
3. Sinha S, Mordohai P, Pollefeys M (2007) Multi-view stereo via graph cuts on the dual of an adaptive tetrahedral mesh. International Conf. Computer Vision, (ICCV2007), Rio de Janeiro
4. Labatut P, Pons JP, Keriven R (2007) Efficient multi-view reconstruction of large-scale scenes using interest points, Delaunay triangulation and graph cuts, International Conf. Computer Vision, (ICCV2007), Rio de Janeiro
5. Faugeras O, Keriven R (1998) Variational principles, surface evolution, PDE's, level set methods, and the stereo problem. IEEE trans image processing, 7(3):336–344
6. Hernández-Esteban C, Schmitt F (2004) Silhouette and stereo fusion for 3d object modeling. CVIU 96(3):367–392
7. Hoppe H, DeRose T, Duchamp T, McDonald J, Stuetzle W (1992) Surface reconstruction from unorganized points. ACM SIGGRAPH '92. Chicago, IL, pp 71–78
8. Amenta N, Choi S, Kolluri R (2001) The power crust. In: Proceedings of 6th ACM symposium on solid modeling. Ann Arbor, MI, pp 249–260
9. Turk G, Levoy M (1994) Zippered polygon meshes from range images. ACM SIGGRAPH '94. Orland, FL, pp 311–318
10. Curless B, Levoy M (1996) A volumetric method for building complex models from range images. ACM SIGGRAPH '96. New Orleans, LA, pp 303–312
11. Kazhdan M, Bolitho M, Hoppe H (2006) Poisson surface reconstruction. Symposium on geometry processing in Proc the fourth eurographics symposium on geometry processing, Aire-la-Ville, Switzerland, pp 61–70
12. Lempitsky V, Boykov Y (2007) Global optimization for shape fitting. IEEE conf computer vision and pattern recognition (CVPR), Minneapois, Minnesota
13. Zach C, Pock T, Bischof H (2007) A globally optimal algorithm for robust TV-L^1 range image integration. International Conf. Computer Vision, Rio de Janeiro
14. Lorenson W, Cline H (1987) Marching Cubes: a high resolution 3d surface construction algorithm. ACM SIGGRAPH '87, New York, pp 163–170
15. Seitz S, Curless B, Diebel J, Scharstein D, Szeliski R (2006) A comparison and evaluation of multi-view stereo reconstruction algorithms. IEEE Conf. Computer Vision and Pattern Recognition (CVPR), New York, pp 519–526
16. Strecha C, von Hansen W, Van Gool L, Fua P, Thoennessen U (2008) On benchmarking camera calibration and multi-view stereo for high resolution imagery. IEEE Conf Computer Vision and Pattern Recognition (CVPR), Anchorage

Depth Distortion

Zhengyou Zhang
Microsoft Research, Redmond, WA, USA

Related Concepts

▸Camera Calibration; ▸Perspective Camera; ▸Active Stereo Vision

Definition

Depth distortion refers to the distortion on the recovered structure from motion or from stereo due to imprecision or errors in the cameras' intrinsic and extrinsic parameters.

Background

In structure from motion, when a camera's intrinsic parameters are imprecise or even erroneous, the estimated motion is not accurate, and the structure inferred from motion will be a distorted version of the true structure in the environment. In stereo, when the cameras' intrinsic and extrinsic parameters are imprecise or even erroneous, the reconstructed 3D scene will be a distorted version of the true scene in the environment.

Theory

As an example, Fig. 1 illustrates a case where the focal length is misestimated. The top view of two cameras is shown. The line segment BE is the true structure in the environment. It is projected to the images according to the true optical center O_1 and O_2, respectively. However, if the estimation of the focal length is not accurate, the reconstructed structure will be distorted. In the figure, the optical centers are estimated to be at O_1' and O_2'. Then, the reconstructed structure will be $B'E'$, which is clearly different from the true one BE.

In general, given N views of a scene, the 3D reconstruction of a point $\boldsymbol{M}$ is a function of its image points $\{\boldsymbol{m}_1, \ldots, \boldsymbol{m}_N\}$, the cameras' intrinsic parameters $\{\boldsymbol{c}_1, \ldots, \boldsymbol{c}_N\}$, and the cameras' extrinsic parameters $\{\boldsymbol{p}_1, \ldots, \boldsymbol{p}_N\}$. That is,

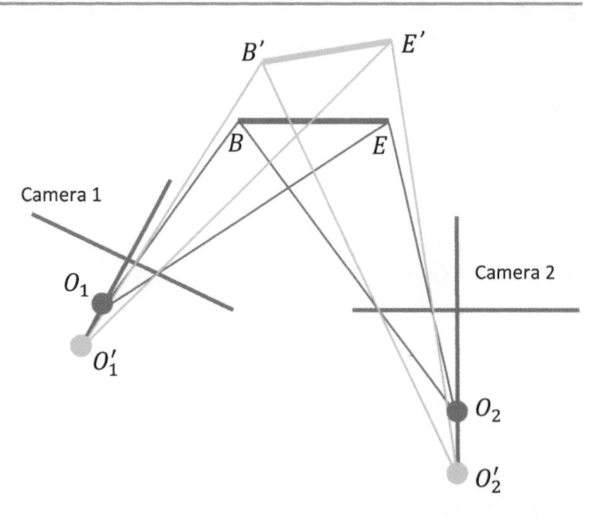

Depth Distortion, Fig. 1 Illustration of depth distortion due to imprecise focal length

$$\boldsymbol{M} = f(\boldsymbol{m}_1, \ldots, \boldsymbol{m}_N; \boldsymbol{c}_1, \ldots, \boldsymbol{c}_N; \boldsymbol{p}_1, \ldots, \boldsymbol{p}_N) . \tag{1}$$

If a parameter, say $\boldsymbol{c}_1$, is misestimated by $\Delta \boldsymbol{c}_1$, i.e., $\boldsymbol{c}_1' = \boldsymbol{c}_1 + \Delta \boldsymbol{c}_1$, then the reconstructed 3D point will be

$$\boldsymbol{M}' = f(\boldsymbol{m}_1, \ldots, \boldsymbol{m}_N; \boldsymbol{c}_1 + \Delta \boldsymbol{c}_1, \ldots, \boldsymbol{c}_N; \boldsymbol{p}_1, \ldots, \boldsymbol{p}_N) , \tag{2}$$

which is a distorted version of $\boldsymbol{M}$ and is different from $\boldsymbol{M}$.

If the error in the parameter is small, we can use Taylor expansion and ignore the high-order terms. The first-order approximation of the error in 3D reconstruction due to $\Delta \boldsymbol{c}_1$ will be

$$\begin{aligned} \boldsymbol{M}' - \boldsymbol{M} = \frac{\partial}{\partial \boldsymbol{c}_1} & (f(\boldsymbol{m}_1, \ldots, \boldsymbol{m}_N; \boldsymbol{c}_1 \\ & + \Delta \boldsymbol{c}_1, \ldots, \boldsymbol{c}_N; \boldsymbol{p}_1, \ldots, \boldsymbol{p}_N)) \\ & \Delta \boldsymbol{c}_1 \equiv \mathbf{J}_{\boldsymbol{c}_1} \Delta \boldsymbol{c}_1 , \end{aligned} \tag{3}$$

where $\mathbf{J}_{\boldsymbol{c}_1}$ is the Jacobian matrix of $f(\cdot)$ with respect to $\boldsymbol{c}_1$. In practice, the exact error in the parameters is usually unknown. If the error can be modeled as a Gaussian, then the error in 3D reconstruction can also be approximated as a Gaussian. Let the error in $\boldsymbol{c}_1'$, $\Delta \boldsymbol{c}_1$, be a Gaussian with mean 0 and covariance matrix $\Sigma_{\boldsymbol{c}_1}$, then the covariance matrix of the 3D reconstruction $\boldsymbol{M}'$ is given by

$$\Sigma_{\boldsymbol{M}} = \mathbf{J}_{\boldsymbol{c}_1} \Sigma_{\boldsymbol{c}_1} \mathbf{J}_{\boldsymbol{c}_1}^T . \tag{4}$$

The reader is referred to [1–3] for details.

The impact of the error in a camera's different intrinsic parameters on 3D reconstruction is not necessarily the same in magnitude. For example, the error in the coordinates of the principal point has little effect in 3D reconstruction [4, 5]. Cheong and Peh [6] provides a more detailed discussion on depth distortion due to calibration uncertainty.

References

1. Ayache N (1991) Artificial vision for mobile robots. MIT, Cambridge
2. Zhang Z, Faugeras OD (1992) 3D dynamic scene analysis: a stereo based approach. Springer, Berlin/Heidelberg
3. Faugeras O (1993) Three-dimensional computer vision: a geometric viewpoint. MIT, Cambridge
4. Triggs B (1998) Autocalibration from planar scenes. In: Proceedings of the 5th European conference on computer vision (ECCV), Freiburg, Germany, pp 89–105
5. Zhang Z (1990) Motion analysis from a sequence of stereo frames and its applications. Ph.D. thesis, University of Paris-Sud, Orsay, Paris, France
6. Cheong LF, Peh CH (2004) Depth distortion under calibration uncertainty. Comput Vis Image Underst 93:221–244

Depth Estimation

Reinhard Koch
Institut für Informatik
Christian-Albrechts-Universität, Kiel, Germany

Synonyms

Depth imaging; Distance estimation; Three dimensional estimation

Related Concepts

▶Plane Sweeping

Definition

Depth estimation describes the process of measuring or estimating distances from sensor data, typically in a 2D array of depth range data. The sensors may be either optical camera configurations (stereo or multiview stereo camera rigs), active projector-camera configurations, or active range cameras.

Background

Depth estimation is one of the fundamental computer vision tasks, as it involves the inverse problem of reconstructing the three-dimensional scene structure from two-dimensional projections. Given a 2D image of a 3D scene, the goal of depth estimation is to recover, for every image pixel, the distance from the camera center to the nearest 3D scene point along the pixel's viewing direction. The resulting 2D array of distance values is called the depth map, which is aligned with the camera coordinate system. The challenge of depth estimation is to recover, with sufficient accuracy, the depth map from the given image. Depth estimation can be accomplished in two principle ways: by triangulation from different viewpoints or by coaxial methods along the camera viewing direction. In both cases, estimation methods may be either passive, based on image data alone, or active, with additional sensor control or scene illumination as supporting data. Binocular stereo, for example, is representative for a passive triangulation method, while laser interferometry is a typical active coaxial method.

As the depth map is organized in a regular 2D image grid and contains the depth value for each pixel, it is sometimes also called 2.5D model [11]. The depth map is not a full 3D shape representation, as it contains depth to the nearest visible 3D scene elements only and does not handle regions that are occluded from the camera's view point. Extensions to layered depth images (LDI) exist that can handle multiple depth values for occluded scene elements [16], but LDIs are still restricted to the camera viewpoint and do not handle full 3D data. Full 3D surface estimation from arbitrary view points is the topic of 3D modeling of surfaces and volumetric 3D reconstruction methods, which can be obtained by merging multiple 2.5D depth maps, or by tomographic methods that actively scan a volume with multiple projections and reconstruct the 3D volume surface density. For tomographic methods, which are common in 3D medical and material science volume analysis, and for 3D modeling, refer to the related concepts. A good overview and recent bibliography on all related concepts to depth estimation can be found in [17].

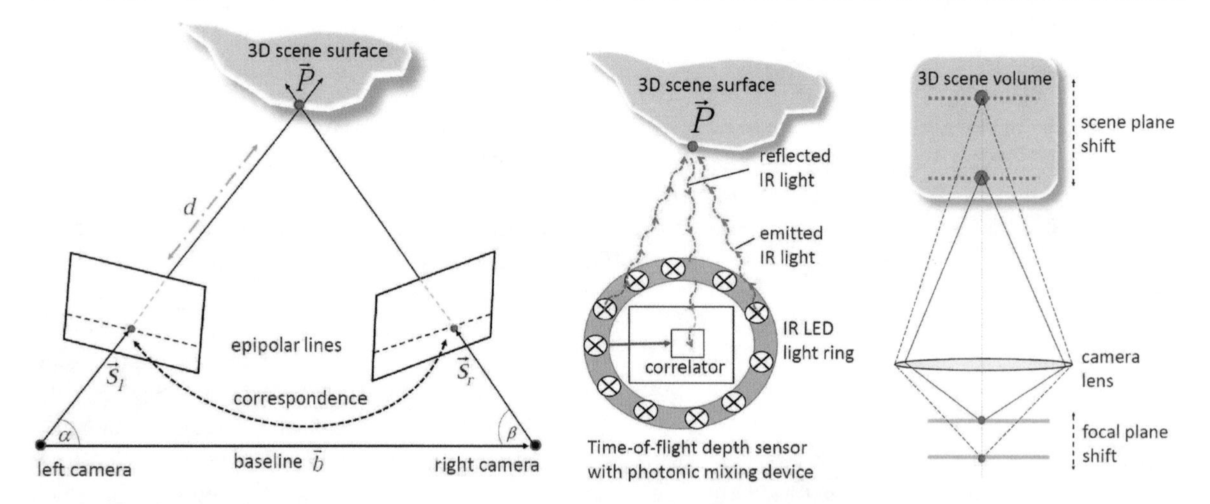

Depth Estimation, Fig. 1 Depth estimation principles. *Left*: depth estimation by stereo triangulation of corresponding projections. *Center*: active time-of-flight depth estimation by phase shift correlation. *Right*: coaxial depth estimation by focal plane shift

Theory

Triangulation

Triangulation is the most common approach for depth estimation. At least two images from different view points observe the same 3D scene, and a 3D scene point is triangulated from the observed 2D projections of the scene point by ray intersection of the viewing rays; see Fig. 1 (left). The depth d is the length of the viewing ray from camera center to ray intersection. It can be computed easily from the triangulation triangle, given the length between the camera centers (base line b) and the two angles α and β between the viewing rays and the base line. This requires that the cameras are calibrated w.r.t. each other in a calibrated stereo rig configuration, and that the corresponding 2D projections of the observed 3D scene point can be related to each other. The latter causes most of the problems in depth estimation, because the 2D search of corresponding projection pairs in the 2D images is difficult.

The correspondence problem can be simplified to a one-dimensional search when the relative orientation between the cameras, the epipolar geometry, is utilized [7]. The spatial triangulation plane or epipolar plane, which is constructed by the camera base line b and the viewing ray $\mathbf{s}_l$ of one camera, intersects the image plane of the other camera in a line, the epipolar line. Hence, the correspondence search is confined to this line and called disparity estimation. Disparity estimation is further simplified if the epipolar line corresponds to horizontal image scan lines. This is the case in standard stereo geometry where both cameras are aligned in identical orientation and shifted in horizontal scanline direction only. Stereo cameras with convergent configurations can be rectified to a virtual alignment by a rectifying homography [1] or with other, more general transformations [13]. A generalization to rectification is the plane sweep, where the images of multiple calibrated cameras are compared by projecting onto a common 3D reference plane. Shifting the plane throughout the depth volume allows multi-camera depth estimation [4].

Passive Correspondence Analysis

Passive correspondence estimation relies on image data alone and requires that the image contains sufficient intensity variations to obtain a unique correspondence match. Correspondences are found by evaluating a cost function of the local intensity data along the search range. There exist a large variety of cost functions and evaluation schemes, ranging from data-driven evaluation of local cost function minimization [6, 18], over semiglobal optimization along epipolar lines [5, 8], to global optimization schemes including smoothness constraints [3]. A good source for algorithms and benchmarking is the Middlebury database and their evaluation test scenario [14].

Active Correspondence Analysis

The main drawback of image-based correspondence analysis is the dependency on sufficient intensity variation in the images. One way out is to replace one of the cameras with an active video projector and to project unique patterns, which are easily identified by the other cameras, into the scene. There exist a variety of approaches. Coded light projectors project a time series of deterministic codes for unique correspondence, or a statistically distributed pattern is projected for optimal correspondence search. The patterns might be projected either in visible or in near infrared light range. Stripe projection systems generate stripes of spectrally colored light [2], or even single laser light stripes are used for easy identification in industrial depth inspection systems. One recent and very successful variant of an active depth triangulation system is the Kinect (a trademark of Microsoft) sensor device that couples an infrared pattern projector and corresponding infrared camera for fast depth estimation with a visual color camera for simultaneous color and depth capture. Such systems deliver reliable dense depth maps for many scenes, but are restricted in depth range to a few meters. Depth estimation in the longer range may be obtained by triangulating laser scanning devices that scan the scene with a rotating mirror laser reflected into the sensing camera.

Coaxial Depth Estimation

One drawback of the triangulation method is that the scene is observed from different viewpoints, and hence, occlusion might occur at object boundaries. If the baseline between the cameras is small, then fewer occlusion boundaries occur, but the depth measurement is not very accurate due to the small triangulation angle. If, on the other hand, a large baseline is chosen, then the triangulation may be accurate but the correspondence problem is more difficult due to the changing perspective. Coaxial methods do not suffer from such problems, as they measure depth from one viewpoint only. Active coaxial depth estimation has its foundations in classical interferometry systems. A coherent laser beam is sent out, reflected back at the 3D scene point, and received at the sensor. The distance the light wave has traveled causes a phase shift between emitted and returning light and hence interference. Due to the short wave length of light, this principle is very accurate but covers only a very small unambiguous depth range. Recently, a novel class of depth sensors has emerged that either exploits the traveling time of a very short light impulse to measure the time of flight directly, or that modulates an infrared LED light source with a periodic signal amplitude and measures the signal phase shift by correlation between outgoing and reflected light. The correlation sensor is a 2D image array of photonic correlation mixers that allows to directly estimate depth in a range of meters, depending on the modulation frequency; see Fig. 1 (center). Typical depth ranges are up to 10 m. Such time-of-flight range cameras [15] are active coaxial depth estimation devices that in principle may deliver dense and accurate depth maps in real-time and can be used for depth estimation of dynamic time-varying scenes [9, 10].

Another option to estimate depth from a coaxial view is to change camera parameters and to relate the observed image changes to depth. This principle is observed by depth from focus algorithms. A series of images with different focal settings and aperture is recorded, and depth is computed from image sharpness variations [12]. This principle has been applied very successfully in depth-of-field microscopy, where a 3D volume of semitransparent objects like biological cells is scanned with shifting focal planes and very narrow depth-of-field settings; see Fig. 1 (right). Thus, a 3D stack of slices is obtained for 3D analysis, very much like 3D tomography.

References

1. Ayache N, Hansen C (1988) Rectification of images for binocular and trinocular stereovision. In: Proceedings of the 9th international conference on pattern recognition. IEEE Computer Society, Rome, Italy, pp 11–16
2. Boyer K, Kak A (1987) Color-encoded structured light for rapid active ranging. IEEE Trans Pattern Anal Mach Intell 9(1):14–28
3. Boykov Y, Kolmogorov V (2004) An experimental comparison of min-cut/max-flow algorithms for energy minimization in vision. IEEE Trans Pattern Anal Mach Intell 26(9):1124–1137
4. Collins RT (1996) A space-sweep approach to true multi-image matching. In: Proceedings of the international conference on computer vision and pattern recognition (CVPR). IEEE Computer Society, San Francisco, CA, USA, pp 358–363

5. Cox IJ, Hingorani SL, Rao SB, Maggs BM (1996) A maximum likelihood stereo algorithm. Comput Vis Image Underst 63:542–567
6. Gong M, Yang R, Wang L, Gong M (2007) A performance study on different cost aggregation approaches used in real-time stereo matching. Int J Comput Vis 75(2): 283–296
7. Hartley R, Zisserman A (2003) Multiple View Geometry in computer vision. Cambridge University Press. ISBN 0-521-54051-8
8. Hirschmueller H (2008) Stereo processing by semi-global matching and mutual information. IEEE Trans Pattern Anal Mach Intell 30(2):328–341
9. Kolb A, Barth E, Koch R, Larsen R (2010) Time-of-flight cameras in computer graphics. Comput Graph Forum 29(1):141–159
10. Kolb A, Koch R (eds) (2009) Dynamic 3D Imaging. Lecture Notes in Computer Science, vol 5742. Springer, Berlin/New York
11. Marr D (1982) Vision. A Computational Investigation into the Human Representation and Processing of Visual Information. W.H. Freeman, San Francisco
12. Nayar SK, Watanabe M, Noguchi M (1996) Real-time focus range sensor. IEEE Trans Pattern Anal Mach Intell 18(12):1186–1198
13. Pollefeys M, Koch R, Gool LV (1999) A simple and efficient rectification method for general motion. In: Proceedings of the 7th international conference on computer vision, vol 1, IEEE Computer Society, Korfu, Greece, pp 496–501
14. Scharstein D, Szeliski R (2002) A taxonomy and evaluation of dense two-frame stereo correspondence algorithms. Int J Comput Vis 47:7–42
15. Schwarte R, Xu Z, Heinol H-G, Olk J, Klein R, Buxbaum B, Fischer H, Schulte J (1997) New electro-optical mixing and correlating sensor: facilities and applications of the photonic mixer device (pmd). In: Proceedings of the SPIE 3100.SPIE
16. Shade J, Gortler SJ, wei He L, Szeliski R (1998) Layered depth images. In: SIGGRAPH, Proceedings of the 25th annual conference on computer graphics and interactive techniques. ACM, New York, pp 231–242
17. Szeliski R (2010) Computer vision, algorithms and applications. Springer, London/New York
18. Tombari F, Mattoccia S, Di Stefano L, Addimanda E (2008) Classification and evaluation of cost aggregation methods for stereo correspondence. In: IEEE conference on computer vision and pattern recognition, 2008 (CVPR 2008). IEEE Computer Society, Anchorage, Alaska, USA, pp 1–8

Depth from Scattering

►Shape from Scatter

Depth Imaging

►Depth Estimation

Descattering

Tali Treibitz
Department of Computer Science and Engineering, University of California, San Diego, La Jolla, CA, USA

Synonyms

Dehazing

Related Concepts

►Dehazing and Defogging; ►Underwater Effects

Definition

The process of descattering refers to enhancing images taken in scattering media.

Background

Imaging in scattering media poses special concerns for computer vision methods. Examples for such media include haze or bad weather, water, blood, and body tissue. Light propagating in such media is scattered and attenuated. When imaging, some light from the source is scattered back from the medium towards the camera, before ever reaching the object. This light is an additive radiance component in the image that veils the object. In air, this additive component is often termed *airlight* and in water and tissue it is termed *backscatter* (this term will be used further on). The backscatter hampers visibility by reducing contrast and signal-to-noise ratio (SNR) in the images [1]. The backscatter increases the measured radiance, which increases the photon noise, without increasing the signal value. Thus, there are methods that aim to *descatter* the image, i.e., reduce the effects of scattering. This sometimes has the effect of extending the visibility range.

Correcting for backscatter in a single image is fundamentally ill posed. When observing a single pixel, it is impossible to know what percentage of its intensity is reflected from the object and how much is contributed by the backscatter. Moreover, the

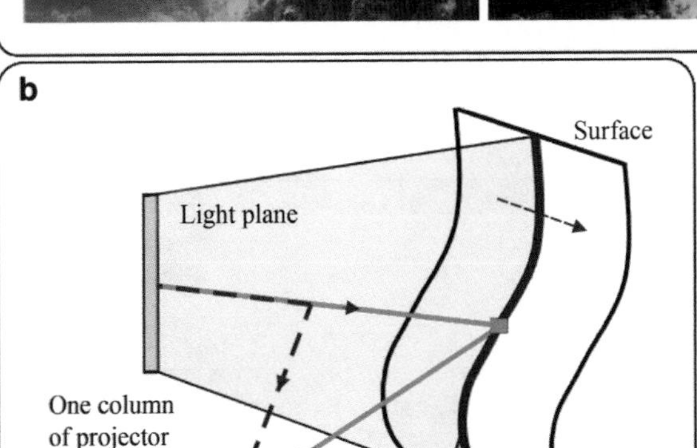

Descattering, Fig. 1 (**a**) Setup for descattering by polarized artificial illumination [5]. (**b**) Setup for the descattering by structured illumination [14]. (**c**) Result from single underwater image descattering [10]. Raw image, descattered image, and range map (*left* to *right*) (Reproduced with permission from the authors)

backscatter value in a pixel is an accumulation of light scattered by the medium between the object and the camera. Thus, the accumulated backscatter increases with the object distance, which generally changes spatially in the image. This is a nonuniform effect which complicates recovery. However, the coupling of backscatter and the 3D structure of the scene also provides an opportunity since recovering either the backscatter or the distance can imply the other. After removal of the estimated backscatter, some methods also try to compensate for light attenuation. Reference [2] provides elaborate physical theory of scattering media. The reader is referred to the "Dehazing and Defogging" entry in this encyclopedia, and to [3] for vision in the atmosphere and to [4, 5] for imaging under artificial illumination.

Methods

Because of the ill-posed nature of the problem, most descattering methods use auxiliary information or specialized hardware. Recently, some methods use a single image with prior assumptions or by applying constraints.

A known range map was used in [6]. In [7] the extremum values of the scene range are used, together with user indication of an airlight area in the image. Instead of range information, Ref. [7] relies on user indication of an area in the image that is not affected by the medium. In [3] the authors use images of the same scene under different weather conditions. Single-image descattering methods relax the need for user input by automatically estimating the airlight location and posing smoothness constraints [8, 9] or applying a dark channel prior [10, 11] (example in Fig. 1a). So far, single-image methods have been demonstrated on images taken under natural ambient illumination.

In highly scattering media, natural illumination has a limited range. Thus, several types of specialized hardware and instrumentation have been deployed. Time-gating [12] exploits the fact that the light from the object arrives to the sensor in a different time than the scattered light, as they travel different optical path lengths. By opening the shutter for a very short period of time, the light reflected off an object can be separated from the scattered light. This requires strong laser illumination and a gated camera, with high temporal resolution. The time of arrival of the object reflectance also reveals its distance.

Rather than scanning the time of flight, various methods perform spatial scanning. Laser line scan methods use a highly collimated laser source that scans the scene. The laser source is imaged by a synchronized narrow field of view. Such systems have been shown to be very efficient, with a range of up to 6 attenuation lengths [13]. Structured light [14–17] methods capture a wide field of view with various illumination patterns. Figure 1b shows an example for a structured light setup.

Another approach modulates light by polarization. It has been shown that high degrees of polarization can be observed in backscatter. Therefore, polarization discrimination has been used to reject scattered light in images taken in haze, underwater, and also through skin. The light can be polarized naturally by scattering [18, 19] or be actively polarized [5, 20]. Figure 1c illustrates a setup of active polarized illumination. Some works assume the light reflected from the object is less polarized than the backscatter. Others assume the light from the object (often metal ones) is more polarized. When polarizing the light source, linear polarization or circular polarization can be chosen to optimize the backscatter degree of polarization in the specific medium.

Often, acoustic waves are less scattered than light in a medium and therefore can penetrate much deeper into the scattering media. However, sonic and ultrasonic images have less resolution than optical images and suffer from speckles. In photoacoustic microscopy, a pulsed laser beam is focused into biological tissue. The laser pulse creates sudden thermal expansion, generating ultrasonic waves, which are then detected to form an image. In this method, the light has to travel only half of the optical path while enabling high resolution.

Fluorescence is emission of light following excitation by light of a different wavelength. This phenomenon is used in microscopy to optically separate (using appropriate filters) the desired signal from scattered light that has the illuminant color.

The process of optical scattering is time reversible. Thus, in [21], the authors have shown that they can backpropagate the scattering by a phase-conjugate mirror.

References

1. Treibitz T, Schechner YY (2009) Recovery limits in pointwise degradation. In: IEEE ICCP, San Francisco
2. Kokhanovsky AA (2004) Light scattering media optics, 3rd edn. Springer, Berlin, p 200
3. Narasimhan SG, Nayar SK (2002) Vision and the atmosphere. Int J Comput Vis 48:233–254
4. Jaffe JS (1990) Computer modelling and the design of optimal underwater imaging systems. IEEE J Ocean Eng 15:101–111
5. Treibitz T, Schechner YY (2009) Active polarization descattering. IEEE Trans Pattern Anal Mach Intell 31:385–399
6. Oakley JP, Satherley BL (1998) Improving image quality in poor visibility conditions using a physical model for contrast degradation. IEEE Trans Image Process 78:167–179
7. Narasimhan SG, Nayar SK (2003) Interactive (de) weathering of an image using physical models. In: IEEE workshop on color and photometric methods in computer vision, Nice
8. Fattal R (2008) Single image dehazing. ACM Trans Graph 27:1–9
9. Tan RT (2008) Visibility in bad weather from a single image. In: Proceedings of the IEEE conference on computer vision pattern recognition (CVPR), Anchorage
10. Carlevaris-Bianco N, Mohan A, Eustice R (2010) Initial results in underwater single image dehazing. In: OCEANS, Seattle
11. He K, Sun J, Tang X (2010) Single image haze removal using dark channel prior. IEEE Trans PAMI 33: 2341–2353
12. Tan C, Sluzek A, Seet G (2005) Model of gated imaging in turbid media. Opt Eng 44:116002
13. Giddings T, Shirron J, Tirat-Gefen A (2005) Eodes-3: An electro-optic imaging and performance prediction model. In: Proceedings of the MTS/IEEE OCEANS 2005, Washington, DC. IEEE, pp 1380–1387
14. Narasimhan SG, Nayar SK, Sun B, Koppal SJ (2005) Structured light in scattering media. In: Proceedings of the IEEE ICCV, Beijing, vol 1, pp 420–427
15. Jaffe J (2010) Enhanced extended range underwater imaging via structured illumination. Opt Expr 18(12):12328–12340
16. Levoy M, Chen B, Vaish V, Horowitz M, McDowall I, Bolas M (2004) Synthetic aperture confocal imaging. ACM Trans Graph 23:825–834
17. Nayar SK, Krishnan G, Grossberg MD, Raskar R (2006) Fast separation of direct and global components of a scene using high frequency illumination. ACM Trans Graph 25:935–944
18. Schechner YY, Karpel N (2005) Recovery of underwater visibility and structure by polarization analysis. IEEE J Ocean Eng 30:570–587
19. Schechner YY, Narasimhan SG, Nayar SK (2003) Polarization-based vision through haze. Appl Opt 42: 511–525

20. Jacques SL, Ramella-Roman J, Lee K (2002) Imaging skin pathology with polarized light. J Biomed Opt 7:329–340
21. Cui M, Yang C (2010) Implementation of a digital optical phase conjugation system and its application to study the robustness of turbidity suppression by phase conjugation. Opt Express 18(4):3444–3455

Description Logics

Bernd Neumann
University of Hamburg, Hamburg, Germany

Synonyms

Concept languages; Terminological logics

Related Concepts

▸Spatiotemporal Reasoning

Definition

Description Logics (DLs) is a family of logic-based knowledge representation formalisms for characterizing object classes and relationships between them.

Background

DLs have been developed to provide well-founded tools for knowledge representation and reasoning. Modern systems offer expressive concept languages with highly optimized reasoners, for example for concept subsumption, instance classification or consistency checks. DLs have gained additional importance due to OWL-DL, the standardized DL for the Semantic Web. Mature software tools exist to support the development of OWL-DL knowledge bases, notably the editor Protégé.

Theory

Knowledge representation in DLs is based on unary predicates called concepts (or concept terms), binary predicates called roles (or role terms), and so-called individuals. A concept is interpreted in a set-theoretical semantics as a set of elements from a domain of discourse (also called universe), a role is interpreted as a set of pairs of elements from the domain, and an individual denotes a particular element of the domain. The elements in the second position of a role pair are called role fillers. Functional roles which map each first argument into at most one role filler are called features.

To construct a knowledge base, one has to fix a set of concept names, a set of role names, and a set of individuals. Names can be used to build complex concept and role terms. This is accomplished with the help of operators whose meaning is precisely defined in terms of the set-theoretical semantics.

Important concept operators supported by DLs include

- Concept union, intersection and negation
- Existential qualification, cardinality restriction
- Role axioms (e.g., reflexivity, disjointness, transitivity, inverse roles)

Definitions of concepts and roles and their relationships to each other form the TBox (terminological box), assertions about individuals in terms of unary and binary relationships form the ABox (assertional box). Typically, the TBox contains taxonomical hierarchies as background knowledge for a domain of discourse, while the ABox describes concrete facts.

Modern DLs are designed to be decidable, i.e., to allow sound and complete reasoning processes. This limits their expressivity to a subset of First Order Logic (FOL). The computational complexity of DL reasoning services has been shown to depend critically on the expressivity of the language.

Concrete domain predicates offer an interesting way to integrate quantitative data from low-level signal processing with symbolic high-level interpretations. Several DL systems allow predicates over natural and real numbers for concept definitions involving numerical restrictions.

One of the first Computer Vision applications developed with knowledge representation using a DL-like formalism was VEIL [1] which employed Loom for deductive reasoning in high-level image interpretation. Loom is a very expressive experimental system, but different from modern DLs, Loom's deductive procedures are not rigorously complete.

The usefulness of DLs for Computer Vision applications is still being explored. In principal, DLs may offer a well-founded formalism for connecting vision to higher-level knowledge, and may provide reasoning facilities for an artificial cognitive system. But there are also aspects which limit the usefulness for Computer Vision.

One problematic aspect is the limited expressivity of DLs which may prohibit intuitive concept formulations for symbolic scene interpretation. For example, in applications such as activity monitoring or situation recognition it is desirable to define compositional hierarchies where concepts describe aggregates of constituents meeting certain constraints. Unfortunately, constraints cannot be expressed except in simple cases, and powerful constraint reasoners are not provided by DL systems.

Another problematic aspect is the limited support for the scene interpretation process which can be provided by the deductive reasoning services of DLs. Scene interpretation is known to be basically equivalent to abduction or logical model construction, hence none of the usual DL reasoning services is immediately applicable.

It has been shown, however, that DLs can be extended or combined with other systems in useful ways. An interesting extension is by a rule system. While the unrestricted use of rules may jeopardize decidability, their employment for ABox reasoning (as "DL-safe rules") can be supported without losing the advantages of a DL system.

In [2] rules are used to model the abduction process for multimedia interpretation, for example to provide deep annotations for multimedia web pages. A DL TBox represents background knowledge, and a DL ABox represents a low-level description of a multimedia document. The rules describe which higher-level hypotheses can be entertained to explain the low-level facts.

In [3] an OWL ontology of activity concepts, extended by SWRL rules (SWRL is the Semantic Web Rule Language proposed for OWL) is used as the DL kernel of a system for activity recognition. The concept definitions of the ontology are translated into compositional hierarchies representing hypothetical activities, and into rules and constraints in JAVA and JESS (the JAVA Expert System Shell). Using rule engines in parallel, the interpretation system tries to instantiate activity hypotheses based on scene observations, this way generating scene descriptions at higher compositional levels.

Application

Applications areas of DL systems include:

- Semantic Web (e.g., logic-based information retrieval)
- Electronic Business (e.g., reasoning about service descriptions)
- Medicine, Bioinformatics (e.g., representing and managing biomedical ontologies)
- Process Engineering (e.g., formal representation of chemical processes)
- Software Engineering (e.g., representing the semantics of UML class diagrams)
- High-level Computer Vision (e.g., defining high-level concepts for scene interpretation)
- Cognitive Robotics (e.g., for representing an activity ontology)

Open Problems

The scalability of DL reasoning services is still a challenge, and new optimization techniques are developed to deal with large knowledge bases. Practical experiences, in particular with OWL-DL, have also revealed the need for more expressivity. SWRL (Semantic Web Rule Language) and its successor RIF (Rule Interchange Format) provide extensions in terms of rule expressions similar to Datalog. Computer Vision applications are mainly restricted to symbolic interpretation processes, DL system support for powerful interfaces to quantitative descriptions has yet to be developed.

References

1. Russ TA, Macgregor RM, Salemi B, Price K, Nevatia R (1996) VEIL: Combining semantic knowledge with image understanding. In: ARPA image understanding workshop, Palm Springs, pp 373–380
2. Gries O, Möller R, Nafissi A, Rosenfeld M, Sokolski K, Wessel M (2010) A probabilistic abduction engine for media interpretation. In: Alferes J, Hitzler P, Lukasiewicz Th (eds) Proceedings of the international conference on web reasoning and rule systems (RR-2010), Bressanone
3. Bohlken W, Neumann B, Hotz L, Koopmann P (2011) Ontology-based realtime activity monitoring using beam

search. In: Crowley JL, Draper BA, Thonnat M (eds) Proc. ICVS 2011, Springer, Berlin/New York, pp 112–121
4. Brachmann RJ, Schmolze JG (1985) An overview of the KL-ONE knowledge representation system. Cogn Sci 9(2):171–216
5. Brachman RJ, McGuinness DL, Patel-Schneider PF, Resnick LA, Borgida A (1991). Living with CLASSIC. When and how to use a KL-ONE-like language. In: Sowa JF (ed) Principles of semantic networks. Morgan Kaufmann, San Mateo, pp 401–456
6. Woods WA, Schmolze JG (1992) The KL-ONE family. In: Lehmann FW (ed) Semantic networks in artificial intelligence. Pergamon, Oxford, pp 133–178
7. Horrocks I (1998) Using an expressive description logic: FaCT or fiction? In: Proceedings of the 6th international conference on principles of knowledge representation and reasoning (KP'98), Trento, pp 636–647
8. Haarslev V, Möller R (2001) Description of the RACER system and its applications. In: Proceedings of the 2001 description logic workshop (DL 2001), Stanford University, pp 132–141
9. Baader F, Calvanese D, McGuinness D, Nardi D, Patel-Schneider P (2003) The description logic handbook. Cambridge University Press, Cambridge
10. Neumann B, Möller RF (2006) On scene interpretation with description logics. In: Nagel HH, Christensen H (eds) Cognitive vision systems. Springer, Berlin/New York, pp 247–275

Dichromatic Reflection Model

Shoji Tominaga
Graduate School of Advanced Integration Science, Chiba University, Inage-ku, Chiba, Japan

Definition

The dichromatic reflection model is a model that describes light reflected from an object's surface as a linear combination of two components. These components are the body (diffuse) reflection and the interface (specular) reflection.

Background

Reflection models are used for image analysis and object recognition in computer vision, as well as image rendering in computer graphics. Shafer proposed the dichromatic reflection model for inhomogeneous dielectric materials [1]. The first body reflection component provides the characteristic object color, and the second interface reflection component has the same spectral composition as illumination. Thus, according to the dichromatic reflection model, all color values on a uniform object surface are described as a linear combination of two color vectors: the illumination color vector and the body color vector. Tominaga and Wandell proposed a method for testing the adequacy of the model. They show that under all illumination and viewing geometries, the spectral reflectance function is described as the weighted sum of two functions of the constant interface reflectance and the body reflectance [2]. This model is called the standard dichromatic reflection model of Type I. Moreover, Tominaga extended this model to describe surface-spectral reflectances of a variety of materials [3].

Theory

Standard Dichromatic Reflection Model

Model for Inhomogeneous Dielectrics (Type I)

The radiance of light reflected from an object's surface, $Y(\theta, \lambda)$, is a function of the wavelength λ and the geometric parameters θ, where λ ranges over a visible wavelength and θ includes the illumination direction angle, the viewing angle, and the phase angle. The dichromatic reflection model for an inhomogeneous dielectric object describes the reflected light as the sum of interface and body reflections, as shown in the following formula:

$$Y(\theta, \lambda) = c_I(\theta)L_I(\lambda) + c_B(\theta)L_B(\lambda), \qquad (1)$$

where the terms $L_I(\lambda)$ and $L_B(\lambda)$ are the spectral power distributions of the interface and body reflection components, respectively. These components are unchanged as the geometric angles vary, and then the weights $c_I(\theta)$ and $c_B(\theta)$ are the geometric scale factors. Let $S_I(\lambda)$ and $S_B(\lambda)$ be the surface-spectral reflectances for the two components, and let $E(\lambda)$ be the spectral power distribution of the incident light to the surface. So we derive

$$Y(\theta, \lambda) = c_I(\theta)S_I(\lambda)\,E(\lambda) + c_B(\theta)S_B(\lambda)E(\lambda). \qquad (2)$$

The total reflectance is defined by dividing the color signal $Y(\theta, \lambda)$ by $E(\lambda)$ as

$$S(\theta, \lambda) = c_I(\theta)S_I(\lambda) + c_B(\theta)S_B(\lambda). \quad (3)$$

The standard dichromatic reflection model incorporates the neutral interface reflection (or constant interface reflection) assumption, which means that the interface reflection component $S_I(\lambda)$ is constant and can be eliminated from Eq. (3). This is shown as follows:

$$S(\theta, \lambda) = c_I(\theta) + c_B(\theta)S_B(\lambda). \quad (4)$$

The neutral interface reflection assumption is applied to most materials, such as plastics and paints, because the interface reflection follows Fresnel's law [4] where the index of refraction is approximately constant for oil and water across the visible spectrum. For such surfaces, the specular reflection appears to have the same color as the illumination.

Extended Dichromatic Reflection Models

Model for Cloth (Type II)

There are many kinds of cloth materials such as silk, wool, rayon, polyester satin, cotton, velveteen, and velour. It is observed that the cotton and velveteen include little specular reflection from the surfaces, while the silk, wool, satin, and velour include gloss on the cloth surfaces, which show dichromatic reflection. The satin, however, is not described in Type I, because surface reflectance for such a material is dichromatic, but the specular reflection at the interface includes no illumination color. Therefore, the surface-spectral reflectance function is described as the generalized dichromatic reflection model of Type II seen below

$$S(\theta, \lambda) = c_I(\theta)\, S_I(\lambda) + c_B(\theta)S_B(\lambda), \quad (5)$$

where the interface reflectance component is not necessarily constant on wavelength.

Model for Metal (Type III)

Metals have quite different reflection properties from dielectric materials, because they only have interface reflection. If the surface is shiny and stainless, body reflection of the reflected light is negligibly small. So a sharp specular highlight is observed only at the viewing angle of the mirror direction. This type of reflection follows Fresnel's law, and the surface-spectral reflectance function depends on the incidence angle of illumination. It is pointed out that the surface-spectral reflectances of some metals are described by two interface reflection components, one of which is the constant spectral reflectance [3]. Therefore, the dichromatic reflection model of Type III is defined (*for* approximating the surface-spectral reflectance function of a metal) as

$$S(\theta, \lambda) = c_{I1}(\theta)\, S_I(\lambda) + c_{I2}(\theta), \quad (6)$$

The right-hand side in Eq. (6) represents two interface reflection components. The first component corresponds to the specular reflection at the normal incidence, which produces the major object color of a metallic surface. The second component is constant on wavelength, which corresponds to the grazing reflection at the horizontal incidence where the spectral reflectance is whitened.

Application

The dichromatic reflection models are useful for object segmentation in image understanding and illuminant estimation in color constancy. Most reflection models in computer graphics are based on the standard dichromatic reflection model [15–19].

References

1. Shafer SA (1985) Using color to separate reflection components. Color Res Appl 10:210–218
2. Tominaga S, Wandell BA (1989) The standard surface reflectance model and illuminant estimation. J Opt Soc Am A 6:576–584
3. Tominaga S (1994) Dichromatic reflection models for a variety of materials. Color Res Appl 19:277–285
4. Born M, Wolf E (1983) Principles of optics. Pergamon, Oxford
5. Klinker GJ, Shafer SA, Kanade T (1988) The measurement of highlights in color images. Int J Comput Vis 2:7–32
6. Lee HC, Breneman EJ, Schulte C (1990) Modeling light reflection for computer color vision. IEEE Trans PAMI 12:402–409
7. Tominaga S, Wandell BA (1990) Component estimation of surface spectral reflectance. J Opt Soc Am A 7(2):312–317
8. Klinker GJ, Shafer SA, Kanade T (1990) A physical approach to color image understanding. Int J Comput Vis 4:7–38

9. Tominaga S (1991) Surface identification using the dichromatic reflection model. IEEE Trans PAMI 13(7):658–670
10. Klinker GJ (1993) A Physical approach to color image understanding. AK Peters, Wellesley
11. Novak CL, Shafer SA (1994) Method for estimating scene parameters from color images. J Opt Soc Am A 11:3020
12. Tominaga S (1996) Multichannel vision system for estimating surface and illumination functions. J Opt Soc Am A 13(11):2163–2173
13. Tominaga S (1996) Surface reflectance estimation by the dichromatic model. Color Res Appl 21(2):104–114
14. Tominaga S (1996) Dichromatic reflection models for rendering object surfaces. J Imaging Sci Technol 40(6): 549–555
15. Phong BT (1975) Illumination for computer-generated pictures. Commun ACM 18:311–317
16. Torrance KE, Sparrow EM (1967) Theory for off-specular reflection from roughened surfaces. J Opt Soc Am 57: 1105–1114
17. Blinn JF (1977) Model of light reflection for computer synthesized pictures. Comput Graph 11:192–198
18. Cook RL, Torrance KE (1981) A reflection model for computer graphics. Comput Graph 15:307–316
19. R Hall (1989) Illumination and color in computer-generated imagery. Springer, Berlin

Differential Geometry of Graph Spaces

▸Isotropic Differential Geometry in Graph Spaces

Differential Geometry of Surfaces in Three-Dimensional Euclidean Space

Jan J. Koenderink
Faculty of EEMSC, Delft University of Technology, Delft, The Netherlands
The Flemish Academic Centre for Science and the Arts (VLAC), Brussels, Belgium
Laboratory of Experimental Psychology, University of Leuven (K.U. Leuven), Leuven, Belgium

Synonyms

Surfaces

Related Concepts

▸Curvature; ▸Differential Invariants; ▸Euclidean Geometry; ▸Osculating Paraboloids; ▸Parametric Curve

Definition

Differential geometry studies spatial entities in local (infinitesimal) neighborhoods. This approach enables one to exploit the power of (multi-)linear algebra. Geometrical entities are differential invariants, the generic example being the curvature of planar curves. The number of relevant differential invariants increases in more complicated settings like – in this entry – that of surfaces in three-dimensional Euclidean space.

Background

The differential geometry of surfaces in three-dimensional Euclidean space is often called "classical differential geometry" as its history goes back to the founding fathers of infinitesimal calculus, Newton and Leibnitz. There exists a huge literature, although novel additions are still forthcoming. This entry reviews the basics. There is ample literature for those needing to delve deeper.

Theory

Surfaces are smooth, two-parameter manifolds immersed into Euclidean three-space $\mathbb{E}^3$. One requires that all partial derivatives are defined, and one often requires some form of "genericity." For instance, many of the constructions discussed here will be undefined for planar surfaces, or spheres, and so forth.

There are many ways to represent surfaces, many of them useful in particular contexts. Perhaps the most commonly useful representation is by means of a parameterization $\mathbf{x}(u, v)$, where $\mathbf{x} \in \mathbb{E}^3$ and $\{u, v\} \in \mathbb{R}^2$. That is to say, the surface is treated as a two-parameter manifold immersed in Euclidean three-space. This representation is used here.

Unlike the case of space curves, there is no obvious way to introduce a parameterization akin to "arc length parameterization," as this would limit the discussion to "developable surfaces," surfaces that are metrically ("can be rolled out into") planes. Most surfaces are intrinsically curved though; the sphere is an example. Thus the parameterization is assumed to be general, although regular, that is to say, the parameter curves are supposed to mesh like the weave and weft of a cloth.

One objective is to describe surfaces purely in terms of their "shape," disregarding their spatial attitude and location. This is akin to the "natural equations" of space curves, which are fully characterized by their curvature and torsion. The case of surfaces is more complicated because one cannot simply characterize them by their curvatures, but needs certain additional constraints. Unlike curves, not anything goes (see below).

The Metrical Description of Surfaces

One naturally attaches a moving frame to a surface, much like the Frenet frame in the case of curves. For a surface one has two tangents, $\mathbf{x}_u$ and $\mathbf{x}_v$, and in $\mathbf{E}^3$ this immediately gives rise to a third vector $\mathbf{x}_u \times \mathbf{x}_v$, thus yielding a complete frame (see Figs. 1 and 2). The third vector is usually normalized as $\mathbf{n}(u, v)$, the surface normal. It is orthogonal to the tangent plane spanned by the two tangents. The metric of the tangent plane is conventionally expressed in terms of the First Principal Form

$$\mathrm{I}(\mathrm{d}u, \mathrm{d}v) = \mathrm{d}\mathbf{x} \cdot \mathrm{d}\mathbf{x} = E\mathrm{d}u^2 + 2F\mathrm{d}u\mathrm{d}v + G\mathrm{d}v^2, \quad (1)$$

where

$$E(u, v) = \mathbf{x}_u \cdot \mathbf{x}_u, \quad (2)$$

$$F(u, v) = \mathbf{x}_u \cdot \mathbf{x}_v, \quad (3)$$

$$G(u, v) = \mathbf{x}_v \cdot \mathbf{x}_v. \quad (4)$$

Because the First Principal Form is positive definite, one necessarily has $E > 0$, $G > 0$, and $EG - F^2 > 0$. Knowing $\mathrm{I}(u, v)$ allows one to calculate the lengths of curves on the surface, the surface area of patches of the surface, angles between directions on the surface, and so forth. Specifically, the area element is

$$\mathrm{d}A(u, v) = \sqrt{EG - F^2}\,\mathrm{d}u\mathrm{d}v = \|\mathbf{x}_u \times \mathbf{x}_v\|\,\mathrm{d}u\mathrm{d}v. \quad (5)$$

The First Principal Form is independent of the representation, that is to say $\mathrm{I}(u, v) = \mathrm{I}'(u', v')$, although the coefficients E, F, and G are not.

The curvature of the surface has to do with the second-order derivatives $\mathbf{x}_{uu}$, $\mathbf{x}_{uv}$, and $\mathbf{x}_{vv}$. One conventionally introduces the Second Fundamental Form

$$\begin{aligned}\mathrm{II}(\mathrm{d}u, \mathrm{d}v) &= -\mathrm{d}\mathbf{x} \cdot \mathrm{d}\mathbf{n} = \mathrm{d}^2\mathbf{x} \cdot \mathbf{n} \\ &= L\mathrm{d}u^2 + 2M\mathrm{d}u\mathrm{d}v + N\mathrm{d}v^2, \end{aligned} \quad (6)$$

where

$$L(u, v) = -\mathbf{x}_u \cdot \mathbf{n}_u = \mathbf{x}_{uu} \cdot \mathbf{n}, \quad (7)$$

$$M(u, v) = -\frac{1}{2}(\mathbf{x}_u \cdot \mathbf{n}_v + \mathbf{x}_v \cdot \mathbf{n}_u) = \mathbf{x}_{uv} \cdot \mathbf{n}, \quad (8)$$

$$N(u, v) = -\mathbf{x}_v \cdot \mathbf{n}_v = \mathbf{x}_{vv} \cdot \mathbf{n}. \quad (9)$$

The Second Principal Form, like the First Principal Form, is independent of the representation, that is, $\mathrm{II}(u, v) = \mathrm{II}'(u', v')$, if the normal has the same direction (otherwise II changes sign), although the coefficients L, M, and N are not.

The Second Principal Form describes the shape of the surface in the sense that the distance of the point $\mathbf{x}(u + \mathrm{d}u, v + \mathrm{d}v)$ to the tangent plane at $\mathbf{x}(u, v)$ is (to the second order in $\{\mathrm{d}u, \mathrm{d}v\}$) equal to $\frac{1}{2}\mathrm{II}_{u,v}(\mathrm{d}u, \mathrm{d}v)$, the surface $\frac{1}{2}\mathrm{II}_{u,v}(\mathrm{d}u, \mathrm{d}v)$ being the osculating paraboloid at $\{u, v\}$.

It is convenient to introduce Dupin's indicatrix (Figs. 3–5)

$$\mathrm{II}_{u,v}(\mathrm{d}u, \mathrm{d}v) = \pm\varepsilon^2, \quad (10)$$

with $\varepsilon \ll 1$, which, generically, is an ellipse or hyperbola. One thinks of the indicatrix as the outline of the wound that is inflicted if you take a chip off the surface with a flat knife. If the indicatrix is an ellipse, the surface is called elliptic; if it is a hyperbola, it is called hyperbolic at that point. In case $LN - M^2 = 0$, one has the degenerate parabolic case.

There are various ways to conceive of the curvature of surfaces. One common way is to consider the curvatures of its normal sections (Fig. 6) that are the curves of intersection with planes that contain the normal direction. The normal curvature depends on the direction $\mathrm{d}u : \mathrm{d}v$, the normal curvature being given by

$$\kappa_n(\mathrm{d}u, \mathrm{d}v) = \frac{\mathrm{II}(\mathrm{d}u, \mathrm{d}v)}{\mathrm{I}(\mathrm{d}u, \mathrm{d}v)} = \frac{L\mathrm{d}u^2 + 2M\mathrm{d}u\mathrm{d}v + N\mathrm{d}v^2}{E\mathrm{d}u^2 + 2F\mathrm{d}u\mathrm{d}v + G\mathrm{d}v^2}. \quad (11)$$

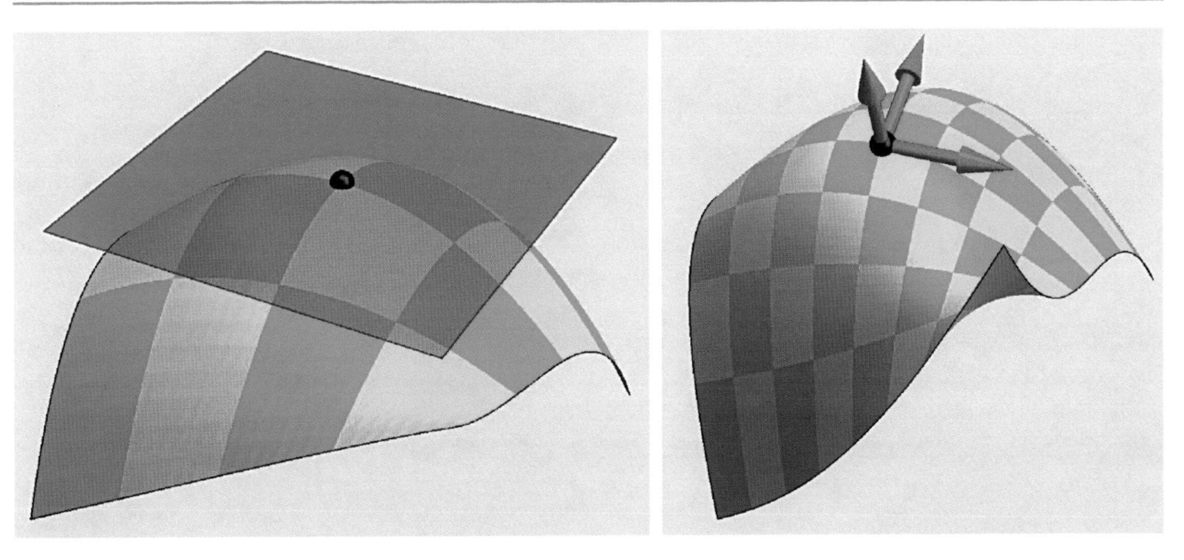

Differential Geometry of Surfaces in Three-Dimensional Euclidean Space, Fig. 1 At *left*, the tangent plane at a point of a smooth surface. At *right*, a frame composed of the two tangents along the parameter directions and the unit surface normal. The tangent vectors span the tangent plane

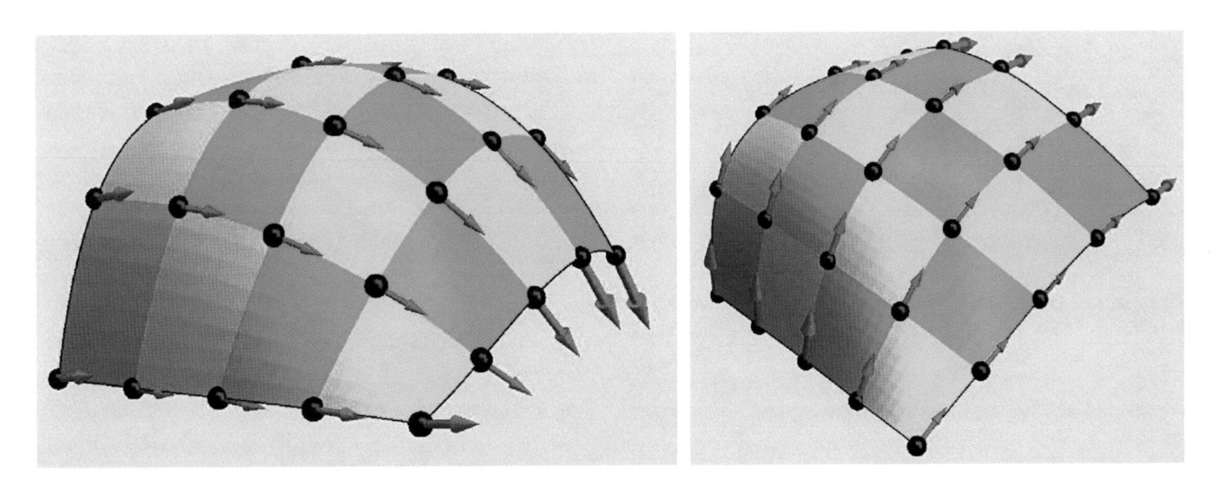

Differential Geometry of Surfaces in Three-Dimensional Euclidean Space, Fig. 2 The two fields of surface tangents $\mathbf{x}_u$ and $\mathbf{x}_v$

The principal curvature varies between two extreme values, the *principal curvatures* $\kappa_{1,2}$. The principal curvatures are the solutions of the quadratic equation

$$\kappa^2 - 2H\kappa + K = 0, \tag{12}$$

where

$$H = \frac{EN + GL - 2FM}{2(EG - F^2)} = \frac{1}{2}(\kappa_1 + \kappa_2), \tag{13}$$

$$K = \kappa_1\kappa_2 = \frac{LN - M^2}{EG - F^2}. \tag{14}$$

The differential invariant $H(u, v)$ is known as the mean curvature, and the invariant $K(u, v)$ as the Gaussian curvature. One easily shows that points with $K > 0$ are elliptic and points with $K < 0$ hyperbolic. For points with $K = 0$, one or both of the principal curvatures vanish. In the latter case, one has planar points (the osculating paraboloid being degenerated to a plane); in the former case, a cylindrical point (the osculating paraboloid being degenerated to a cylinder).

The "mean curvature" is not just the mean of the principal directions. One can easily show that the mean of the normal curvatures of arbitrary mutually perpendicular normal sections also equals the mean curvature.

Another way to understand the Gaussian curvature is through the relation

$$K = \frac{\mathbf{n}_u \times \mathbf{n}_v \, \mathrm{d}u\mathrm{d}v}{\mathbf{x}_u \times \mathbf{x}_v \, \mathrm{d}u\mathrm{d}v} = \frac{\mathrm{d}\Omega}{\mathrm{d}A}, \tag{15}$$

where $\mathrm{d}\Omega$ is the solid angle subtended by the bushel of normals inside the patch of area $\mathrm{d}A$. This is the direct analogon of the definition of the curvature of curves as the rate of change of direction with respect to arc length. One often introduces the Third Principal Form

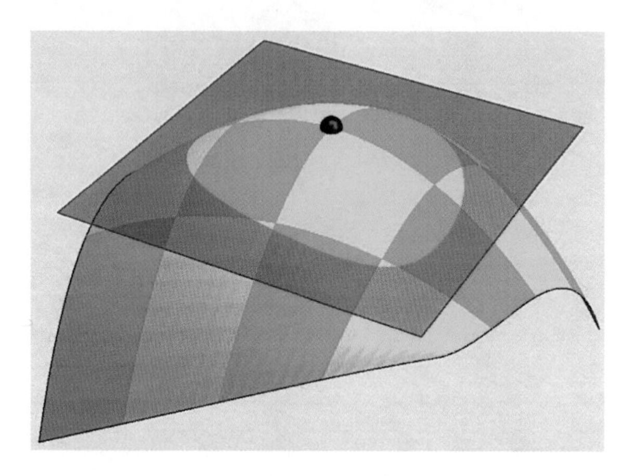

Differential Geometry of Surfaces in Three-Dimensional Euclidean Space, Fig. 3 Dupin's indicatrix can be understood as the shape of the "wound" inflicted to a surface by a cut parallel to the tangent plane. Such curves may have arbitrary shapes, but generically one obtains the cases shown in Fig. 5 for infinitesimally small wounds

$\mathrm{III} = \mathrm{d}\mathbf{n} \cdot \mathrm{d}\mathbf{n}$ through the relation

$$\mathrm{III} - 2H\,\mathrm{II} + K\,\mathrm{I} = 0. \tag{16}$$

Notice that III equals the First Principal Form of the unit sphere. The mapping of the surface $\mathbf{x}(u, v)$ on the unit sphere $\mathbf{n}(u, v)$ is known as the Gauss map, or spherical image. Its area magnification is the Gaussian curvature, and the magnification in an arbitrary direction is the normal curvature in that direction. The Gauss map is a very convenient tool in a great many applications (Figs. 7–11).

The directions $\mathrm{d}u{:}\mathrm{d}v$ of the principal sections are evidently special; they are known as the principal directions of curvature (see Fig. 12). One easily shows them to be mutually orthogonal. Curves for which the tangents coincide with directions of principal curvature are lines of curvature. A patch of the surface free of umbilical points (points where Dupin's indicatrix is a circle, thus where the principal directions are undefined) can be covered with a mesh of two mutually orthogonal families of lines of curvature. For such a parameterization, one has $F = M = 0$, a most "natural" representation of the surface for many purposes. In that case, the principal curvatures are $\kappa_1 = L/E$, $\kappa_2 = N/G$. An intuitive, because geometrical, characterization of the principal directions is $\mathrm{d}\mathbf{n} \times \mathrm{d}\mathbf{x} = 0$. In such a case, one has $\mathrm{d}\mathbf{n} = -\kappa \mathrm{d}\mathbf{x}$, the formula of Rodriguez. On the Gaussian sphere, conjugate directions map on mutually orthogonal directions. Euler's theorem $\kappa_n = \kappa_1 \cos^2\varphi + \kappa_2 \sin^2\varphi$ relates the normal curvature in the direction φ with respect to the first principal direction to the principal curvatures.

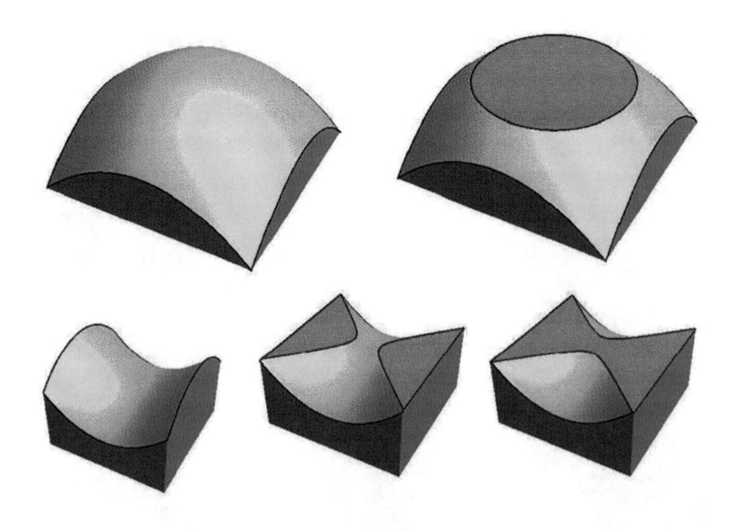

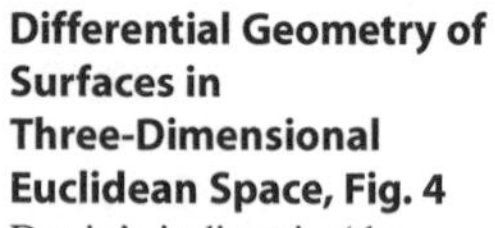

Differential Geometry of Surfaces in Three-Dimensional Euclidean Space, Fig. 4 Dupin's indicatrix (the "wound") in the case of an elliptic (*top*) and a hyperbolic (*bottom*) surface point. In the case of the elliptic point, the indicatrix has only one branch; in the case of the hyperbolic point it has two

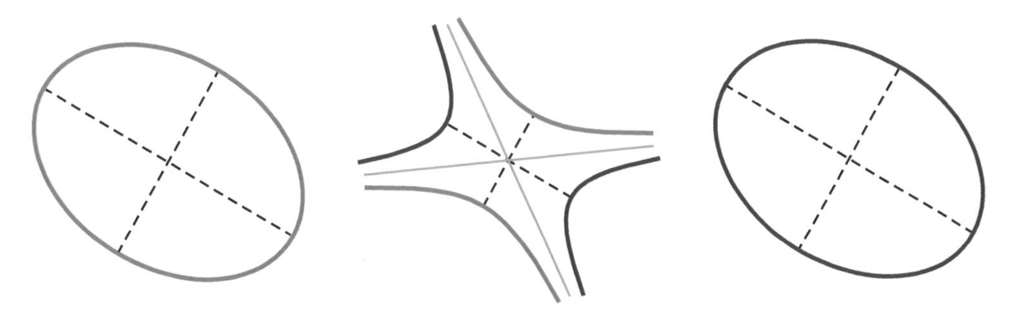

Differential Geometry of Surfaces in Three-Dimensional Euclidean Space, Fig. 5 Dupin's indicatrices for convex and concave elliptic, and for a hyperbolic surface point. The principal directions are dashed, the asymptotic directions drawn in *yellow*. The *red* and *blue* curves represent the two branches of the indicatrix

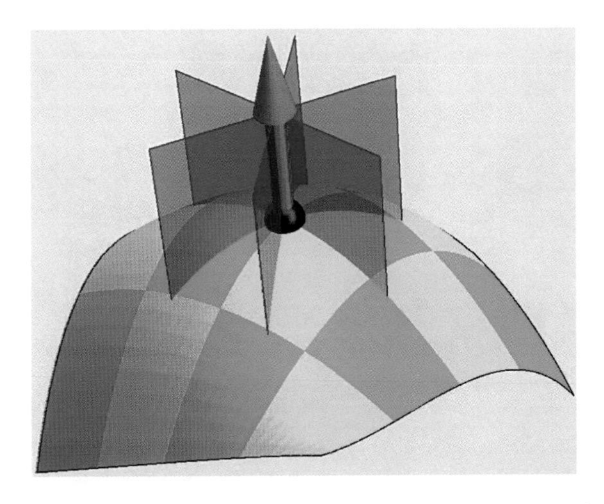

Differential Geometry of Surfaces in Three-Dimensional Euclidean Space, Fig. 6 A series of normal sections. The normal planes contain the surface normal. They cut the surface in planar curves, the "normal sections." The curvature of these normal sections depends on the orientation of the normal plane. The orientations that yield extremes of normal curvature are the principal directions

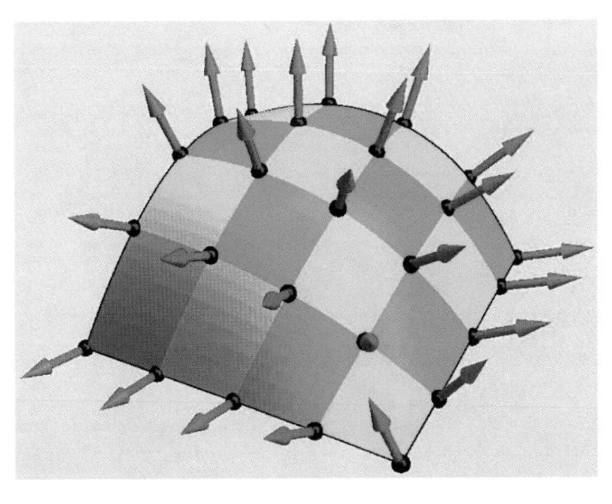

Differential Geometry of Surfaces in Three-Dimensional Euclidean Space, Fig. 7 The field of surface normals. The Gauss map, or spherical image, is obtained by moving all normals to the origin. Hence, one obtains a map from the surface to the unit sphere

Notice that it trivially applies to umbilical and planar points too.

In the case of a hyperbolic point, the principal curvatures are of opposite sign and, by Euler's theorem, there apparently exist directions for which the normal curvature vanishes. These directions are given by $\mathrm{II}(\mathrm{d}u, \mathrm{d}v) = 0$, and are known as asymptotic directions. At generic hyperbolic points, there exist apparently two, mutually transverse, asymptotic directions. They are bisected by the principal directions. A curve whose tangents are asymptotic directions is known as an asymptotic curve. In the neighborhood of a hyperbolic point, one may cover the surface with a mesh of two mutually transverse families of asymptotic lines. In such a case, one has $\mathrm{II}(\mathrm{d}u, \mathrm{d}v) = M\mathrm{d}u\mathrm{d}v$. The curvature vector $\mathbf{k}$ of an asymptotic curve is perpendicular to the normal; typically the osculating plane is tangent to the surface. In such cases, the torsion satisfies $\tau^2 + K = 0$, the formula of Beltrami–Enneper (Fig. 13).

Directions $\mathrm{d}u{:}\mathrm{d}v$ and $\delta u{:}\delta v$ are called mutually conjugate in case one has $\mathrm{d}\mathbf{x} \cdot \delta\mathbf{n} = \delta\mathbf{x} \cdot \mathrm{d}\mathbf{n} = 0$. This condition can also be written in the form $L\mathrm{d}u\delta u + M(\mathrm{d}u\delta v + \delta u\mathrm{d}v) + N\mathrm{d}v\delta v = 0$, which formally looks like a scalar product. At a generic point, each direction has a conjugate mate. The directions of principal curvature are mutually conjugate, whereas

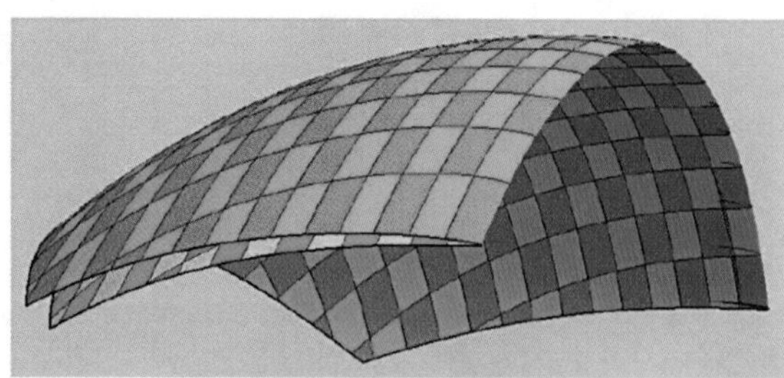

Differential Geometry of Surfaces in Three-Dimensional Euclidean Space, Fig. 8 At *left*, the so-called "shoe surface" $\{u, v, u^3/3 - v^2/2\}$; at *right*, its spherical image. Notice the fold in the spherical image. The fold is image of the parabolic curve $u = 0$. The points with parameters $\{\pm u, v\}$ have parallel normals and thus map on the same point of the unit sphere

the asymptotic directions are self-conjugate. The conjugate directions are important in computer vision because they relate the occluding contour to the viewing direction.

Like in the case of curves, one would like to attach a "moving frame" to the surface (Fig. 14). Classically, one has the Gauss equations

$$\begin{pmatrix} \mathbf{x}_{uu} \\ \mathbf{x}_{uv} \\ \mathbf{x}_{vv} \end{pmatrix} = \begin{pmatrix} \Gamma^1_{11} & \Gamma^2_{11} & L \\ \Gamma^1_{12} & \Gamma^2_{12} & M \\ \Gamma^1_{22} & \Gamma^2_{22} & N \end{pmatrix} \begin{pmatrix} \mathbf{x}_u \\ \mathbf{x}_v \\ \mathbf{n} \end{pmatrix}, \quad (17)$$

and the Weingarten equations

$$\begin{pmatrix} \mathbf{n}_u \\ \mathbf{n}_v \end{pmatrix} = \begin{pmatrix} \beta^1_1 & \beta^2_1 \\ \beta^1_2 & \beta^2_2 \end{pmatrix} \begin{pmatrix} \mathbf{x}_u \\ \mathbf{x}_v \end{pmatrix}, \quad (18)$$

where the Γ^k_{ij} are the Christoffel symbols of the second kind, and

$$\beta^1_1 = \frac{MF - LG}{EG - F^2} \quad (19)$$

$$\beta^2_1 = \frac{LF - ME}{EG - F^2} \quad (20)$$

$$\beta^1_2 = \frac{NF - MG}{EG - F^2} \quad (21)$$

$$\beta^2_2 = \frac{MF - NE}{EG - F^2}. \quad (22)$$

The Christoffel symbols of the second kind depend on the coefficients of the First Principal Form and their

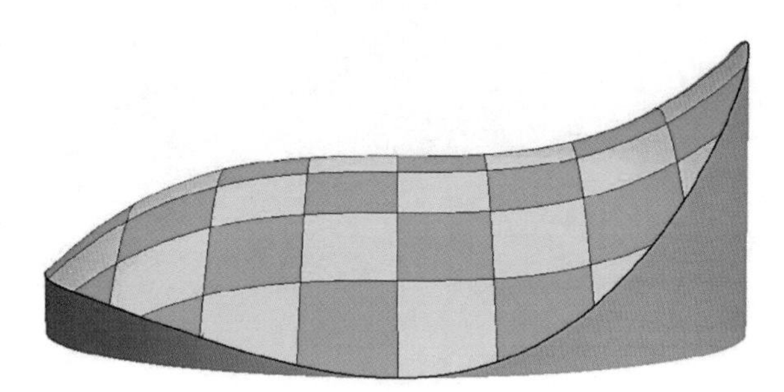

Differential Geometry of Surfaces in Three-Dimensional Euclidean Space, Fig. 9 As seen from a tangent to the parabolic curve, the outline of the shoe surface has an inflection. This is generic

derivatives. One has

$$\Gamma^1_{11} = \frac{GE_u - 2FF_u + FE_v}{2(EG - F^2)} \qquad \Gamma^1_{12} = \frac{GE_v - FG_u}{2(EG - F^2)}$$
$$\Gamma^1_{22} = \frac{2GF_v - GG_u - FG_v}{2(EG - F^2)}$$
$$\Gamma^2_{11} = \frac{2EF_u - EE_v + FE_u}{2(EG - F^2)} \qquad \Gamma^2_{12} = \frac{EG_u - FE_v}{2(EG - F^2)}$$
$$\Gamma^2_{22} = \frac{EG_v - 2FF_v + FG_u}{2(EG - F^2)}. \quad (23)$$

From the Gauss equations, the geometrical interpretation of the Christoffel symbols of the second kind is obvious: the symbol Γ^k_{ij} is the k-component $(k = u, v)$ of the rate of change of the tangent

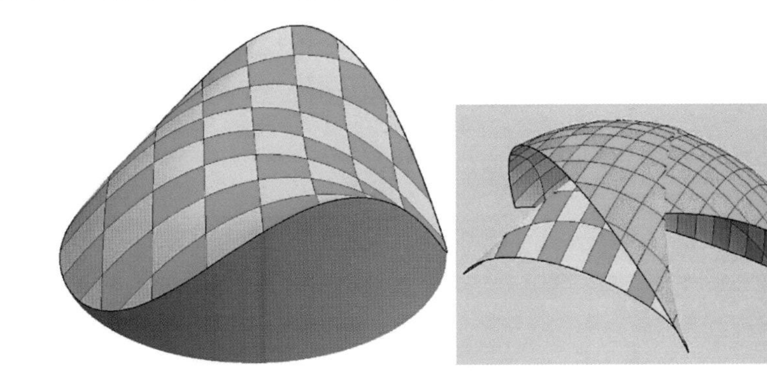

Differential Geometry of Surfaces in Three-Dimensional Euclidean Space, Fig. 10 At *left*, the "Menn's surface" $\{u, v, au^4 + u^2v - v^2\}$; at *right*, its spherical image. Here the Gaussian image has a cusp; in the neighborhood of the cusp point, one has triples of parallel normals at distinct points of the surface

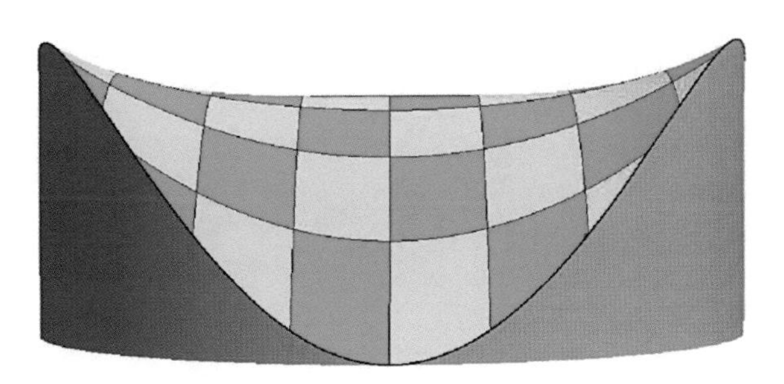

Differential Geometry of Surfaces in Three-Dimensional Euclidean Space, Fig. 11 At the cusp of Menn's surface, the surface is very flat; from this viewpoint, the curvature of the outline is zero. Such points are generically isolated points on parabolic curves

$\mathbf{x}_i$ $(i = u, v)$ when transported over $\mathbf{x}_j$ $(j = u, v)$. Although the notation Γ^k_{ij} might suggest that the geometrical object Γ is a tensorial quantity, this is really not the case; its transformational properties are different.

In analogy with the case of space curves, one might guess that the "natural equations" of a surface are simply the coefficients of the First and Second Principal Forms as a function of $\{u, v\}$. However, this is not the case, the reason being the equality of mixed partial derivatives (i.e., $(\mathbf{x}_{u,v})_{uv} = (\mathbf{x}_{u,v})_{vu}$). One has to impose the additional constraints ("compatibility equations"),

$$L_v - M_u = L\Gamma^1_{12} + M(\Gamma^2_{12} - \Gamma^1_{11}) - N\Gamma^2_{11}, \quad (24)$$

$$M_v - N_u = L\Gamma^1_{22} + M(\Gamma^2_{22} - \Gamma^1_{12}) - N\Gamma^2_{12}, \quad (25)$$

the Codazzi–Mainardi equations, and

$$LN - M^2 = F_{uv} - \frac{1}{2}E_{vv} - \frac{1}{2}G_{uu} + \frac{1}{EG-F^2}\left(\det\begin{vmatrix} 0 & F_v - \frac{1}{2}G_u & \frac{1}{2}G_v \\ \frac{1}{2}E_u & E & F \\ F_u - \frac{1}{2}E_v & F & G \end{vmatrix} - \det\begin{vmatrix} 0 & \frac{1}{2}E_v & \frac{1}{2}G_u \\ \frac{1}{2}E_v & E & F \\ \frac{1}{2}G_u & F & G \end{vmatrix}\right), \quad (26)$$

which, by changing $LN - M^2$ into $K(EG - F^2)$, is known as Gauss' Theorema Egregium. It expresses the Gaussian curvature in terms of the metric and its first-order derivatives, a major result of classical differential geometry. Now we have the result that I and II, satisfying the compatibility equations, and with $E, G, EG - F^2 > 0$, determine the surface up to Euclidean movements.

With special parameterizations, these formulas may simplify greatly. For instance, in terms of a parameterization with lines of curvature, the Codazzi–Mainardi equations can be written as

$$\frac{\partial\kappa_1}{\partial v} = \frac{1}{2}\frac{E_v}{E}(\kappa_2 - \kappa_1), \quad \frac{\partial\kappa_2}{\partial u} = \frac{1}{2}\frac{G_u}{G}(\kappa_1 - \kappa_2). \quad (27)$$

An important special case in many applications is the description of the surface as a Monge patch $x\mathbf{e}_x + y\mathbf{e}_y + z(x, y)\mathbf{e}_z$, where $\mathbf{e}_{x,y,z}$ are an orthonormal Cartesian basis of $\mathbb{E}^3$. One can conventionally write $p = z_x, q = z_y$ for the first- and $r = z_{xx}, s = z_{xy}, t = z_{yy}$ for the second-order partial derivatives

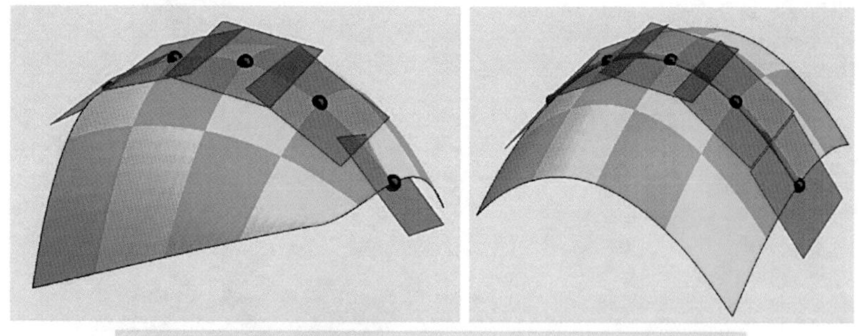

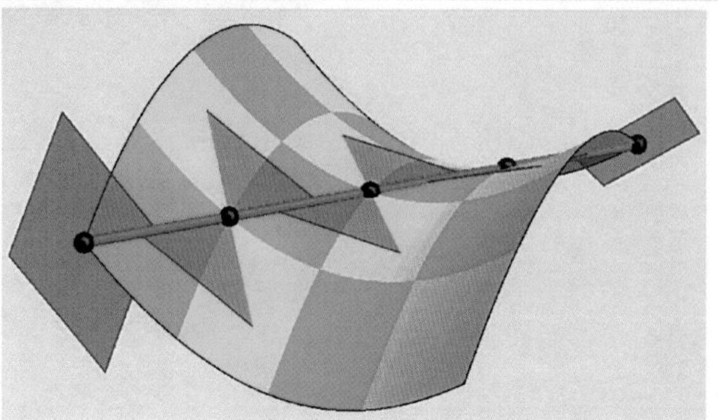

Differential Geometry of Surfaces in Three-Dimensional Euclidean Space, Fig. 12 At *top left*, a field of tangent planes along a curve: notice the rotations of the plane. At *top right*, the tangent planes along a curve of principal curvature. Here the tangent planes rotate about axes at right angles to the curve. At *bottom*, a field of tangent planes along an asymptotic curve. Here the tangent planes rotate about the tangents of the curve

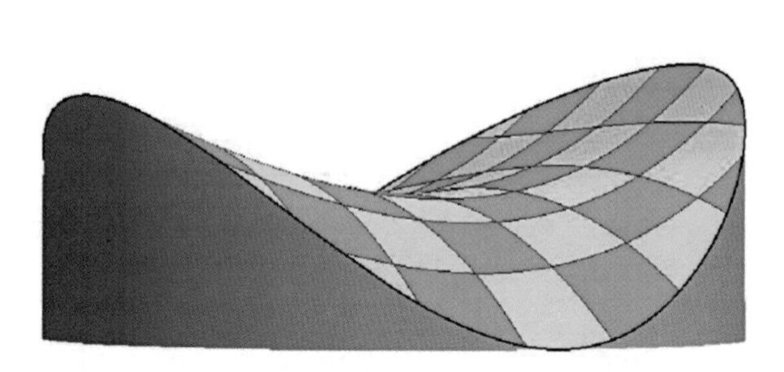

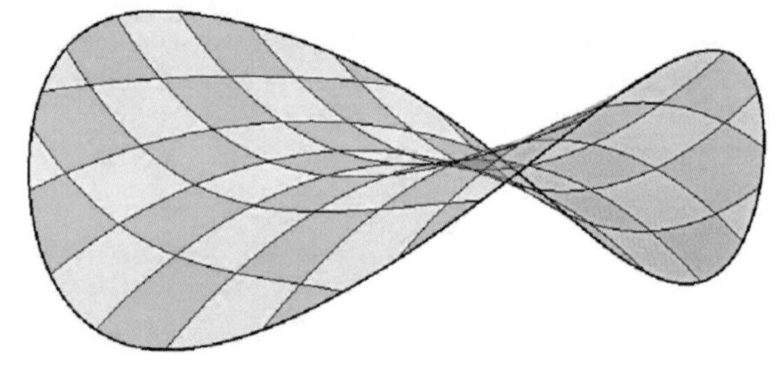

Differential Geometry of Surfaces in Three-Dimensional Euclidean Space, Fig. 13 At *left*, a view of a hyperbolic surface. Notice the singular outline; this is generic for a view along an asymptotic direction. At *right*, a partially transparent hyperbolic surface which allows one to judge the singularity of the outline more easily. It is a cusp. In the opaque view, only one branch of the cusp is visible at any time

of the "height function" $z(x, y)$. Then the tangents are $\mathbf{t}_x = \mathbf{e}_x + p\mathbf{e}_z$ and $\mathbf{t}_y = \mathbf{e}_y + q\mathbf{e}_z$, the normal $\mathbf{n} = (-p\mathbf{e}_x - q\mathbf{e}_y + \mathbf{e}_z)/\sqrt{g}$, where $g = 1 + p^2 + q^2$ is the squared area element $EG - F^2$. The First Fundamental Form is

$$\mathrm{I}(\mathrm{d}x, \mathrm{d}y) = (1+p^2)\mathrm{d}x^2 + 2pq\,\mathrm{d}x\mathrm{d}y + (1+q^2)\mathrm{d}y^2, \tag{28}$$

the Second Principal Form

$$\mathrm{II}(\mathrm{d}x, \mathrm{d}y) = \frac{r\,\mathrm{d}x^2 + 2s\,\mathrm{d}x\mathrm{d}y + t\,\mathrm{d}y^2}{\sqrt{g}}, \tag{29}$$

whereas the Christoffel symbols become

$$\begin{aligned} \Gamma^1_{11} &= \frac{pr}{g}, & \Gamma^1_{12} &= \frac{ps}{g}, & \Gamma^1_{22} &= \frac{pt}{g}, \\ \Gamma^2_{11} &= \frac{qr}{g}, & \Gamma^2_{12} &= \frac{qs}{g}, & \Gamma^2_{22} &= \frac{qt}{g}, \end{aligned} \tag{30}$$

and the coefficients of the Weingarten equations

$$\begin{aligned} \beta^1_1 &= \frac{spq - rq^2 - r}{g^{3/2}}, & \beta^1_2 &= \frac{tpq - sq^2 - s}{g^{3/2}}, \\ \beta^2_1 &= \frac{rpq - sp^2 - s}{g^{3/2}}, & \beta^2_2 &= \frac{spq - tp^2 - t}{g^{3/2}}. \end{aligned} \tag{31}$$

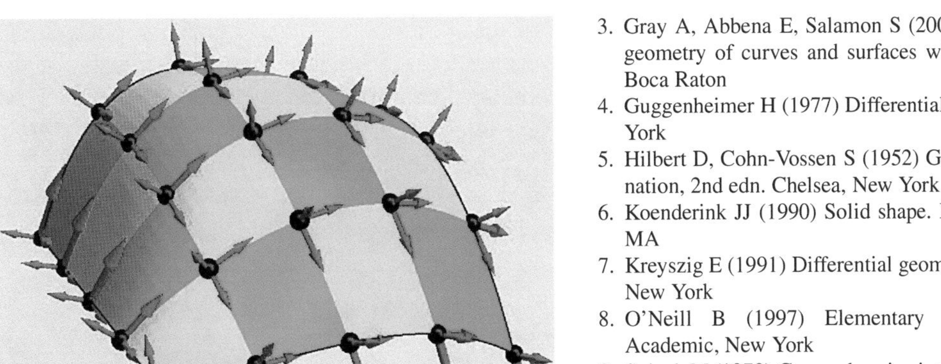

Differential Geometry of Surfaces in Three-Dimensional Euclidean Space, Fig. 14 The "moving frame" of a surface. The rotations and deformations of the frame as it moves over the surface are described by the equations of Gauss and Weingarten

Expression for many types of surface representations (e.g., implicit form of the type $F(x, y, z) = 0$) or special forms (e.g., surfaces of revolution) are readily found in the literature.

Of course, one often has to deal with data that are not in some convenient analytical form. This calls for methods of numerical analysis and computational geometry.

Open Problems

In this section, we described the classical differential theory of surfaces. Many settings in computer vision and image processing do not involve a Euclidean structure though. Cases include affine and projective spaces, as well as (especially important in image processing) "isotropic differential geometry" in "graph spaces." In such cases, one needs to substitute the appropriate differential geometry. Examples abound where authors failed to take this into account and used expressions from Euclidean differential geometry inappropriately. The reader should beware.

References

1. do Carmo MP (1976) Differential geometry of curves and surfaces. Prentice-Hall, Englewood Cliffs
2. Eisenhart LP (2004) A treatise on the differential geometry of curves and surfaces. Dover, New York
3. Gray A, Abbena E, Salamon S (2006) Modern differential geometry of curves and surfaces with mathematica. CRC, Boca Raton
4. Guggenheimer H (1977) Differential geometry. Dover, New York
5. Hilbert D, Cohn-Vossen S (1952) Geometry and the imagination, 2nd edn. Chelsea, New York
6. Koenderink JJ (1990) Solid shape. MIT Press, Cambridge, MA
7. Kreyszig E (1991) Differential geometry, Chapter II. Dover, New York
8. O'Neill B (1997) Elementary differential geometry. Academic, New York
9. Spivak M (1979) Comprehensive introduction to differential geometry, 5 vols. Publish or Perish, Houston
10. Struik DJ (1988) Lectures on classical differential geometry, 2nd edn. Dover, New York

Differential Invariants

Isaac Weiss
Center for Automation Research, University of Maryland at College Park, College Park, MD, USA

Synonyms

Local invariants

Related Concepts

▶Affine Invariants

Basic Concepts

Invariants are entities that do not change under the action of a transformation group, e.g., projective invariants are unchanged under projective transformations. One can distinguish between differential and algebraic invariants. Algebraic invariants involve algebraic forms such as points, lines, conics, etc., while differential invariants involve general differentiable curves and surfaces. Here we concentrate on invariants of curves, mostly in the plane. Differential invariants of surfaces also exist [18]. Also mentioned are differential invariants of space curves and of fields such as optic flow and shading.

Both differential and algebraic invariants, well developed in the mathematical literature, were introduced into computer vision in [11] in order to eliminate the search for the correct viewpoint when trying to recognize an object. Compared to algebraic invariants, the advantages of differential invariants are as follows: (a) Differential invariants can describe any arbitrary curve visible in the image. Algebraic curves in images are quite limited. While ellipses are common as projections of round objects, other algebraic curves are rare. (b) Algebraic invariants require whole curves to be visible and are thus susceptible to occlusions. Differential invariants can be extracted from any visible parts of the curve. The disadvantage is that the differential invariants are harder to extract from the image reliably.

It is useful in projective geometry to replace the normal Cartesian coordinates (x, y) of a point in the plane by a triplet

$$\mathbf{x} = (x_1, x_2, x_3) = \lambda(x, y, 1) \tag{1}$$

with an arbitrary factor λ. To these one adds points with $x_3 = 0$ which have no corresponding Cartesian coordinates but can be thought of as the points at infinity. The point (0,0,0) is excluded from the space. The homogeneous coordinates are also convenient in the affine subgroup of the projective transformation group. In this case, one sets $\lambda = 1$ and $x_3 = 1$.

In these coordinates, a projective transformation (projectivity) can be written as a linear transformation of $\mathbf{x}$ to $\tilde{\mathbf{x}}$:

$$\tilde{\mathbf{x}} = \lambda(\mathbf{x})T\mathbf{x} \tag{2}$$

with T being a constant 3×3 matrix. The factor λ can change arbitrarily from one point to another.

A shape can be described in different ways by various shape descriptors s_k. One can express a curve parametrically as $\mathbf{x}(t)$, in which case the descriptors are usually derivatives with respect to t. It can also be expressed implicitly as a function $f(\mathbf{x}, s_k) = 0$, with the descriptors s_k being constant coefficients, e.g., of a conic. They transform to $\tilde{s}_k$.

A *relative invariant I of weight w* is defined as a function of the shape descriptors s_k that transforms as

$$I(\tilde{s}_k) = J^{-w} I(s_k). \tag{3}$$

J is the Jacobian of the transformation. There are in general different weights for different Js: J of the coordinate transformation is the determinant $|T|$, and there is also J of a parameter change, i.e., $d\tilde{t}/dt$. A change can also result from multiplication of s_k homogeneously by a factor λ. This is of importance in projective homogeneous coordinates. In this case, the invariant can change as

$$I(\lambda s_k) = \lambda^d I(s_k) \tag{4}$$

with d being the *degree* of the invariant.

An invariant of weights and degree zero is *absolute*.

The Jacobians and λ can vary from one point to another, namely, they depend on $\mathbf{x}, t$, but they do not depend on the descriptors s_k of the shape itself.

General Properties

Among the general properties of interest here are whether there is a set of invariants that are necessary and sufficient to describe a shape. For differential invariants, the completeness theorem holds ([3], p. 144):

Theorem 1 *All differential invariants of a (transitive) transformation group in the plane are functions of the two lowest-order invariants and their derivatives.*

This is equivalent to saying that the original curve can be reconstructed from the two lowest-order-independent invariants that exist at each point, up to the relevant transformation. This is because these invariants contain all the information about the curve except for a group transformation.

This completeness makes it possible to create an invariant "signature" of a curve. For example, in the Euclidean case, all invariants can be derived from the curvature and the arc-length at each point, $\kappa(\tau)$. Thus, one can use the signature, namely, the plot of the curvature vs. the arc-length, to recognize the curve up to a Euclidean transformation. Similar quantities can be derived in the affine case as described later. In the projective case, there is no natural arc-length or curvature, but there are still two independent invariants at each curve point that can be plotted against each other. Since this signature is invariant to the viewpoint, the curve can be recognized without a search for the correct viewpoint. This is done without having to find the point correspondence along the curve, a common and difficult problem.

Another general property is that the more general the transformation is, the more descriptors one needs to extract from the image to find invariants. The general projectivity in the plane has eight coefficients that need to be eliminated, so to obtain two invariants, one needs to extract from the image at least ten descriptors per point. The affine subgroup has only six coefficients to eliminate and the Euclidean has three. Thus, more general invariants lead to using higher-order derivatives or higher-order implicit curves.

Overview

Wilczynski's method [6, 17, 18] was the first to obtain closed form formulas for projective invariants of curves and surfaces. While interesting mathematically, it has proved difficult to implement in vision because of the high order of derivatives involved.

Cartan's "moving frame" method [3] is easily applied to transformations that have a natural arc-length, such as Euclidean or unimodular affine transformations (i.e., with $|T| = 1$). In the latter case, one can obtain the affine arc-length and affine curvature. However, it is hard to apply it to transformations (such as projectivities) which do not admit a natural arc-length parameter.

The determinants method takes advantage of the invariant properties of determinants under linear transformations and is also very useful for algebraic invariants. Here it is used to obtain the affine arc-length and curvature more simply than in Cartan's method.

All the above methods depend on the availability of derivatives of the curve $\mathbf{x}(t)$ w.r.t. the parameter t. These can be hard to extract reliably. The canonical method [4, 12] dispenses with both the derivatives and the curve parameter. The parameter is not really a part of the geometry of the curve; it is an artifact introduced for convenience. It turns out that while no method can reduce the number of descriptors that one needs to extract, the canonical method provides descriptors that are more robust and easier to extract. In a nutshell, an auxiliary implicit curve (i.e., a function $f(x, y)$ without a parameter) is fitted around a point of the data curve. Then, the coordinate system is transformed to a canonical, or standard system, in which this auxiliary curve has a particularly simple form. All quantities in this system are invariant.

Another important issue is the connection between 2D images and 3D objects. It has long been known that (in general) the projection from a 3D shape to a single 2D image does not have invariants. There is simply not enough information in a 2D image to make up for the missing depth information by purely geometric means. The invariants discussed above are 2D to 2D (or nD to nD). However, one can find useful invariant *constraints* between invariants of 3D curves and those of their 2D projections [15]. These can identify a shape with the help of information from known models.

Finally, a method is described for obtaining accurate derivatives [13]. The common methods yield incorrect results even in the analytic noiseless case. This systematic error is analyzed and corrected.

A 1D Projective Differential Invariant

First, it is worth mentioning a well-known one-dimensional differential invariant, namely, the Schwarzian derivative [10]. Consider a particle moving along a straight line, with its position at a time t measured by a (nonhomogeneous) coordinate $x(t)$. The Schwarzian derivative S_x is defined as

$$S_x \equiv \left(\frac{x''(t)}{x'(t)}\right)' - \frac{1}{2}\left(\frac{x''(t)}{x'(t)}\right)^2, \tag{5}$$

and it is invariant under projective transformations of the line, given by $\tilde{x} = (ax + b)/(cx + d)$, with arbitrary a, b, c, d, as can be verified directly.

Furthermore, the differential equation

$$S_x = g(t) \tag{6}$$

(where $g(t)$ is given) determines the function $x(t)$ up to 1D projectivity. The Schwarzian derivative is not invariant to change of the parameter t, except by a 1D projectivity similar to that of x.

It is interesting that this invariant can be obtained as an infinitesimal limit of the well-known 1D cross ratio.

Wilczynski's Method

This method finds invariants of the transformation (Eq. 2) in stages. First it finds invariants to the linear part T of the transformation, and from these it derives invariants to λ and then to change in the curve parameter t.

Given a plane curve $\mathbf{x}(t)$, invariants to T can be obtained by solving the linear algebraic system of equations

$$\mathbf{x}''' + 3p_1\mathbf{x}'' + 3p_2\mathbf{x}' + p_3\mathbf{x} = 0 \tag{7}$$

for the three unknowns p_1, p_2, p_3 at point t. It is easy to show, by multiplying the equation through by T, that these solutions p_i are invariant to T. (In fact, p_i are expressible as determinants which fits the determinants method.) However, they are not invariant to change in the arbitrary factor $\lambda(\mathbf{x}(t))$ nor to change in the curve parameter t. From these p_i, one constructs the "semi-invariants":

$$P_2 = p_2 - p_1^2 - p_1' \tag{8}$$

$$P_3 = p_3 - 3p_1p_2 + 2p_1^3 - p_1''. \tag{9}$$

These remain unchanged under multiplication of the coordinates by a factor $\lambda(\mathbf{x})$, but not under change of the parameter t.

The full invariants are

$$\Theta_3 = P_3 - \frac{3}{2}P_2' \tag{10}$$

$$\Theta_8 = 6\Theta_3\Theta_3'' - 7(\Theta_3')^2 - 27P_2\Theta_3^2. \tag{11}$$

Under change of the parameter t, they transform as $\tilde{\Theta}_w = (d\tilde{t}/dt)^{-w}\Theta_w$, and thus w is the weight (Eq. 3). The subscript corresponds to the weight w.

Θ_3 is the only linear invariant, and $\sqrt[3]{\Theta_3}$ can be called the projective arc-length. All other invariants can be derived from Θ_3, Θ_8 and their derivatives. This is a special case of the completeness Theorem 1. The original curve can be derived from these up to a projectivity. One can thus call the above two invariants a *complete* set of independent invariants.

These two invariants still contain the unknown weights, which vary from point to point. Toeliminate them, one can use the invariant $\Theta_{12} = 3\Theta_3\Theta_8' - 8\Theta_3'\Theta_8$. One can now define the two absolute invariants [11]:

$$I_1 = \frac{\Theta_3^8}{\Theta_8^3}, \qquad I_2 = \frac{\Theta_3^4}{\Theta_{12}}. \tag{12}$$

These can be plotted against each other in an invariant plane with coordinates I_1, I_2. One can thus obtain an invariant signature curve identifying the original curve up to a projectivity.

Since these invariants contain the eighth derivative, they are not very practical. The semi-invariant P_2 above contains the fourth derivative only. The other, P_3, contains the fifth, but it can be replaced by

$$P_3^* = P_3 - P_2' \tag{13}$$

which again contain only four derivatives.

One can clearly see the burden that the curve parameter imposes on the method. The semi-invariants P_2, P_3^* are invariant to the projectivity and contain only fourth derivatives. It is the requirement of invariance to the change of parameter t that pushes the number of derivatives needed to eight. Thus, if one can get rid of the parameter, there will be fewer local quantities needed, and the robustness of the invariance will increase.

The Canonical Method

As was seen above, the need to eliminate the parameter t pushes the order of derivatives needed from four to eight. The canonical method [4, 9, 12] does not have a parameter in the first place. It can be applied quite generally, but here the application to projectivities is described.

The basic idea is to transform the given coordinate system to a "canonical," or standard system, which is determined only by the shape itself. Since this canonical system is independent of the original system, it is invariant. All quantities defined in it are thus invariant.

An important simple example is the Euclidean invariants. To find an invariant at a given point $\mathbf{x}_1$ on a curve, one attaches a circle at that point so that the tangent (first derivative) of the circle coincides with that of the curve at $\mathbf{x}_1$ (Fig. 1). Also, the radius of the circle is set so that its second derivative is also equal

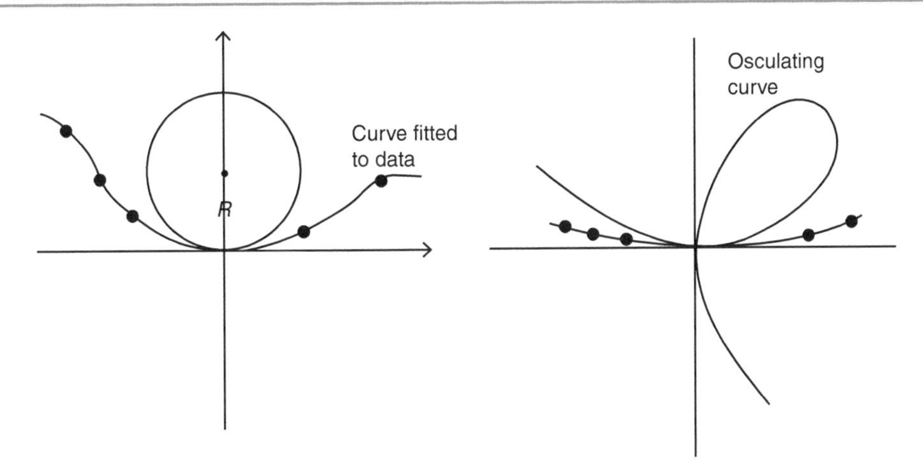

Differential Invariants, Fig. 1 Osculating circle (*left*) and Folium (leaf) of Descartes (*right*)

to the curve's at $\mathbf{x}_1$. Thus, one obtains three so-called points of contact between the given curve and the circle at $\mathbf{x}_1$ that are infinitesimally close and linearly independent.

The next step is to move $\mathbf{x}_1$ to the origin and rotate the curve so that its tangent coincides with the x-axis. Now $\mathbf{x}_1$ and the derivative dy/dx vanish there, Fig. 1. By this, one has exhausted all the Euclidean transformations (translations and rotation) and arrived at a "canonical" Euclidean coordinate system. In this system, all quantities are Euclidean invariants as they cannot be changed by any further Euclidean transformation. In particular, the distance from $\mathbf{x}_1$ to the center of the circle (which now lies on the y-axis) is invariant. This is the radius of curvature which is thus proved invariant.

Although derivatives were mentioned, one does not need to actually find them. Instead, one can fit a circle, $(x - x_0)^2 + (y - y_0) = R^2$, to the curve around point $\mathbf{x}_1$. The coefficients of this circle, R, x_0, y_0, can be found without derivatives or a curve parameter. In principle, one needs only three data points to fit a circle, but in practice, one wants a wider window around $\mathbf{x}_1$. It is convenient to fit a conic in this window. A conic can be expressed as a matrix of coefficients A, satisfying $\mathbf{x}^t A\mathbf{x} = 0$. This is easier to fit because the conic is linear in its coefficients A. One can then find the appropriate circle from this conic algebraically. After moving to the canonical system, the new conic coefficients $\tilde{A}$ are all invariants.

One can generalize the method above to the projective case. Instead of the circle, one uses a more general "osculating curve" with more points of contact with the given curve at $\mathbf{x}_1$. To eliminate the eight projective coefficients, one needs an osculating curve with at least eight contact points. A suitable choice for projectivities is the "nodal cubic" [4] (Fig. 1). After moving to a Euclidean canonical system, this can be expressed as

$$f_{\text{osc}} = c_0x^3 + c_1y^3 + c_2xy^2 + c_3x^2y + c_4y^2 + xy = 0. \tag{14}$$

This curve intersects itself at the origin so it has two tangents there, one lying along the x-axis. The other tangent is called the "projective normal" [6].

The image curve is given as data pixels so one needs to fit a differentiable curve to it. Instead of fitting a conic A around $\mathbf{x}_1$ as before, one fits a higher-order form such as a cubic or a quartic $f(\mathbf{x}) = 0$. In principle, a cubic f will do, having nine coefficients plus the point's position on the curve. In practice, however, it was found [9] that a wide window is necessary for robustness to noise, and this requires a higher-order curve such as a quartic

$$f(x, y) = a_0 + a_1x + \ldots + a_{14}y^4 = 0. \tag{15}$$

(Not all its coefficients need be independent.)

After finding f above by fitting to the data, one finds the osculating nodal cubic (Eq. 14) from it algebraically. The goal is now to transform the coordinates so that this nodal cubic (Eq. 14) takes on the simple coefficient-free form

$$x^3 + y^3 + xy = 0 \tag{16}$$

known as a *folium of Descartes* (Fig. 1) using projective transformations.

In a nutshell, one obtains it as follows. Three of the eight projectivity parameters were already eliminated by moving to a Euclidean canonical frame, namely, moving the origin to $\mathbf{x}_1$ and rotating so that the x-axis is tangent to the curve there. Thus, one already has the x-axis of the canonical system. The canonical y-axis is now chosen as the other tangent of the nodal cubic, the projective normal. One now skews the system so that this projective normal becomes perpendicular to the x-axis. This will eliminate the term with c_4 in the nodal cubic (Eq. 14). Next, the coefficients c_0, c_1 are eliminated by scalings in the x and y directions. One obtains

$$\tilde{x}^3 + \tilde{y}^3 + \tilde{c}_2\tilde{x}\tilde{y}^2 + \tilde{c}_3\tilde{x}^2\tilde{y} + \tilde{x}\tilde{y} = 0. \quad (17)$$

The coefficients in this system, $\tilde{c}_2$ and $\tilde{c}_3$, are *differential affine invariants* because one has reached an affine canonical system – all possible affine transformations (translations, rotation, skewing, scalings) have been used to eliminate all the possible affine transformation coefficients. The remaining coefficients cannot be changed by any further affine transformation so they are affine invariants.

A projective canonical system is obtained by eliminating the last two coefficients $\tilde{c}_2, \tilde{c}_3$ using the remaining projective transformations of tilt and slant. Finally, one transforms the original fitted curve f to this new system and obtains new coefficients $\tilde{a}_i$ for it. Since this system is projectively invariant, these $\tilde{a}_i$ are invariants. One can choose some suitable combinations of them as invariants I_1, I_2.

In summary, the canonical method with implicit curves was used to eliminate the high-order derivatives and the curve parameter. Further details can be found in [9].

The Method of Determinants

The method of determinants is the main tool for deriving invariants of algebraic forms. It can also be used to derive differential affine and Euclidean invariants.

This method takes advantage of the properties of determinants under linear transformations. Many geometric entities can be cast in the form of determinants of points in homogeneous coordinates. In 1D, the distance between points x_1, x_2 is

$$D_{12} = x_1 - x_2 = \begin{vmatrix} x_1 & 1 \\ x_2 & 1 \end{vmatrix}. \quad (18)$$

Similarly, in 2D, the area of a triangle can be written as

$$S_{123} = \frac{1}{2}\begin{vmatrix} x_1 & y_1 & 1 \\ x_2 & y_2 & 1 \\ x_3 & y_3 & 1 \end{vmatrix}. \quad (19)$$

In the projective case, the coordinates can be multiplied by arbitrary factors, e.g., $\mathbf{x}_i = \lambda_i(x_i, y_i, 1)$. Thus, the projective transformation (Eq. 2) transforms a determinant by multiplying it by $|T|$ and λ_i:

$$|\tilde{\mathbf{x}}_1, \tilde{\mathbf{x}}_2, \tilde{\mathbf{x}}_3| = \lambda_1\lambda_2\lambda_3|T||\mathbf{x}_1, \mathbf{x}_2, \mathbf{x}_3|. \quad (20)$$

Thus, a determinant of points in homogeneous coordinates is a *relative projective invariant* of weight 1 in $|T|$ and degree 1 in each λ_i. It is a relative affine invariant with $\lambda_i = 1$.

The above properties are used to find invariants of various algebraic forms. The main trick is to find ratios of various determinants (e.g., lengths, areas) in which all the factors λ_i as well as $|T|$ cancel out, so one gets absolute invariants. When ratios cannot do the trick, then "cross ratios," or ratios of ratios, usually do. Beyond points, determinants of higher-order forms such as conics and cubics are also easily obtained, e.g., for two conics [10]. The duality of points and lines in the projective case makes it possible to interchange their roles in the determinants.

Explicit Differential Affine Invariants

In the canonical method, one obtains implicit (parameter-less) invariants. Here one obtains explicit differential affine invariants by using the determinants of the derivative vectors $\mathbf{x}', \mathbf{x}''$, etc., at point t. The affine case is simpler than the projective case. First, it is assumed that $\lambda = 1$ so λ does not have to be eliminated. Second, the third coordinate is now $x_3 = 1$, and it is involved only with translations. Since the derivatives eliminate any translation coefficients, it is sufficient to deal with a 2D nonhomogeneous coordinate transformation:

$$\begin{pmatrix} \tilde{x} \\ \tilde{y} \end{pmatrix} = T \begin{pmatrix} x \\ y \end{pmatrix} = \begin{pmatrix} a & b \\ c & d \end{pmatrix} \begin{pmatrix} x \\ y \end{pmatrix}. \tag{21}$$

For this transformation, the 2D determinant $|\mathbf{x}', \mathbf{x}''|$ is a relative invariant of weight 1 in $|T|$, analogous to (Eq. 20). It can thus be used to define the *affine arc-length* [3]:

$$\tau = \int_{t_0}^{t} |\mathbf{x}', \mathbf{x}''|^{1/3} dt. \tag{22}$$

It is absolute with respect to $\tilde{t}'$ but relative with respect to T. It will now be used as an invariant parameter for all the differentiations, denoting derivatives as, e.g., $\mathbf{x}_\tau$. Although an arc-length depends on a starting point t_0, the derivatives with respect to it do not. Higher-order relative invariants can now be obtained as further determinants. One immediately obtains the *affine curvature*:

$$\kappa_{\mathrm{af}} = |\mathbf{x}_{\tau\tau}, \mathbf{x}_{\tau\tau\tau}| \tag{23}$$

involving the fourth-order derivative w.r.t. t.

The affine curvature is constant along conics and only conics. It is related to the conic area, a relative invariant, $\pi\kappa_{af}^{3/2}$. Thus, one can interpret the affine curvature geometrically as the area of a conic that osculates the curve at $\mathbf{x}(\tau)$.

For unimodular affine transformations, i.e., with $|T| = 1$, all these relative invariants become absolute. They have been obtained by Cartan in the moving frame method [3].

Explicit Differential Euclidean Invariants

In this case, T is orthonormal, and one has a new invariant, namely, the differential arc-length $\mathbf{x}'^t \cdot \mathbf{x}'$. One can use it to define the integral Euclidean arc-length:

$$\tau = \int (\mathbf{x}'^t \cdot \mathbf{x}')^{1/2} dt. \tag{24}$$

Using derivatives w.r.t. it, this one can define the *Euclidean curvature*

$$\kappa = |\mathbf{x}_\tau, \mathbf{x}_{\tau\tau}| = x_\tau y_{\tau\tau} - y_\tau x_{\tau\tau}. \tag{25}$$

The Euclidean invariants are all absolute as $|T| = 1$.

Projection from 3D to 2D

The above invariants involve the transformation from 2D to 2D shapes. Similar invariants can be derived in nD spaces. In vision, one is more often interested in the projection from 3D objects to 2D images. It has been shown that there are no invariants to such transformations because one looses the depth dimension.

However, it has been shown [15, 16] that there exist invariant *constraints*, i.e., quantities in the image that constrain the corresponding 3D invariants to a subspace of their full invariant space. In particular, for given five 2D image points, the corresponding 3D invariants are constrained to lines in a 3D invariant space. These constraints were obtained by analyzing the projection properties of determinants. Similar methods have been used to obtain constraints in the differential case for the projection of 3D curves to 2D [15].

Such constraints are very useful in combination with given 3D models of the objects. These models provide additional constraints that, together with the invariant constraints, identify an object uniquely. This was applied in [16] to images of vehicles.

In [5], a version of the canonical method has been used to determine whether a space curve lies on some simple surface such as a cylinder or a cone.

Other Invariants

Hybrid invariants, combining differential and algebraic invariants, reduce the number of derivatives at the price of adding points or lines that one needs to match. They are described in [1, 2, 8, 12]. Invariants of fields are described in [7] and are applied to shape from shading in [14].

Accurate Differentiation

Differentiating an image is sensitive to noise and inaccuracies, so one normally convolves the image with a smoothing filter S. Using integration by parts, it is easy to show in theory that the n-th derivative of the image data, $f^{(n)}$, smoothed by S, can be obtained as a convolution with the n-th derivative of the filter S:

$$S \otimes f^{(n)} = S^{(n)} \otimes f. \tag{26}$$

However, the integration by parts is valid only for a *differentiable* filter S. In practice, one commonly uses the truncated Gaussian which is not differentiable at its ends. This is more than a technicality. A quick check with a simple data function such as $f = x^k$ will immediately reveal that the above equation is not valid for the truncated Gaussian filter even in this analytic, noiseless case. The error gets much worse for higher-order derivatives.

To solve the problem, one can replace the truncated Gaussian by a spline approximation of it [13]. For the n-th derivative, one can use an n-th order differentiable spline. In practice, one does not really need to calculate the splines. It is well known that the n-th order derivative of an n-th order spline interpolation of given $n + 1$ points, at the center of the spline, is nothing but the n-th order central *difference* of the given points. One can define a filter for first-order differencing as $D = (-1, 1)$. Then, a general *difference-based* differentiation filter [13] can be defined as

$$D^n \otimes S = D \otimes D \otimes D \otimes \ldots \otimes S. \tag{27}$$

This new filter replaces the erroneous derivative-based filter $S^{(n)}$, and the differentiation (Eq. 26) is replaced by

$$f_s^{(n)} = D^n \otimes S \otimes f \tag{28}$$

where $f_s^{(n)}$ is a smoothed version of $f^{(n)}$. Unlike the old filter, this new filter yields accurate derivatives for polynomials up to order $n + 1$, in which case $f_s^{(n)} = f^{(n)}$. For higher-order polynomials, the derivatives will be smoothed, but there will be no errors from filter truncation.

It is easy to show that the D^n are equal to the standard central difference filters, e.g.,

$$D = (-1, 1) \tag{29}$$

$$D^2 = (1, -2, 1) \tag{30}$$

$$D^3 = (-1, 3, -3, 1). \tag{31}$$

For a pixel distance of h, the above filters are divided by h^n.

References

1. Bruckstein A, Rivlin E, Weiss I (1997) Scale space invariants for recognition. Image Vis Comput 15(5):335–344
2. Doermann D, Rivlin E, Weiss I (1996) Applying algebraic and differential invariants for logo recognition. Mach Vis Appl 9:73–86
3. Guggenheimer H (1963) Differential geometry. Dover, New York
4. Halphen M (1880) Sur les invariants différentiels des courbes gauches. J Ec Polyt 28(1)
5. Keren D, Rivlin E, Shimshoni I, Weiss I (2000) Recognizing 3D objects using tactile sensing and curve invariants. J Math Imaging Vis 12(1):5–23
6. Lane EP (1942) A treatise on projective differential geometry. University of Chicago Press, Chicago
7. Olver PJ (2000) Application of Lie groups to differential equations. Springer, New York
8. Pauwels EJ, Moons T, Van Gool LJ, Oosterlinck A (1995) Recognition of affine planar shapes under affine distortion. Int J Comput Vis 14:49–65
9. Rivlin E, Weiss I (1995) Local invariants for recognition. IEEE Trans Pattern Anal Mach Intell (PAMI) 17(3): 226–238
10. Springer CE (1964) Geometry and analysis of projective spaces. Freeman, San Francisco
11. Weiss I (1988) Projective invariants of shape. Proceedings of computer vision and image processing, Ann Arbor, pp 291–297
12. Weiss I (1993) Noise resistant invariants of curves. IEEE Trans Pattern Anal Mach Intell 15(9):943–948
13. Weiss I (1994) High order differentiation filters that work. IEEE Trans Pattern Anal Mach Intell 16(7):734–739
14. Weiss I (1994) Invariants for recovering shape from shading. In: Mundy J, Zisserman A (eds) Applications of invariance in computer vision II. Springer verlag lecture notes in computer science, vol 825. Springer, Berlin/New York, p 185
15. Weiss I (1999) Model-based recognition of 3D curves from one view. J Math Imaging Vis 10:1–10
16. Weiss I, Ray M (2001) Model-based recognition of 3D objects from single images. IEEE Trans Pattern Anal Mach Intell 23(2):116–128
17. Wilczynski EJ (1906) Projective differential geometry of curves and ruled surfaces. Teubner, Leipzig
18. Wilczynski EJ (1908) Projective differential geometry of curved surfaces (Second Memoir). Am Math Soc Trans 9:79

Diffuse Reflectance

Sanjeev J. Koppal
Harvard University, Cambridge, MA, USA

Synonyms

Diffuse shading; Diffusely scattered reflectance; Lambertian reflectance

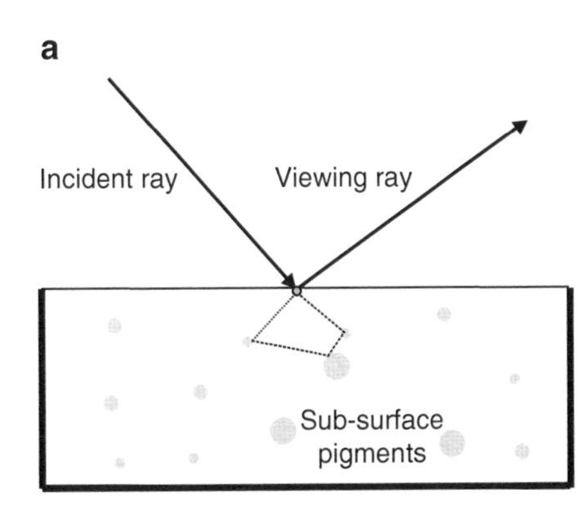

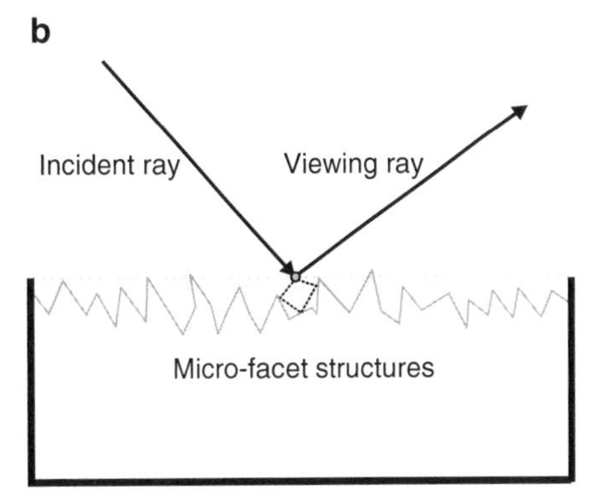

Diffuse Reflectance, Fig. 1 The two diagrams depict diffuse reflection for a scene point with a vertical surface normal. In (**a**), color pigments under a smooth, non-specular surface scatter the incident illumination back into the viewing direction. In (**b**), reflection off microfacets can also cause diffuse reflection, even if the individual microfacets exhibit specular reflection

Related Concepts

►Lambertian Reflectance

Definition

Diffuse reflection is a qualitative term to describe the relative, low-pass responses of different materials to incident illumination.

Background

Diffuse reflection implies that, without scene priors, it is generally impossible to accurately recover the incident illumination on the scene from measured intensities, even if the surface geometry, viewing direction, and object BRDF are known. A large class of materials can be described by diffuse reflectance models which, although usually low parametric, are diverse in nature and do not follow any fixed analytic form.

Theory

Ramamoorthi et al. [1] have described the diffuse reflectance as containing lower frequencies in the appropriatedualspace.Generally,thisimpliesthatalownumber of bases can fully describe the diffuse reflectance function, although the exact number depends on the analytic form of the reflectance function.

Diffuse reflectances from smooth non-specular surfaces are modeled as rays of light that penetrate into the surface of a materials, reflecting off of and being absorbed by flakes of color pigments, as shown in Fig. 1a. Such a physical explanation describes the popular Lambertian model [2] and a variety of extensions such as through microfacets [3]. Rough specular surfaces can also exhibit diffuse reflection and these can be accounted for through microfacet modeling [4] as shown in Fig. 1b.

Since diffuse reflection is a subjective term, many scene effects that exhibit the property of "blurring" the incident illumination are commonly termed as diffuse effects. These include global illumination effects in the scene such as sub-surface scattering and interreflections in skin and marble [5–7]. Strictly speaking, however, such effects are high parametric and global in nature and should not be incorrectly modeled as local, diffuse reflectance.

Computational Tractability

The main popularity of diffuse reflectance models is that they are low parametric and therefore have been used for inferring scene geometry (photometric stereo [8], shape from shading [9]). The high-frequency and global effects of the scene are modeled as noise in these approaches and usually ignored. Diffuse reflection has been shown as a good prior for scene recovery [10, 11]. Psychophysical experiments have also confirmed that humans can accurately estimate scene and illumination properties [12], and much work has been done to replicate this for machine vision, tracing its roots back to retinex theory [13].

References

1. Ramamoorthi R, Hanrahan P (2001) A signal-processing framework for inverse rendering. In: SIGGRAPH, Los Angeles
2. Lambert JH (1760) Photometria sive de mensura de gratibus luminis, colorum et umbrae. Eberhard Klett, Augsburg
3. Oren M, Nayar SK (1995) Generalization of the lambertian model and implications for machine vision. Int J Comput Vis 14(3):227–251
4. Torrance K, Sparrow E (1967) Theory for off-specular reflection from roughened surfaces. J Opt Soc Am 57(9):1105–1112
5. Goesele M, Lensch HPA, Lang J, Fuchs C, Seidel H (2004) Acquisition of translucent objects. In: SIGGRAPH, Los Angeles
6. Jensen WH, Marschner RS, Levoy M, Hanrahan P (2001) A practical model for subsurface light transport. In: SIGGRAPH, Los Angeles
7. Debevec P, Hawkins T, Tchou C, Duiker H, Sarokin W, Sagar M (2000) Acquiring the reflectance field of a human face. In: SIGGRAPH, New Orleans
8. Woodham RJ (1978) Photometric stereo. MIT AI Memo, Cambridge
9. Horn B (1989) Height and gradient from shading. Int J Comput Vis. 5(1):37–75
10. Georghiades A (2003) Incorporating the torrance and sparrow model of reflectance in uncalibrated photometric stereo. In: ICCV, Nice
11. van Ginneken B, Stavridi M, Koenderink JJ (1998) Diffuse and specular reflectance from rough surfaces. Appl opt 37(1):130–139
12. Adelson EH, Pentland AP (1996) The perception of shading and reflectance. Cambridge University Press, New York
13. Land EH, McCann JJ (1971) Lightness and retinex theory. J Opt Soc Am 61(1):1–11

Diffuse Shading

►Diffuse Reflectance

Diffusely Scattered Reflectance

►Diffuse Reflectance

Diffusion

►Diffusion Filtering

Diffusion Filtering

Thomas Brox
Department of Computer Science, University of Freiburg, Freiburg, Germany

Synonyms

Anisotropic diffusion; Diffusion; Diffusion PDEs; Nonlinear diffusion

Related Concepts

►Denoising

Definition

Diffusion filtering is a smoothing technique motivated by equilibration processes in physics that allows for structure-preserving smoothing.

Background

Many problems in computer vision require smoothing or spatial aggregation of information. The most basic problem is image denoising, where noise ought to be removed from an image while preserving its relevant structures. Diffusion filtering is one among several methodologies that are capable of removing noise without blurring and dislocating image edges. The strong relationships to other filters with similar capabilities, such as wavelet shrinkage, the bilateral filter, and variational regularization, have been shown in various papers [1–3].

Theory

Diffusion filtering is motivated by the concept of diffusion in physics. Diffusion leads to equilibration of concentrations, for instance, in fluids. It is by definition a mass preserving process. Mass can move to locally change the concentration levels, but it cannot appear or disappear.

In diffusion, a concentration gradient ∇u creates a flux j as described by Fick's law:

$$j = -D\nabla u, \quad (1)$$

where D is a positive definite, symmetric diffusion matrix, which describes the magnitude and orientation preference of the flux. Combining Fick's law with the mass preservation property leads to the diffusion partial differential equation (PDE)

$$\frac{\partial u}{\partial t} = -\text{div} j = \text{div}(D\nabla u). \quad (2)$$

In image processing, we can interpret image intensities as concentrations and use the same diffusion PDE to smooth these image intensities. The initial condition is set by the input image

$$u(x, y, t = 0) = I(x, y). \quad (3)$$

For $t \to \infty$, diffusion yields the equilibrium state, i.e., a constant value image, where the value is the mean of the input image.

Depending on the choice of the diffusion matrix D, one can distinguish different cases:

- If D is the identity matrix, we obtain homogenous diffusion, where mass is equally distributed in all directions with a magnitude that only depends on the gradient. The corresponding PDE is also known as the heat equation, and there exists an analytic solution, which corresponds to convolution with a Gaussian kernel with variance $\sigma^2 = 2t$.
- If D is a diagonal matrix with identical entries, it degenerates to a scalar g, often called diffusivity, and leads to isotropic diffusion. There is no orientation preference, but the magnitude of the flux may vary from point to point independent of the gradient. The diffusivity can be chosen as a decreasing function of the image gradient, for instance,

$$g(|\nabla I|) = \frac{1}{|\nabla I| + 1}, \quad (4)$$

which leads to a reduction of diffusion across edges and hence edge-preserving smoothing.
- In case of a general matrix D, the diffusion process is called anisotropic diffusion, as there is also a preference with regard to the orientation of smoothing. For instance, it is quite common to smooth along edges, but not across them. There is some confusion about terminology in the literature, as many papers, among them the original work on diffusion filtering by Perona and Malik [4], speak of anisotropic diffusion, although they use a scalar diffusivity rather than the full diffusion tensor.

Diffusion filters can further be separated into linear and nonlinear diffusion filters. In case of linear filters, the diffusion matrix D or diffusivity g does not depend on the smoothed image u but on some external quantity, usually the gradient of the input image I. This leads to PDEs that are linear in u. In case of nonlinear diffusion filters, the diffusion matrix or diffusivity depends on u and thus lead to nonlinear PDEs. Usually, nonlinear diffusion filters should be preferred as they take the enhanced image u into account when defining the areas where smoothing should be reduced, whereas linear diffusion relies on noisy measurements in the input image.

The diffusion PDEs can be solved numerically using iterative schemes. Most common is an explicit

Diffusion Filtering, Fig. 1 *Left*: noisy color image. *Right*: result after applying nonlinear isotropic diffusion

time discretization, which leads to the update equation

$$u^{k+1} = u^k + \tau \text{div}\left(D(u^k)\nabla u^k\right), \quad (5)$$

where τ is a time step size, which must not be too large for the scheme to be stable. Details about discretization, stability, and many other properties of diffusion filters can be found in [5]. For fast implementations, it is also recommended to look into the works on AOS and FED schemes [6, 7].

Application

Diffusion filters are mainly used for image denoising. The corresponding PDEs, however, can be found in numerous other computer vision applications, especially in the context of variational methods. The Euler-Lagrange equation of the regularizer in variational models leads to a diffusion PDE of the general form

$$\text{div}(D\nabla u) = 0, \quad (6)$$

where u in this case is the modeled function that should minimize the variational energy. Very popular is the total variation regularizer [8]

$$\int |\nabla u| dx dy \quad (7)$$

the Euler-Lagrange equation of which is

$$\text{div}\left(\frac{\nabla u}{|\nabla u|}\right) = 0. \quad (8)$$

A (linear) diffusion process is implicitly also part of Laplacian eigenmaps and spectral clustering [9, 10].

Experiment

Diffusion processes can be easily extended to vector-valued images, such as color images. Figure 1 shows a noisy color image and the enhanced version after applying total variation flow

$$\frac{\partial u_k}{\partial t} = \text{div}\left(\frac{\nabla u_k}{\sqrt{\sum_{l=1}^{3} |\nabla u_l|^2}}\right). \quad (9)$$

References

1. Mrázek P, Weickert J, Steidl G (2005) Diffusion-inspired shrinkage functions and stability results for wavelet denoising. Int J Comput Vis 64(2/3):171–186
2. Didas S, Weickert J (2006) From adaptive averaging to accelerated nonlinear diffusion filtering. In: Proceedings of DAGM, Berlin. LNCS, vol 4174. Springer, Berlin/New York, pp 101–110
3. Scherzer O, Weickert J (2000) Relations between regularization and diffusion filtering. J Math Imaging Vis 12(1):43–63
4. Perona P, Malik J (1990) Scale space and edge detection using anisotropic diffusion. IEEE Trans Pattern Anal Mach Intell 12:629–639
5. Weickert J (1998) Anisotropic diffusion in image processing. Teubner, Stuttgart
6. Weickert J, ter Haar Romeny BM, Viergever MA (1998) Efficient and reliable schemes for nonlinear diffusion filtering. IEEE Trans Image Process 7(3):398–410
7. Grewenig S, Weickert J, Bruhn A (2010) From box filtering to fast explicit diffusion. In: Proceedings of DAGM, Darmstadt. LNCS, vol 6376. Springer, pp 533–542
8. Rudin LI, Osher S, Fatemi E (1992) Nonlinear total variation based noise removal algorithms. Phys D 60:259–268
9. Shi J, Malik J (2000) Normalized cuts and image segmentation. IEEE Trans Pattern Anal Mach Intell 22(8):888–905
10. Coifman RR, Lafon S (2006) Diffusion maps. Appl Comput Harmonic Anal 21(1):5–30

Diffusion PDEs

▸Diffusion Filtering

Digital Matting

▸Matte Extraction

Digitization

Peer Stelldinger[1] and Longin Jan Latecki[2]
[1]Computer Science Department, University of Hamburg, Hamburg, Germany
[2]Department of Computer and Information Sciences, Temple University, Philadelphia, PA, USA

Synonyms

Relation between objects and their digital images

Definition

Digitization is a mathematical model of converting continuous subsets of the plane or space (representing real objects) to digital sets in $\mathbb{Z}^2$ or $\mathbb{Z}^3$ or similar grids (representing segmented images of these objects).

Background

A fundamental task of knowledge representation and processing is to infer properties of real objects or situations given their representations. In spatial knowledge representation and, in particular, in computer vision and medical imaging, real objects are represented in a pictorial way as finite and discrete sets of pixels or voxels. The discrete sets result by a quantization process, in which real objects are approximated by discrete sets. In computer vision, this process is called sampling or digitization and is naturally realized by technical devices like computer tomography scanners, CCD cameras or document scanners. Digital images obtained by digitization are suitable to estimate the real object properties like volume and surface area. Therefore, a fundamental question addressed in spatial knowledge representation is: Which properties inferred from discrete representations of real objects correspond to properties of their originals, and under what conditions this is the case? While this problem is well-understood in the 2D case with respect to topology [1–5], it is not as simple in 3D, as shown in [7]. Only recently a first comprehensive answer to this question with respect to important topological and geometric properties of 3D objects has been presented in [8, 9].

The description of geometric and, in particular, topological features in discrete structures is based on graph theory, which is widely accepted in the computer science community. A graph is obtained when a neighborhood relation is introduced into a discrete set, e.g., a finite subset of $\mathbb{Z}^2$ or $\mathbb{Z}^3$, where $\mathbb{Z}$ denotes the integers. On the one hand, graph theory allows investigation into connectivity and separability of discrete sets (for a simple and natural definition of connectivity see [10, 16], for example). On the other hand, a finite graph is an elementary structure that can be easily implemented on computers. Discrete representations are analyzed by algorithms based on graph theory, and the properties extracted are assumed to represent properties of the original objects. Since practical applications, for example in image analysis, show that this is not always the case, it is necessary to relate properties of discrete representations to the corresponding properties of the originals. Since such relations can describe and justify the algorithms on discrete graphs, their characterization contributes directly to the computational investigation of algorithms on discrete structures. This computational investigation is an important part of the research in computer science, and in particular, in computer vision (Marr [11]), where it can contribute to the development of more suitable and reliable algorithms for extracting required shape properties from discrete representations.

It is clear that no discrete representation can exhibit all features of the real original. Thus one has to accept compromises. The compromise chosen depends on the specific application and on the questions which are typical for that application. Real objects and their spatial relations can be characterized using geometric features. Therefore, any useful discrete representation should model the geometry faithfully in order to avoid false conclusions. Topology deals with the invariance of fundamental geometric features like connectivity and separability. Topological properties play an important role, since they are the most primitive object features and human visual system seems to be well-adapted to cope with topological properties.

However, one does not have any direct access to spatial properties of real objects. Therefore, real objects are represented as bounded subsets of the Euclidean space $\mathbb{R}^3$, and their 2D views (projections) as bounded continuous subsets of the plane $\mathbb{R}^2$. Hence, from the theoretical point of view of knowledge representation, the goal is to relate two different pictorial representations of objects in the real world: a discrete and a continuous representation.

Already two of the first books in computer vision deal with the relation between the continuous object and its digital images obtained by modeling a digitization process. Pavlidis [1] and Serra [2] proved independently in 1982 that an r-regular continuous 2D set S (the definition follows below) and the continuous analog of the digital image of S have the same shape in a topological sense. Pavlidis used 2D square grids and Serra used 2D hexagonal sampling grids.

In 3D this problem is much more complicated. In 2005 it has been shown in [7] that the connectivity properties are preserved when digitizing a 3D r-regular object with a sufficiently dense sampling grid, but the preservation of connectivity is much weaker than topology. Stelldinger and Köthe [7] also found out that topology preservation can even not be guaranteed with sampling grids of arbitrary density if one uses the straightforward voxel reconstruction, since the surface of the continuous analog of the digital image may not be a 2D manifold. The question how to guarantee topology preservation during digitization in 3D remained unsolved until 2007.

The solution was provided in [8], where the same digitization model as Pavlidis and Serra is used, also r-regular sets (but in $\mathbb{R}^3$) are used to model the continuous objects. As already shown in [7] the generalization of Pavlidis' straightforward reconstruction method to 3D fails since the reconstructed surface may not be a 2D manifold. For example, Fig. 1a, b shows a continuous object and its digital reconstruction whose surface is not a 2D manifold. However, it is possible to use several other reconstruction methods that all result in a 3D object with a 2D manifold surface. Moreover it is also shown in [8] that these reconstructions and the original continuous object are homeomorphic and their surfaces are close to each other.

The first reconstruction method, Majority interpolation, is a voxel-based representation on a grid with doubled resolution. It always leads to a well-composed set in the sense of Latecki [12], which implies that a lot of problems in 3D digital geometry become relatively simple.

The second method is the most simple one. It just uses balls with a certain radius instead of cubical voxels. When choosing an appropriate radius the topology of an r-regular object will not be destroyed during digitization.

The third method is a modification of the well-known Marching Cubes algorithm [13]. The original Marching Cubes algorithm does not always construct a topologically sound surface due to several ambiguous cases [14, 15]. As shown in [8, 9] most of the ambiguous cases can not occur in the digitization of an r-regular object and that the only remaining ambiguous case always occurs in an unambiguous way, which can be dealt with by a slight modification of the original Marching Cubes algorithm. Thus the generated surface is not only topologically sound, but it also has exactly the same topology as the original object before digitization. Moreover it is shown that one can use trilinear interpolation and that one can even blend the trilinear patches smoothly into each other such that one gets smooth object surfaces with the correct topology. Each of these methods has its own advantages making the presented results applicable to many different image analysis algorithms.

Theory

The (Euclidean) *distance* between two points x and y in $\mathbb{R}^n$ is denoted by $d(x, y)$, and the (Hausdorff) *distance between two subsets of* $\mathbb{R}^n$ is the maximal distance between each point of one set and the nearest point of the other. Let $A \subset \mathbb{R}^n$ and $B \subset \mathbb{R}^m$ be sets. A function $f : A \to B$ is called *homeomorphism* if it is bijective and both it and its inverse are continuous. If f is a homeomorphism, then A and B are *homeomorphic*. Let A, B be two subsets of $\mathbb{R}^n$ (particularly, $n =$ 2 or 3). Then a homeomorphism $f : \mathbb{R}^n \to \mathbb{R}^n$ such that $f(A) = B$ and $d(x, f(x)) \leq r$, for all $x \in \mathbb{R}^n$, is called an r*-homeomorphism* of A to B and A and B are r*-homeomorphic*. A *Jordan curve* is a set $J \subset \mathbb{R}^n$ which is homeomorphic to a circle. Let A be any subset of $\mathbb{R}^n$. The *complement* of A is denoted by A^c. All points in A are *foreground* while the points in A^c are called *background*. The *open ball* in $\mathbb{R}^n$ of radius r and center c is the set $\mathcal{B}_r^0(c) = \{x \in \mathbb{R}^n \mid d(x, c) < r\}$, and the *closed ball* in $\mathbb{R}^n$ of radius r and center c is the set $\overline{\mathcal{B}}_r(c) = \{x \in \mathbb{R}^n \mid d(x, c) \leq r\}$. The *boundary* of A, denoted ∂A, consists of all points $x \in \mathbb{R}^n$ with the property that if B is any open set of $\mathbb{R}^n$ such that $x \in B$, then $B \cap A \neq \emptyset$ and $B \cap A^c \neq \emptyset$.

An open ball $\mathcal{B}_r^0(c)$ is *tangent* to ∂A at a point $x \in \partial A$ if $\partial A \cap \partial \mathcal{B}_r^0(c) = \{x\}$. An open ball $\mathcal{B}_r^0(c)$ is an *osculating open ball of radius* r *to* ∂A *at point* $x \in \partial A$ if $\mathcal{B}_r^0(c)$ is tangent to ∂A at x and either $\mathcal{B}_r^0(c) \subseteq A^0$ or $\mathcal{B}_r^0(c) \subseteq (A^c)^0$.

Definition 1 *A set* $A \subset \mathbb{R}^n$ *is called* r-regular *if, for each point* $x \in \partial A$, *there exist two osculating open balls of radius* r *to* ∂A *at* x *such that one lies entirely in* A *and the other lies entirely in* A^c. *Examples illustrating 2D and 3D cases are shown in Fig. 2.*

Note that the boundary of a 3D r-regular set is a 2D manifold surface. Any set S which is a translated and rotated version of the set $\frac{2 \cdot r'}{\sqrt{3}}\mathbb{Z}^3$ is called a *cubic*

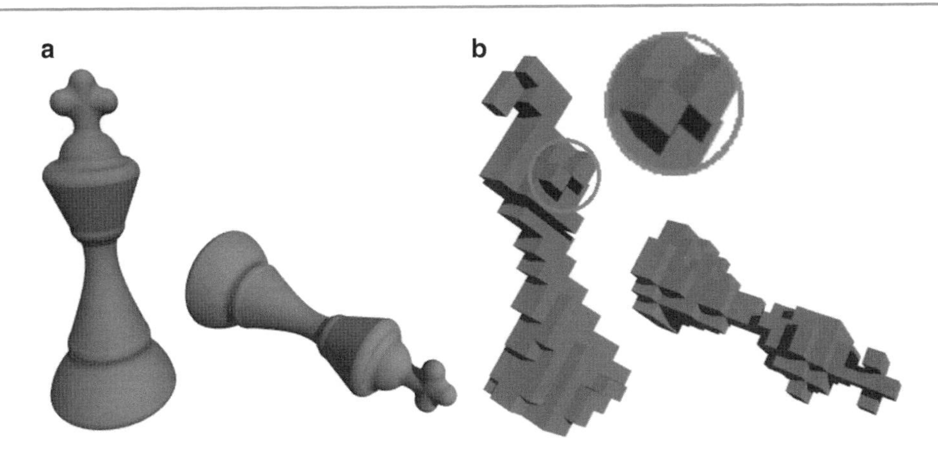

Digitization, Fig. 1 The digital reconstruction (**b**) of an r-regular object (**a**) may not be well-composed, i.e., its surface may not be a 2D manifold as can be seen in the magnification

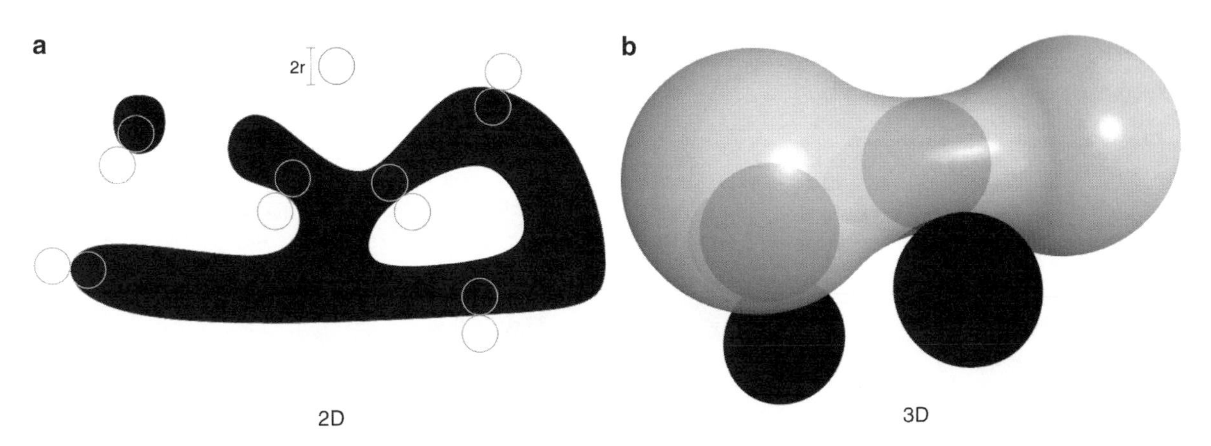

Digitization, Fig. 2 For each boundary point of a **2D/3D** r-regular set exists an outside and an inside osculating open r-disc/ball

r'-*grid* and its elements are called *sampling points.* Note that the distance $d(x,p)$ from each point $x \in \mathbb{R}^3$ to the nearest sampling point $s \in S$ is at most r'. The *voxel* $\mathcal{V}_S(s)$ of a sampling point $s \in S$ is its *Voronoi region* $\mathbb{R}^3$: $\mathcal{V}_S(s) = \{x \in \mathbb{R}^3 \mid d(x,s) \leq d(x,q), \forall q \in S\}$, i.e., $\mathcal{V}_S(s)$ is the set of all points of $\mathbb{R}^3$ which are at least as close to s as to any other point in S. In particular, note that $\mathcal{V}_S(s)$ is a cube whose vertices lie on a sphere of radius r' and center s.

Definition 2 *Let S be a cubic r'-grid, and let A be any subset of $\mathbb{R}^3$. The union of all voxels with sampling points lying in A is the* digital reconstruction of A with respect to S, $\hat{A} = \bigcup_{s \in (S \cap A)} \mathcal{V}_S(s)$.

This method for reconstructing the object from the set of included sampling points is the 3D generalization of the 2D *Gauss digitization* (see [16]) which has been used by Gauss to compute the area of discs and which has also been used by Pavlidis [1] in his sampling theorem.

For any two points p and q of S, $\mathcal{V}_S(p) \cap \mathcal{V}_S(q)$ is either empty or a common vertex, edge or face of both. If $\mathcal{V}_S(p) \cap \mathcal{V}_S(q)$ is a common face, edge, or vertex, then $\mathcal{V}_S(p)$ and $\mathcal{V}_S(q)$ are *face-adjacent, edge-adjacent*, or *vertex-adjacent*, respectively. Two voxels $\mathcal{V}_S(p)$ and $\mathcal{V}_S(q)$ of $\hat{A}$ are *connected in* $\hat{A}$ if there exists a sequence $\mathcal{V}_S(s_1), \ldots, \mathcal{V}_S(s_k)$, with $k \in \mathbb{Z}$ and $k > 1$, such that $s_1 = p$, $s_k = q$, and $s_i \in A$ (or equivalently, $\mathcal{V}_S(s_i) \subset \hat{A}$), for each $i \in \{1, \ldots, k\}$, and $\mathcal{V}_S(s_j)$ and $\mathcal{V}_S(s_{j+1})$ are face-adjacent, for each $j \in \{1, \ldots, k-1\}$. A (connected) *component* of $\hat{A}$ is a maximal set of connected voxels in $\hat{A}$.

Definition 3 *Let S be a cubic r'-grid, and let T be any subset of S. Then $\bigcup_{t \in T} \mathcal{V}_S(t)$ is* well-composed *if*

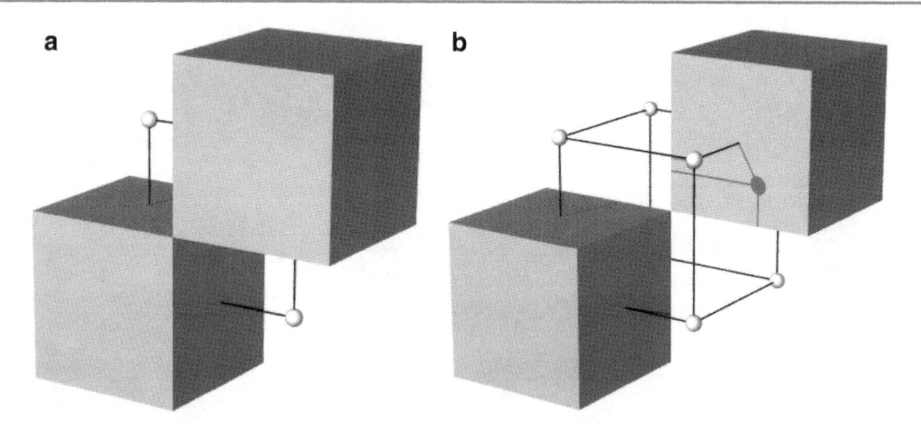

Digitization, Fig. 3 (**a**) Critical configuration (C1). (**b**) Critical configuration (C2). For the sake of clarity, only the voxels of foreground or background points are shown

$\partial(\bigcup_{t\in T} \mathcal{V}_S(t))$ is a surface in $\mathbb{R}^3$, or equivalently, if for every point $x \in \partial(\bigcup_{t\in T} \mathcal{V}_S(t))$, there exists a positive number r such that the intersection of $\partial(\bigcup_{t\in T} \mathcal{V}_S(t))$ and $\mathcal{B}_r^0(x)$ is homeomorphic to the open unit disk in $\mathbb{R}^2$, $\mathbb{D} = \{(x, y) \in \mathbb{R}^2 \mid x^2 + y^2 < 1\}$.

Well-composed digital reconstructions can be characterized by two local conditions depending only on voxels of points of S. Let $s_1, \ldots, s_4$ be any four points of S such that $\bigcap_{i=1}^4 \mathcal{V}_S(s_i)$ is a common edge of $\mathcal{V}_S(s_1), \ldots, \mathcal{V}_S(s_4)$. The set $\{\mathcal{V}_S(s_1), \ldots, \mathcal{V}_S(s_4)\}$ is an instance of the *critical configuration (C1) with respect to* $\bigcup_{t\in T} \mathcal{V}_S(t)$ if two of these voxels are in $\bigcup_{t\in T} \mathcal{V}_S(t)$ and the other two are in $(\bigcup_{t\in T} \mathcal{V}_S(t))^c$, and the two voxels in $\bigcup_{t\in T} \mathcal{V}_S(t)$ (resp. $(\bigcup_{t\in T} \mathcal{V}_S(t))^c$) are edge-adjacent, as shown in Fig. 3a. Now, let $s_1, \ldots, s_8$ be any eight points of S such that $\bigcap_{i=1}^8 \mathcal{V}_S(s_i)$ is a common vertex of $\mathcal{V}_S(s_1), \ldots, \mathcal{V}_S(s_8)$. The set $\{\mathcal{V}_S(s_1), \ldots, \mathcal{V}_S(s_4)\}$ is an instance of the *critical configuration (C2) with respect to* $\bigcup_{t\in T} \mathcal{V}_S(t)$ if two of these voxels are in $\bigcup_{t\in T} \mathcal{V}_S(t)$ (resp. $(\bigcup_{t\in T} \mathcal{V}_S(t))^c$) and the other six are in $(\bigcup_{t\in T} \mathcal{V}_S(t))^c$ (resp. $\bigcup_{t\in T} \mathcal{V}_S(t)$), and the two voxels in $\bigcup_{t\in T} \mathcal{V}_S(t)$ (resp. $(\bigcup_{t\in T} \mathcal{V}_S(t))^c$) are vertex-adjacent, as shown in Fig. 3b. The following theorem from [12] establishes an important equivalence between well-composedness and the (non)existence of critical configurations (C1) and (C2):

Theorem 2 ([12]) *Let S be a cubic r'-grid and let T be any subset of S. Then, $\bigcup_{t\in T} \mathcal{V}_S(t)$ is* well-composed *iff the set of voxels $\{\mathcal{V}(s) \mid s \in S\}$ does not contain any instance of the critical configuration (C1) nor any instance of the critical configuration (C2) with respect to $\bigcup_{t\in T} \mathcal{V}_S(t)$.*

A simple consequence of the 2D digitization theorem by Pavlidis [1] is that the reconstruction of an r'-regular set is well-composed. The main difficulty of 3D digitization as compared to 2D lies in the fact that digital reconstruction $\hat{A}$ of A with respect to S is not guaranteed to be well-composed. An example is provided in Fig. 4. Therefore, it is necessary to repair $\hat{A}$ in order to ensure the topological equivalence between A and repaired $\hat{A}$. The first topology preserving repairing method has been proposed in [8], where also the following theorem is proven. It is an interesting observation that it took 25 years to obtain this 3D theorem.

Theorem 3 ([8]) *If A is an r-regular object and S is a cubic r'-grid with $2r' < r$, then the result of the topology preserving repairing of the reconstruction $\hat{A}$ is r-homeomorphic to A.*

Application

A complete understanding of the loss of information due to the digitization process is fundamental for the justification of any computer vision application. If the relevant information is not contained in the digital image, there is no way to reconstruct it without using context knowledge. Thus, whenever one needs to have guarantees for the correct behavior of some computer vision algorithm one has to be aware of what happens during digitization. This article gives an exemplary

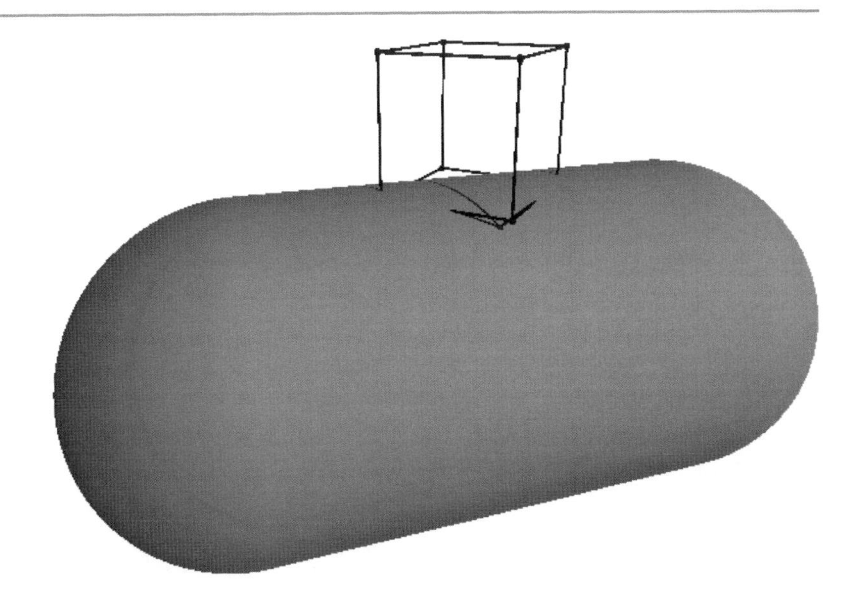

Digitization, Fig. 4 The surface of an object only needs to have an arbitrarily small, but nonzero curvature in order to make occurrences of the critical configuration (C1) possible in the digital reconstruction

insight to the topic, the related problems and the way to solve them.

Open Problems

The analysis of the effect of digitization to the information being extractable from an image is a challenging research area. Newest results approximate real acquisition processes and thus give direct implications for many computer vision algorithms which rely on precise information of the structures being approximated by the digital image. However, in reality the digitization process is still more complicated than the models which are used for topological or geometric sampling theorems. The goal is to derive guarantees for digitization models approximating real digitization processes.

References

1. Pavlidis T (1982) Algorithms for graphics and image processing. Computer Science Press, Cambridge
2. Serra J (1982) Image analysis and mathematical morphology. Academic, San Diego
3. Latecki LJ, Conrad C, Gross A (1998) Preserving topology by a digitization process. J Math Imaging Vis 8:131–159
4. Latecki LJ (1998) Discrete representation of spatial objects in computer vision. Viergever M (ed) Series on computational imaging and vision. Kluwer, Dordrecht
5. Tajine M, Ronse C (2002) Topological properties of hausdorff discretization, and comparison to other discretization schemes. Theor Comput Sci 283(1):243–268
6. Stelldinger P, Köthe U (2003) Shape preservation during digitization: tight bounds based on the morphing distance. In: Pattern recognition, DAGM 2003, Munich, pp 108–115
7. Stelldinger P, Köthe U (2005) Towards a general sampling theory for shape preservation. Image Vis Comput J 23(2):237–248
8. Stelldinger P, Latecki LJ, Siqueira M (2007) Topological equivalence between a 3D object and the reconstruction of its digital image. IEEE Trans Pattern Anal Mach Intell 29(1):126–140
9. Stelldinger P (2008) Image digitization and its influence on shape properties in finite dimensions. DISKI 312, IOS, Amsterdam
10. Kong T, Rosenfeld A (1990) If we use 4- or 8-connectedness for both the objects and the background, the Euler characteristics is not locally computable. Pattern Recognit Lett 11(4):231–232
11. Marr D (1983) Vision. Freeman, San Francisco
12. Latecki LJ (1997) 3D well-composed pictures. Graph Models Image Process 59(3):164–172
13. Lorensen WE, Cline HE (1987) Marching cubes: a high resolution 3D surface construction algorithm. In: SIGGRAPH '87: proceedings of the 14th annual conference on computer graphics and interactive techniques. ACM, New York, pp 163–169
14. Dürst MJ (1988) Letters: additional reference to "marching cubes". SIGGRAPH Comput Graph 22(2): 243
15. Nielson GM, Hamann B (1991) The asymptotic decider: resolving the ambiguity in marching cubes. In: Proceedings of the 2nd IEEE conference on visualization (Visualization'91), San Diego, pp 83–91
16. Klette R, Rosenfeld A (2004) Digital geometry. Morgan Kaufman, Amsterdam
17. Siqueira M, Latecki LJ, Tustison N, Gallier J, Gee J (2008) Topological repairing of 3D digital images. J Math Imaging Vis 30(3):249–274

Dimension Reduction

►Dimensionality Reduction

Dimensional Compression

►Dimensionality Reduction

Dimensional Embedding

►Dimensionality Reduction

Dimensionality Reduction

Eisaku Maeda
NTT Communication Science Laboratories,
Soraku-gun, Kyoto, Japan

Synonyms

Dimensional compression; Dimensional embedding; Dimension reduction

Related Concepts

►Feature Selection

Definition

Dimensionality reduction is the process of reducing the dimension of the vector space spanned by feature vectors (pattern vectors). Various kinds of reduction can be achieved by defining a map from the original space into a dimensionality-reduced space.

Background

The feature space, i.e., the vector space spanned by feature vectors (pattern vectors) defined on d-dimensional space, can be transformed into a vector space of lower-dimension $d'(< d)$ spanned by d'-dimensional feature vectors through linear or nonlinear transformation. This transformation allows feature vectors to be represented by lower-dimensional vectors, and various kinds of vector operations and statistical analysis, such as multivariate analysis, machine learning, clustering, and classification, become less expensive to perform. Moreover, it tackles the "curse of dimensionality," the various problems created by dealing with data of high dimensionality.

Theory and Application

The most common linear approaches for dimensionality reduction are principal component analysis (PCA) and linear discriminant analysis (LDA). PCA, LDA, and related techniques have been used for pattern recognition and computer vision. It is generally considered that PCA offers reduction for information representation while LDA offers reduction for classification (Fig. 1).

PCA find principal axes so as to maximize the variance of the pattern vectors' distribution in a dimensionality-reduced space. It is theoretically identical to a discrete and finite case of Karhunen-Loève transform. In PCA, d' principal axes are obtained as eigenvectors $\Phi_1, \dots, \Phi_{d'}$, which are eigenvectors corresponding to the d' largest eigenvalues of Σ, $\lambda_1, \dots, \lambda_{d'}$. Σ is the covariance matrix of feature vectors and is defined as

$$\Sigma = \frac{1}{n} \sum_{\mathbf{x}} (\mathbf{x} - \mathbf{m})(\mathbf{x} - \mathbf{m})^T \qquad (1)$$

where $\mathbf{x}$ are d-dimensional feature vectors, n is the number of feature vectors, and $\mathbf{m}$ is the mean of feature vectors (i.e., $\mathbf{m} = \frac{1}{n} \sum \mathbf{x}$).

LDA realizes a linear transform into $(c-1)$-dimensional space so as to maximize the between-class distance, which is the distance between the pattern distribution of each class; c is the number of classes (categories). The bases of the transformed space are obtained as the eigenvectors of $\Sigma_W^{-1} \Sigma_B$, where Σ_W is within-class covariance matrix and Σ_B is the between-class covariance matrix. They are written as

$$\Sigma_W = \sum_{i=1}^{c} \left(\frac{1}{n} \sum_{\mathbf{x} \in \mathcal{X}_i} (\mathbf{x} - \mathbf{m}_i)(\mathbf{x} - \mathbf{m}_i)^T \right) \qquad (2)$$

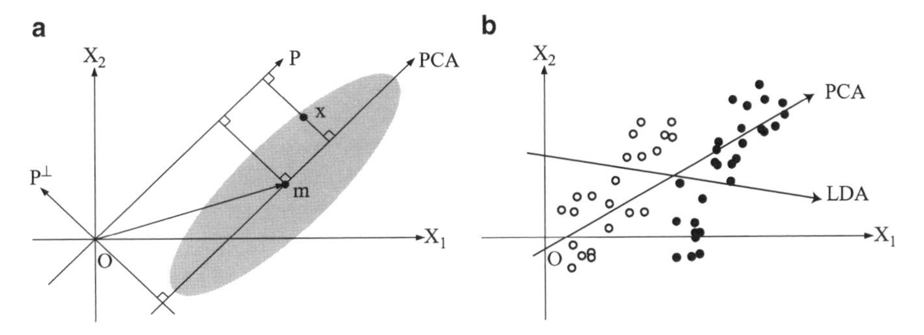

Dimensionality Reduction, Fig. 1 (**a**) Distribution of pattern vectors (*gray region*) and the 1st principal axis. (**b**) Distribution of 2-dimensional feature vectors classified as class 1 (*open circle*) and 2 (*closed circle*) and the principal axes determined by PCA and LDA

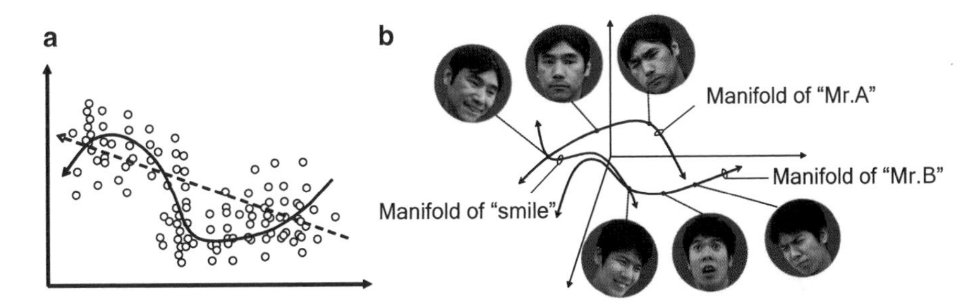

Dimensionality Reduction, Fig. 2 (**a**) Distribution of 2-dimensional pattern vectors and a principal axis determined by linear (*solid line*) and nonlinear (*dotted line*) dimensionality reduction. (**b**) Various embedded manifolds in a face space

$$\Sigma_B = \sum_{i=1}^{c} \frac{n_i}{n}(\mathbf{m}_i - \mathbf{m})(\mathbf{m}_i - \mathbf{m})^T \tag{3}$$

where $\mathbf{m}_i$ is the mean of the feature vectors that belong to ith class $\mathcal{X}_i$ and $\mathbf{m}$ represents the total mean of all feature vectors. As the rank of matrix Σ_B is $(c-1)$, the dimension of the space determined by LDA is not more than $(c-1)$.

PCA- and LDA-based techniques such as class-featuring information compression (CLAFIC) [1] and learning subspace method (LSM) [2] have been studied and widely applied to various pattern recognition tasks like character recognition and phoneme classification. More recently, they have been applied to more complicated tasks, such as face recognition [3, 4] and 3D object recognition [5].

Multidimensional scaling (MDS) is another classic technique for reconstructing a subspace to represent pattern distribution. Given only the dissimilarities (distances) between any pairs of patterns (objects, samples), MDS outputs basis vectors that minimize the distance errors in the subspace spanned by the bases. Distance matrix $\mathbf{D}$, whose (i, j) element is the distance between $\mathbf{x}_i$ and $\mathbf{x}_j$, is transformed by the Young-Householder transformation into matrix $\mathbf{H}$ as follows:

$$\mathbf{H} = -\frac{1}{2}\mathbf{J}_n\mathbf{D}\mathbf{J}_n \tag{4}$$

$$\mathbf{J}_n = \mathbf{I}_n - \frac{1}{n}\mathbf{1}_n \tag{5}$$

where $\mathbf{I}_n$ denotes an n-dimensional identity matrix and $\mathbf{1}_n$ denotes a square matrix, all of whose elements are 1. The principal axes minimizing the distance errors are obtained as the eigenvectors of matrix $\mathbf{H}$.

Nonlinear dimensionality reduction is a powerful tool for pattern recognition and computer vision despite of its higher computational cost than linear methods. Conventional linear dimensionality reduction methods have been extended to nonlinear equivalents through the kernel technique, i.e., kernel PCA [6] and kernel version of LDA [7, 8]. By replacing the inner products of feature vectors $(\mathbf{x}_i, \mathbf{x}_j)$ to $k(\mathbf{x}_i, \mathbf{x}_j)$ in the procedure of linear methods, where k is any

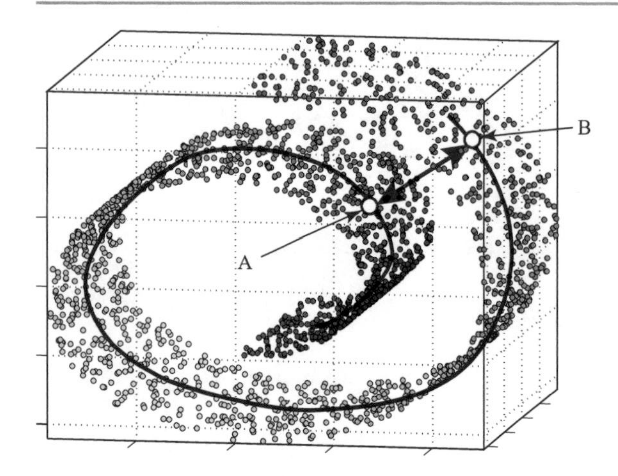

Dimensionality Reduction, Fig. 3 Euclidean distance between A and B (*arrow*) and geodesic distance (*solid line*) in 3-dimensional feature distribution

kernel function, various nonlinear transforms can be achieved depending on the kernel function selected and its parameters.

Many novel techniques of nonlinear dimensionality reduction have been proposed; they are collectively called manifold learning [9, 10]. Pioneering works of manifold learning are Isomap [11] and Locally Linear Embedding (LLE) [12, 13] (Fig. 2).

Isomap is a manifold learning algorithm that preserves the geodesic distance between all feature vectors. The distance is approximately obtained as the shortest path distance by tracking nearest neighbors. In Isomap, these pair-wise geodesic distances between feature vectors are applied to classic MDS (Fig. 3).

LLE, on the other hand, targets low-dimensional manifolds that preserve local geometric relationships between neighbors, so as to minimize cost function $E(\mathbf{w})$,

$$E(\mathbf{w}) = \sum_i \left(\mathbf{x}_i - \sum_j \mathbf{w}_{ij} \mathbf{x}_j \right)^2 \quad (6)$$

where $\mathbf{x}_{ij}$ denote neighbors of $\mathbf{x}_i$ and vw_{ij} denote the linear weights that satisfy $\sum_j \mathbf{w}_{ij} = 1$. Each point $\mathbf{x}_i$ in the d-dimensional space is mapped onto point $\mathbf{y}_i$ in the d'-dimensional space by minimizing the cost function $C(\mathbf{y})$,

$$C(\mathbf{y}) = \sum_i \left(\mathbf{y}_i - \sum_j \mathbf{w}_{ij} \mathbf{y}_j \right)^2 \quad (7)$$

where $\mathbf{w}_{ij}$ are the weights that minimize $E(\mathbf{w})$. Isomap, LLE, and graph Laplacian [14] are variations of kernel PCA with special kernel functions [15].

Another nonlinear method is the feed-forward neural network trained to approximate the identity function [16, 17]. When the trained network has fewer hidden layer units than input and output layer units, the output signals of the hidden layer units can be considered as an encoded vector of the original input pattern vector.

References

1. Watanabe S (1969) Knowing & guessing – quantitative study of inference and information. John Wiley & Sons Inc., Hoboken, NJ, USA
2. Oja E (1983) Subspace methods of pattern recognition. Research Studies Press, Baldock, UK
3. Turk M, Pentland A (1991) Eigenfaces for recognition. J Cogn Neurosci 3(1):71–86
4. Belhumeur P, Hespanha J, Kriegman D (1997) Eigenfaces vs. fisherfaces: recognition using class specific linear projection. IEEE Trans Pattern Anal Mach Intell 19(7): 711–720
5. Murase H, Nayar SK (1995) Visual learning and recognition of 3-d objects from appearance. Int J Comput Vis 14(1): 5–24
6. Schölkopf B, Smola A, Müller KR (1998) Nonlinear component analysis as a kernel eigenvalue problem. Neural Comput 10(5):1299–1319
7. Mika S, Rätsch G, Weston J, Schölkopf B, Müller K (1999) Fisher discriminant analysis with kernels. In: Proceedings of IEEE neural networks for signal processing workshop IX (NNSP'99), Madison, pp 41–48
8. Baudat G, Anouar F (2000) Generalized discriminant analysis using a kernel approach. Neural Comput 12:2385–2404
9. Burges CJ (2005) Geometric methods for feature extraction and dimensional reduction. In: Maimon O, Rokach L (eds) Data mining and knowledge discovery handbook: a complete guide for researchers and practitioners. Springer, NY, USA
10. Pless R, Souvenir R (2009) A survey of manifold learning for images. IPSJ Trans Comput Vis Appl 1:83–94
11. Tenenbaum JB, de Silva V, Langford JC (2000) A global geometric framework for nonlinear dimensionality reduction. Science 290(5500):2319–2323
12. Roweis ST, Saul LK (2000) Nonlinear dimensionality reduction by locally linear embedding. Science 290: 2323–2326
13. Saul LK, Roweis ST, Singer Y (2003) Think globally, fit locally: unsupervised learning of low dimensional manifolds. J Mach Learn Res 4:119–155

14. Belkin M, Niyogi P (2002) Laplacian eigenmaps for dimensionality reduction and data representation. Neural Comput 15(6):1373–1396
15. Ham J, Lee DD, Mika S, Schölkopf B (2004) A kernel view of the dimensionality reduction of manifolds. In: Proceedings of the 21st international conference on machine learning (ICML'04), Banff, pp 369–376
16. DeMers D, Cottrell G (1992) Non-linear dimensionality reduction. In: Advances in Neural Information Processing Systems 5. Morgan Kaufmann Publishers Inc., San Francisco, CA, USA, pp 580–587
17. Hinton GE, Salakhutdinov RR (2006) Reducing the dimensionality of data with neural networks. Science 313(5786):504–507

Dioptrics

►Catadioptric Camera

Discrete AdaBoost

►AdaBoost

Discriminative Random Fields

Sanjiv Kumar
Google Research, New York, NY, USA

Synonyms

Conditional random fields; Two dimensional conditional random fields

Definition

Discriminative random fields (DRFs) are probabilistic discriminative graphical models for classification that allow contextual interactions among labels as well as the observed data using arbitrary discriminative classifiers.

Background

Natural image data shows significant dependencies that should be modeled appropriately to achieve good classification of image entities such as pixels, regions, objects, or the entire image itself. Such dependencies are commonly referred to as *context* in vision. For example, due to local smoothness of natural images, neighboring pixels tend to have similar labels (except at the discontinuities). Different semantic regions in a scene follow plausible spatial configurations, for example, sky tends to occur above water or vegetation. For parts-based object recognition, different parts of an object such as handle, keypad, and front panel in a phone are related to each other through geometric constraints. In fact, the contextual interactions for object recognition are not limited to the parts of an object. These may include interactions among various objects themselves. For example, the presence of a monitor screen increases the probability of having a keyboard or a mouse nearby. The challenge is how to model these different types of context, which include complex dependencies in the observed image data as well as the labels, in a principled manner. Discriminative graphical models provide a solid platform to achieve that.

Traditional generative graphical models, that is, Markov random fields (MRFs) suffer from three main problems, which are overcome by DRFs: First, for computational tractability, the observations are assumed to be conditionally independent in MRFs. This assumption is too restrictive for many natural image analysis applications. Second, interaction among labels in MRFs arises from prior over labeling and hence do not depend on the observed data. This prohibits one from modeling data-dependent interactions in labels that are necessary for a variety of tasks. For example, while implementing local smoothness of labels in image segmentation, it may be desirable to use observed data to modulate the smoothness according to the image intensity gradients. Further, in parts-based object detection to model interactions among object parts, one needs observed data to enforce geometric constraints. DRFs allow interactions among labels based on unrestricted use of observations as necessary. This step is crucial to develop models that can incorporate interactions of different types within the same framework.

Finally, MRFs are used in a probabilistic generative framework that models the joint probability of the observed data and the corresponding labels. However, for classification purposes, one needs to estimate the posterior over labels given the observations, that is,

$P(\boldsymbol{x}|\boldsymbol{y})$, where $\boldsymbol{y}$ is observed data and $\boldsymbol{x}$ are corresponding labels. In a generative MRF framework, one expends efforts to model the joint distribution $p(\boldsymbol{x}, \boldsymbol{y})$, which involves implicit modeling of the observations. In a discriminative framework, one models the distribution $P(\boldsymbol{x}|\boldsymbol{y})$ directly. A major advantage of doing this is that the true underlying generative model may be quite complex even though the class posterior is simple. This means that the generative approach may spend a lot of resources on modeling the generative models which are not particularly relevant to the task of inferring the class labels. Moreover, learning the class density models may become even harder when the training data is limited. The discriminative approach saves one from making simplistic assumptions about the data.

Discriminative Random Field (DRF)

Discriminative random fields (DRFs) are discriminative graphical models based on conditional random fields (CRFs), originally proposed by Lafferty et al. [1] in the context of segmentation and labeling of 1D text sequences. CRFs are discriminative models that directly model the conditional distribution over labels, that is, $P(\boldsymbol{x}|\boldsymbol{y})$ as a Markov random field. This approach allows one to capture arbitrary dependencies between the observations without resorting to any model approximations. DRFs are 2D versions of 1D CRFs, which allow the use of arbitrary discriminative classifiers to model different types of interactions in labels and data, leading to more flexible and powerful generalization of CRFs. These were introduced in vision by Kumar and Hebert [2, 3].

Let the observed data from an input image be given by $\boldsymbol{y} = \{\boldsymbol{y}_i\}_{i \in S}$ where $\boldsymbol{y}_i$ is the data from i^{th} site and $\boldsymbol{y}_i \in \Re^c$. The corresponding labels at the image sites are given by $\boldsymbol{x} = \{x_i\}_{i \in S}$. In a binary classification problem, $x_i \in \{-1, 1\}$. It is easy to extend the formulation to multiclass labeling problems as mentioned later. The random variables $\boldsymbol{x}$ and $\boldsymbol{y}$ are jointly distributed, but in a discriminative framework, a conditional model $P(\boldsymbol{x}|\boldsymbol{y})$ is constructed from the observations and labels, and the marginal $p(\boldsymbol{y})$ is not modeled explicitly. DRFs follow the definition of CRFs given by Lafferty et al. [1].

Definition 1 *CRF: Let $G = (S, E)$ be a graph such that $\boldsymbol{x}$ is indexed by the vertices of G. Then $(\boldsymbol{x}, \boldsymbol{y})$ is said to be a conditional random field if, when conditioned on $\boldsymbol{y}$, the random variables x_i obey the Markov property with respect to the graph: $P(x_i|\boldsymbol{y}, \boldsymbol{x}_{S-\{i\}}) = P(x_i|\boldsymbol{y}, \boldsymbol{x}_{\mathcal{N}_i})$, where $S - \{i\}$ is the set of all the nodes in the graph except the node i, $\mathcal{N}_i$ is the set of neighbors of the node i in G, and $\boldsymbol{x}_\Omega$ represents the set of labels at the nodes in set Ω.*

Thus, a DRF is a random field globally conditioned on the observations $\boldsymbol{y}$. The condition of positivity requiring $P(\boldsymbol{x}|\boldsymbol{y}) > 0, \ \forall\, \boldsymbol{x}$ has been assumed implicitly. Using the Hammersley-Clifford theorem [4] and assuming only up to pairwise clique potentials to be nonzero, the conditional distribution over all the labels $\boldsymbol{x}$ given the observations $\boldsymbol{y}$ in a DRF can be written as

$$P(\boldsymbol{x}|\boldsymbol{y}) = \frac{1}{Z}\exp\left(\sum_{i \in S} A_i(x_i, \boldsymbol{y}) + \sum_{i \in S}\sum_{j \in \mathcal{N}_i} I_{ij}(x_i, x_j, \boldsymbol{y})\right), \tag{1}$$

where Z is a normalizing constant known as the partition function, and $-A_i$ and $-I_{ij}$ are the unary and pairwise potentials, respectively. With a slight abuse of notation, refer A_i as the *association potential* and I_{ij} as the *interaction potential*.

Assuming the random field given in Eq. (1) to be homogeneous, the functional forms of A_i and I_{ij} are independent of the location i. Furthermore, assuming the field to be isotropic implies that the label interactions are nondirectional. In other words, I_{ij} is independent of the relative locations of sites i and j. Thus, subsequently, one can drop the subscripts and simply use the notation A and I to denote the two potentials. In fact, the assumption of isotropy can be easily relaxed at the cost of a few additional parameters. Henceforth, consider the DRF model of the following form:

$$P(\boldsymbol{x}|\boldsymbol{y}) = \frac{1}{Z}\exp\left(\sum_{i \in S} A(x_i, \boldsymbol{y}) + \sum_{i \in S}\sum_{j \in \mathcal{N}_i} I(x_i, x_j, \boldsymbol{y})\right). \tag{2}$$

Figure 1 illustrates a typical DRF for an example image analysis task of man-made structure detection. Given an input image $\boldsymbol{y}$ shown in the bottom layer, suppose the goal is to label each image site (in this

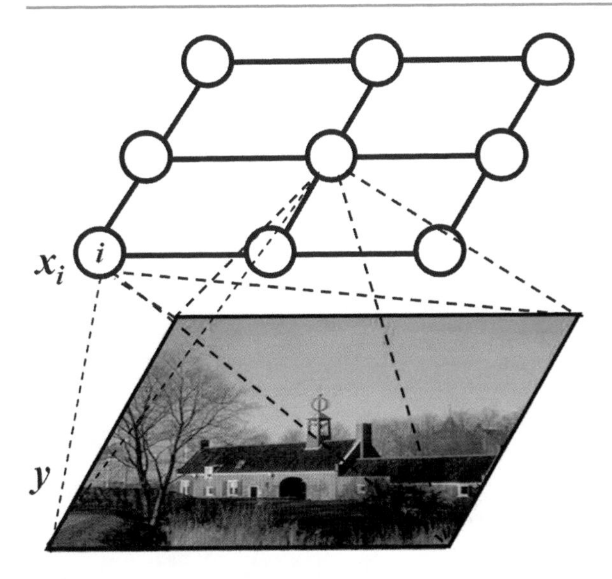

Discriminative Random Fields, Fig. 1 An illustration of a typical DRF for an example task of man-made structure detection in natural images. The aim is to label each site, that is, each 16 × 16 image block whether it is a man-made structure or not. The *top* layer represents the labels on all the image sites. Note that each site i can potentially use features from the whole image y unlike the traditional MRFs

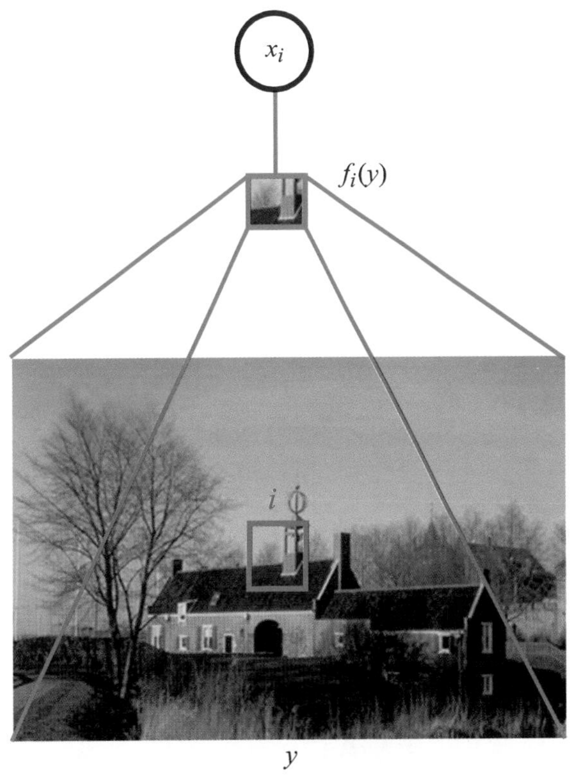

Discriminative Random Fields, Fig. 2 Given a feature vector $f_i(y)$ at site i, the association potential in DRFs can be seen as a measure of how likely the site i will take label x_i, ignoring the effects of other sites in the image. Note that the feature vector $f_i(y)$ can be constructed by pooling arbitrarily complex dependencies in the observed data y

case a 16 × 16 image block) whether it contains a man-made structure or not. The top layer represents the labels $\boldsymbol{x}$ on all the image sites. Note that each site i can potentially use features from the whole image $\boldsymbol{y}$ unlike the traditional MRFs. In addition, DRFs allow to use image data to model interactions between two neighboring sites i and j. The following sections describe how the unary and the pairwise potentials are designed in DRFs.

Association Potential

In the DRF framework, the association potential, $A(x_i, \boldsymbol{y})$, can be seen as a measure of how likely a site i will take label x_i given image $\boldsymbol{y}$, ignoring the effects of other sites in the image (Fig. 2). Suppose $\boldsymbol{f}(.)$ is a function that maps an arbitrary patch in an image to a feature vector such that $\boldsymbol{f} : \mathcal{Y}_p \rightarrow \Re^l$. Here, $\mathcal{Y}_p$ is the set of all possible patches in all possible images. Let $\omega_i(\boldsymbol{y})$ be an arbitrary patch in the neighborhood of site i in image $\boldsymbol{y}$ from which a feature vector $\boldsymbol{f}(\omega_i(\boldsymbol{y}))$ is extracted. Note that the neighborhood used for the patch $\omega_i(\boldsymbol{y})$ need not be the same as the label neighborhood $\mathcal{N}_i$. Indeed, $\omega_i(\boldsymbol{y})$ can potentially be the whole image itself. For clarity, denote the feature vector $\boldsymbol{f}(\omega_i(\boldsymbol{y}))$ at each site i by $\boldsymbol{f}_i(\boldsymbol{y})$. The subscript i indicates the difference just in the feature vectors at different sites, *not* in the functional form of $\boldsymbol{f}(.)$. Then, $A(x_i, \boldsymbol{y})$ is modeled using a local discriminative model that outputs the association of the site i with class x_i as

$$A(x_i, \boldsymbol{y}) = \log P'(x_i | \boldsymbol{f}_i(\boldsymbol{y})), \qquad (3)$$

where $P'(x_i | \boldsymbol{f}_i(\boldsymbol{y}))$ is the local class conditional at site i. This form allows one to use an arbitrary domain-specific probabilistic discriminative classifier for a given task. This can be seen as a parallel to the traditional MRF models where one can use arbitrary local generative classifier to model the unary potential. One possible choice of $P'(.)$ is generalized linear models (GLM), which are used extensively in statistics to model the class posteriors. Logistic function is a commonly used link in GLMs, although other choices such as probit link exist. Using a logistic function, the local class conditional can be written as

$$P'(x_i = 1|\boldsymbol{f}_i(\boldsymbol{y})) = \frac{1}{1+e^{-(w_0+\boldsymbol{w}_1^T \boldsymbol{f}_i(\boldsymbol{y}))}}$$
$$= \sigma(w_0 + \boldsymbol{w}_1^T \boldsymbol{f}_i(\boldsymbol{y})), \qquad (4)$$

where $\boldsymbol{w} = \{w_0, \boldsymbol{w}_1\}$ are the model parameters. This form of $P'(.)$ will yield a linear decision boundary in the feature space spanned by vectors $\boldsymbol{f}_i(\boldsymbol{y})$. To extend the logistic model to induce a nonlinear decision boundary, a transformed feature vector at each site i can be defined as $\boldsymbol{h}_i(\boldsymbol{y}) = [1, \phi_1(\boldsymbol{f}_i(\boldsymbol{y})), \ldots, \phi_R(\boldsymbol{f}_i(\boldsymbol{y}))]^T$ where $\phi_k(.)$ are arbitrary nonlinear functions. These functions can be seen as explicit kernel mapping of the original feature vector into a high dimensional space. The first element of the transformed vector is kept as 1 to accommodate the bias parameter w_0. Further, since $x_i \in \{-1, 1\}$, the probability in Eq. (4) can be compactly expressed as

$$P'(x_i|\boldsymbol{y}) = \sigma(x_i \boldsymbol{w}^T \boldsymbol{h}_i(\boldsymbol{y})). \qquad (5)$$

Finally, for this choice of $P'(.)$, the association potential can be written as,

$$A(x_i, \boldsymbol{y}) = \log(\sigma(x_i \boldsymbol{w}^T \boldsymbol{h}_i(\boldsymbol{y}))) \qquad (6)$$

This transformation ensures that the DRF is equivalent to a logistic classifier if the interaction potential in Eq. (2) is set to zero. Besides GLMs, discriminative classifiers based on SVM, neural network, and boosting have been successfully used in modeling association potential in the literature. Note that in Eq. (6), the transformed feature vector at *each* site i, that is, $\boldsymbol{h}_i(\boldsymbol{y})$ is a function of the whole set of observations $\boldsymbol{y}$. This allows one to pool arbitrarily complex dependencies in the observed data for the purpose of classification. On the contrary, the assumption of conditional independence of the data in the traditional MRF framework allows one to use data only from a particular site, that is, $\boldsymbol{y}_i$, to design the log-density, which acts as the association potential.

Interaction Potential

In the DRF framework, the interaction potential can be seen as a measure of how the labels at neighboring sites i and j interact given the observed image $\boldsymbol{y}$

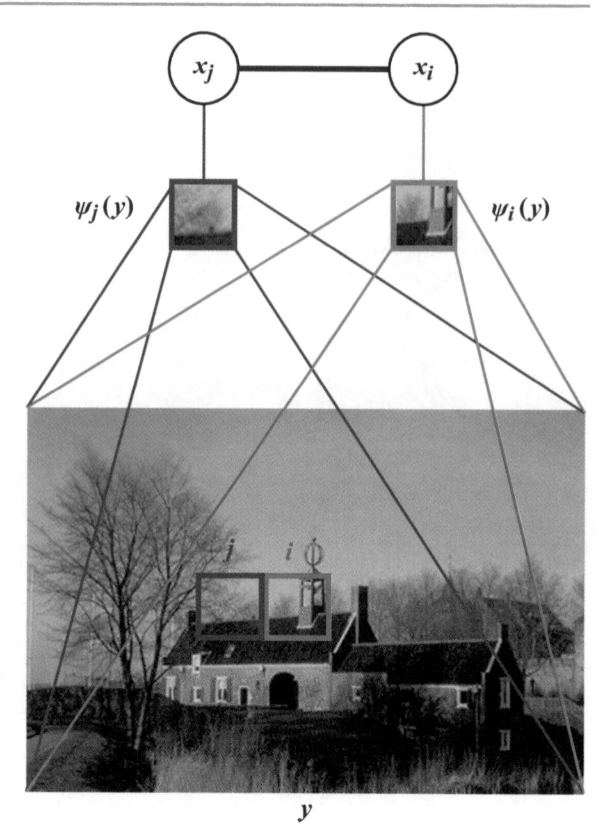

Discriminative Random Fields, Fig. 3 Given feature vectors $\psi_i(\boldsymbol{y})$ and $\psi_j(\boldsymbol{y})$ at two neighboring sites i and j, respectively, the interaction potential can be seen as a measure of how the labels at sites i and j influence each other. Note that such interaction in labels is dependent on the observed image data $\boldsymbol{y}$, unlike the traditional generative MRFs

(Fig. 3). In contrast to generative MRFs, the interaction potential in DRFs is a function of observations $\boldsymbol{y}$. Suppose $\psi(.)$ is a function that maps an arbitrary patch in an image to a feature vector such that $\psi : \mathcal{Y}_p \rightarrow \Re^{\gamma}$. Let $\Omega_i(\boldsymbol{y})$ be an arbitrary patch in the neighborhood of site i in image $\boldsymbol{y}$ from which a feature vector $\psi(\Omega_i(\boldsymbol{y}))$ is extracted. Note that the neighborhood used for the patch $\Omega_i(\boldsymbol{y})$ need not be the same as the label neighborhood $\mathcal{N}_i$. For clarity, denote the feature vector $\psi(\Omega_i(\boldsymbol{y}))$ at site i by $\psi_i(\boldsymbol{y})$. Similarly, denote the feature vector at site j by $\psi_j(\boldsymbol{y})$. Again, to emphasize, the subscripts i and j indicate the difference just in the feature vectors at different sites, *not* in the functional form of $\psi(.)$. Given the features at two different sites, the basic idea is to learn a pairwise discriminative model $P''(x_i = x_j|\psi_i(\boldsymbol{y}), \psi_j(\boldsymbol{y}))$. Note that by choosing the function ψ_i to be different from $\boldsymbol{f}_i$, used in Eq. (4), information different from $\boldsymbol{f}_i$

can be used to model the relations between pairs of sites.

For a pair of sites (i, j), let $\boldsymbol{\mu}_{ij}(\psi_i(\boldsymbol{y}), \psi_j(\boldsymbol{y}))$ be a new feature vector such that $\boldsymbol{\mu}_{ij} : \Re^{\gamma} \times \Re^{\gamma} \rightarrow \Re^{q}$. Denoting this feature vector as $\boldsymbol{\mu}_{ij}(\boldsymbol{y})$ for simplification, the interaction potential is modeled as

$$I(x_i, x_j, \boldsymbol{y}) = x_i x_j \boldsymbol{v}^T \boldsymbol{\mu}_{ij}(\boldsymbol{y}), \qquad (7)$$

where $\boldsymbol{v}$ are the model parameters. Note that the first component of $\boldsymbol{\mu}_{ij}(\boldsymbol{y})$ is fixed to be 1 to accommodate the bias parameter. There are two interesting properties of the interaction potential given in Eq. (7). First, if the association potential at each site and the interaction potentials for all the pairwise cliques except the pair (i, j) are set to zero in Eq. (2), the DRF acts as a logistic classifier which yields the probability of the site pair to have the same labels given the observed data. Of course, one can generalize the form in Eq. (7) as

$$I(x_i, x_j, \boldsymbol{y}) = \log P''(x_i, x_j | \psi_i(\boldsymbol{y}), \psi_j(\boldsymbol{y})), \qquad (8)$$

similar to the association potential and can use arbitrary pairwise discriminative classifier to define this term. The second property of the interaction potential form given in Eq. (7) is that it generalizes the Ising model. The original Ising form is recovered if all the components of vector $\boldsymbol{v}$ other than the bias parameter are set to zero in Eq. (7). A geometric interpretation of interaction potential is that it partitions the space induced by the relational features $\boldsymbol{\mu}_{ij}(\boldsymbol{y})$ between the pairs that have the same labels and the ones that have different labels. Hence, Eq. (7) acts as a data-dependent discontinuity adaptive model that will moderate smoothing when the data from the two sites is "different." The data-dependent smoothing can especially be useful to absorb the errors in modeling the association potential. Anisotropy can be easily included in the DRF model by parameterizing the interaction potentials of different directional pairwise cliques with different sets of parameters $\boldsymbol{v}$.

Parameter Learning and Inference

For 1D sequential CRFs proposed by Lafferty et al. [1], exact maximum likelihood parameter learning is feasible because the induced graph does not contain loops. However, in typical DRFs, the underlying graph contains loops, and it is usually infeasible to exactly maximize the likelihood with respect to the parameters. Therefore, a critical issue in applying DRFs to image-based applications is the design of effective parameter learning techniques that can operate on arbitrary graphs.

Maximum Likelihood Parameter Learning

Let θ be the set of unknown DRF parameters where $\theta = \{\boldsymbol{w}, \boldsymbol{v}\}$. Given M i.i.d. labeled training images, the maximum likelihood estimates of the parameters are given by maximizing the log-likelihood $l(\theta) = \sum_{m=1}^{M} \log P(\boldsymbol{x}^m | \boldsymbol{y}^m, \theta)$, that is,

$$\widehat{\theta} = \underset{\theta}{\operatorname{argmax}} \sum_{m=1}^{M} \left\{ \sum_{i \in S^m} \log \sigma(x_i^m \boldsymbol{w}^T \boldsymbol{h}_i(\boldsymbol{y}^m)) + \sum_{i \in S^m} \sum_{j \in \mathcal{N}_i} x_i^m x_j^m \boldsymbol{v}^T \boldsymbol{\mu}_{ij}(\boldsymbol{y}^m) - \log Z^m \right\}, \qquad (9)$$

where the partition function for the mth image is

$$Z^m = \sum_{\boldsymbol{x}} \exp \left\{ \sum_{i \in S^m} \log \sigma(x_i \boldsymbol{w}^T \boldsymbol{h}_i(\boldsymbol{y}^m)) + \sum_{i \in S^m} \sum_{j \in \mathcal{N}_i} x_i x_j \boldsymbol{v}^T \boldsymbol{\mu}_{ij}(\boldsymbol{y}^m) \right\}.$$

Note that Z^m is a function of the parameters θ and the observed data $\boldsymbol{y}^m$. For learning the parameters using gradient ascent, the derivatives of the log-likelihood are

$$\frac{\partial l(\theta)}{\partial \boldsymbol{w}} = \frac{1}{2} \sum_{m} \sum_{i \in S^m} (x_i^m - \langle x_i \rangle_{\theta; \boldsymbol{y}^m}) \boldsymbol{h}_i(\boldsymbol{y}^m), \qquad (10)$$

$$\frac{\partial l(\theta)}{\partial \boldsymbol{v}} = \sum_{m} \sum_{i \in S^m} \sum_{j \in \mathcal{N}_i} (x_i^m x_j^m - \langle x_i x_j \rangle_{\theta; \boldsymbol{y}^m}) \boldsymbol{\mu}_{ij}(\boldsymbol{y}^m). \qquad (11)$$

Here, $\langle \cdot \rangle_{\theta; \boldsymbol{y}^m}$ denotes expectation with $P(\boldsymbol{x} | \boldsymbol{y}^m, \theta)$. Ignoring $\boldsymbol{\mu}_{ij}(\boldsymbol{y}^m)$, gradient ascent with Eq. (11) resembles the problem of learning in Boltzmann machines.

For arbitrary graphs with loops, the expectations in Eqs. (10) and (11) cannot be computed exactly due to the combinatorial size of the label space.

Sampling procedures such as Markov Chain Monte Carlo (MCMC) can be used to approximate the true expectations. Unfortunately, MCMC techniques have two main problems: a long "burn-in" period (which makes them slow) and high variance in estimates. Although several techniques have been suggested to approximate the expectations, two popular methods are described below (see [5] for other choices and a detailed comparison).

Pseudo-marginal Approximation (PMA)

It is easy to see if true marginal distributions $P_i(x_i | \boldsymbol{y}, \theta)$ at each site, i, and $P_{ij}(x_i, x_j | \boldsymbol{y}, \theta)$ at each pair of sites i and $j \in \mathcal{N}_i$ are known, one can compute exact expectations as

$$\langle x_i \rangle_{\theta;\boldsymbol{y}} = \sum_{x_i} x_i P_i(x_i | \boldsymbol{y}, \theta) \text{ and } \langle x_i x_j \rangle_{\theta;\boldsymbol{y}} = \sum_{x_i, x_j} x_i x_j P_{ij}(x_i, x_j | \boldsymbol{y}, \theta).$$

Since computing exact marginal distributions is in general infeasible, a standard approach is to replace the actual marginals by pseudo-marginals. For instance, one can use loopy belief propagation (BP) to get these pseudo-marginals. It has been shown in practice that for many applications, loopy BP provides good estimates of the marginals.

Saddle Point Approximation (SPA)

In saddle point approximation (SPA), one makes a discrete approximation of the expectations by directly using best estimates of labels at a given setting of parameters. This is equivalent to approximating the partition function (Z) such that the summation over all the label configurations $\boldsymbol{x}$ in Z is replaced by the largest term in the sum, which occurs at the most probable label configuration. In other words, if

$$\hat{\boldsymbol{x}} = \arg\max_{\boldsymbol{x}} P(\boldsymbol{x} | \boldsymbol{y}, \theta),$$

then according to SPA,

$$Z \approx \exp\left\{ \sum_{i \in S} \log \sigma(\hat{x}_i \boldsymbol{w}^T \boldsymbol{h}_i(\boldsymbol{y})) + \sum_{i \in S} \sum_{j \in \mathcal{N}_i} \hat{x}_i \hat{x}_j \boldsymbol{v}^T \boldsymbol{\mu}_{ij}(\boldsymbol{y}) \right\}.$$

This leads to a very simple approximation to the expectation, that is, $\langle x_i \rangle_{\theta;\boldsymbol{y}} \approx \hat{x}_i$. Further assuming a mean-field type decoupling, that is, $\langle x_i x_j \rangle_{\theta;\boldsymbol{y}} = \langle x_i \rangle_{\theta;\boldsymbol{y}} \langle x_j \rangle_{\theta;\boldsymbol{y}}$, it also follows that $\langle x_i x_j \rangle_{\theta;\boldsymbol{y}} \approx \hat{x}_i \hat{x}_j$. Readers familiar with perceptron learning rules can readily see that with such an approximation, the updates in Eq. (10) are very similar to perceptron updates.

However, this discrete approximation raises a critical question: Will the gradient ascent of the likelihood with such gradients converge? It has been shown empirically that while the approximate gradient ascent is not strictly convergent in general, it is weakly convergent in that it oscillates within a set of good parameters or converges to a good parameter with isolated large deviations. In fact, one can show that this weak-convergence behavior is tied to the empirical error of the model [5]. To pick a good parameter setting, one can use any of the popular heuristics used for perceptron learning with inseparable data. For instance, one can let the algorithm run up to some fixed number of iterations and pick the parameter setting that minimizes the empirical error. Even though lack of strict convergence can be seen as a drawback of SPA, the main advantage of these methods is very fast learning of parameters with performance similar to or better than pseudo-marginal methods.

Inference

Given a new test image $\boldsymbol{y}$, the problem of inference is to find the optimal labels $\boldsymbol{x}$ over the image sites, where optimality is defined with respect to a given cost function. Maximum a posteriori (MAP) solution is a widely used estimate that is optimal with respect to the zero-one cost function defined as

$$C(\boldsymbol{x}, \boldsymbol{x}^*) = 1 - \delta(\boldsymbol{x} - \boldsymbol{x}^*), \tag{12}$$

where $\boldsymbol{x}^*$ is the true label configuration, and $\delta(\boldsymbol{x} - \boldsymbol{x}^*)$ is 1 if $\boldsymbol{x} = \boldsymbol{x}^*$, and 0 otherwise. The MAP solution is defined as

$$\hat{\boldsymbol{x}} = \arg\max_{\boldsymbol{x}} P(\boldsymbol{x} | \boldsymbol{y}, \theta).$$

For binary classifications, the MAP estimate can be computed exactly for an undirected graph using the max-flow/min-cut type of algorithms if the probability distribution meets certain conditions [6]. For the DRF model, since max-flow algorithms do not allow negative interaction between the sites, the data-dependent smoothing for each clique is set to be $\mathbf{v}^T\boldsymbol{\mu}_{ij}(\mathbf{y}) = \max\{0, \mathbf{v}^T\boldsymbol{\mu}_{ij}(\mathbf{y})\}$, yielding an approximate MAP solution.

An alternative to the MAP solution is the maximum posterior marginal (MPM) solution which is optimal for the sitewise zero-one cost function defined as

$$C(\mathbf{x}, \mathbf{x}^*) = \sum_{i \in S} (1 - \delta(x_i - x_i^*)), \quad (13)$$

where x_i^* is the true label at the ith site. The MPM solution at each site is defined as

$$\hat{x_i} = \arg\max_{\mathbf{x}_i} P_i(\mathbf{x}_i | \mathbf{y}, \theta), \quad \text{where} \quad P_i(\mathbf{x}_i | \mathbf{y}, \theta) = \sum_{x - x_i} P(\mathbf{x} | \mathbf{y}, \theta),$$

and $x - x_i$ denotes all the node variables except for node i. The MPM computation requires marginalization over a large number of variables which is generally NP-hard. However, as discussed before, one can use loopy BP to obtain an estimate of the MPM solution.

Application

DRFs and related models have been applied to a wide variety of problems in computer vision to incorporate context in, for example, image denoising, scene segmentation and region classification, object recognition, and video tagging.

Experimental Results

Here is an application of binary DRFs to the problem of man-made structure detection in natural scenes. It is a difficult problem due to presence of significant clutter and wide variability in the appearance of man-made structure as well as the background. The training and the test set contained 108 and 129 images, respectively, from the Corel image database. Each image is divided into nonoverlapping 16×16 pixels blocks. Each block forms a site in the graph. The whole training set contained 36,269 blocks from the *nonstructured* class and 3,004 blocks from the *structured* class. Histograms over gradient orientations were used to extract heaved central-shift moment-based 14-dimensional feature vector. In the unary classifier, an explicit quadratic kernel is used to map the feature vector into a 119-dimensional space. The pairwise data vector is obtained by concatenating the unary vectors at two sites. The parameters of the DRF model were learned using the maximum likelihood framework as described before.

For an input test image given in Fig. 4a, the *structure* detection results from three methods are shown in Fig. 4. The blocks identified as *structured* have been shown enclosed within an artificial boundary. It can be noted that for similar detection rates, the number of false positives have significantly reduced for the DRF-based detection. Locally, different branches may yield features similar to those from the man-made structures. The logistic classifier does not enforce smoothness in labels, which led to increased isolated false positives. However, the MRF solution with Ising model simply smooths the labels without taking observations into account resulting in a smoothed false-positive region around the tree branches.

To carry out the quantitative evaluations, the detection rates and the number of false positives per image for each technique are compared. At first data, interactions are not allowed for any method by extracting features from individual sites. The comparative results for the three methods are given in Table 1 next to "MRF", "Logistic$^-$" and "DRF$^-$". For comparison purposes, the false-positive rate of the logistic classifier is fixed to be the same as the DRF in all the experiments. For similar false positives, the detection rates of the traditional MRF and the DRF are higher than the logistic classifier due to the label interaction. However, the higher detection rate of the DRF in comparison to the MRF indicates the gain due to the use of discriminative models in the association and interaction potentials. In the next experiment, to take advantage of the power of the DRF framework, data interaction was allowed for both the logistic classifier as well as the DRF ("Logistic" and "DRF" in Table 1). The DRF detection rate increases substantially, and the false positives decrease further indicating the importance of allowing the data interaction in addition to the label interaction.

Discriminative Random Fields, Fig. 4 Structure detection results on a test example for different methods. For similar detection rates, DRF reduces the false positives considerably. (**a**) Input image, (**b**) logistic, (**c**) MRF, and (**d**) DRF

Discriminative Random Fields, Table 1 Detection rates (DR) and false positives (FP) for the test set containing 129 images (49, 536 sites). FP for logistic classifier were kept to be the same as for DRF for DR comparison. Superscript '—' indicates no neighborhood data interaction was used

	MRF	Logistic^{-}	DRF^{-}	Logistic	DRF
DR (%)	58.35	47.50	61.79	60.80	72.54
FP (per image)	2.44	2.28	2.28	1.76	1.76

Extensions and Related Work

A large number of extensions of the basic binary DRFs have been proposed. For instance, *multiclass* DRFs allow label sets consisting of more than two labels and *hierarchical DRFs* incorporate context at multiple levels [7]. For example, for parts-based object detection, local context is the geometric relationship among parts of an object, while the relative spatial configurations of different objects (e.g., monitor, keyboard, and mouse) provides the global context. Learning in DRFs was extended to a semi-supervised paradigm by Lee et al. [8], while a Hidden-DRF model with latent variables was proposed in [9].

Open Problems

Hierarchical DRFs are of great interest since such models are quite powerful and can parse scenes at multiple levels of granularity. Robust parameter learning and inference in such models is a very challenging open problem.

References

1. Lafferty J, McCallum A, Pereira F (2001) Conditional random fields: probabilistic models for segmenting and labeling sequence data. In: Proceedings of the international conference on machine learning. Morgan Kaufmann Publishers, San Francisco
2. Kumar S, Hebert M (2003) Discriminative random fields: a discriminative framework for contextual interaction in classification. In: Proceedings of the IEEE international conference on computer vision (ICCV), Nice, vol 2, pp 1150–1157
3. Kumar S, Hebert M (2004) Discriminative fields for modeling spatial dependencies in natural images. In: Thrun S, Saul L, Schölkopf B (eds) Advances in neural information processing systems 16, The MIT Press, Boston
4. Hammersley JM, Clifford P Markov field on finite graph and lattices. Unpublished
5. Kumar S, August J, Hebert M (2005) Exploiting inference for approximate parameter learning in discriminative fields: an empirical study. In: Fourth international workshop on energy minimization methods in computer vision and pattern recognition (EMMCVPR), St. Augustine
6. Greig DM, Porteous BT, Seheult AH (1989) Exact maximum a posteriori estimation for binary images. J R Stat Soc 51(2):271–279
7. Kumar S, Hebert M (2005) A hierarchical field framework for unified context-based classification. In: IEEE international conference on computer vision (ICCV), Beijing
8. Lee C, Wang S, Jiao F, Schuurmans D, Greiner R (2006) Learning to model spatial dependency: semi-supervised discriminative random fields. In: Neural information processing systems conference (NIPS), Vancouver
9. Quattoni A, Collins M, Darrell T (2004) Conditional random fields for object recognition. In: Neural information processing systems (NIPS), Vancouver

Disocclusion

►Inpainting

Distance Estimation

►Depth Estimation

Dual Differential Geometry

►Isotropic Differential Geometry in Graph Spaces

Dynamic Optimization

►Dynamic Programming

Dynamic Programming

Boris Flach and Vaclav Hlavac
Department of Cybernetics, Czech Technical University in Prague, Faculty of Electrical Engineering, Prague 6, Czech Republic

Synonyms

Dynamic optimization

Definition

Dynamic Programming (DP) is a paradigm used in algorithms for solving optimization problems. It relies on the construction of nested subproblems such that the solution of the main problem can be obtained from the solutions of the subproblems.

Background

The paradigm was introduced by the mathematician Richard Bellman in the 1940s and applied in control theory [1].

Theory

The applicability of the paradigm relies on the following two assumptions:

Optimal substructure means that a system of nested subproblems can be constructed in such a way that the solution of the main problem can be obtained from the solutions of these subproblems.

Overlapping subproblems means that the smaller subproblems in the next level are only slightly smaller and moreover the set of these subproblems is small as well. This distinguishes DP from "divide and conquer" methods.

The method starts by solving the smallest subproblems directly. The obtained results are stored and used for solving the bigger subproblems in the next higher level. Applying this principle for each step avoids expensive recursions.

The following two algorithms illustrate applications of the DP paradigm.

Viterbi Algorithm

The task is to find the maximal value of a function F which depends on n discrete variables $x_1, x_2, \ldots, x_n \in K$ in a finite domain K. The function has the form

$$F(x_1, \ldots, x_n) = f_1(x_1, x_2) + f_2(x_2, x_3) + \ldots + f_{n-1}(x_{n-1}, x_n). \quad (1)$$

An appropriate nested system of subproblems is defined by

$$\phi_i(k) = \max_{x_1,\ldots,x_i} \left[f_1(x_1, x_2) + \ldots + f_i(x_i, k) \right] \quad (2)$$

for all $i = 1, \ldots, n-1$ and all $k \in K$. It is quite obvious that the solutions for $\phi_{i+1}()$ can be obtained from those for $\phi_i()$ by

$$\phi_{i+1}(k) = \max_{k' \in K} \left[\phi_i(k') + f_{i+1}(k', k) \right]. \quad (3)$$

The solution of the simplest subproblems $\phi_1(k)$ is found simply by enumeration and similarly, the solution of the main problem is eventually obtained by solving $\max_k \phi_{n-1}(k)$.

Additional memory space is required if the task is to calculate a maximizer $(x_1^*, \ldots, x_n^*)$ of F rather than its maximal value. In such a case a "pointer"

$$p_{i+1}(k) = \operatorname*{argmax}_{k' \in K} \left[\phi_i(k') + f_{i+1}(k', k) \right] \quad (4)$$

has to be stored for each $i = 1, \ldots, n-1$ and each $k \in K$. The required maximizer is found simply by backtracking in this list, that is, $x_i^* = p_{i+1}(x_{i+1}^*)$.

Floyd Warshall Algorithm

The task is to calculate the shortest paths for all pairs of vertices in a weighted graph (V, E, w). It is assumed for simplicity that all nonexistent edges are included but have infinite length. Let $S(i, j)$, $i, j \in V$, denote the length of the shortest path connecting vertex i with vertex j. The nested system of subproblems is defined as follows: Let $S(i, j, k)$ denote the length of the shortest path connecting vertex i with vertex j which is passing only through vertices of the subset $V_k = \{1, 2, \ldots, k\} \subset V$. In other words, $S(i, j, k)$ is the length of the shortest path connecting i and j in the subgraph induced by the vertex set $\{i, j\} \cup V_k$. Again, it is easy to see that $S(i, j, k+1)$ can be calculated from $S(.,.,k)$. The sought-after shortest path either passes through vertices from V_k only or otherwise, is composed from a part connecting vertices i and $k+1$ and a second part connecting vertices $k+1$ and j. This leads to

$$S(i, j, k+1) = \min\left\{ S(i, j, k), S(i, k+1, k) + S(k+1, j, k) \right\}. \quad (5)$$

The solutions of the smallest subproblems are obviously $S(i, j, 0) = w(i, j)$.

The algorithm has time complexity $\mathcal{O}(|V|^3)$ and space complexity $\mathcal{O}(|V|^2)$ and provides the lengths of the shortest paths between all pairs of vertices. A simple modification of the algorithm allows one to reconstruct the shortest path for each pair of vertices. The key idea is to store an additional matrix of size $|V| \times |V|$, whose elements represent the largest of the indices of intermediate vertices in the shortest path connecting vertices i and j. This matrix is initialized and updated along with matrix S. Once calculated, it allows one to reconstruct the shortest path between each pair of vertices recursively.

A more comprehensive analysis of DP and other examples can be found in [2, 3].

References

1. Bellman R (2003) Dynamic programming. Dover, Mineola
2. Denardo EV (2003) Dynamic programming: models and applications. Dover, Mineola
3. Sniedovich M (2010) Dynamic programming: foundations and principles. Taylor and Francis, Boca Raton

E

Edge Detection

James H. Elde
Centre for Vision Research, York University, Toronto, ON, Canada

Related Concepts

►Boun-dary Detection; ►Scale Selection

Definition

Edge detection is the process of label the image pixels that lie on the boundaries where abrupt intensity discontinuity occur.

Background

The light projected from a visual scene into an eye or camera is typically piecewise smooth as a function of visual angle. Since nearby points on a surface tend to have similar attitude, reflectance, and illumination, the pixels to which these surface points project tend to have similar intensity. This rule is broken when two adjacent pixels project from points on either side of an occlusion boundary, since the points now project from different surfaces that may well have different attitude, reflectance, and illumination, and typically an abrupt change in image intensity results. Intensity edges also arise when neighboring pixels project from points on the same surface that happens to straddle a surface crease, pigment change, or shadow boundary.

Since these abrupt changes in image intensity correspond to significant physical events in the scene, the problem of reliably detecting and localizing these edges is an important and fundamental early vision problem. While detection of object boundaries (occlusion edges) is sometimes seen as the main goal, reliable detection of surface creases and reflectance edges also has clear importance for shape estimation and object recognition, and even cast shadows provide important information about relief, surface contact, and scene layout.

Since edges are sparse, edge detection is also motivated from a differential image coding viewpoint: a large fraction of the information in an image can be captured by coding just the locations, 2D orientations, and intensity changes of these edges.

The problem of edge detection dates back to the first days of computer vision: back to Roberts' thesis at least [20]. In its simplest form, the goal is to label the image pixels that lie on (or very near) a step discontinuity in the image. This definition has been generalized in a number of useful ways: to allow for possible blurring of the step discontinuity, due to shading or defocus, for example, and to require an estimate of the local 2D orientation of the edge as well as its location.

- Biological basis [9]
- History [20]
- Problem definition [1, 2, 5, 6, 13, 15, 18]

Theory and Application

Roberts based his edge detector on a simple 2 pixel × 2 pixel discrete differential operator related to the gradient magnitude (Fig. 1). Since the operator relies

K. Ikeuchi (ed.), *Computer Vision*, DOI 10.1007/978-0-387-31439-6,

Edge Detection, Fig. 1 Edge detection example [4–6]. *Top left*: Original greyscale image. *Top right*: Edge map. *Bottom left*: Reconstruction of original image from brightness and contrast stored only at edge locations. *Bottom right*: Reconstruction including edge blur information

upon only 4 pixels, the results are highly sensitive to noise and are dominated by the fine structure of the image. Over the intervening years, most edge detection algorithms have continued to rely on a first stage of local differential filtering but have innovated in the design of smoothing kernels to increase signal-to-noise ratio, in the order of the differential operators used, and in how they are combined. This local differential filtering approach aligns well with physiological data showing that neurons early in the primate visual pathway can to some degree be approximated as stabilized low-order differential operators [9].

By far the most popular smoothing kernel has been the 2D Gaussian $G(x, y) = \frac{1}{2\pi\sigma^2} \exp\left(-\frac{x^2+y^2}{2\sigma^2}\right)$. Marr and Hildreth [17] proposed the use of second-order isotropic Laplacian-of-Gaussian ("Mexican hat") filters $\nabla^2 G(x, y)$. The $\nabla^2 G$ filter produces a signed response that crosses zero precisely at the location of an (ideal) step edge, and Marr and Hildreth proposed that edges thus be identified with such zero crossings. Due to the isotropy of the $\nabla^2 G$ filter, the response is invariant to the orientation of the edge, a desirable property, since edges can occur at any orientation. Marr and Hildreth also observed that different edges occur at different scales, and so employed $\nabla^2 G$ filters over a range of scales σ. Observing that spurious zero crossings at a single scale can occur due to interference between multiple distinct edges, Marr and Hildreth proposed a *scale combination rule*: edges are deemed valid only if zero crossings at the same location and orientation are found at more than one scale.

Marr and Hildreth obtained orientation invariance by using isotropic filters and approximate scale invariance by combining information across scales. However, these invariance properties came at the price detection and localization performance. This became clear partly through the work of John Canny [2], who used a variational approach to determine an edge detector that would be (nearly) optimal precisely in terms of detection rate and localization performance, while avoiding multiple responses to the same edge. The outcome was a an edge detection filter that is well approximated by a first-derivative-of-Gaussian function. Canny used two such filters to estimate stabilized partial derivatives and hence the local gradient vector ∇G:

$$\nabla G = \begin{pmatrix} G_x \\ G_y \end{pmatrix}, \quad \text{where } G_x = \frac{\partial G}{\partial x} \text{ and } G_y = \frac{\partial G}{\partial y}$$

The first-derivative-of-Gaussian filter can be shown to be steerable [7]: the filter can be synthesized at any orientation from a linear combination of the two basis filters G_x and G_y. Thus, Canny's approach amounts to computing the first derivative in the gradient direction at every pixel of the smoothed image. In this sense, the approach still enjoys the orientation invariance property emphasized by Marr and Hildreth: a rotation of the edge in the image will not change the estimated gradient magnitude. Canny's approach, however, delivers higher signal-to-noise because the filter more closely approximates the shape of an extended edge in the tangent direction. While Marr and Hildreth's $\nabla^2 G$

operator computes the second derivative along the edge, Canny's ∇G operator locally integrates along the edge over the support of the Gaussian kernel.

While Marr and Hildreth localized edges at the zero crossings of the $\nabla^2 G$ response, Canny localized edges at maxima of the gradient magnitude, taken in the gradient direction, found using a process called *non-maximum suppression*. Since not all of the resulting maxima correspond to significant edges, a threshold on the gradient magnitude must be applied to reduce the false positive rate. Canny devised a clever heuristic technique, dubbed *thresholding with hysteresis*, that removes many false positives without unduly affecting the hit rate (proportion of correct detections). The technique exploits the fact that real weak edges are often chain-connected along a contour to stronger edges, while false positives are more likely to be isolated. The technique employs two thresholds: all edges that are above the low threshold and chain-connected to edges above the high threshold are identified as edges.

The Canny edge detector may well be the most widely used algorithm in the history of computer vision. Its early adoption is likely derived in part from the open availability of the source code, but it's continuing widespread use reflects the fact that it continues to perform well in comparison to more recent algorithms.

While most versions of Canny's algorithm in use detect edges only at a single scale, in fact both Marr and Hildreth and Canny recognized the problem of scale, but dealt with it in slightly different ways. While Marr and Hildreth conjunctively combined responses across scale (zero crossings of the $\nabla^2 G$ response must be found at multiple scales at the same location and orientation to signal an edge), Canny proposed to *disjunctively* combine edges detected at different scales. This raises the *multiple response problem*: how do we know whether two extremal responses, at different scales but similar locations and orientations, signal two distinct edges in the image, or a single edge corrupted by noise? Canny proposed a *feature synthesis* method to deal with this problem, that signals the larger scale edge only if it could not be satisfactorily explained by the smaller scale responses in a local neighborhood.

Canny's proposed technique localizes edges using the smallest scale at which they are detected, an approach that Canny justified based upon his variational optimization, and this motivated the later development of *edge focusing* techniques [1] in which edges are detected at coarse scales and then tracked through scale space to finer scales for better localization. As it turns out, Canny's one-dimensional analysis does not generalize to two dimensions: Elder and Zucker [5, 6] later showed that for an ideal step or blurred step edge corrupted by noise, localization improves monotonically with increasing scale. This result highlights the difference between theory and practice: since edges in real images are always finite in extent, often curved, and surrounded by other kinds of image structure, neither detection nor localization is likely to be optimized by maximizing scale.

Elder and Zucker pointed out that while it is difficult to model all of the real-world effects that limit the effectiveness of large filters, the small-scale problem is tractable, since performance is limited by noise that can be modeled as white, and for which parameters can be estimated. Based on this observation, they proposed a method for selecting the scale of differential operators called *local scale control*, in which scales are search from fine to coarse to determine the *minimum reliable scale*, that is the smallest scale at which the sign of the derivative measurement is statistically reliable. This approach avoids spurious responses to noise, while minimizing bias due to the finite extent of the edge, curvature, and interference from neighbouring image structure. They also demonstrated methods for subpixel localization down to roughly $1/20$ of a pixel precision [5].

A different scale selection approach for edge detection was proposed at roughly the same time by Lindeberg [15]. Lindeberg's method selects the scale maximizing the response of so-called γ-normalized differential operators. This approach has the property that the scales selected for an ideal, noise-free, isolated, blurred straight edge of infinite extent will be proportional to the blur scale of the edge, and independent of the edge contrast. Thus, small scales should be selected for sharp edges, and larger scales for blurred edges. Unlike the local scale control approach of Elder and Zucker, the selection of scale is not related to the signal-to-noise ratio: in general a large-scale filter will have a smaller response to an ideal edge than a small-scale filter, even when the signal-to-noise for the two filters is the same. As a result of this bias to smaller scales, many noisy edges are detected, and these must be eliminated using some form of post hoc thresholding.

These studies also raised issues around the order of differential operators to employ. While Canny localized edges at gradient maxima, Elder and Zucker localized edges at zero crossings of the second derivative, steered in the gradient direction. Specifically edges are localized at the boundary between two pixels where the sign of the second derivative changes from positive to negative in the gradient direction (this avoids detections of minima in the gradient magnitude). Where the same scales used for these two operators, the results would be identical. However, the Elder and Zucker approach tunes the scales of the two operators independently, based upon the signal-to-noise properties of the operators, and as a result the gradient maxima and second derivative zero crossings are decoupled. While the statistical testing method cannot be used to distinguish real and spurious maxima, it can be used to test for response sign, and hence to detect zero crossings. Thus, the use of the second derivative here is critical. Lindeberg [15] has argued for explicit calculation and testing of the sign of the third derivative, but this is equivalent to checking the sign of the second derivative on either side of the zero crossing. However, Elder and Zucker do show that explicit computation of the third derivative is useful for estimating the blur of an edge, which may be useful for discriminating different types of edges, and for recovering depth from defocus.

In their 1980s paper [17], Marr and Hildreth speculated about the possibility that the locations and gradient magnitudes at oriented zero crossings over multiple scales might constitute a complete encoding of the image, allowing, in principle, for the image to be perfectly reconstructed from the edge representation. While this may seem to be going in the wrong direction from a computer vision point of view, the question is important because it addresses whether an edge code could serve as a *complete* early visual representation, providing sufficient information for all higher-level algorithms.

Since Marr and Hildreth's original conjecture, there have been numerous theoretical and empirical studies exploring the possible completeness of an edge code. Mallat and Zhong [16] demonstrated excellent reconstruction results based upon such a code. However, since edges are represented at many scales, and at the finest scales edge density is very high, this code is highly overcomplete. More compact codes can be derived, but at the expense of noticeable artifact in the reconstruction.

Elder [4] explored an alternative reconstruction approach based upon their edge representation. Rather than storing information from all scales at each edge point, only the location, 2-bit orientation, contrast, brightness, and blur of each edge point were stored, resulting in a far more compact code. Reconstruction is excellent as long as both the intensity and the blur information are employed in the reconstruction.

Since edge detection is often just the first computational step in a computer vision pipeline, and applications may have real-time requirements, efficiency has always been a factor in the design of edge detection algorithms. Both Marr and Hildreth [17] and Canny [2] deliberately employed filters that were $x-y$ separable, allowing 2D convolutions to be computed by composing much cheaper 1D convolutions. Rachid Deriche [3] further improved upon these efficiency by developing recursive filters based upon Canny's original design criteria, allowing very fast implementation on sequential hardware. More recent algorithms have tended to rely upon steerable filters [7] that allow oriented operators to be implemented using only a few basis functions.

While linear filtering forms the front end of most edge detection algorithms, a number of interesting non-linear techniques have been studied. In fact this has a long tradition, going back to early work that sees the problem as model fitting [10] and includes active contour ("snake") methods for fitting semi-local deformable contour models to visual data [12]. While effective in many applications, active contour methods tend to have more parameters to tune, and results are sensitive to these and to initial conditions.

Still to include:

- Scale [13]
- Color edge detection [14]
- Nonlinear filtering [11]
- Nonlocal methods [19]
- Evaluation [8, 13, 18]
- Redefinition in terms of "salient" boundaries [13, 18]

References

1. Bergholm F (1987) Edge focusing. IEEE Trans Pattern Anal Mach Intell 9(6):726–741
2. Canny J (1986) A computational approach to edge-detection. IEEE Trans Pattern Anal Mach Intell 8(6): 679–698
3. Deriche R (1987) Using Canny's criteria to derive a recursively implemented optimal edge detector. Int J Comput Vis 1(2):167–187

4. Elder JH (1999) Are edges incomplete? Int J Comput Vis 34(2–3):97–122
5. Elder JH, Zucker SW (1996) Scale space localization, blur and contour-based image coding. In: Proceedings of the IEEE conference on computer vision and pattern recognition (CVPR). IEEE Computer Society, Los Alamitos, pp 27–34
6. Elder JH, Zucker SW (1998) Local scale control for edge detection and blur estimation. IEEE Trans Pattern Anal Mach Intell 20(7):699–716
7. Freeman WT, Adelson EH (1991) The design and use of steerable filters. IEEE Trans Pattern Anal Mach Intell 13(9):891–906
8. Heath M, Sarkar S, Sanocki T, Bowyer K (1998) Comparison of edge detectors – a methodology and initial study. Comput Vis Image Underst 69(1):38–54
9. Hubel DH, Wiesel TN (1968) Receptive fields and functional architecture of monkey striate cortex. J Physiol 195:215–243
10. Hueckel MH (1971) An operator which locates edges in digitized pictures. J Assoc Comput Mach 18:113–125
11. Iverson LA, Zucker SW (1995) Logical/linear operators for image curves. IEEE Trans Pattern Anal Mach Intell 17(10):982–996
12. Kass M, Witkin A, Terzopoulos D (1987) Snakes – active contour models. Int J Comput Vis 1(4):321–331
13. Konishi S, Yuille AL, Coughlan JM, Zhu SC (2003) Statistical edge detection: learning and evaluating edge cues. IEEE Trans Pattern Anal Mach Intell 25(1):57–74
14. Lee HC, Cok DR (1991) Detecting boundaries in a vector field. IEEE Trans Signal Process 39(5):1181–1194
15. Lindeberg T (1998) Edge detection and ridge detection with automatic scale selection. Int J Comput Vis 30(2):117–154
16. Mallat S, Zhong S (1992) Characterization of signals from multiscale edges. IEEE Trans Pattern Anal Mach Intell 14(7):710–732
17. Marr D, Hildreth E (1980) Theory of edge-detection. Proc R Soc Lond B 207(1167):187–217
18. Martin DR, Fowlkes CC, Malik J (2004) Learning to detect natural image boundaries using local brightness, color, and texture cues. IEEE Trans Pattern Anal Mach Intell 26(5):530–549
19. Perona P, Malik J (1990) Scale-space and edge-detection using anisotropic diffusion. IEEE Trans Pattern Anal Mach Intell 12(7):629–639
20. Roberts L (1965) Machine perception of 3-dimensional solids. In: Tippett J (ed) Optical and electro-optical information processing. MIT, Cambridge, MA

Eigenspace Methods

Tomokazu Takahashi and Hiroshi Murase
Faculty of Economics and Information, Gifu Shotoku Gakuen University, Gifu, Japan

Synonyms

Methods of image recognition in a low-dimensional eigenspace

Related Concepts

▸Principal Component Analysis (PCA)

Definition

The eigenspace method is an image recognition technique that achieves object recognition, object detection, and parameter estimation from images using the distances between input and gallery images in a low-dimensional eigenspace. Here, the eigenspace is constructed based on a statistical method, such as principal component analysis or Karhunen-Loève transform, so that the variation in the appearances of target objects can be represented in a low-dimensional space efficiently. In particular, a technique called the parametric eigenspace method represents the rotation and translation of a target object or a light source as a manifold in an eigenspace. Accordingly, this method performs object recognition and parameter estimation using distances in the manifold.

Background

Appearance-based object recognition is a technique that recognizes a target object by matching between input and preregistered gallery images. One of the simplest methods to achieve this image matching calculates the distances between the pixel values of these images. However, it is difficult for many applications to apply this method due to two problems: (1) processing time needed to calculate the distance between images increases depending on the size of the images and (2) memory space needed to store the gallery images of target objects grows depending on the number of the objects.

Principal component analysis (PCA) can be used as one of the unsupervised dimensionality reduction techniques that transform a sample set of high-dimensional vectors to the set of low-dimensional vectors with minimum information loss. This technique first calculates eigenvectors that are at right angles to each other and maximizes the variances in their directions; it then constructs a low-dimensional eigenspace defined by a small number of eigenvectors. The low-dimensional vectors are obtained by projecting the high-dimensional vectors to the eigenspace. An eigenspace method constructs a low-dimensional

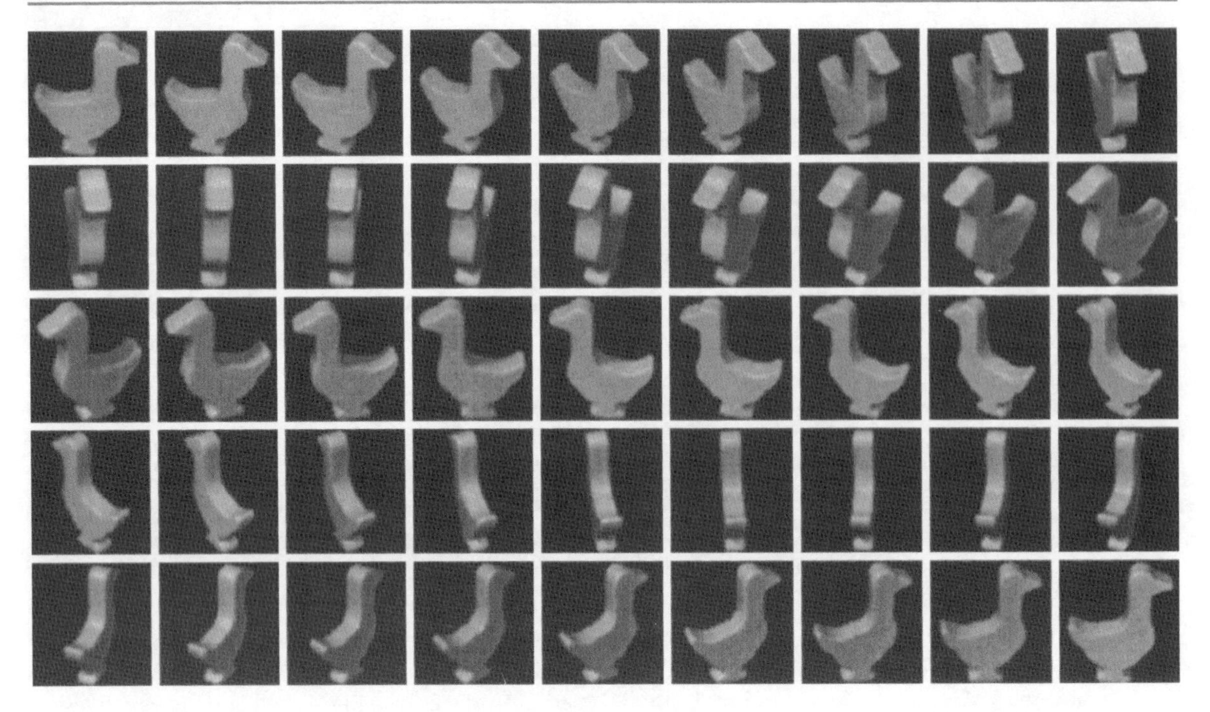

Eigenspace Methods, Fig. 1 Image samples in an object image set containing object images with different horizontal poses

eigenspace from preregistered gallery images of target objects and calculates the distances between input and gallery images in the low-dimensional space. Thus, the eigenspace method can reduce the processing time and memory space efficiently without degrading the recognition performance.

Theory

An image recognition procedure using an eigenspace method is divided into learning and recognition stages. The learning stage, which is performed beforehand, constructs a low-dimensional eigenspace from a large number of learning images of target objects and then projects each learning image to the eigenspace. On the other hand, the recognition stage projects an input image of a target object to the eigenspace and then recognizes the input object by matching between input and learning images in the low-dimensional space.

Learning Stage

For each learning image, the method first normalizes the image size so that the number of pixels can be N, and it represents them as N-dimensional vectors $\boldsymbol{x}_1, \ldots, \boldsymbol{x}_M$. M represents the number of learning images. Here, we assume that each target object has one learning image; therefore, M also represents the number of target objects. A matrix X is constructed as follows:

$$X = \left[\frac{\boldsymbol{x}_1 - \boldsymbol{c}}{||\boldsymbol{x}_1 - \boldsymbol{c}||} \cdots \frac{\boldsymbol{x}_M - \boldsymbol{c}}{||\boldsymbol{x}_M - \boldsymbol{c}||} \right], \tag{1}$$

$$\boldsymbol{c} = \frac{1}{M} \sum_{m=1}^{M} \boldsymbol{x}_m. \tag{2}$$

Eigenvectors are obtained by solving the following eigenequation:

$$XX^{\mathrm{T}}\boldsymbol{u}_i = \lambda_i \boldsymbol{u}_i, \tag{3}$$

where $\boldsymbol{u}_i$ represents an eigenvector of XX^{T} corresponding to an eigenvalue λ_i. Eigenvectors $\boldsymbol{u}_1, \ldots, \boldsymbol{u}_N$ are arranged in descending order of their eigenvalues. The method constructs a low-dimensional eigenspace from $k(<< N)$ eigenvectors corresponding to the k largest eigenvalues and then obtains k-dimensional vectors $\boldsymbol{f}_m$ by projecting N-dimensional vectors $\boldsymbol{x}_m$ to the eigenspace using the following equation:

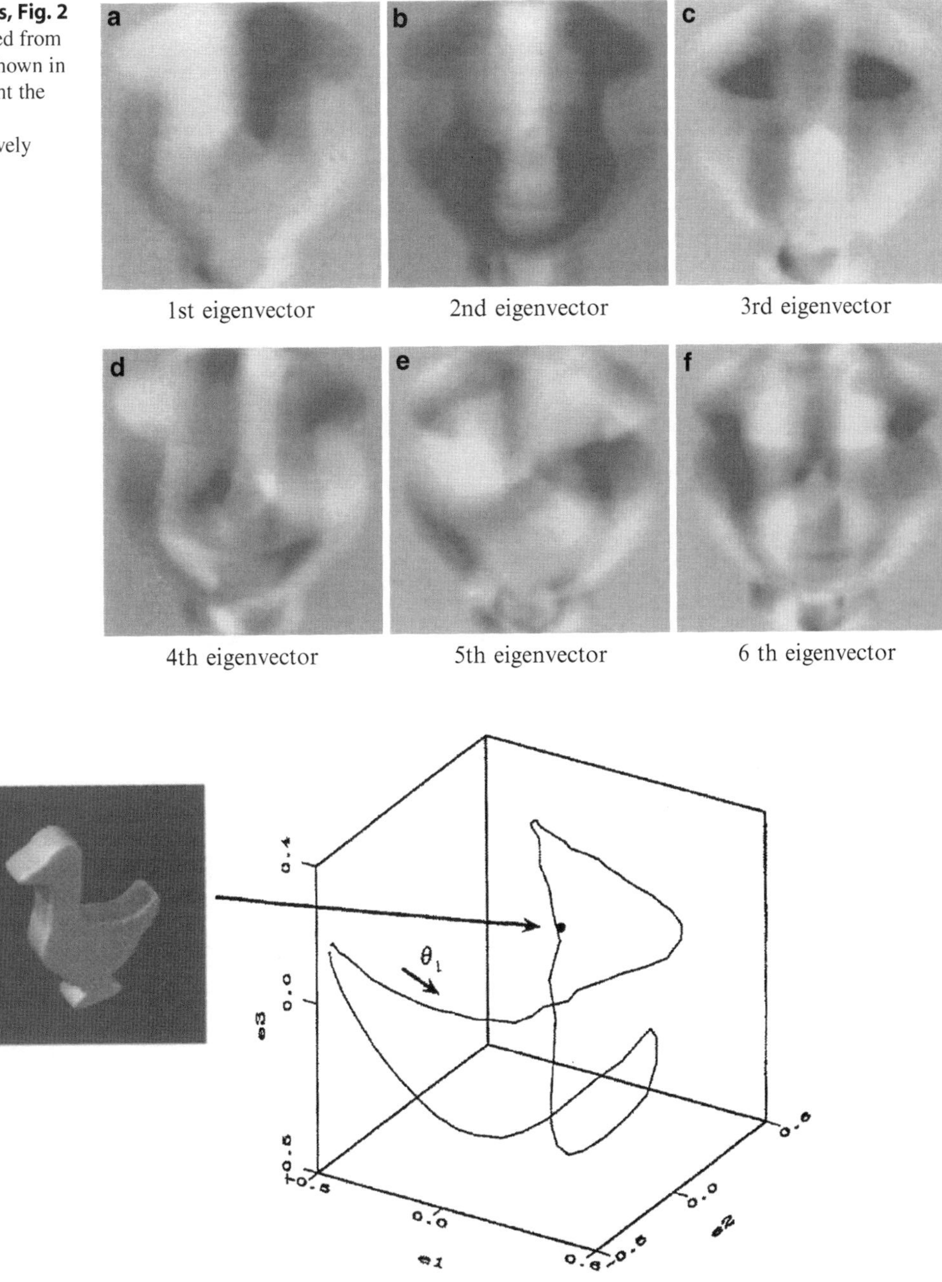

Eigenspace Methods, Fig. 2 Eigenvectors calculated from the object image set shown in Fig. 1. (**a**)–(**f**) represent the first through sixth eigenvectors, respectively

Eigenspace Methods, Fig. 3 A manifold in the object eigenspace constructed from the eigenvectors shown in Fig. 2. Transition of object appearances according to the horizontal pose parameter θ_1 draws a closed smooth curve in the eigenspace

$$\boldsymbol{f}_m = [\boldsymbol{u}_1 \cdots \boldsymbol{u}_k]^{\mathrm{T}} \frac{\boldsymbol{x}_m - \boldsymbol{c}}{||\boldsymbol{x}_m - \boldsymbol{c}||}. \tag{4}$$

Consequently, the memory space needed to store the learning images can be greatly reduced.

Recognition Stage

In the same manner as in the learning stage, the method first normalizes the image size of an input image and represents it as an N-dimensional vector $\boldsymbol{y}$. The method then obtains a k-dimensional vector $\boldsymbol{g}$ by projecting $\boldsymbol{y}$ to the low-dimensional eigenspace constructed in the learning stage using the following equation:

$$g = [u_1 \cdots u_k]^{\mathrm{T}} \frac{y - c}{||y - c||}. \quad (5)$$

The method recognizes the input object as an object $\hat{m}$ that minimizes the distance between the input image g and the learning image f_m using the following equation:

$$\hat{m} = \arg\min_m ||g - f_m||^2. \quad (6)$$

Accordingly, the processing time needed to calculate the distance between the images can be greatly reduced.

Based on an efficient representation of human face images using PCA as proposed by Sirovich and Kirby [1], Turk and Pentland [2] proposed a method using "eigenfaces" to detect and recognize human faces in images. This method calculates a small number of eigenvectors from a large number of learning face images and measures the distances between input and learning face images in a low-dimensional eigenspace. The eigenvectors calculated from the face images are called eigenfaces. Since this method was reported, image recognition using the eigenspace method has become an active area of research.

Murase and Nayar [3] proposed a "parametric eigenspace method" for recognizing three-dimensional objects and estimating their poses simultaneously. This method constructs a manifold as a smooth curved line approximated from a point set in an eigenspace for each target object. The point set is obtained by projecting learning images with various poses to the eigenspace. Object recognition and pose estimation are achieved by calculating the distances between an input image and manifolds. In order to achieve object recognition and pose estimation simultaneously, two types of eigenspaces are constructed: a universal eigenspace from the learning images of all target objects and an object eigenspace for each target object. The method first recognizes an object in the universal eigenspace and then estimates a pose of the object in the object eigenspace. Figure 1 shows samples in an object image set containing object images with different horizontal poses, and Fig. 2 shows eigenvectors that were calculated from the image set. On the other hand, the closed smooth curve shown in Fig. 3 represents a manifold in the object eigenspace that was constructed from the eigenvectors. The manifold was approximated from the points that were obtained by projecting the image in the image set shown in Fig. 1 to the eigenspace.

Ohba and Ikeuchi [4] proposed a method using "eigen window" to recognize partially occluded objects accurately and estimate their poses. The method extracts multiple local regions with high detectability, uniqueness, and reliability from an object region in each learning image. Eigen windows are constructed from the extracted local regions by PCA. Object recognition and pose estimation are achieved by matching between the eigen windows and local regions extracted from an input image in a similar way.

In addition to the approaches described above, there have been a number of techniques related to the eigenspace method. These include techniques of density estimation of samples in a high-dimensional space [5], nonlinear expansion of PCA [6, 7], image recognition using two-dimensional PCA [8], and image recognition using high-order tensors [9].

Application

The eigenspace method is used as a fundamental technique in arbitrary computer vision applications, such as object recognition, detection, and tracking, because the method works effectively despite its algorithm being quite simple.

References

1. Sirovich L, Kirby M (1987) Low-dimensional procedure for the characterization of human faces. J Opt Soc Am A 4(3):519–524
2. Turk M, Pentland A (1991) Eigenfaces for recognition. J Cognit Neurosci 3(1):71–86
3. Murase H, Nayar S (1995) Visual learning and recognition of 3-d objects from appearance. Int J Comput Vis 14:5–24
4. Ohba K, Ikeuchi K (1997) Detectability, uniqueness, and reliability of eigen window for stable verification of partially occluded objects. IEEE Trans Pattern Anal Mach Intell 19(9):1043–1048
5. Moghaddam B, Pentland A (1997) Probabilistic visual learning for object representation. IEEE Trans Pattern Anal Mach Intell 19(7):696–710
6. Schölkopf B, Smola A, Müller K (2006) Nonlinear component analysis as a kernel eigenvalue problem. Neural Comput 10(5):1299–1319

7. Belkin M, Niyogi P (2006) Laplacian eigenmaps for dimensionality reduction and data representation. Neural Comput 15(6):1373–1396
8. Yang J, Zhang D, Frangi A, Yang J (2004) Two-dimensional pca: a new approach to appearance-based face representation and recognition. IEEE Trans Pattern Anal Mach Intell 26(1):131–137
9. Vasilescu M, Terzopoulos D (2002) Multilinear analysis of image ensembles: tensorfaces. Proc Eur Conf Comput Vis (ECCV) 1:447–460

Eight-Point Algorithm

Zhengyou Zhang
Microsoft Research, Redmond, WA, USA

Related Concepts

▶Eight-Point Algorithm; ▶Epipolar Geometry; ▶Essential Matrix; ▶Fundamental Matrix

Definition

The 8-point algorithm is a linear technique to estimate the essential matrix or the fundamental matrix from eight or more point correspondences.

Background

When dealing with multiple images, it is essential to determine the relative geometry between them, which is known as the epipolar geometry. Between two images of a scene, given a pair of corresponding image points $(\mathbf{m}_i, \mathbf{m}'_i)$, the following epipolar constraint must be satisfied:

$$\widetilde{\mathbf{m}}_i'^T \mathbf{M} \widetilde{\mathbf{m}}_i = 0 , \tag{1}$$

where $\widetilde{\mathbf{m}}_i = \begin{bmatrix} \mathbf{m}_i \\ 1 \end{bmatrix}$ is point $\mathbf{m}_i$ in homogeneous coordinates. Similarly, $\widetilde{\mathbf{m}}'_i$ is point $\mathbf{m}'_i$ in homogeneous coordinates. Matrix $\mathbf{M}$ is a 3×3 matrix. If the images are calibrated with known intrinsic camera parameters and the image points are expressed in the normalized image coordinate system, then matrix $\mathbf{M}$ is known as the *essential matrix*, and is usually denoted by $\mathbf{E}$; otherwise, matrix $\mathbf{M}$ is known as the *fundamental matrix* and is usually denoted by $\mathbf{F}$. Both essential matrix and fundamental matrix must satisfy certain properties, but the common property is that it is a rank-2 matrix, i.e., the determinant of matrix $\mathbf{M}$ is equal to zero.

Theory

The 8-point algorithm ignores the constraints on the elements of matrix $\mathbf{M}$ and treats them independently. Let us define a 9-D vector $\mathbf{x}$ using the elements of matrix $\mathbf{M}$ such that

$$\mathbf{x} = [M_{11}, M_{21}, M_{31}, M_{12}, M_{22}, M_{32}, M_{13}, M_{22}, M_{33}]^T , \tag{2}$$

where M_{ij} is the (i, j) element of matrix $\mathbf{M}$. Let $\mathbf{m} = [u, v]^T$. Then, the epipolar constraint (Eq. 1) can be rewritten as

$$\mathbf{a}_i^T \mathbf{x} = 0 , \tag{3}$$

where

$$\mathbf{a}_i = [u_i \widetilde{\mathbf{m}}_i'^T, v_i \widetilde{\mathbf{m}}_i'^T, \widetilde{\mathbf{m}}_i'^T]^T . \tag{4}$$

It is clear that $\mathbf{x}$ can only be determined up to a scale factor.

Given eight or more point correspondences, $\mathbf{x}$ can be determined as follows. Each point correspondence yields one epipolar equation (Eq. 1). With N ($N \geq 8$) point correspondences, we can stack them together into a vector equation as follows:

$$\mathbf{A}^T \mathbf{x} = \mathbf{0} , \tag{5}$$

with $\mathbf{A} = [\mathbf{a}_1, \ldots, \mathbf{a}_N]$. The solution to $\mathbf{x}$ can be obtained by minimizing the least squares, subject to $\|\mathbf{x}\| = 1$. With Lagrange multiplier, this is equivalent to minimizing

$$\|\mathbf{A}^T \mathbf{x}\|^2 + \lambda(1 - \|\mathbf{x}\|^2) . \tag{6}$$

With simple algebra, it can be found that the solution to $\mathbf{x}$ is the eigenvector of the 9×9 matrix $\mathbf{A}\mathbf{A}^T$ associated with the smallest eigenvalue.

When the image coordinates (u_i, v_i) are in pixels, the elements of matrix $\mathbf{A}$ have orders of difference in value, and matrix $\mathbf{A}\mathbf{A}^T$ may not be well conditioned. One remedy is to pre-normalize the image points, and several solutions are examined in [1]. One simplest approach is to perform a scaling and translation such that all image coordinates are within [−1, 1]. Compared with using directly the pixel coordinates, significant improvement in accuracy has been observed.

The matrix **M** estimated above is obtained by ignoring its property. For example, the estimated **M** is usually not rank-2. To obtain the closest rank-2 matrix, "closest" in terms of Fronobius norm, we perform a singular value decomposition on **M**, i.e.,

$$\mathbf{M} = \mathbf{USV}\,, \tag{7}$$

where $\mathbf{S} = \operatorname{diag}(s_1, s_2, s_3)$ with $s_1 \geq s_2 \geq s_3 \geq 0$. Replacing the smallest singular value by zero, i.e., $\hat{\mathbf{S}} = \operatorname{diag}(s_1, s_2, 0)$, then

$$\mathbf{U}\hat{\mathbf{S}}\mathbf{V} \equiv \hat{\mathbf{M}} \tag{8}$$

is the optimal rank-2 matrix.

For more details about the epipolar geometry, the reader is referred to [2, 3]. The reader is referred to [4] for a review of various methods for determining the epipolar geometry (the essential matrix and the fundamental matrix), to [5] for a study of the relationship between various optimization criteria, and to [6] for how to obtain a more robust Euclidean motion and structure estimation via the estimation of fundamental matrix.

References

1. Hartley R (1997) In defense of the eight-point algorithm. IEEE Trans Pattern Anal Mach Intell 19(6):580–593
2. Faugeras O, Luong QT, Papadopoulo T (2001) The geometry of multiple images. MIT, Cambridge
3. Hartley R, Zisserman A (2000) Multiple view geometry in computer vision. Cambridge University Press, Cambridge/New York
4. Zhang Z (1998) Determining the epipolar geometry and its uncertainty: a review. Int J Comput Vis 27(2):161–195
5. Zhang Z (1998) On the optimization criteria used in two-view motion analysis. IEEE Trans Pattern Anal Mach Intell 20(7):717–729
6. Zhang Z (1997) Motion and structure from two perspective views: from essential parameters to euclidean motion via fundamental matrix. J Opt Soc Am A 14(11): 2938–2950

Ellipse Fitting

Zhi-Yong Liu
Institute of Automation, Chinese Academy of Sciences, Beijing, P. R. China

Synonyms

Ellipse matching

Definition

Fit one or more ellipses to a set of image points.

Background

Fitting geometric primitives to image data is a basic task in pattern recognition and computer vision. The fitting allows reduction and simplification of image data to a higher level with certain physical meanings. One of the most important primitive models is ellipse, which, being a projective projection of a circle, is of great importance for a variety of computer vision-related applications.

Ellipse fitting methods can be roughly divided into two categories: least square fitting and voting/clustering. Least square fitting, though usually fast to implement, requires the image data presegmented and is sensitive to outliers. On the other hand, voting techniques can detect multiple ellipses at once and exhibit some robustness against noise but suffer from a heavier computational and memory load. Furthermore, most of standard ellipse-fitting methods cannot be directly used in real-world applications involving a noisy environment. Thus, the arc finding-based techniques will also be described in the entry, with emphasis on ellipse fitting in complicated images, though, strictly speaking, they cannot be taken as a counterpart of the above two categories.

Due to space limitation, there are some other techniques that cannot be covered by the entry but with some references listed in the recommended reading, such as the moment [16] and genetic algorithm [12]-based methods.

Theory

Least Square Fitting

Least square fitting is realized by minimizing some distance measure between ellipse and image data as follows:

$$\mathbf{\Theta} = \arg\min_{\mathbf{\Theta}} \sum_{i=1}^{N} \mathcal{D}(\mathbf{\Theta}, \mathbf{x}_i)^2, \tag{1}$$

where $\mathcal{D}(\mathbf{\Theta}, \mathbf{x}_i)$ denotes some distance between pixel $\mathbf{x}_i = [x_i, y_i]^T$ and the ellipse specified by $\mathbf{\Theta}$.

There are two main distance measures for ellipse fitting: geometric distance and algebraic distance. In efficiency, the algebraic distance-based techniques outperform its counterpart because a direct solution of Eq. (1) is obtainable by using algebraic distance. However, the algebraic fitting suffers from the drawback of bias estimation and unclear physical interpretations on the estimated errors and fitting parameters. For instance, some algebraic fitting results are not invariant to the coordinates transformation of image data.

Geometric Fitting

An ellipse takes the following equation:

$$\frac{[(x - x_0)\cos\phi + (y - y_0)\sin\phi]^2}{a^2} + \frac{[(x - x_0)\sin\phi + (y - y_0)\cos\phi]^2}{b^2} = 1, \tag{2}$$

where x_0, y_0 are coordinates of center, ϕ is the orientation, and a, b are the semiaxis, as shown in Fig. 1. Geometric distance, also known as shortest or orthogonal distance, is defined by the distance between one image point and its orthogonal projection upon the ellipse, as illustrated by $D_1(\mathbf{x}_i) = \|\mathbf{x}_i - \mathbf{x}_{i1}\|$ in Fig. 1. In such a setting, the ellipse fitted becomes actually a principal curve, which is best in the sense of mean square reconstruction error minimization. However, the distance D_1 is analytically intractable. There exist some techniques to tackle the problem, unavoidably involving some iterative numerical algorithm.

Ahn et al. [1] use a Gaussian-Newton algorithm for geometric fitting, by implicitly describing the orthogonal projection point $\mathbf{x}_1$. A temporary coordinate system

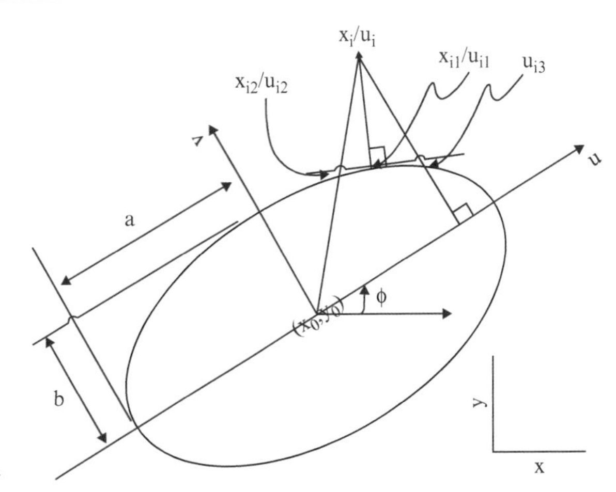

Ellipse Fitting, Fig. 1 An ellipse with center at (x_0, y_0), orientation ϕ, and semiaxis a and b. The projection of point $\mathbf{x}_i$, or $\mathbf{u}_i$ in coordinate system uv, upon the ellipse is the point with the shortest distance on the ellipse, denoted by $\mathbf{x}_{i1}$ ($\mathbf{u}_{i1}$)

uv is introduced as follows:

$$\mathbf{u} = \mathbf{R}(\mathbf{x} - \mathbf{x}_0), \tag{3}$$

where $\mathbf{R} = \begin{pmatrix} \cos\phi, & \sin\phi \\ -\sin\phi, & \cos\phi \end{pmatrix}$ is a rotation matrix. In coordinate system uv, the ellipse is transformed with center at the origin and without rotation, i.e., with the simple formulation as

$$f_1(u, v) = \frac{u^2}{a^2} + \frac{v^2}{b^2} - 1 = 0. \tag{4}$$

For a data point $\mathbf{u}_i$, the tangent line through its projection point $\mathbf{u}_{i1}$ on the ellipse is perpendicular to the line connecting the two points $\mathbf{u}_i$ and $\mathbf{u}_{i1}$:

$$f_2(u, v) = b^2 u(v_i - v) - a^2 v(u_i - u) = 0. \tag{5}$$

Then, a generalized Newton method is employed to find $\mathbf{u}_{i1}$ as follows:

$$\begin{aligned} \mathbf{Q}_n \Delta\mathbf{u} &= -\mathbf{f}(\mathbf{u}_n) \\ \mathbf{u}_{n+1} &= \mathbf{u}_n + \Delta\mathbf{u} \end{aligned} \tag{6}$$

where $\mathbf{f} \triangleq (f_1, f_2)^T$, and the Jacobian matrix

$$\mathbf{Q} = \begin{pmatrix} b^2 u, & a^2 v \\ (a^2 - b^2)v + b^2 v_i, & (a^2 - b^2)u - a^2 u_i \end{pmatrix}. \quad (7)$$

It is noted that several solutions (maximum to 4) could be found by the iteration above. In order to get the right one, one can choose a proper initial value as

$$\mathbf{u}_0 = 0.5(\mathbf{u}_{i2} + \mathbf{u}_{i3}),$$

where $\mathbf{u}_{i2} = \mathbf{u}_i \frac{ab}{\sqrt{b^2 u_i^2 + a^2 v_i^2}}$ and $\mathbf{u}_{i3} = \begin{cases} (u_i, \mathrm{sign}(v_i)\frac{b}{a}\sqrt{a^2 - u_i^2})^T & \text{if } |u_i| < a, \\ (\mathrm{sign}(u_i)a, 0)^T & \text{else.} \end{cases}$, as illustrated in Fig. 1.

Once $\mathbf{u}_1$ has been found, through a backward transformation of Eq. (3), the error distance vector becomes

$$\mathbf{D}_1(\mathbf{x}_i) = \mathbf{x}_i - \mathbf{x}_{i1} = \mathbf{R}^{-1}(\mathbf{u}_i - \mathbf{u}_{i1}). \quad (8)$$

By implicitly describing the orthogonal projection $\mathbf{u}_1$ through f_1 and f_2 defined by Eqs. 4 and 5, the Jacobian matrix with respect to the five parameters $\theta = (x_0, y_0, a, b, \phi)^T$ is as follows:

$$\mathbf{J}_{\mathbf{x}_{i1},\theta} = (\mathbf{R}^{-1}\mathbf{Q}^{-1}\mathbf{B})|_{\mathbf{u}=\mathbf{u}_1}, \quad (9)$$

where $\mathbf{Q}$ is given by Eq. 7 and $\mathbf{B} = (\mathbf{B}_1, \mathbf{B}_2, \mathbf{B}_3, \mathbf{B}_4, \mathbf{B}_5)$ with

$$\begin{aligned}
\mathbf{B}_1 &= (b^2 u \cos\phi - a^2 v \sin\phi, b^2(v_i - v)\cos\phi \\
&\quad + a^2(u_i - u)\sin\phi)^T, \\
\mathbf{B}_2 &= (b^2 u \sin\phi + a^2 v \cos\phi, b^2(v_i - v)\sin\phi \\
&\quad - a^2(u_i - u)\cos\phi)^T, \\
\mathbf{B}_3 &= (a(b^2 - v^2), 2av(u_i - u))^T, \\
\mathbf{B}_4 &= (b(a^2 - u^2), -2bu(v_i - b))^T, \\
\mathbf{B}_5 &= ((a^2 - b^2)uv, (a^2 - b^2)(u^2 - v^2 - uu_i + vv_i))^T.
\end{aligned}$$

Finally, based on Eqs. (8) and (9), the fitting can be accomplished by a Gaussian-Newton iteration as

$$\theta_{n+1} = \theta_n + \lambda \mathbf{J}^{-1}_{\mathbf{x}_{i1},\theta}\mathbf{D}_1(\mathbf{x}_i), \quad (10)$$

where λ denotes a step-size.

Instead of the shortest distance, an alternative approach was proposed to use radial distance, as illustrated by $D_2(\mathbf{x}_i) = \|\mathbf{x}_i - \mathbf{x}_{i2}\|$ in Fig. 1, where $\mathbf{x}_{i2}$ is the intersection of the ellipse and the radial line passing through $\mathbf{x}_i$. Distance D_2 is analytically obtained as [9]

$$\mathcal{D}_2(\mathbf{\Theta}, \mathbf{x}) = |\frac{ab}{\sqrt{(\kappa)}} - 1|\sqrt{(x - x_0)^2 + (y - y_0)^2}, \quad (11)$$

with $\kappa \triangleq a^2(\cos\phi(y - y_0) - \sin\phi(x - x_0))^2 + b^2(\cos\phi(x - x_0) + \sin\phi(y - y_0))^2$. However, solution of Eq. 1 can still not be directly reached. Usually, some iterative technique is still required to get the fitting results, such as stochastic gradient algorithm.

Some other developments on geometric fitting are referred to [2, 13].

Algebraic Fitting

In fact, a more commonly used distance measure for ellipse fitting is algebraic distance, which defines the distance by the second-order polynomial

$$\mathcal{D}(\mathbf{\Theta}, \mathbf{u}) = ax^2 + bxy + cy^2 + dx + ey + f, \quad (12)$$

with the constraint

$$b^2 - 4ac < 0 \quad (13)$$

and with $\mathbf{\Theta} = [a, b, c, d, e, f]^T$. The problem of Eq. (1) with the distance measure given by Eq. (12) has been extensively studied in the name of conic fitting. In order to fit an ellipse, though the general problem of conic fitting could be solved directly, the constraint of Eq. (13) generally makes the methods iterative. By constraining $b^2 - 4ac = -1$ and by transferring the problem to solving a generalized eigensystem, Fitzgibbon et al. [3] propose a fast and direct ellipse fitting method. Specifically, the problem is reformulated as

$$\min_{\mathbf{\Theta}} \|\mathbf{D}\mathbf{\Theta}\|^2 \text{ subject to } \mathbf{\Theta}^T \mathbf{C}\mathbf{\Theta} = 1, \quad (14)$$

where $\mathbf{D} = [\mathbf{u}_1, \mathbf{u}_2, \ldots \mathbf{u}_N]^T$ with $\mathbf{u}_i = [x^2, xy, y^2, x, y, 1]$, and the constraint matrix $\mathbf{C}$ is defined as

$$\mathbf{C} = \begin{pmatrix} 0 & 0 & 2 & 0 & 0 & 0 \\ 0 & -1 & 0 & 0 & 0 & 0 \\ 2 & 0 & 0 & 0 & 0 & 0 \\ & & \mathbf{0} & & & \end{pmatrix}, \qquad (15)$$

where $\mathbf{0} \in \mathbb{R}^{3\times 6}$ is a null matrix. By introducing Lagrange multipliers λ and differentiating, the conditions for the optimal $\boldsymbol{\Theta}$ become

$$\begin{aligned} \mathbf{S}\Theta &= \lambda\mathbf{C}\Theta, \\ \Theta^T\mathbf{C}\boldsymbol{\Theta} &= 1, \end{aligned} \qquad (16)$$

where $\mathbf{S} = \mathbf{D}^T\mathbf{D}$. Thus, problem (14) can be solved by finding the eigenvector Θ_i of the generalized eigensystem defined by Eq. (16). It is also noted that there are six solution pairs (λ_i, Θ_i). By definition, the one corresponding to the minimal positive eigenvalue λ_i is chosen as the solution because

$$\|\mathbf{D}\boldsymbol{\Theta}\|^2 = \boldsymbol{\Theta}^T\mathbf{D}^T\mathbf{D}\boldsymbol{\Theta} = \boldsymbol{\Theta}^T\mathbf{S}\boldsymbol{\Theta} = \lambda\boldsymbol{\Theta}^T\mathbf{C}\boldsymbol{\Theta} = \lambda. \qquad (17)$$

Some other developments on algebraic fitting are referred to [5, 14].

Voting-Based Techniques

Different from the least square fitting techniques which can fit only one primitive, the voting-based methods, consisting of mainly the Hough transform, can detect multiple primitives at once. Even for single ellipse fitting, the Hough transform-based techniques are more robust against outliers. On the other hand, however, Hough transform usually suffers from a larger computational and memory load, due to the fact that an ellipse involves five parameters. To alleviate the problem, there exist some modifications of the original method.

One popular way to tackle the problem is to decompose the five-dimensional space into several subspace with lower dimensions, thanks to some special geometric features of ellipse. A common decomposition method is first to locate the center of the ellipse, and then to specify the remaining three parameters. The center of an ellipse can be located in several ways, based on different geometric features of ellipse.

One geometric feature of ellipse frequently used is described as follows [18]. Let $\mathbf{x}_i$ and $\mathbf{x}_j$ be two points on an ellipse, and find their midpoint $\mathbf{m}_{ij} = \frac{1}{2}(\mathbf{x}_i + \mathbf{x}_j)$

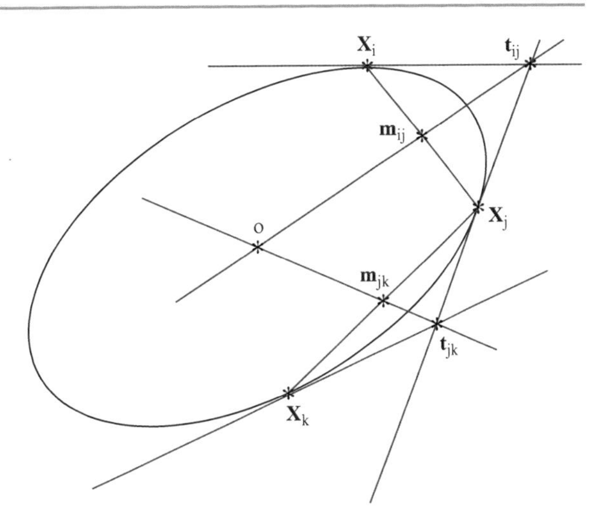

Ellipse Fitting, Fig. 2 Intersection of lines $\overline{t_{ij}m_{ij}}$ and $\overline{t_{jk}m_{jk}}$ is center of the ellipse

and the intersection point $\mathbf{t}_{ij}$ of their tangent lines, as illustrated in Fig. 2. Then, the line connecting the two points $\mathbf{m}_{ij}$ and $\mathbf{t}_{ij}$ is a radial line that passes through center of the ellipse. That is, any point pairs on an ellipse with unparallel tangents can produce a radial line. Consequently, the center candidates can be located in the following way:

1. Generate the radial lines based on every point pairs in the image.
2. Define a 2-dimensional histogram on $x - y$ plane of the image, and record each radial line by incrementing the histogram bin through which the line passes.
3. Find all local maxima in the 2-dimensional histogram as the center candidates.

Another interesting feature of ellipse arises from its symmetry [6]. For two horizontal lines h_i and h_j, their intersections with an ellipse are denoted by $\mathbf{x}_{li}$, $\mathbf{x}_{ri}$ and $\mathbf{x}_{lj}$, $\mathbf{x}_{rj}$, respectively, as illustrated in Fig. 3. The line $\overline{\mathbf{x}_{mi}\mathbf{x}_{mj}}$ will pass through center of the ellipse, where the midpoint $\mathbf{x}_{mi} \triangleq \frac{1}{2}(\mathbf{x}_{li} + \mathbf{x}_{ri})$ and $\mathbf{x}_{mj} \triangleq \frac{1}{2}(\mathbf{x}_{lj} + \mathbf{x}_{rj})$. Similarly, for two vertical lines v_i and v_j, the line $\overline{\mathbf{y}_{mi}\mathbf{y}_{mj}}$ also passes through center of the ellipse, where line $\overline{\mathbf{y}_{mi}\mathbf{y}_{mj}}$ is gotten in a similar way as $\overline{\mathbf{x}_{mi}\mathbf{x}_{mj}}$. Thus, intersection of the two lines $\overline{\mathbf{x}_{mi}\mathbf{x}_{mj}}$ and $\overline{\mathbf{y}_{mi}\mathbf{y}_{mj}}$ gives center of the ellipse. Based on the geometric feature, a center detection method is given as follows:

1. Fully horizontally scan the image, and find the midpoints $\mathbf{m}_{hi}$ of every possible point pairs on each scanning line.

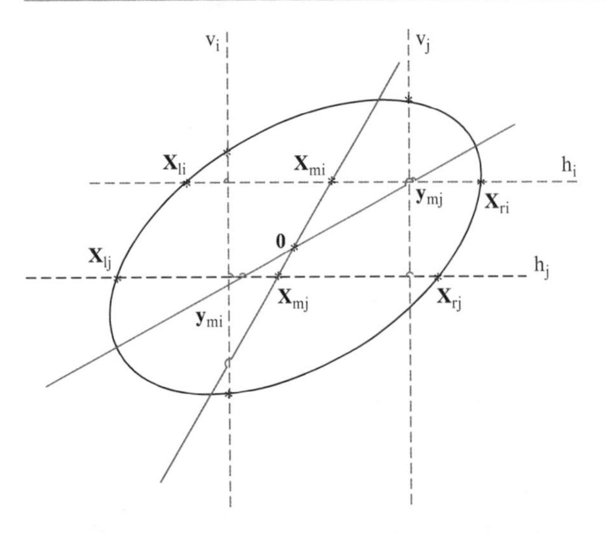

Ellipse Fitting, Fig. 3 Intersection of lines $\overline{x_{mi}x_{mj}}$ and $\overline{y_{mi}y_{mj}}$ is center of the ellipse

2. A 2-dimensional Hough transform is applied to find all of the lines l_h formed by the midpoints $\mathbf{m}_{hi}$ $(i = 1, 2, \ldots)$.
3. All lines l_v are found in a similar way as above on vertical scanning.
4. Every intersection of l_h and l_v gives a center candidate.

Once the center candidates have been extracted, detection of the three remaining parameters can be reached through a standard voting process in an accumulator space or can be further decomposed into detection of orientation and semiaxis, respectively. A typical decomposition strategy for the three parameters is to find first axis ratio $\frac{a}{b}$ and orientation ϕ and then axis length [4]. Axis ratio and orientation satisfy the following relationship:

$$\frac{a^2}{b^2} = \frac{(x_i - x_0)\cos\phi + (y_i - y_0)\sin\phi}{(x_i - x_0)\sin\phi - (y_i - y_0)\cos\phi} \tan(\theta_i - \phi + \frac{\pi}{2}), \tag{18}$$

where θ_i denotes the angle of the tangent line through $\mathbf{x}_i$ and x_0, y_0 have been extracted in the previous stage. Thus, a two-dimensional plane can be built based on Eq. (18) for each point on the ellipse, and consequently the axis ratio and orientation can be estimated by finding the maxima in the plane. It is noted that, for each one center candidate, only the data points that participate in the extraction of center are used in the above voting process.

Finally, ellipse fitting can be accomplished by specifying a or b. Taking a as an example as follows, which satisfies

$$a = \sqrt{(x_i - x_0)^2 + \gamma(y_i - y_0)^2}, \tag{19}$$

where $\gamma \triangleq \frac{a}{b}$ is gotten in the previous step. Hence, voting on a one-dimensional accumulator by the data points pertaining each candidate will retrieve the axis length of the target.

There are also some improvements of the Hough transform that are used to alleviate the computational load, such as the randomized Hough transform. Because discussion on the variants of Hough transform is out of the scope of current entry, readers are referred to some other related entries or some recommended readings such as [11].

Some other developments on voting-based fitting are referred to [15, 17].

Arc Finding-Based Techniques

Arc finding-based techniques accomplish ellipse fitting by finding elliptic arcs in the edge map. Thus, they in general require edge detection as a preprocessing. Arc finding is of particular interest for ellipse fitting, thanks to the fact that any fragment/elliptic arc (longer than five pixels) of an ellipse can retrieve the whole one. This makes such techniques suitable for ellipse fitting in complicated image because it is usually easier to extract arcs than a whole ellipse from an edge map. Arc finding-based ellipse fitting generally takes the following three steps: arc finding, arc fitting and grouping, and ellipse fitting.

In order to extract ellipse, the arc found should be a *neat* curve without any branches. A simple approach to get neat curve is described as follows [9]: first, a thinning operator is applied to make the edge with one pixel width; second, delete those pixels that have more than two pixels in its 8-neighbor field; finally, a standard chain finding algorithm is used to find neat curves.

Each curve is further fitted by an ellipse, by using a geometric fitting technique mentioned above. The reason for choosing geometric fitting lies in that, first, geometric fitting provides a proper criterion

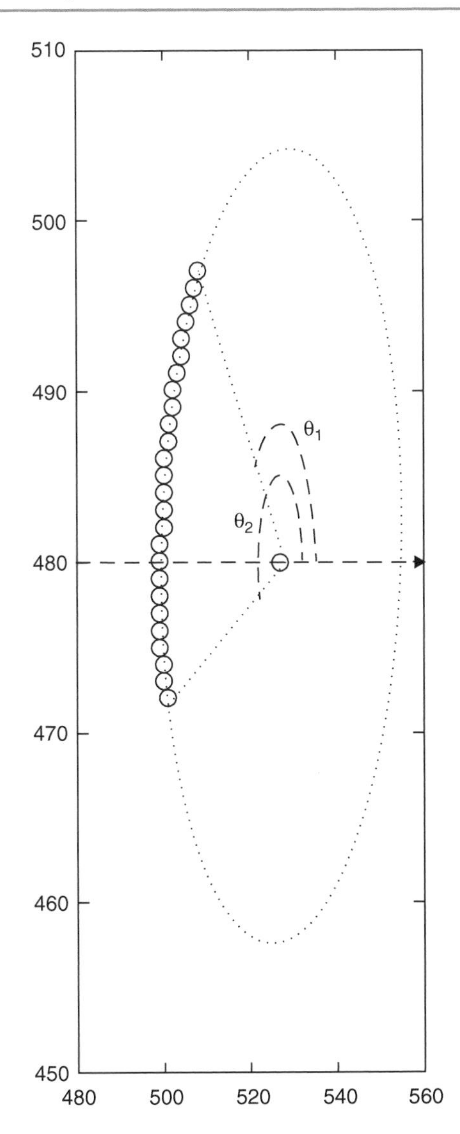

Ellipse Fitting, Fig. 4 The beginning and ending angles of an elliptic arc

(error measure) to evaluate the curve to be a qualified elliptic arc or not and, second, geometric fitting gives a more accurate result than algebraic fitting and Hough transform on a small elliptic fragment, which is frequently encountered since the neat curve finding tends to break the curves with branches into several shorter ones. Finally, each curve is characterized by a 8-dimensional vector

$$[x_0, y_0, a, b, \phi, e, \theta_1, \theta_2]^T, \tag{20}$$

where $e = \frac{1}{N}\sum_{i=1}^{N} \mathcal{D}(\boldsymbol{\Theta}, \mathbf{x}_i)^2$ denotes the mean square fitting error and θ_1 and θ_2 denote its beginning

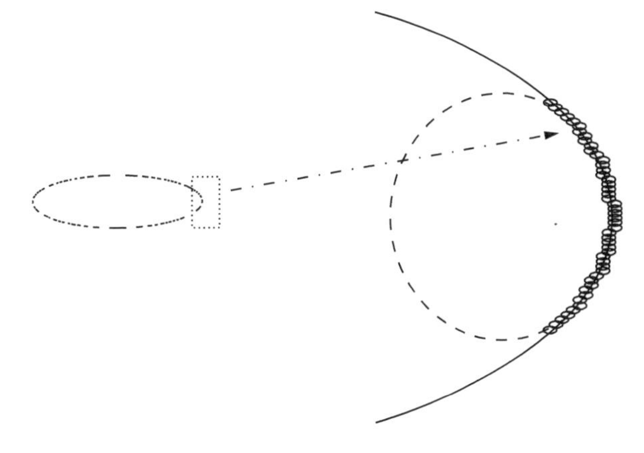

Ellipse Fitting, Fig. 5 Fitting results of a small elliptic fragment: in the image on *right* side, dash ellipse is gotten by algebraic fitting and solid ellipse by geometric fitting

and ending angles, i.e., its direction, as illustrated in Fig. 4. The curves whose fitting error is smaller than a predefined threshold are chosen as qualified elliptic arcs. The elliptic arcs belonging to one same ellipse are then grouped together according to the following two rules:

- Their positions and shapes are close to each other, by the first five parameters.
- They are mutually complementary in direction to form an entire ellipse, by the last two parameters.

During the arc grouping, the data points belonging to each single ellipse are also segmented. Consequently, each ellipse can be finally fitted by least square fitting on these segmented point sets individually.

It is noticed that the arc finding methods employ an integrated framework since its fitting technique comes from least square fitting. However, such integration seems to be unavoidable in real applications, especially in unstructured environments that usually involve heavy noises.

Some other developments on arc finding-based techniques are referred to [8, 10].

Open Problems

Statistical bias of algebraic fitting make the fitting results tend to shrink, especially on small fragments or on data points with heavier noises. Although there are some works on general conic fitting to remove

Ellipse Fitting, Fig. 6 Real-world application: ellipse fitting results on gear image

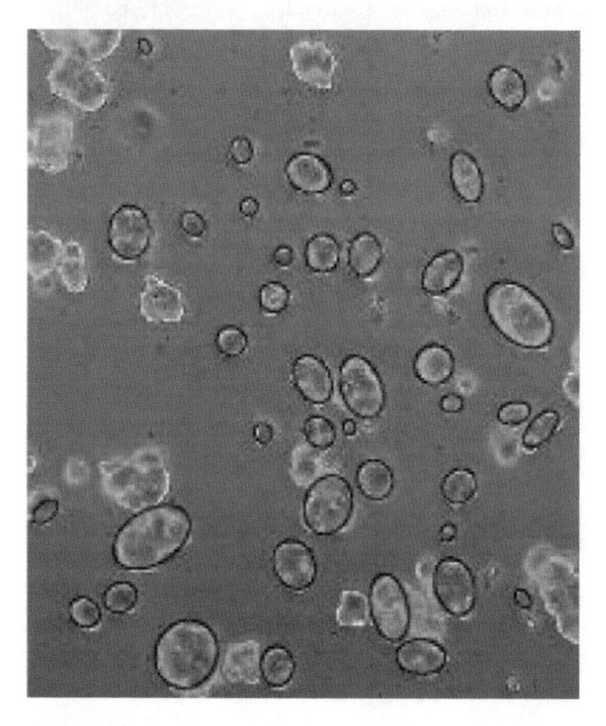

Ellipse Fitting, Fig. 7 Real-world application: ellipse fitting results on blood cell image

the bias [7], discussion on direct ellipse fitting is lacked. Since direct algebraic fitting seems to be the fastest algorithm for ellipse fitting, it is of theoretical and practical significance to make some progress on removing bias in the fitting process.

On the other hand, although there are some methods proposed for ellipse fitting in complicated images, it still remains a big challenge to fit ellipse in uncontrolled real-world environments.

Experimental Results

It is not intended to demonstrate all of the algorithms described above. Only two experimental results are given, with the first one to demonstrate the bias of algebraic fitting on a small elliptic arc and the second one to give one glimpse of state-of-art real-world application in noisy environments of ellipse fitting.

Image points used in the first experiments are fetched from a small elliptic fragment, as shown in Fig. 5. The dash ellipse is the fitting result of algebraic fitting, which shrinks to be smaller than the right one as denoted by the bigger solid ellipse that is resulted from geometric fitting.

Two real-world applications of ellipse fitting on holes of gear and blood cells are shown in Figs. 6 and 7, respectively.

References

1. Ahn SJ, Rauh W, Warnecke HJ (2001) Least-squares orthogonal distances fitting of circle, sphere, ellipse, hyperbola, and parabola. Pattern Recognit 34:2283–2302
2. Cui Y, Weng J, Reynolds H (1996) Estimation of ellipse parameters using optimal minimum variance estimator. Pattern Recognit Lett 17:309–316
3. Fitzgibbon AW, Pilu M, Fisher RB (1999) Direct least-squares fitting of ellipses. IEEE Trans Pattern Anal Mach Intell 21(5):476–480
4. Guil N, Zapata EL (1997) Lower order circle and ellipse hough transform. Pattern Recognit 30:1729–1744
5. Halir R, Flusser V (1998) Numerically stable direct least squares fitting of ellipses. In: WSCG'98 conference proceedings, Plzen-Bory
6. Ho CT, Chen LH (1995) A fast ellipse/circle detector using geometric symmetry. Pattern Recognit 28:117–124
7. Kanatani K (1994) Statistical bias of conic fitting and renormalization. IEEE Trans Pattern Anal Mach Intell 16(3): 320–326
8. Kim E, Haseyama V, Kitajima H (2002) Fast and robust ellipse extraction from complicated images. In: Proceedings of IEEE international conference on information technology and applications, Bathurst, NSW, Australia
9. Liu ZY, Qiao H (2009) Multiple ellipses detection in noisy environments: a hierarchical approach. Pattern Recognit 42:2421–2433

10. Mai F, Hung YS, Zhong H, Sze WF (2008) A hierarchical approach for fast and robust ellipse extraction. Pattern Recognit 8(41):2512–2524
11. McLaughlin RA (1998) Randomized hough transform: improved ellipse detection with comparison. Pattern Recognit Lett 19:299–305
12. Roth G, Levine MD (1994) Geometric primitive extraction using a genetic algorithm. IEEE Trans Pattern Anal Mach Intell 16(9):901–905
13. Spath H (1997) Orthogonal distance fitting by circles and ellipses with given data. Comput Stat 12:343–354
14. Taubin G (1991) Estimation of planar curves, surfaces and non-planar space curves defined by implicit equations with applications to edge and range image segmentation. IEEE Trans Pattern Anal Mach Intell 13(11):1115–1138
15. Tsuji S, Matsumoto F (1978) Detection of ellipses by a modified hough transformation. IEEE Trans Comput 25: 777–781
16. Voss K, Suesse H (1997) Invariant fitting of planar objects by primitives. IEEE Trans Pattern Anal Mach Intell 19(1):80–84
17. Yipa KK, Tama KS, Leung NK (1992) Modification of hough transform for circles and ellipses detection using a 2-dimensional array. Pattern Recognit 25:1007–1022
18. Yuen HK, Illingworth J, Kittler J (1989) Detecting partially occluded ellipses using the hough transform. Image Vis Comput 7:31–37

Ellipse Matching

▸Ellipse Fitting

EM-Algorithm

▸Expectation Maximization Algorithm

Environment Mapping

▸Image-Based Lighting

Epipolar Constraint

Zhengyou Zhang
Microsoft Research, Redmond, WA, USA

Synonyms

Coplanarity constraint

Related Concepts

▸Epipolar Geometry

Definition

Epipolar constraint states that in stereovision with two cameras, given a point in one image, its corresponding point in the other image must lie on a line, known as the *epipolar line*. This constraint arises from the fact that the pair of corresponding image points and the optical centers of the two cameras must lie on a plane (known as *coplanarity constraint*), and the intersection of this plane with the image plane is the epipolar line.

Background

See entry ▸Epipolar Geometry for details.

Epipolar Geometry

Zhengyou Zhang
Microsoft Research, Redmond, WA, USA

Synonyms

Multiple view geometry; Multiview geometry

Related Concepts

▸Epipolar Constraint; ▸Essential Matrix; ▸Fundamental Matrix

Definition

Epipolar geometry describes the geometric relationship between two camera systems. It is captured by a 3×3 matrix known as *essential matrix* for calibrated cameras and as *fundamental matrix* for uncalibrated cameras. It states that for a point observed in one camera, its corresponding point in the other camera must lie on a line. This is known as the *epipolar constraint*.

It reduces the search space of correspondences from two dimensions to one dimension. In motion and structure from motion, this constraint is also known as *coplanarity constraint* because the optical centers of the cameras and a pair of corresponding image points must lie in a single plane.

Background

The epipolar geometry exists between any two camera systems. Consider the case of two cameras as shown in Fig. 1. Let C and C' be the optical centers of the first and second cameras, respectively. Given a point $\mathbf{m}$ in the first image, its corresponding point in the second image is constrained to lie on a line called the *epipolar line* of $\mathbf{m}$, denoted by $\mathbf{l}'_{\mathbf{m}}$. The line $\mathbf{l}'_{\mathbf{m}}$ is the intersection of the plane $\boldsymbol{\Pi}$, defined by $\mathbf{m}$, C, and C' (known as the *epipolar plane*), with the second image plane $\mathcal{I}'$. This is because image point $\mathbf{m}$ may correspond to an arbitrary point on the semi-line $C\mathtt{M}$ ($\mathtt{M}$ may be at infinity) and that the projection of $C\mathtt{M}$ on $\mathcal{I}'$ is the line $\mathbf{l}'_{\mathbf{m}}$. Furthermore, one observes that all epipolar lines of the points in the first image pass through a common point $\mathbf{e}'$, which is called the *epipole*. Epipole $\mathbf{e}'$ is the intersection of the line CC' with the image plane $\mathcal{I}'$. This can be easily understood as follows. For each point $\mathbf{m}_k$ in the first image $\mathcal{I}$, its epipolar line $\mathbf{l}'_{\mathbf{m}_k}$ in $\mathcal{I}'$ is the intersection of the plane $\boldsymbol{\Pi}^k$, defined by $\mathbf{m}_k$, C, and C', with image plane $\mathcal{I}'$. All epipolar planes $\boldsymbol{\Pi}^k$ thus form a pencil of planes containing the line CC'. They must intersect $\mathcal{I}'$ at a common point, which is $\mathbf{e}'$. Finally, one can easily see the symmetry of the epipolar geometry. The corresponding point in the first image of each point $\mathbf{m}'_k$ lying on $\mathbf{l}'_{\mathbf{m}_k}$ must lie on the epipolar line $\mathbf{l}_{\mathbf{m}'_k}$, which is the intersection of the same plane $\boldsymbol{\Pi}^k$ with the first image plane $\mathcal{I}$. All epipolar lines form a pencil containing the epipole $\mathbf{e}$, which is the intersection of the line CC' with the image plane $\mathcal{I}$. The symmetry leads to the following observation. If $\mathbf{m}$ (a point in $\mathcal{I}$) and $\mathbf{m}'$ (a point in $\mathcal{I}'$) correspond to a single physical point $\mathtt{M}$ in space, then $\mathbf{m}$, $\mathbf{m}'$, C, and C' must lie in a single plane. This is the well-known *coplanarity constraint* in solving motion and structure from motion problems when the intrinsic parameters of the cameras are known [1].

The computational significance in matching different views is that for a point in the first image, its correspondence in the second image must lie on the epipolar line in the second image, and then the search space for a correspondence is reduced from 2 dimensions to 1 dimension. This is called the *epipolar constraint*.

If the line linking the two optical centers is parallel to one or both of the image planes, then the epipole in one or both of the images goes to infinity, and the epipolar lines are parallel to each other. Additionally, if the line linking the two optical centers is parallel with the horizontal scanlines of the cameras, then the epipolar lines become horizontal, too. This is the assumption of many stereo algorithms which have horizontal epipolar lines.

Theory

Before proceeding further, the reader is referred to the entry ▸Camera Parameters (Intrinsic, Extrinsic) for the description of camera perspective projection matrix and intrinsic and extrinsic parameters. It is assumed that the reader is familiar with the notation used in that entry.

Assume that the second camera is brought from the position of the first camera through a rotation $\mathbf{R}$ followed by a translation $\mathbf{t}$. Thus, any point (X, Y, Z) in the first camera coordinate system has coordinates (X', Y', Z') in the second camera coordinate system such that

$$\begin{bmatrix} X \\ Y \\ Z \end{bmatrix} = \mathbf{R} \begin{bmatrix} X' \\ Y' \\ Z' \end{bmatrix} + \mathbf{t} \tag{1}$$

where

$$\mathbf{R} = \begin{bmatrix} r_{11} & r_{12} & r_{13} \\ r_{21} & r_{22} & r_{23} \\ r_{31} & r_{32} & r_{33} \end{bmatrix} \quad \text{and} \quad \mathbf{t} = \begin{bmatrix} t_X \\ t_Y \\ t_Z \end{bmatrix}.$$

$\mathbf{R}$ has nine components but there are only three degrees of freedom. There are six constraints on $\mathbf{R}$. Indeed, a rotation matrix $\mathbf{R}$ must satisfy

$$\mathbf{R}\mathbf{R}^T = \mathbf{I}, \tag{2}$$

and

$$\det(\mathbf{R}) = 1. \tag{3}$$

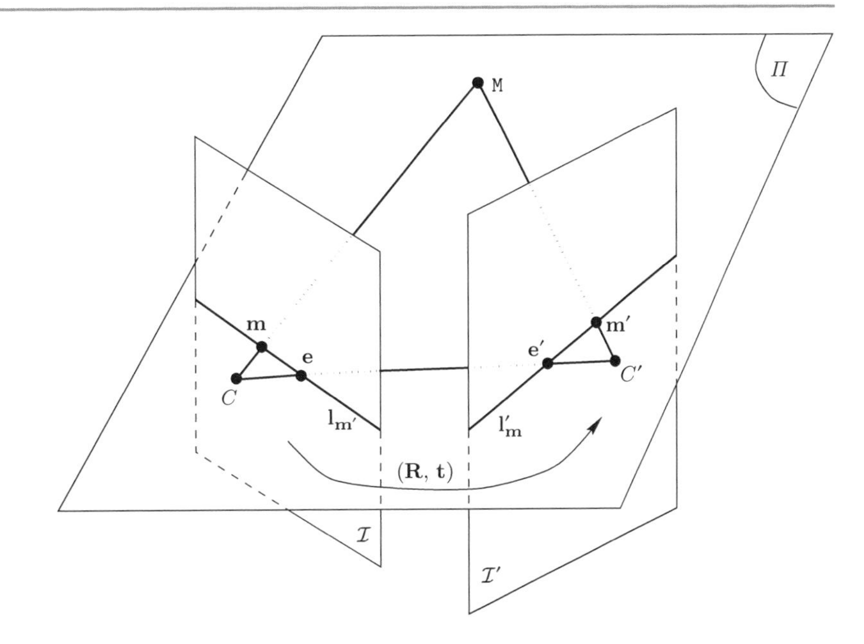

Epipolar Geometry, Fig. 1
The epipolar geometry

See, for example [2], for more details on the different representations of the rotation and its properties.

In the following, we first derive the epipolar equation with the normalized image coordinates, then extend it to include the pixel image coordinates, and finally formulate in terms of camera perspective projection matrices.

Working with Normalized Image Coordinates

The two images of a space point $\mathrm{X} = [X, Y, Z]^T$ are $[x, y, 1]^T$ and $[x', y', 1]^T$ in the first and second normalized images, respectively. They are denoted by $\tilde{\mathbf{x}}$ and $\tilde{\mathbf{x}}'$. Let $\mathrm{X}' = [X', Y', Z']^T$ be the coordinates of the same space point in the second camera coordinate system. From the pinhole model, we have

$$\tilde{\mathbf{x}} = \mathrm{X}/Z,$$
$$\tilde{\mathbf{x}}' = \mathrm{X}'/Z'.$$

Eliminating the structure parameters X and X′ using Eq. (1), we obtain

$$\tilde{\mathbf{x}} = \frac{1}{Z}(Z'\mathbf{R}\tilde{\mathbf{x}}' + \mathbf{t}),$$

which contains still two unknown structure parameters Z and Z'. The cross product of the above equation with vector $\mathbf{t}$ yields

$$\mathbf{t} \times \tilde{\mathbf{x}} = \frac{Z'}{Z}\mathbf{t} \times \mathbf{R}\tilde{\mathbf{x}}'.$$

Its dot product (or inner product) with $\tilde{\mathbf{x}}$ gives

$$\tilde{\mathbf{x}}^T \mathbf{t} \times (\mathbf{R}\tilde{\mathbf{x}}') = 0. \tag{4}$$

Here, the quantity Z'/Z has been removed.

Equation (4) is very important in solving motion and structure from motion. Geometrically, it is very clear. The three vectors CC', $C\tilde{\mathbf{x}}$, and $C'\tilde{\mathbf{x}}'$ are coplanar. When expressed in the first camera coordinate system, they are equal to $\mathbf{t}$, $\tilde{\mathbf{x}}$, and $\mathbf{R}\tilde{\mathbf{x}}'$, respectively. The coplanarity of the three vectors implies that their mixed product should be equal to 0, which gives Eq. (4).

Let us define a mapping $[\cdot]_\times$ from a 3D vector to a 3×3 *antisymmetric matrix* (also called *skew symmetric matrix*):

$$\begin{bmatrix} x_1 \\ x_2 \\ x_3 \end{bmatrix}_\times = \begin{bmatrix} 0 & -x_3 & x_2 \\ x_3 & 0 & -x_1 \\ -x_2 & x_1 & 0 \end{bmatrix}. \tag{5}$$

It is clear that

$$[\mathbf{t}]_\times = -[\mathbf{t}]_\times^T. \tag{6}$$

Using this mapping, we can express the cross product of two vectors by the matrix multiplication of a 3×3 matrix and a three-vector: $\mathbf{t} \times \tilde{\mathbf{x}} = [\mathbf{t}]_\times \tilde{\mathbf{x}}, \ \forall \tilde{\mathbf{x}}$. Equation (4) can then be rewritten as

$$\tilde{\mathbf{x}}^T \mathbf{E}\tilde{\mathbf{x}}' = 0, \tag{7}$$

where

$$\mathbf{E} = [\mathbf{t}]_\times \mathbf{R}. \tag{8}$$

We call this equation the *epipolar equation*.

Matrix $\mathbf{E}$ is known under the name of the *essential matrix*. It was first proposed by Longuet-Higgins [1] for structure from motion. The essential matrix is determined completely by the rotation and translation between the two cameras. Because $[\mathbf{t}]_\times$ is antisymmetric, we have $\det([\mathbf{t}]_\times) = 0$. Thus, we have

$$\det(\mathbf{E}) = \det([\mathbf{t}]_\times)\det(\mathbf{R}) = 0. \tag{9}$$

For more properties of the essential matrix, see [3, 4].

Before gaining an insight of Eq. (7), we recall how to represent a line in a plane. Any line can be described by an equation of the form

$$ax + by + c = 0. \tag{10}$$

Thus, the line can be represented by a three-vector $\mathbf{l} = [a, b, c]^T$ such that a point $\tilde{\mathbf{x}} = [x, y, 1]^T$ on it must satisfy

$$\tilde{\mathbf{x}}^T \mathbf{l} = 0. \tag{11}$$

Of course, the three-vector $\mathbf{l}$ is only defined up to a scale factor. Multiplying $\mathbf{l}$ by any nonzero scalar λ gives $\lambda\mathbf{l}$, which describes exactly the same line. If a line goes through two given points $\tilde{\mathbf{x}}_1$ and $\tilde{\mathbf{x}}_2$, we have

$$\tilde{\mathbf{x}}_1^T \mathbf{l} = 0 \quad \text{and} \quad \tilde{\mathbf{x}}_2^T \mathbf{l} = 0,$$

and it is easy to see that the line is represented by

$$\mathbf{l} = \tilde{\mathbf{x}}_1 \times \tilde{\mathbf{x}}_2, \tag{12}$$

that is, the cross product of the two point vectors.

For point $\tilde{\mathbf{x}}' = [x', y', 1]^T$ in the second image, its corresponding point X' in space must be on the semi-line $C'\mathrm{X}'_\infty$ passing through $\tilde{\mathbf{x}}'$, where X'_∞ is a point at infinity. From the pinhole model, point X' can be represented as

$$\mathrm{X}' = \lambda\tilde{\mathbf{x}}' = \lambda \begin{bmatrix} x' \\ y' \\ 1 \end{bmatrix}, \quad \lambda \in (0, \infty).$$

This is in fact the parametric representation of the semi-line $C'\mathrm{X}'_\infty$. If we express this point in the coordinate system of the first camera, we have

$$\mathrm{X} = \mathbf{R}\mathrm{X}' + \mathbf{t} = \lambda\mathbf{R}\tilde{\mathbf{x}}' + \mathbf{t}, \quad \lambda \in (0, \infty).$$

The projection of the semi-line $C'\mathrm{X}'_\infty$ on the first camera is still a line, denoted by $\mathbf{l}_{\mathbf{x}'}$, on which the corresponding point in the first image of point $\mathbf{x}$ must lie. The line $\mathbf{l}_{\mathbf{x}'}$ is known as the *epipolar line* of $\mathbf{x}'$. The epipolar line can be defined by two points. The first point can be obtained by projecting X with $\lambda = 0$, which gives $\tilde{\mathbf{e}} = \frac{1}{t_Z}\mathbf{t}$, where t_Z is the Z-component of the translation vector $\mathbf{t}$. This is in fact the projection of the optical center C of the second camera on the first camera and is called the *epipole* in the first image. The second point can be obtained by projecting X with $\lambda = \infty$, which gives $\tilde{\mathbf{x}}_\infty = \frac{1}{\mathbf{r}_3^T\tilde{\mathbf{x}}'}\mathbf{R}\tilde{\mathbf{x}}'$, where $\mathbf{r}_3$ is the third row of the rotation matrix $\mathbf{R}$. As described in the last paragraph, the epipolar line $\mathbf{l}_{\mathbf{x}'}$ is represented by

$$\mathbf{l}_{\mathbf{x}'} = \tilde{\mathbf{e}} \times \tilde{\mathbf{x}}_\infty = \mathbf{t} \times \mathbf{R}\tilde{\mathbf{x}}' = \mathbf{E}\tilde{\mathbf{x}}'. \tag{13}$$

Here we have multiplied the original vector by t_Z and $\mathbf{r}_3^T\tilde{\mathbf{x}}'$ because, as we said, a three-vector for a line is only defined up to a scalar factor.

If now we reverse the role of the two camera, we find that the epipolar geometry is symmetric for the two cameras. For a given point $\mathbf{x}$ in the first image, its corresponding epipolar line in the second image is

$$\mathbf{l}'_{\mathbf{x}} = \mathbf{E}^T\tilde{\mathbf{x}}.$$

It is seen that the transpose of matrix $\mathbf{E}$, $\mathbf{E}^T$, defines the epipolar lines in the second image.

From the above discussion, Eq. (7) says nothing more than that point $\mathbf{x}$ is on the epipolar line $\mathbf{l}_{\mathbf{x}'}$, that is,

$$\tilde{\mathbf{x}}^T\mathbf{l}_{\mathbf{x}'} = 0 \quad \text{with } \mathbf{l}_{\mathbf{x}'} = \mathbf{E}\tilde{\mathbf{x}}',$$

or that point $\tilde{\mathbf{x}}'$ is on the epipolar line $\mathbf{l}'_{\mathbf{x}}$, that is,

$$\mathbf{l}'^T_{\mathbf{x}}\tilde{\mathbf{x}}' = 0 \quad \text{with } \mathbf{l}'_{\mathbf{x}} = \mathbf{E}^T\tilde{\mathbf{x}}.$$

The epipoles are intersections of all epipolar lines. That is, epipoles satisfy all the epipolar line equations.

Let the normalized coordinates of the epipole in the first image be **e**. Then, from Eq. (7), **e** satisfies

$$\tilde{\mathbf{e}}^T \mathbf{E} \tilde{\mathbf{x}}' = 0 \tag{14}$$

regardless of $\mathbf{x}'$. This means,

$$\tilde{\mathbf{e}}^T \mathbf{E} = \mathbf{0}^T \tag{15}$$

at anytime. That is,

$$\tilde{\mathbf{e}}^T \mathbf{E} = \tilde{\mathbf{e}}^T [\mathbf{t}]_\times \mathbf{R} = \mathbf{0}^T. \tag{16}$$

Since **R** is an orthonomal matrix, we have

$$\tilde{\mathbf{e}}^T [\mathbf{t}]_\times = \mathbf{0}. \tag{17}$$

The solution is

$$\tilde{\mathbf{e}} = \left[\frac{t_X}{t_Z}, \frac{t_Y}{t_Z}, 1\right]^T. \tag{18}$$

This is exactly the projection of the optical center of the second camera onto the first image plane, as we have already explained geometrically. For the second image, we have

$$\mathbf{E}\tilde{\mathbf{e}}' = [\mathbf{t}]_\times \mathbf{R}\tilde{\mathbf{e}}' = \mathbf{0}. \tag{19}$$

Thus,

$$\mathbf{R}\tilde{\mathbf{e}}' = \mathbf{t}. \tag{20}$$

The position of the epipole can then be determined as

$$\tilde{\mathbf{e}}' = \frac{1}{\mathbf{r}_3' \cdot \mathbf{t}} \mathbf{R}^T \mathbf{t} = \left[\frac{\mathbf{r}_1' \cdot \mathbf{t}}{\mathbf{r}_3' \cdot \mathbf{t}}, \frac{\mathbf{r}_1' \cdot \mathbf{t}}{\mathbf{r}_3' \cdot \mathbf{t}}, 1\right]^T, \tag{21}$$

where $\mathbf{r}_i' = [r_{1i}, r_{2i}, r_{3i}]^T$, and $i = 1,2,3$ are the column vectors of **R**.

For the epipole in the first image to go to infinity, we must have

$$t_Z = 0. \tag{22}$$

This means that the translation of the camera has to be within the focal plane of the first camera. For both epipoles in the two images to go to infinity, then

$$\mathbf{r}_3' \cdot \mathbf{t} = 0. \tag{23}$$

This implies that the optical center of the first camera lies in the focal plane of the second camera. Furthermore, if we require the two focal planes to be a single one, then besides $t_Z = 0$, r_{13} and r_{23} have to be 0. Since $\|\mathbf{r}_i'\| = 1$, we have

$$r_{33} = 1. \tag{24}$$

Thus **R** is in the form of

$$\mathbf{R} = \begin{bmatrix} \cos\theta & \sin\theta & 0 \\ -\sin\theta & \cos\theta & 0 \\ 0 & 0 & 1 \end{bmatrix}. \tag{25}$$

This means that the rotation can be only around the optical axis of the cameras.

Substituting Eq. (22) and (25) for (8), we have

$$\begin{aligned} \mathbf{E} &= [\mathbf{t}]_\times \mathbf{R} \\ &= \begin{bmatrix} 0 & 0 & t_Y \\ 0 & 0 & -t_X \\ -t_Y\cos\theta - t_X\sin\theta & -t_Y\sin\theta + t_X\cos\theta & 0 \end{bmatrix}. \end{aligned} \tag{26}$$

If we expand the above equation, it is clear that there is only linear terms of the image coordinates, rather than quadric terms in the original form. That means the epipolar lines are parallel in both images and the orientations are independent of the image points.

Working with Pixel Image Coordinates

If two points **m** and $\mathbf{m}'$, expressed in pixel image coordinates in the first and second camera, are in correspondence, they must satisfy the following equation

$$\tilde{\mathbf{m}}^T \mathbf{F} \tilde{\mathbf{m}}' = 0, \tag{27}$$

where

$$\mathbf{F} = \mathbf{A}^{-T} \mathbf{E} \mathbf{A}'^{-1}, \tag{28}$$

and **A** and $\mathbf{A}'$ are respectively the of the first and second camera. Equation (27) is easily verified. From the pinhole camera model, the **x** are related to the pixel coordinates **m** by $\tilde{\mathbf{x}} = \mathbf{A}^{-1}\tilde{\mathbf{m}}$. Plunging it into Eq. (7) yields Eq. (27). This is a fundamental constraint for two pixels to be in correspondence between two images.

As with the normalized image coordinates, the above Eq. (27) can also be derived from the pinhole model. Without loss of generality, we assume that the world coordinate system coincides with the second camera coordinate system. From the camera perspective projection model, we have

$$s\tilde{\mathbf{m}} = \mathbf{A}\,[\mathbf{R}\ \mathbf{t}]\begin{bmatrix}\mathtt{M}'\\1\end{bmatrix}$$

$$s'\tilde{\mathbf{m}}' = \mathbf{A}'\,[\mathbf{I}\ \mathbf{0}]\begin{bmatrix}\mathtt{M}'\\1\end{bmatrix}.$$

Eliminating $\mathtt{M}'$, s, and s' in the above two equations, we obtain, not at all surprising, Eq. (27).

The 3×3 matrix $\mathbf{F}$ is called the *fundamental matrix*. With this fundamental matrix, we can express the epipolar equation for two unnormalized images in the same form as for the normalized images. Since $\det(\mathbf{E}) = 0$,

$$\det(\mathbf{F}) = 0. \tag{29}$$

$\mathbf{F}$ is of rank 2. Besides, it is only defined up to a scalar factor. If $\mathbf{F}$ is multiplied by an arbitray scalar, Eq. (27) still holds. Therefore, a fundamental matrix has only seven degrees of freedom. There are only seven independent parameters among the nine elements of the fundamental matrix.

We now derive the expression of the epipoles. The epipole $\mathbf{e}$ in the first image is the projection of the optical center C' of the second camera. Since $C' = \mathbf{0}$, from the pinhole model, we have

$$s_e\tilde{\mathbf{e}} = \mathbf{A}\,[\mathbf{R}\ \mathbf{t}]\begin{bmatrix}0\\1\end{bmatrix} = \mathbf{A}\mathbf{t}, \tag{30}$$

where s_e is a scale factor. Thus, the epipole $\tilde{\mathbf{e}}$ is equal to $\mathbf{At}$ divided by its third element. Similarly, the epipole $\mathbf{e}'$ in the second image is the projection of the optical center C of the first camera. The optical center is determined by

$$\mathbf{A}\,[\mathbf{R}\ \mathbf{t}]\begin{bmatrix}C\\1\end{bmatrix} = \mathbf{0},$$

which gives

$$C = -\mathbf{R}^{-1}\mathbf{t}.$$

Therefore, the epipole $\mathbf{e}'$ is given by

$$s'_e\tilde{\mathbf{e}}' = \mathbf{A}'\,[\mathbf{I}\ \mathbf{0}]\begin{bmatrix}C\\1\end{bmatrix} = -\mathbf{A}'\mathbf{R}^{-1}\mathbf{t}, \tag{31}$$

that is, it is equal to $-\mathbf{A}'\mathbf{R}^{-1}\mathbf{t}$ divided by the third element of the vector.

Now we show, for a given point $\mathbf{m}'$ in the second image, how to compute the corresponding epipolar line $\mathbf{l}_{\mathbf{m}'}$ in the first image. It is determined by two points. Besides the epipole $\mathbf{e}$, we need another point. This point can be the projection of any point $\widehat{\mathtt{M}}'$ on the optical ray $\langle C', \tilde{\mathbf{m}}'\rangle$. In particular, we can choose $\widehat{\mathtt{M}}'$ such that

$$\tilde{\mathbf{m}}' = \mathbf{A}'\ [\mathbf{I}\ \mathbf{0}]\begin{bmatrix}\widehat{\mathtt{M}}'\\1\end{bmatrix} = \mathbf{A}'\widehat{\mathtt{M}}',$$

that is, the scale factor is equal to 1. This gives $\widehat{\mathtt{M}}' = \mathbf{A}'^{-1}\tilde{\mathbf{m}}'$. The projection of this point in the first camera, $\widehat{\mathbf{m}}$, is given by

$$s_m\tilde{\widehat{\mathbf{m}}} = \mathbf{A}\ [\mathbf{R}\ \mathbf{t}]\begin{bmatrix}\widehat{\mathtt{M}}'\\1\end{bmatrix} = \mathbf{A}(\mathbf{R}\mathbf{A}'^{-1}\tilde{\mathbf{m}}' + \mathbf{t}),$$

where s_m is the scale factor. As already described in Eq. (10) on page 250Working with Normalized Image Coordinatesequation.9.10.80, a line can be represented by a three-vector defined *up to a scale factor*. According to Eq. (12), the epipolar line $\mathbf{l}_{\mathbf{m}'}$ is given by

$$\begin{aligned}\mathbf{l}_{\mathbf{m}'} &= s_e s_m \tilde{\mathbf{e}} \times \tilde{\widehat{\mathbf{m}}}\\ &= (\mathbf{At}) \times [\mathbf{A}(\mathbf{R}\mathbf{A}'^{-1}\tilde{\mathbf{m}}' + \mathbf{t})]\\ &= (\mathbf{At}) \times (\mathbf{A}\mathbf{R}\mathbf{A}'^{-1}\tilde{\mathbf{m}}').\end{aligned}$$

It can be shown that $(\mathbf{Ax}) \times (\mathbf{Ay}) = \det(\mathbf{A})\,\mathbf{A}^{-T}(\mathbf{x} \times \mathbf{y})$ for all vectors $\mathbf{x}$ and $\mathbf{y}$ if matrix $\mathbf{A}$ is invertible. Therefore, we have

$$\mathbf{l}_{\mathbf{m}'} = \mathbf{A}^{-T}[\mathbf{t} \times (\mathbf{R}\mathbf{A}'^{-1}\mathbf{m}')] = \mathbf{F}\tilde{\mathbf{m}}'$$

with $\mathbf{F}$ given by Eq. (28). Then any point $\mathbf{m}$ on the epipolar line of $\mathbf{m}'$ satisfies $\tilde{\mathbf{m}}^T\mathbf{F}\tilde{\mathbf{m}}' = 0$, and this is exactly Eq. (34). Therefore, we obtain geometrically the same equation.

Now we reverse the role of the two images and consider the epipolar line $\mathbf{l}'_{\mathbf{m}}$ in the second image for a given point $\mathbf{m}$ in the first image. Line $\mathbf{l}'_{\mathbf{m}}$ goes through the epipole $\mathbf{e}'$. We choose the projection of a point $\widehat{\mathtt{M}}'$ on the optical ray $\langle C, \tilde{\mathbf{m}}\rangle$ such that

$$\tilde{\mathbf{m}} = \mathbf{A}\,[\mathbf{R}\ \mathbf{t}]\begin{bmatrix}\widehat{\mathtt{M}}'\\1\end{bmatrix},$$

that is, the scale factor is chosen to be 1. This gives

$$\widehat{\mathrm{M}}' = (\mathbf{AR})^{-1}(\tilde{\mathbf{m}} - \mathbf{At}).$$

Its projection in the second camera gives

$$\begin{aligned} s'_m \tilde{\hat{\mathbf{m}}}' &= \mathbf{A}'\,[\mathbf{I}\ \mathbf{0}] \begin{bmatrix} \widehat{\mathrm{M}}' \\ 1 \end{bmatrix} \\ &= \mathbf{A}'(\mathbf{AR})^{-1}(\tilde{\mathbf{m}} - \mathbf{At}) \\ &= \mathbf{A}'\mathbf{R}^{-1}\mathbf{A}^{-1}\tilde{\mathbf{m}} - \mathbf{A}'\mathbf{R}^{-1}\mathbf{t}. \end{aligned}$$

The epipolar line $\mathbf{l}'_{\mathbf{m}}$ is thus represented by

$$\begin{aligned} \mathbf{l}'_{\mathbf{m}} &= s'_e s'_m \tilde{\mathbf{e}}' \times \tilde{\hat{\mathbf{m}}}' \\ &= -(\mathbf{A}'\mathbf{R}^{-1}\mathbf{t}) \times (\mathbf{A}'\mathbf{R}^{-1}\mathbf{A}^{-1}\tilde{\mathbf{m}}) \\ &= -(\mathbf{A}'\mathbf{R}^{-1})^{-T}(\mathbf{t} \times \mathbf{A}^{-1}\tilde{\mathbf{m}}) \\ &= -\mathbf{A}'^{-T}\mathbf{R}^{T}[\mathbf{t}]_{\times}\mathbf{A}^{-1}\tilde{\mathbf{m}} \\ &= \mathbf{F}^{T}\tilde{\mathbf{m}}. \end{aligned}$$

In the above, we have used the following properties:

- $(\mathbf{Ax}) \times (\mathbf{Ay}) = \det(\mathbf{A})\mathbf{A}^{-T}(\mathbf{x} \times \mathbf{y}),\ \forall \mathbf{x}, \mathbf{y}$ if matrix $\mathbf{A}$ is invertible.
- $(\mathbf{AB})^{-1} = \mathbf{B}^{-1}\mathbf{A}^{-1}$ if matrices $\mathbf{A}$ and $\mathbf{B}$ are invertible.
- $\mathbf{R}^{T} = \mathbf{R}^{-1}$ if $\mathbf{R}$ is a rotation matrix.
- $[\mathbf{t}]_{\times}^{T} = -[\mathbf{t}]_{\times}$ if $[\mathbf{t}]_{\times}$ is an antisymmetric matrix.

It is thus clear that if $\mathbf{F}$ describes epipolar lines in the first image for points given in the second image, then $\mathbf{F}^{T}$ describes epipolar lines in the second image for points given in the first image. The two images play a symmetric role in the epipolar geometry.

We now compute the epipoles from a different point of view. By definition, all epipolar lines in the first image go through the epipole $\mathbf{e}$. This implies

$$\tilde{\mathbf{e}}^{T}\mathbf{F}\tilde{\mathbf{m}}' = 0, \quad \forall \mathbf{m}',$$

or in vector equation form

$$\mathbf{F}^{T}\tilde{\mathbf{e}} = \mathbf{0}. \tag{32}$$

Pludging Eq. (28) into the above equation gives

$$\mathbf{A}'^{-T}\mathbf{R}^{T}[\mathbf{t}]_{\times}\mathbf{A}^{-1}\tilde{\mathbf{e}} = \mathbf{0}.$$

Because $[\mathbf{t}]_{\times}\mathbf{t} = \mathbf{0}$, up to a scale factor, we have $s_e\mathbf{A}^{-1}\tilde{\mathbf{e}} = \mathbf{t}$, and thus $s_e\tilde{\mathbf{e}} = \mathbf{At}$. This is exactly Eq. (30). Similarly, for the epipole $\mathbf{e}'$ in the second image, we have

$$\mathbf{F}\tilde{\mathbf{e}}' = \mathbf{0} \tag{33}$$

or

$$\mathbf{A}^{-T}[\mathbf{t}]_{\times}\mathbf{R}\mathbf{A}'^{-1}\tilde{\mathbf{e}}' = \mathbf{0}.$$

This gives $s'_e\mathbf{R}\mathbf{A}'^{-1}\tilde{\mathbf{e}}' = -\mathbf{t}$, or $s'_e\tilde{\mathbf{e}}' = -\mathbf{A}'\mathbf{R}^{-1}\mathbf{t}$. This is exactly Eq. (31).

Working with Camera Perspective Projection Matrices

In several applications, for example, in the case of calibrated stereo, the camera perspective projection matrices are given, and we want to compute the epipolar geometry. Let $\mathbf{P}$ and $\mathbf{P}'$ be the projection matrices of the first and second camera. Furthermore, the 3×4 matrix $\mathbf{P}$ is decomposed as the concatenation of a 3×3 submatrix $\mathbf{B}$ and a three-vector $\mathbf{b}$, that is, $\mathbf{P} = [\mathbf{B}\ \mathbf{b}]$. Similarly, $\mathbf{P}' = [\mathbf{B}'\ \mathbf{b}']$.

From the pinhole model, we have

$$s\tilde{\mathbf{m}} = [\mathbf{B}\ \mathbf{b}] \begin{bmatrix} \mathrm{M}' \\ 1 \end{bmatrix}$$

$$s'\tilde{\mathbf{m}}' = [\mathbf{B}'\ \mathbf{b}'] \begin{bmatrix} \mathrm{M}' \\ 1 \end{bmatrix}.$$

Assume that $\mathbf{B}$ and $\mathbf{B}'$ are invertible, we can compute M' from each of the above equations:

$$\mathrm{M}' = s\mathbf{B}^{-1}\tilde{\mathbf{m}} - \mathbf{B}^{-1}\mathbf{b}$$
$$\mathrm{M}' = s'\mathbf{B}'^{-1}\tilde{\mathbf{m}}' - \mathbf{B}'^{-1}\mathbf{b}'.$$

The right sides of the above equations must be equal, which gives

$$s\mathbf{B}^{-1}\tilde{\mathbf{m}} = s'\mathbf{B}'^{-1}\tilde{\mathbf{m}}' + \mathbf{B}^{-1}\mathbf{b} - \mathbf{B}'^{-1}\mathbf{b}'.$$

Multiplying both sides by $\mathbf{B}$ gives

$$s\tilde{\mathbf{m}} = s'\mathbf{B}\mathbf{B}'^{-1}\tilde{\mathbf{m}}' + \mathbf{b} - \mathbf{B}\mathbf{B}'^{-1}\mathbf{b}'.$$

Performing a cross product with $\mathbf{b} - \mathbf{B}\mathbf{B}'^{-1}\mathbf{b}'$ yields

$$s(\mathbf{b} - \mathbf{B}\mathbf{B}'^{-1}\mathbf{b}') \times \tilde{\mathbf{m}} = s'(\mathbf{b} - \mathbf{B}\mathbf{B}'^{-1}\mathbf{b}') \times \mathbf{B}\mathbf{B}'^{-1}\tilde{\mathbf{m}}'.$$

Eliminating the arbitrary scalars s and s' by multiplying $\tilde{\mathbf{m}}^{T}$ from the left (i.e., dot product) gives

$$\tilde{\mathbf{m}}^T \mathbf{F} \tilde{\mathbf{m}}' = 0, \tag{34}$$

where

$$\mathbf{F} = [\mathbf{b} - \mathbf{B}\mathbf{B}'^{-1}\mathbf{b}']_{\times} \mathbf{B}\mathbf{B}'^{-1}. \tag{35}$$

We thus obtain the epipolar equation in terms of the perspective projection matrices. Again, it is clear that the roles of $\mathbf{m}$ and $\mathbf{m}'$ are symmetric, and we have $\tilde{\mathbf{m}}'^T \mathbf{F}^T \tilde{\mathbf{m}} = 0$.

Now let us show how to compute the epipoles. The epipole $\mathbf{e}$ in the first image is the projection of the optical center C' of the second camera, and the optical center C' is given by

$$C' = -\mathbf{B}'^{-1}\mathbf{b}'.$$

We thus have

$$s_e \tilde{\mathbf{e}} = \mathbf{P} \begin{bmatrix} C' \\ 1 \end{bmatrix} = \mathbf{b} - \mathbf{B}\mathbf{B}'^{-1}\mathbf{b}', \tag{36}$$

where s_e is a scale factor. Thus, epipole $\mathbf{e}$ is equal to $(\mathbf{b} - \mathbf{B}\mathbf{B}'^{-1}\mathbf{b}')$ divided by its third element. Similarly, the epipole in the second image, $\mathbf{e}'$, is equal to $(\mathbf{b}' - \mathbf{B}'\mathbf{B}^{-1}\mathbf{b})$ divided by its third element.

Next, we show how, for a given point $\mathbf{m}'$ in the second image, to compute the corresponding epipolar line $\mathbf{l}_{\mathbf{m}'}$ in the first image. The epipolar line must go through the epipole $\mathbf{e}$. We thus need another point to determine it. This point can be the projection of any point $\widehat{\mathrm{M}}'$ on the optical ray $\langle C', \tilde{\mathbf{m}}' \rangle$. In particular, we can choose $\widehat{\mathrm{M}}'$ such that

$$\tilde{\mathbf{m}}' = \mathbf{P}' \begin{bmatrix} \widehat{\mathrm{M}}' \\ 1 \end{bmatrix} = \mathbf{B}'\widehat{\mathrm{M}}' + \mathbf{b}',$$

that is, the scale factor is equal to 1. This gives $\widehat{\mathrm{M}}' = \mathbf{B}'^{-1}(\tilde{\mathbf{m}}' - \mathbf{b}')$. According to the pinhole model, the image $\hat{\mathbf{m}}$ of this point is given by

$$s_m \tilde{\hat{\mathbf{m}}} = \mathbf{P} \begin{bmatrix} \widehat{\mathrm{M}}' \\ 1 \end{bmatrix} = \mathbf{B}\mathbf{B}'^{-1}\tilde{\mathbf{m}}' + (\mathbf{b} - \mathbf{B}\mathbf{B}'^{-1}\mathbf{b}'),$$

where s_m is the scale factor. As already described in Eq. (10) on page 250Working with Normalized Image Coordinatesequation.9.10.80, a line can be represented by a three-vector defined *up to a scale factor*. According to Eq. (12), the epipolar line $\mathbf{l}_{\mathbf{m}'}$ is given by

$$\begin{aligned} \mathbf{l}_{\mathbf{m}'} &= s_e s_m \tilde{\mathbf{e}} \times \tilde{\hat{\mathbf{m}}} \\ &= (\mathbf{b} - \mathbf{B}\mathbf{B}'^{-1}\mathbf{b}') \times [\mathbf{B}\mathbf{B}'^{-1}\tilde{\mathbf{m}}' + (\mathbf{b} - \mathbf{B}\mathbf{B}'^{-1}\mathbf{b}')] \\ &= (\mathbf{b} - \mathbf{B}\mathbf{B}'^{-1}\mathbf{b}') \times (\mathbf{B}\mathbf{B}'^{-1}\tilde{\mathbf{m}}'), \end{aligned}$$

or

$$\mathbf{l}_{\mathbf{m}'} = \mathbf{F}\tilde{\mathbf{m}}', \tag{37}$$

where $\mathbf{F}$ is given by Eq. (35). Then any point $\mathbf{m}$ on the epipolar line of $\mathbf{m}'$ satisfies $\tilde{\mathbf{m}}^T \mathbf{F} \tilde{\mathbf{m}}' = 0$, and this is exactly Eq. (34). Therefore, we obtain geometrically the same equation. Because of symmetry, for a given point $\mathbf{m}$ in the first image, its corresponding epipolar line in the second image is represented by the vector $\mathbf{F}^T \tilde{\mathbf{m}}$.

Now, we show that if the images are calibrated, then the $\mathbf{F}$ is reduced to the $\mathbf{E}$. Since the images are calibrated, the points $\mathbf{m}$ can be expressed in normalized coordinates, that is, $\mathbf{m} = \mathbf{x}$. Without loss of generality, the world coordinate system is assumed to coincide with the second camera coordinate system. From the perspective projection model, we have the following camera projection matrices:

$$\mathbf{P} = [\mathbf{R}\ \mathbf{t}] \quad \text{and} \quad \mathbf{P}' = [\mathbf{I}\ \mathbf{0}].$$

This implies that $\mathbf{B} = \mathbf{R}$, $\mathbf{b} = \mathbf{t}$, $\mathbf{B}' = \mathbf{I}$, and $\mathbf{b}' = \mathbf{0}$. Pludging them into Eq. (35) gives $\mathbf{F} = [\mathbf{t}]_{\times}\mathbf{R}$, which is exactly the essential matrix Eq. (8).

In the above derivation of the fundamental matrix, a camera perspective projection matrix $\mathbf{P}$ is decomposed into a 3×3 matrix $\mathbf{B}$ and a three-vector $\mathbf{b}$, and $\mathbf{B}$ must be invertible. Later, we provide a more general derivation directly in terms of the camera projection matrices $\mathbf{P}$ and $\mathbf{P}'$.

Fundamental Matrix and Epipolar Transformation

We examine the relationship between the fundamental matrix and the (i.e., the transformation of the epipoles and the epipolar lines between the two images).

For any point $\mathbf{m}'$ in the second image, its epipolar line $\mathbf{l}_{\mathbf{m}'}$ in the first image is given by $\mathbf{l}'_{\mathbf{m}} = \mathbf{F}\tilde{\mathbf{m}}'$. It must go through the $\tilde{\mathbf{e}} = [e_1, e_2, e_3]^T$ and a point $\tilde{\mathbf{m}} = [u, v, s]^T$, that is, $\mathbf{l}_{\mathbf{m}'} = \tilde{\mathbf{e}} \times \tilde{\mathbf{m}} = \mathbf{F}\tilde{\mathbf{m}}'$. Here, we use the for the image points. Symmetrically, the epipolar line in the second image $\mathbf{l}'_{\mathbf{m}}$ of point $\mathbf{m}$ is given by $\mathbf{l}'_{\mathbf{m}} = \mathbf{F}^T \tilde{\mathbf{m}}$ and must go through the epipole $\tilde{\mathbf{e}}' = [e'_1, e'_2, e'_3]^T$ and a point $\tilde{\mathbf{m}}' = [u', v', s']^T$, that is, $\mathbf{l}'_{\mathbf{m}} = \tilde{\mathbf{e}}' \times \tilde{\mathbf{m}}' = \mathbf{F}^T \tilde{\mathbf{m}}$. In other words, the epipole $\mathbf{e}'$ is *on* the epipolar line $\mathbf{l}'_{\mathbf{m}}$ for any point $\mathbf{m}$; that is,

$$\tilde{\mathbf{e}}'^T \mathbf{l}'_{\mathbf{m}} = \tilde{\mathbf{e}}'^T \mathbf{F}^T \tilde{\mathbf{m}} = 0 \quad \forall \mathbf{m},$$

which yields

$$\mathbf{F}\tilde{\mathbf{e}}' = \mathbf{0}. \tag{38}$$

Let $\mathbf{c}_1$, $\mathbf{c}_2$, and $\mathbf{c}_3$ be the column vectors of $\mathbf{F}$, and we have $e'_1\mathbf{c}_1 + e'_2\mathbf{c}_2 + e'_3\mathbf{c}_3 = \mathbf{0}$; thus the rank of $\mathbf{F}$ is at most two. The solution to the epipole $\tilde{\mathbf{e}}'$ is given by

$$\begin{aligned} e'_1 &= F_{23}F_{12} - F_{22}F_{13} \\ e'_2 &= F_{13}F_{21} - F_{11}F_{23} \\ e'_3 &= F_{22}F_{11} - F_{21}F_{12}, \end{aligned} \tag{39}$$

up to, of course, a scale factor. Similarly, for the epipole in the first image, we have

$$\mathbf{F}^T\tilde{\mathbf{e}} = \mathbf{0}, \tag{40}$$

which gives

$$\begin{aligned} e_1 &= F_{32}F_{21} - F_{22}F_{31} \\ e_2 &= F_{31}F_{12} - F_{11}F_{32} \\ e_3 &= F_{22}F_{11} - F_{21}F_{12}, \end{aligned} \tag{41}$$

also up to a scale factor.

Now let us examine the relationship between the epipolar lines. Once the epipole is known, an epipolar line can be parameterized by its direction vector. Consider $\mathbf{l}'_{\mathbf{m}} = \tilde{\mathbf{e}}' \times \tilde{\mathbf{m}}'$; its direction vector $\mathbf{u}'$ can be parameterized by one parameter τ' such that $\mathbf{u}' = [1, \tau', 0]^T$. A particular point on $\mathbf{l}'_{\mathbf{m}}$ is then given by $\tilde{\mathbf{m}}' = \tilde{\mathbf{e}}' + \lambda'\mathbf{u}'$, where λ' is a scalar. Its epipolar line in the first image is given by

$$\begin{aligned} \mathbf{l}_{\mathbf{m}'} &= \mathbf{F}\tilde{\mathbf{m}}' = \mathbf{F}\tilde{\mathbf{e}}' + \lambda'\mathbf{F}\mathbf{u}' = \lambda'\mathbf{F}\mathbf{u}' \\ &= \lambda' \begin{bmatrix} F_{11} + F_{12}\tau' \\ F_{21} + F_{22}\tau' \\ F_{31} + F_{32}\tau' \end{bmatrix}. \end{aligned} \tag{42}$$

This line can also be parameterized by its direction vector $\mathbf{u} = [1, \tau, 0]^T$ in the first image, which implies that

$$\begin{aligned} \mathbf{l}_{\mathbf{m}'} &\cong \tilde{\mathbf{e}} \times (\tilde{\mathbf{e}} + \lambda\mathbf{u}) = \lambda\tilde{\mathbf{e}} \times \mathbf{u} \\ &= \lambda \begin{bmatrix} -(F_{11}F_{22} - F_{21}F_{12})\tau \\ F_{11}F_{22} - F_{21}F_{12} \\ (F_{32}F_{21} - F_{22}F_{31})\tau - F_{31}F_{12} + F_{11}F_{32} \end{bmatrix}, \end{aligned} \tag{43}$$

where $\cong$ means "equal" up to a scale factor and λ is a scalar. By requiring that Eq. (42) and Eq. (43) represent the same line, we have

$$\tau = \frac{a\tau' + b}{c\tau' + d}, \tag{44}$$

where

$$\begin{aligned} a &= F_{12} \\ b &= F_{11} \\ c &= -F_{22} \\ d &= -F_{21}. \end{aligned} \tag{45}$$

Writing in matrix form gives

$$\rho \begin{bmatrix} \tau \\ 1 \end{bmatrix} = \begin{bmatrix} a & b \\ c & d \end{bmatrix} \begin{bmatrix} \tau' \\ 1 \end{bmatrix},$$

where ρ is a scale factor. This relation is known as the *homography* between τ and τ', and we say that *there is a homography between the epipolar lines in the first image and those in the second image*. The above is, of course, only valid for the epipolar lines having the nonzero first element in the direction vector. If the first element is equal to zero, we should parameterize the direction vector as $[\tau, 1, 0]^T$, and similar results can be obtained.

At this point, we can see that the epipolar transformation is defined by the coordinates $\tilde{\mathbf{e}} = [e_1, e_2, e_3]^T$ and $\tilde{\mathbf{e}}' = [e'_1, e'_2, e'_3]^T$ of the epipoles and the four parameters a, b, c, and d of the homography between the two pencils of the epipolar lines. The coordinates of each epipole are defined up to a scale factor, and the parameters of the homography, as can be seen in Eq. (44), are also defined up to a scale factor. Thus, we have in total seven free parameters. This is exactly the number of parameters of the fundamental matrix.

If we have identified the parameters of the epipolar transformation, that is, the coordinates of the two epipoles and the coefficients of the homography, then we can construct the fundamental matrix, from Eq. (39), (41), and (45), as

$$\begin{aligned}
F_{11} &= be_3e'_3 \\
F_{12} &= ae_3e'_3 \\
F_{13} &= -(ae'_2 + be'_1)e_3 \\
F_{21} &= -de_3e'_3 \\
F_{22} &= -ce_3e'_3 \\
F_{23} &= (ce'_2 + de'_1)e_3 \\
F_{31} &= (de_2 - be_1)e'_3 \\
F_{32} &= (ce_2 - ae_1)e'_3 \\
F_{33} &= -(ce'_2 + de'_1)e_2 + (ae'_2 + be'_1)e_1.
\end{aligned} \tag{46}$$

The determinant $ad - bc$ of the homography is equal to the determinant of the first 2×2 submatrix of $\mathbf{F}$, $F_{11}F_{22} - F_{12}F_{21}$, which is zero when the epipoles are at infinity.

General Form of Epipolar Equation for Any Projection Model

In this section, we will derive a which does not assume any particular projection model.

Intersecting Two Optical Rays

The projections for the first and second cameras are represented respectively as

$$s\tilde{\mathbf{m}} = \mathbf{P}\tilde{\mathtt{M}}, \tag{47}$$

and

$$s'\tilde{\mathbf{m}}' = \mathbf{P}'\tilde{\mathtt{M}}', \tag{48}$$

where $\tilde{\mathbf{m}}$ and $\tilde{\mathbf{m}}'$ are augmented image coordinates and $\tilde{\mathtt{M}}$ and $\tilde{\mathtt{M}}'$ are augmented space coordinates of a single point in the two camera coordinate systems. Here both projection matrices *do not include the extrinsic parameters*.

The same point in the two camera coordinate systems can be related by

$$\tilde{\mathtt{M}} = \mathbf{D}\tilde{\mathtt{M}}', \tag{49}$$

where

$$\mathbf{D} = \begin{bmatrix} \mathbf{R} & \mathbf{t} \\ \mathbf{0}_3^T & 1 \end{bmatrix}$$

is the Euclidean transformation matrix compactly representing both rotation and translation. Now substituting Eq. (49) for (47), we have

$$s\tilde{\mathbf{m}} = \mathbf{P}\mathbf{D}\tilde{\mathtt{M}}'. \tag{50}$$

For an image point $\tilde{\mathbf{m}}'$, Eq. (48) actually defines an optical ray on which every space point $\tilde{\mathtt{M}}'$ projects onto the second image at $\tilde{\mathbf{m}}'$. This optical ray can be written in pamametric form as

$$\tilde{\mathtt{M}}' = s'\mathbf{P}'^{+}\tilde{\mathbf{m}}' + \mathbf{p}'^{\perp}, \tag{51}$$

where $\mathbf{P}'^{+}$ is the pseudoinverse matrix of $\mathbf{P}'$:

$$\mathbf{P}'^{+} = \mathbf{P}'^{T}(\mathbf{P}'\mathbf{P}'^{T})^{-1}, \tag{52}$$

and $\mathbf{p}'^{\perp}$ is a four-vector that is perpendicular to all the row vectors of $\mathbf{P}'$, that is,

$$\mathbf{P}'\mathbf{p}'^{\perp} = \mathbf{0}. \tag{53}$$

There are an infinite number of matrices that satisfy $\mathbf{P}'\mathbf{P}'^{+} = \mathbf{I}$. Thus, $\mathbf{P}'^{+}$ is not unique. See [5, 6] for how to derive this particular pseudoinverse matrix.

It remains to determine $\mathbf{p}'^{\perp}$. First note that such a vector does exist because the difference between the row dimension and column dimension is one, and the row vectors are generally independent of each other. Actually, one way to obtain $\mathbf{p}'^{\perp}$ is

$$\mathbf{p}'^{\perp} = (\mathbf{I} - \mathbf{P}'^{+}\mathbf{P}')\boldsymbol{\omega}, \tag{54}$$

where $\boldsymbol{\omega}$ is an arbitrary four-vector. To show that it is perpendicular to every row vector of $\mathbf{P}'$, we multiply $\mathbf{P}'$ and $\mathbf{p}'^{\perp}$:

$$\mathbf{P}'(\mathbf{I} - \mathbf{P}'^{+}\mathbf{P}')\boldsymbol{\omega} = (\mathbf{P}' - \mathbf{P}'\mathbf{P}'^{T}(\mathbf{P}'\mathbf{P}'^{T})^{-1}\mathbf{P}')\boldsymbol{\omega} = \mathbf{0}$$

which is indeed a zero vector.

Actually, the following equation always stands, as long as the of the 3×4 matrix $\mathbf{P}'$ is 3:

$$\mathbf{I} - \mathbf{P}'^{+}\mathbf{P}' = \mathbf{I} - \mathbf{P}'^{T}(\mathbf{P}'\mathbf{P}'^{T})^{-1}\mathbf{P}' = \frac{\mathbf{p}'^{\perp}\mathbf{p}'^{\perp T}}{\|\mathbf{p}'^{\perp}\|^2}. \tag{55}$$

The effect of matrix $\mathbf{I} - \mathbf{P}'^{+}\mathbf{P}'$ is to transform an arbitrary vector to a vector that is perpendicular to every row vector of $\mathbf{P}'$. If $\mathbf{P}'$ is of rank 3 (which is usually the case), then $\mathbf{p}^{\perp}$ is unique up to a scale factor.

Equation (51) is easily justified by projecting $\mathtt{M}'$ onto the image using Eq. (48), which indeed gives $\tilde{\mathbf{m}}'$. If we look closely at the equation, we can find that $\mathbf{p}'^{\perp}$ actually defines the optical center, which always projects onto the origin, and $\mathbf{P}'^{+}\tilde{\mathbf{m}}'$ defines the direction of the optical ray corresponding to image point $\tilde{\mathbf{m}}'$. For a particular value s', Eq. (51) corresponds to a point on the optical ray defined by $\mathbf{m}'$.

Similarly, an image point $\tilde{\mathbf{m}}$ in the first image also defines an optical ray. Requiring the two rays to intersect in space means that a point $\tilde{\mathrm{M}}'$ corresponding to a particular s' in Eq. (51) must project onto the first image at $\tilde{\mathbf{m}}$. That is,

$$s\tilde{\mathbf{m}} = s'\mathbf{PDP}'^{+}\tilde{\mathbf{m}}' + \mathbf{PDp}'^{\perp}, \tag{56}$$

where $\mathbf{PDp}'^{\perp}$ is the $\mathbf{e}$ in the first image.

Performing a cross product with $\mathbf{PDp}'^{\perp}$ yields

$$s(\mathbf{PDp}'^{\perp}) \times \tilde{\mathbf{m}} = (\mathbf{PDp}'^{\perp}) \times (s'\mathbf{PDP}'^{+}\tilde{\mathbf{m}}').$$

Eliminating s and s' by multiplying $\tilde{\mathbf{m}}^T$ from the left (equivalent to an inner product), we have

$$\tilde{\mathbf{m}}^T\mathbf{F}\tilde{\mathbf{m}}' = 0, \tag{57}$$

where

$$\mathbf{F} = \left[\mathbf{PDp}'^{\perp}\right]_{\times} \mathbf{PDP}'^{+} \tag{58}$$

is the general form of fundamental matrix. It is evident that the roles that the two images play are symmetrical.

Note that Eq. (58) will be the essential matrix $\mathbf{E}$ if $\mathbf{P}$ and $\mathbf{P}'$ do not include the intrinsic parameters, that is, if we work with normalized cameras.

We can also include all the intrinsic and extrinsic parameters in the two projection matrices $\mathbf{P}$ and $\mathbf{P}'$, so that for a 3D point $\tilde{\mathrm{M}}'$ in a world coordinate system, we have

$$s\tilde{\mathbf{m}} = \mathbf{P}\tilde{\mathrm{M}}', \tag{59}$$

$$s'\tilde{\mathbf{m}}' = \mathbf{P}'\tilde{\mathrm{M}}'. \tag{60}$$

Similarly we get

$$s\tilde{\mathbf{m}} = s'\mathbf{PP}'^{+}\tilde{\mathbf{m}}' + \mathbf{Pp}'^{\perp}, \tag{61}$$

The same line of reasoning will lead to the general form of epipolar equation

$$\tilde{\mathbf{m}}^T\mathbf{F}\tilde{\mathbf{m}}' = 0,$$

where

$$\mathbf{F} = \left[\mathbf{Pp}'^{\perp}\right]_{\times} \mathbf{PP}'^{+}. \tag{62}$$

It can also be shown that this expression is equivalent to Eq. (35) for the full perspective projection (see next subsection), but it is more general. Indeed, Eq. (35) assumes that the 3×3 matrix $\mathbf{B}'$ is invertible, which is the case for full perspective projection but not for affine cameras, while Eq. (62) makes use of the pseudoinverse of the projection matrix, which is valid for both full perspective projection as well as affine cameras. Therefore the equation does not depend on any specific knowledge of projection model. Replacing the projection matrix in the equation by specific projection matrix for each specific projection model produces the epipolar equation for that specific projection model.

The Full Perspective Projection Case

Here we work with normalized cameras. Under the full perspective projection, the projection matrices for the two cameras are the same:

$$\mathbf{P}_p = \mathbf{P}'_p = \begin{bmatrix} 1 & 0 & 0 & 0 \\ 0 & 1 & 0 & 0 \\ 0 & 0 & 1 & 0 \end{bmatrix}. \tag{63}$$

It is not difficult to obtain

$$\mathbf{P}'^{+}_p = \mathbf{P}^T_p,$$

and

$$\mathbf{p}'^{\perp}_p = \begin{bmatrix} 0 \\ 0 \\ 0 \\ 1 \end{bmatrix}.$$

Now substituting the above equations for Eq. (58), we have obtained the essential matrix

$$\mathbf{E}_p = \left[\mathbf{P}_p\mathbf{Dp}'^{\perp}_p\right]_{\times} \mathbf{P}_p\mathbf{DP}'^{+}_p = [\mathbf{t}]_{\times}\mathbf{R}, \tag{64}$$

which is exactly the same as what we derived in the last section.

For the full perspective projection, we can prove that Eq. (62) is equivalent to Eq. (35). From definitions, we have

$$\left[\mathbf{B}' \quad \mathbf{b}'\right]\mathbf{p}'^{\perp}_p = \mathbf{0},$$

$$[\mathbf{B}' \quad \mathbf{b}']\mathbf{P}'^{+}_{p} = \mathbf{I}.$$

It is easy to show

$$\mathbf{P}\mathbf{p}'^{\perp}_{p} = \lambda(\mathbf{b} - \mathbf{B}\mathbf{B}'^{-1}\mathbf{b}'),$$
$$\mathbf{P}'^{+}_{p} = \begin{bmatrix} \mathbf{B}'^{-1} - \mathbf{B}'^{-1}\mathbf{b}'\mathbf{q}^{T} \\ \mathbf{q}^{T} \end{bmatrix},$$

where λ is a nonzero scalar and $\mathbf{q}$ is a nonzero arbitrary three vector. Substituting them for Eq. (62) yields

$$\mathbf{F} = \lambda[\mathbf{b} - \mathbf{B}\mathbf{B}'^{-1}\mathbf{b}']_{\times}\left(\mathbf{B}\mathbf{B}'^{-1} - (\mathbf{b} - \mathbf{B}\mathbf{B}'^{-1}\mathbf{b}')\mathbf{q}^{T})\right)$$
$$= \lambda[\mathbf{b} - \mathbf{B}\mathbf{B}'^{-1}\mathbf{b}']_{\times}\mathbf{B}\mathbf{B}'^{-1}, \quad (65)$$

which completes the proof as $\mathbf{F}$ is defined up to a scale factor.

The reader is referred to [5, 6] for the epipolar geometry between affine cameras and for a general expression of the fundamental matrix for both perspective and affine cameras. The reader is referred to [7, 8] for various algorithms of determining essential matrix and fundamental matrix from point correspondences. The reader is referred to [9, 10] for a general treatment of geometry across multiple cameras.

References

1. Longuet-Higgins H (1981) A computer algorithm for reconstructing a scene from two projections. Nature 293:133–135
2. Zhang Z, Faugeras OD (1992) 3D dynamic scene analysis: a stereo based approach. Springer, Berlin/Heidelberg
3. Maybank S (1992) Theory of reconstruction from image motion. Springer, Berlin/New York
4. Faugeras O (1993) Three-dimensional computer vision: a geometric viewpoint. MIT, Cambridge
5. Xu G, Zhang Z (1996) Epipolar geometry in stereo, motion and object recognition. Kluwer Academic, Dordrecht/Boston
6. Zhang Z, Xu G (1998) A unified theory of uncalibrated stereo for both perspective and affine cameras. J Math Imaging Vis 9:213–229
7. Zhang Z (1997) Motion and structure from two perspective views: from essential parameters to euclidean motion via fundamental matrix. J Opt Soc Am A 14(11):2938–2950
8. Zhang Z (1998) Determining the epipolar geometry and its uncertainty: a review. Int J Comput Vis 27(2):161–195
9. Hartley R, Zisserman A (2000) Multiple view geometry in computer vision. Cambridge University Press, Cambridge/New York
10. Faugeras O, Luong QT (2001) In: Papadopoulo T (ed) The geometry of multiple images. MIT, Cambridge/London

Error Concealment

▶Inpainting

Error-Correcting Graph Matching

▶Many-to-Many Graph Matching

Error-Tolerant Graph Matching

▶Many-to-Many Graph Matching

Essential Matrix

Zhengyou Zhang
Microsoft Research, Redmond, WA, USA

Related Concepts

▶Epipolar Geometry; ▶Fundamental Matrix

Definition

Essential matrix is a special 3×3 matrix which captures the geometric relationship between two calibrated cameras or between two locations of a single moving camera.

Background

See entry ▶Epipolar Geometry for details.

Theory

Because the cameras are calibrated, we use the normalized image coordinates. The two images of a space point $\boldsymbol{X} = [X, Y, Z]^T$ are $[x, y, 1]^T$ and $[x', y', 1]^T$ in the first and second images and are denoted by $\widetilde{\boldsymbol{x}}$ and $\widetilde{\boldsymbol{x}}'$. Assume that the second camera is brought from the position of the first camera through a rotation $\mathbf{R}$

followed by a translation t. From the epipolar geometry, we have the following equation

$$\widetilde{x}^T \mathbf{E} \widetilde{x}' = 0 , \tag{1}$$

where

$$\mathbf{E} = [t]_\times \mathbf{R} . \tag{2}$$

And $[t]_\times$ is a 3×3 antisymmetric matrix defined by vector t. This equation is called the *epipolar equation*, and matrix $\mathbf{E}$ is known as the *essential matrix*.

Essential matrix has a number of properties, including:

1. $\det \mathbf{E} = 0$, because $[t]_\times$ is an antisymmetric matrix.
2. $\mathbf{E}^T t = \mathbf{0}$, because $([t]_\times \mathbf{R})^T t = \mathbf{R}^T [t]_\times^T t = -\mathbf{R}^T (t \times t) = \mathbf{0}$.
3. $\mathbf{E}\mathbf{E}^T = (t^t t)\mathbf{I} - tt^T$, because $\mathbf{E}\mathbf{E}^T = ([t]_\times \mathbf{R})([t]_\times \mathbf{R})^T = [t]_\times \mathbf{R}\mathbf{R}^T [t]_\times^T = -[t]_\times^2$.
4. $\|\mathbf{E}\|^2 = 2\|t\|^2$. Here, $\|\mathbf{E}\|$ is the Frobenius norm of matrix $\mathbf{E}$, i.e., $\|\mathbf{E}\|^2 = \sum_{i,j} E_{ij}^2$.

It can be shown that a real 3×3 matrix can be decomposed into the multiplication of an antisymmetric matrix $\mathbf{T}$ and a rotation matrix $\mathbf{R}$ if and only if one of $\mathbf{E}$'s singular values is 0 while the other two are equal.

Euclidean Geometry

Gunnar Sparr
Centre for Mathematical Sciences, Lund University, Lund, Sweden

Related Concepts

▸Algebraic Curve; ▸Camera Calibration

Definition

Euclidean geometry deals with properties of geometric configurations that are preserved under isometric (or *length preserving*) transformations. Alternatively, it may be characterized as a mathematical theory based on an axiom system (that can be traced back to Euclid) expressing, in modern terminology, incidence, order, congruence, continuity, and parallelity. Euclidean geometry is today a special case of many geometric theories (projective, affine, and Riemannian geometries, Hilbert spaces . . .).

Historical Background

Euclidean geometry has a long and glorious history (cf. [1–3]), having lived at the core of the development of science and culture since antiquity. It is today not an area of research per se, but still plays an important role in many contexts.

Euclidean geometry is one of the oldest manifestations of humans in science. The latter part of the word *geometry* originates from the Greek word *metri'a* for *measure*, and the subject developed in the antiquity as an empirical science for surveying. It was given a scientific formulation by the Greek mathematician Euclid in Alexandria, about 300 B.C. Starting from a small set of intuitively appealing *axioms*, in his monumental treatise, the *Elements*, he deduced a large number of *propositions* about geometrical figures (cf. [8]). In the Elements, plane geometry was presented in essentially the way it is today taught in secondary school. Also the solid geometry of three dimensions was addressed. Two other big contributors to ancient geometry, both active in the third century B.C., were Archimedes, with equations for, e.g., the circumference of the circle and the area of the sphere, and Apollonius, with investigations of conic sections.

For over 2,000 years, the attribute *Euclidean* was unnecessary, because no other kind of geometry was conceived of. Early, however, the fifth of Euclid's axioms, the *parallel axiom*, was met with challenge, and many unsuccessful efforts were made to deduce it from the other axioms. However, it took until the nineteenth-century before its independence was settled by the construction of consistent geometric models with other parallelity concepts. Some contributors to such non-Euclidean geometries were Gauss, Bolyai, and Lobachevsky.

By today's standard of rigor, the treatment of Euclid is not without objections, using some assumptions and concepts not explicitly accounted for. Beginning in the nineteenth-century, several categorical axiom systems were presented, e.g., by Hilbert in 1899 (cf. [6, 9]).

Also other geometric systems were discovered and developed, like affine geometry, beginning with Euler in the eighteenth-century, and projective geometry, through Poncelet, von Staudt, a.o. in the nineteenth-century. While in affine geometry, parallelity plays a prominent role; in projective geometry, the concept does not exist, in that every pair of lines intersect.

In contrast to the then prevalent *synthetic* approach to geometry, based on geometric constructions, in the seventeenth century, Descartes introduced coordinate systems and founded the *analytic geometry*. Also de Moivre made significant contributions. The use of coordinates enabled the study of geometry by means of algebra and calculus. In this formalism, in a natural way, Euclidean geometry can be generalized to higher dimensions, by means of the notion of Euclidean space; see below.

From a more modern mathematical point of view, geometry in some space is the study of properties of configurations that are preserved under some group of one-to-one transformations of the space in question (cf. [1–3]). This viewpoint was first formulated by Klein 1872 in the so-called Erlanger program, cf. [6]. It has been the key to a fruitful interplay between geometry, algebra, analysis, and topology. For Euclidean geometry, the characterizing transformation group is the group of isometric transformations, reflecting, e.g., the crucial property of congruence, that two triangles are congruent if one of them can be moved rigidly onto the other one.

Theory and Applications

The topic of Euclidean geometry today has branched out in many directions. Below, we focus on a few of relevance to computer vision.

Synthetic Euclidean Geometry

The synthetic approach to Euclidean geometry (the one used by Euclid) starts from the basic concepts of *point, line, and surface*. The rules for interaction between these are described by *axioms*. Euclid used five axioms (cf. [8]):

1. Through any two points, there is exactly one line.
2. A finite line segment can be extended to a line.
3. A circle can be drawn with any center point and any radius.
4. All right angles are equal to another.
5. The parallel postulate: If a straight line falling on two straight lines make the interior angles on the same side less than two right angles, the two straight lines, if produced indefinitely, meet on that side on which are the angles less than the two right angles.

For a rigorous exposition of Euclidean geometry, see [9], built on Hilbert's system of 20 axioms in five main groups: combination, order, parallelity, congruence, and continuity.

From the beginning, Euclidean geometry mostly dealt with squares and rectangles, right-angled triangles, trapezia, and circles. Some well-known examples of results from Euclidean geometry are the theorem of Pythagoras, the theorem stating that the sum of angles in a triangle equals two right angles, and the theorem stating that the periphery angle corresponding to an arch of a circle equals half the angle at the center.

The proofs in this tradition were constructive. Questions were raised about the possibility of generally solving geometric problems by ruler and compass. Using advances in algebra, the inherent impossibility of such constructions for some famed problems was proved, e.g., the trisectioning of an angle, the doubling of the cube, and the squaring of a circle.

Analytic Geometry: The Space $\mathbf{R}^n$

In the analytic geometry, as learned at school, points in the plane are represented by coordinates as (x_1, x_2), and points in space as (x_1, x_2, x_3). This inspires to consider n-tuples of real numbers $(x_1, x_2, \ldots, x_n)$, which together build up $\mathbf{R}^n$. This space forms the scene for geometry, not only in 1, 2, and 3 dimensions, but also in higher dimensions. The elements are called points $X : (x_1, x_2, \ldots, x_n)$.

The set $\mathbf{R}^n$ can be provided with different structures:

Linear Space. Equip $\mathbf{R}^n$ with the operations of addition, $\boldsymbol{x} + \boldsymbol{y} = (x_1, \ldots, x_n) + (y_1, \ldots, y_n) = (x_1 + y_1, \ldots, x_n + y_n)$, and multiplication with a scalar $\lambda \boldsymbol{x} = (\lambda x_1, \ldots, \lambda x_n)$. Then, in a way known from introductory courses in linear algebra, the notion of linear space is introduced. The elements of $\mathbf{R}^n$ are called *vectors*. The notion of *dimension* is defined, giving $\mathbf{R}^n$ the dimension n.

Affine Space. Combining the point and vector interpretations of $\mathbf{R}^n$, affine spaces are defined in a way such that, loosely speaking, it is possible to subtract points to get vectors and add a vector to a point to get another point. (On the contrary, it is not possible to add points.) For a thorough presentation, see [1]. For affine spaces, the notation A^n will be used below.

Euclidean Space. Known from linear algebra is also the notion of *scalar product* on $\mathbf{R}^n$, being a function $\boldsymbol{x} \cdot \boldsymbol{y}$ such that for all vectors $\boldsymbol{x}$, $\boldsymbol{y}$, and all scalars λ:

- $\boldsymbol{x} \cdot \boldsymbol{y} = \boldsymbol{y} \cdot \boldsymbol{x}$.
- $\boldsymbol{x} \cdot (\lambda_1 \boldsymbol{y}_1 + \lambda_2 \boldsymbol{y}_2) = \lambda_1 \boldsymbol{x} \cdot \boldsymbol{y}_1 + \lambda_2 \boldsymbol{x} \cdot \boldsymbol{y}_2$.
- $\boldsymbol{x} \cdot \boldsymbol{x} \geq 0$ with equality if and only if $\boldsymbol{x} = 0$.

Given a scalar product, a *norm* on the linear space $\mathbf{R}^n$ is defined by $\|\boldsymbol{x}\| = \sqrt{\boldsymbol{x} \cdot \boldsymbol{x}}$. Two vectors $\boldsymbol{x}$ and $\boldsymbol{y}$ are orthogonal if $\boldsymbol{x} \cdot \boldsymbol{y} = 0$. An orthonormal basis $\boldsymbol{e}_1, \ldots \boldsymbol{e}_n$ is characterized by $\boldsymbol{e}_i \cdot \boldsymbol{e}_j = 0$ for $i \neq j$ and $= 1$ for $i = j$.

Having a scalar product, the affine space A^n can be provided with a *metric*, by which the distance between points is given by $d(X, Y) = \|\boldsymbol{u}\|$, where $\boldsymbol{u}$ is the vector from X to Y. In particular, in an orthonormal basis, the distance between the points $X : (x_1, \ldots, x_n)$ and $Y = (y_1, \ldots, y_n)$ is

$$|XY| = ((x_1 - y_1)^2 + \ldots (x_n - y_n)^2)^{1/2}$$

(in agreement with the theorem of Pythagoras in the planar case).

The scalar product also makes it possible to define the *angle* θ between two vectors $\boldsymbol{x}$ and $\boldsymbol{y}$ as

$$\boldsymbol{x} \cdot \boldsymbol{y} = \|\boldsymbol{x}\| \|\boldsymbol{y}\| \cos \theta .$$

Provided with a scalar product for vectors, A^n becomes a Euclidean space, below denoted E^n.

Euclidean Geometry and Transformation Groups

Given a set S, consider the group $\mathrm{Bij}(S)$ consisting of all one-to-one transformations $f : S \rightarrow S$. According to the view of the Erlanger program, to impose a geometry on S is the same as to specify a subgroup G of $\mathrm{Bij}(S)$ and to say that two subsets A, B of S are equivalent if there is an $f \in G$ such that $fA = B$. See [1–3].

Two important such transformation groups on $\mathbb{R}^n$ are $\mathrm{GL}(n)$, consisting of all non-singular $n \times n$-matrices, and $\mathrm{O}(n)$, consisting of all orthogonal $n \times n$-matrices.

For planar Euclidean geometry, S is the affine space A^2 and G is the group of all *isometric transformations*, i.e., transformations such that for any points $X, Y \in \mathbb{R}^2$, $d(TX, TY) = d(X, Y)$. In fact, it can be proven that every mapping T with this property can be written as

$$T : X \rightarrow QX + b \text{ , with } Q \in \mathrm{O}(n) .$$

In particular, T is an affine map, mapping lines onto lines. Thus, it also maps triangles onto triangles, after which the isometric property guarantees congruence. In higher dimensions holds the analogous formula for T.

In Euclidean geometry, also the notion of *similarity* plays a prominent role, which could motivate the use of similarity transformations above (where Q is replaced by $cQ, c \neq 0$). Note that the isometric transformations form a subgroup of the similarity transformations.

The example of planar Euclidean geometry is so crucial that it is worthwhile to comment further on the structure of its transformation group. The formula above shows that an isometric transformation is composed by a rotation around the origin (expressed by Q), followed by a translation (by b). Furthermore, it is possible to prove that any such transformation is equal to either the identity, or a translation, or a rotation (around some point), or a reflection in some line, or a so-called glide reflection (a translation followed by a reflection in a line parallel to the translation). Also in $\mathbb{R}^3$, it is possible to make an analogous characterization, where a type screw (composition of a rotation and a translation) is added. Note in particular that in this way, isometries on $\mathbb{R}^2$ and $\mathbb{R}^3$ are characterized by geometric constructions only, without use of any particular coordinate system.

On the Stratum of Euclidean, Affine and Projective Geometry

Euclidean geometry is one of the layers in a hierarchy of structures that can be imposed on $\mathbb{R}^n$ and which are of high relevance to computer vision. With a term introduced by Faugeras [5], these form a *geometric stratum*. For more details on the respective geometries, see [1–3, 10].

Affine Geometry. For the linear space $\mathbf{R}^n$, a particular role is played by linear subspaces of the form $a_1 x_1 + \ldots + a_n x_n = 0$, called *hyperplanes* (always passing through the origin). As a linear space, its dimension is $n-1$. Analogously, in the affine space A^n, one considers *affine hyperplanes* $a_1 x_1 + \ldots +$

$a_n x_n = a$ (which for $a \neq 0$ do not pass through the origin). In the case $n = 2$, one talks about *lines* instead of hyperplanes.

For affine geometry on $\mathbf{R}^n$, the characterizing transformations are

$$T : X \to AX + b \,, \text{ with } A \in \text{GL(n)} \,.$$

By such transformations, affine hyperplanes are mapped onto affine hyperplanes. Two nonintersecting hyperplanes are mapped onto two nonintersecting hyperplanes. Hence, the property of parallelity is preserved and is thus a concept within affine geometry. On the contrary, distance is not preserved and is an alien concept for affine geometry. However, it is possible to compare distances along a line (e.g., saying that M is the midpoint of a line segment PQ), but it is not possible to measure and compare distances on nonparallel lines. Neither does the concept of angle live in affine geometry, in lack of a scalar product.

Also affine geometry can be built up from axioms. Compared to Euclidean geometry, one way of doing this is by replacing the parallel axiom by an axiom "for any point P and any line ℓ, not through P, there is at least one line through P which does not meet ℓ," plus another one, involving seven-point configurations, named after Desargues.

Projective Geometry. By a point in the so-called projective space $\mathbb{P}^n$ is meant a line through the origin in $\mathbb{R}^{n+1}$. In other words, a point is represented by any of the vectors $\lambda(x_1, x_2, \ldots x_{n+1})$, $\lambda \neq 0$. This is called *homogeneous coordinates* for the point in $\mathbb{P}^n$. By a *projective transformation* T is meant a mapping represented by $\text{GL}(n + 1)$ in homogeneous coordinates, i.e., $y = \lambda T x$ for some λ.

Looking in particular at plane projective geometry, where points are represented by lines through the origin in $\mathbb{R}^3$, a *projective line* is represented by a plane through the origin in $\mathbb{R}^3$. Since every pair of such planes intersect in a line through the origin, i.e., a point in $\mathbb{P}^2$, the notion of parallelity does not exist in projective geometry.

The projective space $\mathbb{P}^n$, embedded in $\mathbb{R}^{n+1}$, can also be visualized by means of an affine hyperplane in $\mathbb{R}^{n+1}$, e.g., $\Pi : x_{n+1} = 1$. Every line not parallel to Π intersects Π in a unique point. Besides these, $\mathbb{P}^n$ contains points corresponding to lines parallel to Π. The latter points of $\mathbb{P}^n$ are called *points at infinity* and may be identified with directions in Π. The set of points at infinity is called the *line at infinity*. To sum up, the projective space $\mathbb{P}^n$ can be visualized by a model formed by an n-dimensional affine space extended with points at infinity.

Perspective Transformations: Multiple View Geometry

For computer vision, *perspective transformations* play a crucial role. Specializing to three dimensions, consider two affine 3-dimensional hyperplanes Π and Π' in $\mathbb{R}^4$. If C is a point outside Π and Π', a mapping $P : \Pi \longrightarrow \Pi'$ is defined by considering the intersections of lines through C with Π and Π'. Then C is called the *focal point* of the *perspective transformation* P. In particular, one notes that if Π and Π' is nonparallel, then point at infinity of Π is mapped onto *ordinary* points of Π' and vice versa. If Π and Π' are parallel, then points at infinity are mapped onto points at infinity, and the perspective transformation is an affine transformation.

To sum up, Euclidean geometry is a special case of affine geometry, as follows from the fact that the group of isometries is a subgroup of the affine group. Affine geometry is a special case of projective geometry, since the affine transformations form a subgroup of the projective ones, characterized by leaving the line at infinite invariant.

Of particular importance to computer vision is the case of perspective transformations from three dimensions to two. These are singular, i.e., not one to one, contrary to the ones discussed above. They may be visualized by letting C tend to the plane Π', which in the limit gives a mapping $\Pi \longrightarrow \Pi \cap \Pi'$, from three to two dimensions. Moreover, the ambient space $\mathbb{R}^4$ becomes superfluous and can be replaced by the three-dimensional affine space Π.

Multiple view geometry deals with the situation where a number of such two-dimensional perspective images are known of a common three-dimensional object (cf. [4, 7]). With no further information, reconstruction is possible up to projective transformations. Having more geometric structure available, e.g., Euclidean information on the image planes and focal points, more structure can be obtained for the reconstruction, ideally making it Euclidean. This situation is termed *camera calibration*; see [7].

Some Other Geometries Embracing Euclidean Geometry

Algebraic Geometry. Besides hyperplanes, crucial roles in the study of Euclidean (as well as projective and affine) geometry are played by *conics*. While hyperplanes have equations involving first-order polynomials, conics are described by second-order polynomials. In two dimensions, this leads to the analytic geometry of conic sections, representing ellipses, hyperbolas, and parabolas. Algebraic geometry is the generalization of this to higher dimensions and higher-order polynomials, where Euclidean geometry thus falls out as a special case.

Riemannian geometry. Riemann geometry lives on a so-called Riemannian manifold, which, loosely speaking, is a space constructed by deforming and patching together Euclidean spaces according to certain rules, guaranteeing, e.g., *smoothnes*. Such a space enjoys notions of distance and angle but behaves in a curved, non-Euclidean manner. The simplest Riemannian manifold consists of $\mathbb{R}^n$ with a constant scalar product, leading to classical Euclidean geometry. Riemannian geometry plays a prominent role in general relativity.

Hilbert Spaces. Considering infinite sequences $(x_1, x_2, \ldots)$ instead of finite ones, $(x_1, x_2, \ldots, x_n)$, will lead to convergence problems whenever forming sums. Restricting oneself to sequences with $\sum_1^\infty x_i^2$ finite, a prototype of so-called Hilbert spaces is obtained. In a natural way, a scalar product is defined, leading to a distance measure, by means of which it is possible to prove, e.g., the infinite-dimensional analogue of the theorem of Pythagoras and many other theorems of Euclidean geometry.

References

1. Berger M (1987) Geometry I and II. Springer Universitext. Springer, Berlin/Heidelberg/New York/London/Paris/Tokyo
2. Coxeter HSM (1989) Introduction to geometry, 2nd edn. Wiley Classics Library, Wiley/New York
3. Eves H (1972) A survey of geometry. Allyn and Bacon, Boston
4. Faugeras O (1993) Three-dimensional computer vision. A geometric viewpoint. MIT, Cambridge/London
5. Faugeras O (1995) Stratification of three-dimensional vision: projective, affine, and metric representations. J Opt Soc Am A 12:465–484
6. Greenberg MJ (2008) Euclidean and non-Euclidean geomtries: Development and history. W.H. Freeman
7. Hartley R, Zisserman A (2003) Multiple view geometry in computer vision, 2nd edn. Cambridge University Press, Cambridge
8. Heath TL (1956) The thirteen books of Euclid's elements, 3 vols. Dover, New York
9. Hilbert D, Cohn-Vossen S (1952) Geometry and the imagination. Chelsea, New York
10. Veblen O, Young JW (1910–1918) Projective geometry I and II. Ginn and Co., Boston

Evolution of Robotic Heads

Michael R. M. Jenkin
Department of Computer Science and Engineering, York University, Toronto, ON, Canada

Related Concepts

▸Camera Calibration

Definition

Robotic heads are actively controlled camera platforms, typically designed to mimic the head and camera (eye) motions associated with humans. Early designs were built to study the role of eye and head motion in active vision systems [1]. Later designs are also used in the study of human-robot interaction.

Background

Cameras have a finite field of view and thus must be actively controlled in order to bring out of view portions of the scene into view and to track dynamic scene events. The need for active control of camera geometry becomes even more critical when multiple cameras are involved as changes in relative camera geometry can simplify stereo image processing and can be used to bring specific scene features within the tuning range of multiple-camera scene reconstruction algorithms.

The development of single camera robotic heads can be traced back to the Stanford Cart [2]. Binocular systems began to appear in the late 1980s and early 1990s (e.g., [3–7]). By the late 1990s, head designs were beginning to be driven by research interest in human-robot interaction (e.g., [8, 9]). Current head designs are typically designed to be anthropomorphic

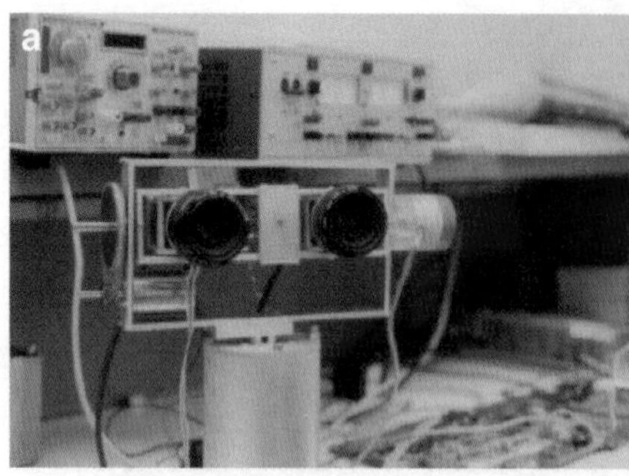

Harvard

b

TRISH 1

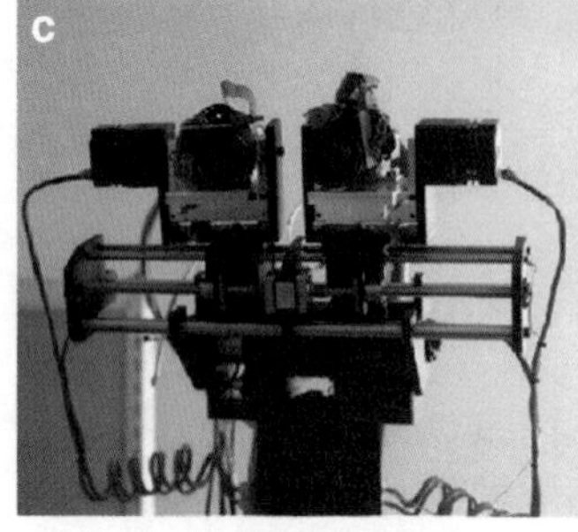

KTH

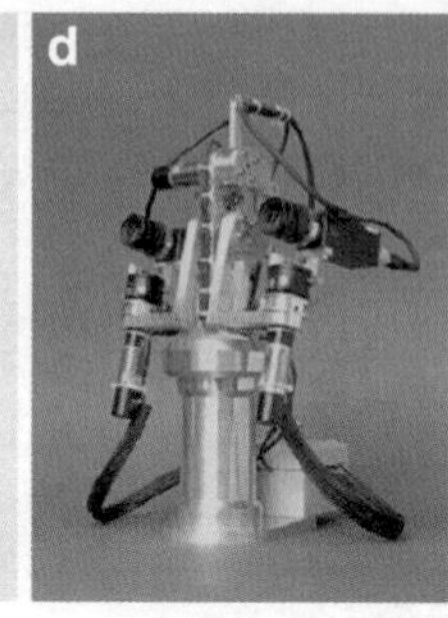

Yorick

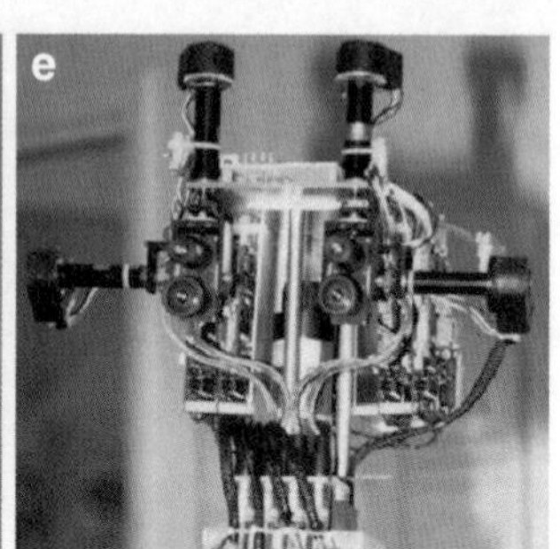

Cog

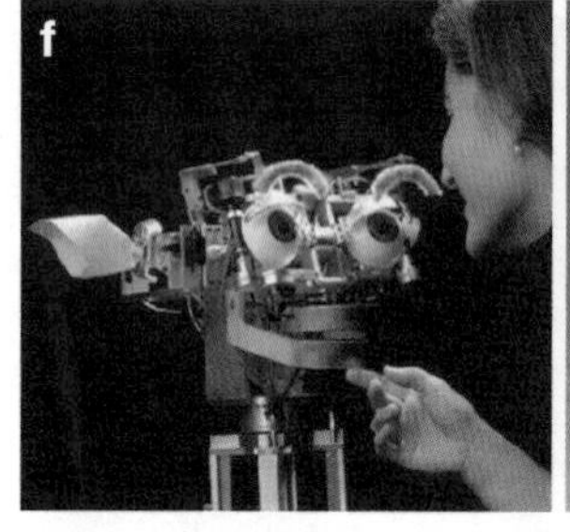

Kismet

MER camera mast

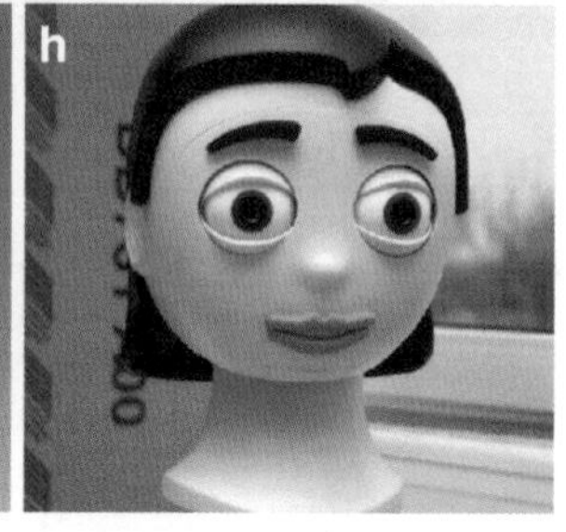

Flobi

Evolution of Robotic Heads, Fig. 1 The evolution of robotic heads. (**a**) Appears with the kind permission of J. Clark (**b**) and (**c**) Appear courtesy of M. Jenkin. (**d**) Appears with the kind permission of P. Sharkey. (**e**) Appears courtesy NAS/JPL-Caltech. (**f**) Appears with the kind permission of Blefeld University

(e.g., [10]) or to meet specific requirements of the sensors or the application (e.g., [11]). See Fig. 1 for examples.

Theory

Although there is a wide range of different head designs, stereo robotic heads are perhaps the most common. Such systems either have fixed relative geometry between the two cameras or the relative geometry is controllable. Individual cameras may also be equipped with controllable intrinsic camera settings. Heads are typically mounted on pan and tilt "necks" that drive the entire head to look at different portions of the scene. Controllable parameters can include:

- *Head pan.* Pan angle introduced by a robotic neck.
- *Head tilt.* Tilt angle introduced by a robotic neck.
- *Baseline.* The displacement between the left and right cameras.
- *Fixation point.* The point at which the optical axes of the left and right cameras intersect. The fixation point can be defined in various ways including the individual pan directions of the left and right cameras and the vergence/version angles.

– *Cyclotorsion*. The roll of each camera about its optical axis.
– *Zoom/focus/aperture*. Intrinsic settings of each camera.

Being able to control the relative pose of the two cameras provides the sensor the ability to bring different portions of space into alignment so that objects appear at the same position in both cameras. As stereo image processing is typically performed only over a limited range of disparities (image differences), changing the relative geometry between the cameras leads to the camera system attending to different regions of space. The region of space that is brought into alignment is known as the horopter (see [12] for a review of the geometry).

Early designs were strongly constrained by the mass of available camera housings and especially the mass of controllable focus/zoom lenses. This is particularly evident in the early designs shown in Fig. 1. More recent designs have taken advantage of the decreased size and mass of camera systems.

Many modern robotic heads have been developed to help study aspects of human-robot interaction. Such "social robots" are designed to not only use multiple sensors to reconstruct 3D environments but they often also provide a range of actuators to simulate human facial responses.

References

1. Ballard DH (1991) Animate vision. Artif Intell J 48:57–86
2. Moravec HP (1983) The Stanford Cart and the CMU rover. Proc IEEE 71:872–884
3. Krotkov E (1989) Active computer vision by cooperative focus and stereo. Springer, New York
4. Ferrier NJ, Clark JJ (1993) The Harvard binocular head. Int J Pattern Recognit Artif Intell 7(1):9–32
5. Milios E, Jenkin M, Tsotsos J (1993) Design and performance of TRISH, a binocular robot head with torsional eye movements. Int J Pattern Recognit Artif Intell 7(1): 51–68
6. Madden BC, von Seelen U (1995) PennEyes: a binocular active vision system. Technical report MS-CIS-95-37, University of Pennsylvania
7. Sharkey PM, Murray DW, McLauchlan PF, Brooker JP (1998) Hardware development of the Yorick series of active vision systems. Microprocess Microsyst 21:363–375
8. Scassellati B (1998) Eye finding via face detection for a foveated, active vision system. In: Proceedings of the AAAI-98, Madison, pp 969–976
9. Breazeal C, Scassellati B (1999) How to build robots that make friends and influence people. In: IEEE/RSJ international conference on robots and autonomous systems, Kyongju
10. Lütkebohle I, Hegel F, Schulz S, Hackel M, Wrede B, Wachsmuth S, Sagerer G (2010) The Bielefeld anthropomorphic robot head "Flobi". In: IEEE international conference on robotics and automation, Anchorage
11. Goldberg SB, Maimone MW, Matthies L (2002) Stereo vision and rover navigation software for planetary exploration. In: 2002 aerospace conference, Big Sky
12. Hansard M, Horaud R (2008) Cyclopean geometry of binocular vision. J Opt Soc Am A 25(9):2357–2369

Expectation Maximization Algorithm

Boris Flach and Vaclav Hlavac
Department of Cybernetics, Czech Technical University in Prague, Faculty of Electrical Engineering, Prague 6, Czech Republic

Synonyms

EM-algorithm

Related Concepts

▸Maximum Likelihood Estimation

Definition

The Expectation Maximization algorithm iteratively maximizes the likelihood of a training sample with respect to unknown parameters of a probability model under the condition of missing information. The training sample is assumed to represent a set of independent realizations of a random variable defined on the underlying probability space.

Background

One of the main paradigms of statistical pattern recognition and Bayesian inference is to model the relation between the observable features $x \in \mathcal{X}$ of an object and its hidden state $y \in \mathcal{Y}$ by a joint probability measure $p(x, y)$. This probability measure is often known only up to some parameters $\theta \in \Theta$. It is thus necessary to estimate these parameters from a training sample, which is assumed to represent a sequence of independent realizations of a random variable. If, ideally, these are realizations of pairs (x, y), then the corresponding estimation methods are addressed as *supervised* learning. It is, however, quite common that some of those variables describing the hidden state are latent. These latent variables are never observed in the training data. Therefore, it is necessary to marginalize over them in order to estimate the unknown parameters θ. Corresponding estimation methods are known as *unsupervised* learning. Moreover, especially in computer vision, the observation x *and* the hidden state y both may have a complex structure. The latter can be, e.g., a segmentation, a depth map, or a similar object. Consequently, it is often not feasible to provide the complete information y for the realizations in the sample. This means to estimate the parameters in the situation of missing information. The EM algorithm is a method searching for maximum likelihood estimates of the unknown parameters under such conditions.

Theory

All the situations described in the previous section can be treated in a uniform way by assuming the training sample as a set of independent realizations of a random variable.

Let Ω be a finite sample space, $\mathcal{F}$ be its power set, and $p_\theta : \mathcal{F} \to \mathbb{R}_+$ be a probability measure defined up to unknown parameters $\theta \in \Theta$. Let $X : \Omega \to \mathcal{X}$ be a random variable and $T = (x_1, x_2, \ldots, x_n)$ be a sequence of independent realizations of X (see, e.g., [1, 2] for a formal definition of independent realizations). The maximum likelihood estimator provides estimates of the unknown parameters θ by maximizing the probability of T:

$$\theta^* = \underset{\theta}{\operatorname{argmax}} \prod_{i=1}^{n} p_\theta(\Omega_i), \tag{1}$$

where Ω_i denotes the pre-image $\{\omega \in \Omega \mid X(\omega) = x_i\}$. If the logarithm is taken, the task reads equivalently:

$$\theta^* = \underset{\theta}{\operatorname{argmax}}\, L(x_1, \ldots, x_n, \theta)$$

$$= \underset{\theta}{\operatorname{argmax}} \sum_{i=1}^{n} \log \sum_{\omega \in \Omega_i} p_\theta(\omega). \tag{2}$$

Remark 1. It is often assumed that Ω is a Cartesian product $\Omega = \mathcal{X} \times \mathcal{Y}$ and that X simply projects onto the first component $X(x, y) = x$. Then the probability $p_\theta(\Omega_i) = \sum_{y \in \mathcal{Y}} p_\theta(x_i, y)$ is nothing but the marginalization over all possible y. This special case will be considered in an example below. □

The optimization task (Eq. 2) is often complicated and hardly solvable by standard optimization methods – either because the objective function is not concave or because θ represents a set of parameters of different natures. Suppose, however, that the task of parameter estimation is feasible if complete information, i.e., a set of realizations of $\omega \in \Omega$, is available. This applies in particular if the corresponding simpler objective function $\sum_i \log p_\theta(\omega_i)$ is concave with respect to θ or if the task decomposes into simpler, independent optimization tasks with respect to individual components of a parameter collection.

The key idea of the Expectation Maximization algorithm is to exploit this circumstance and to solve the optimization task (Eq. 2) by iterating the following two feasible tasks:

1. Given a current estimate of θ, determine the missing information, i.e., $p_\theta(\omega | \Omega_i)$, for each element $x_i \in T$.
2. Given the complete information, solve the corresponding estimation task, resulting in an improved estimate of θ.

To further substantiate this idea of "iterative splitting" of task (Eq. 2), it is convenient to introduce nonnegative auxiliary variables $\alpha_i(\omega)$, $\omega \in \Omega_i$, for each element x_i of the learning sample T such that they fulfil:

$$\sum_{\omega \in \Omega_i} \alpha_i(\omega) = 1, \quad \forall i = 1, 2, \ldots, n. \tag{3}$$

These variables α_i can be seen as (so far arbitrary) posterior probabilities $p(\omega | \Omega_i)$ for $\omega \in \Omega_i$, given a realisation x_i. The log-likelihood of a realization x_i can be written by their use as:

$$\log p_\theta(\Omega_i) = \sum_{\omega \in \Omega_i} \alpha_i(\omega) \log p_\theta(\Omega_i)$$

$$= \sum_{\omega \in \Omega_i} \alpha_i(\omega) \log p_\theta(\omega) - \sum_{\omega \in \Omega_i} \alpha_i(\omega) \log \frac{p_\theta(\omega)}{p_\theta(\Omega_i)}, \quad (4)$$

where the first equality follows directly from (Eq. 3). The log-likelihood of the training sample can be therefore expressed equivalently as:

$$L(x_1, \ldots, x_n, \theta) = \sum_{i=1}^{n} \sum_{\omega \in \Omega_i} \alpha_i(\omega) \log p_\theta(\Omega_i) = \sum_{i=1}^{n} \sum_{\omega \in \Omega_i} \alpha_i(\omega) \log p_\theta(\omega) - \sum_{i=1}^{n} \sum_{\omega \in \Omega_i} \alpha_i(\omega) \log p_\theta(\omega|\Omega_i). \quad (5)$$

The expression as a whole does not depend on the specific choice of the auxiliary variables α, whereas the minuend and subtrahend do. Moreover, note that the minuend is nothing but the likelihood of a sample of complete data, if the α are interpreted as the missing information, i.e., posterior probabilities for $\omega \in \Omega_i$ given the observation x_i.

Starting with some reasonable choice for the initial $\theta^{(0)}$ the likelihood is iteratively increased by alternating the following two steps. The (E)xpectation step calculates new α such that whatever new θ will be chosen subsequently, the subtrahend will not increase. The (M)aximization step relies on this and maximizes the minuend only, avoiding to deal with the subtrahend:

$$\text{E-step} \quad \alpha_i^{(t)}(\omega) = p_{\theta^{(t)}}\big(\omega|\Omega_i\big) \quad (6)$$

$$\text{M-step} \quad \theta^{(t+1)} = \operatorname*{argmax}_{\theta} \sum_{i=1}^{n} \sum_{\omega \in \Omega_i} \alpha_i^{(t)}(\omega) \log p_\theta(\omega). \quad (7)$$

From the conceptual point of view the E-step can be seen as inference – it calculates the missing data, i.e., the posterior probabilities $p_{\theta^{(t)}}\big(\omega|\Omega_i\big)$ for each element x_i in the training sample. The M-step utilizes these posterior probabilities for a supervised learning step. The names themselves stem from a rather formal view: the E-step calculates the α and therefore the objective function in (Eq. 7) which has the form of an expectation of $\log p_\theta(\omega)$. The computation of this objective function is sometimes considered to be a part of the E-step. The name for the M-step is obvious.

It is easy to see that the likelihood is monotonically increasing: The choice (Eq. 6) for α guarantees that the subtrahend in (Eq. 5) can only decrease whatever new θ will be chosen in the subsequent M-step. This follows from the inequality:

$$\sum_{\omega \in \Omega_i} p_\theta(\omega|\Omega_i) \log p_{\theta'}(\omega|\Omega_i) \leqslant \sum_{\omega \in \Omega_i} p_\theta(\omega|\Omega_i) \log p_\theta(\omega|\Omega_i) \ \ \forall \theta' \neq \theta. \quad (8)$$

Since the M-step chooses the new θ so as to maximize the minuend, the likelihood can only increase (or stay constant). Another convenient way to prove monotonicity of the EM algorithm can be found in [3, 4]. These tutorials consider the EM algorithm as the maximization of a lower bound of the likelihood.

It remains unclear whether the global optimum of the likelihood is reached in a fix-point of the algorithm. Moreover, it happens quite often that the M-step is infeasible for complex models p_θ. Then a weaker form of the EM algorithm is used by choosing $\theta^{(t+1)}$ so as to guarantee an increase of the objective function of the M-step.

The derivation of the concept of the EM algorithm was given here for a discrete probability space and discrete random variables. It can be however generalized for uncountable probability spaces and random variables X with continuous probability density.

Example 1. The EM algorithm is often considered for the following special case. The sampling space Ω is a Cartesian product $\Omega = \mathcal{X} \times \mathcal{Y}$ and the random variable X simply projects onto the first component $X(x, y) = x$. The parameters $\theta \in \Theta$ of the probability $p_\theta(x, y)$ are to be estimated given a sequence of independent realizations of x. In this special case, the log-likelihood has the form:

$$L = \sum_{i=1}^{n} \log \sum_{y \in \mathcal{Y}} p_\theta(x_i, y). \quad (9)$$

Its decomposition (Eq. 5) is:

$$L = \sum_{i=1}^{n} \sum_{y \in \mathcal{Y}} \alpha_i(y) \log p_\theta(x_i, y) - \sum_{i=1}^{n} \sum_{y \in \mathcal{Y}} \alpha_i(y) \log p_\theta(y|x_i). \quad (10)$$

The EM algorithm itself then reads:

$$\text{E-step } \alpha_i^{(t)}(y) = p_{\theta^{(t)}}(y|x_i) \quad (11)$$

$$\text{M-step } \theta^{(t+1)} = \operatorname*{argmax}_\theta \sum_{i=1}^{n} \sum_{y \in \mathcal{Y}} \alpha_i^{(t)}(y) \log p_\theta(x_i, y). \quad (12)$$

History and Applications

The classic paper [5] is often cited as the first one introducing the EM algorithm in its general form. It should be noted, however, that the method was introduced and analyzed substantially earlier for a broad class of pattern recognition tasks in [6] and for exponential families in [7].

A comprehensive discussion of the EM algorithm can be found in [8] and in the context of pattern recognition in [9, 10]. Standard application examples are parameter estimation for mixtures of Gaussians [8] and the Mean Shift algorithm [11]. Another important application is parameter estimation for Hidden Markov Models. This model class is extensively used for automated speech recognition. The corresponding EM algorithm is known as Baum-Welch algorithm in this context [12]. Rather complex applications of the EM algorithm arise in the context of parameter estimation for Markov Random Fields [13].

References

1. Meintrup D, Schäffler S (2005) Stochastik. Springer, Berlin
2. Papoulis A (1990) Probability and Statistics. Prentice-Hall, Englewood Cliffs
3. Minka T (1998) Expectation-maximization as lower bound maximization. Tutorial, MIT, Cambridge, MA
4. Dellaert F (2002) The expectation maximization algorithm. Technical report GIT-GVU-02-20, Georgia Institute of Technology
5. Dempster AP, Laird NM, Rubin DB (1977) Maximum likelihood from incomplete data via the EM algorithm. J R Stat Soci Ser B 39(1):1–38
6. Schlesinger MI (1968) The interaction of learning and self-organization in pattern recognition. Kibernetika 4(2):81–88
7. Sundberg R (1974) Maximum likelihood theory for incomplete data from an exponential family. Scand J Stat 1(2): 49–58
8. McLachlan GJ, Krishnan T (1997) The EM algorithm and extensions. Wiley, New York
9. Schlesinger MI, Hlavac V (2002) Ten lectures on statistical and structural pattern recognition. Kluwer Academic Publishers, Dordrecht
10. Bishop CM (2006) Pattern recognition and machine learning. Springer, New York
11. Cheng Y (1995) Mean shift, mode seeking, and clustering. IEEE Trans Pattern Anal Mach Intell 17(8):790–799
12. Jelinek F (1998) Statistical methods for speech recognition. MIT, Cambridge, MA
13. Li SZ (2009) Markov random field modeling in image analysis. 3rd edn. Advances in pattern recognition. Springer, London

Exploration: Simultaneous Localization and Mapping (SLAM)

Samunda Perera[1], Nick Barnes[2] and Alexander Zelinsky[3]
[1]Canberra Research Laboratory, NICTA, Canberra, Australia
[2]Australian National University, Canberra, Australia
[3]CSIRO, Information Sciences, Canberra, Australia

Synonyms

Concurrent mapping and localization (CML); Visual SLAM

Related Concepts

►Structure-from-Motion (SfM)

Definition

Exploration refers to gathering data about an environment through sensors in order to discover its structure. A fundamental technique for exploration of an unknown environment is Simultaneous Localization and Mapping (SLAM). SLAM is a technique that supports the incremental building of a 3-D map representation of an environment while also using the same incremental map to accurately localize the observer in order to minimize errors in the map building.

SLAM techniques can be implemented with laser range finders, monocular vision, stereo vision, and RGB-D cameras.

Background

An accurate representation of an environment is highly useful for autonomous and human assistive applications. Such a representation can be a high level minimalistic map derived from a sparse set of landmarks or a detailed dense 3-D model (e.g., Fig. 1). An accurate map can be used to perform useful tasks such as answering the question, "Where am I ?" and enables navigation and guiding tasks such as "How to get there?." When dealing with unknown environments, the map representation must be acquired through an exploration phase. With SLAM the map can be built through a data gathering or map building phase, by either online or off-line methods. In online methods, the map is incrementally built, and location within the map is simultaneously calculated. With off-line methods the data is collected and then post-processed to yield an accurate map representation.

The key process of SLAM is an optimization algorithm. Noisy observations are made about the landmarks from different viewpoints (poses), and the task is to solve the optimization problem of estimating the position of each landmark and the set of camera viewpoints which best explain the observations. In off-line mapping this is a global optimization problem where all observations are available (e.g., Bundle Adjustment (BA) [2] applied to a full image dataset). In online mapping this is performed sequentially, using observations available up to and including the current observation. Thus, in online mapping, an incremental version of the map is available as the map is being constructed.

Two different methods are used in online mapping, namely, filtering-based and keyframe-based methods. Filtering-based methods are based on the following nature of sequential mapping. As camera observations/measurements are noisy, there is induced error in the newly identified landmarks. Later when one localizes with respect to the current map, again there is induced error in this new pose estimate due to error in observation and error in these new landmarks. So in subsequent new landmark additions to the map, there will be error in landmarks due to both measurement errors and the camera viewpoint estimation error.

Exploration: Simultaneous Localization and Mapping (SLAM), Fig. 1 A 3-D map generated by RGB-D Mapping by Henry et al. [1] (Copyright © 2012 SAGE. Reprinted by Permission of SAGE)

Due to the common error associated with camera viewpoint, estimates of the landmarks are all necessarily correlated with each other [3]. Therefore a consistent full solution to the combined localization and mapping problem requires a joint state composed of the camera pose and every landmark position, to be updated following each landmark observation and hence the name Simultaneous Localization And Mapping.

In contrast keyframe-based methods do not propagate a probability distribution over time but employ a subset of image frames, called keyframes, in the map to perform efficient global optimization of the map and keyframe poses within a practical time limit. Although non keyframes are discarded in the optimization, it permits a large number of image observations per keyframe to be efficiently processed.

Visual SLAM is possible with both monocular and stereo cameras, and these systems are termed monoSLAM and stereoSLAM, respectively. In addition, RGB-D Mapping [1] which generates a dense 3-D map with RGB-D cameras has recently been introduced. It should be noted that, despite the computational effort, the use of cameras provides several unique advantages over other sensor types such as laser rangefinders and sonars. They are inexpensive, low power, compact, and capture scene information up to very large distances.

Theory

A complete SLAM system performs following operations in a cycle. First a set of new landmarks are identified in the environment, initialized (*new landmark/feature initialization*), and inserted into the map. Upon moving to a new location, *observations* are made and matched with the map landmarks, and this is referred to as *data association*. Here, *relocalization* methods are used to recover from possible tracking failures. Next, a prior estimate of the observer motion is obtained (*motion estimation*). Then the estimates of landmark positions and sensor pose that best explains the data are sought (*optimization*). If reobserving a previously mapped portion of the environment, care should be taken to ensure correct closure of the trajectory loop and consistency of the map (*loop closure*). Algorithm 1 illustrates this process (Note that as online mapping is an incremental process, Algorithm 1 assumes for generality some initial map is available. This may be the result of a single observation and new landmark initialization from the starting position).

Input: Initial set of landmarks and initial state of observer
Output: SLAM Map and observer state at each time instant
repeat
(1) Observation
(2) Data association
if tracking failure detected **then**
(3) Relocalization
(4) Motion estimation
(5) Optimization
if loop closure detected **then**
(6) Loop closure correction
(7) New landmark initialization

Algorithm 1: SLAM operation

An example situation is illustrated by Fig. 2. At time k, the true state of the observer is given by $\mathbf{x}_k$. The observer makes observations $\mathbf{z}_{k,i}, \mathbf{z}_{k,j}, \mathbf{z}_{k,p}$ of static landmarks whose true states are given by $\mathbf{m}_i, \mathbf{m}_j$, and $\mathbf{m}_p$, respectively. In the landmark selection process, it also discovers a new landmark whose true state is given by $\mathbf{m}_q$. Based on the corresponding measurement $\mathbf{z}_{k,q}$, the state of this new landmark is then initialized and inserted into the map. At time $k+1$, the observer has moved to a state $\mathbf{x}_{k+1}$ under the control input $\mathbf{u}_{k+1}$. The observer has been able to reobserve two existing landmarks in the map, namely, $\mathbf{m}_p$ and $\mathbf{m}_q$. Both the corresponding measurements $\mathbf{z}_{k+1,p}, \mathbf{z}_{k+1,q}$ and control input $\mathbf{u}_{k+1}$ can be utilized to obtain a prior estimate of the incremental motion of the observer between time k and $k+1$.

At time $k+n$, the observer has travelled along a long trajectory loop and is revisiting a previously mapped part of the environment, namely, landmarks $\mathbf{m}_i$ and $\mathbf{m}_j$. However, as can be seen from Fig. 2, the estimated state of the observer might not show a closed loop. Loop closure methods detect this situation and correct the state of the observer and the states of landmarks in the map.

The following subsections details the key stages of the SLAM system operation as given in Algorithm 1.

Observation

A new image frame is fetched from the camera and features (e.g., interest points, lines) which can be potential landmarks are extracted.

Data Association

Data association refers to finding correspondences between observations and landmarks. There are a number of methods in this regard. The active search/covariance-driven gating approach projects individual covariance predictions of landmarks into the observation image and limits the observation area based on suitable Mahalanobis distance. The Joint Compatibility Branch and Bound (JCBB) method [4] uses the fact that observation prediction errors are correlated with each other and hence calculates the joint probability of any set of correspondences. Active matching [5], on the other hand, considers the joint distribution and uses information theory to guide which landmarks to measure and when to measure them.

Relocalization

In case of rapid (unmodelled) motion, blur, and occlusion, camera pose tracking can fail and will result in subsequent corruption of the map. Relocalization refers to automatically detecting and recovering from such tracking failures to preserve the integrity of the map. Tracking failure can be detected by considering the percentage of unsuccessful observations and large uncertainty in the camera pose. The pose of the camera is recovered by establishing correspondence between the image and the map and solving the resulting perspective pose estimation problem [6].

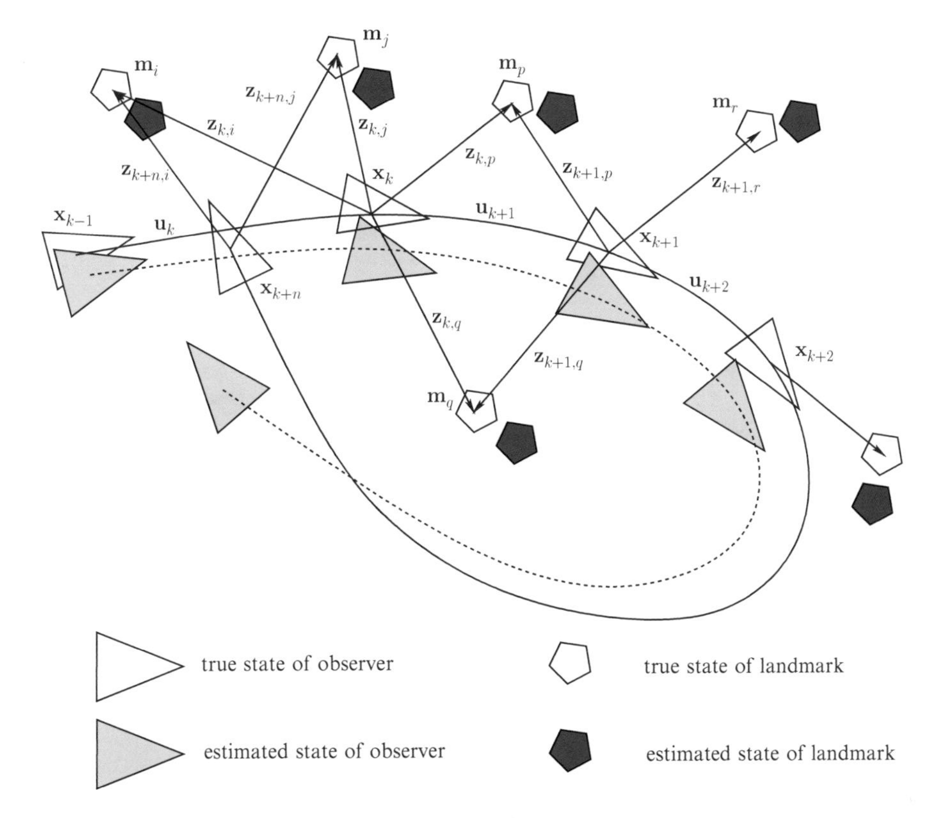

Exploration: Simultaneous Localization and Mapping (SLAM), Fig. 2 SLAM problem

Motion Estimation

The observer's movement can be predicted based on odometry information or by making smooth motion assumptions (e.g., a constant velocity motion model). This information stands as a prior estimate of the observer state and covariance which is used later in the optimization stage. An example of such observer-state covariance computation is given by Eq. 5.

Optimization: Filtering-Based SLAM

Probabilistically the SLAM problem requires the computation of the joint posterior density of the landmark locations and observer (camera) state at time k, given the available observations and control inputs up to and including time k, together with the initial state of the observer [7–9].

This probability distribution can be stated as $P(\mathbf{x}_k, \mathbf{m}|\mathbf{Z}_{0:k}, \mathbf{U}_{0:k}, \mathbf{x}_0)$ where the quantities are defined as below (see Fig. 2).

$\mathbf{x}_k$ = observer state vector

$\mathbf{m}$ = vector describing the set of all landmark locations, i.e., landmark states $\mathbf{m}_i$

$\mathbf{Z}_{0:k}$ = set of all landmark observations $\mathbf{z}_{i,k}$

$\mathbf{U}_{0:k}$ = history of control inputs $\mathbf{u}_k$

Given an estimate for the problem at time $k-1$ (i.e., $P(\mathbf{x}_{k-1}, \mathbf{m}|\mathbf{Z}_{0:k-1}, \mathbf{U}_{0:k-1})$) a recursive solution to the problem can be obtained by employing a state transition model and an observation model which characterize the effect of the control input and observation, respectively.

Motion model: $P(\mathbf{x}_k|\mathbf{x}_{k-1}, \mathbf{u}_k)$
Observation model: $P(\mathbf{z}_k|\mathbf{x}_k, \mathbf{m})$

Therefore, the SLAM algorithm can now be expressed as a standard two-step recursive prediction (time update) correction (measurement update) form as below:

Time Update

$$P(\mathbf{x}_k, \mathbf{m}|\mathbf{Z}_{0:k-1}, \mathbf{U}_{0:k}, \mathbf{x}_0) = \int P(\mathbf{x}_k|\mathbf{x}_{k-1}, \mathbf{u}_k) \times P(\mathbf{x}_{k-1}, \mathbf{m}|\mathbf{Z}_{0:k-1}, \mathbf{U}_{0:k-1}, \mathbf{x}_0) d\mathbf{x}_{k-1} \quad (1)$$

Measurement Update

$$P(\mathbf{x}_k, \mathbf{m}|\mathbf{Z}_{0:k}, \mathbf{U}_{0:k}, \mathbf{x}_0) = \frac{P(\mathbf{z}_k|\mathbf{x}_k, \mathbf{m}) P(\mathbf{x}_k, \mathbf{m}|\mathbf{Z}_{0:k-1}, \mathbf{U}_{0:k}, \mathbf{x}_0)}{P(\mathbf{z}_k|\mathbf{Z}_{0:k-1}, \mathbf{U}_{0:k})} \quad (2)$$

Two main solution methods to the probabilistic SLAM problem include Extended Kalman Filter (EKF)-based SLAM [3] and Rao-Blackwellized Filter (RBF)-based SLAM [10, 11].

Extended Kalman Filter (EKF)-Based SLAM

As noted, the solution to the SLAM problem requires a joint state composed of the observer state and the state of every landmark, to be updated following each landmark observation. This motivates to use the Kalman filter which provides a recursive solution to the discrete data linear filtering problem. When the motion model and measurement errors/noise are independent white Gaussian, the Kalman filter computes an optimal gain which minimizes mean-square error of the posterior estimate of the system. The Gaussian distribution is completely characterized by its mean and covariance. Here correlations between observer state and different landmarks can be properly represented in the total system covariance matrix of the Kalman filter. As the system would generally be nonlinear, the Extended Kalman Filter (EKF) is used.

Propagation of Uncertainty A fundamental problem in estimating covariance matrices is how to propagate uncertainty through a function. Consider a nonlinear function f which acts on the vector $\mathbf{a}$ of input variables to produce the output vector $\mathbf{b}$:

$$\mathbf{b} = f(\mathbf{a}) \quad (3)$$

Under affine approximation, the error covariance of the output vector $P_{\mathbf{bb}}$ is obtained in terms of the error covariance of the input vector $P_{\mathbf{aa}}$ by Eq. 4 (The covariance between two arbitrary vectors $\mathbf{a}$ and $\mathbf{b}$ is defined as $P_{\mathbf{ab}} = E[(\mathbf{a} - E[\mathbf{a}])(\mathbf{b} - E[\mathbf{b}])^T]$):

$$P_{\mathbf{bb}} = \frac{\partial f}{\partial \mathbf{a}} P_{\mathbf{aa}} \frac{\partial f}{\partial \mathbf{a}}^T \quad (4)$$

State Prediction The general form of the EKF prediction steps can be customized for SLAM to yield faster computation. Note that there is no need to perform a time update for landmark states $\mathbf{m}_i : i = 1 \ldots n$ if the landmarks are stationary. Hence, only the observer's state, $\mathbf{x}_k$, is predicted ahead. The new error covariance of the observer's predicted state should also be computed. In addition, though there is no change in error covariance of landmarks, there will be change in error covariance between each landmark and observer state (since observer state changed). Given below are the new time update equations for the covariance items that require an update.

Notation:

$\{.\}^-$ = time update (prior) estimate
$\mathbf{n}$ = observer motion model noise
$P_{\mathbf{nn}}$ = observer motion model noise covariance
f_v = observer state transfer function $\mathbf{x}_{k+1} = f_v(\mathbf{x}_k, \mathbf{n})$ such that $\mathbf{x}_{k+1}^- = f_v(\mathbf{x}_k, \mathbf{0})$

$$P_{\mathbf{x}_{k+1}\mathbf{x}_{k+1}}^- = \frac{\partial f_v}{\partial \mathbf{x}_k} P_{\mathbf{x}_k\mathbf{x}_k} \frac{\partial f_v}{\partial \mathbf{x}_k}^T + \frac{\partial f_v}{\partial \mathbf{n}} P_{\mathbf{nn}} \frac{\partial f_v}{\partial \mathbf{n}}^T \quad (5)$$

$$P_{\mathbf{x}_{k+1}\mathbf{m}_i}^- = \frac{\partial f_v}{\partial \mathbf{x}_k} P_{\mathbf{x}_k\mathbf{m}_i} \quad (6)$$

The computational complexity of the EKF-SLAM solution grows quadratically with the number of landmarks. This is due to the calculations involved in the EKF update steps.

Rao-Blackwellized Filter (RBF)-Based SLAM

Rao-Blackwellized filter based SLAM methods include FastSLAM [10] and FastSLAM 2 [11] algorithms. The algorithms utilize a key point in SLAM, i.e., given the observer path/trajectory (if the observer path is assumed correct), different landmark measurements are independent of each other. So a particle filter is used to estimate the observer's path and for each particle a map is maintained. Since landmarks within this map are independent/uncorrelated, they can be represented with separate low-dimensional EKFs. Therefore, it has linear complexity rather than the quadratic complexity of EKF-SLAM.

Landmark Representation As noted, probabilistic SLAM (EKF/RBF SLAM) requires the uncertainty of landmarks to be modelled. A landmark's uncertainty modelling depends upon the parameterization used to represent its state. The simplest way to parametrize a landmark is by the common three-dimensional Cartesian coordinates $\mathbf{m}_i^{car} = (X_i, Y_i, Z_i)^T$. This is applicable in stereo camera SLAM systems where the landmark state can be obtained by triangulation, and the uncertainty follows the common Gaussian assumption. However, in single camera SLAM systems, in which it is not possible to estimate a landmark's state from a single measurement, there is large uncertainty in the depth direction which cannot be modelled as Gaussian. Therefore, alternative parameterizations are proposed to alleviate this issue.

Civera et al. [12] represent a feature using inverse depth parameterization as

$$\mathbf{m}_i^{idp} = (x_i, y_i, z_i, \theta_i, \phi_i, \rho_i)^T, \tag{7}$$

where x_i, y_i, z_i represent the camera position from which the feature was first observed, θ_i, ϕ_i represent the azimuth and elevation of the corresponding ray, and ρ_i represents the inverse depth to the landmark along this ray. The relationship of $\mathbf{m}_i^{idp}$ to $\mathbf{m}_i^{car}$ is as follows:

$$\mathbf{m}_i^{car} = \begin{pmatrix} X_i \\ Y_i \\ Z_i \end{pmatrix} = \begin{pmatrix} x_i \\ y_i \\ z_i \end{pmatrix} + \frac{1}{\rho_i} \begin{pmatrix} \cos\phi_i \sin\theta_i \\ -\sin\phi_i \\ \cos\phi_i \cos\theta_i \end{pmatrix}. \tag{8}$$

Marzorati et al. [13] use the homogenous coordinates

$$\mathbf{m}_i^{isp} = (x_i, y_i, z_i, \omega_i)^T, \tag{9}$$

which relates with $\mathbf{m}_i^{car}$ as

$$\mathbf{m}_i^{car} = \begin{pmatrix} X_i \\ Y_i \\ Z_i \end{pmatrix} = \frac{1}{\omega_i} \begin{pmatrix} x_i \\ y_i \\ z_i \end{pmatrix} \tag{10}$$

where ω_i is the inverse scale.

Although the number of parameters per landmark has increased from three of the minimal representation ($\mathbf{m}_i^{car}$), $\mathbf{m}_i^{idp}$ and $\mathbf{m}_i^{isp}$ are able to represent the landmark uncertainty as Gaussian and make the measurement equation more linear.

Optimization: Keyframe-Based SLAM

Keyframe-based SLAM splits camera pose tracking and mapping into separate processes. The accuracy of camera pose estimation is increased by using measurements from large numbers of landmarks per image frame. The map is usually estimated by performing bundle adjustment/optimization over only a set of frames (keyframes) since measurements from nearby image frames provides redundant and therefore less information for a given computational budget [14]. Here a new keyframe can be selected and inserted to the map based on a sliding window of most recent camera poses or by ensuring a minimum distance to the pose of the nearest keyframe already in the map.

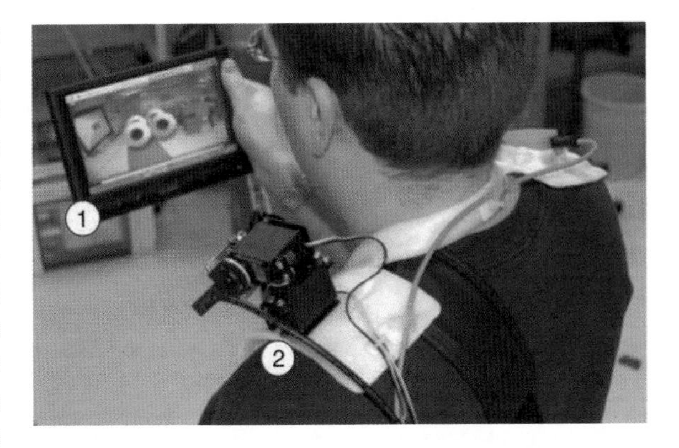

Exploration: Simultaneous Localization and Mapping (SLAM), Fig. 3 Wearable SLAM system for augmented reality by Castle et al. [18] – (1) handheld display with camera mounted on the back (2) active camera capable of pan and tilt (Reprinted from [18], with permission from Elsevier)

Loop Closure

Ideally when a SLAM system reobserves a previously mapped region of an environment, it should correctly recognize the corresponding landmarks in the map. However, particularly in long trajectory loops, landmark states in two such regions of the map can be incompatible given their uncertainty estimates. Therefore, loop closure methods are utilized to align such regions and make the map globally coherent. These include image-to-image matching [15], image-to-map matching [16], and hybrid methods [17].

New Landmark Initialization

Significant and distinguishable parts of the environment are selected as landmarks/features. In visual SLAM, these typically include interest points and edges/lines in the observed image. Upon selecting suitable landmarks, their position states and covariances are initialized and inserted into the map. Adding a new

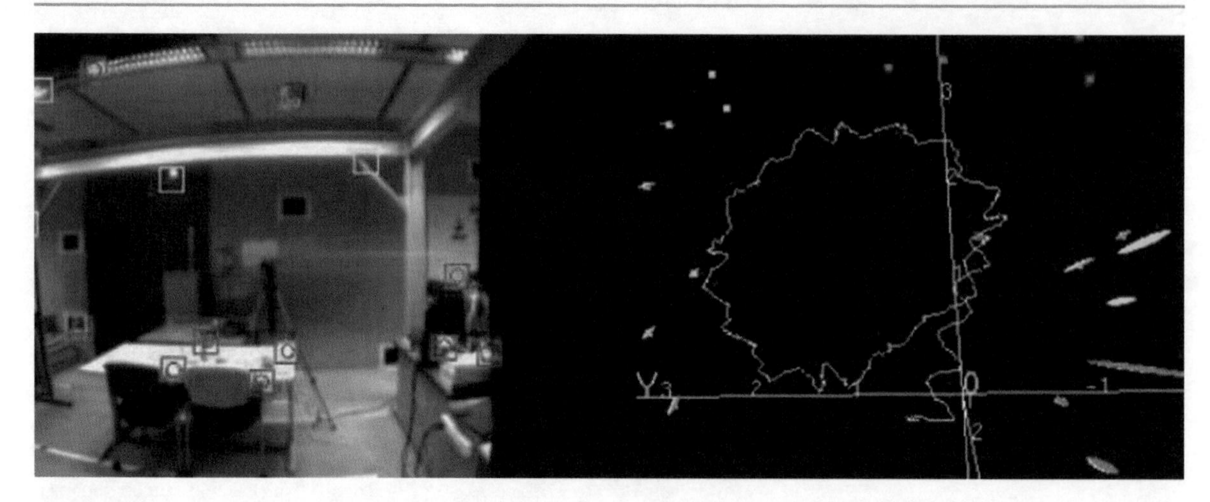

Exploration: Simultaneous Localization and Mapping (SLAM), Fig. 4 A snapshot from MonoSLAM [20] as a humanoid robot walks in a circular trajectory. Here rectangular patches on the camera image (*left*) denote landmarks which are tracked interest points. In the 3-D map (*right*), *yellow* trace is the estimated robot trajectory and ellipsoids show landmark location uncertainties (Copyright © 2007 IEEE. All rights reserved. Reprinted, with permission, from [20])

landmark $\mathbf{m}_{new}$ to the map is a function of observer pose $\mathbf{x}_k$ and new landmark observation $\mathbf{z}$ which is given by $\mathbf{m}_{new} = f_{new}(\mathbf{x}_k, \mathbf{z})$. To add a new landmark, one needs to know the error covariance of the new landmark with other landmarks $\mathbf{m}_{oth}$ and the observer $\mathbf{x}_k$. This can be calculated using the system uncertainty at that time as below:

$$P_{\mathbf{m}_{new}\mathbf{m}_{new}} = \frac{\partial f_{new}}{\partial \mathbf{x}_k} P_{\mathbf{x}_k\mathbf{x}_k} \frac{\partial f_{new}}{\partial \mathbf{x}_k}^T + \frac{\partial f_{new}}{\partial \mathbf{z}} R \frac{\partial f_{new}}{\partial \mathbf{z}}^T \tag{11}$$

$$P_{\mathbf{x}_k\mathbf{m}_{new}} = P_{\mathbf{x}_k\mathbf{x}_k} \frac{\partial f_{new}}{\partial \mathbf{x}_k}^T \tag{12}$$

$$P_{\mathbf{m}_{oth}\mathbf{m}_{new}} = P_{\mathbf{m}_{oth}\mathbf{x}_k} \frac{\partial f_{new}}{\partial \mathbf{x}_k}^T \tag{13}$$

Note R denotes the error covariance of the observation $\mathbf{z}$.

Application

Visual SLAM approaches have been successfully used in large scale outdoor environment mappings, and due to the human-like visual sensing of the camera, it has found applications in augmented reality applications [18] (see Fig. 3).

Open Problems

Despite recent work [19], the performance of visual SLAM is not sufficiently robust in dynamic environments where multiple moving objects are present in the scene. This is similar to the motion segmentation problem in computer vision. In addition, mapping with RGB-D cameras [1] is an emerging field with the recent introduction of affordable RGB-D cameras.

Experimental Results

Figure 4 gives a snapshot of one of the first real-time MonoSLAM system in operation [20].

References

1. Henry P, Krainin M, Herbst E, Ren X, Fox D (2012) RGB-D mapping: using Kinect-style depth cameras for dense 3D modeling of indoor environments. Intern J Robot Res 31(5):647–663
2. Hartley RI, Zisserman A (2004) Multiple view geometry in computer vision, 2nd edn. Cambridge University Press, Cambridge. ISBN: 0521540518
3. Smith R, Self M, Cheeseman P (1990) Autonomous robot vehicles. In: Estimating uncertain spatial relationships in robotics. Springer, New York, pp 167–193
4. Neira J, Tardos JD (2001) Data association in stochastic mapping using the joint compatibility test. IEEE Trans Robot Autom 17(6):890–897

5. Chli M, Davison AJ (2008) Active matching. In: Proceedings of 10th European conference on computer vision (ECCV'08), Marseille
6. Williams B, Klein G, Reid I (2007) Real-time SLAM relocalisation. In: Proceedings of IEEE 11th international conference on computer vision ICCV 2007, Rio de Janeiro, pp 1–8
7. Durrant-Whyte H, Bailey T (2006) Simultaneous localization and mapping: part i. IEEE Robot Autom Mag 13(2):99–110
8. Bailey T, Durrant-Whyte H (2006) Simultaneous localization and mapping (SLAM): part ii. IEEE Robot Autom Mag 13(3):108–117
9. Thrun S, Burgard W, Fox D (2005) Probabilistic robotics. In: Intelligent robotics and autonomous agents. MIT, Cambridge
10. Montemerlo M, Thrun S, Koller D, Wegbreit B (2002) FastSLAM: a factored solution to the simultaneous localization and mapping problem. In: Proceedings of the AAAI national conference on artificial intelligence, Edmonton. AAAI, pp 593–598
11. Montemerlo M, Thrun S, Koller D, Wegbreit B (2003) FastSLAM 2.0: an improved particle filtering algorithm for simultaneous localization and mapping that provably converges. In: Proceeding of the international conference on artificial intelligence (IJCAI), Acapulco, pp 1151–1156
12. Civera J, Davison AJ, Montiel J (2008) Inverse depth parametrization for monocular SLAM. IEEE Trans Robot 24(5):932–945
13. Marzorati D, Matteucci M, Migliore D, Sorrenti DG (2009) On the use of inverse scaling in monocular SLAM. In: Proceedings of IEEE international conference on robotics and automation ICRA'09, Kobe, pp 2030–2036
14. Strasdat H, Montiel JMM, Davison AJ (2010) Real-time monocular SLAM: why filter? In: Proceedings of IEEE international robotics and automation (ICRA) conference, Anchorage, pp 2657–2664
15. Cummins M, Newman P (2008) FAB-MAP: probabilistic localization and mapping in the space of appearance. Intern J Robot Res 27(6):647–665
16. Williams B, Cummins M, Neira J, Newman P, Reid I, Tardos J (2008) An image-to-map loop closing method for monocular SLAM. In: Proceedings of IEEE/RSJ international conference on intelligent robots and systems IROS 2008, Nice, pp 2053–2059
17. Eade E, Drummond T (2008) Unified loop closing and recovery for real time monocular SLAM. In: BMVC 2008, Leeds
18. Castle RO, Klein G, Murray DW (2010) Combining monoSLAM with object recognition for scene augmentation using a wearable camera. J Image Vis Comput 28(11): 1548–1556
19. Migliore D, Rigamonti R, Marzorati D, Matteucci M, Sorrenti DG (2009) Use a single camera for simultaneous localization and mapping with mobile object tracking in dynamic environments. In: Proceedings of international workshop on safe navigation in open and dynamic environments application to autonomous vehicles, Kobe
20. Davison AJ, Reid ID, Molton ND, Stasse O (2007) MonoSLAM: real-time single camera SLAM. IEEE Trans Pattern Anal Mach Intell 29(6):1052–1067

Extended Gaussian Image (EGI)

Sing Bing Kang[1] and Berthold K. P. Horn[2]
[1]Microsoft Research, Redmond, WA, USA
[2]Department of Electrical Engineering and Computer Science, MIT, Cambridge, MA, USA

Synonyms

Surface orientation histogram (discrete version of EGI)

Related Concepts

▸Extended Gaussian Image (EGI)

Definition

The extended Gaussian image (EGI) of a 3-D object is a function defined on a unit sphere. The value of the EGI at a point on the unit sphere is the inverse of the curvature at the corresponding point on the object – where the corresponding point is the one that has the same surface orientation. In the case of a polyhedral object, the value on the sphere is zero except for impulses at points on the sphere corresponding to faces of the polyhedron. The "size" or volume of each impulse is equal to the area of the corresponding face. The mapping from a 3-D polyhedron to the EGI is illustrated in Fig. 1. Figure 2 shows the mapping for a 3-D piecewise smooth object.

Background

Object recognition and determination of object pose (orientation and translation) are two fundamental, interlinked tasks in 3-D computer vision and robotics. The shape of an object can be measured directly using, for example, a rangefinder or binocular stereo techniques or indirectly using, say, photometric stereo. A representation for that shape is needed that makes recognition and attitude determination relatively easy. One requirement is that the representation transforms in a simple way when the object is rotated.

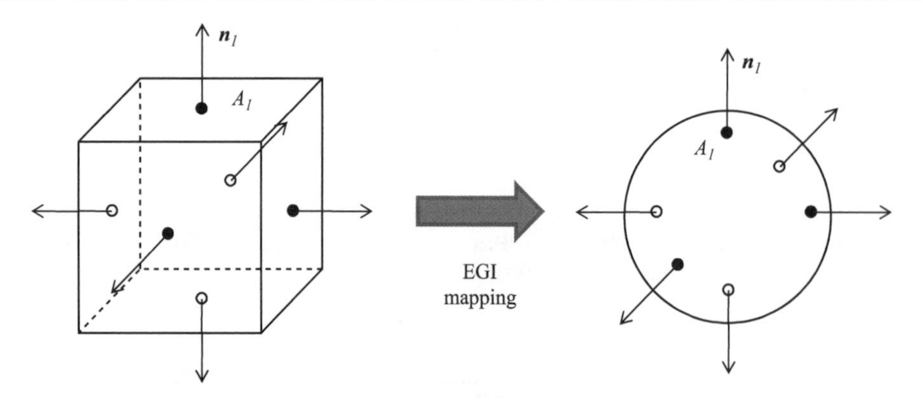

Extended Gaussian Image (EGI), Fig. 1 The EGI of a polyhedron. *Left*: cube with surface normals. *Right*: *arrows* on the unit sphere represent impulses corresponding to faces of the cube

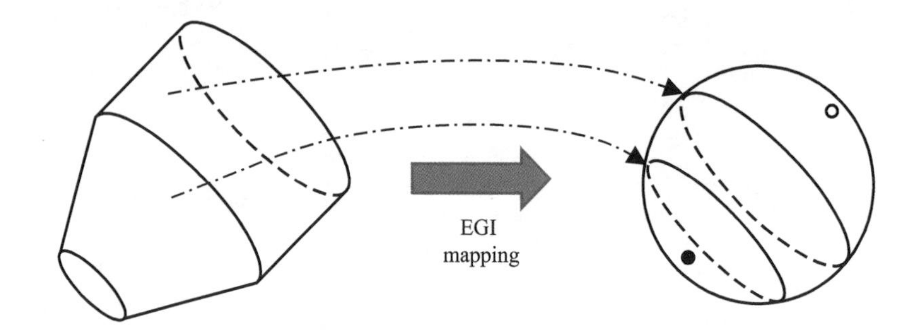

Extended Gaussian Image (EGI), Fig. 2 The EGI of a piecewise smooth object. *Left*: piecewise smooth object. *Right*: the two flat ends map to two impulses while the conical surface maps to a *small circle*, and the cylindrical surface maps to a *large circle* on the sphere

Theory

EGIs can be used to represent both smoothly curved objects as well as polyhedra. For ease of explanation, the convex polyhedra are discussed first. Minkowski [11] showed in 1897 that a convex polyhedron is fully specified (up to translation) by the areas and orientations of its faces (see also [14] and [10]). The area and orientation of faces can be represented conveniently by point masses (i.e., impulses of mass density) on a sphere. The extended Gaussian image of a polyhedron is obtained by placing a mass equal to the area of each face at a point on the sphere with the same surface orientation as the face.

More specifically, the Gaussian image of an object is obtained by mapping from a point on the surface of the object to the point on the unit sphere that has the same surface normal. In the case of a convex object, this mapping is invertible – only one point on the object's surface corresponds to a point on the unit sphere. This mapping extends in a natural way to curves and surface patches. The Gaussian curvature is the limit of the ratio of the (signed) area of a patch on the unit sphere to the area of the corresponding patch on the surface of the object, as the size of the patches becomes smaller and smaller. The Gaussian curvature is everywhere nonnegative in the case of a convex object. The extended Gaussian image (EGI) associates a value with a point on the unit sphere equal to the inverse of the Gaussian curvature at the corresponding point on the object.

A.D. Alexandrov showed that a smoothly curved convex object is fully specified by curvature given as a function of surface orientation [1]. So the EGI representation is unique for both smoothly curved convex objects and convex polyhedra. Note, however, that neither of the proofs is constructive and thus they do not provide a basis for reconstructing an object from its EGI. Importantly, reconstruction is *not* required for recognition and orientation determination.

The discrete and continuous versions of the EGI can be related as follows: An object with planar faces can be thought of as the limit of an object with patches of spherical surfaces as the radius of the curvature of the patches get larger and larger. A spherical patch of radius R contributes a constant value R^2 (its Gaussian curvature) to an area of the unit sphere corresponding to all of the surface orientations at points in that patch. As the radius tends to infinity, the value increases, at the same time that the area on the sphere shrinks. So the EGI of a polyhedron consists of a set of impulses, each of which has "volume" equal to the area of the corresponding face. It is convenient to treat these impulses as weights (or arrows of varying length) placed on the unit sphere. This is the discrete EGI. Conversely, the continuous EGI can be viewed as the limit of point mass density on the sphere as an approximation of a smoothly curved surface using a tessellation consisting of planar facets is made finer and finer.

The EGI has the following properties:

- *Translation invariance.* The EGI is not affected by object translation.
- *Rotation tracking.* Rotation of the object induces an equal rotation of the EGI since the unit surface normals rotate with the object.
- *Total mass equals total surface area.* Follows directly from the definition.
- *Center of mass at origin.* If the object is closed and bounded, the center of mass of the EGI is at the center of the Gaussian sphere.
- *Uniqueness.* There is only one closed convex object that corresponds to a particular EGI [1, 15].

If the object is not convex, more than one point on the object will contribute to a given point on the Gaussian sphere. One way of extending the definition of the EGI in the non-convex case is to use the sum of the inverses of the absolute values of the Gaussian curvature of all points on the surface that have the same orientation. While still useful in recognition and orientation determination, multiple non-convex objects may have the same EGI. For example, the EGI of all tori with the same surface area ($4\pi^2 R\rho$) and axis orientation is $2R\rho \sec \eta$, while there is but one *convex* object with that EGI. (Here R and ρ are the major axis and minor axis, respectively, while η is the angle between the surface normal and the plane perpendicular to the axis of the torus.)

In generating an EGI from an object, explicit information on shapes of faces and their adjacency relationships is not kept. Interestingly, a convex polyhedron can be recovered from its EGI [6, 9] as well as face adjacency and edge length information [12]. For a much more detailed treatment on EGIs, see [4]. The EGI has also been examined more broadly in the context of orientation-based representations [8].

The original version of the EGI is translation invariant. While it allows orientation to be extracted without regard to translation, a separate step is required to compute the translation. A variant of the EGI, the Complex EGI [7], uses complex mass to represent area (magnitude) and distance (phase). This decouples translation from rotation, allowing rotation to be determined in the usual fashion using the magnitudes while phase is subsequently used to compute translation. A support-function-based representation described in [13] encodes descriptors that are both local (tangent plane) and global (position and orientation of tangent plane).

Other extensions of the EGI address another inherent feature of not explicitly coding structural information. For example, the CSG-EESI (Constructive Solid Geometry-Enhanced Extended Spherical Image) [17] has two levels. The higher level contains the CSG tree which describes who the various subparts form the body, while the lower level describes the subparts using enhanced spherical images similar to that of [15]. Another tree-like representation is that of Hierarchical Extended Gaussian Image (HEGI) [18]. An HEGI description can be constructed as a tree where each leaf node corresponds to an Extended Gaussian Image (EGI) description of a convex part resulting from the recursive convex hull decomposition of an object. The COSMOS system described in [3] uses support functions based on Gaussian curvature and mean curvedness; connectivity information is maintained as a list.

Application

The EGI has been used for recognition and bin-picking of topmost objects (e.g., [2, 5, 6, 19]). It has also been used for symmetry detection [16] (for both reflectional and rotational symmetries).

In practice, the continuous EGI of a smoothly curved object is numerically approximated. The surface of the object may be divided into many patches, each of which is approximated by a planar facet of known area and orientation. The unit sphere may also be tessellated into cells, in each of which a total is accumulated of the areas of facets that have orientations falling within the range of orientations of that cell. Regular and semiregular tessellations of the sphere can be obtained by projecting regular polyhedra (Platonic solids) or semiregular polyhedra (Archimedean solids). Unfortunately there are only 5 regular solids, with the icosahedron having 20 facets, which is not a fine enough tessellation. There are 13 semiregular solids, but again there are relatively few faces in these tessellations, the truncated icosahedron, for example, having only 32 faces. Finer, but less regular, tessellations can be obtained by subdividing faces of the regular and semiregular tessellations.

In typical practical applications, the object is only partially observed. Assuming the object is not occluded by other objects, the EGI of an observed convex object has information on at most one hemisphere. For a non-convex object, there is the issue of self-occlusion, so that the weight distribution for a set of tracked surface normal may change with object orientation. The most obvious solution is that of brute force search through the space of possible discrete orientations (e.g., as was done in [7]), but this is computationally expensive. Ikeuchi [6] reduces the search space by computing the ratio of the surface area to the projected area and constraining the freedom of rotation (since this ratio is independent of planar rotation for a given line of sight).

References

1. Alexandrov AD (1942) Existence and uniqueness of a convex surface with a given integral curvature. C R (Doklady) Acad Sci URSS (NS) 35(5):131–134
2. Brou P (1983) Finding objects in depth maps. PhD thesis, Department of Electrical Engineering and Computer Science, MIT
3. Dorai C, Jain AK (1997) COSMOS – a representation scheme for 3D free-form objects. PAMI 19(10):1115–1130
4. Horn BKP (1984) Extended Gaussian images. Proc IEEE 72(12):1671–1686
5. Horn BKP, Ikeuchi K (1984) The mechanical manipulation of randomly oriented parts (picking parts out of a bin of parts). Sci Am 251(2):100–111
6. Ikeuchi K (1981) Recognition of 3-D objects using the extended Gaussian image. In: IJCAI, Vancouver, pp 595–608
7. Kang SB, Ikeuchi K (1993) The complex EGI: a new representation for 3-D pose determination. PAMI 15(7): 707–721
8. Liang P, Taubes CH (1994) Orientation-based differential geometric representations for computer vision applications. PAMI 16(3):249–258
9. Little JJ (1983) An iterative method for reconstructing convex polyhedra from extended Gaussian images. In: Proceedings of the national conference on artificial intelligence, Washington, DC, pp 247–254
10. Lyusternik LA (1963) Convex figures and polyhedra. Dover, New York
11. Minkowski H (1897) Allgemeine Lehrsatze über die konvexen Polyeder. In: Nachrichten von der Königlichen Gesellschaft der Wissenschaften, mathematisch-physikalische Klasse, Göttingen, pp 198–219
12. Moni S (1990) A closed-form solution for the reconstruction of a convex polyhedron from its extended Gaussian image. In: ICPR, Atlantic City, NJ, vol 1, pp 223–226
13. Nalwa VS (1989) Representing oriented piecewise C^2 surfaces. IJCV 3(2):131–153
14. Pogorelov AV (1956) Differential geometry. Noordhoff, Groningen
15. Smith DA (1979) Using enhanced spherical images. Technical report, MIT AI Lab Memoo No. 530
16. Sun C, Sherrah J (1997) 3D symmetry detection using the extended Gaussian image. PAMI 19(2):164–168
17. Xie SE, Calvert TW (1988) CSG-EESI: a new solid representation scheme and a conversion expert system. PAMI 10(2):221–234
18. Xu JZ, Suk M, Ranka S (1996) Hierarchical EGI: a new method for object representation. In: 3rd international conference on signal processing, Beijing, vol 2
19. Yang HS, Kak AC (1986) Determination of the identification, position and orientation of the topmost object in a pile. CVGIP 36(2/3):229–255

Extrinsic Parameters

►Camera Extrinsic Parameters

F

Face Identification

Xiaogang Wang
Department of Electronic Engineering, Chinese University of Hong Kong, Shatin, Hong Kong

Related Concepts

▶Face Identification

Definition

Face identification is to automatically identify a person by computers based on a query face image. In order to determine the identity of the query face image, the face images of all the registered persons in the database are compared against the query face image and are reranked based on the similarities.

Background

Face identification is a powerful technology with wide applications to biometrics, surveillance, law enforcement, human-computer interaction, and image and video search. It is often confused with another research topic called face verification. Face verification is to validate a claimed identity based on the query image. It compares the query image against the face images whose identity is claimed and decides to either accept or reject the claimed identity. Face identification involves one-to-many matches, while face verification involves one-to-one matches. As the number of registered persons in the database increases, both the accuracy and the efficiency of face identification decreases. However, the performance of the face verification is not much affected by the size of the database.

Face identification is one of the most important biometric technologies. Compared with other biometric attributes, face images are easy to be captured in a nonintrusive way. Face identification has higher compatibility with other systems. As the fast growth of digital cameras as well as images and videos on the Internet is fast, face identification has become one of the hottest research topics in computer vision.

Face identification has some major challenges, such as large intrapersonal variations, small interpersonal variations, and the small sample size problem, to be solved. Face images of the same person may appear very differently because of variations of poses, illuminations, aging, occlusions, makeups, hair styles, and expressions. On the other hand, if the database is large, some faces of different persons may be quite similar. It is critical for face identification algorithms to effectively extract the interpersonal variations and depress the intrapersonal variations. In face identification, the visual features to be compared are usually in very high-dimensional spaces, while each person registered in the database only has a few face images for training. Therefore, it is easy for the classifier to overfit the training set. This is called the small sample size problem. Many developed face identification algorithms are to address the three challenges motioned above. Detailed surveys of face identification methods can be found in [1, 2].

K. Ikeuchi (ed.), *Computer Vision*, DOI 10.1007/978-0-387-31439-6,

Theory and Applications

The face identification pipeline is shown in Fig. 1. Given an input image, the face region is first located by a face detection algorithm [3]. Then the landmarks on the face are automatically located by a face alignment algorithm [4] through matching the face region with a predefined face graph. An example of a face graph model is shown in Fig. 2. Face alignment is critically important for depressing the variations caused by poses and expressions. It is an important pre-step for extracting visual features, especially local features. At the preprocessing step, both geometric rectification and photometric rectification are applied. For example, given the face alignment result, face images are usually cropped and transformed such that the two eye centers and the mouth center of all the face images are at fixed positions. In order to reduce the effect of lighting variations, the face region can be normalized by histogram equalization or gamma correction. In some approaches [5], a difference-of-Gaussian (DoG) filter was used to remove both high-frequency variations caused by noise and low-frequency illumination variations.

Feature Extraction

There are four types of features used for face identification: geometric features, holistic features, local features, and semantic features. Geometric features measure the sizes of facial components and the distances between components. They were used in some early face recognition systems developed in 1980s and 1990s. Because of the lack of robustness and discriminative power, they are not widely used by current face identification approaches. Holistic features and local features are most popularly used for face identification. Holistic features are simply high-dimensional vectors by concatenating all the pixels within the face regions. Local features are extracted by characterizing the texture patterns within local regions. The local regions can be defined by a regular grid or the neighborhoods around detected landmarks. Compared with holistic features, extracting local features requires higher-resolution face images. Gabor responses [6] and local binary patterns (LBP) [7] are two types of local features most widely used.

Gabor responses are obtained by the convolution of local face regions with a set of Gabor kernels. Gabor kernels are characterized as localized, orientation selective, and frequency selective, and they are similar to the receptive field profiles in cortical simple cells. A Gabor kernel is the product of a Gaussian envelope and a plane wave:

$$\Psi_k(\boldsymbol{x}) = \frac{\|\boldsymbol{k}\|}{\sigma^2} \cdot e^{-\frac{\|k\|^2\|x\|^2}{2\sigma^2}} \cdot \left[e^{i k \cdot x} - e^{-\frac{\sigma^2}{2}}\right]. \quad (1)$$

$\boldsymbol{x} = (x, y)$ is the variable in the spatial domain and k is the frequency vector, which determines the scale and the orientation of the Gabor kernel,

$$k = k_s e^{i\phi_d}, \quad (2)$$

where $k_s = \frac{k_{\max}}{f^s}$, $f = 2$, $s = 0, \ldots, L-1$ and $\phi_d = (\pi \cdot d)/8$, for $d = 0, \ldots, \boldsymbol{M}-1$. There are totally $L \times \boldsymbol{M}$ Gabor kernels at L different scales along $\boldsymbol{M}$ different orientations. The number of oscillations under the Gaussian envelope function is determined by $\delta = 2\pi$. The term $\exp(-\sigma^2)$ is subtracted to make the kernel DC-free, and thus, the Gabor responses are insensitive to illumination. Figure 3 shows the real parts of Gabor kernels at 5 scales along 8 orientations.

Given an image $I(\boldsymbol{x})$, the Gabor response at location x_0 is computed by convolution:

$$(\Psi_k * I)(\boldsymbol{x}_0) = \int \Psi_k(\boldsymbol{x}_0 - \boldsymbol{x}) I(x) d^2(\boldsymbol{x}). \quad (3)$$

Since the phases of Gabor responses changes drastically with translation, usually only the magnitudes of the responses at landmarks or on a grid are used as local features for face identification. Gabor responses are robust to illumination variations and can tolerate the misalignment caused by variations of poses and expression to some extent.

LBP is a powerful texture descriptor introduced by Ojala et al. [8]. As shown in Fig. 4a, it defines the neighborhood of a pixel i by uniformly sampling P pixels along the circle centered at i with radius R. The pixels in the neighborhood are assigned with binary numbers by thresholding against the value of pixel i, as shown in Fig. 4b. These binary numbers are converted to a decimal number, which indicates the local binary pattern of pixel i. A local binary pattern is called uniform if it contains at most two bitwise transitions from 0 to 1 or vice versa when the binary string is considered circular. For example, 00011110 is a uniform pattern

Face detection → Face alignment → Preprocessing → Feature extraction → Classification

Face Identification, Fig. 1 Face identification pipeline

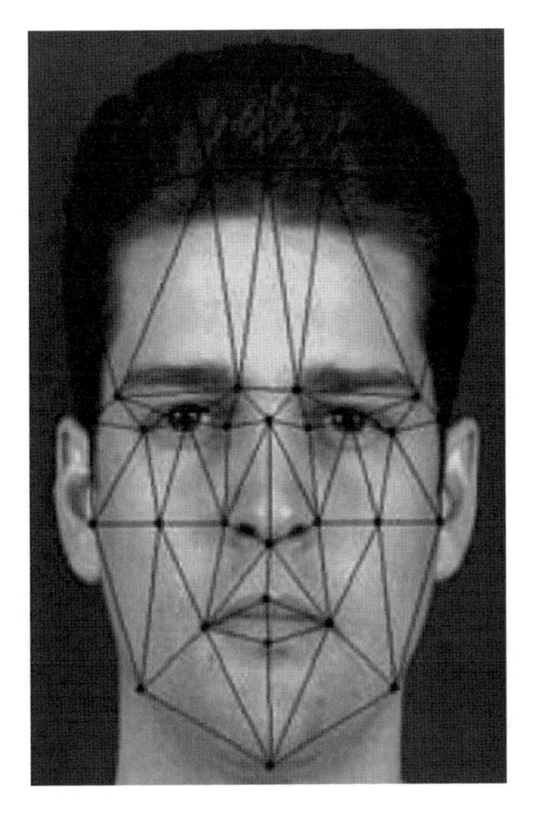

Face Identification, Fig. 2 Face graph model

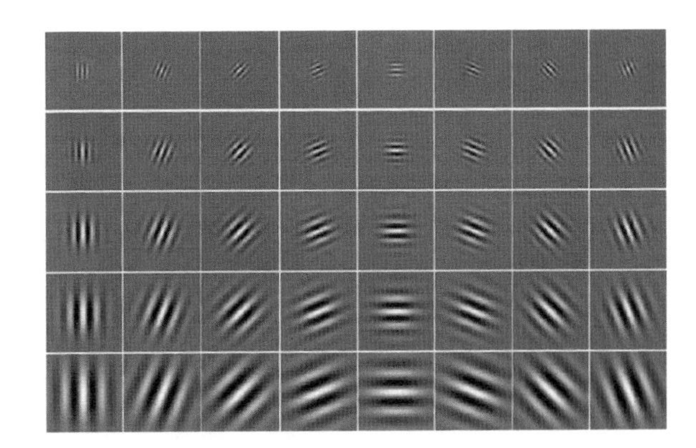

Face Identification, Fig. 3 Real parts of Gabor kernels at 5 scales and 8 orientations

and 00010100 is not. It is observed that uniform patterns appear much more frequently than nonuniform patterns. Both accuracy and computational efficiency of face identification can be improved if only uniform patterns are kept and all the nonuniform patterns are mapped to a single label.

Local binary patterns are treated as words of a codebook. The face region is divided into local regions by a grid as shown in Fig. 4c. The histograms of uniform patterns inside the local regions are used as features for face recognition. LBP characterizes a very large set of edges and has high discrimination power. If the values of centered pixels and their neighborhoods are under the same monotonous transformation, local binary patterns do not change. Therefore, LBP is robust to lighting variations. Since the histograms within local regions are used as features, they are robust to misalignment and pose variations.

The idea of LBP has been extended to develop other types of local descriptors [9, 10] for face identification. For example, a learning-based descriptor was proposed in [11]. At each pixel, a low-level feature vector is extracted by sampling its neighboring pixels in the ring-based pattern as shown in Fig. 5. $r \times 8$ pixels are sampled at even intervals on the ring of r. After sampling, the feature vector is normalized into unit length. Then an encoding method is applied to encode the normalized feature vectors into discrete codes using the PCA tree. Histograms of codes within local regions are used as local features. Since the descriptor is unsupervised learned from the training set instead of being handcrafted, the learned codes are more uniformly distributed and the resulting code histogram can achieve better discriminative power and robustness trade-off than LBP.

Semantic features were proposed for face identification in very recent years. Two types of semantic features, "attribute" and "simile," were proposed in [12]. To extract the attribute features, a set of binary classifiers are trained to recognize the presence or absence of describable aspects of visual appearance (e.g., "does the face has double chin?" and "does the face have high

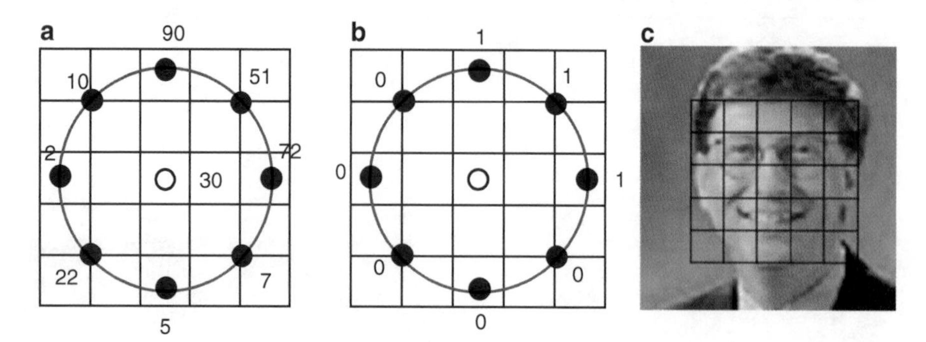

Face Identification, Fig. 4 LBP operator. (**a**): the neighborhood of a pixel. (**b**): the local binary pattern is labeled as $11100000 = 224$. (**c**) The face region is divided into local regions

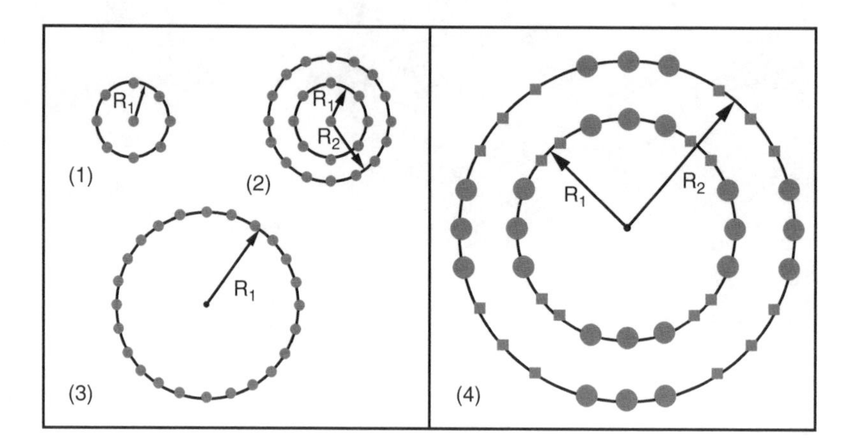

Face Identification, Fig. 5 Different sampling methods used in [11] for the learning-based descriptor. (*1*) $R_1 = 1$, with center; $R_1 = 1$, $R_2 = 2$, with center; (*3*) $R_1 = 3$, no center; and (*4*) $R_1 = 4$, $R_2 = 7$, no center (The sampling dots on the *green square*-labeled arcs are omitted for better visuality)

cheekbones?"). The output of each classifier is a score between 1 and −1. The raw outputs of attribute classifiers are concatenated to form the feature vector for face identification. The simile features of a face image are the similarities between this face image and a set of reference people. Semantic features are high-level visual features. They are inspired by the process of face identification by human beings, since human beings often describe a face using some attribute terms or using its similarities with some known people. Semantic features are more insensitive to the variations of poses, illuminations, and expressions.

The resolution of face images is an important factor affecting feature extraction. Local features usually require higher resolutions than holistic features. As the development of digital cameras, extraction of features from very high resolution images [13] is an important topic to be explored.

Classification

The extracted features are usually in very high-dimensional spaces and include large intrapersonal variations and noise. It is inaccurate and inefficient to directly use them for face identification. They are typically projected into low-dimensional spaces or fed into trained classifiers in order to compare two face images. PCA [14] is widely used for dimension reduction. In face identification, intrapersonal variations are the main factor deteriorating the identification accuracy and may be larger than interpersonal variations. However, they cannot be depressed by PCA. Therefore, although PCA can improve the identification efficiency, it cannot effectively improve the accuracy.

Bayesian face recognition [15] effectively models of the intra-personal variations using the intrapersonal subspace, which is spanned by the eigenvectors $\{\boldsymbol{e}_k\}$ and eigenvalues $\{\lambda_k\}$ of matrix,

$$C = \sum_{\ell(\boldsymbol{x}_i)=\ell(\boldsymbol{x}_j)} (\boldsymbol{x}_i - \boldsymbol{x}_j)(\boldsymbol{x}_i - \boldsymbol{x}_j)^T, \qquad (4)$$

where $\boldsymbol{x}_i$ is the feature vector and $\ell(\boldsymbol{x}_i)$ is its identity. The distance between two feature vectors $\boldsymbol{x}$ and $\boldsymbol{x}'$ is computed as

$$d(\boldsymbol{x},\boldsymbol{x}') = \sum y_k^2/\lambda_k, \qquad (5)$$

where y_k is the projection of $\boldsymbol{x} - \boldsymbol{x}'$ on eigenvector $\boldsymbol{e}_k$. Since the intrapersonal variations concentrate on the first few eigenvectors with the largest eigenvalues λ_k, they are effectively depressed by weighting the inverse of λ_i. Therefore, Bayesian face recognition can effectively improve the accuracy by reducing the intrapersonal variations.

LDA [16] is another widely used subspace method. LDA tries to find the subspace that best discriminates different face classes by maximizing the between-class scatter matrix S_b, while minimizing the within-class scatter matrix S_w in the projective subspace. Suppose there are L persons in the training set and X_c is the set of feature vectors of person c. S_w and S_b are defined as

$$S_w = \sum_{c=1}^{L} \sum_{\boldsymbol{x}_i \in X_c} (\boldsymbol{x}_i - \boldsymbol{m}_c)(\boldsymbol{x}_i - \boldsymbol{m}_c)^T, \qquad (6)$$

$$S_b = \sum_{c=1}^{L} n_c (\boldsymbol{m}_c - \boldsymbol{m})(\boldsymbol{m}_c - \boldsymbol{m})^T, \qquad (7)$$

where $\boldsymbol{m}_c$ is the mean of feature vectors in X_c, n_c is the number of feature vectors in X_c, and $\boldsymbol{m}$ is the mean of all the feature vectors in the training set. LDA subspace is spanned by a set of vectors W satisfying

$$W = \arg\max \left| \frac{W^T S_b W}{W^T S_w W} \right|. \qquad (8)$$

Feature vectors are first projected into the LDA subspace before being used to compare the distance between two face images. W can be computed as the eigenvectors of $S_w^{-1} S_b$. However, because of the small sample size problem, S_w^{-1} is often singular and it is easy for W to overfit the training set. Many improvements of LDA, such as dual-space LDA [17], null-space LDA [18], and direct LDA [19], have been made to address this problem.

In [20, 21], the three popular subspace methods, PCA, Bayesian face recognition, and LDA, were unified under one framework. Based on this framework, a unified subspace analysis method was developed using PCA, Bayes, and LDA as three steps. Fist PCA is used to reduce noise and data dimensionality. Then within the PCA subspace, Bayesian analysis is used to learn the intrapersonal subspace and to depress the intrapersonal variations. In the last step of LDA, interpersonal variations are extracted and the noise is further reduced. A 3D parameter space is constructed using the three subspace dimensions as axes. The choice of the three parameters greatly affects the performance of face identification. The original PCA, LDA, and Bayes methods only occupy some local lines or areas in the 3D parameter space and therefore cannot achieve the best performance. Searching through this parameter space, an optimal subspace for face identification is obtained.

When reducing the data dimensionality by PCA to solve the small sample size problem, some discriminative features have to be discarded in order to train a stable LDA classifier. A random sampling subspace method [22, 23] was proposed to effective solve the small sample size problem and keep nearly all the discriminative information in the meanwhile by integrating multiple classifiers. Many works [24] have shown that integrating an ensemble of classifiers can effectively improve the performance of face identification. The multiple classifiers can be constructed based on different features, within different local face regions, using different training examples or even under different parameter settings.

PCA and LDA are linear subspace methods and assume a Euclidean structure of the face space. Manifold learning [25] assumes that face images reside on a manifold structure and obtains a face subspace using locality preserving projections. It is used to reduce the variations of poses, expressions, and illuminations.

Besides subspace methods, other classifiers such as SVM [26] and boosting [27] were also used for face identification. In recent years, sparse representation [28] was applied to face identification and drew a lot of attentions. Using ℓ^1 regularization, it effectively handles errors due to occlusion and corruption by exploiting the fact that these errors are often sparse. LDA was extended to non-parametric discriminant analysis assuming that the distribution of face data is non-Gaussian [29].

Face Identification Based on Other Types of Data

Most of the face identification methods assumed that the queries are 2D face photos. However, other types of face data, such as 3D data, video sequences, infrared images, and sketch drawings, can also be used queries in some specific application scenarios. Compared to extensive research work on 2D face photos, relatively less work has been on these topics.

3D face identification [30] measures geometry of rigid features on the face. Compared with 2D face recognition, it is more robust to the variations of illuminations, makeups, and poses. Its major technological limitation is the acquisition of 3D face data, which requires range cameras. With the development and popularity of 3D cameras, the research on 3D face identification is drawing more and more attentions.

Compared with single images, video sequences provide extra dynamic information and multiple face instances and therefore can effectively improve both the robustness and accuracy of face identification [31, 32]. The head pose may vary significantly in a video sequence, and therefore video-based face identification requires robust face tracking and pose estimation. Faces in videos are usually at lower resolutions with blurring effects. These factors add extra challenges for video-based face identification.

In order to be robust to lighting variations, infrared imaging was used for face identification in recent years [33, 34]. Infrared face images were used as queries. Infrared cameras can capture clear images at nighttime and in hazy conditions. An external illumination source is not required since the face emits thermal energy.

In law enforcement, automatically identifying suspects from the police mug-shot database can help the police narrow down potential suspects quickly. However, in some cases, the photo image of a suspect for query is not available. The best substitute is often a sketch drawing based on the recollection of an eyewitness. Therefore, face identification using sketch drawings as queries become an interesting research topic. Photos and sketches are in different modalities and have significant differences on shape and texture. It requires matching sketches with photos in the database and is more challenging. It also involves the unknown psychological mechanism of sketch generation. A few works [35, 36] have been done on face sketch recognition. However, this problem is far from being well solved yet.

Benchmark Databases

In the past decades, many databases have been built as benchmarks to evaluate the performance of face identification and face verification. Among existing databases, FRGC [37] and Multi-PIE [38] are in larger scales and are widely used. Face images in most existing databases were collected in controlled environments. In recent years, there has been an increasing interest in studying face recognition in uncontrolled environments. It was shown that the performance of many existing face recognition approaches dropped significantly in such uncontrolled environments. The LFW [39] database collected 13,233 face images of 1,040 from the Internet and these face images were collected in uncontrolled environments. LFW has drawn a lot of attentions in the field face recognition in recent years. Besides the 2D face photo images, some databases include other types of face data. For examples, FRGC includes both 2D and 3D face data. XM2VTS [40] includes face video sequences for the evaluation of face identification using videos as queries. The CUFS database [36] includes face photos and face sketch drawings of 606 subjects and is used for the research of face sketch recognition.

References

1. Li SZ, Jain A (eds) (2004) Handbook of face recognition. Springer, New York
2. Zhao W (2003) Face recognition: a literature survey. ACM Comput Surv 35:399–458
3. Viola P, Jones MJ (2004) Robust real-time face detection. Int J Comput Vis 57:137–154
4. Cootes TF, Taylor CJ (2004) Statistical models of appearance for computer vision. Technical report, imaging science and biomedical engineering, University of Manchester
5. Hua G, Akbarzadeh A (2009) A robust elastic and partial matching metric for face recognition. In: Proceedings of the IEEE international conference on computer vision, Kyoto
6. Wiskott L, Fellous J, Kruger N, Malsburg C (1997) Face recognition by elastic bunch graph. IEEE Trans PAMI 19:775–779
7. Ahonen T, Hadid A, Pietikainen M (2006) Face description with local binary patterns: application to face recognition. IEEE Trans PAMI 28:2037–2041
8. Ojala T, Peitikainen M, Maenpaa T (2002) Multiresolution gray-scale and rotation invariant texture classification with local binary patterns. IEEE Trans PAMI 24:971–987
9. Zhang W, Shan S, Gao W, Chen X, Zhang H (2005) Local gabor binary pattern histogram sequence (LGBPHS): a novel non-statistical model for face representation and recognition. In: Proceedings of the IEEE international conference on computer vision, Beijing
10. Tan X, Triggs B (2010) Enhanced local texture feature sets for face recognition under difficult lighting conditions. IEEE Trans Image Process 19(6):1635–1650

11. Cao Z, Yin Q, Tang X, Sun J (2010) Face recognition with learning-based descriptor. In: Proceedings of the IEEE international conference on computer vision and pattern recognition (CVPR), Providence
12. Kumar N, Berg AC, Belhumeur P (2009) Attribute and simile classifiers for face verification. In: Proceedings of the IEEE international conference on computer vision, Kyoto
13. Lin D, Tang X (2006) Recognizing high resolution faces: from macrocosm to microcosm. In: Proceedings of the European conference on computer vision (ECCV), Graz
14. Turk M, Pentland A (1991) Face recognition using eigenfaces. J Cognit Neurosci 3:71–86
15. Moghaddam B, Jebara T, Pentland A (2000) Bayesian face recognition. Pattern Recognit 33:1771–1782
16. Belhumeur P, Hespanda J, Kriegman D (1997) Eigenfaces vs. fisherfaces: recognition using class specific linear projection. IEEE Trans PAMI 19:711–720
17. Wang X, Tang X (2004) Dual-space linear discriminant analysis for face recognition. In: Proceedings of the IEEE international conference on computer vision and pattern recognition (CVPR), Washington, DC
18. Chen L, Liao H, Ko M, Lin J, Yu G (2000) A new lda-based face recognition system which can solve the small sample size problem. J Pattern Recognit 33:1713–1726
19. Yu H, Yang J (2001) A direct lda algorithm for high-dimensional data – with application to face recognition. Pattern Recognit 34:2067–2070
20. Wang X, Tang X (2003) Unified subspace analysis for face recognition. In: Proceedings of the IEEE international conference on computer vision, Nice
21. Wang X, Tang X (2004) A unified framework for subspace face recognition. IEEE Trans PAMI 26:1222–1228
22. Wang X, Tang X (2004) Random sampling lda for face recognition. In: Proceedings of the IEEE international conference on computer vision and pattern recognition (CVPR), Washington, DC
23. Wang X, Tang X (2006) Random sampling for subspace face recognition. Int J Comput Vis 70:91–104
24. Su Y, Shan S, Chen X, Gao W (2007) Hierarchical ensemble of global and local classifiers for face recognition. In: Proceedings of the IEEE international conference on computer vision, Rio de Janeiro
25. He X, Yan S, Hu Y, Niyogi P, Zhang H (2005) Face recognition using laplacianfaces. IEEE Trans PAMI 27:328–340
26. Heisele B, Ho P, Poggio T (2001) Face recognition with support vector machines: Global versus component-based approach. In: Proceedings of the IEEE international conference on computer vision, Vancouver
27. Wang X, Zhang C, Zhang Z (2009) Boosted multi-task learning for face verification with applications to web images and videos search. In: Proceedings of the IEEE international conference on computer vision and pattern recognition (CVPR), Miami
28. Wright J, Yang AY, Ganesh A, Sastry SS, Ma Y (2009) Robust face recognition via sparse representation. IEEE Trans PAMI 31:210–227
29. Li Z, Lin D, Tang X (2009) Nonparametric discriminant analysis for face recognition. IEEE Trans PAMI 31:755–761
30. Flynn P, Chang K, Bowyer K (2006) A survey of approaches and challenges in 3d and multi-modal 3d+2d face recognition. Comput Vis Image Underst 101:1–5
31. Tang X, Li Z (2004) Frame synchronization and multi-level subspace analysis for video-based face recognition. In: Proceedings of the IEEE international conference on computer vision and pattern recognition (CVPR), Washington, DC
32. Lee K, Ho J, Yang M, Kriegman D (2003) Video-based face recognition using probabilistic appearance manifolds. In: Proceedings of the IEEE international conference on computer vision and pattern recognition (CVPR), Madison
33. Li S, Chu R, Liao S, Zhang L (2007) Illumination invariant face recognition using near-infrared images. IEEE Trans PAMI 29:627–639
34. Lin D, Tang X (2006) Inter-modality face recognition. In: Proceedings of the European conference on computer vision (ECCV), Graz
35. Tang X, Wang X (2003) Face sketch synthesis and recognition. In: Proceedings of the IEEE international conference on computer vision, Nice
36. Wang X, Tang X (2009) Face photo-sketch synthesis and recognition. IEEE Trans PAMI 31(11):1955–1967
37. Phillips, PJ, Flynn PJ, Bowyer KW (2005) Overview of the face recognition grand challenge. In: Proceedings of the IEEE international conference on computer vision and pattern recognition (CVPR), San Diego
38. Gross R, Matthews I, Cohn J, Kanade T, Baker S (2008) Multi-pie. Image Vis Comput 26:15–26
39. Huang GB, Manu R, Berg T, Learned-Miller E (2007) Labeled faces in the wild: A database for studying face recognition in unconstrained environments. Technical report, University of Massachusetts, Amherst
40. Messer K, Matas J, Kittler J, Luettin J, Maitre G (1999) Xm2vtsdb: the extended m2vts database. In: Proceedings of the IEEE international conference on audio and video-based biometric person authentication, Washington, DC

Face Modeling

Zicheng Liu[1] and Zhengyou Zhang[2]
[1]Microsoft Research, Microsoft Corporation, Redmond, WA, USA
[2]Microsoft Research, Redmond, WA, USA

Synonyms

Face reconstruction; Three dimensional face modeling

Related Concepts

▸Structure-from-Motion (SfM)

Definition

Face modeling usually refers to the problem of recovering 3D face geometries from one or more images, though the recovery of lighting and skin reflectance is sometimes considered as face modeling as well.

Background

Because face is a special type of object which people are all familiar with, there has been tremendous interests among researchers in the problem of 3D face reconstruction. The most reliable and accurate way to obtain face geometries is by using active sensors such as laser scanners and structured light systems. So far, laser scanners are the most commonly used and most accurate active sensors. Structured light systems are becoming more popular because they are capable of capturing continuous motions. Some structured light systems use visible light sources, while others use invisible light sources such as infrared lights. Visible light systems give better signals, but they are intrusive.

The disadvantage of active sensors is that they are usually expensive and not widely available. Recently, Microsoft released a depth camera, called Kinect, which uses active sensors. The device works with both game consoles and PCs, and has popularized the use of depth cameras in many applications including face modeling.

Given that cameras are everywhere, it is not surprising that researchers have been fascinated by the problem of 3D face reconstruction from images. This can be thought of as a special case of the classical computer vision problem of 3D structure recovery from images. One could directly apply a generic 3D reconstruction technique to face images. It usually does not work very well because face skins are usually smooth making it difficult to find accurate matchings across images. Since all human faces are similar in terms of their rough shape and topology, a lot of the research has been devoted to developing techniques that leverage the prior knowledge on faces. For example, one can start from a generic face mesh and try to adjust the vertex positions of the mesh to fit the image observations. To reduce the number of degrees of freedoms involved in the fitting process, people have proposed to use a linear space of face geometries to constrain the parameter space. One can use a set of pre-captured or hand-designed face geometries as the examples. Any face is assumed to be a linear combination of the example faces.

Theory

Face modeling techniques can be divided into two categories based on whether the illumination effects are modeled or not. The methods that belong to the first category do not model illumination effects [1–4]. They have origins from structure from motion or stereovision. The methods in the second category takes illumination into account [5–8]. In fact, they leverage the shading information for the geometry reconstruction. These techniques can be traced to shape from shading. For a more detailed and systematic descriptions of face modeling techniques and applications, the readers are referred to the book [20].

Regardless of which method is used, the first question is how to represent a face. As mentioned earlier, linear space representation has been shown to be an effective way to constrain the parameter space. There are two different ways to construct a linear space representation.

The first, called a morphable model, was proposed by Blanz and Vetter in their seminar paper [5]. They used a set of existing face meshes obtained by laser scanners. These meshes must be aligned so that there is a correspondence between the vertices of different meshes.

Let $\mathrm{v}_i = (X_i, Y_i, Z_i)^T$, $i = 1, \ldots, n$, denote the vertices of a face mesh. Its geometry is represented by a vector

$$\mathcal{S} = (\mathrm{v}_1^T, \ldots, \mathrm{v}_n^T)^T = (X_1, Y_1, Z_1, \ldots, X_n, Y_n, Z_n)^T. \tag{1}$$

Suppose there are $m + 1$ face meshes which are obtained by using laser scanner or some other means. Let $\mathcal{S}^j$ denote the geometry of the jth mesh, $j = 1, \ldots, m + 1$. These faces generate a linear space of face geometries:

$$F = \left\{ \sum_{j=1}^{m+1} \alpha_j \mathcal{S}^j : \sum_{j=1}^{m+1} \alpha_j = 1 \right\}. \tag{2}$$

Let $\mathcal{S}^0$ denote the average face geometry, that is, $\mathcal{S}^0 = \frac{1}{m+1} \sum_{j=1}^{m+1} \mathcal{S}^j$. Denote $\delta\mathcal{S}^j = \mathcal{S}^j - \mathcal{S}^0$, $j = 1, \ldots, m + 1$. Note that these vectors are linearly dependent. Principal component analysis can be performed on the vectors $\delta\mathcal{S}^1, \ldots, \delta\mathcal{S}^{m+1}$. Let $\sigma_1^2, \ldots, \sigma_m^2$ denote the eigenvalues with $\sigma_1^2 \geq \sigma_2^2 \geq \ldots \geq \sigma_m^2$. Let

$\mathcal{M}^1, \ldots, \mathcal{M}^m$ denote the corresponding eigenvectors. Then any face $\mathcal{S} \in F$ can be represented as the average face $\mathcal{S}^0$ plus a linear combination of the eigenvectors, that is,

$$\mathcal{S} = \mathcal{S}^0 + \sum_{j=1}^{m} c_j \mathcal{M}^j. \tag{3}$$

c_j are the geometry coefficients. The prior probability for geometry coefficients $c_1, \ldots, c_m$ is given by

$$Pr(c_1, \ldots, c_m) = e^{-\frac{1}{2}\sum_{j=1}^{m} \frac{c_j^2}{\sigma_j^2}}. \tag{4}$$

The second approach is to manually design the average mesh $\mathcal{S}^0$ and a set of deformation vectors which act as $\delta\mathcal{S}^j$ as proposed in [2, 9]. Each deformation vector corresponds to an intuitive way of deforming the face. For face modeling purpose, the deformation vectors are used in almost the same way as the eigenvectors. The only difference is that for eigenvectors, there is a prior probability for the geometry coefficients (see (Eq. 4)). For deformation vectors, one can predefine a valid range for each model coefficient.

Given the face representation as shown in (Eq. 3), the problem of face reconstruction becomes searching for the model coefficients c_j. As mentioned earlier, the method of solving for the model coefficients can be divided into two categories depending on whether they model illumination effects or not.

The methods that belong to the first category typically assume there are two or more input images corresponding to different views of a face. If the camera motions corresponding to the views are not known, one can use structure-from-motion techniques to estimate the camera motion and obtain a set of 3D points, which are usually quite sparse and noisy. After that, one can solve for the model coefficients and the head pose by minimizing the total distances of the 3D points to the model. Numerically, one can solve it through an Iterative Closest Point (ICP) procedure. For details, the readers are referred to [2].

If the camera motions corresponding to the views are known such as in a stereo rig, one can perform dense stereo matching and obtain a depth map of the face. One such system was developed by Chen and Medioni [10]. The obtained depth map is usually quite noisy and has spikes. One could use a face model to fit the obtained depth map or inject a face model representation in the stereo reconstruction process in a way similar to the model-based bundle adjustment formulation in [11], but the camera motions are no longer variables anymore in this case.

Another approach is to use two orthogonal views [12–14]: one frontal view and one side view. The frontal view provides the information relative to the horizontal and vertical axis, while the side view provides depth information. Since the number of feature points that can be detected on the two views is usually quite small, one could use a linear space face representation to fit the detected feature points and obtain a complete face geometry.

The methods that belong to the second category only require a single view. Since the publication of Blanz and Vetter's paper [5] which assumed a single light source, researchers have extended their technique to handle more general lighting conditions. These techniques leverage shading information and they typically recover the shape, reflectance, and lighting simultaneously. Most techniques assume the face skin is Lambertian. Similar to linear space representation for face geometry, the diffuse reflectance components (also called albedo or texture) can also be represented as a linear combination of example face albedos. In this way, the unknowns for the reflectance are the albedo (also called texture) coefficients. One effective tool to model the lighting is the spherical harmonics representation which was proposed by Ramamoorthi and Hanrahan [15] and Basri and Jacobs [16]. The basic idea is that the irradiance can be well approximated by using a linear combination of a small number of spherical harmonic basis functions. They showed that with nine spherical harmonic basis functions, the average approximation error is no more than 1%. The coefficients of the spherical harmonic basis are called lighting coefficients. Given the pose, geometry model coefficients, albedo coefficients, and lighting coefficients, one can synthesize an image of the face. The objective is to solve for these parameters so that the synthesized image matches the input image. The optimization problem can be solved through an iterative procedure. For details, the readers are referred to [8].

Application

Face modeling has many applications. The techniques allow people to create personalized avatars which can be used in chatting rooms, e-mails, greeting cards, and games. Many human-machine dialog systems use realistic-looking avatars as visual representation of the computer agent that interacts with the human user. Face models are useful for 3D head pose tracking and facial expression tracking. In teleconferencing, face modeling techniques can be used for eye-gaze correction to improve video conferencing experience [17]. Face modeling techniques are useful for face recognition to handle pose variations [18] and lighting variations [19].

References

1. Fua P, Miccio C (1998) From regular images to animated heads: a least squares approach. In: European conference on computer vision (ECCV), Freiburg, pp 188–202
2. Zhang Z, Liu Z, Adler D, Cohen M, Hanson E, shan Y (2004) Robust and rapid generation of animated faces from video images: a model-based modeling approach. Int J Comput Vision 58(2):93–120
3. Dimitrijevic M, Ilic S, Fua P (2004) Accurate face models from uncalibrated and ill-lit video sequences. In: Computer vision and pattern recognition (CVPR), II, Washington, DC, pp 1034–1041
4. Amberg B, Blake A, Fitzgibbon A, Romdhani S, Vetter T (2007) Reconstructing high quality face-surfaces using model based stereo. In: International conference on computer vision, Rio de Janeiro, Brazil
5. Blanz V, Vetter T (1999) A morphable model for the synthesis of 3d faces. In: Proceedings of the Annual conference series on computer graphics, SIGGRAPH, Los Angeles, pp 187–194
6. Zhang L, Wang S, Samaras D (2005) Face synthesis and recognition from a single image under arbitrary unknown lighting using a spherical harmonic basis morphable model. In: Computer vision and pattern recognition II (CVPR), San Diego, pp 209–216
7. Lee J, Moghaddam B, Pfister H, Machiraju R (2005) A bilinear illumination model for robust face recognition. In: International conference on computer vision, II, Beijing, pp 1177–1184
8. Wang Y, Zhang L, Liu Z, Hua G, Wen Z, Zhang Z, Samaras D (2009) Face re-lighting from a single image under arbitrary unknown lighting conditions. IEEE Trans Pattern Recognit Mach Intell 31(11):1968–1984
9. Liu Z, Zhang Z, Jacobs C, Cohen M (2000) Rapid modeling of animated faces from video. In: Proceedings of visual 2000, Lyon, pp 58–67
10. Chen Q, Medioni G (2001) Building 3-d human face models from two photographs. J VLSI Signal Process 27(1–2): 127–140
11. Shan Y, Liu Z, Zhang Z (2001) Model-based boundle adjustment with application to face modeling. In: International conference on computer vision, Vancouver, pp 644–651
12. Akimoto T, Suenaga Y, Wallace RS (1993) Automatic 3d facial models. IEEE Comput Graph Appl 13(5):16–22
13. Ip HH, Yin L (1996) Constructing a 3d individualized head model from two orthogonal views. Vis Comput 12:254–266
14. Dariush B, Kang SB, Waters K (1998) Spatiotemporal analysis of face profiles: detection, segmentation, and registration. In: Proceeding of the 3rd international conference on automatic face and gesture recognition, IEEE, pp 248–253
15. Ramamoorthi R, Hanrahan P (2001) A signal-processing framework for inverse rendering. In: SIGGRAPH, Los Angeles, pp 117–128
16. Basri R, Jacobs D (2003) Lambertian reflectance and linear subspaces. Pattern Anal Mach Intell 25(2):218–233
17. Yang R, Zhang Z (2002) Eye gaze correction with stereovision for video tele-conferencing. In: Proceedings of the 7th European conference on computer vision (ECCV), Copenhagen, vol II, pp 479–494
18. Romdhani S, Blanz V, Vetter T (2002) Face identification by fitting a 3d morphable model using linear shape and texture error functions. In: European conference on computer vision (ECCV'2002), IV, Copenhagen, pp 3–19
19. Qing L, Shan S, Gao W, Du B (2005) Face recognition under generic illumination based on harmonic relighting. Int J Pattern Recognit Artif Intell 19(4):513–531
20. Zicheng L, Zhengyou Z (2011) Face geometry and appearance modeling: concepts and applications. Cambridge University Press

Face Reconstruction

►Face Modeling

Factorization

Hanno Ackermann
Leibniz University Hannover, Hannover, Germany

Synonyms

Structure-from-motion (SfM); Three dimensional reconstruction

Definition

Given arbitrary many images and 3D-points, the so-called *factorization* algorithm [1] is a noniterative technique for simultaneously estimating 3D-structure along with orientations and positions of the cameras which observed the 3D-scene or object. It is based on the factorization of a matrix consisting of all measurements.

The original algorithm requires a particular affine camera. It was later extended to more general affine cameras and even to the full projective camera model. Other variants can handle lines, triangles, or ellipses instead of 2D-point correspondences. It can be further extended to nonrigid scenes or objects, not merely rigid ones. It can also minimize geometric error criteria instead of algebraic errors and even allow unknown entries in the measurement matrix.

Background

Let the 3×3 upper triangular matrix K_i denote the calibration of the ith camera, the 3×3 rotation matrix R_i its orientation and the 3-vector t_i its position, and X_j^H the jth 3D-feature in homogeneous coordinates. The perspective projection x_{ij} of X_j^H by the ith camera can be modeled by

$$x_{ij} \simeq K_i \left[R_i \mid t_i \right] X_j^H. \tag{1}$$

The symbol $\simeq$ implies that every scalar multiple of the observed coordinates x_{ij} is a solution to this equation. Estimating all the camera parameters and the vectors X_j^H from the measurements x_{ij} at the same time is one of the oldest and most important problems in computer vision. The vectors X_j usually represent 3D-points in homogeneous coordinates, but they can also express lines, triangles, ellipses, or other features.

These days, the most widely used algorithm for structure-from-motion estimation, *bundle adjustment*, iteratively minimizes the reprojection error (The Euclidean distance between the projections of the estimated 3D-points X_j^H into the images and the measured 2D-points x_{ij}). It is regarded as the most accurate method, but it requires a good initialization and can handle only rigid scenes or rigid objects [2]. In this sense, the factorization algorithm is more flexible and better suited to problems that do not demand so high accuracy.

Theory

If the nonhomogeneous 3D-vectors X_j are projected into the images by an affine camera (Any pinhole camera is reasonably approximated by an affine camera if the object is very far away from the camera. This implies that the depth variation within the scene is neglectable small compared with the distance between camera and scene), (Eq. 1) can be simplified to

$$x_{ij} = K_i \left[R_i \mid t_i \right] X_j^H = \underbrace{K_i R_i}_{P_i} X_j + K_i t_i. \tag{2}$$

As opposed to the case of a projective camera in (Eq. 1), matrix K_i is 2×3 in (Eq. 2).

The vectors x_{ij} are collected into a measurement matrix for all m images and the n 3D-features

$$W = \begin{bmatrix} x_{11} & \cdots & x_{1n} \\ \vdots & \ddots & \vdots \\ x_{m1} & \cdots & x_{mn} \end{bmatrix} = \underbrace{\begin{bmatrix} P_1 \\ \vdots \\ P_m \end{bmatrix}}_{P} \underbrace{\begin{bmatrix} X_1 & \cdots & X_n \end{bmatrix}}_{X} + \underbrace{\begin{bmatrix} K_1 t_1 \\ \vdots \\ K_m t_m \end{bmatrix}}_{t} \begin{bmatrix} 1 & \cdots & 1 \end{bmatrix}. \tag{3}$$

Since matrices P and X both have rank 3 in general (Matrix P only has three columns and X only three rows; cf. (Eq. 2).) and the matrix given by the product of t and $\begin{bmatrix} 1 & \cdots & 1 \end{bmatrix}$ has rank 1, W cannot have more than rank 4, because each column of W is a linear combination of the columns of P and t.

The world coordinate system can be arbitrarily defined, so let us place it in such a way that the centroid of all X_i is at the coordinate origin, i.e., $\sum_{j=1}^{n} X_j = 0$. Then, the translation $K_i t_i$ is the centroid of all the observations of the ith image:

$$K_i t_i = \frac{1}{n} \sum_{j=1}^{n} x_{ij}, \tag{4}$$

Hence, if $K_i t_i$ is subtracted from x_{ij} to define

$$x'_{ij} = x_{ij} - K_i t_i, \tag{5}$$

the resulting measurement matrix is written as $W' = PX$, which has rank 3. The matrix W' can be factorized using the singular value decomposition in the form

$$W' = U \Sigma V^\top = U \left(\Sigma V^\top \right). \tag{6}$$

The original factorization algorithm in [1] decomposes W' in the form $\left(U\Sigma^{1/2}\right)\left(\Sigma^{1/2}V^\top\right)$, but the decomposition in (Eq. 6) is preferred for numerical reasons and simplicity of notation. Due to the rank-3 constraint, only the first three singular values on the diagonal of the matrix Σ are nonzero in the absence of noise. It therefore suffices to take only the corresponding three vectors in U and V and truncate Σ accordingly. The matrices U and $\left(\Sigma V^\top\right)$ define distorted camera parameters and a distorted 3D-structure, respectively, since singular value decomposition does not impose that each row-tuple of U are rows of a rotation matrix. Therefore, the computed camera parameters and the reconstructed 3D-structure are affinely distorted and need to be corrected in a subsequent step.

Since any nonsingular 3×3 matrix A may be inserted into the decomposition of W',

$$W' = UAA^{-1}\left(\Sigma V^\top\right), \tag{7}$$

without altering W', matrix A can be determined as second step of the factorization algorithm. Denote by $I_{3\times3}$ the 3×3 identity matrix. As in the case of self-calibrating a perspective camera by means of the dual absolute quadric, matrix A is determined from (Eq. 2) such that the two rows U_i of U corresponding to image i equal $K_i R_i$

$$\begin{aligned} K_i R_i &= U_i A \\ (K_i R_i)(K_i R_i)^\top &= (U_i A)(U_i A)^\top \\ K_i \underbrace{R_i R_i^\top}_{I_{3\times3}} K_i^\top &= U_i \underbrace{AA^\top}_{T} U_i^\top \\ K_i K_i^\top &= U_i T U_i^\top . \end{aligned} \tag{8}$$

For the simplest affine camera, the *orthographic* camera, the matrices K_i consist of the first two rows of a 3×3 identity matrix, i.e., $K_i K_i^\top$ equals the 2×2 identity matrix for all images. Hence, if $u_{i1}^\top$ and $u_{i2}^\top$ are the first and the second rows of U_i, respectively, the following three equations are obtained:

$$1 = u_{i1}^\top T u_{i1}, \tag{9}$$

$$1 = u_{i2}^\top T u_{i2}, \tag{10}$$

$$0 = u_{i1}^\top T u_{i2} = u_{i2}^\top T u_{i1}. \tag{11}$$

Equations 9–Eq. 11 constitute three linear equations in the six unknown entries of the symmetric matrix T per image. If we define $t_6 = \left[T_{11} \quad T_{22} \quad T_{33} \quad T_{12} \quad T_{13} \quad T_{23}\right]^\top$ where T_{ab} denotes the entry of T in the ath row and the bth column, and $H^{11} = u_{i1}u_{i1}^\top$, $H^{22} = u_{i2}u_{i2}^\top$, $H^{12} = u_{i1}u_{i2}^\top$, (Eq. 9)–(Eq. 11) can be written as

$$1 = \left[\ H_{11}^{11} \quad H_{22}^{11} \quad H_{33}^{11} \quad 2H_{12}^{11} \quad 2H_{13}^{11} \quad 2H_{23}^{11}\ \right] t_6, \tag{12}$$

$$1 = \left[\ H_{11}^{22} \quad H_{22}^{22} \quad H_{33}^{22} \quad 2H_{12}^{22} \quad 2H_{13}^{22} \quad 2H_{23}^{22}\ \right] t_6, \tag{13}$$

$$0 = \left[\ H_{11}^{12} \quad H_{22}^{12} \quad H_{33}^{12} \quad (H_{12}^{12} + H_{21}^{12}) \quad (H_{13}^{12} + H_{31}^{12}) \quad (H_{23}^{12} + H_{32}^{12})\ \right] t_6. \tag{14}$$

Here, H_{ab} denotes the (a, b)th entry of matrix H. For simplicity, the subscript i denoting the image is omitted.

After estimating t_6 from the three equations (Eq. 12)–(Eq. 14) per image using normal equations, the correcting matrix A can be determined by eigenvalue decomposition $T = UDU^\top$ as $A = UD^{1/2}$. The matrix of camera parameters P can be taken as $P = UA$ and the 3D-structure as $X = A^{-1}\Sigma V^\top$.

Extensions

The original factorization algorithm was extended to more sophisticated affine camera models. For the projective camera model, noniterative [3] and iterative [4, 5] extensions exist. The latter alternate matrix decomposition with refinement of the *projective depths* λ_{ij}, scalar variables are chosen so that

$$\lambda_{ij} x_{ij} = K_i \left[R_i \mid t_i\right] X_j^H \tag{15}$$

holds true under perspective projection. In [6], a projective extension is proposed which minimizes the reprojection error to obtain projectively distorted estimates of shape and motion parameters. The reprojection error is a geometric error measure instead of the algebraic error which is minimized by factorization schemes. There also exist algorithms which can handle missing observations, i.e., not all observations x_{ij} need to be known.

A generalization of the affine factorization algorithm to the *kinematic chain* model was proposed in [7]. An algorithm for scenes or objects which generally deform nonrigidly was introduced in [8]. Using prior knowledge on the shape and pose variety of human bodies, a projective factorization algorithm was proposed in [9] which can also handle missing observations.

Open Problems

Many unsolved problems remain in all variants of the factorization scheme. One of the most prominent difficulties stems from the requirement that all observations of the measurement matrix must be known. For the affine factorization scheme, the matrix decomposition step can be solved by an alternating projection method [10], yet such algorithms turn out to be unstable for the projective camera model and can therefore handle only small amounts of unknown observations. More robust algorithms combine different error metrics [11, 12] or use strong prior knowledge [9].

For iterative extensions, the problem of initialization reappears. If arbitrarily initialized, convergence to a reasonable local minimum is not guaranteed, so the resulting reconstructions can be strongly distorted.

Algorithms on nonrigid scenes or objects can handle small nonrigid deformations, unless restrictive prior knowledge is imposed. For the projective camera model without further prior knowledge, no successful 3D-reconstruction methods have been proposed which can reconstruct nonrigidly deforming scenes unless special devices, such as stereo cameras, are used or partially rigid scenes are assumed.

Lastly, the decomposition of matrix T is not possible if some eigenvalues are negative.

References

1. Tomasi C, Kanade T (1992) Shape and motion from image streams under orthography: a factorization method. Int J Comput Vis 9(2):137–154
2. Frahm J, Fite-Georgel P, Gallup D, Johnson T, Raguram R, Wu C, Jen Y, Dunn E, Clipp B, Lazebnik S, Pollefeys M (2010) Building rome on a cloudless day. In: Proceedings of the 11th European conference on computer vision (ECCV), Heraklion, Crete, Greece, pp 368–381
3. Sturm PF, Triggs B (1996) A factorization based algorithm for multi-image projective structure and motion. In: The proceedings of the fourth European conference on computer vision (ECCV), vol 2. Springer, London, pp 709–720
4. Heyden A, Berthilsson R, Sparr G (1999) An iterative factorization method for projective structure and motion from image sequences. Int J Comput Vis 17(13):981–991
5. Mahamud S, Hebert M (2000) Iterative projective reconstruction from multiple views. In: The proceedings of the IEEE conference on computer vision and pattern recognition (CVPR), vol 2. Hilton Head, SC, USA, pp 430–437
6. Hung YS, Tang WK (2006) Projective reconstruction from multiple views with minimization of 2d reprojection error. Int J Comput Vis 66(3):305–317
7. Yan J, Pollefeys M (2008) A factorization-based approach for articulated nonrigid shape, motion and kinematic chain recovery from video. IEEE Pattern Anal Mach Intell (PAMI) 30(5):865–877
8. Bregler C, Hertzmann A, Biermann H (2000) Recovering non-rigid 3d shape from image streams. In: IEEE computer vision and pattern recognition (CVPR), Hilton Head, pp 690–696
9. Hasler N, Ackermann H, Rosenhahn B, Thormählen T, Seidel H (2010) Multilinear pose and body shape estimation of dressed subjects from image sets. In: IEEE conference on computer vision and pattern recognition (CVPR), San Francisco
10. Wold H (1966) Estimation of principal components and related models by iterative least squares. In: Krishnaiah PR (ed) Multivariate analysis. Academic, New York, p p 391–420
11. Martinec D, Pajdla T (2002) Structure from many perspective images with occlusions. In 7th European conference on computer vision (ECCV), Copenhagen, pp 542–544
12. Ackermann H, Rosenhahn B (2010) A linear solution to 1-dimensional subspace fitting under incomplete data. In: Asian conference on computer vision (ACCV) Queenstown, NZ

Feature Reduction

►Feature Selection

Feature Selection

Rama Chellappa[1] and Pavan Turaga[2]
[1]Department of Electrical Engineering and Computer Engineering and UMIACS, University of Maryland, College Park, MD, USA
[2]Department of Electrical and Computer Engineering, Center for Automation Research University of Maryland, College Park, MD, USA

Synonyms

Dimensionality reduction; Feature reduction; Variable selection

Definition

Feature selection refers to a set of techniques for automatically extracting important features from raw observations, often in a task-dependent manner.

Background

Since early works in pattern analysis [1], there has been interest in extracting parsimonious and meaningful features from raw data for tasks such as recognition, compression, etc. While features based on domain knowledge prove useful, automatic methods for feature selection that can find optimal transformations of raw data are of particular interest. Toward this, linear and nonlinear methods for feature selection have been suggested. Invertible transforms are of particular interest when one is interested in obtaining the original signal, or its approximation, from features. Task specific criteria used for feature selection range from minimizing the reconstruction error to maximizing the separability between classes for classification tasks.

Theory

The basic intuition behind feature selection methods is to extract lower-dimensional information from high-dimensional data such as images, videos, etc., in a domain-dependent or task-specific manner. This is approached by restricting the complexity of the feature selection operator, such as linear or nonlinear feature selection, and then formulating a task-dependent criterion function that measures the quality of the obtained features. Then, the problem is cast as optimizing the criterion function over the set of admissible feature selection operators.

Given a set of data points denoted by $\mathcal{X} = \{x_1, x_2, \ldots, x_k\}$ where $x_i \in \mathbb{R}^n$, optionally along with lower-dimensional attributes or labels $l_i = \mathcal{L}(x_i)$ associated to each x_i, a set of admissible operators Φ, where for any $\phi \in \Phi$, $\phi : \mathbb{R}^n \rightarrow \mathbb{R}^d$, is a mapping from the raw data space to a feature space of dimension $d < n$, and a task-specific criterion function $\mathcal{E}(\mathcal{X}, \mathcal{L}, \phi)$, the goal of feature selection is to find a mapping $\phi \in \Phi$ such that the criterion function $\mathcal{E}$ is optimized. In computer vision, the x_i's typically are images or other statistics of images; the labels l_i's can be discrete-valued such as names of objects and identities of people, etc., or continuous-valued such as the pose of an object, the age of a person, etc.

Linear methods restrict ϕ to the set of linear transforms, which can be represented as a matrix multiplication given by $y = \phi(x) = \mathbf{W}^T x$, where $\mathbf{W}$ is a $n \times d$ matrix. Examples of linear feature selection methods include principal component analysis (PCA) [1], independent component analysis (ICA) [2], Fisher's linear discriminant analysis (FLDA) [3], support vector machines (SVM) [4], partial least squares (PLS) [5], boosting [6], etc. Nonlinear extensions of linear techniques are commonly achieved by means of the kernel method [7].

Examples

Principal Component Analysis: In PCA, the goal is to obtain linear projections that allow optimal reconstruction of data as measured in terms of the reconstruction error. Thus, given a set of data points $\mathcal{X}$, the goal is to obtain a $n \times d$ orthonormal matrix $\mathbf{W}$, such that $\mathcal{E}(\mathcal{X}, \mathbf{W}) = \sum_{i=1}^{k} \left\| x_i - \mu - \mathbf{W}\mathbf{W}^{\mathrm{T}} x_i \right\|^2$, is minimized, where μ is the mean of the set $\mathcal{X}$. For this case, the optimal $\mathbf{W}$ has columns which are the top d eigenvectors of the data covariance matrix [8, 9].

Fisher's Discriminant Analysis: In FLDA, along with a set of points $\{x_i\}$, one is given a set of discrete class labels $l_i \in \{1, 2, \ldots, c\}$. The goal is to obtain linear projections of the data so that separation between classes is increased with respect to the spread within each class. This is measured in terms of within-class and between-class scatter matrices. Denote $D_i = \{x \in \mathcal{X} | \mathcal{L}(x) = i\}$, the subset of points belonging to the ith class, and m and m_i as the mean of entire set $\mathcal{X}$ and the ith class, respectively. The withinclass scatter is defined as $S_W = \sum_{i=1}^{c} S_i$, where $S_i = \sum_{x \in D_i} (x - m_i)(x - m_i)^T$, and the between-class scatter is defined as $S_B = \sum_{i=1}^{c} n_i (m_i - m)(m_i - m)^T$, where n_i is the cardinality of D_i. Then, the criterion function to be maximized is $\mathcal{E}(\mathcal{X}, \mathcal{L}, \mathbf{W}) = \frac{|\mathbf{W}^{\mathrm{T}} S_B \mathbf{W}|}{|\mathbf{W}^{\mathrm{T}} S_W \mathbf{W}|}$. For this case, the optimal $\mathbf{W}$ has columns which are the top d generalized eigenvectors of $S_B \mathbf{w}_i = \lambda_i S_W \mathbf{w}_i$ [9].

Projection Pursuit: Projection pursuit is a broad class of techniques for exploratory data analysis where the goal is to find projections of data along interesting directions. This is typically stated as searching for directions of non-Gaussianity. Criterion functions to measure non-Gaussianity include entropy and higher-order moments such as the kurtosis. For the case of entropy, the Gaussian distribution has the largest entropy among the class of zero mean and unit variance densities. Thus, to maximize non-Gaussianity, the criterion function to be maximized is $E[f(\mathbf{W}^T x) \log f(\mathbf{W}^T x)]$ (the negative entropy), where f is the estimated probability density function of the projections [10].

Application

Feature selection has found very wide applications in face recognition [11–13]. Kernelized versions of PCA, LDA [14], and SVMs [15] have been successful in face recognition. Similar applications have been proposed in numerous object recognition tasks. Recently, the method of partial least squares has found successful application in human detection and face recognition [16, 17].

Open Problems and Recent Trends

Feature-Selection for Non-Euclidean Manifolds: The techniques for feature selection described above are implicitly designed to operate on Euclidean spaces. The definitions of various entities used – L_2 distortion measure, the mean and covariances – are valid only for Euclidean spaces. In computer vision, there are several applications where the data space is a non-Euclidean manifold embedded in a larger ambient Euclidean space. Examples of analytical manifolds that appear in vision include the hypersphere, Hilbert sphere, Grassmann manifold, covariance matrices, etc. Applications of these manifolds in vision include shape analysis [18], human detection and tracking [19], and activity analysis [20]. Often, it is possible to extend feature-selection concepts to manifolds by taking recourse to Riemannian geometry. The significant modification lies in redefining the notions of distances in terms of manifold geodesics and statistical quantities such as means and covariances in terms of intrinsic statistics [21, 22]. For example, the counterpart of PCA for manifolds is called principal geodesic analysis (PGA) [23]. However, these procedures are often approximations, and it is very hard in general to quantify their degree of accuracy.

Further, in many cases, the geometry of the manifold is not known analytically, but one has access to several samples from the manifold. In this setting, one first takes recourse to manifold learning techniques such as LLE [24], Isomaps [25], Laplacian eigenmaps [26], etc. Manifold learning techniques aim to find an embedding of the manifold to a Euclidean space. Since any such embedding results in distortions to the geometric properties of the manifold, different algorithms aim to preserve different properties of the original manifold. Once an acceptable embedding into a Euclidean space is found, feature selection can proceed using one of the existing algorithms mentioned before. However, the number and density of samples available on the manifold strongly affects the quality of features obtained.

Feature Selection and Sensing: There has been recent interest in designing sensors and cameras that can directly sense the required task-dependent features instead of sensing first and then performing feature selection. By directly sensing features, one can potentially reduce the load on the sensor without loss of performance in the given application. This is achieved in practice by modifying camera elements, such as by introducing a programmable micro mirror array [27] or optical masks [28], in a way that the required transformations of the raw data are directly sensed. However, different designs are needed to preserve different properties of the images. In parallel, there has been recent interest in sensing "universal" features by random projections. The Johnson-Lindenstrauss lemma is used as motivation for using random projections directly for inference tasks [29]. Cameras to sense random projections of data have been designed using micro-mirror arrays [30]. Though these are application agnostic, the dimensionality required to preserve the universality of these features keeps increasing with the number of data points.

References

1. Hotelling H (1933) Analysis of a complex of statistical variables into principal components. J Educ Psychol 24:417–441
2. Hyvärinen A, Oja E (2000) Independent component analysis: algorithms and applications. Neural Netw 13:411–430

3. Fisher RA (1936) The use of multiple measurements in taxonomic problems. Ann Eugen 7:179–188
4. Cortes C, Vapnik V (1995) Support-vector networks. Mach Learn 20(3):273–297
5. Wold H (1985) Partial least squares. In: Kotz S, Johnson NL (eds) Encyclopedia of statistical sciences. Wiley, New York, pp 581–591
6. Freund Y, Schapire RE (1995) A decision-theoretic generalization of on-line learning and an application to boosting. In: Proceedings of the Second European Conference on Computational Learning Theory. Barcelona, Spain, pp 23–37
7. Shawe-Taylor J, Cristianini N (2004) Kernel methods for pattern analysis. Cambridge University Press, New York
8. Jolliffe IT (1986) Principal component analysis. Springer, New York
9. Duda RO, Hart PE, Stork DG (2001) Pattern classification, 2nd edn. Wiley, New York
10. Jones MC, Sibson R (1987) What is projection pursuit? J R Stat Soc 150(1):1–37
11. Turk M, Pentland A (1991) Eigenfaces for recognition. J Cognit Neurosci 3(1):71–86
12. Etemad K, Chellappa R (1997) Discriminant analysis for recognition of human face images. J Opt Soc Am A 14:1724–1733
13. Bartlett MS, Movellan JR, Sejnowski TJ (2002) Face recognition by independent component analysis. IEEE Trans Neural Netw 13:1450–1464
14. Yang MH (2002) Kernel Eigenfaces vs. Kernel Fisherfaces: face recognition using Kernel methods. In: Proceedings of the fifth IEEE international conference on automatic face and gesture recognition, Washington DC, p 215
15. Guo G, Li SZ, Chan K (2000) Face recognition by support vector machines. In: Proceedings of the fourth IEEE international conference on automatic face and gesture recognition, Grenoble, p 196
16. Schwartz WR, Kembhavi A, Harwood D, Davis LS (2009) Human detection using partial least squares analysis. In: International conference on computer vision, Kyoto
17. Schwartz WR, Guo H, Davis LS (2010) A robust and scalable approach to face identification. In: European Conference on Computer Vision (ECCV), Crete
18. Srivastava A, Joshi SH, Mio W, Liu X (2005) Statistical shape analysis: clustering, learning, and testing. IEEE Trans Pattern Anal Mach Intell 27:590–602
19. Tuzel O, Porikli F, Meer P (2008) Pedestrian detection via classification on Riemannian manifolds. IEEE Trans Pattern Anal Mach Intell 30:1713–1727
20. Veeraraghavan A, Srivastava A, Roy Chowdhury AK, Chellappa R (2009) Rate-invariant recognition of humans and their activities. IEEE Trans Image Process 18: 1326–1339
21. Bhattacharya R, Patrangenaru V (2003) Large sample theory of intrinsic and extrinsic sample means on manifolds-I. Ann Stat 31(1):1–29
22. Pennec X (2006) Intrinsic statistics on Riemannian manifolds: basic tools for geometric measurements. J Math Imaging Vis 25:127–154
23. Fletcher PT, Lu C, Pizer SM, Joshi S (2004) Principal geodesic analysis for the study of nonlinear statistics of shape. IEEE Trans Med Imaging 23:995–1005
24. Roweis ST, Saul LK (2000) Nonlinear dimensionality reduction by locally linear embedding. Science 290: 2323–2326
25. Tenenbaum JB, de Silva V, Langford JC (2000) A global geometric framework for nonlinear dimensionality reduction. Science 290:2319–2323
26. Belkin M, Niyogi P (2003) Laplacian eigenmaps for dimensionality reduction and data representation. Neural Comput 15(6):1373–1396
27. Nayar SK, Branzoi V, Boult TE (2006) Programmable imaging: towards a flexible camera. Int J Comput Vis 70:7–22
28. Veeraraghavan A, Raskar R, Agrawal A, Mohan A, Tumblin J (2007) Dappled photography: mask enhanced cameras for heterodyned light fields and coded aperture refocusing. ACM Trans Graph 26:69
29. Johnson WB, Lindenstrauss J (1984) Extensions of Lipschitz mappings into a Hilbert space. In: Conference on modern analysis and probability. American Mathematical Society, Providence, RI, pp 189–206
30. Duarte MF, Davenport MA, Takhar D, Laska JN, Sun T, Kelly KF, Baraniuk RG (2008) Single-pixel imaging via compressive sampling. IEEE Signal Process Mag 25:83–91
31. Candes EJ, Tao T (2006) Near-optimal signal recovery from random projections: universal encoding strategies? IEEE Trans Inf Theory 52(12):5406–5425

Field of View

Srikumar Ramalingam
Mitsubishi Electric Research Laboratories, Cambridge, MA, USA

Synonyms

Field of vision

Related Concepts

►Center of Projection

Definition

Field of view refers to the angular volume of 3D space sampled by the light rays of a camera.

Background

The pinhole camera is one of the most successful mathematical models in the field of computer vision, image processing, and graphics. People naturally accepted this imaging model because of its extreme simplicity and its closeness to an image perceived by the human visual system. In a pinhole camera, the projection rays from scene points (light rays) intersect at a single point (optical center). Typically, these conventional cameras have a very small field of view, around 50°. Omnidirectional cameras have a larger field of view and have been extremely useful in several applications like videoconferencing, augmented reality, surveillance, and large-scale 3D reconstruction. These cameras can be constructed in a simple manner, for they can be made from conventional cameras by using additional lenses or mirrors. For example, Fig. 1a shows an E8 Nikon Coolpix camera appended with a fish-eye lens having a field of view of 183° × 360°. Another possibility is to use mirrors in addition to lenses to increase the field of view. These configurations are referred to as *catadioptric*, where "cata" comes from mirrors (reflective) and "dioptric" comes from lenses (refractive). Figure 1b, c show two catadioptric configurations with hyperbolic and parabolic mirrors, respectively.

Theory and Camera Models

Fish-Eye Camera Model

Fish-eye lenses have a short focal length and a very large field of view (cf. Fig. 2a). However, when the field of view is greater than 180°, the concept of focal length is not defined. For example, the focal length is not defined for the E8 fish-eye lens of Nikon which has a field of view of 183° × 360°. Several works have used fish-eye lenses for creating omnidirectional images [1–3]. Geometrically, omnidirectional cameras can be either single viewpoint or noncentral. Single viewpoint configurations are preferred to noncentral systems because they permit the generation of geometrically correct perspective images from the image(s) captured by the camera. In addition, most theories and algorithms developed for conventional cameras hold good for single-center omnidirectional cameras. In theory, fish-eye lenses do not provide a single viewpoint imaging system [4]. The projection rays pass through a small disk in space rather than a single point. Nevertheless, in practice, it is usually a good assumption to consider these cameras as single viewpoint cameras [5]. Perspective images synthesized from fish-eye images are visibly very accurate without any distortions. Several distortion functions can be used to model fish-eye and central catadioptric images [6, 7]. Some of them are mentioned below.

- *Stereographic projection*: Several radially symmetric models [8] were used for fish-eye images. One of them is the stereographic projection. This model gives a relation between θ, the angle made by a scene point, the optical center and the optical axis, and the distance r between the associated image point and the distortion center

$$r = k \tan \frac{\theta}{2}$$

  where k is the only parameter to be estimated.
- *Equidistant projection*:

$$r = k\theta$$

- *Equisolid angle projection*:

$$r = k \sin \frac{\theta}{2}$$

- *Sine law projection*:

$$r = k \sin \theta$$

  On fitting the Nikon FC-E8 fish-eye lens with the four radially symmetric models (stereographic, equidistant, equisolid angle, and sine law), it was found that the stereographic projection gave the lowest error [9]. The error is the Euclidean distance between the original image pixels and the projected image pixels using the models.
- *Combined stereographic and equisolid angle model*: In [9], Bakstein and Pajdla followed a model-fitting approach to identify the right projection model for fish-eye cameras. Their model is a combination of stereographic and equisolid angle models. The following relation was obtained with four parameters:

$$r = a \tan \frac{\theta}{b} + c \sin \frac{\theta}{d}$$

  On the whole, they used 13 camera parameters: six external motion parameters $(\mathsf{R}, \mathbf{t})$, one aspect

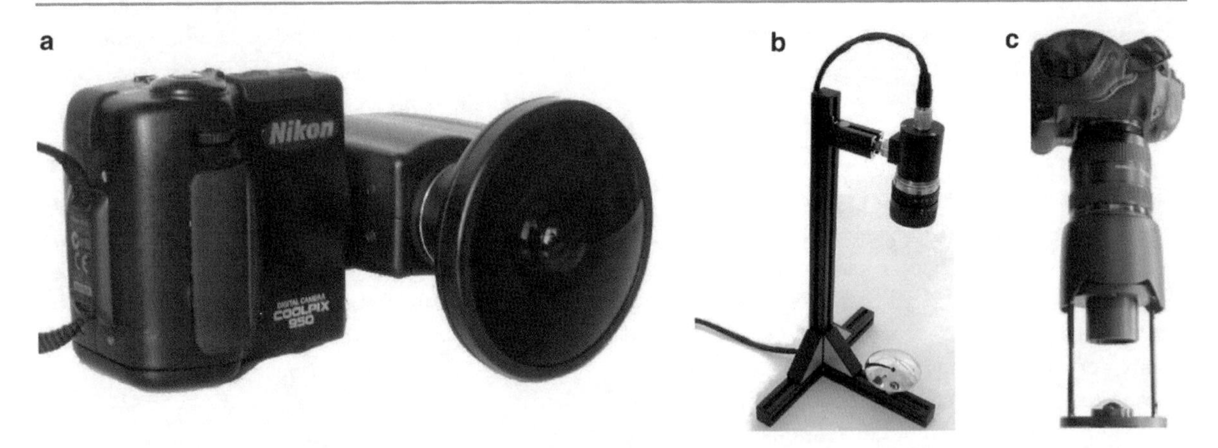

Field of View, Fig. 1 (**a**) Fisheye camera (**b**) Catadioptric configuration using a hyperbolic mirror and a pinhole camera. (**c**) Catadioptric configuration using a parabolic mirror and an orthographic camera

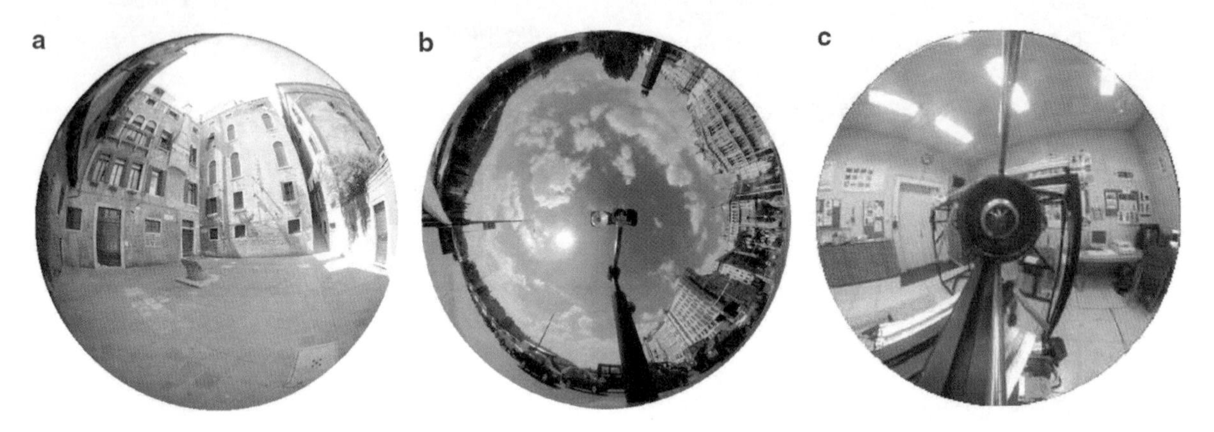

Field of View, Fig. 2 Omnidirectional images captured by three different cameras are shown: (**a**) Image of a fish-eye camera (**b**) image of a catadioptric camera using a hyperbolic mirror and a pinhole camera. (**c**) image of a catadioptric camera using a parabolic mirror and an orthographic camera. Courtesy of Branislav Micusik and Tomas Pajdla

ratio (β), two parameters for the principal point (u_0, v_0), and the four parameters of the above projection model (a, b, c, d).

- *Polynomial lens distortion model*: Most distortion corrections assume the knowledge of the distortion center. Let r_d refer to the distance of an image point from the distortion center. The distance of the same image point in the undistorted image is given by

$$r_u = r_d(1 + k_1 r_d^2 + k_2 r_d^4 + \ldots)$$

where k_1 and k_2 are distortion coefficients [10].

- *Field of view (FOV)*: The distortion function is given by

$$r_u = \frac{\tan(r_d w)}{2 \tan \frac{w}{2}}$$

The above distortion correction function is based on a single parameter w. It is a good idea to correct the distortion using the polynomial model followed by the field of view model [6].

- *Division model (DM)*: The distortion correction function is given by

$$r_u = \frac{r_d}{(1 + k_1 r_d^2 + k_2 r_d^4 + \ldots)}$$

where the k_i are the distortion coefficients.

Catadioptric Camera Model

Vision researchers have been interested in catadioptric cameras [4, 11–17] because they allow numerous possibilities in constructing omnidirectional cameras. The possibilities arise from the differences in size, shape, orientation, and positioning of the mirrors with respect to the camera. Please refer to the encyclopedia entry on catadioptric camera for more details.

Application

- *Larger field of view*: Fig. 2 shows images captured by three different omnidirectional cameras. The first image is captured by a fish-eye camera, the second is captured by a catadioptric system constructed using a hyperbolic mirror and a perspective camera, and finally, the third is captured by another catadioptric camera constructed using a parabolic mirror and an orthographic camera. One can make the following observation from the omnidirectional images. A very large scene, which usually requires several pinhole images, can be captured in a single omnidirectional image although of course at a lower resolution.
- *Stable motion estimation*: Motion estimation is a challenging problem for pinhole images, especially when a larger number of images are involved. On the other hand, omnidirectional cameras stabilize the motion estimation and improves its accuracy [18–20]. In the case of small rigid motions, two different motions can yield nearly identical motion fields for classical perspective cameras. However, this is impossible in the case of omnidirectional cameras. By improving the stability of motion estimation, omnidirectional cameras also contribute to a stable 3D reconstruction.

A detailed survey of various camera models, calibration, and 3D reconstruction algorithms is given in [21, 22].

References

1. Miyamoto K (1964) Fish eye lens. J Opt Soc Am 54(8):1060–1061
2. Slater J (1932) Photography with the whole sky lens. Am Photogr
3. Wood R (1902) Fish-eye view, and vision under water. Philos Mag 12:159–162
4. Nalwa V (1996) A true omnidirectional viewer. Technical report, Bell Laboratories, Holmdel
5. Ying X, Hu Z (2004) Can we consider central catadioptric cameras and fisheye cameras within a unified imaging model. European conference on computer vision (ECCV), Prague, pp 442–455
6. Devernay F, Faugeras O (2001) Straight lines have to be straight. Mach Vis Appl 13:14–24
7. Claus D, Fitzgibbon A (2005) A rational function lens distortion model for general cameras. International conference on computer vision, vol 1, pp 213–219
8. Fleck M (1995) The wrong imaging model. Technical Report TR 95-01, University of Iowa
9. Bakstein H, Pajdla T (2002) Panoramic mosaicing with a 180 field of view lens. IEEE workshop on omnidirectional vision, Copenhagen
10. Brown D (1971) Close-range camera calibration. Photogr Eng 37(8):855–866
11. Bogner S (1995) Introduction to panoramic imaging. IEEE SMC, vol 54. pp 3100–3106
12. Charles J, Reeves R, Schur C (1987) How to build and use an all-sky camera. Astron Mag
13. Hong J (1991) Image based homing. International conference on robotics and automation
14. Murphy JR (1995) Application of panoramic imaging to a teleoperated lunar rover. IEEE SMC conference, pp 3117–3121
15. Rees DW (1970) Panoramic television viewing system. United States Patent (3,505,465)
16. Yagi Y, Kawato S (1990) Panoramic scene analysis with conic projection. International conference on robots and systems (IROS)
17. Yamazawa K, Yagi Y, Yachida M (1995) Obstacle avoidance with omnidirectional image sensor hyperomni vision. International conference on robotics and automation, pp 1062–1067
18. Brodsky T, Fernmuller C, Aloimonos Y (1996) Directions of motion fields are hardly ever ambiguous. European conference on computer vision (ECCV), vol 2, 110–128
19. Koch O, Teller S (2007) Wide-area egomotion estimation from known 3d structure. IEEE conference on computer vision pattern recognition (CVPR)
20. Ramalingam S, Bouaziz S, Sturm P, Brand M (2010) Skyline2gps: geo-localization in urban canyons using omni-skylines. IROS
21. Sturm P, Ramalingam S, Tardif JP, Gasparini S, Barreto J (2011) Camera models and fundamental concepts used in geometric computer vision. Found Trends Comput Graph Vis 6(1–2):1–183
22. Ramalingam S (2006) Generic imaging models: calibration and 3d reconstruction algorithms. PhD Thesis, INRIA Rhone Alpes

Field of Vision

▸Field of View

Filling In

▸Inpainting

Fisher-Rao Metric

Stephen J. Maybank
Department of Computer Science and Information Systems, Birkbeck College University of London, London, UK

Synonyms

Rao Metric

Related Concepts

►Fisher-Rao Metric; ►Maximum Likelihood Estimation

Definition

The Fisher-Rao metric is a particular Riemannian metric defined on a parameterized family of conditional probability density functions (pdfs). If two conditional pdfs are near to each other under the Fisher-Rao metric, then the square of the distance between them is approximated by twice the average value of the log likelihood ratio of the conditional pdfs.

Background

Suppose that a parameterized family of conditional pdfs is given and it is required to find the parameter value corresponding to the conditional pdf that best fits a given set of data. It is useful to have a distance function defined on pairs of conditional pdfs, such that if a given conditional pdf is a close fit to the data, then all the conditional pdfs near to it are also close fits to the data. Any such distance function should be independent of the choice of parameterization of the family of conditional pdfs and independent of the choice of parameterization of the data. The Fisher-Rao metric is the only known Riemannian metric which yields a distance function with both the required independence properties [1].

Theory

Let X be an open subset of a Euclidean space $\mathbb{R}^n$ and let T be an open subset of a Euclidean space $\mathbb{R}^d$. Let x, θ be vectors in X and T, respectively, and let $p(x|\theta)$ be a probability density function defined for x in X and conditional on θ in T. The set T is a parameter space for the family of conditional pdfs $\theta \mapsto p(x|\theta)$. Let θ_i for $1 \leq i \leq d$ be the components of θ. With these choices of parameterization for X and T, the Fisher-Rao metric on T is defined by the following family of $d \times d$ matrices:

$$J_{ij}(\theta) = -\int_X \left(\frac{\partial^2}{\partial\theta_i \partial\theta_j} \ln p(x|\theta)\right) p(x|\theta)\, dx,$$
$$1 \leq i, j \leq d, \theta \in T. \quad (1)$$

The same matrix $J(\theta)$ is also defined by the formulae:

$$J_{ij}(\theta) = \int_X \left(\frac{\partial}{\partial\theta_i} \ln p(x|\theta)\right)\left(\frac{\partial}{\partial\theta_j} \ln p(x|\theta)\right) p(x|\theta)\, dx,$$
$$1 \leq i, j \leq d, \theta \in T. \quad (2)$$

Rao notes in [6] that $J(\theta)$ defines a Riemannian metric on T. For information about (Eq. 1) and (Eq. 2), see [1], Sect. 2.3.

Let $\theta + \Delta\theta$ be a point in T near to θ, and let $\mathrm{dist}(\theta, \theta + \Delta\theta)$ be the length of the shortest path in T from θ to $\theta + \Delta\theta$, as measured using the Fisher-Rao metric. This shortest path is called the geodesic between θ and $\theta + \Delta\theta$. It can be shown that:

$$\Delta\theta^\top J(\theta)\Delta\theta = \mathrm{dist}(\theta, \theta + \Delta\theta)^2 + O(\|\Delta\theta\|^3). \quad (3)$$

The Fisher-Rao metric is also defined when X is a finite set. Let X have n elements and let θ_i be the probability of the ith element of X. The θ_i for $1 \leq i \leq n$ are not independent because they sum to 1. Let θ be the vector formed from the probabilities θ_i for $1 \leq i \leq n-1$. The parameter space T is the open subset of $\mathbb{R}^{n-1}$ consisting of vectors θ with components θ_i such that:

$$0 < \theta_i < 1, \quad 1 \leq i \leq n-1,$$
$$0 < \sum_{i=1}^{n-1} \theta_i < 1.$$

The Fisher-Rao metric is defined on T by the following $(n-1) \times (n-1)$ matrix:

$$J_{ij}(\theta) = \theta_n^{-1}, \quad 1 \le i, j \le n-1, i \ne j,$$
$$J_{ii}(\theta) = \theta_n^{-1} + \theta_i^{-1}, \quad 1 \le i \le n-1.$$

Let θ and $\theta + \Delta\theta$ be nearby points in T. Then the approximation (Eq. 3) to the square of the distance between the pdfs $p(x|\theta)$ and $p(x|\theta + \Delta\theta)$ is:

$$\sum_{i=1}^{n} \theta_i^{-1} (\Delta\theta_i)^2.$$

Let $D(\theta(1)\|\theta(2))$ be the Kullback-Liebler divergence between the conditional pdfs $p(x|\theta(1))$ and $p(x|\theta(2))$ [1]. In the continuous case, in which X is an open subset of $\mathbb{R}^n$, the Kullback-Leibler divergence is given by:

$$D(\theta(1)\|\theta(2)) = \int_X \ln\left(\frac{p(x|\theta(1))}{p(x|\theta(2))}\right) p(x|\theta(1))\,dx. \tag{4}$$

On setting $\theta = \theta(1)$, $\theta + \Delta\theta = \theta(2)$, it follows from (Eq. 4) that:

$$\Delta\theta^\top J(\theta)\Delta\theta = \frac{1}{2} D(\theta\|\theta + \Delta\theta) + O(\|\Delta\theta\|^3). \tag{5}$$

An equation similar to (Eq. 5) holds when X is a finite set.

The matrix $J(\theta)$ used to define the Fisher-Rao metric appears in the theory of maximum likelihood estimation [2]. Suppose that N points $x(i)$ for $1 \le i \le N$ are sampled independently from X using the conditional pdf $p(x|\theta)$. Let $\hat{\theta}$ be the maximum likelihood estimate of θ:

$$\hat{\theta} = \operatorname{argmax} \phi \mapsto \prod_{i=1}^{N} p(x(i)|\phi).$$

If N is large, then the distribution of $\theta - \hat{\theta}$ is closely approximated by a Gaussian distribution with expected value 0 and covariance:

$$N^{-1} J(\theta)^{-1}.$$

Let N be any fixed positive integer, let $\phi \equiv \phi(x(1), \ldots, x(N))$ be any unbiased estimator of θ, and let C be the covariance of ϕ. Then the matrix:

$$C - N^{-1} J(\theta)^{-1}$$

is positive semi-definite. The matrix $N^{-1} J(\theta)^{-1}$ is known as the Cramér-Rao lower bound for C [1].

In the continuous case, in which X is an open subset of $\mathbb{R}^n$, it is rare to find a closed form expression for the Fisher-Rao metric. An example of a closed form expression is provided by the family of Gaussian densities for which $X = \mathbb{R}$. Let $\theta = (\mu, t)$, such that μ, t are points in $\mathbb{R}$ with $t > 0$. The parameter space T is the upper half plane in $\mathbb{R}^2$. Let $p(x|\theta)$ be the Gaussian pdf for x in $\mathbb{R}$ with expected value μ and standard deviation $t/\sqrt{2}$. In this case, the scaled Fisher-Rao metric, $(1/2)J(\theta)$, coincides with the Poincaré metric on T:

$$\frac{1}{2} J(\theta) = \frac{1}{t^2}\begin{pmatrix} 1 & 0 \\ 0 & 1 \end{pmatrix}, \quad \theta \in T.$$

It is shown in [3] that under certain conditions the Fisher-Rao metric can be closely approximated by a simpler metric.

Application

The Fisher-Rao metric provides a theoretical basis for the Hough transform which is used to detect geometrical structures such as lines and circles. The size and shape of the Hough transform accumulators and the number of accumulators can be calculated using the Fisher-Rao metric [3]. Peter and Rangarajan [5] describe planar shapes using weighted sums of Gaussian pdfs. The associated Fisher-Rao metric is used to define geodesics in the parameter manifold for the shape pdfs. Each segment of a geodesic specifies a continuous family of shapes which interpolate between the two shapes represented by the end points of the segment. An algorithm to find geodesics is described in [4].

References

1. Amari S-I (1985) Differential-geometric methods in statistics. Lecture notes in statistics, vol. 28. Springer, New York
2. Fisher RA (1922) On the mathematical foundations of theoretical statistics. Phil Trans R Soc London Ser A 222:309–368
3. Maybank SJ (2004) Detection of image structures using Fisher Information and the Rao metric. IEEE Trans Pattern Anal Mach Intell 26(12):49–62
4. Mio W, Liu X (2006) Landmark representations of shapes and Fisher-Rao geometry. In: Proceedings of the IEEE

international conference on image processing, Atlanta. IEEE, pp 2113–2116
5. Peter A, Rangarajan A (2009) Information geometry for landmark shape analysis: unifying shape representation and deformation. IEEE Trans Pattern Anal Mach Intell 31(2):337–350
6. Rao C (1945) Information and accuracy attainable in the estimation of statistical parameters. Bull Calcutta Math Soc 37:81–91

Fisheye Camera

▸Omnidirectional Camera

Fisheye Lens

Peter Sturm
INRIA Grenoble Rhône-Alpes, St Ismier Cedex, France

Related Concepts

▸Omnidirectional Vision

Definition

A fisheye lens is a lens giving a field of view of about 180° or larger.

Background

The terms fisheye camera and fisheye view seem to have been introduced by Wood in 1906 [1]. Wood was interested in the way a fish perceives objects outside the water. Besides studying the problem theoretically, he also built a camera that mimics the fisheye view. To do so, he immersed a pinhole camera in a casing filled with water and that had a glass plate as one of its faces, through which the camera could acquire images of the outside world. One basic observation Wood made is that since the camera looks from a denser medium (water) into a lighter one (air), its effective field of view is larger than its native one, due to the refraction happening at the interface between the media. This effect is related to the so-called Snell's window. In particular, when looking from water into air and supposing that the water surface is still, the entire hemisphere above the water can be seen within a circular cone-shaped field of view with an opening angle of about 96°.

In the 1920s, Bond and Hill independently invented purely glass-based fisheye lenses that, like Wood's water-based camera, gave hemispherical fields of view [2, 3]. Many improvements were made subsequently, for instance, on chromatic aberrations; see, for example, [4–6]. Fisheye lenses can achieve larger than hemispheric fields of view, for example, Martin reports a design with a 310° field of view [6]. An example of a fisheye lens and an image acquired using one is given in Fig. 1.

Theory

Since fisheye cameras capture an entire hemisphere or more in a single image, the image is bound to show strong distortions. These cameras can thus not be modeled using a pinhole model and even classical polynomial distortion models are insufficient. Better suited models have been proposed [4, 7, 8], as follows. Let ϕ be the angle between the optical axis and an incoming light ray and θ the angle between the optical axis and the ray leaving the lens towards the image plane. Then, we can make the following definitions:

$$\begin{aligned}
\text{Perspective projection: } & \theta = \phi \\
\text{Stereographic projection: } & \tan\theta = k \tan\frac{\phi}{2} \\
\text{Equidistant projection: } & \tan\theta = k\phi \\
\text{Equi-solid angle projection: } & \tan\theta = k \sin\frac{\phi}{2} \\
\text{Sine-law projection: } & \tan\theta = k \sin\phi,
\end{aligned}$$

where k is a free parameter.

It is sometimes useful to express these projection models with respect to the distance r of an image point from the principal point:

$$\begin{aligned}
\text{Perspective projection: } & r = m \tan\phi \\
\text{Stereographic projection: } & r = m \tan\frac{\phi}{2} \\
\text{Equidistant projection: } & r = m\phi \\
\text{Equi-solid angle projection: } & r = m \sin\frac{\phi}{2} \\
\text{Sine-law projection: } & r = m \sin\phi,
\end{aligned}$$

Fisheye Lens, Fig. 1 Two images of a fisheye conversion lens and an image taken with a fisheye lens. The image is necessarily heavily distorted since it "contains" an entire hemispheric field of view

where m is a free parameter, proportional to the camera's focal length.

Various other models for fisheye lenses and other omnidirectional cameras have been proposed in the literature, for instance [5, 9–12]. These and other models are described in [13], which also provides references to calibration methods. It seems that most fisheye lenses are designed to approach the equidistant model. In practice, an accurate calibration of a fisheye camera may require to add a classical polynomial distortion model "on top" of a specific fisheye projection model.

Application

Among the first applications of fisheye lenses were meteorology, via the study of cloud formations, and forest management, via the assessment of leaf coverage via fisheye images of forest canopies. Other applications are the same as those of other omni-directional cameras, where a wide field of view is beneficial, for instance in mobile robotics or video surveillance.

References

1. Wood R (1906) Fish-eye views, and vision under water. Philos Mag 6 12(68):159–162
2. Bond W (1922) A wide angle lens for cloud recording. Philos Mag 44(263):999–1001
3. Hill R (1924) A lens for whole sky photographs. Q J R Meteorol Soc 50(211):227–235
4. Miyamoto K (1964) Fish eye lens. J Opt Soc Am 54(8):1060–1061
5. Kumler J, Bauer M (2000) Fish-eye lens designs and their relative performance. In: Proceedings of SPIE conference on current developments in lens design and optical systems engineering, San Diego, USA, pp 360–369
6. Martin C (2004) Design issues of a hyper-field fisheye lens. In: Sasián J, Koshel R, Manhart P, Juergens R (eds) Proceedings of SPIE conference on novel optical systems design and optimization VII, Bellingham, WA, vol 5524, pp 84–92
7. Beck C (1925) Apparatus to photograph the whole sky. J Sci Instrum 2(4):135–139
8. Fleck M (1995) Perspective projection: The wrong imaging model. Technical Report TR 95–01, Department of Computer Science, University of Iowa, Iowa City, IA 52242, USA
9. Herbert T (1986) Calibration of fisheye lenses by inversion of area projections. Appl Opt 25(12):1875–1876
10. Gennery D (2006) Generalized camera calibration including fish-eye lenses. Int J Comput Vis 68(3):239–266
11. Bakstein H, Pajdla T (2002) Panoramic mosaicing with a 180° field of view lens. In: Proceedings of the workshop on omnidirectional vision, Copenhagen, Denmark, pp 60–68
12. Devernay F, Faugeras O (2001) Straight lines have to be straight. Mach Vis Appl 13(1):14–24
13. Sturm P, Ramalingam S, Tardif JP, Gasparini S, Barreto J (2011) Camera models and fundamental concepts used in geometric computer vision. Found Trends Comput Graph Vis 6(1–2):1–183
14. Wikipedia (2011) Fisheye lens. http://en.wikipedia.org/wiki/Fisheye_lens. Accessed 3 Aug 2011

Fluorescent Lighting

Stephen Lin
Microsoft Research Asia, Beijing Sigma Center, Beijing, China

Definition

Fluorescent lighting is the illumination of a scene by the fluorescence of phosphor in a gas-discharge lamp.

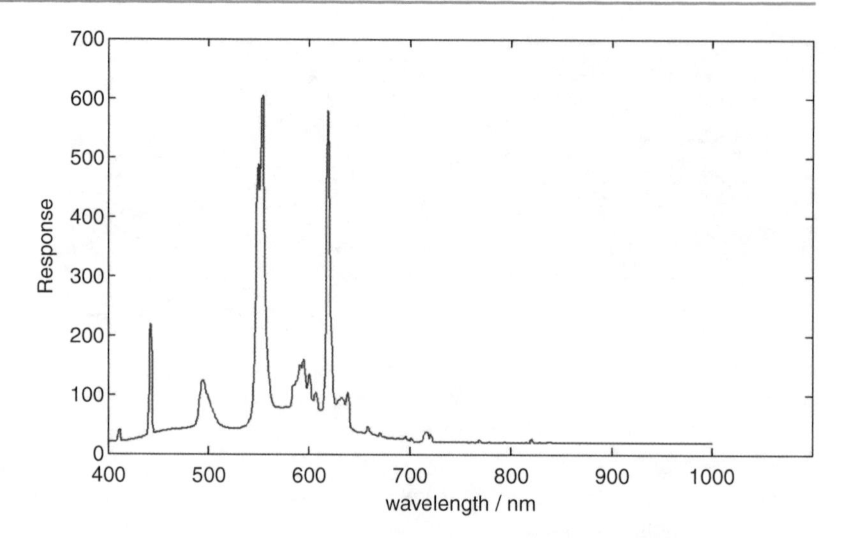

Fluorescent Lighting, Fig. 1 A typical color spectrum of fluorescent light

Background

In fluorescent lighting, electrical energy is converted into radiant energy by a physical process in the fluorescent lamp. The lamp consists of a sealed glass tube containing mercury vapor, an inert gas (such as argon) at low pressure, and a phosphor coating on the inside surface. After ionizing the gas, electrical current flows through the gas between electrodes at the ends of the tube. The current excites the mercury atoms, which then emit ultraviolet light. Ultraviolet (UV) light is not visible to the human eye, but is used to cause fluorescence of the phosphor, which absorbs the UV radiation and produces light in the visible range of the color spectrum. Additional details on fluorescent lamp operation can be found in [3].

Fluorescent light has certain properties that distinguish it from other forms of illumination. These include a color spectrum with sharp peaks that correspond to the chemical composition of the phosphor, and flicker at twice the frequency of the alternating current. Some computer vision algorithms are designed to take advantage of these properties for particular purposes.

Application

The color spectrum of fluorescent light contains peaks as exemplified in Fig. 1. These peaks originate from energy emission of the phosphor, and lie at specific wavelengths according to the phosphor composition. The correspondence of spectrum peaks to particular wavelengths has been utilized in computer vision for calibration of multispectral sensing devices based on dispersive optics [1]. In [1], the spectrum peaks are also used to identify the presence of fluorescent lighting in a scene, as the characteristic peaks are detectable even in light reflected from object surfaces.

Fluorescent lighting also exhibits a regular, high-frequency flicker due to the cycles of electrical current flow in fluorescent lamps. This stroboscopic effect is used in the Mova Contour™ system for markerless motion tracking of human faces [2, 4]. The face is covered with a phosphorescent makeup and then illuminated by the flicker of a blacklight fluorescent lamp, which uses a phosphor that converts the shortwave UV radiation of the mercury vapor to a long-wave UV light that stimulates the makeup. At intervals of the flicker when the fluorescent lighting is off, emission from the phosphorescent makeup is recorded by multiple cameras. The random patterns of makeup formed by a rough applicator sponge are tracked and also triangulated to form 3D models.

References

1. Du H, Tong X, Cao X, Lin S (2009) A prism-based system for multispectral video acquisition. In: Proceedings of the international conference on computer vision. IEEE, Kyoto, Japan
2. Geller T (2008) Overcoming the uncanny valley. IEEE Comput Graph Appl 28(4):11–17
3. Henkenius M (1991) How it works: fluorescent lamp. Pop Mech 10:59–60
4. King BA, Paulson LD (2007) Motion capture moves into new realms. IEEE Comput 40(9):13–16

Focal Length

Peter Sturm
INRIA Grenoble Rhône-Alpes, St Ismier Cedex, France

Synonyms

Principal distance

Related Concepts

▶Center of Projection; ▶Camera Calibration; ▶Image Plane; ▶Optical Axis; ▶Pinhole Camera Model

Definition

The focal length has different, related, meanings. In optics, the focal length of a lens or optical system is the distance from the center to the point on the optical axis where a bundle of incoming rays parallel to the optical axis get focused to. In geometric image formation models such as the pinhole model, the focal length usually represents the distance between the center of projection and the image plane.

Background

The concept of focal length stems from the area of optics. The focal length of a lens is generally defined as the distance from the center of the lens to the point where a set of incoming parallel light rays that are parallel to the optical axis get focused. It is thus related to the magnification operated by an optical system.

In computer vision, the image formation process carried out by the optics and electronics of a camera is usually modeled by simple geometric models.

In the following, the thin lens model and the pinhole model are discussed. More general background is given, for example, in [1–3].

Theory

Thin lens model. Consider first a simple "physical" model for a lens, the thin lens model, for the case of a lens whose two outer surfaces are convex and spherical. For simplicity, we assume here that the two spheres have the same radius R. Let n be the index of refraction of the lens material and n_0 that of the surrounding medium (for a vacuum $n_0 = 1$, for air $n_0 \approx 1.0008$).

Incoming rays get bent by the lens, due to the successive refractions in the two surfaces of the lens. For lenses with spherical surfaces, even incoming rays that are parallel to the optical axis do not converge to the same point on the optical axis after these refractions. Rather, they hit the optical axis in a segment; this is known as spherical aberration; see Fig. 1.

Let us now make two common approximations. First, under the thin lens approximation, one assumes that both refractions happen at the same point, on the lens' plane instead of on the two spherical surfaces. Second, we make the so-called paraxial approximation by assuming that along the path of the refracted light ray, all angles it forms with the optical axis and the normals of the spherical surfaces are small (leading, e.g., to the approximation $\sin\alpha \approx \alpha$ for such angles). Under these two simplifying assumptions, all incoming rays that are parallel to the optical axis are focused by the lens to a point on the optical axis that is at the following distance from the lens center:

$$f = \frac{R\, n_0}{2(n - n_0)} \ .$$

This distance is the focal length of the lens, under the paraxial thin lens approximation. This formula can be generalized to lenses with two spherical surfaces of different radii, in the form of the so-called lensmaker's equation [4, 5]. Lenses with other bounding shapes than spherical ones exist of course and can be studied similarly [2].

Under the above assumptions, if one wishes to take a sharp picture of a distant object, one would put the image plane at a distance of f from the lens. However, if the object is at a close distance s from the lens, then the light rays emerging from it converge at a distance s' from the lens that is different from f and that is given by [2, 6]:

$$\frac{1}{s} + \frac{1}{s'} = \frac{1}{f} \ .$$

This classical relationship can be easily derived by considering similar triangles and the following two light rays emitted from a point P at a distance s from the lens; cf. the lower part of Fig. 1. The ray parallel

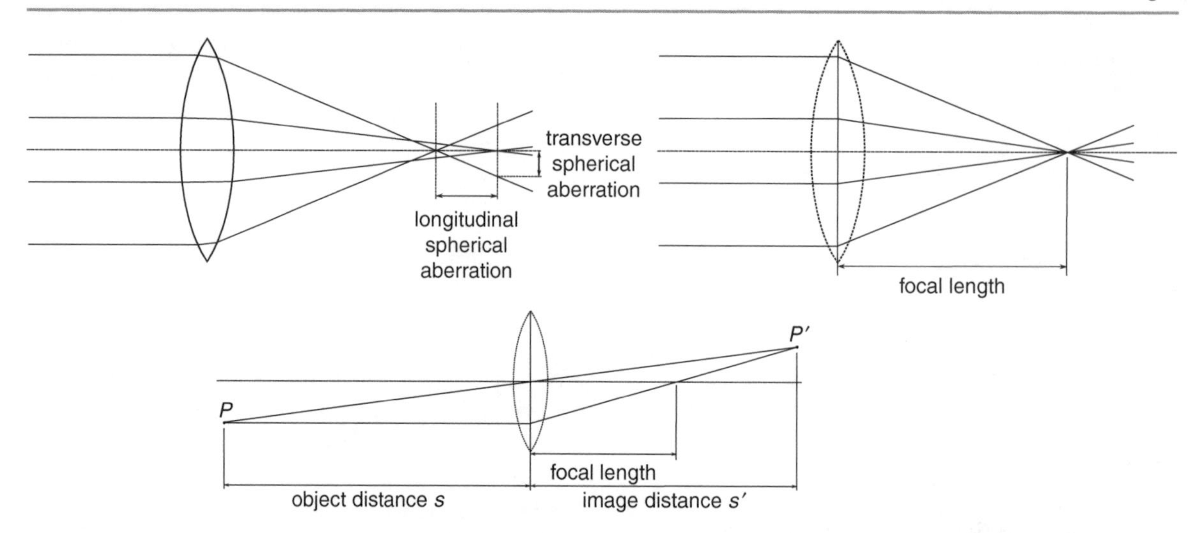

Focal Length, Fig. 1 *Upper left:* spherical aberration of a lens with two spherical surfaces. *Upper right:* focal length in the paraxial thin lens approximation. *Bottom:* image of an object point P at finite distance from the lens, under the paraxial thin lens model

to the optical axis (supposing here it enters the aperture) intersects the optical axis at a distance f from the lens, after being bent by it, as explained above. The ray going through the lens center does not get bent under the thin lens assumption. The two rays thus converge in a point P' at a distance of s' from the lens plane. To get a sharp picture, one would thus put the image plane at distance s'.

Pinhole model. In the pinhole model and other camera models, the focal length is defined differently than in optics, as the distance between the center of projection and the image plane. As seen above, this is in general different from the focal length of optical systems, even in the very simple case of the paraxial thin lens model thereof. The two definitions coincide however if a camera is focused to infinity and if the center of the thin lens model is considered as center of projection in the pinhole model.

The focal length of the pinhole model is often expressed in the nonmetric unit "number of pixels" (true focal length divided by the density of pixels). For real cameras, the focal length is usually given in millimeters. An often used convention is to characterize a camera by the focal length that an equivalent 35-mm-format camera would have: the focal length of a lens that, if used with a 35-mm-format image area, would have the same field of view as the camera under consideration.

Application

The focal length of the pinhole or other camera models typically used in computer vision is part of the camera's intrinsic parameters, which can be computed by camera calibration. Camera calibration is an important requirement in most applications where geometric information about the scene or the camera movement is to be determined from images.

The difference between the meaning of the pinhole model's focal length and that of a true optical system made of lenses (the pinhole model is lensless) is stressed again here. A consequence of what is explained above is that when calibrating a camera using the pinhole model, that model's focal length is affected by both a change in focus and zoom of the actual camera [7].

References

1. Wikipedia (2011) Focal length. http://en.wikipedia.org/wiki/Focal_length. Accessed 3 August 2011
2. Hecht E (2001) Optics. 4th edn. Addison, Wesley
3. Geissler P (2000) Imaging optics. In: Jähne B, Haußecker H (eds) Computer vision and applications. Academic, San Diego, pp 53–84
4. HyperPhysics (2011) Light and vision. http://hyperphysics.phy-astr.gsu.edu/hbase/ligcon.html#c1. Accessed 4 August 2011
5. Wikipedia (2011) Lens (optics). http://en.wikipedia.org/wiki/Lens_(optics). Accessed 4 August 2011

6. Wikipedia (2011) Thin lens. http://en.wikipedia.org/wiki/Thin_lens. Accessed 4 August 2011
7. Willson R, Shafer S (1993) Modeling and calibration of zoom lenses. In: Gruen A, Huang T (eds) Camera calibration and orientation determination. Springer, Verlag, pp 137–161

Focus Bracketing

▸Multi-focus Images

Foreground-background Assignment

▸Occlusion Detection

Fresnel Conditions

▸Fresnel Equations

Fresnel Equations

Daisuke Miyazaki
Graduate School of Information Sciences, Hiroshima City University, Asaminami-ku, Hiroshima, Japan

Synonyms

Fresnel Conditions; Fresnel's law; Light Transmission and Reflection Coefficients

Related Concepts

▸Polarized Light in Computer Vision; ▸Polarizer; ▸Polarization

Definition

The four Fresnel equations express the reflection and transmission coefficients of light components whose electric-field vector is either parallel or perpendicular to the plane of incidence.

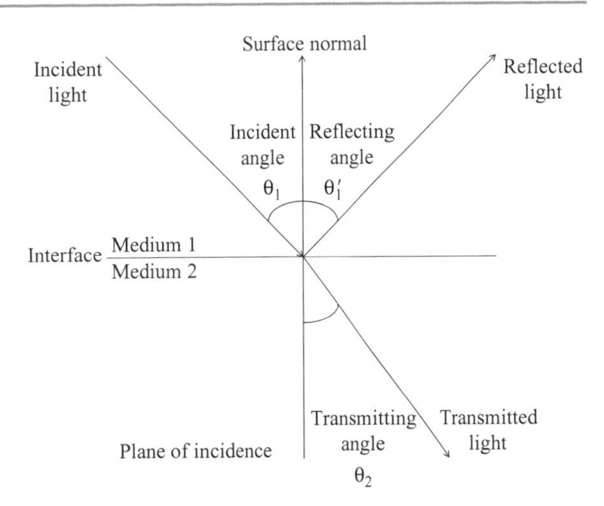

Fresnel Equations, Fig. 1 Reflection, refraction, and transmission at the interface of two materials. The surface is assumed to be optically smooth

Background

Based on the theory of electromagnetism, the Fresnel equations express the reflection and transmission coefficients of light that hits an interface between two media. This entry introduces amplitude reflectivity, amplitude transmissivity, intensity reflectivity, and intensity transmissivity.

Theory

Figure 1 illustrates a light ray that hits the interface between two materials, the refractive indices of which are denoted by n_1 and n_2, respectively. Part of the light is reflected from the interface, while another part penetrates the surface and refracts as it enters the second material. The plane including the surface normal and the incident light ray is called the plane of incidence (POI). The incident light, the reflected light, and the transmitted light are denoted as the subscripts i, r, and t, respectively.

Being an electromagnetic wave, light carries an oscillating electric field. The oscillating field (called E-vector) has amplitude components that are parallel or perpendicular to the POI. These components are denoted by p and s, respectively. Here, p is associated with the term "parallel," while s is associated with the word "senkrecht," which means "perpendicular" in German. The incidence, reflection, and transmission angles are defined as θ_1, θ_1', and θ_2, respectively, as illustrated in Fig. 1.

For optically smooth objects, the incidence and reflection angles are equal, $\theta_1 = \theta_1'$, while θ_1 and θ_2 are related by Snell's law (cf. Sect. 1.5.1 in the 5th edition of Born and Wolf [1], Sect. 4.4.1 in the 4th edition of Hecht [2]):

$$n_1 \sin\theta_1 = n_2 \sin\theta_2 . \tag{1}$$

The ratio of the amplitude of the reflected light to that of the incident light is called reflection coefficient (or, amplitude reflectivity), r. The ratio of the amplitude of the transmitted light to that of the incident light is called transmission coefficient (or, amplitude transmissivity), t. These coefficients are generally different for the two E-vector components. The coefficients for the p-component and s-component are derived from the theory of optics as (cf. Sect. 1.5.2 in the 5th edition of Born and Wolf [1], Sect. 4.6.2 in the 4th edition of Hecht [2]):

$$r_p = \frac{\tan(\theta_1 - \theta_2)}{\tan(\theta_1 + \theta_2)} \tag{2}$$

$$r_s = -\frac{\sin(\theta_1 - \theta_2)}{\sin(\theta_1 + \theta_2)} \tag{3}$$

$$t_p = \frac{2\sin\theta_2\cos\theta_1}{\sin(\theta_1 + \theta_2)\cos(\theta_1 - \theta_2)} \tag{4}$$

$$t_s = \frac{2\sin\theta_2\cos\theta_1}{\sin(\theta_1 + \theta_2)} . \tag{5}$$

These coefficients are plotted in Fig. 2. These equations are called Fresnel equations.

The intensity reflectivity of the p-component R_p and that of the s-component R_s, and the intensity transmissivity of the p-component T_p and that of the s-component T_s are (cf. Sect. 1.5.3 in the 5th edition of Born and Wolf [1]):

$$R_p = \frac{\tan^2(\theta_1 - \theta_2)}{\tan^2(\theta_1 + \theta_2)} \tag{6}$$

$$R_s = \frac{\sin^2(\theta_1 - \theta_2)}{\sin^2(\theta_1 + \theta_2)} \tag{7}$$

$$T_p = \frac{\sin 2\theta_1 \sin 2\theta_2}{\sin^2(\theta_1 + \theta_2)\cos^2(\theta_1 - \theta_2)} \tag{8}$$

$$T_s = \frac{\sin 2\theta_1 \sin 2\theta_2}{\sin^2(\theta_1 + \theta_2)} . \tag{9}$$

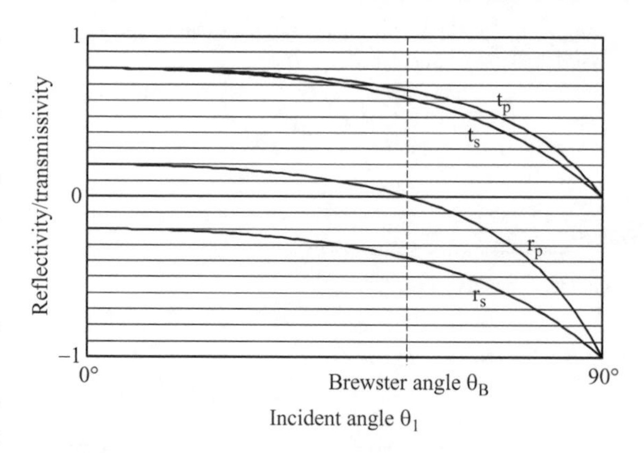

Fresnel Equations, Fig. 2 Amplitude reflectivity of the p-component r_p and s-component r_s. Amplitude transmissivity of the p-component t_p and s-component t_s. The plots correspond to the case where $n_2/n_1 = 1.5$

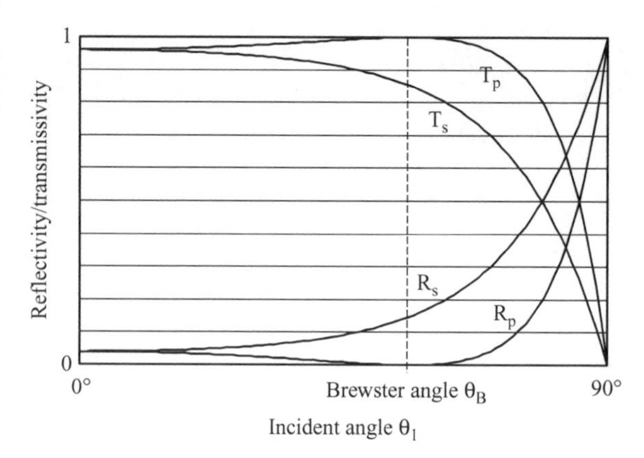

Fresnel Equations, Fig. 3 Intensity reflectivity of the p-component R_p and s-component R_s. Intensity transmissivity of the p-component T_p and s-component T_s. The plots correspond to the case where $n_2/n_1 = 1.5$

They are plotted in Fig. 3.

From the above equations, $R_p = 0$ can be obtained at a special incidence angle. This angle is referred to as the Brewster angle, θ_B. The Brewster angle is obtained by substituting $\theta_1 + \theta_2 = \pi/2$ (namely, $R_p = 0$) into (Eq. 1), yielding (cf. Sect. 1.5.3 in the 5th edition of Born and Wolf [1]):

$$\tan\theta_B = \frac{n_2}{n_1} . \tag{10}$$

From (Eq. 1) to (Eq. 6)–(Eq. 9), and defining $\theta = \theta_1$, $n = n_2/n_1$, the following can be derived (cf. Appendix A.6 in Miyazaki [3]):

$$R_p = \frac{1 + n^2 - (n^2 + 1/n^2)\sin^2\theta - 2\cos\theta\sqrt{n^2 - \sin^2\theta}}{1 + n^2 - (n^2 + 1/n^2)\sin^2\theta + 2\cos\theta\sqrt{n^2 - \sin^2\theta}} \quad (11)$$

$$R_s = \frac{1 + n^2 - 2\sin^2\theta - 2\cos\theta\sqrt{n^2 - \sin^2\theta}}{1 + n^2 - 2\sin^2\theta + 2\cos\theta\sqrt{n^2 - \sin^2\theta}} \quad (12)$$

$$T_p = \frac{4\cos\theta\sqrt{n^2 - \sin^2\theta}}{1 + n^2 - (n^2 + 1/n^2)\sin^2\theta + 2\cos\theta\sqrt{n^2 - \sin^2\theta}} \quad (13)$$

$$T_s = \frac{4\cos\theta\sqrt{n^2 - \sin^2\theta}}{1 + n^2 - 2\sin^2\theta + 2\cos\theta\sqrt{n^2 - \sin^2\theta}} . \quad (14)$$

Application

Fresnel equations are mainly used to model and analyze light transmission and specular reflection. Intensity transmissivity ((Eq. 8) and (Eq. 9)) is also used for analyzing diffuse reflection and thermal radiation, since they are caused by radiation from beneath the object surface.

References

1. Born M, Wolf E (1974) Principles of optics. Pergamon, New York
2. Hecht E (2002) Optics. Pearson, San Francisco
3. Miyazaki D (2005) Shape estimation of transparent objects by using polarization analyses. PhD thesis, The University of Tokyo

Fresnel's Law

▶Fresnel Equations

Fundamental Matrix

Zhengyou Zhang
Microsoft Research, Redmond, WA, USA

Related Concepts

▶Epipolar Geometry; ▶Essential Matrix

Definition

Fundamental matrix is a special 3×3 matrix which captures the geometric relationship between two cameras or between two locations of a single moving camera.

Background

See entry ▶Epipolar Geometry for details.

Theory

If two points $\boldsymbol{m}$ and $\boldsymbol{m}'$, expressed in pixel image coordinates in the first and second camera, are in correspondence, they must satisfy the following equation

$$\widetilde{\boldsymbol{m}}^T \mathbf{F} \widetilde{\boldsymbol{m}}' = 0 , \quad (1)$$

where

$$\mathbf{F} = \mathbf{A}^{-T}\mathbf{E}\mathbf{A}'^{-1} , \quad (2)$$

$$\mathbf{E} = [\boldsymbol{t}]_\times \mathbf{R} , \quad (3)$$

$\mathbf{A}$ and $\mathbf{A}'$ are respectively the intrinsic matrix of the first and second camera, and $(\mathbf{R}, \boldsymbol{t})$ is the rigid transformation between the first and second camera. This is a fundamental constraint for two pixels to be in correspondence between two images. The 3×3 matrix $\mathbf{F}$ is called the *fundamental matrix*, and the 3×3 matrix $\mathbf{E}$ is known as the *essential matrix* (see entry ▶Essential Matrix).

As can be seen in Eq. (2), the fundamental matrix and the essential matrix are related. If the cameras are calibrated, i.e., if $\mathbf{A}$ and $\mathbf{A}'$ are known, we can use the normalized image coordinates, and the fundamental matrix becomes the essential matrix.

Because $\det \mathbf{E} = 0$, we have $\det \mathbf{F} = 0$. Thus, the fundamental matrix is singular (rank 2). From Eq. (1), it is also clear that $\mathbf{F}$ is defined up to a scale factor, which depends on the translation magnitude that cannot be determined from image alone. Therefore, a fundamental matrix only has 7 degrees of freedom. Given one pair of corresponding image points, we have one constraint on $\mathbf{F}$ as expressed by Eq. (1). We thus need at least seven or more point correspondences in order to determine the fundamental matrix between two images. The reader is referred to [1] for various algorithms of determining the fundamental matrix from point correspondences.

References

1. Zhang Z (1998) Determining the epipolar geometry and its uncertainty: a review. Int J Comput Vis 27(2):161–195

G

Gait Analysis

▶Gait Recognition

Gait Biometrics

▶Gait Recognition

Gait Recognition

Darko S. Matovski, Mark S. Nixon and John N. Carter
School of Electronics and Computer Science,
University of Southampton, Southampton,
Hampshire, UK

Synonyms

Automatic gait recognition; Gait analysis; Gait biometrics

Related Concepts

▶Face Identification; ▶Optical Flow; ▶Principal Component Analysis (PCA)

Definition

The way a person walks (or runs) combined with their posture is known as gait. Recognizing individuals by their particular gait using automated vision-based algorithms is known as gait recognition.

Background

Gait has few important advantages over other forms of biometric identification. It can be acquired at a distance when other biometrics are obscured or the resolution is insufficient. It does not require subject cooperation and can be acquired in a noninvasive manner. It is easy to observe and hard to disguise as walking is necessary for human mobility. Gait can be acquired from a single still image or from a temporal sequence of images (e.g., a video).

Shakespeare made several references to the individuality of gait, e.g., in *The Tempest* [Act 4 Scene 1], Cares observes "*High'st Queen of state, Great Juno comes; I know her by her gait*" and in Henry IV Part II [Act 2, Scene 3], "*To seem like him: so that, in speech, in gait, in diet, in affections of delight, in military rules, humors of blood, he was the mark and glass, copy and book.*"

The aim of medical research has been to classify the components of gait for the treatment of pathologically abnormal patients. Murray et al. [17] created standard movement patterns for pathologically normal people. Those patterns were then used to identify pathologically abnormal patients.

The biomechanics literature makes observations concerning identity: "A given person will perform his or her walking pattern in a fairly repeatable and characteristic way, sufficiently unique that it is possible to recognize a person at a distance by their gait" [27].

Psychophysiological studies such as [5, 11] have shown that humans can recognize friends and the sex of a person solely by their gait with 70–80 % accuracy. These and similar studies have inspired the use of gait as a biometric trait.

K. Ikeuchi (ed.), *Computer Vision*, DOI 10.1007/978-0-387-31439-6,

Recently, there has been a rapid growth in the number of surveillance systems, aimed to improve safety and security. These systems are yet to include recognition capabilities, and gait recognition could be a most suitable choice. The primary aim of surveillance videos is to monitor people. However, the video data can be of a low quality (poor resolution, time lapse, etc.), and the subject can try to conceal the more conventional biometrics. Nevertheless, such video can provide sufficient data for gait recognition technology, and there is already research in using gait biometrics as a forensic tool [4]. Gait recognition could be employed at a border crossing or any high-throughput environment. Gait contains very rich information and is considered to be unique. Studies have shown that gait can also be used to reveal a person's identity, gender, emotional state, etc.

Recognition by gait is one of the newest biometrics since its development only started when computer memory and processing speed became sufficient to process sequences of image data with reasonable performance. The potential for gait recognition is great, and hence there is a vast interest in computer vision research in extracting gait features.

Theory

A gait recognition system primarily consists of a computer vision system. A gait signature is created by extracting images of a walking subject which is then compared to the signatures of known subjects. Figure 1 shows an example of some of the basic steps in a gait recognition system.

Step 1: Data can be acquired using a single or multiple cameras. If data is acquired using a single camera, recognition can be performed using a 2D gait signature such as the Gait Energy Image (GEI – shown in step 4). However, if multiple but synchronized cameras are used, the number of possibilities is greater. Examples of the usage of multiple-synchronized cameras include:

- Producing a 3D gait model and using it for recognition.
- 3D information can be used to improve recognition for a 2D approach by producing a non-normalized version of a 2D signature.
- Achieving a view-invariant recognition. A gait signature from any view can be re-created using 3D data which can be mapped to a signature acquired in an outdoor environment where typically only a single and nonoptimal view is available.

Step 2: An example of preprocessing step is background subtraction or background segmentation. The subject can be acquired easily and reliably by using chroma-keying if there are clear color difference between the subject and the background. Background subtraction can measure the naturally occurring scene behind the walking subject using one of the plethora of computer vision techniques.

Step 3: As human gait is periodic, a gait sequence (sample) can consist of multiple gait cycles. Identifying the most suitable cycle can lead to better recognition rates. Signal processing techniques can be applied to the foreground signal (sum of foreground pixels) in the case of binary image.

Step 4: There are number of approaches to produce a gait signature, some of which are described later. A baseline gait signature was proposed in [22]. An example of a signature is shown in step 4.

Step 5: A gait signature can be used directly within a classifier. Alternatively, features can be extracted from a signature, and those features can be used for classification. Again, there is a selection of classification techniques; in the simplest case, a classifier such as k-nearest neighbor (using Euclidian distance) can be used.

Databases

A database can be collected for various purposes. Primary concerns include uniqueness and practicality. A database should contain enough subjects to allow for an estimate of inter- and intra-subject variation. The current databases contain smaller number of subjects compared to databases used to evaluate performance of other biometrics (e.g., face, fingerprint). However, there are databases that include covariate factors and application potential. Some of the most well-known databases together with some of their characteristics are shown in Table 1.

Approaches to Gait Recognition

The approaches to gait recognition can be divided in two main groups: *model-based* and *model-free*

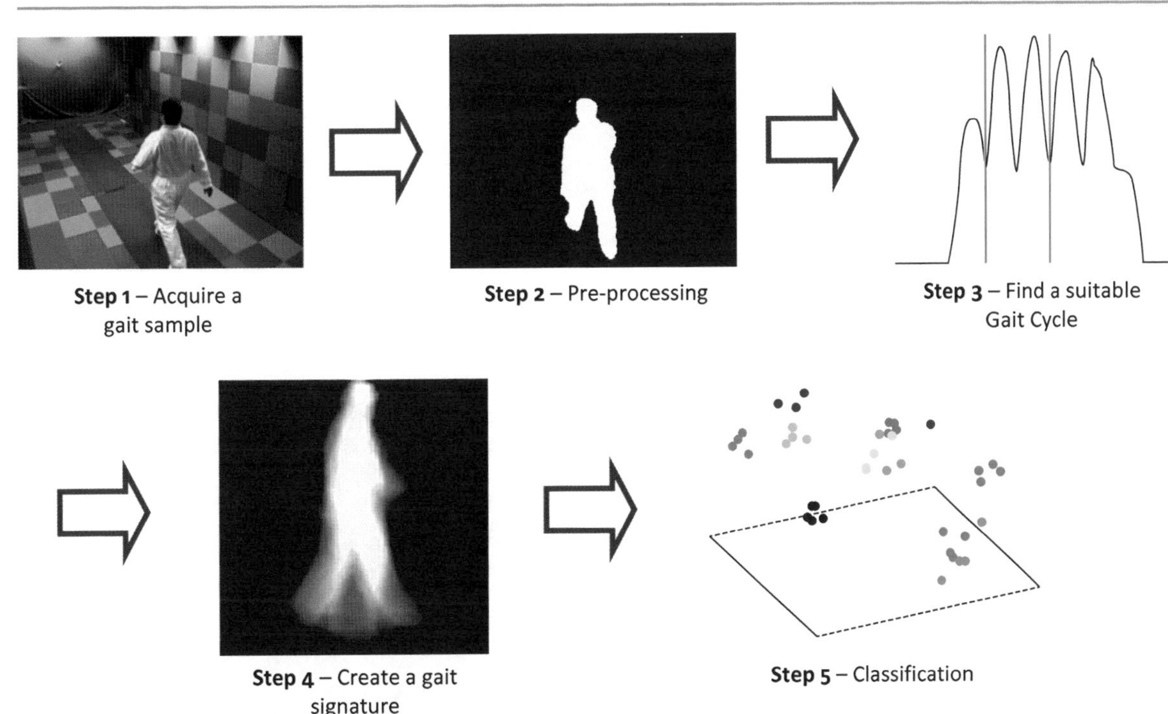

Gait Recognition, Fig. 1 General steps of a gait recognition system

Gait Recognition, Table 1 Details of some of the well-known gait databases

Name	Subjects	Sequences	Covariates	Viewpoints	Indoor(I)/outdoor(O)
HumanID (USF)	122	1,870	Y	2	O
SOTON 2002	114	>2,500	Y	2	I/O
CMU MoBo	100	600	Y	6	I (treadmill)
MIT 2002	24	194	Y	1	I
UMD 2002	44	176	N	1	O
CASIA 2006	124	1,240	Y	11	I
SOTON multimodal [21]∗1	>300	>5,000	Y	12	I
Osaka University	1,035	2,070	N	2	I

(see Table 2). Model-based approaches use the human body structure, and model-free methods use the whole motion pattern of the human body. Which approach is adopted depends on the acquisition conditions. Model-free (appearance-based) approaches use the input images directly to produce a gait signature without fitting a model. These approaches can perform recognition at lower resolutions which makes them suitable for outdoor applications, where a subject can be at a large distance from the camera. Model-based approaches typically require higher resolution images of a subject to be able to fit the model accurately.

The table is taken from [18, 19]. Example papers for all of the approaches can be found in the original sources.

Model-Free Approaches

The model-free approaches derive the human silhouette by separating the moving object from the background. The subject can then be recognized by measurements that reflect shape and/or movement. The simplest approach is to simply form an average of the silhouettes over a complete gait cycle [15]. The approach is called the Gait Energy Image (GEI), and it is shown in Fig. 2. Motion Silhouette Image (MSI) is a similar representation to the GEI. The value of each pixel is computed as a function of motion of that pixel in the temporal dimension over all silhouettes that are part of a single gait cycle. Both the GEI and MSI are easy to compute, but they are vulnerable to appearance

Gait Recognition, Table 2 Approaches to gait recognition

Model-free analysis		Model-based analysis	
Moving shape	Shape + motion	Structural	Modeled
Unwrapped silhouette; silhouette similarity; relational statistics; self-similarity; key frame analysis; frieze patterns; area; symmetry; point distribution models; key poses	Eigenspace sequences; hidden Markov model; average silhouette; moments; ellipsoidal fits; kinematic features; gait style and content	Stride parameters; human parameters; joint trajectories	Articulated model; dual oscillator; linked feature trajectories

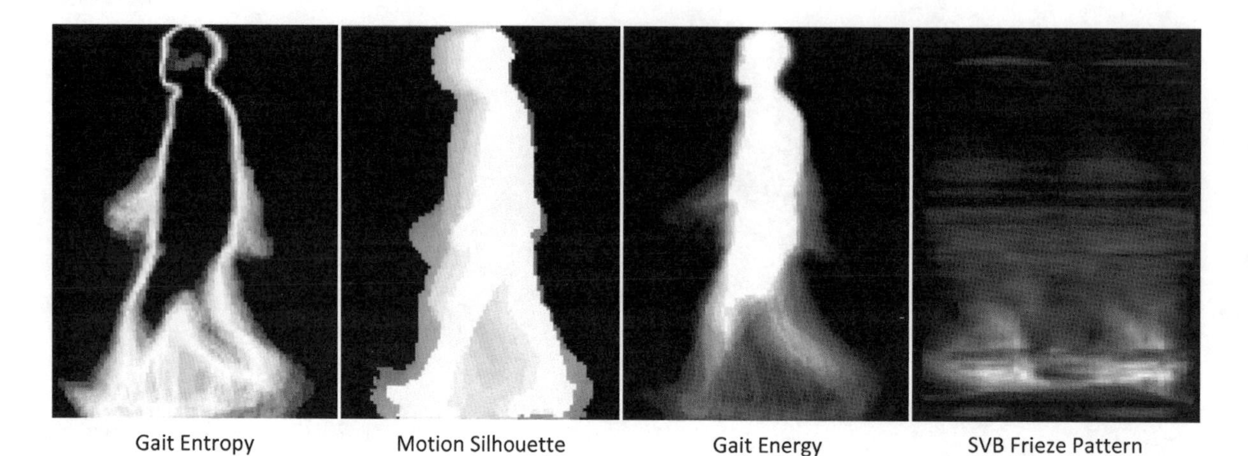

Gait Recognition, Fig. 2 Examples of model-free gait signatures

changes of the human silhouette. Frieze pattern represents the information contained in a gait sequence by horizontal and vertical projections of the silhouettes. Its extension, SVB Frieze patterns, use key frame subtraction in order to mitigate the effects of appearance changes on the silhouette (see Fig. 2). The Gait Entropy Image (GEnI) is another example of a compact gait representation (signature). GEnI is computed by calculating the Shannon entropy for each pixel. Shannon entropy measures the uncertainty associated with a random variable.

The gait signatures for the approaches shown in Fig. 2 are usually used directly for classification. There are additional ways of extracting gait signatures without using a model. Some examples are described below:

- Little and Boyd [14] derive a dense optical flow for each image sequence. Scale-independent scalar features of each flow, based on moments of the moving point, characterize the spatial distribution of the flow. The periodic structure of these sequences of scalars is analyzed. The scalar sequences for an image sequence have the same fundamental period but differ in phase, which is used as a feature for recognition of individuals by the shape of their motion.
- BenAbdelkader et al. [2] use background modeling to track the subject for a number of frames and extract a sequence of segmented images of the person. A self-similarity plot is computed via correlation of each pair of images in this sequence. For recognition, PCA (principal component analysis) is used to reduce the dimensionality of the plots. A k-nearest neighbor rule is used on the reduced space for classification. Another silhouette-based gait recognition technique using PCA has been proposed by Liang et al. [13]. Eigenspace transformation based on principal component analysis (PCA) is applied to time-varying distance signals derived from a sequence of silhouette images to reduce the dimensionality of the input feature space. Supervised pattern classification techniques are performed in the lower-dimensional eigenspace for recognition.

- Hayfron-Acquah et al. [8] proposes a method for automatic gait recognition based on analyzing the symmetry of human motion. The Generalised Symmetry Operator is used to locate features according to their symmetrical properties rather than relying on the boarders of a shape. The symmetry operator is used on the optical flow image to produce a gait signature. For purposes of classification, the similarity differences between the Fourier descriptions of the gait signatures are calculated using Euclidean distance.
- Gait is a temporal sequence and can be modeled using hidden Markov models (HMM). The statistical nature makes the model relatively robust. The postures that an individual adopts are regarded as states of the HMM and are typical to that individual and provide means of discrimination [24].
- Kale et al. [10] use two different image features to directly train a HMM: the width of the outer contour of a binary silhouette and the entire binary silhouette itself.

Model-Based Approaches

The advantages of the previous approaches (silhouette or features derived from it) are speed and simplicity. However, model-based approaches are better at handling occlusion, noise, scale, and rotation. Model-based approaches require a high resolution therefore not very suitable for outdoor surveillance.

Model-based approaches incorporate knowledge of the shape and dynamics of the human body into the extraction process. These approaches extract features that fit a physical model of the human body. A gait model consists of shapes of various body parts and how those shapes move relative to each other (motion model). The shape model for a human subject can use ellipse to describe the head and the torso, quadrilaterals to describe the limbs, and rectangles to describe the feet. Alternatively, arbitrary shapes could be used to describe the edges of the body parts. The motion model describes the dynamics of the motion of the different body parts. Using a model ensures that only image data corresponding to allowable human shape and motion is extracted, reducing the effect of noise. The models can be 2- or 3-dimensional. Most of the current models are 2-dimensional, but deliver good results on databases of more than 100 subjects.

m = mass
θ = angular displacement
l = length of the limb
Subscripts T and K denote high and lower leg respectively.

Gait Recognition, Fig. 3 Example of a gait model – the dynamically coupled pendulum model [28]

Some examples of model-based approaches are described below:

- Yam et al. [28] have used pendular motion and the understanding of biomechanics of human locomotion to develop two models: a bilateral symmetric and analytical model (employs the concept of forced couple oscillator). See Fig. 3. The gait signature is the phase-weighted magnitude of the Fourier description of both the thigh and knee rotation.
- Bouchrika and Nixon [3] have proposed a new approach to extract human joints. Spatial model templates for human motion are derived from the analysis of gait data collected from manual labeling. Motion templates describing the motion of the joints are parameterized using the elliptic Fourier descriptors

$$\begin{bmatrix} x(t) \\ y(t) \end{bmatrix} = \begin{bmatrix} a_0 \\ b_0 \end{bmatrix} + \begin{bmatrix} \cos(\alpha) - \sin(\alpha) \\ \sin(\alpha)\cos(\alpha) \end{bmatrix} \begin{bmatrix} X(t) * S_x \\ Y(t) * S_y \end{bmatrix}$$

where α is the rotation angle, s_x and s_y are the scaling factors across the horizontal and vertical axes, respectively, and $X(t)$ and $Y(t)$ are the Fourier summation. Hough transform is used in the feature extraction process.
- Wang et al. [26] have proposed an algorithm based upon the fusion of static and dynamic body information. The static body information is in a form of a compact representation obtained by Procrustes shape analysis. The dynamic information is obtained by a model-based approach which tracks the subject and recovers joint-angle trajectories of lower limbs. A fusion at the decision level is used to improve recognition results. Figure 1 shows an example of the results obtained.

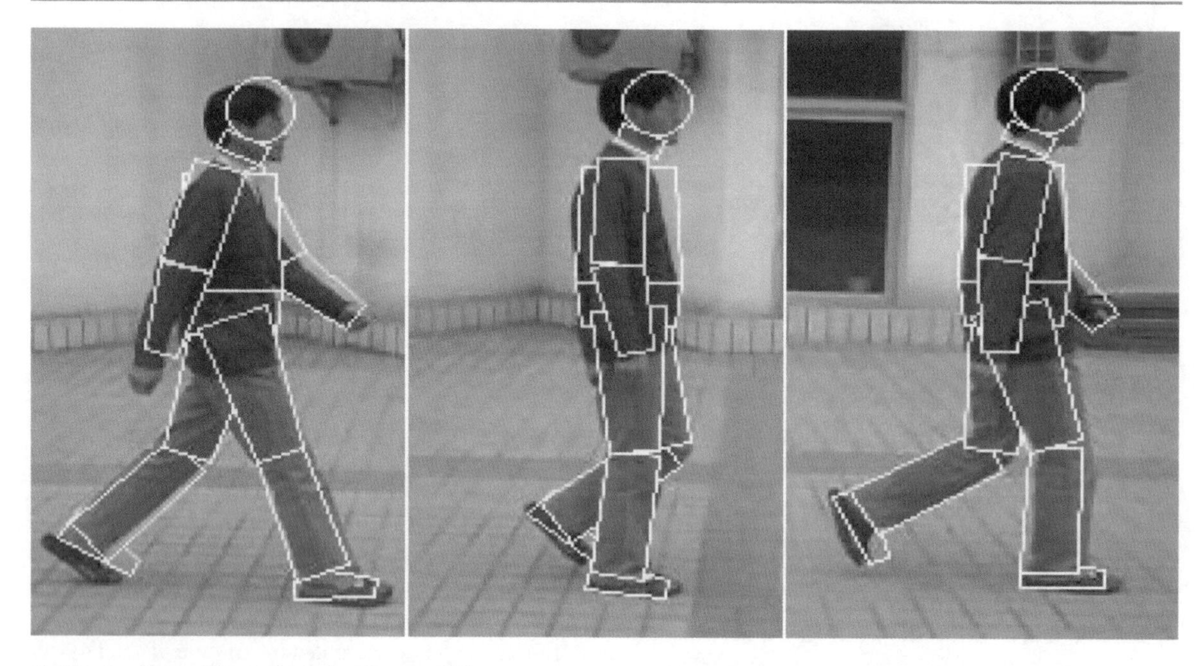

Gait Recognition, Fig. 4 Example of results [26]

Gait Recognition, Table 3 Progression of gait recognition systems

Time period	No of subjects	Source	Recognition rate	Notes
1990s	~10	USC	95.2	
2000s	~120	HiD, CASIA, Southampton	75–99 %	Recognition rate depends on covariates
Recent (2010)	>300	Southampton multimodal	95–100 %	Includes time-dependent covariates
Recent (2010)	>1000	Osaka University	90 %	No covariates

There have been moves towards developing 3D gait models. Examples of work in this fields are [7, 25]. Guoying et al. [7] use video sequences from multiple cameras to construct 3D human models. The motion is tracked by applying a local optimization algorithm. The length of key segments is extracted as static parameters, and the motion trajectories of lower limbs are used as dynamic features. Linear time normalization is used for matching and recognition. Three-dimensional approaches are robust to changes in viewpoint and have a great potential. However at present, experiments only on small databases are possible mainly due to high computational requirements.

Gait is dependent on large number of parameters (joint angles and body segment size) which leads to complex models with many free parameters. Finding the best fit model for a subject leads to searching a high-dimensional parameter space. Therefore, there is a trade off between the accuracy of the model (complexity) and computational cost. The models are often simplified based on certain acceptable assumptions, e.g., a system could assume constant walking speed. However, as computing power increases, the problems arising of high complexity can be mitigated.

Experimental Results

The current state of the art achieves very high recognition rates (close to 100 %) on relatively large databases (>300 subjects) when the training and test data are recorded under similar conditions. An example of progression in performance over time is shown in Table 2. However, recognition rate can drop with change of clothing, shoes, walking surface, and pose. Many current studies focus on solving these problems. Recent major achievements in gait recognition are

Gait Recognition, Table 4 Some experiments comprising the HumanID gait challenge problem

Experiment	Probe	# of subjects	Difference
A	Different camera view than gallery	122	View
B	Subjects wore different shoes	54	Shoe
C	Different camera view and different shoes	54	Shoe, view
D	Subjects walked on a different surface	121	Surface
E	Different shoes and different walking surface	60	Surface, shoes
F	Different walking surface and different camera view	121	Surface, view
G	Different walking surface, different shoes, and different camera view	60	Surface, shoe, view

described in [16, 20]. Matovski et al. [16] have shown that elapsed time does not affect gait recognition and that gait can be used as a reliable biometric over time and at a distance. The world's largest gait database of more than 1,000 people has been constructed to enable statistically reliable performance evaluation of gait recognition performance [20] (Table 3).

The HumanID gait challenge problem [22] was set up to outline a baseline algorithm for gait recognition and propose a number of difficult experiments for the existing gait matchers. The gallery set consists of 122 subjects walking on a grass surface recorded by a single camera.

Table 4 shows the differences of the probe set compared to the gallery set for each of the challenge experiments.

The results in Fig. 5 show the progress in gait recognition over a period of 2 years for the experiments shown in Table 4.

Application

Gait research is currently at an evaluation stage rather than an application stage. However, the potential for gait recognition is great. The complete unobtrusiveness without any subject cooperation or contact for data acquisition makes gait particularly attractive for identification purposes. It could be used in applications including forensics, security, immigration, and surveillance.

Many surveillance systems capture only a low-resolution video at varying lighting conditions, and gait recognition might be the only plausible choice for automatic recognition. A bank robber may wear a mask so you cannot see his face, wear gloves so you cannot get fingerprints, and wear a hat so you cannot get DNA evidence – but they have to walk or run into the bank, and they could be identified from their gait.

Gait recognition has been used as evidence for conviction in some criminal cases. A man in Bolton (UK) was convicted based on his distinctive gait. A CCTV footage of the burglar captured near the crime scene was compared to a video captured at the police station by a podiatrist specializing in gait analysis. In 2004, a perpetrator robbed a bank in Denmark. The Institute of Forensic Medicine in Copenhagen was contacted by the police to perform gait analysis, as they thought the perpetrator had a unique gait. The institute instructed the police to establish a covert recording of the suspect from the same angles as the surveillance recordings for comparison. The gait analysis revealed several characteristic matches between the perpetrator and the suspect. For example, both the perpetrator (to the left) and the suspect showed inverted left ankle (white arrow) during left leg's stance phase and markedly outward rotated feet (see Fig. 6). The suspect was convicted of robbery, and the court found that gait analysis is a very valuable tool [12].

One system named the Biometric Tunnel [23] has led to the first live demonstration of gait as a biometric and could indicate a possible route for future deployment of the technology. The left side of Fig. 7 depicts the system. It consists of a simple corridor with 12 synchronized and fixed cameras. The subjects are asked to walk through the middle, and the lighting and background are controlled to facilitate analysis. The right side of Fig. 7 shows the details of the arrangement. The system is designed with a high-throughput environment in mind.

Open Problems

Although a large number of gait recognition algorithms have been reported, it is important to note that gait biometrics is still in its infancy. The majority of

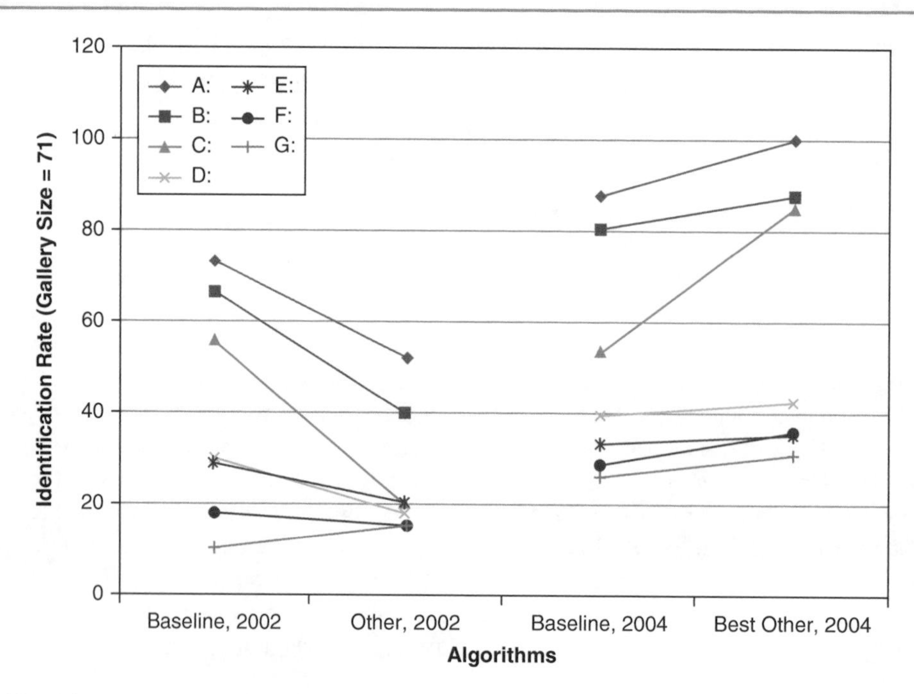

Gait Recognition, Fig. 5 The progress from the baseline over 2 years for the various experiments shown in Table 4 [22]

Gait Recognition, Fig. 6 Bank robbery

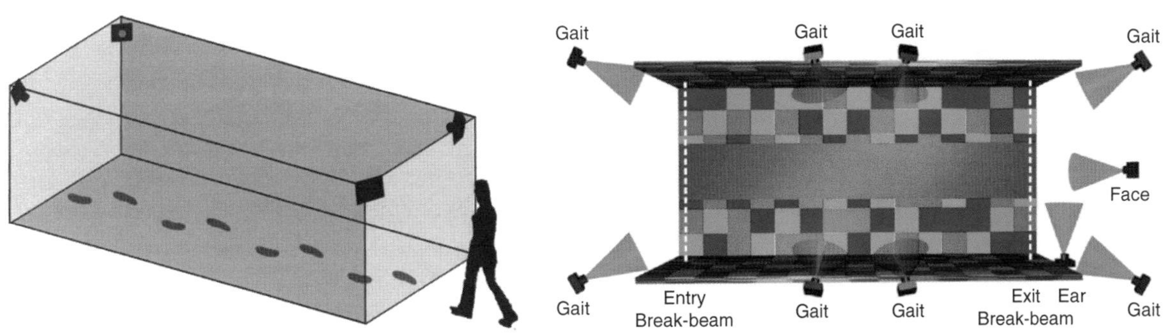

Gait Recognition, Fig. 7 The biometric tunnel

studies achieve good recognition rates on gallery and probe acquired in similar conditions. However, it is very challenging to extract gait features that are invariant to change in appearance as well as to conditions that affect a person's gait. Examples of things that can change and negatively affect the effectiveness of current gait algorithms are change of clothing, shoe type, carrying a load, and injuries/medical conditions. Clothing for instance can change the observed pattern of motion and make it difficult to accurately locate joint position. Furthermore, there are certain factors that are related to the environment and not the subjects themselves that can cause difficulties for current gait matchers. Examples of environmental confounding factors are camera viewing angle, background, and illumination.

Recently, studies have reported progress in solving some of the issues outlined above. A study by Goffredo et al. [6] describes a self-calibrating view-invariant gait recognition algorithm. Hossain et al. [9] have developed a clothing invariant gait matcher. Aqmar et al. [1] are the most recent approach focussed on speed variation.

Currently, gait recognition can deliver very high recognition rates in a constrained environment and if certain factors are controlled. A move towards developing algorithms invariant to change over time is needed. Furthermore, additional work is required to translate the research to outside environment and to explore how scalable it is. Attempts so far suggest that developing highly reliable gait-based human identification system in a real-world application is, and will continue to be, very challenging. In the short term, some of the challenges associated with gait recognition can be addressed by fusing gait with other biometrics [29].

References

1. Aqmar MR, Shinoda K, Furui S (2010) Robust gait recognition against speed variation. In: Proceedings of 20th IEEE international conference on pattern recognition (ICPR), Istanbul, Turkey
2. Benabdelkader C, Cutler R, Davis L (2002) Motion-based recognition of people in eigengait space. In: Proceedings of IEEE international conference automatic face and gesture recognition (AFGR). IEEE, Piscataway, pp 267–272
3. Bouchrika I, Nixon M (2008) Exploratory factor analysis of gait recognition. In: 8th IEEE international conference on automatic face and gesture recognition (AFGR), Amsterdam, The Netherlands
4. Bouchrika I, Goffredo M, Carter JN, Nixon MS (2010) On using gait in forensic biometrics. Journal of Forensic Sciences 56(4):882–889 July 2011
5. Cutting JE, Kozlowski LT (1977) Recognizing friends by their walk: gait perception without familiarity cues. Bull Psychon Soc 9(5):353–356
6. Goffredo M, Bouchrika I, Carter JN, Nixon MS (2010) Self-calibrating view-invariant gait biometrics. IEEE Trans Syst Man Cybern Part B Cybern 40(4):997–1008
7. Guoying Z, Guoyi L, Hua L, Pietikainen M (2006) 3D gait recognition using multiple cameras. In: Proceedings of 7th IEEE international conference automatic face and gesture recognition (AFGR), Southampton, pp 529–534
8. Hayfron-Acquah J, Nixon M, Carter J (2003) Automatic gait recognition by symmetry analysis. Pattern Recognit Lett 24(13):2175–2183
9. Hossain MA, Makihara Y, Junqui W, Yagi Y (2010) Clothing-invariant gait identification using part-based clothing categorization and adaptive weight control. Pattern Recognit 43(6):2281–2291
10. Kale A, Sundaresan A, Rajagopalan A, Cuntoor N, Roy-Chowdhury A, Kruger V, Chellappa R (2004) Identification of humans using gait. IEEE Trans Image Process 13(9):1163–1173
11. Kozlowski LT, Cutting JE (1977) Recognizing the sex of a walker from a dynamic point-light display. Percept Psychophys 21(6):575–580
12. Larsen PK, Simonsen EB, Lynnerup N (2008) Gait analysis in forensic medicine. J Forensic Sci 53(5): 1149–1153

13. Liang W, Tieniu T, Huazhong N, Weiming H (2003) Silhouette analysis-based gait recognition for human identification. IEEE Trans Pattern Anal Mach Intell 25(12): 1505–1518
14. Little J, Boyd J (1998) Recognizing people by their gait: the shape of motion. Videre: J Comput Vis Res 1(2): 1–32
15. Liu Z, Sarkar S (2004) Simplest representation yet for gait recognition: averaged silhouette. In: Proceedings of 17th IEEE international conference pattern recognition (ICPR), Cambridge, UK, pp 211–214
16. Matovski DS, Nixon MS, Mahmoodi S, Carter JN (2010) The effect of time on the performance of gait biometrics. In: 4th IEEE international conference biometrics: theory, applications and systems (BTAS), Washington, DC, USA
17. Murray M, Drought A, Kory R (1964) Walking patterns of normal men. J Bone Jt Surg 46(2):335
18. Nixon MS, Carter JN (2006) Automatic recognition by gait. Proc IEEE 94(11):2013–2024
19. Nixon MS, Tan TN, Chellappa R (2005) Human identification based on gait. Springer, New York
20. Okumura M, Iwama H, Makihara Y, Yagi Y (2010) Performance evaluation of vision-based gait recognition using a very large-scale gait database. In: 4th IEEE international conference biometrics: theory, applications and systems (BTAS), Washington, DC
21. Samangooei S, Bustard J, Nixon MS, Carter JNN (2010) On acquisition and analysis of a dataset comprising of gait, ear and semantic data. In: Bhanu B, Govindaraju V (eds) Multibiometrics for human identification. Cambridge University Press, Cambridge. (in press)
22. Sarkar S, Phillips PJ, Zongyi L, Isidro Robledo V, Grother P, Bowyer KW (2005) The HumanID gait challenge problem: data sets, performance, and analysis. IEEE Trans Pattern Anal Mach Intell 27(2):162–177
23. Seely R, Samangooei S, Lee M, Carter J, Nixon M (2008) University of Southampton multi-biometric tunnel and introducing a novel 3d gait dataset. In: Proceedings of 2nd IEEE international conference biometrics: theory, applications and systems (BTAS), Washington, DC
24. Sundaresan A, Roychowdhury A, Chellappa R (2003) A hidden markov model based framework for recognition of humans from gait sequences. In: Proceedings of IEEE international conference image processing (ICIP), Barcelona, pp 93–96
25. Urtasun R, Fua P (2004) 3D tracking for gait characterization and recognition. In: Proceedings of 6th IEEE international conference automatic face and gesture recognition (AFGR), Seoul, pp 17–22
26. Wang L, Ning H, Tan T, Hu W (2004) Fusion of static and dynamic body biometrics for gait recognition. IEEE Trans Circuits Syst Video Technol 14(2):149–158
27. Winter DA (2009) Biomechanics and motor control of human movement. Wiley, Ottawa
28. Yam C, Nixon M, Carter J (2004) Automated person recognition by walking and running via model-based approaches. Pattern Recognit 37(5):1057–1072
29. Zhou X, Bhanu B (2008) Feature fusion of side face and gait for video-based human identification. Pattern Recognit 41(3):778–795

Gamut Mapping

Rajeev Ramanath[1] and Mark S. Drew[2]
[1]DLP® Products, Texas Instruments Incorporated, Plano, TX, USA
[2]School of Computing Science, Simon Fraser University, Vancouver, BC, Canada

Synonyms

Color management

Related Concepts

▶Gamut Mapping

Definition

Gamut Mapping refers to the process of translating colors in one device's color space to that of another. This process is performed on colors in images and video so as to create a rendition of a source image (typically in a capture device's color space) in an output device's color space while meeting several rendering intents: absolute and relative colorimetric fidelity, perceptual accuracy, and the problem of saturation – each of which trades off one color property at the expense of another.

Background

Different media – cameras, printers, displays – have different achievable gamuts, depending on the manner in which color is either captured or reproduced. This typically means that one medium can have colors that may not be reproducible on another. The color gamut of a device may be displayed as a volume of achievable colors, and typically this is shown in a CIELAB or CIELUV (see Fig. 1), or as a projection on the CIE xy or u"v" chromaticity diagram (see Fig. 2). The three-dimensional representation is far more informative than the two-dimensional projection as it captures the nuances of the color space, specifically around the luminance of the primaries and the associated black

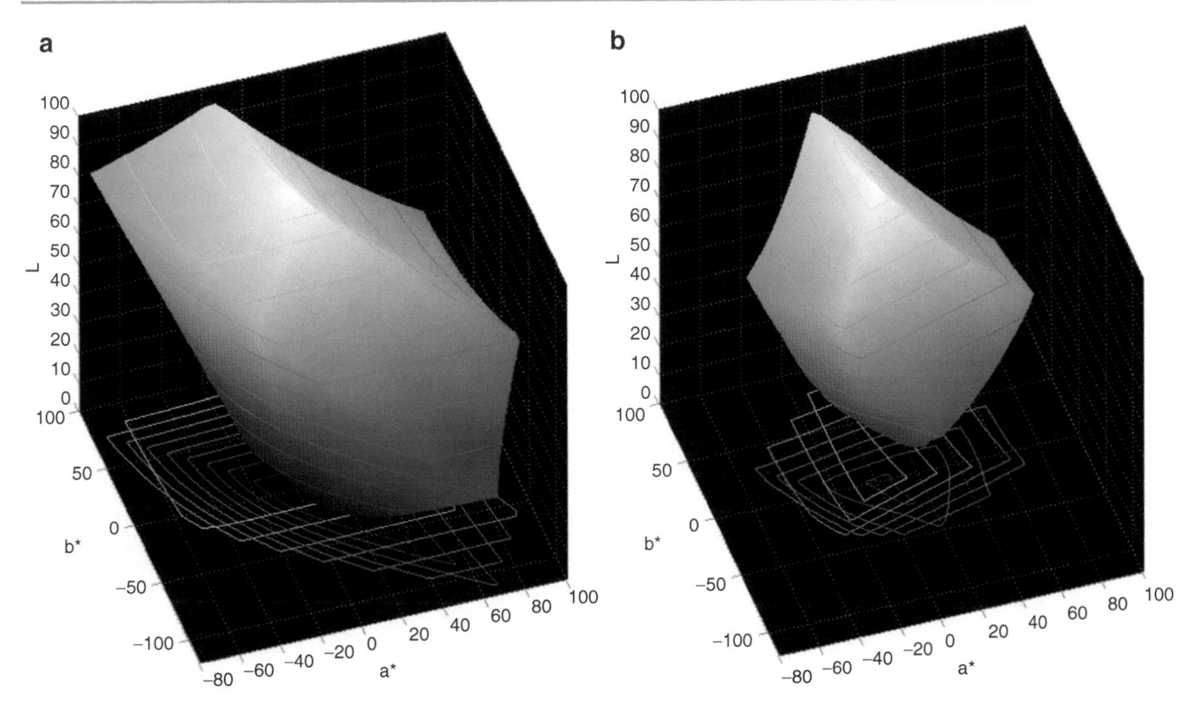

Gamut Mapping, Fig. 1 Three-dimensional renderings of color gamuts: (**a**) device with sRGB color space, (**b**) device with SWOP color space

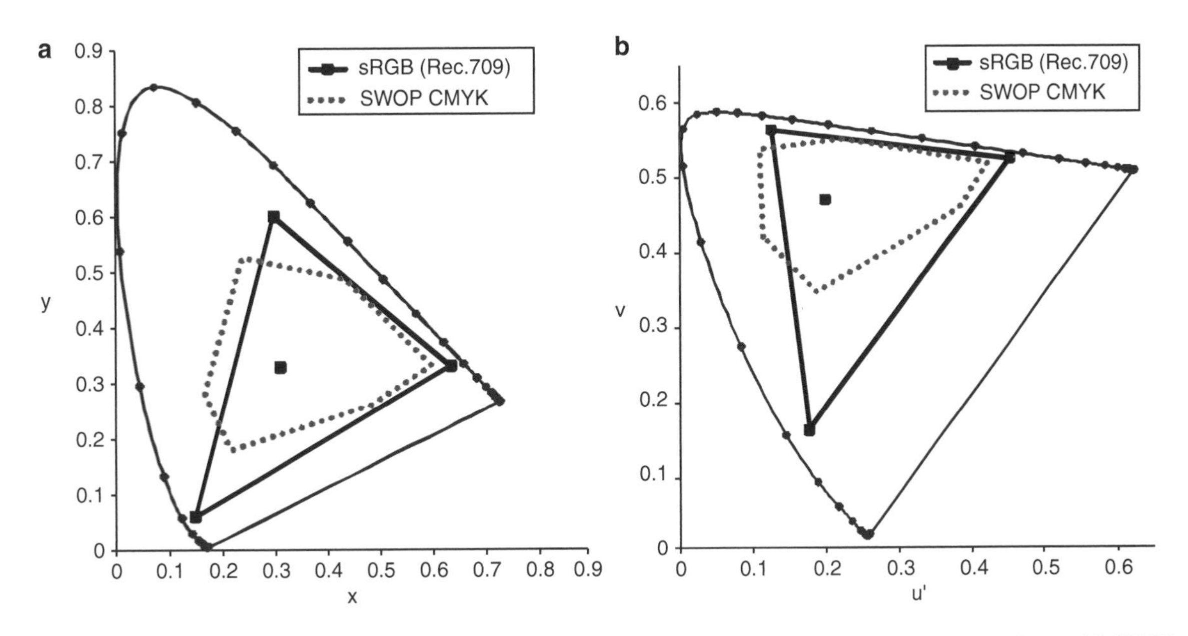

Gamut Mapping, Fig. 2 Two-dimensional renderings of color gamuts of a device with sRGB color space and one with SWOP color space in (**a**) CIE xy chromaticity diagram, (**b**) CIE u"v" chromaticity diagram

and white levels of the medium. The two-dimensional projection requires multiple luminance slices to be plotted for it to be comparably informative.

Theory

The challenge of gamut mapping may be described by means of an example of a case where an image stored within an sRGB color space needs to be reproduced in a print medium with SWOP colors: Fig. 3 shows two color gamuts – the sRGB color gamut (wireframe) almost enclosing a SWOP color space (solid). It is interesting to note that there are regions of the color space that are represented by the sRGB color space but are not represented by the SWOP color space and vice versa. In performing such a mapping, the following challenges arise:

- How should black and white be mapped?
- Should colors in the intersection of the two gamut volumes be reproduced as such, or should they be compressed?
- How should input colors that are outside the output color space be reproduced?
- What must be done with colors that can be created with the output color space but cannot be represented in the input color space?

The objective of gamut mapping algorithms is to translate colors in the input color space to achievable colors in the output color space so as to meet certain key criteria that are referred to as rendering intents (per ICC guidelines [3]):

- *ICC-absolute colorimetric intent*: "Chromatically adapted tristimulus values of in-gamut colors are unchanged." This intent preserves the relationship among in-gamut colors at the expense of out-of-gamut colors while maintaining accuracy. Again, nothing is specifically stated about the mapping of out-of-gamut colors.
- *Media-relative colorimetric intent*: "The use of media-relative colorimetry enables colour reproductions to be defined which maintain highlight detail, while keeping the medium white, even when the original and reproduction media differ in colour." This intent also preserves the relationship between in-gamut colors at the expense of out-of-gamut colors while maintaining accuracy. Nothing is specifically stated about the mapping of out-of-gamut colors.

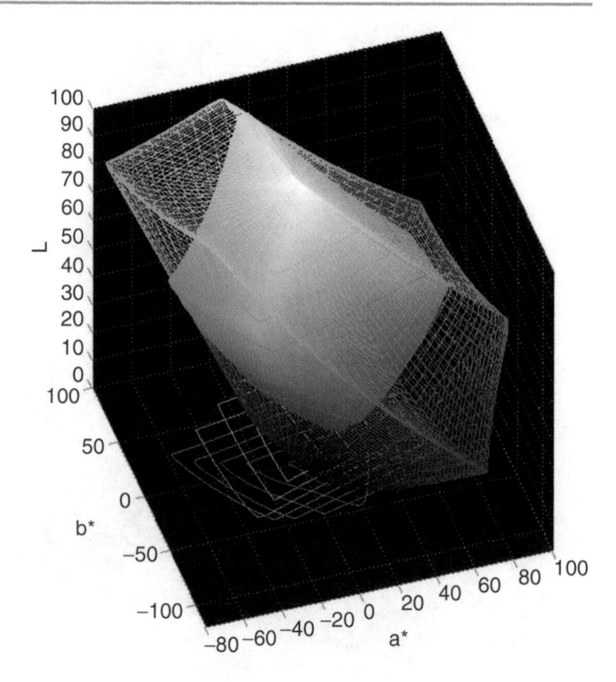

Gamut Mapping, Fig. 3 Three-dimensional renderings of color gamuts of an sRGB color space (wireframe) and a SWOP color space (solid)

- *Perceptual intent*: "The exact gamut mapping of the perceptual intent is vendor specific and is useful for general reproduction of pictorial images, typically includes tone scale adjustments to map the dynamic range of one medium to that of another, and gamut warping to deal with gamut mismatches." This objective of this intent is clear from the definition, and typically involves proprietary algorithms to perform the gamut mapping for general reproduction of images, particularly pictorial or photographic-type images.
- *Saturation intent*: "The exact gamut mapping of the saturation intent is vendor specific and involves compromises such as trading off preservation of hue in order to preserve the vividness of pure colours." This objective of this intent is also clear from the definition, and also typically involves proprietary algorithms to perform the gamut mapping for images and video that contains charts and diagrams.

Most gamut mapping algorithms are performed in either the device space or in a perceptual space: the choice of which space to use is highly application and preference dependent. For the sake of simplicity, in this

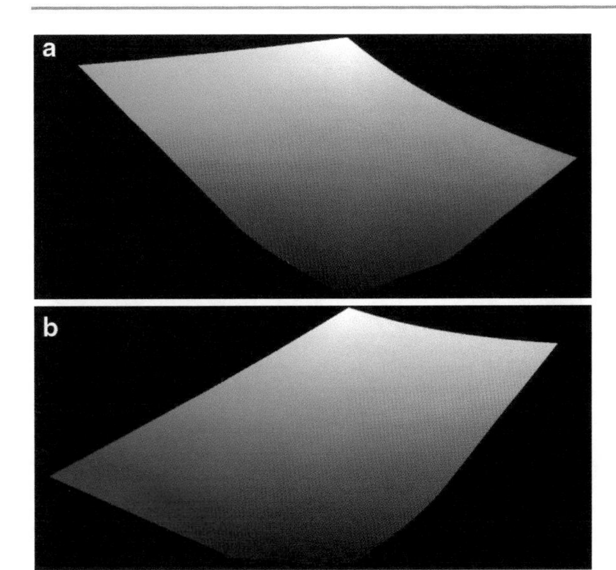

Gamut Mapping, Fig. 4 A set of slices of the color gamut of an sRGB color space (in a CIELAB representation) showing the (**a**) *red–green* slice and (**b**) a *yellow–blue* slice

article, we will restrict ourselves to a CIELAB space along with its perceptual correlates: CIELAB Lightness, Chroma, and Hue. However it should be noted that gamut mapping algorithms could be defined in a variety of spaces, making use of each space's general perceptual correlates of hue, chroma/colorfulness, and lightness/brightness.

Typical color gamuts are shown in Fig. 3 and particular Hue slices for red–green, and yellow–blue are shown in Fig. 4. That is, these slices each show a vertical slice through a color solid, each at a particular position around the surrounding circle we consider to be Hue – i.e., is a color red, is it green, etc. The Chroma increases from zero in the center of the solid to a maximum value on the outside of the solid, and Lightness (the correlate of brightness) goes from minimum (black) at the bottom to maximum (white) at the top. To ease illustration, Hue slices will be abstracted using figures like those in Fig. 5 composed of highly simplified shapes (triangles in this case) – real-world color gamuts can be far more complicated and not included in these illustrations. It is to be noted that the point $L_{M\text{-out}}$ is often referred to as the *cusp* for that Hue slice.

Multiple variations of different scenarios of Hue slices are shown in Fig. 6 to illustrate the variety of problems one might encounter with gamut mapping even when the shape of the gamut is simplified using triangles. These four cases show the input gamut to be larger (in the Lightness scale) than the output gamut, but it is straightforward to envision the opposite set of cases (swapping input and output color spaces) to be possible as well for different gamut pairs. It is also straightforward to envision gamut pairs for which the Lightness scales match. Clearly, even for one gamut pair, one chosen approach for "mapping" the gamuts may work for most Hue slices, but might fail for a different Hue slice.

In general, gamut mapping algorithms may be classified based on which of four approaches is taken:

- Gamut Clipping Algorithms
- Gamut Compression/Expansion Algorithms
- Spatial Gamut Mapping Algorithms
- Memory-Color-Aware Gamut Mapping Algorithms

Gamut Clipping

Gamut clipping approaches may be described as those that address those colors that are outside the gamut of the output color space. Some of the most common approaches include those that are *nearest neighbor* and those that are *Hue-preserving*. Nearest-neighbor approaches map the color in the input gamut's color space to a color in the output gamut's color space that is nearest as defined by one of the following criteria:

- Nearest color in the output color space along a vector of constant Lightness: Mapping along the vector of constant Lightness tends to address the mapping issue as a simple one-dimensional problem of changing only the Chroma of a color. This is shown as approach (a) in Fig. 7, with the mapping occurring along the dashed blue lines.
- Nearest color in the output color space along a vector to the point in the Hue slice with a Lightness of 50. Mapping along the vector pointing to a Lightness of 50 has the advantage that both Lightness and Chroma are adjusted, which helps maintain contrast between colors. This is shown as approach (b) in Fig. 7. This approach is sometimes further divided as a mapping to a different Lightness value for colors above and below the *cusp* of the input gamut, shown as approach (c) in Fig. 7.
- Nearest color in the output color space along a vector to the point in the Hue slice with a Lightness of the *cusp* ($L_{M\text{-out}}$ in Fig. 5): Mapping to the cusp has the advantage of maintaining the general trend of

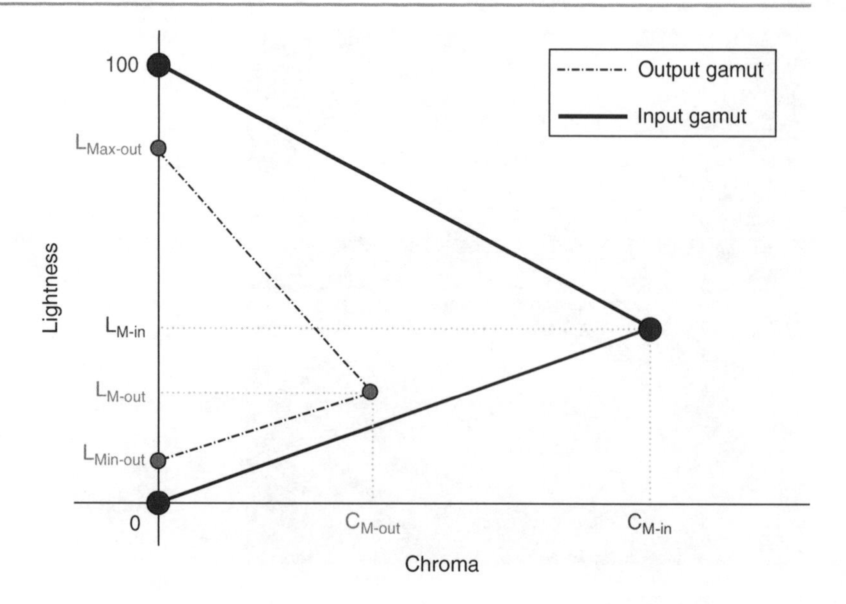

Gamut Mapping, Fig. 5 A highly simplified Hue slice of color gamut of a gamut pair showing the min-max and most chromatic points in the two gamuts

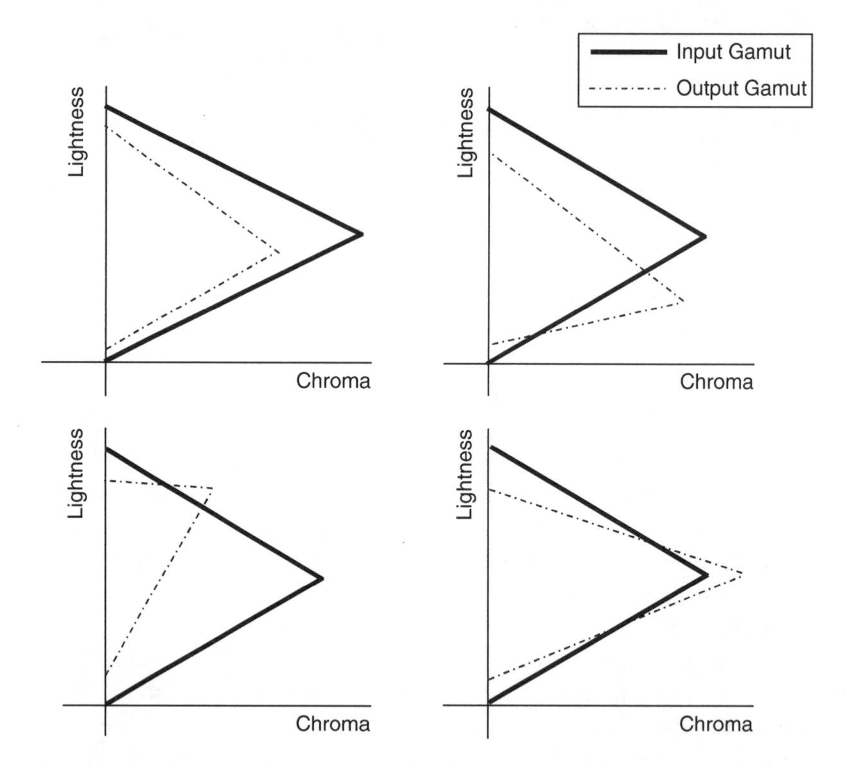

Gamut Mapping, Fig. 6 Four variations of Hue slices showing challenges that gamut mapping algorithms have to address. The complement of these approaches also exist wherein the input and output gamuts are reversed in these pictures

the Lightness in the input and output color spaces. This is shown as approach (d) in Fig. 7.

- Nearest color in the output color space based on a color difference measure: If one were to use the CIELAB difference measure ΔE^*_{ab}, which is approximately perceptually uniform, a color would be mapped to an output color that is perceptually closest to the input color as defined by the ΔE^*_{ab} measure. A slightly different variant of this approach would be to use a different color difference measure, such as ΔE_{94}, or ΔE^*_{00} (CIEDE2000), or some other preferred color

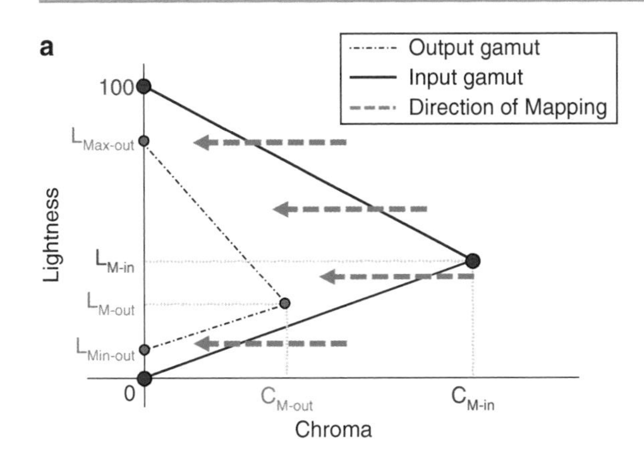

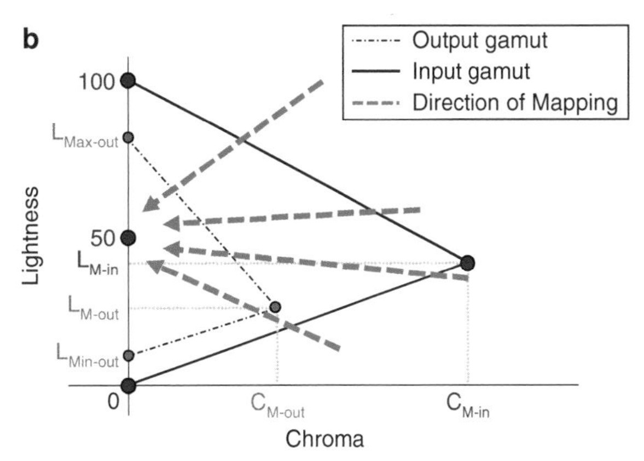

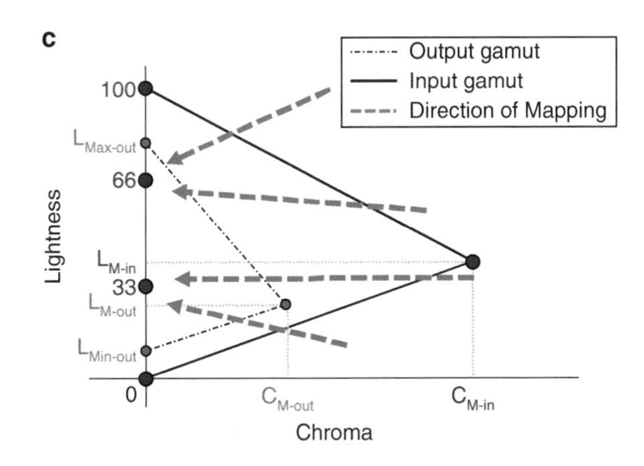

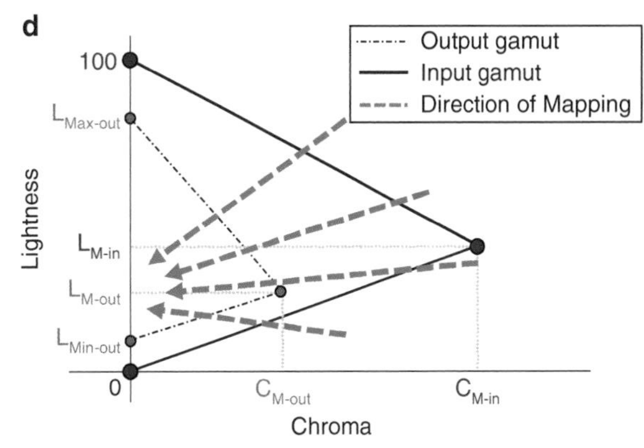

Gamut Mapping, Fig. 7 Different directions for mapping colors showing the (**a**) mapping along lines of constant Lightness, (**b**) mapping along lines of fixed Lightness, (**c**) mapping along lines of varying lightness with a simple example of two different Lightnesses, (**d**) mapping along the vectors to the output gamut's cusp

difference measure. *It is to be noted that this can result in shifts in the Hue descriptor for a color.*

Gamut Compression/Expansion

One of the challenges that arises with gamut clipping is that this invariably results in loss of detail in images when multiple input colors map to the same output color, especially when large differences in color gamuts are to be overcome. If this behavior is not desired, gamut compression (or alternatively expansion) is typically performed. The approaches for gamut compression/expansion may be classified as belonging to one of three classes:

- *Lightness compression/expansion* approaches operate only along the Lightness/luminance axis when the largest difference is along this axis of the color space. The resulting Chroma (after Lightness mapping) could either be clipped or get compressed/expanded appropriately.
- *Chroma compression/expansion* approaches operate only along the Chroma axis. The resulting Lightness (after Chroma mapping) could either be clipped or get compressed/expanded appropriately.
- *Lightness and chroma compression* approaches operate on both Lightness and Chroma at the same time by mapping colors along various directions, as for those shown in Fig. 7a–d.

In all these approaches, independent of which dimension of the color space is compressed/expanded, or which direction of compression/expansion is chosen, the choice of the input–output relationship impacts the outcome significantly. Figure 8 shows various input–output relationships for the two parameters – Lightness and Chroma – that may be used for the above approaches. All these approaches are typically performed on a hue-by-hue basis.

Spatial Gamut Mapping Algorithms

The gamut mapping approaches discussed thus far consider individual colors in an image in isolation, aside from their relationship in the color space. This disregards key factors that are known to impact the appearance of colors to human observers, such as spatial frequency of the image/color, surround colors, and other color appearance phenomena that can greatly change the perception of a color. Discounting these factors typically results in loss of detail in images or a degradation in the perception of the image – although specific colors may have been "accurately" mapped. A generic class of approaches called spatial gamut mapping algorithms take the spatial relationships of colors into account when mapping colors. A simple, yet robust, framework proposed by Balasubramanian et al. [1] is shown in Fig. 9 where the two gamut mappers (a) and (b) are tuned for the specific application. The objective of the spatial high-pass filter is to allow for Lightness variations in the signal when high-frequency content in the image is encountered. This is based on the observation that humans possess the most visual acuity in the brightness dimension of color, and so high spatial frequency content is more perceived along the Lightness axis.

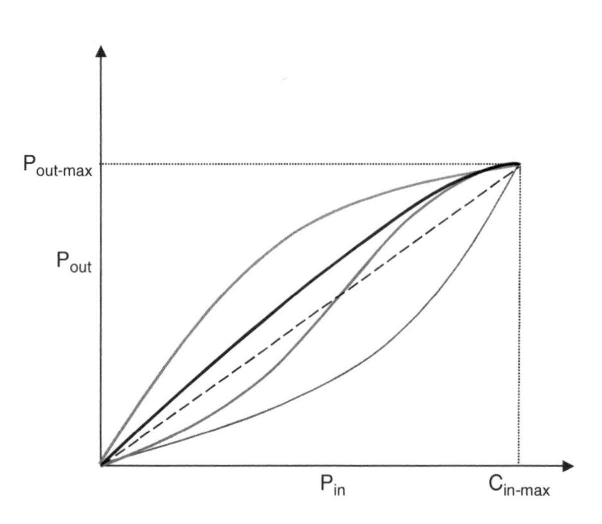

Gamut Mapping, Fig. 8 Different input–output relationships for gamut compression/expansion, represented for parameter P which denotes Lightness and/or Chroma

Memory-Color-Aware Gamut Mapping Algorithms

Although this class of algorithms may be combined with the three other classes of gamut mapping solutions, it is important to address this issue separately as this is at the core of high-quality gamut mapping algorithms. One of the most challenging tasks of a gamut mapping algorithm arises from the fact that its performance is typically evaluated by a human observer. Human observers have visual systems that, apart from being highly complex in terms of the appearance phenomena that determine their behavior, have preferences: the sky has to be a certain shade of blue, human skin tone rendition has to be a "certain" desired color, the color of green grass has to be a very specific shade of green, etc. This requires gamut mapping algorithms to comprehend these colors, deemed *memory colors*: colors that the human observer "knows" to be a certain shade, especially in the context in which they are presented. Approaches that attempt to maintain the rendition of memory colors tend to be proprietary to the vendors. These approaches tend to be mostly probabilistic in nature, taking into account a very large database of memory colors as rendered in the output media and deemed suitable by the solution providers.

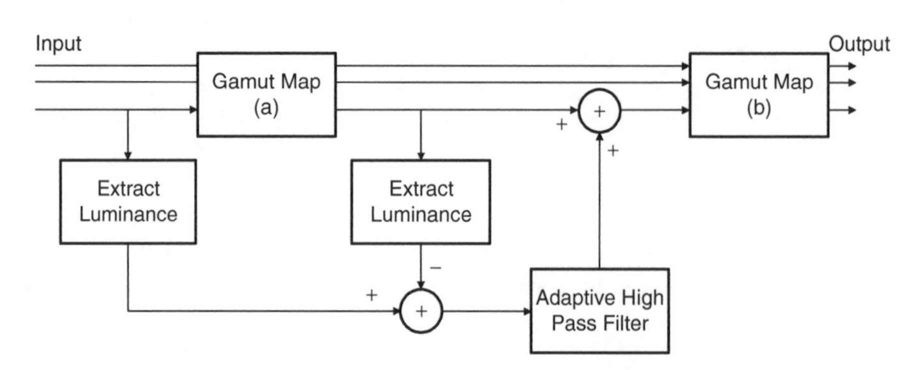

Gamut Mapping, Fig. 9 A generic framework for spatial gamut mappers

Open Problems

The rendering intents as specified by the ICC say little about how the mapping needs to be performed [5]. Open questions like the following still remain unanswered:

- *What must be done when there is a combination of the rendering intents?* This is a relatively common problem when a pictorial image containing the logo of a corporation needs to be gamut mapped – the logo needs to be accurate while the image may need to be perceptually mapped.
- *What attributes of the appearance of colors must be used?* Colors may be defined by their Lightness, Chroma, and Hue; or by their brightness, colorfulness, and Hue. The choice of which triplet needs to be used will help determine the color space that is to be chosen for gamut mapping.
- *What color space is best suited to perform gamut mapping?* In the case of colorimetric intents, it is relatively clear that the tristimulus values need to be maintained/adapted, and the chromatic-adaptation transform is specified to be the linear Bradford model. Even if one were to know clearly which appearance attributes to use, for each set of appearance attributes, and application, different color spaces lend themselves differently. More importantly, different color spaces provide different "predictors" for these appearance attributes and the gamut mapping algorithm needs to rely on the accuracy of these predictors.

Interested readers are referred to works by various authors in the general field of color imaging and reproduction that address not just the immediate needs of gamut mapping algorithms but the more important aspect of their interaction with the larger image and color processing chain in typical capture, display, and print systems [2, 4, 6].

References

1. Balasubramanian R, deQueiroz R, Eschbach R, Wu W (2000) Gamut mapping to preserve spatial luminance variations. J Imaging Sci Technol, 45(Part 5): 436–443
2. Hunt RWG (2004) Reproduction of colour, 6th edn. Wiley, Chichester
3. International Color Consortium® (2004) Specification ICC.1:2004–10 (Profile Version 4.2.0.0) Image technology colour management – Architecture, profile format, and data structure. International Color Consortium, Reston, VA
4. Morovic J (2008) Color gamut mapping. Wiley, Chichester
5. Morovic J, Luo MR (2001) The fundamentals of gamut mapping: a survey. J Imaging Sci Technol 45(3):283–290
6. Sharma G (eds) (2002) Digital color imaging handbook. CRC, Hoboken

Gaze Control

▸Active Sensor (Eye) Movement Control

Generalized Bas-Relief (GBR) Transformation

▸Bas-Relief Ambiguity

Generic Object Recognition

Svetlana Lazebnik
Department of Computer Science, University of Illinois at Urbana-Champaign, Urbana, IL, USA

Synonyms

Object class recognition (categorization)

Related Concepts

▸Object Class Recognition (Categorization) ▸Object Detection

Definition

Generic object recognition is the problem of identifying the category membership of an object contained in a photograph. This term is largely synonymous with object class recognition (categorization), though it places a greater emphasis on approaches aimed at recognizing a broad range of natural categories, as opposed to object instances or specialized categories (e.g., faces).

Background

Historically, much of the effort in the field of object recognition has focused on identifying particular object instances or members of a few select categories, such as faces, cars, and pedestrians. However, in the last few years, the community has started moving towards the goal of building *general-purpose* recognition systems whose ability to identify everyday categories matches that of humans. This is a big challenge: according to cognitive scientists, human beings can identify around 3,000 entry-level categories and 30,000 visual categories overall [1], and the number of categories distinguishable with domain expertise may be on the order of 10^5 [2].

Theory

For a summary of the major strategies for modeling general visual categories, refer to the entry on object class recognition (categorization). Apart from the representational issues discussed in that entry, computational scalability issues are central to generic object recognition (see, e.g., [3]). To achieve scalable and accurate recognition in the many-class setting, it is important to have a good model of the semantic, visual, and structural relationships between the different categories. In particular, organizing classes into visual or semantic hierarchies [4, 5] allows the complexity of recognition to scale sublinearly in the number of classes, makes the system less prone to confusion between unrelated classes, and enables it to perform cross-category generalization (e.g., an unfamiliar type of a car may still be recognized as a vehicle). Another promising strategy for cross-category generalization is the learning of attributes [6]. For example, the system may not have been trained to recognize a tapir, but it may still be able to infer that it is a four-legged herbivorous mammal. More generally, *transfer learning*, or the sharing of knowledge between related categories, is likely to be a key component of an effective generic recognition system, enabling it to acquire new models from very few images (see, e.g., [7]).

Progress in generic object recognition crucially depends on the availability of diverse and large-scale datasets. Fortunately, beginning with the Caltech-101 dataset [8], the field has seen a rapid succession of datasets containing ever increasing numbers of images and categories. Notable current datasets include Caltech-256 [9], LabelMe [10], Tiny Images [11], the PASCAL Visual Object Classes Challenge [12], and ImageNet [13]. In particular, ImageNet contains over 11 million images and over 15,000 concepts organized in a hierarchy, making it a good testbed for new general-purpose recognition algorithms.

References

1. Biederman I, Mezzanotte R, Rabinowitz J (1982) Scene perception: Detecting and judging object undergoing relational violations. Cogn Psychol 14:143–147
2. Perona P (2009) Visual recognition circa 2008. In: Dickinson S, Leonardis A, Schiele B, Tarr M (eds) Object categorization: computer and human vision perspectives. Cambridge University Press, Cambridge/New York, pp 55–68
3. Deng J, Berg AC, Li K, Li FF (2010) What does classifying more than 10,000 image categories tell us? European conference on computer vision (ECCV), Heraklion
4. Marszalek M, Schmid C (2008) Constructing category hierarchies for visual recognition. European conference on computer vision (ECCV), Marseille, pp 479–491
5. Marszalek M, Schmid C (2007) Semantic hierarchies for visual object recognition. IEEE conference on computer vision and pattern recognition (CVPR), Minneapolis, pp 1–7
6. Farhadi A, Endres I, Hoiem D, Forsyth D (2009) Describing objects by their attributes. IEEE conference on computer vision and pattern recognition (CVPR), Miami, pp 1778–1785
7. Fei-Fei L, Fergus R, Perona P (2006) One-shot learning of object categories. IEEE Trans Pattern Anal Mach Intell 28(4):594–611
8. Fei-Fei L, Fergus R, Perona P (2004) Learning generative visual models from few training examples: an incremental Bayesian approach tested on 101 object categories. IEEE CVPR workshop on generative-model based vision. http://www.vision.caltech.edu/Image_Datasets/Caltech101
9. Griffin G, Holub AD, Perona P (2006) The Caltech-256. Caltech technical report. http://www.vision.caltech.edu/Image_Datasets/Caltech256/
10. Russell B, Torralba A, Murphy K, Freeman WT (2008) LabelMe: a database and web-based tool for image annotation. Int J Comput Vis 77(1–3):157–173
11. Torralba A, Fergus R, Freeman WT (2008) 80 million tiny images: a large dataset for non-parametric object and scene recognition. IEEE Trans Pattern Anal Mach Intell 30(11):1958–1970
12. Everingham M, Van Gool L, Williams CKI, Winn J, Zisserman A (2010) The Pascal visual object classes (VOC) challenge. Int J Comput Vis 88(2):303–338. http://pascallin.ecs.soton.ac.uk/challenges/VOC/
13. Deng J, Dong W, Socher R, Li LJ, Li K, Fei-Fei L (2009) ImageNet: a large-scale hierarchical image database. IEEE conference on computer vision and pattern recognition (CVPR), Miami, pp 248–255

Generically Describable Situation

►Situation Graph Trees

Geodesic Active Contour

►Geodesics, Distance Maps, and Curve Evolution

Geodesics, Distance Maps, and Curve Evolution

Ron Kimmel
Department of Computer Science, Technion – Israel Institute of Technology, Haifa, Israel

Synonyms

Geodesic active contour; Segmentation; Variational methods

Related Concepts

►Edge Detection; ►Semantic Image Segmentation

Definition

Geodesics are locally the shortest paths in some space. Equivalently, geodesics are curves for which small perturbation increases their length according to some measure. A minimal geodesic corresponds to the shortest geodesic connecting two points. In computer vision and pattern recognition, the way distance is measured and the resulting geodesics define the specific application, see [1] for a pedagogical introduction to the field of numerical geometry of images.

Background

Edge detectors can be defined from a variational perspective where an edge is a curve along which the image gradient aligns with the curve's normal [2–6]. These types of edges are geodesics in the sense of integrating the inner product between the curve's normal and the image gradient and by Green's Theorem can be shown to be the Marr-Hildreth edge detectors [7], also known as LZC or Laplacian zero crossings. In fact, the more sophisticated Canny-Haralick [8, 9] edge detector can be viewed in a similar geometric-variational manner [3].

The first to formulate computer vision problems from a variational axiomatic view point was probably Horn [10]. For example, shape from shading, the problem of shape reconstruction from an intensity image, can be casted into a minimal geodesic computation problem. In a similar fashion, the image domain can be separated into meaningful regions by treating the image either as a set of pixels, each tries to place itself in the right segment, or as contours that attempt to locate themselves along the boundaries between objects. These kinds of contours live somewhere between histogram segmentation and edge detection and could be referred to as *edge integrators*. The *geodesic active contour* (GAC) model was first presented in [11]; see also [12] for a link between vector quantization and the GAC model. It is a computational framework to find local geodesics in a domain weighted by some edge indicator function. Its first implementation was via the Osher-Sethian level set framework [13]. Minimal geodesics in that metric were shown to be the solution of an eikonal equation [14] that could be efficiently solved by the *fast marching method* [15, 16]; see also its extension to surfaces [17].

Eikonal solvers compute distance maps in a specific metric. Shape morphology, which defines algebra of shapes via Minkowski operations on sets of points, can also be related to distance maps. For an intuitive explanation, one can think of adding two shapes as the process of adding two numbers A and B. The result can be thought of as adding n times the number $\frac{1}{n}B$ to A. As $n \to \infty$, addition can be considered as a differential process. In a similar manner, erosion and dilation operations in mathematical morphology of shapes have a differential interpretations. Its continues formulation was put forward in a number of publications [18–20] relating to level sets and eikonal solvers.

The *geodesic active contour* model evolves a given contour into a local geodesic via a heat flow in the space of weighted contours. Other heat flows evolve the image level sets themselves towards geodesics in Euclidean, affine, or perspective geometries. For example, the curvature scale space [21, 22] and the affine scale space [23, 24] can simplify the image structure in a topology preserving manner, where topology refers to the connectivity of the image level sets.

Theory

Consider the *geodesic active contour* functional

$$E(C) = \int g(C(s))ds, \qquad (1)$$

where g is some edge indicator function, e.g., $g(x, y) = (|\nabla I| + \epsilon)^{-1}$, and $C(s) = \{x(s), y(s)\}$ is an arclength parametrized planar contour, for which the first variation is given by

$$\frac{\delta E}{\delta C} = -(\kappa g - \langle \nabla g, \boldsymbol{n} \rangle)\boldsymbol{n}, \qquad (2)$$

where $\boldsymbol{n}$ is the curve's normal and κ its curvature. One could either find geodesics under that metric by a gradient descent process $C_t = -\frac{\delta E}{\delta C}$, that in a level set formulation reads

$$\phi_t = \text{div}\left(g\frac{\nabla \phi}{|\nabla \phi|}\right)|\nabla \phi|.$$

Here, ϕ is an implicit representation of the curve C. The way to find minimal geodesics under the same metric would be by solving the eikonal equation $|\nabla u| = g$ with $u(p) = 0$ at some source point p in the image domain, and then extract the minimal geodesic by gradient descent flow $C_t = -\nabla u$ from some target point q to the source point p.

Interesting Measures

In the shape from shading problem, the surface $z(x, y)$ is reconstructed from the shading image $I(x, y) = \langle N, l \rangle$, where N is the surface normal and l is the light source direction. For $l = \hat{z}$, one needs to solve for z in the eikonal equation

$$|\nabla z| = \sqrt{\frac{1 - I^2}{I^2}}$$

The Marr-Hildreth edge detector is given by the extremal curves of

$$E(C) = \int \langle n, \nabla I \rangle ds,$$

for which the first variation vanishes for $\Delta I = 0$, while the Canny-Haralick edge detector integrates the extremal curves of

$$E(C) = \int \langle n, \nabla I \rangle ds - \iint_{\Omega_C} \kappa(I)|\nabla I| dx dy.$$

In this case the first variation w.r.t. C (provided by Euler-Lagrange equation) vanishes for $I_{\xi\xi} = 0$. Equivalently, edges are found where the second derivative of the image along the image gradient direction $\xi = \frac{\nabla I}{|\nabla I|}$ is equal to zero.

Minimizing the Euclidean arclength $s = \int |C_p| dp$ leads to the curvature flow, $C_t = C_{ss}$ or equivalently $C_t = \kappa \boldsymbol{n}$, that for an image simultaneous evolution of all its level sets reads

$$I_t = \text{div}\left(\frac{\nabla I}{|\nabla I|}\right)|\nabla I|.$$

The Equi-Affine arclength $v = \int |\det(C_p, C_{pp})|^{1/3} dp$ leads to the affine heat flow $C_t = C_{vv}$, or equivalently $C_t = \kappa^{1/3}\boldsymbol{n}$, that for all image level sets simultaneous evolution is given by

$$I_t = \left(\text{div}\left(\frac{\nabla I}{|\nabla I|}\right)\right)^{1/3} |\nabla I|.$$

Application

Total variation [25] methods in image processing, edge detection and integration, segmentation, structure reconstruction SFS [26, 27], and shape morphology (algebra of shapes) are all applications of computing geodesics and geodesic distances. Each application has its own distance measure and computational preferred flavor.

Open Problems

The theoretical machinery and concepts that motivated the usage of differential geometry and geodesics were adopted to the field of shape matching, processing, and analysis. Further links to metric geometry in general and open problems in this field can be found in [28].

References

1. Kimmel R (2003) Numerical geometry of images: theory, algorithms and applications. Springer, Boston. ISBN:0-387-95562-3
2. Vasilevskiy A, Siddiqi K (2002) Flux maximizing geometric flows. IEEE Trans Pattern Anal Mach Intell 24(12): 1565–1578
3. Kimmel R, Bruckstein AM (2003) On regularized Laplacian zero crossings and other optimal edge integrators. Int J Comput Vis 53(3):225–243
4. Kimmel R (2003) Fast edge integration. In Geometric level set methods in imaging, vision, and graphics. Springer, New York, pp 59–77
5. Desolneux A, Moisan L, Morel JM (2003) Variational snake theory. In: Osher S, Paragios N (eds) Geometric level set methods in imaging, vision and graphics. Springer, New York, pp 79–99
6. Fua P, Leclerc YG (1990) Model driven edge detection. Mach Vis Appl 3:45–56
7. Marr D, Hildreth E (1980) Theory of edge detection. Proc R Soc Lond B 207:187–217
8. Haralick R (1984) Digital step edges from zero crossing of second directional derivatives. IEEE Trans Pattern Anal Mach Intell 6(1):58–68
9. Canny J (1986) A computational approach to edge detection. IEEE Trans Pattern Anal Mach Intell 8(6):679–698
10. Horn BKP (1986) Robot vision. MIT press. McGraw-Hill Higher Education, New York
11. Caselles V, Kimmel R, Sapiro G (1997) Geodesic active contours. Int J Comput Vis 22(1):61–79
12. Chan T, Vese L (1999) An active contour model without edges. In: Scale-space theories in computer vision. Springer, Berlin, pp 141–151
13. Osher SJ, Sethian JA (1988) Fronts propagating with curvature dependent speed: algorithms based on Hamilton-Jacobi formulations. J Comput Phys 79:12–49
14. Cohen LD, Kimmel R (1997) Global minimum for active contours models: a minimal path approach. Int J Comput Vis 24(1):57–78
15. Tsitsiklis JN (1995) Efficient algorithms for globally optimal trajectories. IEEE Trans Autom Control 40(9): 1528–1538
16. Sethian JA (1996) A marching level set method for monotonically advancing fronts. Proc Natl Acad Sci 93(4): 1591–1595
17. Kimmel R, Sethian JA (1998) Computing geodesic paths on manifolds. Proc Natl Acad Sci USA 95(15):8431–8435
18. Brockett RW, Maragos P (1992) Evolution equations for continuous-scale morphology. In: Proceedings IEEE international conference on acoustics, speech, and signal processing, San Francisco, pp 1–4
19. Sapiro G, Kimmel R, Shaked D, Kimia B, Bruckstein AM (1993) Implementing continuous-scale morphology via curve evolution. Pattern Recognit 26(9):1363–1372
20. Kimmel R, Sethian JA (2001) Optimal algorithm for shape from shading and path planning. J Math Imaging Vis 14(3):237–244
21. Gage M, Hamilton RS (1986) The heat equation shrinking convex plane curves. J Diff Geom 23:69–96
22. Grayson MA (1987) The heat equation shrinks embedded plane curves to round points. J Diff Geom 26:285–314
23. Alvarez L, Guichard F, Lions PL, Morel JM (1993) Axioms and fundamental equations of image processing. Arch Ration Mech 123:199–257
24. Sapiro G, Tannenbaum A (1993) Affine invariant scale-space. Int J Comput Vis 11(1):25–44
25. Rudin L, Osher S, Fatemi E (1992) Nonlinear total variation based noise removal algorithms. Physica D 60:259–268
26. Bruckstein AM (1988) On shape from shading. Comput Vis Graph Image Process 44:139–154
27. Horn BKP, Brooks MJ (eds) (1989) Shape from shading. MIT, Cambridge
28. Bronstein A, Bronstein M, Kimmel R (2008) Numerical geometry of non-rigid shapes. Springer, New York

Geometric Algebra

Leo Dorst
Intelligent Systems Laboratory Amsterdam,
Informatics Institute, University of Amsterdam,
Amsterdam, The Netherlands

Synonyms

Clifford algebra

Definition

Geometric algebra is an algebra based on a geometric product of vectors in an inner product space (a.k.a. the Clifford product). It naturally extends common geometrical techniques from linear algebra, to process subspaces and operations between them in a direct, coordinate-free manner. It permits division by subspaces (notably vectors), which leads to much more compact expression of common operations and algorithms in linear algebra, and more direct solutions to equations. Geometric algebra includes quaternions, in a real construction as a ratio of vectors, and extends them to encode rigid body motions. Geometric algebra also extends vector calculus to permit direct differentiation with respect to geometrical quantities and operators, thus allowing classically scalar optimization techniques to be transferred directly to geometric settings.

Background

Geometric algebra (GA) (or Clifford algebra) and its predecessor Grassmann algebra date from the 1870s and 1840s and are therefore older than linear algebra (LA). They formed the first formalization of how to compute with linear subspaces. Simplification of the ideas by Gibbs and others with the intention to construct a compact algebraic system for 3D computations in engineering around 1900 then led to the neglect of these more general frameworks. Now that parts of their structure have effectively been patched onto LA in an *ad hoc* manner with ideas like homogeneous coordinates and quaternions, it may pay to reconsider the use of GA to unify geometric computations. Hestenes was the first to realize this in the 1980s [4], initially for physics. In the 1990s, his ideas percolated into robotics, computer vision, and computer graphics.

Theory

Geometric algebra assumes a vector space V^n with inner product, denoted "·" (aka dot product) between vectors. GA views this dot product as the symmetric part of an underlying *geometric product* denoted " " (a half space, or juxtaposition); the other, antisymmetric part being the outer product, denoted "$\wedge$", signifying the *span* of vectors. The axioms of the geometric product are surprisingly few: it is bilinear, associative, distributive over vector addition, scalars commute, and the geometric product of a vector $\mathbf{v}$ with itself is a scalar equal to $\mathbf{v} \cdot \mathbf{v}$.

From this, it follows that for a (non-null) vector $\mathbf{v}$, the geometric product is invertible with $\mathbf{v}^{-1} = \mathbf{v}/(\mathbf{v} \cdot \mathbf{v})$. This makes division by a vector a permissible operation. That in turn simplifies many geometric expressions. For instance, a reflection of a vector in a plane with normal $\mathbf{n}$ at the origin can be rewritten in a much simpler form than the classical linear algebra expression:

$$\mathbf{x} \mapsto \mathbf{x} - 2\,\frac{\mathbf{x} \cdot \mathbf{n}}{\mathbf{n} \cdot \mathbf{n}}\,\mathbf{n} = \mathbf{x} - 2\left(\frac{\mathbf{x}\,\mathbf{n} + \mathbf{n}\,\mathbf{x}}{2}\right)\mathbf{n}^{-1} = -\mathbf{n}\,\mathbf{x}\,\mathbf{n}^{-1}.$$

Planar reflection is an orthogonal transformation, and concatenating such reflections leads to the general expression of an orthogonal transformation as being characterized as a product of invertible vectors – an algebraic and computationally practical consequence of the famous Cartan-Dieudonné theorem. In particular, a rotation in 3D may be represented as the product of two reflections, i.e., an element $\mathbf{m}\,\mathbf{n}$ applied in a "sandwiching" manner to a vector:

$$\mathbf{x} \;\mapsto\; (\mathbf{mn})\,\mathbf{x}\,(\mathbf{mn})^{-1}.$$

Taking the normal vectors to be normalized to unity, this element $(\mathbf{mn})$ has the same algebraic properties as a *unit quaternion*. Quaternions in GA are therefore simply products (or ratios) of real vectors denoting double planar reflections. Parametrizing this product of vectors by their angle $\mathbf{I}\phi/2$ (i.e., $\phi/2$ from $\mathbf{n}$ to $\mathbf{m}$ to produce a rotation over ϕ in that sense), we may write

$$\mathbf{m}\,\mathbf{n} = \mathbf{n}\cdot\mathbf{m} - \mathbf{n}\wedge\mathbf{m} = \cos(\phi/2) - \mathbf{I}\sin(\phi/2) = e^{-\mathbf{I}\phi/2},$$

where $\mathbf{I}$ is the unit "bivector" $\mathbf{n} \wedge \mathbf{m}$ of the plane span$(\mathbf{n}, \mathbf{m})$, with an orientation determined by the order of the spanning vectors. One may verify (by the axioms of the geometric product) that $\mathbf{I}^2 = -1$, so that the "complex" nature of the quaternions is merely a consequence of the axioms of the geometric product of a real vector space. The final expression as an exponent then follows from the usual Taylor series definition of the exponential, extended to GA elements.

This parametrization of an orthogonal transformation as the exponent of a bivector uses the minimal $\frac{1}{2}n(n-1)$ parameters in an n-D vector space. In applications estimating an orthogonal transformation from data, this is more convenient than the usual over-parametrization as an $n \times n$ matrix M, needing to impose orthogonality as the $\frac{1}{2}n(n+1)$ constraints $M^T M = I$.

An element that can be made as the product of unit vectors, or expressed as the exponent of a bivector, is called a *rotor* in GA. In multiplying operators, this rotor representation is more efficient than matrices (a generalization of the well-known advantage of unit quaternions over rotation matrices). For applying the operator to vectors only, matrices are faster; but rotors can be applied, without modification, to *any* geometric element. The reason behind this is that an orthogonal transformation represented by a rotor V trivially preserves the geometric product in a covariant manner:

$$V\,(AB)\,V^{-1} = (VAV^{-1})\,(VBV^{-1}).$$

As long as we make any construction in GA from a linear combination of geometric products, and as long as we can represent the transformations of interest as rotors, we have a *structure-preserving* implementation of our geometry. As we will see below, this may require choosing a clever embedding relationship between the space in which our geometric elements reside, and the geometric algebra that represents them.

As one multiplies elements of geometric algebra, elements of various *grades* appear. These can be decomposed on a basis consisting of *blades*, which are elements factorizable by the outer product. Such elements represent subspaces of a dimension equal to their grade. For instance, starting from a basis $\{\mathbf{e}_1, \mathbf{e}_2, \mathbf{e}_3\}$ for the vector space $\mathbb{R}^3$, one obtains eight basis blades:

$$\left\{ \underbrace{1}_{\text{scalar basis}}, \quad \underbrace{\mathbf{e}_1, \mathbf{e}_2, \mathbf{e}_3}_{\text{vector basis}}, \quad \underbrace{\mathbf{e}_2 \wedge \mathbf{e}_3, \mathbf{e}_3 \wedge \mathbf{e}_1, \mathbf{e}_1 \wedge \mathbf{e}_3}_{\text{2-blade basis}}, \quad \underbrace{\mathbf{e}_1 \wedge \mathbf{e}_2 \wedge \mathbf{e}_3}_{\text{3-blade basis}} \right\}.$$

Geometrically, these can be interpreted as the point at the origin, three vector directions, three 2-directions of planar elements, and a volume element. A geometric algebra of an n-dimensional vector space has a basis of 2^n elements (like its Grassmann algebra). The high dimensionality of a geometric algebra is not an argument against its use, since the possibility to generate all structure-preserving interactions between such elements automatically at compile time offsets much run time complexity [3]; also, the geometrically meaningful elements tend to be obtained by multiplication of basis elements, and those are sparse on this basis. (Incidentally, such multiplicative generation of meaningful elements may be taken as a distinction between geometric algebra and Clifford algebra, since the latter permits arbitrary addition in its theoretical construction.) In fact, the "extra" elements are mostly geometric objects one would make anyway in classical LA, but now integrated in one consistent framework.

Scalar and vector *calculus* is naturally extended to GA elements. This makes it possible to optimize orientation estimations by differentiating a well-chosen cost criterion with respect to rotors, and demanding that the result be zero. Such capabilities are beginning to provide powerful coordinate-free geometrical optimization methods (see, e.g., [10]), in which specific parametrizations can be chosen conveniently afterward.

The Conformal Model

In linear algebra, homogeneous coordinates are an embedding trick that permits us to use the linear transformations of an $(n+1)$-dimensional space to represent *projective transformations* of an n-dimensional space computationally. In GA, there is an embedding which allows computing with *conformal transformations* in an n-dimensional space using the rotors (i.e., orthogonal transformations) from the algebra of an $(n+2)$-dimensional space [1]. This space has an inner product with signature $(n+1, 1)$, i.e., one can provide it with an orthonormal basis of which $n+1$ vectors square to $+1$, and 1 vector squares to -1. A more commonly used basis is a "null basis" involving the vector n_∞ representing the point at infinity, and n_o, an arbitrarily chosen finite point (which can serve as an origin, similarly to the extra dimensions in homogeneous coordinates). The products of the two non-Euclidean dimensions obey $n_o^2 = n_\infty^2 = 0$, and $n_o \cdot n_\infty = -1$. This, combined with a well chosen embedding of points, permits the space to be an isometric image of Euclidean space (which homogeneous coordinates are not). The resulting geometric algebra is often referred to as CGA (for conformal geometric algebra).

The embedding of Euclidean space into CGA is that a point P at Euclidean location $\mathbf{p}$ relative to n_o

is represented by a vector $p \in \mathbb{R}^{n+1,1}$ given by

$$p = \alpha\,(n_o + \mathbf{p} + \frac{1}{2}\mathbf{p}^2 n_\infty), \quad \text{with } \alpha \in \mathbb{R}$$

(which may be viewed as an extension of the homogeneous coordinate representation $\alpha(n_o + \mathbf{p})$). Note that p is a null vector: $p^2 = 0$. A simple computation now shows that the squared Euclidean distance between two normalized points is effectively the inner product between their representative vectors:

$$-\frac{1}{2}\|\mathbf{p} - \mathbf{q}\|^2 = (\frac{p}{-n_\infty \cdot p}) \cdot (\frac{q}{-n_\infty \cdot p}).$$

The factor $-n_\infty \cdot p$ is the homogeneous weight α of the point representation, so dividing by it provides a normalization. It follows that orthogonal transformations of $\mathbb{R}^{n+1,1}$ which preserve the point at infinity n_∞ preserve Euclidean distances, and hence represent rigid body motions of $\mathbb{R}^n$. Their rotors (sometimes called *motors*) form a more general form of unit quaternion which can represent rotations around a general axis, combined with a shift along that axis. When n_∞ is not preserved, a rotor of the algebra represents a general conformal transformation.

As to the blades in CGA, a general vector of the space $\mathbb{R}^{4,1}$ (to be specific) represents a Euclidean 3D sphere in a dual manner (with a hyperplane as a special case). A point is merely a dual sphere of zero radius. A 4-blade (i.e., the outer product of four points) represents the oriented sphere through those points, unless one of them is n_∞, in which case it represents an oriented plane. Similarly, 3-blades are oriented circles or lines, and 2-blades are oriented point pairs. Other useful geometric elements obtain an algebraic definition, such as tangent vectors, tangent planes, and purely directional elements. In CGA, therefore, classical primitives already used effectively in the specification of geometrical algorithms finally obtain an algebraic expression that permits automatic generation of their transformation properties under the rotors representing the operators of interest. The generating capabilities of CGA are quite powerful: e.g., $\mathbf{P} = (\mathbf{X} \cdot \mathbf{A})/\mathbf{A}$ represents the projection of a blade $\mathbf{X}$ onto a blade $\mathbf{A}$; if $\mathbf{X}$ is a line and $\mathbf{A}$ is a sphere, then $\mathbf{P}$ is a great circle. Such geometrical nuggets may be found in tutorial texts like [3].

Application

While on paper GA may look more involved than LA, its structure-preserving properties make for cleaner software for geometrical applications; the algebra playing the role – at compiler time if required – of generating the transformation methods for the various geometric data objects automatically [3]. The formalism presumably allows for the expression of all desired geometry, and helpfully hampers the construction of elements without geometric meaning (if certain simple construction rules are followed).

The novel representation of rigid body motions and the associated geometric primitives has already permitted compact treatment of such geometrical computer visions tasks as camera calibration [6, 10], pose estimation [8], and pose interpolation [2]. Image transformations in space and value can be given their own geometric algebra [5] and signal processing [9]. Applications to color processing or quaternionic Fourier transforms are so far less compelling.

Open Problems

Currently, the GA equivalents of a suite of techniques from applied linear algebra are being uncovered and turned into practical tools. These include eigenproblems, singular value decomposition, spectral analysis, calculus (Lagrange optimization) and statistics (geometric noise characterization [7]), and more. Since linear algebra and vector calculus are contained within geometric algebra, one can always defer to known techniques, but the employment of the full suite of GA techniques in sample problems is showing promise.

Moreover, other useful GA models are being identified: for 3D projective geometry, the geometric algebra of 3D lines involving the space $\mathbb{R}^{3,3}$ can represent projective transformations as orthogonal transformations, and hence perform them through structure-preserving rotor computations.

References

1. Anglès P (1980) Construction de revêtements du group conform d'un espace vectorial muni d'une "métrique" de type (p,q). Annales de l'Institut Henri Poincaré Sect A XXXIII:33–51

2. Dorst L, Valkenburg RJ (2011) Square root and logarithm of rotors in 3D conformal geometric algebra using polar decomposition. In: Dorst L, Lasenby J (eds) Guide to geometric in practice. Springer, New York, pp 81–104
3. Dorst L, Fontijne D, Mann S (2009) Geometric algebra for computer science: an object-oriented approach to geometry. Morgan Kaufman, San Francisco
4. Hestenes D, Sobczyk G (1984) Clifford algebra to geometric calculus. Reidel, Dordrecht/Boston
5. Koenderink JJ (2002) A generic framework for image geometry. In: Lasenby J, Dorst L, Doran C (eds) Applications of geometric algebra in computer science and engineering. Birkhäuser, Boston, pp 319–332
6. Lasenby J, Stevenson A (2000) Using geometric algebra for optical motion capture. In: Bayro-Corrochano E, Sobczyk G (eds) Applied clifford algebras in computer science and engineering. Birkhäuser, Boston
7. Perwass C (2009) Geometric algebra with applications in engineering. Springer, Berlin
8. Rosenhahn B, Sommer G (2005) Pose estimation in conformal geometric algebra. part i: The stratification of mathematical spaces; part ii: Real-time pose estimation using extended feature concepts. J Math Imaging Vis (JMIV) 22:47–70
9. Sommer G, Zang D (2007) Parity symmetry in multidimensionals signals. Commun pure Appl Anal 6(3): 829–852
10. Valkenburg RJ, Dorst L (2011) Estimating motors from a variety of geometric data in 3D conformal geometric algebra. In: Dorst L, Lasenby J (eds) Guide to geometric in practice. Springer, London, pp 25–46

Geometric Calibration

Zhengyou Zhang
Microsoft Research, Redmond, WA, USA

Related Concepts

▶Calibration; ▶Calibration of Projective Cameras; ▶Camera Calibration

Definition

Geometric calibration is the process of determining the geometric property of a camera such as its intrinsic and extrinsic parameters. It is often referred to as simply *camera calibration* in computer vision. The reader is referred to entry ▶Camera Calibration for a discussion on other camera calibration tasks.

Background

Much work has been done, starting in the photogrammetry community (see [1, 2] to cite a few) and more recently in computer vision ([3–10] to cite a few). According to the dimension of the calibration objects, we can classify those techniques roughly into four categories.

3D reference object-based calibration. Camera calibration is performed by observing a calibration object whose geometry in 3-D space is known with very good precision. Calibration can be done very efficiently [11]. The calibration object usually consists of two or three planes orthogonal to each other. Sometimes, a plane undergoing a precisely known translation is also used [5], which equivalently provides 3D reference points. This approach requires an expensive calibration apparatus and an elaborate setup.

2D plane-based calibration. Techniques in this category requires to observe a planar pattern shown at a few different orientations [12, 13]. Different from Tsai's technique [5], the knowledge of the plane motion is not necessary. Because almost anyone can make such a calibration pattern by himself/herself, the setup is easier for camera calibration.

1D line-based calibration. Calibration objects used in this category are composed of a set of collinear points [14, 15]. As will be shown, a camera can be calibrated by observing a moving line around a fixed point, such as a string of balls hanging from the ceiling.

Self-calibration. Techniques in this category do not use any calibration object and can be considered as 0D approach because only image point correspondences are required. Just by moving a camera in a static scene, the rigidity of the scene provides in general two constraints [9, 16] on the cameras' internal parameters from one camera displacement by using image information alone. Therefore, if images are taken by the same camera with fixed internal parameters, correspondences between three images are sufficient to recover both the internal and external parameters which allow us to reconstruct 3-D structure up to a similarity [17, 18]. Although no calibration objects are necessary, a large number of parameters need to be estimated, resulting in a much harder mathematical problem.

Other techniques exist: vanishing points for orthogonal directions [19, 20] and calibration from pure rotation [21, 22].

Theory

What follows is an excerpt of the review on camera calibration [23]. The reader is referred to [23] for a complete review of various geometric calibration techniques.

Geometric calibration depends on which camera model is used. The reader is referred to entry ▸Camera Model for a presentation of various camera models. In the sequel, we use the perspective camera model.

Camera Calibration with 3D Objects

The traditional way to calibrate a camera is to use a 3D reference object such as those shown in Fig. 1. In Fig. 1a, the calibration apparatus used at INRIA [11] is shown, which consists of two orthogonal planes, on each a checker pattern is printed. A 3D coordinate system is attached to this apparatus, and the coordinates of the checker corners are known very accurately in this coordinate system. A similar calibration apparatus is a cube with checker patterns painted in each face, so in general three faces will be visible to the camera. Figure 1b illustrates the device used in Tsai's technique [5], which only uses one plane with checker pattern, but the plane needs to be displaced at least once with known motion. This is equivalent to knowing the 3D coordinates of the checker corners.

According to the perspective camera model, the relationship between the 3D point $\boldsymbol{M}$ and its image projection $\boldsymbol{m}$ is given by

$$s\tilde{\mathbf{m}} = \underbrace{\mathbf{A}\big[\mathbf{R} \quad \boldsymbol{t}\big]}_{\mathbf{P}}\tilde{\mathbf{M}} \equiv \mathbf{P}\tilde{\mathbf{M}}\,, \tag{1}$$

$$\text{with } \mathbf{A} = \begin{bmatrix} \alpha & \gamma & u_0 \\ 0 & \beta & v_0 \\ 0 & 0 & 1 \end{bmatrix} \tag{2}$$

$$\text{and} \quad \mathbf{P} = \mathbf{A}\big[\mathbf{R} \quad \boldsymbol{t}\big] \tag{3}$$

where s is an arbitrary scale factor; $(\mathbf{R}, \boldsymbol{t})$, called the extrinsic parameters, is the rotation and translation which relates the world coordinate system to the camera coordinate system; and $\mathbf{A}$ is called the camera intrinsic matrix, with (u_0, v_0) the coordinates of the principal point, α and β the scale factors in image u and v axes, and γ the parameter describing the skew of the two image axes. The 3×4 matrix $\mathbf{P}$ is called the camera projection matrix, which mixes both intrinsic and extrinsic parameters. Matrix $\mathbf{P}$ is defined up to a scale factor, so it only has 11 free parameters.

Given one 2D-3D correspondence $(\boldsymbol{m}_i, \boldsymbol{M}_i)$, we can write down two equations based on Eq. (1). There are in total 11 unknowns (5 intrinsic parameters and 6 extrinsic parameters). If we have 6 or points that do not lie on a single plane, a unique solution is available.

A popular technique in this category consists of four steps [11]:

1. Detect the corners of the checker pattern in each image.
2. Estimate the camera projection matrix $\mathbf{P}$ using linear least squares.
3. Recover intrinsic and extrinsic parameters $\mathbf{A}$, $\mathbf{R}$, and $\mathbf{t}$ from $\mathbf{P}$.
4. Refine $\mathbf{A}$, $\mathbf{R}$, and $\mathbf{t}$ through a nonlinear optimization.

Note that it is also possible to first refine $\mathbf{P}$ through a nonlinear optimization and then determine $\mathbf{A}$, $\mathbf{R}$, and $\mathbf{t}$ from the refined $\mathbf{P}$.

It is worth noting that using corners is not the only possibility. We can avoid corner detection by working directly in the image. In [24], calibration is realized by maximizing the gradients around a set of control points that define the calibration object. Figure 2 illustrates the control points used in that work.

Camera Calibration with 2D Objects: Plane-Based Technique

This technique only requires the camera to observe a planar pattern from a few different orientations [12, 13]. An example is shown in Fig. 3.

Given an image of the model plane, a homography between the known model plane and the image can be estimated. Let us denote it by $\mathbf{H} = [\boldsymbol{h}_1 \quad \boldsymbol{h}_2 \quad \boldsymbol{h}_3]$. Without loss of generality, we assume the model plane is on $Z = 0$ of the world coordinate system. This yields

$$[\boldsymbol{h}_1 \quad \boldsymbol{h}_2 \quad \boldsymbol{h}_3] = \lambda\mathbf{A}\,[\boldsymbol{r}_1 \quad \boldsymbol{r}_2 \quad \boldsymbol{t}\,]\,,$$

Geometric Calibration, Fig. 1 3D apparatus for calibrating cameras

a

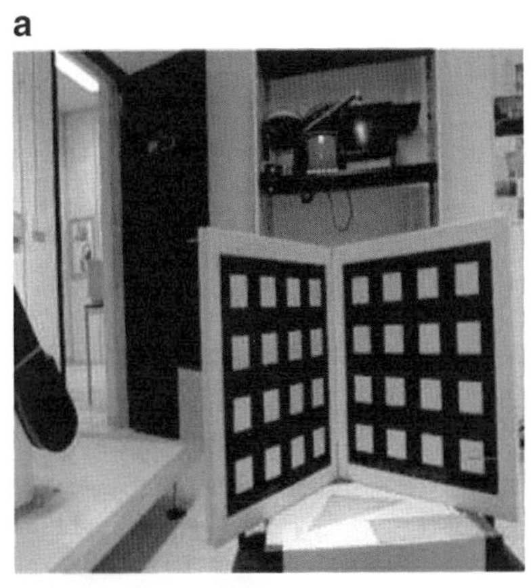

b

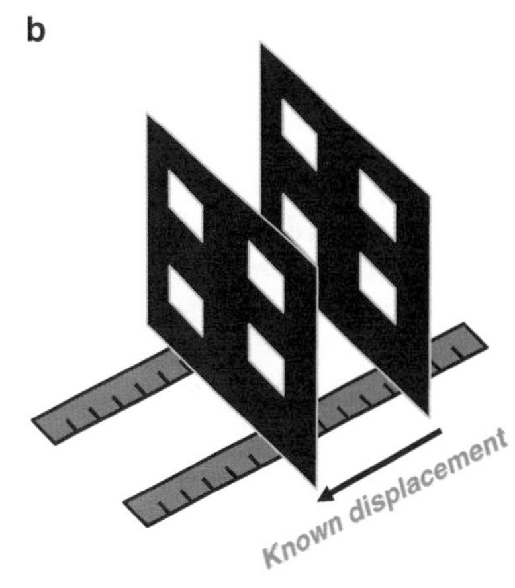

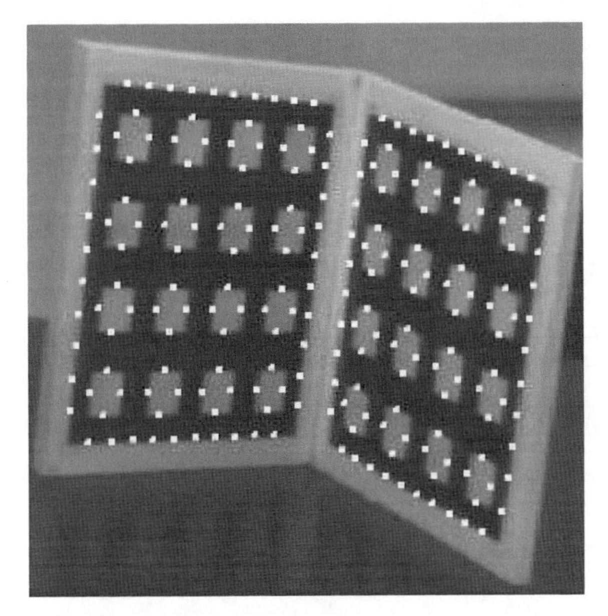

Geometric Calibration, Fig. 2 Control points used in a gradient-based calibration technique

where λ is an arbitrary scalar. The reader is referred to [12, 25] for the derivation. Using the knowledge that $\boldsymbol{r}_1$ and $\boldsymbol{r}_2$ are orthonormal, we have

$$\boldsymbol{h}_1^T \mathbf{A}^{-T} \mathbf{A}^{-1} \boldsymbol{h}_2 = 0 \tag{4}$$

$$\boldsymbol{h}_1^T \mathbf{A}^{-T} \mathbf{A}^{-1} \boldsymbol{h}_1 = \boldsymbol{h}_2^T \mathbf{A}^{-T} \mathbf{A}^{-1} \boldsymbol{h}_2 \,. \tag{5}$$

These are the two basic constraints on the intrinsic parameters, given one homography. Because a homography has 8 degrees of freedom and there are 6 extrinsic parameters (3 for rotation and 3 for translation), we can only obtain 2 constraints on the intrinsic parameters.

If the camera is shown with the model plane at N general orientations, we have $3N$ constraints on five intrinsic parameters. With $N \geq 3$, a unique solution is available. There is a degenerate configuration in this technique when planes are parallel to each other. See [15] for a more detailed description.

A recommended calibration procedure is as follows:

1. Print a pattern and attach it to a planar surface.
2. Take a few images of the model plane under different orientations by moving either the plane or the camera.
3. Detect the feature points in the images.
4. Estimate the five intrinsic parameters and all the extrinsic parameters using the closed-form solution.
5. Estimate the coefficients of the radial distortion.
6. Refine all parameters, including lens distortion parameters, through maximum likelihood estimation.

Solving Camera Calibration with 1D Objects

A 1D object consisting of three or more points on a line. An example is shown in Fig. 4.

As discussed in [14, 26], calibration is impossible with a free moving 1D calibration object, no matter how many points on the object. It is, however, possible if the 1D object moves around a fixed point. In the sequel, let the fixed point be point $\boldsymbol{A}$, and $\boldsymbol{a}$ is the corresponding image point. We need three parameters, which are unknown, to specify the coordinates of $\boldsymbol{A}$ in the camera coordinate system, while image point $\boldsymbol{a}$ provides two scalar equations according to (Eq. 1).

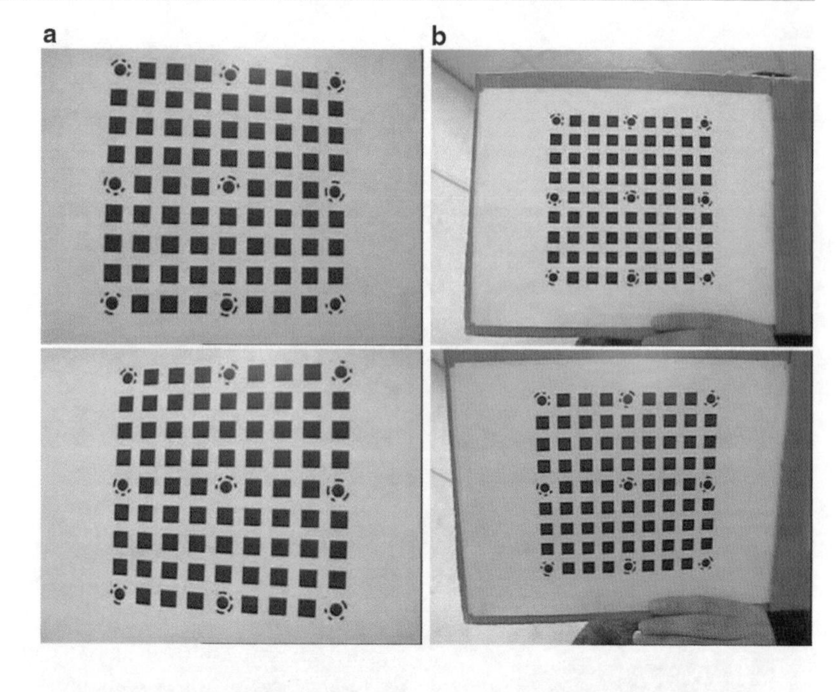

Geometric Calibration, Fig. 3 Two sets of images taken at different distances to the calibration pattern. Each set contains five images. On the *left*, three images from the set taken at a close distance are shown. On the *right*, three images from the set taken at a larger distance are shown

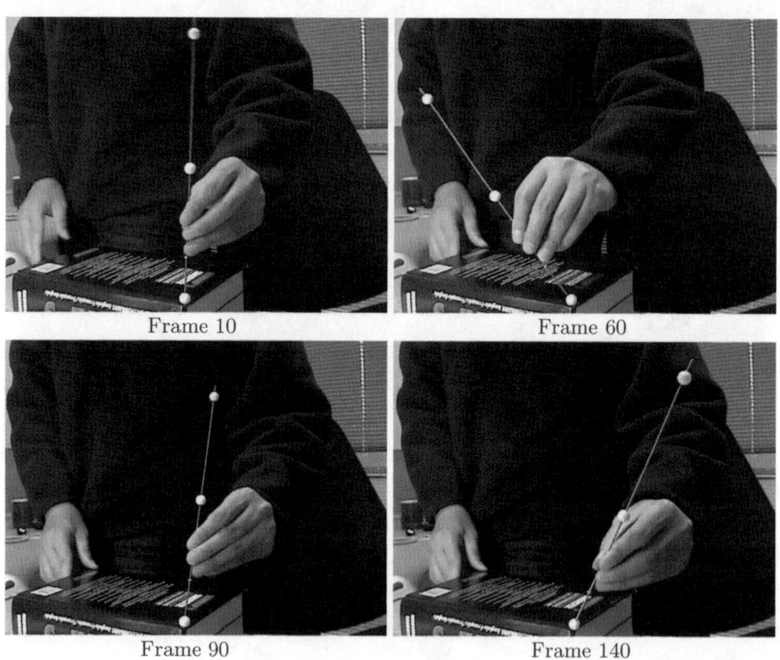

Geometric Calibration, Fig. 4 Sample images of a 1D object used for camera calibration

Two Points with Known Distance

They could be the endpoints of a stick, and we move the stick around the endpoint that is fixed. Let $\boldsymbol{B}$ be the free endpoint and $\boldsymbol{b}$, its corresponding image point. For each observation, we need two parameters to define the orientation of the line $\boldsymbol{AB}$ and therefore the position of $\boldsymbol{B}$ because the distance between $\boldsymbol{A}$ and $\boldsymbol{B}$ is known. Given N observations of the stick, we have five intrinsic parameters, three parameters for $\boldsymbol{A}$ and $2N$ parameters for the free endpoint positions to estimate, that is, the total number of unknowns is $8 + 2N$. However, each observation of $\boldsymbol{b}$ provides two equations, so together with $\boldsymbol{a}$ we only have in total $2+2N$ equations. Camera calibration is thus impossible.

Three Collinear Points with Known Distances

By adding an additional point, say C, the number of unknowns for the point positions still remains the same, that is, $8+2N$. For each observation, $\boldsymbol{b}$ provides two equations, but $\boldsymbol{c}$ only provides one additional equation because of the collinearity of $\boldsymbol{a}$, $\boldsymbol{b}$, and $\boldsymbol{c}$. Thus, the total number of equations is $2 + 3N$ for N observations. By counting the numbers, we see that if we have six or more observations, we should be able to solve camera calibration, and this is the calibration technique as developed in [14, 26].

Four or More Collinear Points with Known Distances

Again, the number of unknowns and the number of independent equations remain the same because of invariance of cross ratios. This said, the more collinear points we have, the more accurate camera calibration will be in practice because data redundancy can combat the noise in image data.

Self-calibration

Self-calibration is also called auto-calibration. Techniques in this category do not require any particular calibration object. They can be considered as 0D approach because only image point correspondences are required. Just by moving a camera in a static scene, the rigidity of the scene provides in general two constraints [9, 16, 17] on the cameras' internal parameters from one camera displacement by using image information alone. Absolute conic is an essential concept in understanding these constraints. Therefore, if images are taken by the same camera with fixed internal parameters, correspondences between three images are sufficient to recover both the internal and external parameters which allow us to reconstruct 3-D structure up to a similarity [17, 18]. Although no calibration objects are necessary, a large number of parameters need to be estimated, resulting in a much harder mathematical problem. The reader is referred to two books [27, 28] which provide an excellent recount of those techniques.

Discussions

Although many calibration techniques exist, no single calibration technique is the best for all. It really depends on the situation a user needs to deal with. Following are my few recommendations:

- Calibration with apparatus vs. self-calibration. Whenever possible, if we can pre-calibrate a camera, we should do it with a calibration apparatus. Self-calibration cannot usually achieve an accuracy comparable with that of pre-calibration because self-calibration needs to estimate a large number of parameters, resulting in a much harder mathematical problem. When pre-calibration is impossible (e.g., scene reconstruction from an old movie), self-calibration is the only choice.
- Partial vs. full self-calibration. Partial self-calibration refers to the case where only a subset of camera intrinsic parameters are to be calibrated. Along the same line as the previous recommendation, whenever possible, partial self-calibration is preferred because the number of parameters to be estimated is smaller. Take an example of 3D reconstruction with a camera with variable focal length. It is preferable to pre-calibrate the pixel aspect ratio and the pixel skewness.
- Calibration with 3D vs. 2D apparatus. Highest accuracy can usually be obtained by using a 3D apparatus, so it should be used when accuracy is indispensable and when it is affordable to make and use a 3D apparatus. From the feedback I received from computer vision researchers and practitioners around the world in the last couple of years, calibration with a 2D apparatus seems to be the best choice in most situations because of its ease of use and good accuracy.
- Calibration with 1D apparatus. This technique is relatively new, and it is hard for the moment to predict how popular it will be. It, however, should be useful especially for calibration of a camera network. To calibrate the relative geometry between multiple cameras as well as their intrinsic parameters, it is necessary for all involving cameras to simultaneously observe a number of points. It is hardly possible to achieve this with 3D or 2D calibration apparatus if one camera is mounted in the front of a room while another in the back. An exception is when those apparatus are made transparent; then the cost would be much higher. This is not a problem for 1D objects. We can, for example, use a string of balls hanging from the ceiling.

In this entry, we have only considered the linear projective projection. With real cameras, lens distortion sometimes has to be considered. The reader is referred to the entry ▸Calibration of Projective Cameras.

References

1. Brown DC (1971) Close-range camera calibration. Photogramm Eng 37(8):855–866
2. Faig W (1975) Calibration of close-range photogrammetry systems: mathematical formulation. Photogramm Eng Remote Sens 41(12):1479–1486
3. Gennery D (1979) Stereo-camera calibration. In: Proceedings of the 10th image understanding workshop, Los Angeles, pp 101–108
4. Ganapathy S (1984) Decomposition of transformation matrices for robot vision. Pattern Recognit Lett 2:401–412
5. Tsai RY (1987) A versatile camera calibration technique for high-accuracy 3D machine vision metrology using off-the-shelf tv cameras and lenses. IEEE J Robot Autom 3(4): 323–344
6. Faugeras O, Toscani G (1986) The calibration problem for stereo. In: Proceedings of the IEEE conference on computer vision and pattern recognition (CVPR), Miami Beach, FL. IEEE, Washington, DC, pp 15–20
7. Weng J, Cohen P, Herniou M (1992) Camera calibration with distortion models and accuracy evaluation. IEEE Trans Pattern Anal Mach Intell 14(10):965–980
8. Wei G, Ma S (1993) A complete two-plane camera calibration method and experimental comparisons. In: Proceedings of the fourth international conference on computer vision, Berlin, pp 439–446
9. Maybank SJ, Faugeras OD (1992) A theory of self-calibration of a moving camera. Int J Comput Vis 8(2): 123–152
10. Faugeras O, Luong T, Maybank S (1992) Camera self-calibration: theory and experiments. In: Sandini G (ed) Proceedings of the 2nd European conference on computer vision (ECCV). Lecture notes in computer science, vol 588, Santa Margherita Ligure, Italy. Springer, Berlin, pp 321–334
11. Faugeras O (1993) Three-dimensional computer vision: a geometric viewpoint. MIT, Cambridge
12. Zhang Z (2000) A flexible new technique for camera calibration. IEEE Trans Pattern Anal Mach Intell 22(11): 1330–1334
13. Sturm P, Maybank S (1999) On plane-based camera calibration: a general algorithm, singularities, applications. In: Proceedings of the IEEE conference on computer vision and pattern recognition (CVPR), Fort Collins, Colorado. IEEE Computer Society, Los Alamitos, pp 432–437
14. Zhang Z (2002) Camera calibration with one-dimensional objects. In: Proceedings of the European conference on computer vision (ECCV'02), vol IV, Copenhagen, Denmark, pp 161–174
15. Zhang Z (2004) Camera calibration with one-dimensional objects. IEEE Trans Pattern Anal Mach Intell 26(7): 892–899
16. Luong QT (1992) Matrice Fondamentale et Calibration Visuelle sur l'Environnement-Vers une plus grande autonomie des systèmes robotiques. Ph.D. thesis, Université de Paris-Sud, Centre d'Orsay
17. Luong QT, Faugeras O (1997) Self-calibration of a moving camera from point correspondences and fundamental matrices. Int J Comput Vis 22(3):261–289
18. Hartley RI (1994) An algorithm for self calibration from several views. In: Proceedings of the IEEE conference on computer vision and pattern recognition (CVPR), Seattle, WA. IEEE, Los Alamitos, pp 908–912
19. Caprile B, Torre V (1990) Using vanishing points for camera calibration. Int J Comput Vis 4(2):127–140
20. Liebowitz D, Zisserman A (1998) Metric rectification for perspective images of planes. In: Proceedings of the IEEE conference on computer vision and pattern recognition (CVPR), Santa Barbara, California. IEEE Computer Society, Los Alamitos, pp 482–488
21. Hartley R (1994) Self-calibration from multiple views with a rotating camera. In: Eklundh JO (ed) Proceedings of the 3rd European conference on computer vision. Lecture notes in computer science, Stockholm, Sweden, vol 800–801. Springer, New York, pp 471–478
22. Stein G (1995) Accurate internal camera calibration using rotation, with analysis of sources of error. In: Proceedings of the fifth international conference on computer vision, Cambridge, MA, pp 230–236
23. Zhang Z (2004) Camera calibration. In: Medioni G, Kang S (eds) Emerging topics in computer vision. Prentice Hall Professional Technical Reference, Upper Saddle River, pp 4–43
24. Robert L (1995) Camera calibration without feature extraction. Comput Vis Graph Image Process 63(2):314–325. Also INRIA technical report 2204
25. Zhang Z (1998) A flexible new technique for camera calibration. Technical Report MSR-TR-98-71, Microsoft Research available together with the software at http://research.microsoft.com/~zhang/Calib/
26. Zhang Z (2001) Camera calibration with one-dimensional objects. Technical report MSR-TR-2001-120, Microsoft Research
27. Hartley R, Zisserman A (2000) Multiple view geometry in computer vision. Cambridge University Press, Cambridge
28. Faugeras O, Luong QT (2001) In: Papadopoulo T (ed) The geometry of multiple images. MIT, Cambridge/London

Geometric Fusion

▸Three dimensional View Integration

Geons

Sven J. Dickinson[1] and Irving Biederman[1,2]
[1]Department of Computer Science, University of Toronto, Toronto, ON, Canada
[2]Departments of Psychology, Computer Science, and the Neuroscience Program, University of Southern California, Los Angeles, CA, USA

Synonyms

Recognition-by-components (RBC) theory

Related Concepts

▶Object Class Recognition (Categorization)

Definition

Geons are a set of less than 50 qualitative 2-D or 3-D part classes derived from permuting a set of four dichotomous and trichotomous properties of a generalized cylinder (GC). The values of these properties are nonaccidental in that they can be resolved from a general viewpoint, e.g., whether the axis of a cylinder is straight or curved. Geons were originally introduced by Biederman [9, 10] as the foundation for his recognition-by-components (RBC) theory for human shape perception, whereby object-centered models are represented as concatenations of geons, and object recognition from a 2-D image proceeds by matching recovered parts, typically segmented at regions of matched concavity, and their relations to object models.

Background

The concept of modeling an object as a composition of generalized cylinders dates back to Binford [18], who spawned a generation of object-recognition systems based on generalized cylinders, e.g., [20, 44, 45, 51, 54]. Generalized cylinders suffered from unbounded complexity, for arbitrarily complex functions could be used to define the axis, cross section, and sweep functions. As a result, it became popular to restrict the complexity of these functions, e.g., straight axis, constant or linear sweep, rotationally symmetric cross section, in order to facilitate their overconstrained recovery from sparse image data.

In the mid-1980s, two alternative restrictions on generalized cylinders emerged from the computer vision and human vision communities, respectively. Pentland [46] introduced the superquadric ellipsoid to the computer vision community – a 3-D, symmetry-based part representation that afforded a large degree of descriptive power with a small number of parameters. Around the same time, Biederman [9, 10] introduced geons to the human vision community as part of his recognition-by-components (RBC) theory. Like superquadric ellipsoids, geons exploited symmetry to reduce the complexity of a generalized cylinder. However, while the superquadric ellipsoid was a metric shape representation, the geon was a qualitative shape categorization. Thus, when a superquadric ellipsoid was recovered from an image, the recovered parameters defined a specific shape (a generative model), whereas when a geon was recovered from an image, it defined a symbolic part class (non-generative category) with only coarse (rather than exact) metric specification.

The appeal of the geon was twofold: (1) its properties were based on the sorts of judgments that humans are very good at, e.g., judging whether a line was straight or curved rather than estimating its exact curvature; and (2) each geon class afforded a high degree of within-class shape deformation, offering great potential for shape categorization and invariance over orientation in depth. Given extensive experiments with humans and primates that lent strong support of his RBC theory, the computer vision community quickly set out to develop computational models for geon recovery from 2-D images.

Theory

Geons define a partitioning of a subspace of the generalized cylinders. Like generalized cylinders, each geon is defined by its axis function, its cross-section function, and its sweep function. Biederman noted that humans are (1) much better at distinguishing between straight and curved lines than they are at estimating curvature; (2) much better at distinguishing parallelism from nonparallel symmetry than they are at estimating the angle between two causally related lines; and (3) good at distinguishing between various types of vertices produced by a cotermination of contours, such as a fork from an arrow from a L-junction. Drawing on these properties of the human visual system, Biederman mapped the spaces of the three generalized cylinder parameters to dichotomous and trichotomous values (Fig. 1):

- *Axis shape*: the axis takes on two possible values: straight or curved.
- *Cross-section shape*: the cross-section shape takes on two possible values: straight-edged or curved-edged.

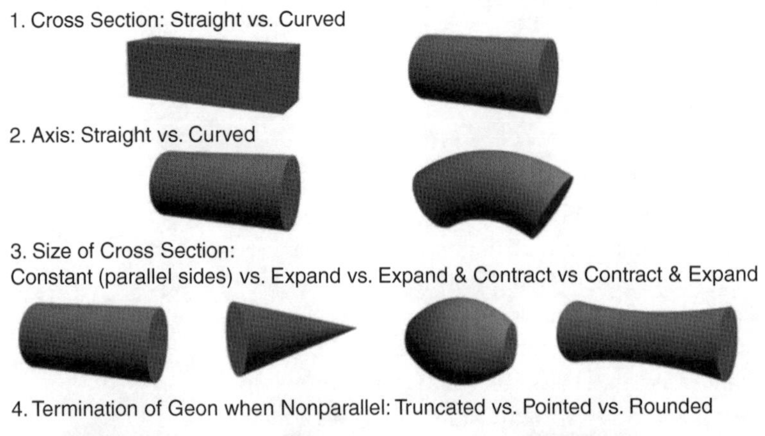

The set of geons is generated by variations in the production function for generalized cylinders that produce viewpoint-invariant (= nonaccidental) shape differences
1. Cross Section: Straight vs. Curved
2. Axis: Straight vs. Curved
3. Size of Cross Section:
Constant (parallel sides) vs. Expand vs. Expand & Contract vs Contract & Expand
4. Termination of Geon when Nonparallel: Truncated vs. Pointed vs. Rounded

© Irving Biederman

Geons, Fig. 1 The space of approximately 50 geons is defined by permuting the dichotomous and trichotomous properties of a restricted space of generalized cylinders

- *Sweep function*: the cross-section sweep function takes on four possible values: constant, monotonically increasing (or decreasing), monotonically increasing and then decreasing, or monotonically decreasing and then increasing.
- *Termination*: given a nonconstant sweep function, the termination of a geon could be truncated, end in a point (projects into an L-vertex), or end as a curved surface.

Originally, Biederman [10] posited cross-section symmetry as another attribute (with three possible values: rotationally symmetric, possessing an axis of reflective symmetry, or asymmetry) but that attribute was dropped as experiments showed that people assume symmetrical cross-sections, even when the cross-section is asymmetrical (as with an airplane wing).

Permuting the possible values of these four functions defines a space of $2 \times 2 \times 4 \times 3 = 48$ 3-D geons, as illustrated in Fig. 1. Adding 2D geons, e.g., circle, quadrilateral, and triangle, and subtracting the eight instances of constant sweep (2 axis shape $\times$ 2 cross-section shape $\times$ 2 point and curved terminations) when the sweep function is constant brings the total to about 50.

Related Work

Hummel et al. [37, 38] first proposed a connectionist model for recovering geons from line drawings that achieved invariance to viewpoint. In the computer vision community, Bergevin and Levine were the first to propose a computational model for geon recovery and geon-based recognition [4–8]. Dickinson et al. [27–29] introduced a hybrid object representation combining 3-D object-centered volumetric parts and 2-D viewer-centered aspects modeling the parts. While the framework was applicable to any vocabulary of volumetric parts, it was demonstrated on a qualitative shape vocabulary very similar to geons. Many geon-based frameworks followed, including probabilistic approaches [39], logic-based approaches [32], parametric geon recovery from range data [26, 48, 53], deformable contour-based approaches [47], deformable volume-based approaches [25], and active vision approaches [24, 31]. See [23] for a panel discussion on the strengths and weakness of geons and the challenges that lie ahead.

Open Problems

Geons have tremendous potential as a part representation in support of object categorization. They are qualitative and can support a high degree of within-class deformation, they (like generalized cylinders) map to the natural part structure of objects (when such elongated part structure exists), they are viewpoint-invariant 3-D parts that support object-centered 3-D models (which, in turn, better support scaling to large databases), and there is psychophysical support for them (the human is still, by far, the best example of an object categorization system). Despite these advantages, geons declined as a subject of study in the computer vision community in the late 1990s, in part due to the advent of appearance-based recognition and a general movement away from shape features.

The main reason for their decline was not necessarily a shortcoming of the representation, i.e., geons, but rather the community's inability to extract qualitative shape from real images of real objects. Except for those approaches operating on range images, the work reviewed above operated on either line drawings or uncluttered scenes containing simple, textureless objects. The key assumption made by these systems was that a salient contour in the image maps one-to-one to a salient surface discontinuity (or occluding contour) on a geon. Unfortunately, in a real scene, objects contain texture, shadows, reflectance contours, and structural "noise" (surface discontinuities that are not salient with respect to the geon class), all of which introduce unwanted contours. Moreover, images of contours (both good and bad) may be broken or noisy, requiring complex perceptual grouping and multiscale analysis to restore and capture the salient shape of the contours. Yet despite these conditions, humans and primates have absolutely no trouble distinguishing (or abstracting) those contours that mark orientation and depth discontinuities – the critical contours for geon extraction – from contours reflecting variations in surface texture, color, lighting, shadows, etc.

As discussed in Dickinson [22], the recognition community's gradual movement from shape toward appearance, coupled with the community's interest in engineering practical systems, drew attention away from basic research on shape modeling in support of object categorization. However, the community is once again realizing that over the set of exemplars belonging to an object category, shape is far more invariant than appearance. As a result, shape-based object categorization systems (mainly using contours) are beginning to reemerge, e.g., [33]. But a return to local contour-based features is not sufficient, as local shape features are still too exemplar-specific. Rather, such features must be perceptually grouped and abstracted to form more generic shape structures that offer the within-class deformation invariance required for effective categorization. Geons offer a powerful shape abstraction with great categorization potential, but only when more progress has been made on the mid-level challenges of perceptual grouping and intermediate-level shape abstraction. Some early work along these lines has started to appear [50].

Experimental Results: Computer Vision

Figure 2 illustrates three examples of geon recovery systems in the computer vision community. In Fig. 2a, the system of Bergevin and Levine [7] recovers geons from line drawings. In Fig. 2b, the system of Dickinson et al. [24] recovers geon-like volumetric parts from real images of simple objects, as does the system of Pilu and Fisher [47], as shown in Fig. 2c.

Experimental Results: Human Vision

There is now substantial neural and behavioral evidence for the representation of objects as an arrangement of geons, as specified by the recognition-by-components theory. This evidence can be summarized in terms of six independent assumptions. Any one (or several) of these assumptions can be made independent of RBC but, to date, RBC is the only theory from which all six derive.

> The representation of an object is largely edge-based – specifically, those edges specifying orientation and depth discontinuities – rather than surface-based (i.e., color, texture).

Reaction times (RTs) and error rates for naming briefly presented images of objects are as fast for line drawings as they are for full, color photography [16]. This is also true of verification in which the observer verifies whether a name ("chair"), provided prior to an image of an object, matches the object. The equivalence in performance for identifying line drawings and photography is evident even when the objects have a diagnostic color/texture, such as a fish, fork, or banana, as opposed to objects with nondiagnostic surface properties, e.g., a chair or a lamp, which can be any color or texture.

The equivalence of photography and line drawings is also witnessed in fMRI activity where the adaptation (i.e., the reduction) of the BOLD signal that is evident with a repetition of a stimulus, fMRI-a, is the same when the images have the same format, i.e., identical photographs or line drawings, as when they have different formats, one a photograph and the other a line drawing [34]. This invariance to surface properties is

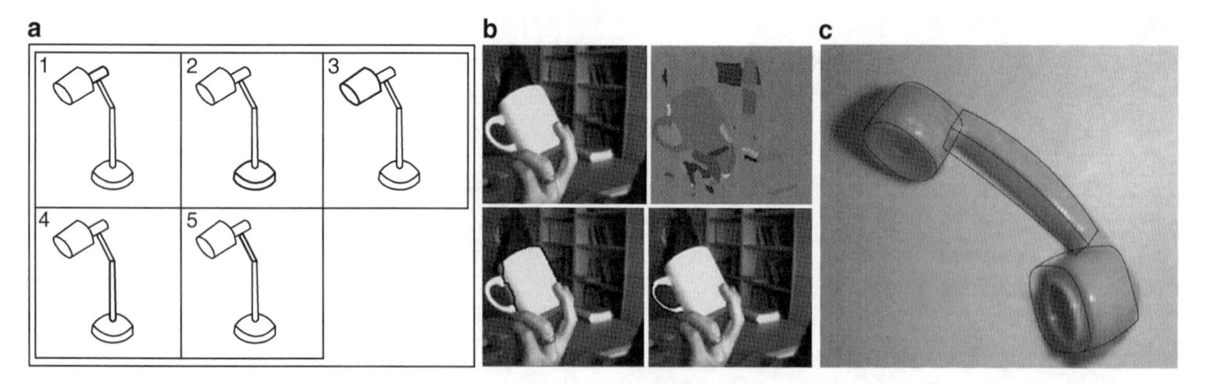

Geons, Fig. 2 Three examples of geon recovery in the computer vision community: (**a**) decomposing a line drawing of a lamp into its constituent geon parts, with bold contours indicating parts: 1 - rear shade, 2 - base, 3 - front shade, 4 - lower neck, and 5 - upper neck (Bergevin and Levine [7]); (**b**) from a region segmentation (*upper right*) of the image of an occluded cup (*upper left*), the two recovered constituent qualitative volumetric parts (with matched contours highlighted in black) are shown in *lower left* (body cylinder) and *lower right* (handle bent cylinder) (Dickinson et al. [24]); and (**c**) decomposing a phone into its constituent geons parts (Pilu and Fisher [47])

also seen in the response of many single neurons in object-sensitive areas in the macaque [41]. In fMRI, the processing of surface properties, color and texture, activates different cortical areas than those activated when processing shape [21].

There are few transformations to appearance as dramatic as rendering a line drawing from a photograph yet the readily achieved invariance to this transformation poses a major challenge to appearance-based theories of object recognition.

Objects are represented by parts rather than local features, templates, or concepts.

Object priming is the facilitation that ensues as a consequence of a prior perception of an object. It can be readily evidenced by a reduction in RTs and error rates in the naming of brief, masked presentations of objects and has been documented over a 14-month period from the first to second presentation of the images. (The reduction in the magnitude of the BOLD response to a repeated stimulus, termed fMRI adaptation, is generally attributed to more efficient coding and is interpreted as a neural correlate of priming.) Almost all of this priming is visual (i.e., perceptual) rather than lexical (easier access to the name itself) in that an object with the same name but a substantially different shape, e.g., a grand piano followed by an upright piano, evidences almost no facilitation.

Studies with complementary, contour-deleted line drawings document that all the priming can be attributed to the repetition of the parts (in their appropriate relations) as opposed to local features, i.e., the specific lines and vertices in the image [14]. Thus, if every other vertex and line from each geon is deleted from one image of an object and the deleted contour composes the other member of a complementary pair, as in the two images of a flashlight on the left side of Fig. 3a (so if the two were superimposed they would comprise an intact image with no overlap of contour), the degree of priming between members of a complementary pair – which depict the same parts though with different local contours – is equal to the priming between identical images. This implies that none of the priming can be attributable to the local contours (i.e., the local lines and vertices). Presumably, the local contours are required to activate a representation of the part, but once that part (in its appropriate relations) is activated there is no contribution of the initial local image features.

Instead of deletion of local features, if the deletion is of half the parts of a complex object, as shown in Fig. 3b, then there is no visual priming between members of a complementary pair. Thus the priming is completely dependent on the overlap in the parts in the two images. These effects on behavioral priming have their exact counterpart in fMRI-a. Here, local feature complements show the same reduction in the BOLD response as when the identical images are repeated, suggesting equivalent representations, but repetition of part complements show a complete loss of adaptation

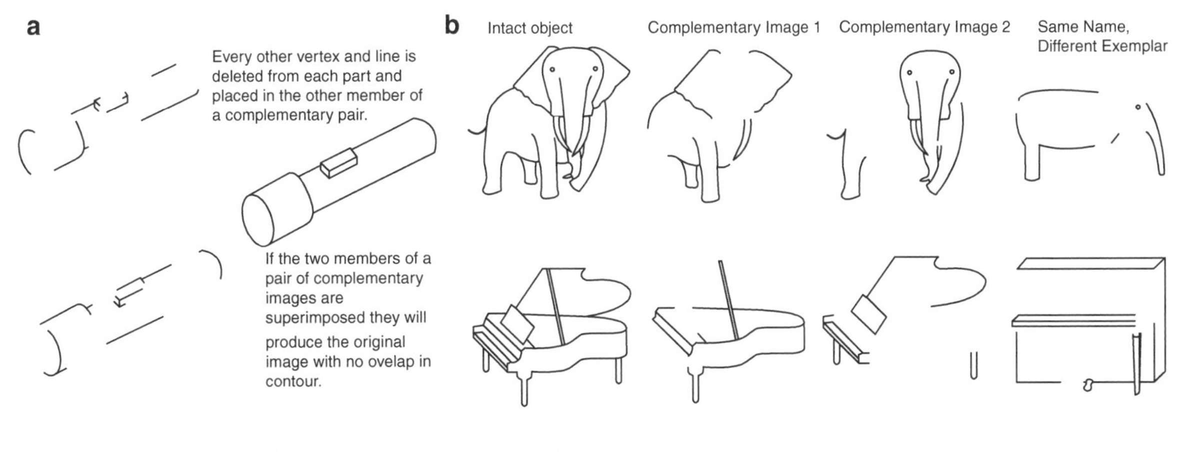

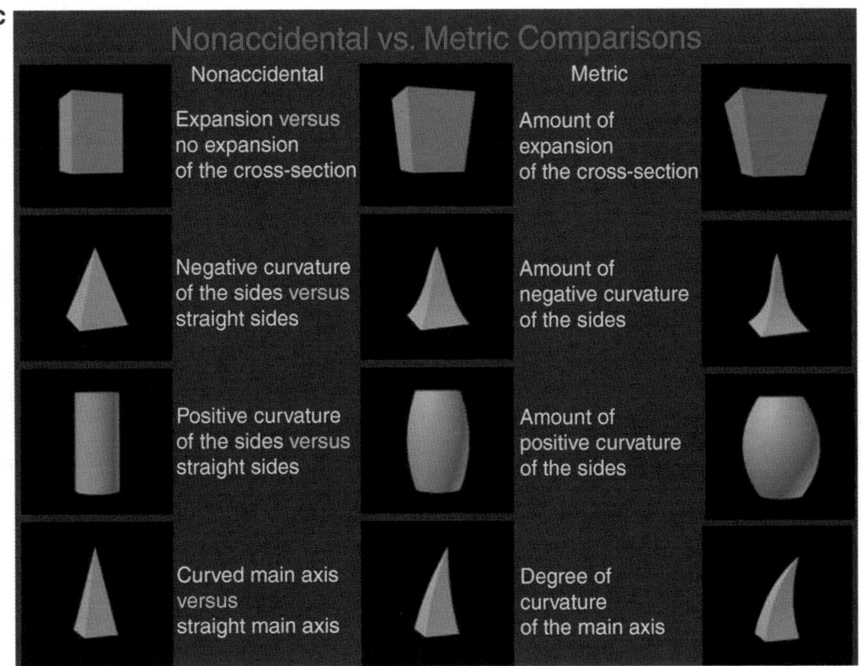

Geons, Fig. 3 Psychophysical evidence in support of Geons: (**a**) members of a local contour-deleted complementary pair, which have the same parts but different local features, prime each other as much as they do themselves; priming is not attributable to local contours; (**b**) there is no visual priming between members of a complementary pair when they have no parts in common, as between the images of the second and third columns [14]; and (**c**) equal image differences between nonaccidental (between center and left columns) and metric properties (between center and right columns). Geons are distinguished by nonaccidental properties. Discrimination is much faster and more accurate for differences in nonaccidental than metric properties

thus indicating that there is no overlap in visual representations when the images are composed of different parts, even though they are of the same subordinate concept, e.g., both grand pianos [35].

Evidence against a template representation derives from studies of the priming of depth-rotated stimuli. As long as the same parts can be readily extracted in two different images of the same object, recognition or matching of a rotated object will be achieved with virtually no cost. However, if because of self-occlusion some parts disappear and other parts emerge, then priming is reduced or object matching is impaired [15].

Single cell recordings in the inferior temporal lobe (IT) of the macaque, the area generally accepted to mediate object recognition, generally fire as strongly to one or two of the parts of an object as they do to the complete object [40].

Parts are distinguished by nonaccidental properties (NAPs) and only coarsely by metric properties (MPs).

Values of various dimensions of geons can be regarded as singular or nonsingular. A singular value, such as 0 curvature (i.e., a straight contour), retains that value as the object is rotated in depth. A nonsingular value, such as a nonzero value of curvature (i.e., a curved contour), can vary with the orientation in depth of that contour. In addition to curvature, parallelism of two contours can have a singular value of zero convergence (or divergence) or a nonzero value. Two or three contours that coterminate can be regarded as a singular value of zero separation between their terminations, forming vertices, such as Ls, arrows, or forks. This framework can define NAP differences as the difference between singular and nonsingular values as, e.g., a difference between a curved and a straight contour produced by the parallel sides of the cylinder on the left in the third row of Fig. 3c and the middle barrel. Metric differences are differences in non-singular values, such as two contours with unequal nonzero curvatures, as with the slightly curved and more curved barrels in the third row.

The aforementioned invariance to rotation in depth holds only if the objects that are to be discriminated differ in NAPs [13, 15]. Objects differing only in metric properties incur high costs when they are encountered at a different orientation in depth. At equal orientations, the discrimination of two shapes as being same or different is markedly easier if the shapes differ in NAPs than MPs [17]. Cells in the IT region of the macaque modulate (i.e., vary their firing rate) much more to a change in a NAP compared to an MP [41, 52]. Even pigeons show greater sensitivity to differences in NAPs than MPs [1]. In these comparisons of the sensitivity of NAPs and MPs, the physical differences are equated according to a model of V1 [42], the first stage of cortical shape coding.

Dimensions of generalized cylinders (GCs) are independently coded and have psychophysical and neural reality.

The set of geons is generated by combinations of the values of the independent dimensions shown in Fig. 1. (In addition, as noted previously, there can be coarse variation in the metric of these geons, such as their aspect ratio or degree of axis curvature.) Are simple object parts actually coded by independent combinations of these dimensions (vs. just being nondimensionalized variations in shape templates)? One measure of independent coding of perceptual dimensions is whether human observers can selectively attend to one dimension without any cost from variations in another, to-be-ignored, dimension. For example, the speed and accuracy of discriminating different shapes is unaffected by whether the colors of those shapes are held constant or varied. It might seem plausible that shape could be attended while ignoring a surface feature such as color. Would efficient selective attention be manifested when observers are attending to one shape dimension, say axis curvature, while ignoring variations in another shape dimension, say aspect ratio. The answer is clearly yes [43]. Moreover, a multidimensional analysis of the firing of a population of IT cells to a set of stimuli similar to that depicted in Fig. 3c shows that 95% of the variance of the spike rates can be modeled in terms of independent coding of the GC dimensions [40].

Low sensitivity for discriminating complex, irregular shapes (= texture?) compared to simple shapes but high sensitivity for distinguishing regular from irregular.

Geons are simple and regular. What about complex, highly irregular objects, such as a bush or a crumpled sweater? It would be highly unlikely that people are employing geons for the precise representation of such objects. Interestingly, the evidence is that people do not represent such variation in any detail beyond the fact that the shapes are irregular and some simple nonaccidental characterizations, e.g., whether the surfaces are round or pointed. This is also true of IT cells [2]. Essentially, objects with irregular parts are treated as texture, rather than shape.

There is a more general point to be made here. GCs (and geons) were criticized for their unwieldiness for modeling objects such as bushes. But this is confusing a graphics system, in which the goal is to achieve an exact replica of the image, with a biological recognition system designed to do basic- and subordinate-level classification in which irrelevant variation is best ignored.

Objects are represented by a structural description that specifies simple parts and relations.

Geons are the representation of the parts of an object, but objects are typically composed of more than one part. In the same manner that people are sensitive to the order of phonemes, so "rough" and "fur" have the same phonemes but in different order, people are sensitive to the arrangement of parts of an object, so they can say, e.g., that a vertical cylinder is attached end-to-middle and perpendicular to the top of a larger horizontal brick. That geons and their relations may be coded independently is documented by a remarkable patient with a left inferior temporal lesion who had no problem distinguishing objects differing in their geons but could not distinguish objects that differed in the relations among the same geons [3]. Recent neuroimaging studies show that such relations are specified explicitly at the same cortical locus, the lateral occipital complex, that object shape is specified [36].

References

1. Lazareva OF, abd Wasserman EA, Biederman I (2008) Pigeons and humans are more sensitive to nonaccidental than to metric changes in visual objects. Behav Processes 77:199–209
2. Kayaert G, Biederman I, Vogels R (2005) Representation of regular and irregular shapes in macaque inferotemporal cortex. Cereb Cortex 15:1308–1321
3. Behrmann M, Peterson MA, Moscovitch M, Suzuki S (2006) Integrative agnosia: deficit in encoding relations between parts. J Exp Psychol Human Percept Perf 32(5):1169–1184
4. Bergevin R, Levine MD (1988) Recognition of 3-D objects in 2-D line drawings: an approach based on geons. Technical report TR-CIM-88, Center for intelligent machines, McGill University, Nov 1988
5. Bergevin R, Levine MD (1989) Generic object recognition: building coarse 3D descriptions from line drawings. In: Proceedings, IEEE workshop on interpretation of 3D scenes, Austin, pp 68–74
6. Bergevin R, Levine MD (1992) Extraction of line drawing features for object recognition. Pattern Recognit 25(3): 319–334
7. Bergevin R, Levine MD (1992) Part decomposition of objects from single view line drawings. CVGIP: Image Underst 55(1):73–83
8. Bergevin R, Levine MD (1993) Generic object recognition: building and matching coarse 3D descriptions from line drawings. IEEE Trans Pattern Anal Mach Intell 15:19–36
9. Biederman I (1985) Human image understanding: recent research and a theory. Comput Vis Graph Image Process 32:29–73
10. Biederman I (1987) Recognition-by-components: a theory of human image understanding. Psychol Rev 94:115–147
11. Biederman I (1990) Higher level vision. In: An invitation to cognitive science: visual cognition and action, vol 2. MIT, Cambridge, MA, pp 41–72
12. Biederman I (2000) Recognizing depth-rotated objects: a review of recent research and theory. Spat Vis 13:241–253
13. Biederman I, Bar M (1999) One-shot viewpoint invariance in matching novel objects. Vis Res 39:2885–2899
14. Biederman I, Cooper EE (1991) Priming contour-deleted images: evidence for intermediate representations in visual object recognition. Cognit Psychol 23:393–419
15. Biederman I, Gerhardstein PC (1993) Recognizing depth-rotated objects: evidence and conditions for 3D viewpoint invariance. J Exp Psychol Hum Percept Perform 19: 1162–1182
16. Biederman I, Ju G (1988) Surface vs. edge-based determinants of visual recognition. Cognit Psychol 20:38–64
17. Biederman, I, Yue, X, Davidoff J (2009) Representation of shape in individuals from a culture with minimal exposure to regular simple artifacts: sensitivity to nonaccidental vs. metric properties. Psychol Sci 20:1437–1442
18. Binford TO (1971) Visual perception by computer. In: Proceedings, IEEE conference on systems and control, Miami
19. Borges D, Fisher R (1997) Class-based recognition of 3d objects represented by volumetric primitives. Image Vis Comput 15(8):655–664
20. Brooks R (1983) Model-based 3-D interpretations of 2-D images. IEEE Trans Pattern Anal Mach Intell 5(2): 140–150
21. Cant JS, Goodale MA (2009) Asymmetric interference between the perception of shape and the perception of surface properties. J Vis 9(13):11–20
22. Dickinson S (2009) The evolution of object categorization and the challenge of image abstraction. In: Dickinson S, Leonardis A, Schiele B, Tarr M (eds) Object categorization: computer and human vision perspectives. Cambridge University Press, New York, pp 1–37
23. Dickinson S, Bergevin R, Biederman I, Eklundh J-O, Jain A, Munck-Fairwood R, Pentland A (1997) Panel report: the potential of geons for generic 3-D object recognition. Image Vis Comput 15(4):277–292
24. Dickinson S, Christensen H, Tsotsos J, Olofsson G (1997) Active object recognition integrating attention and viewpoint control. Comput Vis Image Underst 67(3): 239–260
25. Dickinson S, Metaxas D (1994) Integrating qualitative and quantitative shape recovery. Int J Comput Vis 13(3): 1–20
26. Dickinson S, Metaxas D, Pentland A (1997) The role of model-based segmentation in the recovery of volumetric parts from range data. IEEE Trans Pattern Anal Mach Intell 19(3):259–267
27. Dickinson S, Pentland A, Rosenfeld A (1990) A representation for qualitative 3-D object recognition integrating object-centered and viewer-centered models. In: Leibovic K (ed) Vision: a convergence of disciplines. Springer, New York, pp 398–421
28. Dickinson S, Pentland A, Rosenfeld A (1992) From volumes to views: an approach to 3-D object recognition. CVGIP: Image Underst 55(2):130–154

29. Dickinson S, Pentland A, Rosenfeld A (1992) 3-D shape recovery using distributed aspect matching. IEEE Trans Pattern Anal Mach Intell 14(2):174–198
30. Du L, Munck-Fairwood R (1993) A formal definition and framework for generic object recognition. In: Proceedings, 8th Scandinavian conference on image analysis, University of Tromsø, Norway
31. Eklundh J-O, Olofsson G (1992) Geon-based recognition in an active vision system. In: ESPRIT-BRA 3038, vision as process. Springer-Verlag ESPRIT Series
32. Fairwood R (1991) Recognition of generic components using logic-program relations of image contours. Image Vis Comput 9(2):113–122
33. Ferrari V, Jurie F, Schmid C (2010) From images to shape models for object detection. Int J Comput Vis 87(3): 284–303
34. Grill-Spector K, Kourtzi Z, Kanwisher N (2001) The lateral occipital complex and its role in object recognition. Vis Res 41(10–11):1409–1422
35. Hayworth KJ, Biederman I (2006) Neural evidence for intermediate representations in object recognition. Vis Res 46:4024–4031
36. Hayworth KJ, Lescroart MD, Biederman I (2011) Visual relation encoding in anterior LOC. J Exp Psychol Human Percept Perf 37(4):1032–1050
37. Hummel JE, Biederman I (1992) Dynamic binding in a neural network for shape recognition. Psychol Rev 99: 480–517
38. Hummel JE, Biederman I, Gerhardstein P, Hilton H (1988) From edges to geons: a connectionist approach. In: Proceedings, connectionist summer school, Carnegie Mellon University, pp 462–471
39. Jacot-Descombes A, Pun T (1992) A probabilistic approach to 3-D inference of geons from a 2-D view. In: Proceedings, SPIE applications of artificial intelligence X: machine vision and robotics, Orlando pp 579–588
40. Kayaert G, Biederman I, Op de Beeck H, Vogels R (2005) Tuning for shape dimensions in macaque inferior temporal cortex. Eur J Neurosci 22:212–224
41. Kayaert G, Biederman I, Vogels R (2003) Shape tuning in macaque inferior temporal cortex. J Neurosci 23: 3016–3027
42. Lades M, Vorbruggen JC, Buhmann J, Lange J, von der Malsburg C, Wurtz RP, Konen W (1993) Distortion invariant object recognition in the dynamic link architecture. IEEE Trans Comput 42:300–311
43. Lescroart MD, Biederman I, Yue X, Davidoff J (2010) A cross-cultural study of the representation of shape: sensitivity to underlying generalized-cone dimensions. Vis Cogn 18(1):50–66
44. Marr D, Nishihara H (1978) Representation and recognition of the spatial organization of three-dimensional shapes. R Soc Lond B 200:269–294
45. Nevatia R, Binford TO (1977) Description and recognition of curved objects. Artif Intell 8:77–98
46. Pentland A (1986) Perceptual organization and the representation of natural form. Artif Intell 28:293–331
47. Pilu M, Fisher RB (1996) Recognition of geons by parametric deformable contour models. In: Proceedings, European conference on computer vision (ECCV), LNCS, Springer, Cambridge, UK, April 1996, pp 71–82
48. Raja N, Jain A (1992) Recognizing geons from superquadrics fitted to range data. Image Vis Comput 10(3):179–190
49. Raja N, Jain A (1994) Obtaining generic parts from range images using a multi-view representation. CVGIP: Image Underst 60(1):44–64
50. Sala P, Dickinson S (2010) Contour grouping and abstraction using simple part models. In: Proceedings, European conference on computer vision (ECCV) Crete, Greece, Sept 2010
51. Ulupinar F, Nevatia R (1993) Perception of 3-D surfaces from 2-D contours. IEEE Trans Pattern Anal Mach Intell 15:3–18
52. Vogels R, Biederman I, Bar M, Lorincz A (2001) Inferior temporal neurons show greater sensitivity to nonaccidental than metric differences. J Cogn Neurosci 13:444–453
53. Wu K, Levine MD (1997) 3-D shape approximation using parametric geons. Image Vis Comput 15(2):143–158
54. Zerroug M, Nevatia R (1996) Volumetric descriptions from a single intensity image. Int J Comput Vis 20(1/2):11–42

Gesture Recognition

Matthew Turk
Computer Science Department and Media Arts and Technology Graduate Program, University of California, Santa Barbara, CA, USA

Synonyms

Human motion classification

Definition

Vision-based gesture recognition is the process of recognizing meaningful human movements from image sequences that contain information useful in human-human interaction or human-computer interaction. This is distinguished from other forms of gesture recognition based on input from a computer mouse, pen or stylus, sensor-based gloves, touch screens, etc.

Background

Automatic image-based gesture recognition is an area of computer vision motivated by a range of application areas, including the analysis of human-human communication, sign language interpretation, human-robot interaction, multimodal human-computer interaction, and gaming. Human gesture has a long history of interdisciplinary study by psychologists, linguists, anthropologists, and others in the context of human communication [9], exploring the role of gesture in face-to-face conversation, universal and cultural aspects of gesture, the influence of gesture in human evolution and in child development, and other topics, going back at least to the work of Charles Darwin with *The Expression of the Emotions in Man and Animals* (1872). Research in computer vision-based gesture recognition began primarily in the 1990s as computers began to be capable of supporting real-time (or *interactive time*, fast enough to support human interaction) processing and recognition of video streams.

Several gesture taxonomies or categorizations have been developed by different researchers that underscore the breadth of the problem in general. Cadoz [4] described three functional roles of human gesture: semiotic (gestures to communicate meaningful information), ergodic (gestures to manipulate the environment), and epistemic (gestures to discover the environment through tactile experience). Most work in automated gesture recognition is concerned with the first role (semiotic gestures), whereas the area of human activity analysis tends to focus on the latter two. Kendon [11] described a gesture continuum, defining five types of gestures: gesticulation, language-like gestures, pantomimes, emblems, and sign languages. Each of these has a varying association with verbal speech, language properties, spontaneity, and social regulation, indicating that human gesture is indeed a complex phenomenon. Gesticulation, defined as spontaneous, speech-associated gesture, makes up a large portion of human gesture and is further characterized by McNeill [14] into four types:

- Iconic – representational gestures depicting some feature of the object, action, or event being described
- Metaphoric – gestures that represent a common metaphor, rather than the object or event directly
- Beat – small, formless gestures, often associated with word emphasis
- Deictic – pointing gestures that refer to people, objects, or events in space or time

These gesture types modify the content of accompanying speech and often help to disambiguate speech, similar to the role of spoken intonation. Cassell et al. [5] described early research in conversational agents that models the relationship between speech and gesture and generates interactive dialogs between three-dimensional animated characters that gesture as they speak.

Vision-Based Gesture Recognition

Vision-based gesture recognition must detect human movements from image sequences, ideally in real time and independent of the specific user, the imaging condition, the camera viewpoint, clothing and other confusing factors, and the significant variation in how people gesture. Aspects of a gesture that may be critical to its interpretation include spatial information (where the gesture occurs and/or refers to), pathic information (the path a gesture takes), symbolic information (sign(s) made during a gesture), and affective information (the emotional quality of a gesture, which may be related to the speed and magnitude of a gestural act, as well as to facial expression).

A gesture may be considered as a continuous set of movements or as a sequence of discrete poses or postures. Gestures are inherently dynamic and time varying, while postures are specific – and static – configurations; recognizing specific configurations (such as making a "victory sign") should properly be referred to as *posture recognition*. Analyzing movement (such as dance or general behaviors in a social situation) is generally referred to as *activity analysis* or *activity recognition* [1].

Unless the gestures are constrained to a particular point in time (e.g., with a "push to gesture" functionality), it is necessary to determine when a dynamic gesture begins and ends. This temporal segmentation/detection of gesture is a challenging problem, particularly in less constrained environments where several kinds of spontaneous gestures are possible amidst other non-gestural movement. While temporal detection and segmentation of gestures may be attempted as a first step, other approaches combine spatial (or spatiotemporal) segmentation with recognition [2, 12].

A typical approach to human gesture recognition involves detecting and tracking component body parts, such as hands, arms, head, torso, legs, and feet, based on an articulated body model, and subsequently classifying the movement into one of a set of known gestures (e.g., [19]). The output of the tracking stage is a time-varying sequence of parameters describing (2D or 3D) positions, velocities, and angles of the relevant body parts and features, possibly including a representation of uncertainty that indicates limitations of the sensor and the algorithms. An alternative is to take a view-based approach, which computes parameters directly from image motion, generally bypassing human body modeling (e.g., [6, 7]).

Hand gestures have received particular attention in gesture recognition, as hands provide the opportunity for a wide range of meaningful gestures, as evidenced by the rich history of human sign languages such as American Sign Language (ASL) (e.g., [18, 20]), and may be convenient for quickly and naturally conveying information in vision-based interfaces (e.g., [3, 8]). Video-only approaches have had limited success, however, due to the complexities of highly articulated hands and skin-on-skin occlusions.

Recently, there has been a significant amount of work in gesture recognition from depth imagery or combinations of video (RGB) and depth data, largely driven by the availability of the Microsoft Kinect sensor (and SDK/toolkit) and the use of body modeling, tracking, and gesture recognition in consumer applications using the Kinect [16]. Other companies are developing new technologies for gesture recognition (e.g., [17] and [13]), as well as for spatial operating environments that leverage tracking and gesture technologies (e.g., [15]).

Open Problems

Gesture recognition is a broadly defined set of problems and challenges, for which there are some domain-specific solutions that are adequate for commercial use; however, the general problems are largely unsolved. At the low level, there are limitations to any choice of sensor type, and work remains to be done on integrating data from multiple sensors. There is no agreement on how to best represent the sensed spatial and temporal information and its relationship to human movement. Temporal segmentation of natural dynamic gestures is unlikely to be solved without a deep understanding of the gesture semantics – i.e., the high-level context in which the gestures take place. Despite the recent impact of depth sensors on this area, the field is still wide open for solutions that can provide precise and robust gesture recognition in a wide range of environments.

Research in vision-based gesture recognition can be stimulated by the creation and sharing of thorough, annotated data sets that capture a wide range of spontaneous gestures and imaging conditions and by apples-to-apples comparisons such as the recent ChaLearn Gesture Challenges [10].

References

1. Aggarwal JK, Ryoo MS (2011) Human activity analysis: a review. ACM Comput Surv 43(3):1–43. ACM
2. Alon J, Athitsos V, Yuan Q, Sclaroff S (2009) A unified framework for gesture recognition and spatiotemporal gesture segmentation. IEEE Trans Pattern Anal Mach Intell 31:1685–1699
3. Bretzner L, Laptev I, Lindeberg T (2002) Hand gesture recognition using multi-scale colour features, hierarchical models and particle filtering. In: IEEE conference on automatic face and gesture recognition. IEEE Computer Society, Los Alamitos
4. Cadoz C (1994) Les réalités virtuelles. Dominos/Flammarion, Paris
5. Cassell J, Steedman M, Badler N, Pelachaud C, Stone M, Douville B, Prevost S, Achorn B (1994) Modeling the interaction between speech and gesture. In: Proceedings of the sixteenth conference of the cognitive science Society. Lawrence Erlbaum Associates, Hillsdale
6. Cutler R, Turk M (1998) View-based interpretation of real-time optical flow for gesture recognition. In: Proceedings of the 1998 IEEE conference on automatic face and gesture recognition, Nara, Japan, 14–16 Apr
7. Darrell TJ, Penland AP (1993) Space-time gestures. In: IEEE conference on vision and pattern recognition (CVPR), New York, NY
8. Freeman WT, Tanaka K, Ohta J, Kyuma K (1996) Computer vision for computer games. In: 2nd international conference on automatic face and gesture recognition, Killington, VT, USA, pp 100–105
9. http://www.gesturestudies.com/
10. http://gesture.chalearn.org/
11. Kendon A (1972) Some relationships between body motion and speech. In: Siegman AW, Pope B (eds) Studies in dyadic communication. Pergamon Press, New York
12. Kim D, Song J, Kim D (2007) Simultaneous gesture segmentation and recognition based on forward spotting accumulative HMMs. Pattern Recognit 40:2012–2026
13. Leap Motion. http://www.leapmotion.com/
14. McNeill D (1992) Hand and mind: what gestures reveal about thought. University of Chicago Press, Chicago
15. Oblong Industries. http://www.oblong.com/
16. Shotton J, Fitzgibbon A, Cook M, Sharp T, Finocchio M, Moore R, Kipman A, Blake A (2011) Real-time human pose

recognition in parts from a single depth image. In: IEEE conference on computer vision pattern recognition (CVPR). IEEE, Piscataway
17. SoftKinetic. http://www.softkinetic.com/SoftKinetic.aspx
18. Starner T, Weaver J, Pentland A (1998) Real-time American sign language recognition using desk and wearable computer-based video. IEEE Trans Pattern Anal Mach Intell 20(12):1371–1375
19. Turk M (2001) In: Stanney K (ed) Handbook of virtual environment technology. Lawrence Erlbaum Associates, Inc.
20. Vogler C, Goldenstein S (2008) Toward computational understanding of sign language. Technol Disabil 20(2): 109–119

Gradient Vector Flow

Chenyang Xu[1] and Jerry L. Prince[2]
[1]Siemens Technology-To-Business Center, Berkeley, CA, USA
[2]Electrical and Computer Engineering, Johns Hopkins University, Baltimore, MD, USA

Synonyms

GVF

Related Concepts

▶Edge Detection

Definition

Gradient vector flow is the vector field that is produced by a process that smooths and diffuses an input vector field and is usually used to create a vector field that points to object edges from a distance.

Background

Finding objects or homogeneous regions in images is a process known as image segmentation. In many applications, the locations of object edges can be estimated using local operators that yield a new image called an edge map. The edge map can then be used to guide a deformable model, sometimes called an active contour or a snake, so that it passes through the edge map in a smooth way, therefore defining the object itself.

A common way to encourage a deformable model to move toward the edge map is to take the spatial gradient of the edge map, yielding a vector field. Since the edge map has its highest intensities directly on the edge and drops to zero away from the edge, these gradient vectors provide directions for the active contour to move. When the gradient vectors are zero, the active contour will not move, and this is the correct behavior when the contour rests on the peak of the edge map itself. However, because the edge itself is defined by local operators, these gradient vectors will also be zero far away from the edge, and therefore the active contour will not move toward the edge when initialized far away from the edge.

Gradient vector flow (GVF) is the process that spatially extends the edge map gradient vectors, yielding a new vector field that contains information about the location of object edges throughout the entire image domain. GVF is defined as a diffusion process operating on the components of the input vector field. It is designed to balance the fidelity of the original vector field, so it is not changed too much, with a regularization that is intended to produce a smooth field on its output.

Although GVF was designed originally for the purpose of segmenting objects using active contours attracted to edges, it has been since adapted and used for many alternative purposes. Some newer purposes including defining continuous medial axis representation [1], extracting scale-invariant image features [2], regularizing image anisotropic diffusion algorithms [3], finding the centers of ribbon-like objects [4], and much more.

Theory

The theory of GVF was originally described in [5]. Let $f(x, y)$ be an edge map defined on the image domain. For uniformity of results, it is important to restrict the intensities to lie between 0 and 1, and by convention $f(x, y)$ takes on larger values (close to 1) on the object edges. The gradient vector flow (GVF) field is given by the vector field $\mathbf{v}(x, y) = [u(x, y), v(x, y)]$ that minimizes the energy functional

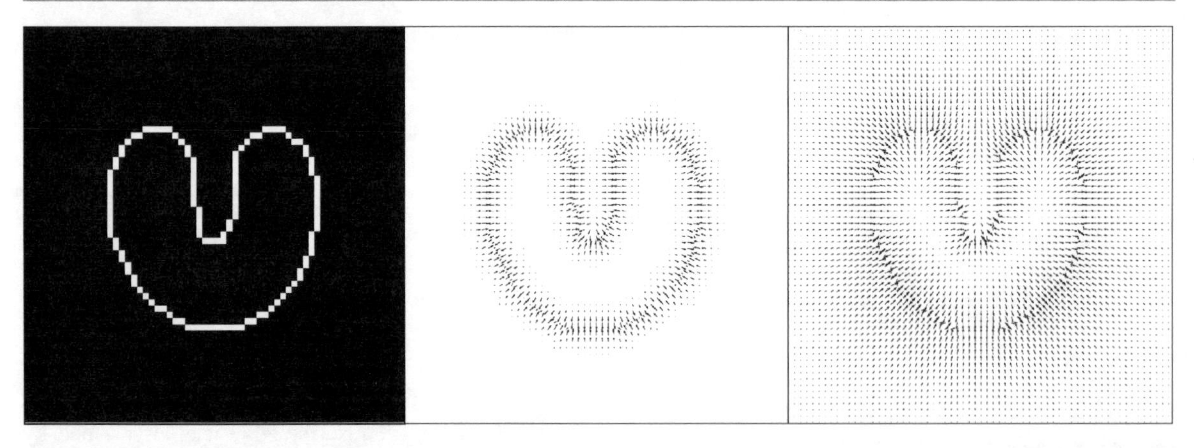

Gradient Vector Flow, Fig. 1 An edge map (*left*) describes the boundary of an object. The gradient of the (blurred) edge map (*center*) points toward the boundary but is very local. The gradient vector flow (GVF) field (*right*) also points toward the boundary but has a much larger capture range

$$\mathcal{E} = \iint_{\mathbb{R}^2} |\nabla f|^2 |\mathbf{v} - \nabla f|^2 + \mu (u_x^2 + u_y^2 + v_x^2 + v_y^2)\, dx\, dy\,. \tag{1}$$

In this equation, subscripts denote partial derivatives, and the gradient of the edge map is given by the vector field $\nabla f = (f_x, f_y)$. Figure 1 shows an edge map, the gradient of the (slightly blurred) edge map, and the GVF field generated by minimizing $\mathcal{E}$.

Equation (1) is a variational formulation that has both a data term and a regularization term. The first term in the integrand is the data term. It encourages the solution $\mathbf{v}$ to closely agree with the gradients of the edge map since that will make $\mathbf{v} - \nabla f$ small. However, this only needs to happen when the edge map gradients are large since $\mathbf{v} - \nabla f$ is multiplied by the square of the length of these gradients. The second term in the integrand is a regularization term. It encourages the spatial variations in the components of the solution to be small by penalizing the sum of all the partial derivatives of $\mathbf{v}$. As is customary in these types of variational formulations, there is a regularization parameter $\mu > 0$ that must be specified by the user in order to trade off the influence of each of the two terms. If μ is large, for example, then the resulting field will be very smooth and may not agree as well with the underlying edge gradients.

Theoretical Solution. Finding $\mathbf{v}(x, y)$ to minimize (Eq. 1) requires the use of calculus of variations since $\mathbf{v}(x, y)$ is a function, not a variable. Accordingly, the Euler equations, which provide the necessary conditions for $\mathbf{v}$ to be a solution, can be found by calculus of variations, yielding

$$\mu \nabla^2 u - (u - f_x)(f_x^2 + f_y^2) = 0\,, \tag{2a}$$

$$\mu \nabla^2 v - (v - f_x)(f_x^2 + f_y^2) = 0\,, \tag{2b}$$

where ∇^2 is the Laplacian operator. It is instructive to examine the form of the equations in (Eq. 2). Each is a partial differential equation that the components u and v of $\mathbf{v}$ must satisfy. If the magnitude of the edge gradient is small, then the solution of each equation is guided entirely by Laplace's equation, for example, $\nabla^2 u = 0$, which will produce a smooth scalar field entirely dependent on its boundary conditions. But the boundary conditions are effectively provided by the locations in the image where the magnitude of the edge gradient is large, where the solution is driven to agree more with the edge gradients.

Computational Solutions. There are several ways to compute GVF. First, the energy function $\mathcal{E}$ itself (Eq. 1) can be directly discretized and minimized, for example, by gradient descent. Second, the partial differential equations in (Eq. 2) can be discretized and solved numerically. The original GVF paper used such an iterative approach, while later papers introduced considerably faster implementation such as an octree-based method [6] and a multi-grid method [7].

Extensions and Advances. GVF is easily extended to higher dimensions. The energy function is readily written in a vector form as

$$\mathcal{E} = \int_{\mathbb{R}^n} \mu|\nabla \mathbf{v}|^2 + |\nabla f|^2 |\mathbf{v} - \nabla f|^2 d\mathbf{x}\,, \quad (3)$$

which can be solved by gradient descent or by finding and solving its Euler equation. Figure 2 shows an illustration of a three-dimensional GVF field on the edge map of a simple object (see [8]).

The data and regularization terms in the integrand of the GVF functional can also be modified. A modification described in [9], called *generalized gradient vector flow* (GGVF), defines two scalar functions and reformulates the energy as

$$\mathcal{E} = \int_{\mathbb{R}^n} g(|\nabla f|)|\nabla \mathbf{v}|^2 + h(|\nabla f|)|\mathbf{v} - \nabla f|^2 d\mathbf{x}\,. \quad (4)$$

While the choices $g(\nabla f|) = \mu$ and $h(|\nabla f|) = |\nabla f|^2$ reduce GGVF to GVF, the alternative choices $g(|\nabla f|) = \exp\{-|\nabla f|/K\}$ and $h(\nabla f|) = 1 - g(|\nabla f|)$, for K a user-selected constant, can improve the tradeoff between the data term and its regularization in some applications.

The variational formulation of GVF has also been modified in *motion GVF* (MGVF) to incorporate object motion in an image sequence [10]. Whereas the diffusion of GVF vectors from a conventional edge map acts in an isotropic manner, the formulation of MGVF incorporates the expected object motion between image frames.

An alternative to GVF called vector field convolution (VFC) provides many of the advantages of GVF, has superior noise robustness, and can be computed faster [11]. The VFC field $\mathbf{v}_{\mathrm{VFC}}$ is defined as the convolution of the edge map f with a vector field kernel $\mathbf{k}$:

$$\mathbf{v}_{\mathrm{VFC}}(x, y) = f(x, y) * \mathbf{k}(x, y)\,, \quad (5)$$

where

$$\mathbf{k}(x, y) = \begin{cases} m(x, y)\left(\dfrac{-x}{\sqrt{x^2+y^2}}, \dfrac{-y}{\sqrt{x^2+y^2}}\right) & (x, y) \neq (0, 0) \\ (0, 0) & \text{otherwise} \end{cases} \quad (6)$$

The vector field kernel $\mathbf{k}$ has vectors that always point toward the origin, but their magnitudes, determined in detail by the function m, decrease to zero with increasing distance from the origin.

The beauty of VFC is that it can be computed very rapidly using a fast Fourier transform (FFT), a multiplication, and an inverse FFT. The capture range can be large and is explicitly given by the radius R of the vector field kernel. A possible drawback of VFC is that weak edges might be overwhelmed by strong edges, but that problem can be alleviated by the use of a hybrid method that switches to conventional forces when the snake gets close to the boundary.

Properties. GVF has characteristics that have made it useful in many diverse applications. It has already been noted that its primary original purpose was to extend a local edge field throughout the image domain, far away from the actual edge in many cases. This property has been described as an extension of the *capture range* of the external force of an active contour model. It is also capable of moving active contours into concave regions of an object's boundary. These two properties are illustrated in Fig. 3.

Previous forces that had been used as external forces (based on the edge map gradients and simply related variants) required pressure forces in order to move boundaries from large distances and into concave regions. Pressure forces, also called balloon forces, provide continuous force on the boundary in one direction (outward or inward) and tend to have the effect of pushing through weak boundaries. GVF can often replace pressure forces and yield better performance in such situations.

Because the diffusion process is inherent in the GVF solution, vectors that point in opposite directions tend to compete as they meet at a central location, thereby defining a type of geometric feature that is related to the boundary configuration but not directly evident from the edge map. For example, *perceptual edges* are gaps in the edge map which tend to be connected visually by human perception [12]. GVF helps to connect them by diffusing opposing edge gradient

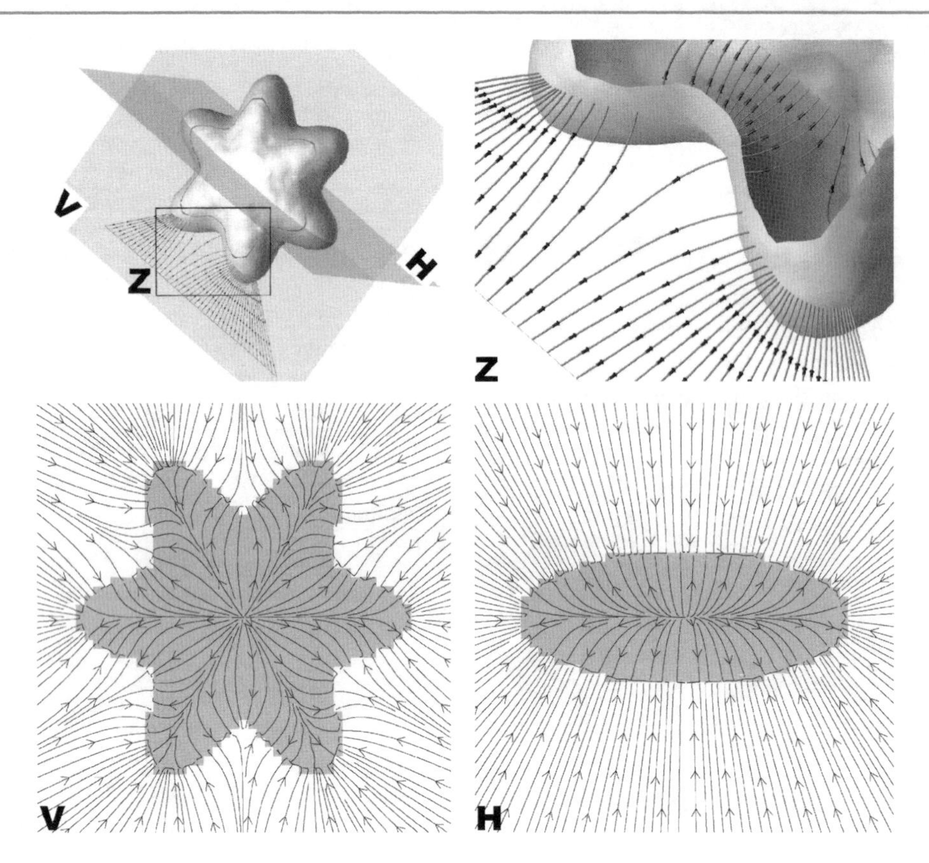

Gradient Vector Flow, Fig. 2 The object shown in the *top left* is used as an edge map to generate a three-dimensional GVF field. Vectors and streamlines of the GVF field are shown in the (*Z*) zoomed region, (*V*) vertical plane, and (*H*) horizontal plane

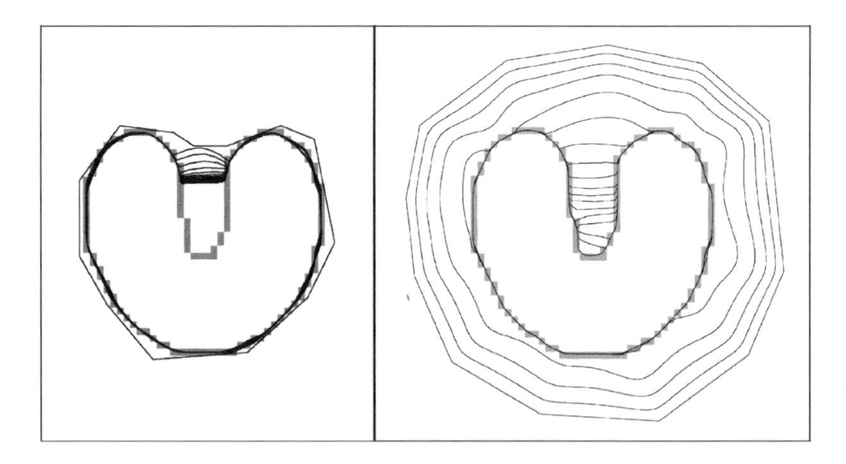

Gradient Vector Flow, Fig. 3 An active contour with traditional external forces (*left*) must be initialized very close to the boundary, and it still will not converge to the true boundary in concave regions. An active contour using GVF external forces (*right*) can be initialized farther away, and it will converge all the way to the true boundary, even in concave regions

vectors across the gap, and even though there is no actual edge map, active contour will converse to the perceptual edge because the GVF vectors drive them there (see [13]). This property carries over when there are so-called *weak edges* identified by regions of edge maps having lower values.

GVF vectors also meet in opposition at central locations of objects, thereby defining a type of medialness. This property has been exploited as an alternative definition of the skeleton of objects [1] and also as a way to initialize deformable models within objects such that convergence to the boundary is more likely.

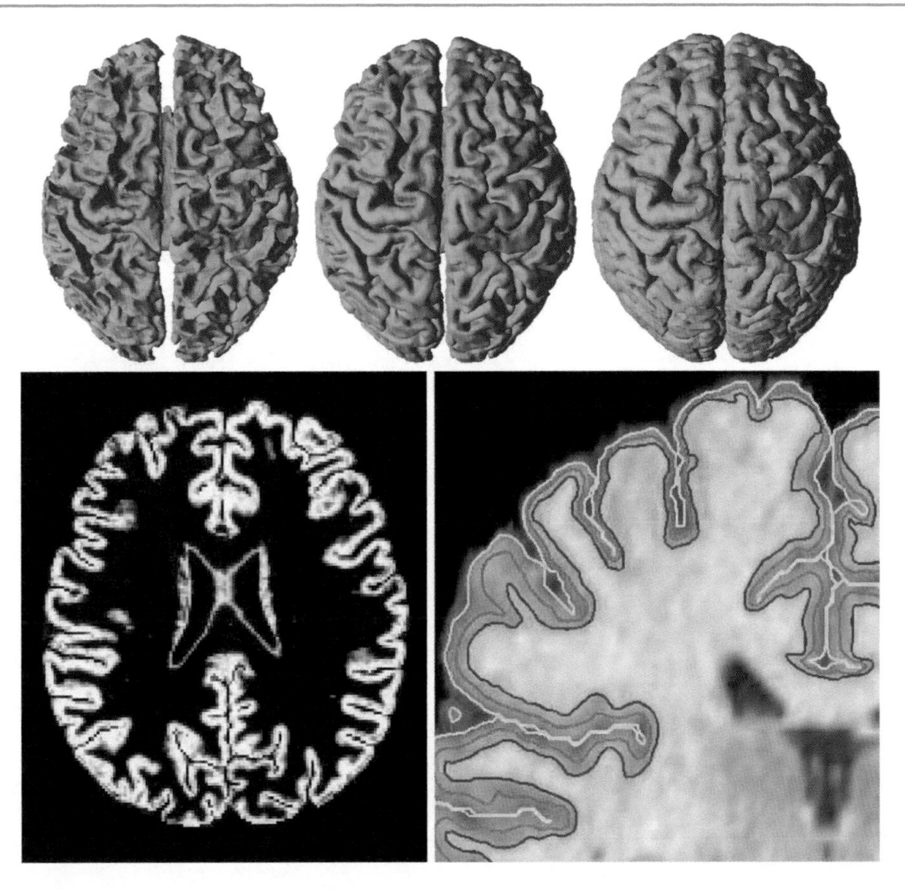

Gradient Vector Flow, Fig. 4 The inner, central, and outer surfaces of the human brain cortex (*top*) are found sequentially using GVF forces in three geometric deformable models. The central surface uses the gray matter membership function (*bottom left*) as an edge map itself, which draws the central surface to the central layer of the cortical gray matter. The positions of the three surfaces are shown as nested surfaces in a coronal cutaway (*bottom right*)

Application

The most fundamental application of GVF is as an external force in a deformable model. A typical application considers an image $I(\mathbf{x})$ with an object delineated by intensity from its background. Thus, a suitable edge map $f(\mathbf{x})$ could be defined by

$$f(\mathbf{x}) = \frac{|\nabla(I(\mathbf{x}) * G_\sigma(\mathbf{x}))|}{\max_{\mathbf{x}} |\nabla(I(\mathbf{x}) * G_\sigma(\mathbf{x}))|}, \qquad (7)$$

where G_σ is a Gaussian blurring kernel with standard deviation σ and $*$ is convolution. This definition is applicable in any dimension and yields an edge map that falls in the range $[0, 1]$. Gaussian blurring is used primarily so that a meaningful gradient vector can always be computed, but σ is generally kept fairly small so that true edge positions are not overly distorted. Given this edge map, the GVF vector field $\mathbf{v}(\mathbf{x})$ can be computed by solving (Eq. 2).

The deformable model itself can be implemented in a variety of ways including parametric models such as the original snake [12] or active surfaces and implicit models including geometric deformable models [14]. In the case of parametric deformable models, the GVF vector field $\mathbf{v}$ can be used directly as the external forces in the model. If the deformable model is defined by the evolution of the (two-dimensional) active contour $\mathbf{X}(s, t)$, then a simple parametric active contour evolution equation can be written as

$$\gamma \mathbf{X}_t = \alpha \mathbf{X}_{ss} - \mathbf{v}(\mathbf{X}). \qquad (8)$$

Here, the subscripts indicate partial derivatives, and γ and α are user-selected constants.

In the case of geometric deformable models, then the GVF vector field $\mathbf{v}$ is first projected against the normal direction of the implicit wavefront, which defines an additional speed function. Accordingly, then the evolution of the signed distance function $\phi_t(\mathbf{x})$ defining a simple geometric deformable contour can be written as

$$\gamma\phi_t = [\alpha\kappa - \mathbf{v} \cdot \frac{\nabla\phi}{|\nabla\phi|}]|\nabla\phi|, \qquad (9)$$

where κ is the curvature of the contour and α is a user-selected constant.

A more sophisticated deformable model formulation that combines the geodesic active contour flow with GVF forces was proposed in [15]. This paper also shows how to apply the Additive Operator Splitting schema [16] for rapid computation of this segmentation method. The uniqueness and existence of this combined model were proven in [17].

GVF has been used to find both inner, central, and central cortical surfaces in the analysis of brain images [4], as shown in Fig. 4. The process first finds the inner surface using a three-dimensional geometric deformable model with conventional forces. Then the central surface is found by exploiting the central tendency property of GVF. In particular, the cortical membership function of the human brain cortex, derived using a fuzzy classifier, is used to compute GVF as if itself were a thick edge map. The computed GVF vectors point toward the center of the cortex and can then be used as external forces to drive the inner surface to the central surface. Finally, another geometric deformable model with conventional forces is used to drive the central surface to a position on the outer surface of the cortex.

References

1. Hassouna MS, Farag AY (2009) Variational curve skeletons using gradient vector flow. IEEE Trans Pattern Anal Mach Intell 31(12):2257–2274
2. Engel D, Curio C (2008) Scale-invariant medial features based on gradient vector flow fields. In: 19th IEEE international conference on pattern recognition (ICPR), Tampa, pp 1–4
3. Yu H, Chua CS (2006) GVF-based anisotropic diffusion models. IEEE Trans Image Process 15(6): 1517–1524
4. Han X, Pham DL, Tosun D, Rettmann ME, Xu C, Prince JL et al (2004) CRUISE: cortical reconstruction using implicit surface evolution. NeuroImage 23(3):997–1012
5. Xu C, Prince JL (1998) Snakes, shapes, and gradient vector flow. IEEE Trans Image Process 7(3):359–369
6. Esteban CH, Schmitt F (2004) Silhouette and stereo fusion for 3D object modeling. Comput Vis Image Underst 96(3):367–392
7. Han X, Xu C, Prince JL (2007) Fast numerical scheme for gradient vector flow computation using a multigrid method. IET Image Process 1(1):48–55
8. Xu C, Han X, Prince JL (2008) Gradient vector flow deformable models. In: Bankman I (ed) Handbook of medical image processing and analysis, 2nd edn. Academic, Burlington, MA, USA, pp 181–194
9. Xu C, Prince JL (1998) Generalized gradient vector flow external forces for active contours. Signal Process 71(2):131–139
10. Ray N, Acton ST (2004) Motion gradient vector flow: an external force for tracking rolling leukocytes with shape and size constrained active contours. IEEE Trans Med Imaging 23(12):1466–1478
11. Li B, Acton ST (2007) Active contour external force using vector field convolution for image segmentation. IEEE Trans Image Process 16(8):2096–2106
12. Kass M, Witkin A, Terzopoulos D (1988) Snakes: active contour models. Int J Comput Vis 1:321–331
13. Xu C, Prince JL (2012) Active contours, deformable models, and gradient vector flow. Online document: GVF online resource including code download. http://www.iacl.ece.jhu.edu/static/gvf/
14. Xu C, Yezzi A, Prince JL (2000) On the relationship between parametric and geometric active contours and its applications. In: 34th Asilomar conference on signals, systems and computers, Pacific Grove, vol 1, pp 483–489
15. Paragios N, Mellina-Gottardo O, Ramesh V (2004) Gradient vector flow fast geometric active contours. IEEE Trans Pattern Anal Mach Intell 26(3):402–407
16. Goldenberg R, Kimmel R, Rivlin E, Rudzsky M (2001) Fast geodesic active contours. IEEE Trans Image Process 10(10):1467–1475
17. Guilot L, Bergounioux M (2009) Existence and uniqueness results for the gradient vector flow and geodesic active contours mixed model. Commun Pure Appl Anal 8(4):1333–1349

GVF

►Gradient Vector Flow

H

Hand-Eye Calibration

Songde Ma[1] and Zhanyi Hu[2]
[1]Ministry of Science & Technology, Beijing, China
[2]National Laboratory of Pattern Recognition, Institute of Automation, Chinese Academy of Sciences, Beijing, China

Synonyms

Robot-camera calibration; Tracker-camera calibration

Related Concepts

▶Camera Calibration

Definition

The hand-eye calibration problem first appeared and got its name from the robotics community, where a camera (eye) was mounted on the gripper (hand) of a robot. The camera was calibrated using a calibration pattern. Then the problem of identifying the unknown transformation from the camera to the hand coordinate system is known as the hand-eye calibration.

Background

There is a strong need for an accurate hand-eye calibration. The reasons are twofold: (i) to map sensor-centered measurements into the robot-world coordinate and (ii) to allow for an accurate prediction of the pose of the sensor on the basis of the arm motion – in fact these are often complementary aspects of the same problem [1].

Theory

A typical hand-eye structure is shown in Fig. 1, where a camera is rigidly mounted on the gripper of the robot. In Fig. 1, $\mathbf{X}$, a 4×4 matrix, denotes the Euclidean transformation between the hand coordinate system and the camera coordinate system, i.e., the hand-eye calibration. Roughly speaking, the hand-eye calibration methods can be divided into two categories: one is to decompose the matrix $\mathbf{X}$ into its rotational and translational parts then optimize the rotation at first, followed by an optimization for the translational part. The other is to optimize the rotation and translation simultaneously.

The standard approach to hand-eye calibration relies on (i) a known reference object and (ii) the possibility to reliably track points on this reference object in order to obtain corresponding points between pairs of images. As shown in Fig. 2, $\mathbf{A}$, $\mathbf{B}$, $\mathbf{C}$, and $\mathbf{D}$ are Euclidean transformation matrices between different coordinate systems, which can be represented by a 4×4 homogeneous matrix as

$$\begin{bmatrix} \mathbf{R} & \mathbf{t} \\ \mathbf{0}^t & 1 \end{bmatrix} \qquad (1)$$

where $\mathbf{R}$ is a 3×3 rotation matrix and $\mathbf{t}$ is a 3×1 vector. Matrix $\mathbf{A}$ denotes the transformation between the first and second position of the robot hand (also known as tool center point motion, TCP motion), which could be

K. Ikeuchi (ed.), *Computer Vision*, DOI 10.1007/978-0-387-31439-6,

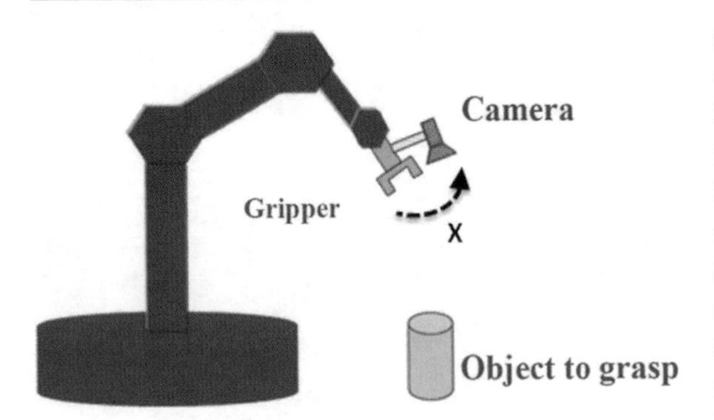

Hand-Eye Calibration, Fig. 1 A camera (eye) mounted at the gripper (hand) of a robot

read out from the robot controller. Since the camera is calibrated beforehand, the transformation matrices **C** and **D** are assumed known, then the motion of the camera, **B**, can be expressed as

$$\mathbf{B} = \mathbf{C}\mathbf{D}^{-1} \tag{2}$$

From this we have the basic equation of the hand-eye calibration as

$$\mathbf{AX} = \mathbf{XB} \tag{3}$$

A number of approaches have been proposed for the determination of the hand-eye calibration matrix **X** from Eq. (3). Here is a brief historical development of the approaches.

1980s

Early solutions decoupled the rotational part of **X** from the translational one, yielding some simple, fast, but error-prone formulations, since rotation estimation errors could propagate to the translational part. Shiu et al. [2] used angle-axis representation and least-squares fitting to calculate the rotation then the translation. Tsai et al. [3] gave a closed-form solution using a more efficient linear algorithm. The number of unknowns in their method is unchanged no matter how many measurements are available. Wang [4] compared [2] and [3] with real data and showed that [3] is slight better than [2].

1990s

Zhuang et al. [5] extended and simplified part of the results in [2] by reformulating the solutions for the rotational part of the homogeneous transform equation as a quaternion equation. Chou et al. [6] presented another quaternion-based approach where a closed-form solution is obtained using the generalized inverse method with singular value decomposition (SVD). Based on the concept of *screw motion*, Chen [7] did not decouple rotational and translational terms for the first time. Zhuang et al. [8] applied nonlinear optimization directly for **X** estimation by minimizing a similar expression to the Frobenius norms of homogeneous matrices of transformation errors. Park et al. [9] performed nonlinear optimization in the same way but again under the decomposed formulation. Lu et al. [10] introduced the eight-space formulation based on quaternions and derived a closed-form least-squares solution using Schur decomposition and Gaussian elimination. Horaud et al. [11] solved simultaneously for rotation and translation using Levenberg–Marquardt technique. Ma [12] gave a linear approach for camera self-calibration and head–eye calibration by controlling the camera platform to undertake at least three orthogonal motions. It is the first approach to combine both camera self-calibration and hand-eye calibration based on active vision concept. Wei et al. [13] nonlinearly minimized algebraic distances and performed a fully automatic hand-eye and camera calibration. Daniilidis [14] introduced the dual quaternions – an algebraic representation of the screw theory to describe motions, which makes it possible to find a fast SVD-based joint solution for rotation and translation.

2000s

Bayro-Corrochano et al. [15] gave an SVD-based linear solution of the coupled problem by the use of motors within the geometric algebra framework. Andreff et al. [16] combined structure from motion with known robot motions for the calibration. They did not enforce the nonlinear orthogonality constraint by increasing the dimensionality of the rotational part and managed to formulate the problem as a single homogeneous linear system. Fassi et al. [17] investigated the standard equation using a geometrical approach and gave some new properties of the equation. Fassi et al. highlighted the reason of over-constrained system when multiple instances of the equation are to be solved simultaneously. Schmidt et al. [18] presented a calibration approach which, in contrast to the standard method, does not require a calibration pattern for determining camera position and orientation.

Hand-Eye Calibration, Fig. 2 Standard approach of hand-eye calibration

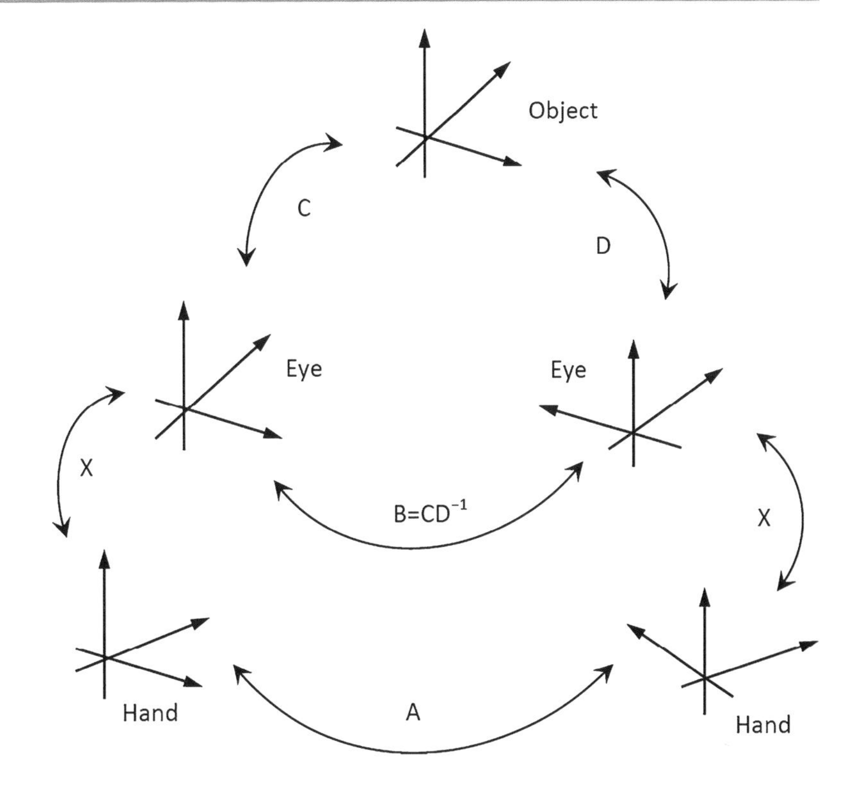

Application

Hand-eye calibration is useful in many industrial applications, for instance, grasping objects, visual servoing, robot navigation, et al. For example, in robot-assisted endoscopic surgery [19], the hand–eye transformation has to be estimated every time when the camera head is mounted anew on the endoscope optics, which is done before each operation because it has to be sterilized. Therefore, an automatic and robust hand-eye calibration algorithm is both desirable and welcome.

References

1. Strobl KH, Hirzinger G (2006) Optimal hand-eye calibration. In: Proceedings of the IEEE/RSJ international conference on intelligent robots and systems, Beijing, China, pp 4647–4653
2. Shiu YC, Ahmad S (1989) Calibration of wrist-mounted robotic sensors by solving homogeneous transform equations of the form AX=XB. IEEE Trans Robot Autom 5(1):16–29
3. Tsai RY, Lenz RK (1989) A new technique for fully autonomous and efficient 3D robotics hand/eye calibration. IEEE Trans Robot Autom 5(3):345–358
4. Wang CC (1992) Extrinsic calibration of a vision sensor mounted on a robot. IEEE Trans Robot Autom 2(8): 161–175
5. Zhuang H, Roth ZS (1991) Comments on 'calibration of wrist-mounted robotic sensors by solving homogenous transform equations of the form AX = XB'. IEEE Trans Robot Autom 7(6):877–878
6. Chou JCK, Kamel M (1991) Finding the position and orientation of a sensor on a robot manipulator using quaternions. Int J Robot Res 10(3):240–254
7. Chen HH (1991) A screw motion approach to uniqueness analysis of head–eye geometry. In: Proceedings of the IEEE conference of computer vision and pattern recognition (CVPR), Hawaii, USA, pp 145–151
8. Zhuang H, Shiu YC (1993) A noise–tolerant algorithm for robotic hand–eye calibration with or without sensor orientation measurement. IEEE Trans Syst Man Cybern 23(4):1168–1175
9. Park FC, Martin BJ (1994) Robot sensor calibration: solving AX=XB on the euclidean group. IEEE Trans Robot Autom 10(5):717–721
10. Lu YC, Chou JC (1995) Eight-space quaternion approach for robotic hand–eye calibration. In: Proceedings of the IEEE international conference on systems, man and cybernetics, Vancouver, BC, Canada, pp 3316–3321
11. Horaud R, Dornaika F (1995) Hand–eye calibration. Int J Robot Res 14(3):195–210
12. Ma SD (1996) A self-calibration technique for active vision systems. IEEE Trans Robot Autom 12(1):114–120
13. Wei GQ, Arbter K, Hirzinger G (1998) Active self–calibration of robotic eyes and hand–eye relationships with

model identification. IEEE Trans Robot Autom 14(1): 158–166
14. Daniilidis K (1999) Hand–Eye Calibration Using Dual Quaternions. International Journal of Robotics Research 18(3) 286–298
15. Bayro–Corrochano E, Daniilidis K, Sommer G (2000) Motor algebra for 3D kinematics: the case of the hand-eye calibration. J Math Imaging Vis 13(2):79–100
16. Andreff N, Horaud R, Espiau B (2001) Robot hand–eye calibration using structure–from–motion. Int J Robot Res 20(3):228–248
17. Fassi I, Legnani G (2005) Hand to sensor calibration: a geometrical interpretation of the matrix equation AX = XB. J Robot Syst 22(9):497–506
18. Schmidt J, Vogt F, Niemann H (2005) Calibration–free hand–eye calibration: a structure–from–motion approach. Lect Notes Comput Sci 3663:67–74
19. Schmidt J, Vogt F, Niemann H (2003) Robust hand-eye calibration of an endoscopic surgery robot using dual quaternions. Lect Notes Comput Sci 2781(DAGM 2003):548–556

Hazard Detection

►Obstacle Detection

Heterogeneous Parallel Computing

►High-Performance Computing in Computer Vision

High-Performance Computing in Computer Vision

Guna Seetharaman
Air Force Research Lab RIEA, Rome, NY, USA

Synonyms

Computing architectures for machine perception; Heterogeneous parallel computing

Definition

The central objective of a computer vision task is to perceive visual data and develop a response. Various processes take place in the *data to decision* chain. High-performance computing refers to the capacity to achieve a computational task at the required fidelity with minimal resources including time and endurance. Recent explosive growth in video cameras and the resulting ubiquity of visual data broaden the scope of high-performance computer vision well beyond robotics and automation. It is difficult to crisply separate the processing and perception mechanisms underlying this chain. Inspired by neural information processing in the visual system, they are often broadly grouped as low-, intermediate-, and higher-level vision. Low-level vision entails very simple computations applied at each pixel and its immediate neighborhood. Examples include edge detection, texture, order statistics, and optical flow. Intermediate level analysis is concerned with local and regional consistencies including coherence. Examples include contour tracing, connected component labeling, and Hough transforms. These accumulative methods do not fully capture certain evidence-driven assimilative processes, where one or more salient features detected in the scene steer the evidence gathering low-level primitives over the rest of the image. Under special circumstances, the low-level and intermediate-level vision are coupled in an iterative style, giving rise to a recursive model of perceptive processing. Examples include optimal placement of edges, deformable templates, subjective contours, and extraction plus grouping of gestalts. Higher-level processes are defined and analyzed from a broader perspective, facilitating a net perception sufficient to achieve detection, recognition, localization, association, prediction, etc. They drive complex applications such as super resolution, segmentation, reconstruction, object recognition, content-based image retrieval, and tracking etc.

Background

Abstract model of computation is comprised of an arithmetic and logic unit (ALU) to perform a primitive operation, an access mechanism to fetch or modify the operands without imposing limits on the size of the data-set or program, ability to concurrently maintain multiple independent intermediate results impacting the course of execution, and the richness of its instruction set. Parallel computing, pipelined processing, and application specific hardware including heterogeneous and reconfigurable processors have all been demonstrated as a means to achieve high performance. We will focus our discussion on approaches to achieve high performance in computer vision.

Distinct types of parallel computing models are used to characterize the overall computation required for computer vision: data-level parallelism, cooperating concurrent processes, and competing concurrent threads with and without mutual preemption. Computations for low-level vision would follow data parallelism and are efficiently implemented using specialized designs including reconfigurable and pipelined processors to perform vector, block, and stream operations. Modern general purpose graphic processing units (GPUs) and stream and vector processing CPUs are good at this. Intermediate vision processes are efficiently performed using concurrent/multi-threaded processing with multi-core and multi-threaded processors and sufficient cache memory. Higher-level processes [1–3] are well suited for computing clusters consisting of super-scalar processors with node-level capabilities to efficiently process vector and regular gridded data. Most clusters employ message passing mechanisms to move data between processors; interconnection topologies such as bi-orthogonal linear buses, meshes, tori, trees, and hypercubes have all been tried in practice. The need for heterogeneous parallel computing for vision and all aspects of artificial intelligence was envisioned early on [4]. Low-level vision algorithms were developed to take advantage of early parallel processing architectures [5]. Parallel computing algorithms at all levels of the vision hierarchy have been adapted to suit prevailing architectures [6] including the modern CMOS image sensors [7]. Modern day desktop computers make a powerful combination of heterogenous processor, multi-core, and multi-processor systems more accessible and affordable. This has further broadened high-performance computing and shifted the focus on software methods, tools, and versatility of application scenarios.

Images acquired through physical cameras suffer a loss of depth. Computational methods aimed at recovering depth information involve iterative optimization of certain regularized functions which are compute-intensive. They are highly uniform, if not identical, across the entire image, making them a candidates for parallel processing. Vision algorithms are often designed to follow a divide-and-conquer approach. For example, the image is partitioned into four quadrants, each processed independently and the results grouped by a merging strategy. Each quadrant will in turn be recursively split into sub quadrants for further processing, until it is not necessary to split further. A problem specific preprocessing or postprocessing step is used at the split and merge stages, respectively, and each partition can be processed independently in parallelized fashion. The computation is said to follow an adaptive quad-tree control structure. A seminal algorithm in computer graphics also known as the area-subdivision (Warnock) algorithm [18] and a widely known quad-tree-based image segmentation algorithm [19] follow this structure. Specific aspects of each computation will determine the type of data-flow: either top-down or bottom-up. Certain complex visual tasks involve bidirectional inferencing where one or more salient objects present in the scene steer the way the rest of the scene is processed. Such tasks will sift up and down the quad-tree in cyclical fashion. In general, they will include intra-block computations and adjacent-block data access to apply spatial operators. The interdependence between incremental computations is captured by a hybrid topological structure, a pyramid, which combines a quad-tree with a two-dimensional mesh. Basic computations such as convolution are performed as a systolic or vector operation across the mesh, using single instruction multiple data (SIMD) primitives [8]. They can also be achieved by carefully (de)composed sequence of partial sums and efficiently implemented (on a standard von Neumann machine) as pipelined primitives exploiting the order in which the pixels will be visited. Piecewise affine computations can be used to model generalized geometric deformations and motion fields [20, 21] with efficient implementation using triangular patches. Integral histogram is a powerful emerging low-level processing method for speeding up multi-resolution sliding window-based searching and other computer vision tasks using block-level pipelined computations. Multi-core parallel implementations of integral histograms [9] can significantly scale up computation to large image and video collections.

Classic space filling curves such as Hilbert curves, Peano scans, and Morton curves have been used as a locality preserving pixel organization [10, 11] and an alternative to the standard raster scan which scans the image left to right and top to bottom visiting each pixel exactly once. The spatially compact nature of space filling scans tend to cohere highly correlated pixels as maximal subsequences, producing a significant improvement on aggregative operators such as the run-length coding and compressive sensing.

A recent paper on using spatially coherent data structures for representing matrix equations combines the advantages of a binary-tree control structure and the spatially compact nature of Peano scans to achieve cache-oblivious computations [12] and algorithms on sequential machines. Z-tree is built by inserting a Hilbert curve-scanned sequence of pixel at the terminal level of a binary tree and then building the tree upward. Such a tree represents very large images as a collection of tiles, permitting mesh-algorithms at any given level of the Z-tree and SIMD processing over two or more aligned tiles. Spatio-temporal visualization of gigapixel-sized images and videos can be extended to incorporate such tiled representations [13–15, 22].

Our understanding of human visual perception suggests multiple competing chains of evidence accumulation and incremental inferencing neural processes. The basic unit of such inferences are known as gestalts [16]. They cohere and coalesce into a candidate interpretation, which are indirectly influenced by context and cues about the scene. They account for subjective processes associated with observer expectations. Evidence gathered in one location across an image may at times add to, concur with, be neutral to, or compete against the interpretation arising from elsewhere within the image. A consistent interpretation is achieved through a constraint propagation and resolution method known as relaxation labeling and belief propagation techniques [17]. Such methods have been studied thoroughly in the perception of line drawings of a complex scene of trihedral blocks. Animals that blend with the background such as the well-known dalmatian dog camouflaged against a background of rocks are sometimes used to illustrate the inherent difficulties of separating foreground (the dog) from background based on local contrast or edge structures only. Any cue about the existence of a dog immediately contributes to the coalescing of an otherwise disconnected set of patches into a meaningful grouping in the shape of dog. From a computational standpoint, two approaches have been proposed each with significant success [1]: model-driven computations including hypothesize and verify methods and rule-based methods. The former is built around procedural programs and models that evoke a model driven sequence of tasks and operations. In the latter, a collection of associations between physical scene/object situations and resulting observations across images are combined effectively using rule-based computations to achieve one of more stable interpretations. These methods lend themselves to parallel computations. Light-weighted threads with carefully orchestrated spatial- and feature-space-based interdependencies will be required to efficiently implement these processes. Modern multi-core heterogeneous processors are more suitable for this.

From a computer engineering perspective, the granularity of the parallelism is of importance for constructing high-throughput dedicated machines. However, from a computer science and computing architecture perspective, one will see that parallel computation for computer vision requires multiple architectures across scale and level of information processing. Topics not covered but deserving of further exploration is focal plane and light-field processing using optical parallel computing and quantum computing for vision algorithms. There is a diversity in the intrinsic parallelism that is seen at each stage of the data-to-decision chain.

References

1. Cantoni V, Levialdi S, Zavidovique B (2011) 3C vision: cues, context and channels. Elsevier, Amsterdam
2. Scientific American (1998) Science's vision: the mechanics of sight.
3. Scientific American (2008) Special report on PERCEPTION: 105 mind-bending illusions
4. Reddy R (1985) Super chips for artificial intelligence. IEEE Int Conf on Solid-State Circuits XXVIII:54–55
5. Bader DA, JáJá J (1996) Parallel algorithms for image histogramming and connected components with an experimental study. J Parallel Distrib Comput 35(2):173–190
6. Gross T, OĤallaron D (1998) iWarp: anatomy of a parallel computing system. MIT, Cambridge
7. Gamal A, Eltoukhy H (2005) CMOS image sensors. IEEE Circuits Devices Mag 21(3):6–20
8. Seetharaman G (1995) A simplified design strategy for mapping image processing algorithms on a simd torus. J Theor Comput Sci 140(2):319–331
9. Bellens P, Palaniappan K, Badia RM, Seetharaman G, Labarta J (2011) Parallel implementation of the integral histogram. Lecture notes in computer science (ACIVS), vol 6915. Springer, Berlin/New York, pp 586–598
10. Seetharaman G, Zavidovique B (1997) Image processing in a tree of peano coded images. In: Proceedings computer architectures for machine perception (CAMP '97), IEEE computer society, Cambridge, MA, pp 229–234
11. Asano T, Ranjan D, Roos T, Welzl E, Widmayer P (1997) Space-filling curves and their use in the design of geometric data structures. Theor Comput Sci 181(1):3–15
12. Heinecke A, Bader M (2009) Towards many-core implementation of LU-decomposition decomposition using peano curves. In: Proceedings of the combined workshop on

unconventional high performance computing plus memory access workshop. Association for Computing Machinary, New York, pp 21–30
13. Hasler AF, Palaniappan K, Manyin M, Dodge J (1994) A high performance interactive image spreadsheet (IISS). Comput Phys 8(4):325–342
14. Palaniappan K, Fraser J (2001) Multiresolution tiling for interactive viewing of large datasets. In: 17th international AMS conference on interactive information and processing systems (IIPS) for meteorology, oceanography and hydrology, American meteorological society, Albuquerque, pp 338–342
15. Ponto K, Doerr K, Kuester F (2010) Giga-stack: a method for visualizing giga-pixel layered imagery on massively tiled displays. Future Gener Comput Systems 26(5):693–700
16. Rock I, Palmer S (1990) The legacy of gestalt psychology. Sci Am 263(6): 84–90
17. Grauer-Gray S, Kambhamettu C, Palaniappan K (2008) GPU implementation of belief propagation using CUDA for cloud tracking and reconstruction. In: 5th IAPR workshop on pattern recognition in remote sensing (ICPR), Tampa, FL, pp 1–4
18. Warnock J.E, (1969) A hidden surface algorithm for computer generated halftone pictures. Ph.D. dissertation. The University of Utah. AAI6919002
19. Horowitz S.L, Theodosios Pavlidis (1976) Picture segmentation by a tree traversal algorithm. J. ACM 23(2): 368–388
20. Seetharaman G, Gasperas G, Palaniappan K (2000) A piecewise affine model for image registration in nonrigid motion analysis. IEEE Int. Conf. Image Processing, 1:561–564
21. Zhou L, Kambhamettu C, Goldgof D, Palaniappan K, Hasler A. F, (2001) Tracking non-rigid motion and structure from 2D satellite cloud images without correspondences. IEEE Trans. Pattern Analysis and Machine intelligence, 23(11):1330–1336
22. Fraser J, Haridas A, Seetharaman G, Rao R, Palaniappan K (2013) KOLAM: A cross-platform architecture for scalable visualization and tracking in wide-area motion imagery. Proc. SPIE Conf. Geospatial InfoFusion III (Defense, Security and Sensing: Sensor Data and Information Exploitation), 8747:87470N

Histogram

Ying Nian Wu
Department of Statistics, UCLA, Los Angeles, CA, USA

Definition

Histogram is a graphical display of the distribution of observed values of some random variable.

Theory and Applications

Histogram was introduced by Karl Pearson. In a histogram, the range of the observed values is divided into a number of bins of equal length. A rectangle is erected on top of each bin so that the area of the rectangle equals the frequency that the observed values fall into this bin. A histogram is a more detailed summary of a distribution than mean and variance. It can be considered a nonparametric estimate of the probability density function of the random variable.

In image analysis and computer vision, histograms are often obtained by spatial pooling and they serve as image features. For a texture image, histograms of responses from a bank of filters are pooled over the image domain, and these histograms serve as features that characterize the texture pattern. Histograms can also be pooled within local windows as local texture features. Heeger and Bergen [3] proposed an algorithm for texture synthesis by matching histograms of filter responses. Zhu, Wu, and Mumford [7] proposed a Markov random field model for stochastic textures. The model is the maximum entropy distribution that matches observed marginal histograms of filter responses. The spatially pooled histograms discard the position information. This is appropriate for characterizing texture patterns, which are spatially stationary.

The local orientation histograms of image intensity gradients are key components of two of the most successful image features, namely, SIFT (scale-invariant feature transform) [5] and HoG (histogram of oriented gradients) [2]. Such histograms are pooled within local cells, where each pixel contributes a weighted vote to the histogram bin that corresponds to the orientation of the intensity gradient at this pixel. The weight can be the magnitude of the gradient or some nonlinear transformation of it. Such histograms are very informative descriptions of local image patches, and they are partially invariant to local geometric distortions or shape deformations, because they are spatially pooled within local cells where information about the exact positions of the gradients is discarded.

A more general version of histogram is also used in the bag-of-words method for image classification [1]. The basic idea is to obtain a codebook of "words" by quantizing local image features via clustering. The histogram is in the form of the frequencies of the occurrences of the codewords. The histogram can be pooled over the entire image. One can also divide the image into subregions and pool the histograms within these subregions [4]. Such histograms can then be used for image classification, usually by SVM with histogram intersection kernel [4, 6]. Again, because the spatial pooling of the histograms discards the exact position information, such histograms are invariant to shape deformations.

References

1. Csurka G, Dance C, Fan L, Willamowski J, Bray C (2004) Visual categorization with bags of keypoints. In: Workshop of ECCV, Prague
2. Dalal N, Triggs B (2005) Histograms of oriented gradients for human detection. In: IEEE conference on computer vision pattern recognition (CVPR), San Diego
3. Heeger DJ, Bergen JR (1995) Pyramid-based texture analysis/synthesis. In: SIGGRAPH'95, Los Angeles, pp 229–238
4. Lazebnik S, Schmid C, Ponce J (2006) Beyond bags of features: spatial pyramid matching for recognizing natural scene categories. In: IEEE conference on computer vision pattern recognition (CVPR), New York
5. Lowe DG (1999) Object recognition from local scale-invariant features. In: ICCV, Kerkyra
6. Maji S, Berg AC, Malik J (2008) Classification using intersection kernel support vector machines is efficient. In: IEEE conference on computer vision pattern recognition (CVPR), Anchorage
7. Zhu SC, Wu YN, Mumford DB (1997) Filter, random field, and maximum entropy (FRAME): towards a unified theory for texture modeling. Int J Comput Vis 27:107–126

Human Appearance Modeling and Tracking

▶Appearance-Based Human Tracking

Human Motion Classification

▶Gesture Recognition

Human Pose Estimation

Leonid Sigal
Disney Research, Pittsburgh, PA, USA

Synonyms

Articulated pose estimation; Body configuration recovery

Related Concepts

▶Human Pose Estimation

Definition

Human pose estimation is the process of estimating the configuration of the body (pose) from a single, typically monocular, image.

Background

Human pose estimation is one of the key problems in computer vision that has been studied for well over 15 years. The reason for its importance is the abundance of applications that can benefit from such a technology. For example, human pose estimation allows for higher-level reasoning in the context of human-computer interaction and activity recognition; it is also one of the basic building blocks for marker-less motion capture (MoCap) technology. MoCap technology is useful for applications ranging from character animation to clinical analysis of gait pathologies.

Despite many years of research, however, pose estimation remains a very difficult and still largely unsolved problem. Among the most significant challenges are the following: (1) variability of human visual appearance in images, (2) variability in lighting conditions, (3) variability in human physique, (4) partial occlusions due to self-articulation and layering of objects in the scene, (5) complexity of human skeletal structure, (6) high dimensionality of the pose, and (7) the loss of 3D information that results from observing the pose from 2D planar image projections. To date, there is no approach that can produce satisfactory results in general, unconstrained settings while dealing with all of the aforementioned challenges.

Theory and Application

Human pose estimation is typically formulated probabilistically to account for ambiguities that may exist in the inference (though there are notable exceptions, e.g., [11]). In such cases, one is interested in estimating the posterior distribution, $p(\mathbf{x}|\mathbf{z})$, where $\mathbf{x}$ is the pose of the body and $\mathbf{z}$ is a feature set derived from the image. The key modeling choices that affect the inference are:

- The representation of the pose – $\mathbf{x}$
- The nature and encoding of image features – $\mathbf{z}$
- The inference framework required to estimate the posterior – $p(\mathbf{x}|\mathbf{z})$

Next, the primary lines of research in pose estimation with respect to these modeling choices are reviewed. It is worth noting that these three modeling choices are not always independent. For example, some inference frameworks are specifically designed to utilize a given representation of the pose.

Representation: The configuration of the human body can be represented in a variety of ways. The most direct and common representation is obtained by parameterizing the body as a *kinematic tree* (see Fig. 1), $\mathbf{x} = \{\tau, \theta_\tau, \theta_1, \theta_2, \ldots, \theta_N\}$, where the pose is encoded using position of the root segment (to keep the kinematic tree as short as possible, the pelvis is typically; used as the root segment), τ; orientation of the root segment in the world, θ_τ and a set of relative joint angles, $\{\theta_i\}_{i=1}^N$, that represent the orientations of body parts with respect to their parents along the tree (e.g., the orientation of the thigh with respect to the pelvis, shin with respect to the thigh).

Kinematic tree representation can be obtained for 2D, 2.5D, and 3D body models. In 3D, $\tau \in \mathbb{R}^3$ and $\theta_\tau \in SO(3)$; $\theta_i \in SO(3)$ for spherical joints (e.g., neck), $\theta_i \in \mathbb{R}^2$ for saddle joints (e.g., wrist), and $\theta_i \in \mathbb{R}^1$ for hinge joints (e.g., knee) and represents the pose of the body in the world. Note, the actual representation of the rotations in 3D is beyond the scope of this entry. In 2D, $\tau \in \mathbb{R}^2$ and $\theta_\tau \in \mathbb{R}^1$; $\theta_i \in \mathbb{R}^1$ corresponds to pose of the *cardboard* person in the image plane. 2.5D representations are the least common and are extensions of the 2D representation such that the pose, $\mathbf{x}$, is augmented with (typically discrete) variables encoding the relative depth (layering) of body parts with respect to one another in the 2D *cardboard* model. In all cases, be it in 2D or 3D, this representation results in a high-dimensional pose vector, $\mathbf{x}$, in $\mathbb{R}^{30} - \mathbb{R}^{70}$, depending on the fidelity and exact parameterization of the skeleton and joints. Alternatively, one can parameterize the pose of the body by 2D or 3D locations of the major joints [6]. For example, $\mathbf{x} = \{\mathbf{p}_1, \mathbf{p}_2, \ldots, \mathbf{p}_N\}$, where $\mathbf{p}_i$ is the joint location in the world, $\mathbf{p}_i \in \mathbb{R}^3$, or, in an image, $\mathbf{p}_i \in \mathbb{R}^2$. This latter representation is less common, however, because it is not invariant to the morphology (body segment lengths) of a given individual.

A typical alternative to the kinematic tree models is to model the body as a set of parts, $\mathbf{x} = \{\mathbf{x}_1, \mathbf{x}_2, \ldots, \mathbf{x}_M\}$, each with its own position and orientation in space, $\mathbf{x}_i = \{\tau_i, \theta_i\}$, that are connected by a set of statistical or physical constraints that enforce skeletal (and sometimes image) consistency. Because part-based parameterization is redundant, it results in an even higher-dimensional representation. However, it does so in a way that makes it efficient to infer the pose, as will be discussed in a later section. Methods that utilize such a parameterization are often called *part-based*. As in kinematic tree models, the parts can be defined in 2D [2, 4, 7, 8, 11, 16] or in 3D [21], with 2D parameterizations being significantly more common. In 2D, each part's representation is often augmented with an additional variable, s_i, that accounts for uniform scaling of the body part in the image, i.e., $\mathbf{x}_i = \{\tau_i, \theta_i, s_i\}$ with $\tau_i \in \mathbb{R}^2$, $\theta_i \in \mathbb{R}^1$, and $s_i \in \mathbb{R}^1$.

Image features: Performance of any pose estimation approach depends substantially on the observations, or image features, that are chosen to represent salient parts of the image with respect to the human pose. A related and equally important issue is one of how these features are encoded. In addition to using different encodings, some approaches propose to reduce the dimensionality of the resulting feature vectors through vector quantization or *bag-of-words* representations. These coarser representations simplify feature matching but at the expense of losing spatial structure in the image. Common features and encoding methods are illustrated in Fig. 2.

Over the years, many features have been proposed by various authors. The most common features include image silhouettes [1], for effectively separating the person from background in static scenes; color [16], for modeling unoccluded skin or clothing; edges [16], for modeling external and internal contours of the body; and gradients [5], for modeling the texture over the body parts. Less common features include shading and focus [14]. To reduce dimensionality and increase robustness to noise, these raw features are often encapsulated in image descriptors, such as shape context [1, 2, 6], SIFT [6], and histogram of oriented gradients [5]. Alternatively, hierarchical multilevel image encodings can be used, such as HMAX [12], spatial pyramids [12], and vocabulary trees [12]. The effectiveness of different feature types on pose estimation has been studied in the context of several inference architectures; see [2] and [12] for discussions and quantitative analyses.

H

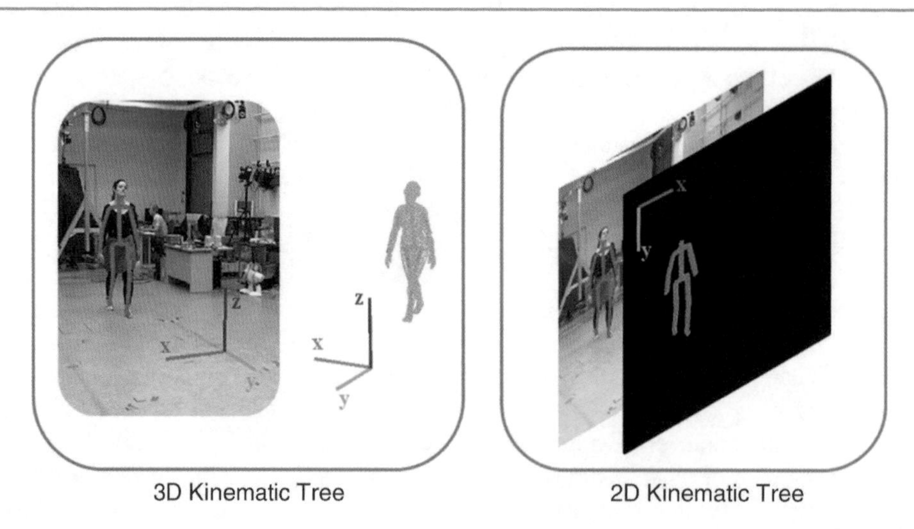

Human Pose Estimation, Fig. 1 *Skeleton representation*: Illustration of the 3D and 2D kinematic tree skeleton representation on the *left* and *right*, respectively

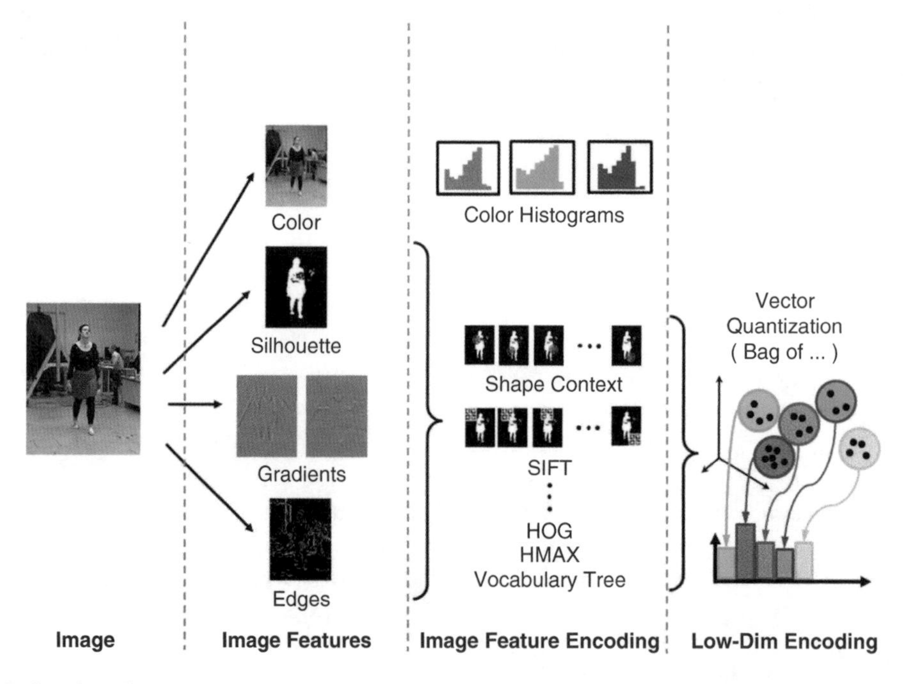

Human Pose Estimation, Fig. 2 *Image features*: Illustration of common image features and encoding methods used in the literature

Inference (regression models): Characterizing the posterior distribution, $p(\mathbf{x}|\mathbf{z})$, can be done in a number of ways. Perhaps the most intuitive way is to define a parametric [1, 6, 12] or nonparametric [15, 18, 22, 25] form for the conditional distribution $p(\mathbf{x}|\mathbf{z})$ and learn the parameters of that distribution from a set of training exemplars. This class of models is more widely known as *discriminative methods*, and they have been shown to be very effective for pose estimation. Such methods directly learn $p(\mathbf{x}|\mathbf{z})$ from a labeled dataset of poses and corresponding images, $\mathcal{D} = \{(\mathbf{x}_i, \mathbf{z}_i)\}_{i=1}^{N}$, which can be produced artificially using computer graphics software packages (e.g., Poser) [1, 12, 18]. The inference takes a form of probabilistic regression. Once a regression function is learned, a scanning window approach is typically used at test time to detect a portion of the image (bounding box) where the person resides; $p(\mathbf{x}|\mathbf{z})$ is then used to characterize the configuration of the person in that target window.

The simplest method in this category is the one of linear regression [1], where the body configuration, $\mathbf{x}$, is assumed to be a linear combination of the image features, $\mathbf{z}$, with additive Gaussian noise,

$$\mathbf{x} = A[\mathbf{z} - \mu_z] + \mu_x + \nu; \qquad \nu \sim \mathcal{N}(0, \Sigma);$$

$\mu_x = \frac{1}{N}\sum_{i=1}^{N}\mathbf{x}_i$, and $\mu_z = \frac{1}{N}\sum_{i=1}^{N}\mathbf{z}_i$ are means computed over the training samples to center the data. Alternatively, this can be written as

$$p(\mathbf{x}|\mathbf{z}) = \mathcal{N}(A[\mathbf{z} - \mu_z] + \mu_x, \Sigma). \tag{1}$$

The regression coefficients, A, can be learned easily from paired training samples, $\mathcal{D} = \{(\mathbf{x}_i, \mathbf{z}_i)\}_{i=1}^{N}$, using the least squares formulation (see [1] for details).

Parametric vs. nonparametric: Parametric discriminative methods [1, 6, 12] are appealing because the model representation is fixed with respect to the size of the training dataset $\mathcal{D}$. However, simple parametric models, such as linear regression [1] or relevance vector machine [1], are unable to deal with complex non-linear relationships between image features and poses. Nonparametric methods, such as nearest neighbor regression [18] or kernel regression [18], are able to model arbitrary complex relationships between input features and output poses. The disadvantage of these nonparametric methods is that the model and inference complexity are both functions of the training set size. For example, in kernel regression,

$$p(\mathbf{x}|\mathbf{z}) = \sum_{i=1}^{N} \mathcal{K}_x(\mathbf{x}, \mathbf{x}_i)\frac{\mathcal{K}_z(\mathbf{z}, \mathbf{z}_i)}{\sum_{k=1}^{N}\mathcal{K}_z(\mathbf{z}, \mathbf{z}_k)}, \tag{2}$$

where $\mathcal{K}_x(\cdot,\cdot)$ and $\mathcal{K}_z(\cdot,\cdot)$ are kernel functions measuring the similarity of the arguments (e.g., Gaussian kernels) and the inference complexity is $O(N)$ (where N is the size of the training dataset). More sophisticated nonparametric methods, such as Gaussian Process Latent Variable Models (GPLVMs), can have even higher complexity; GPLVMs have $O(N^3)$ learning and $O(N^2)$ inference complexity. In practice, nonparametric methods tend to perform better but are slower.

Dealing with ambiguities: If one assumes that $p(\mathbf{x}|\mathbf{z})$ is unimodal [1], conditional expectation can be used to characterize the plausible configuration of the person in an image given the learned model. For example, for linear regression in Eq. (1),
$E[\mathbf{x}|\mathbf{z}] = A[\mathbf{z} - \mu_z] + \mu_x$; for kernel regression in Eq. (2),

$$E[\mathbf{x}|\mathbf{z}] = \sum_{i=1}^{N}\mathbf{x}_i\frac{\mathcal{K}_z(\mathbf{z}, \mathbf{z}_i)}{\sum_{k=1}^{N}\mathcal{K}_z(\mathbf{z}, \mathbf{z}_k)}. \tag{3}$$

In practice, however, most features under standard imaging conditions are ambiguous, resulting in multimodal distributions. Ambiguities naturally arise in image projections, where multiple poses can result in similar, if not identical, image features (e.g., front- and back-facing poses yield nearly identical silhouette features). To account for these ambiguities, parametric mixture models were introduced in the form of Mixture of Experts [6, 12]. Nonparametric alternatives, such as Local Gaussian Process Latent Variable Models (LGPLVM) [25], cluster the data into convex local sets and make unimodal predictions within each cluster, or search for prominent modes in $p(\mathbf{x}|\mathbf{z})$ [15, 22].

Learning: Obtaining the large datasets that are required for learning discriminative models that can generalize across motions and imaging conditions is challenging. Synthetic datasets often do not exhibit the imaging characteristics present in real images, and real fully labeled datasets are scarce. Furthermore, even if large datasets could be obtained, learning from vast amounts of data is not a trivial task [6]. To address this issue, two solutions were introduced: (1) learning from small datasets by discovering an intermediate low-dimensional latent space for regularization [15, 22] and (2) learning in semisupervised settings, where a relatively small dataset of paired samples is accompanied by a large amount of unlabeled data [12, 15, 22].

Limitations: Despite popularity and lots of successes, discriminative methods do have limitations. First, they are only capable of recovering a relative 3D configuration of the body and not its position in 3D space. The reason for this is practical, as reasoning about position in 3D space would require prohibitively large training datasets that span the entire 3D volume of the space visible from the camera. Second, their performance tends to degrade as the distributions of test and training data start to diverge; in other words, generalization remains one of the key issues. Lastly, learning discriminative models efficiently from large datasets that cover wide range of realistic activities and postures remains a challenging task.

Inference (generative): Alternatively, one can take a generative approach and express the desired posterior, $p(\mathbf{x}|\mathbf{z})$, as a product of a likelihood and a prior:

$$p(\mathbf{x}|\mathbf{z}) \propto \underbrace{p(\mathbf{z}|\mathbf{x})}_{likelihood} \underbrace{p(\mathbf{x})}_{prior}. \tag{4}$$

Characterizing this high-dimensional posterior distribution is typically hard; hence, most approaches rely on a posteriori (MAP) solutions that look for the most probable configurations that are both typical (have high *prior* probability) and can explain the image data well (have high *likelihood*):

$$\mathbf{x}_{MAP} = \arg\max p(\mathbf{x}|\mathbf{z}). \tag{5}$$

Searching for such configurations, however, in the high-dimensional (40+) articulation space is very challenging and most approaches frequently get stuck in local optima. Global hierarchical search methods, such as Annealed Particle Filter [10], have shown some promising results for simple skeletal configurations, where body is mostly upright and when observations from multiple cameras are available. For the more general articulations and monocular observations that are often the focus of pose estimation algorithms, this class of methods has not been very successful to date.

Inference (part-based models): To battle the inference complexity of generative models, part-based models have been introduced. These methods originate in the object recognition community with formulation of Fischler and Elschlager (1973) and assume that a body can be represented as an assembly of parts that are connected by constraints imposed by the joints within the skeletal structure (and, sometimes, by the image constraints imposed by projections onto an image plane that account for occlusions). This formulation reduces the inference complexity because likely body part locations can be searched for independently, only considering the nearby body parts that constrain them, which significantly prunes the total search space.

Among the earliest successes along this line of research is the work of Lee and Cohen [13]. Their approach focused on obtaining proposal maps for the locations of individual joints within an image. These proposal maps were obtained based on a number of features that were computed densely over the image. For example, face detection was used to obtain hypotheses for the location of the head; head-shoulder contour matching, obtained using a deformable contour model and gradient descent, was used as evidence for shoulder joint locations; elliptical skin regions, obtained using skin-color segmentation, were used to determine the locations of the lower arms and lower legs. In addition, second-derivative (ridge) observations were used as evidence for other limbs of the body. Given proposals for the different joints, weighted by the confidence of corresponding detectors, a data-driven Markov Chain Monte Carlo (MCMC) approach was used to recover 3D configurations of the skeleton. This inference relied on direct inverse kinematics (IK) obtained from 2D proposal maps. To further improve the results, a kinematic jump proposal process was also introduced. The kinematic jump proposal process involves flipping a body part or a set of parts (i.e., the head, a hand, or an entire arm) in the depth direction around its pivotal joint.

Other part-based approaches try to assemble regions of an image into body parts and successively construct those parts into a body. Prime examples of such methods are introduced by Mori et al. [14] and Ren et al. [17]. In [14] superpixels were first assembled into body parts based on the evaluation of low-level image cues, including contour, shape, shading, and focus. The part proposals were then pruned and assembled together using length, body part adjacency, and clothing symmetry. A similar approach was taken in [17], but line segments were used instead of assembling superpixels. Parallel lines were assembled into candidate parts using a set of predefined rules, and the candidate parts were in turn assembled into the body with a set of joint, scale, appearance, and orientation consistency constraints. Unlike [14], the search for the most probable body configurations was formulated as a solution to an Integer Quadratic Programming (IQP) problem.

The most traditional and successful approach, however, is to represent the body using a Markov Random Field (MRF) with body parts corresponding to the nodes and constraints between parts encoded by potential functions that account for physical and statistical dependencies (see Fig. 3). Formally, the posterior,

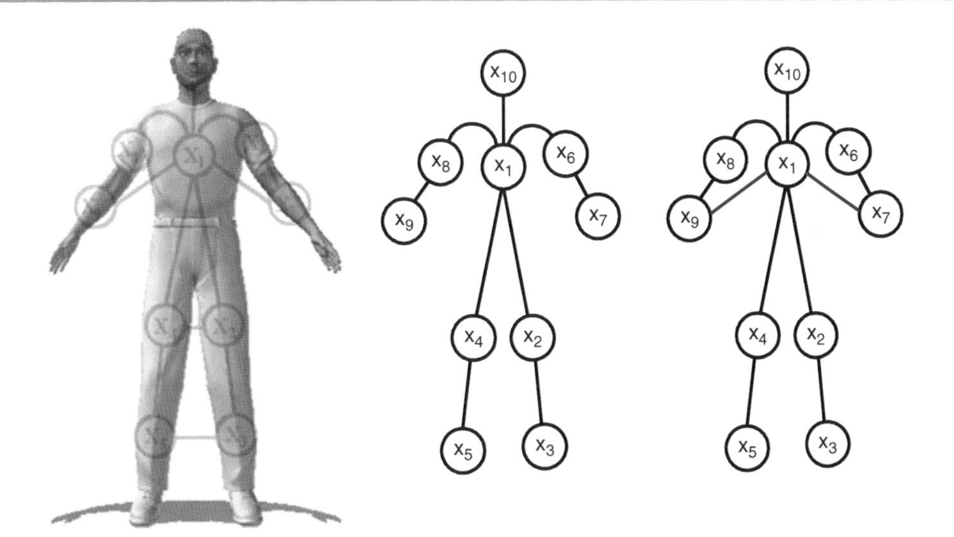

Human Pose Estimation, Fig. 3 *Pictorial structures model*: Illustrated is the depiction of the 10-part tree-structured pictorial structures model (*middle*) and a non-tree-structured (loopy) pictorial structures model (*right*). In the non-tree-structured model, additional constraints encoding occlusions are illustrated in *blue*

$p(\mathbf{x}|\mathbf{z})$, can be expressed as

$$\begin{aligned} p(\mathbf{x}|\mathbf{z}) &\propto p(\mathbf{z}|\mathbf{x})p(\mathbf{x}) \\ &= p(\mathbf{z}|\{\mathbf{x}_1, \mathbf{x}_2, ..., \mathbf{x}_M\})p(\{\mathbf{x}_1, \mathbf{x}_2, ..., \mathbf{x}_M\}) \\ &\approx \underbrace{\prod_{i=1}^{M} p(\mathbf{z}|\mathbf{x}_i)}_{likelihood} \underbrace{p(\mathbf{x}_1) \prod_{(i,j)\in E} p(\mathbf{x}_i, \mathbf{x}_j)}_{prior}. \end{aligned} \quad (6)$$

In this case, pose estimation takes the form of inference in a general MRF network. The inference can be solved efficiently using message-passing algorithms, such as Belief Propagation (BP). BP consists of two distinct phases: (1) a set of message-passing iterations are executed to propagate consistent part estimates within a graph, and (2) marginal posterior distributions are estimated for every body part [2, 8, 16]. A typical formulation looks at the configuration of the body in the 2D image plane and assumes discretization of the pose for each individual part, e.g., $\mathbf{x}_i = \{\tau_i, \theta_i, s_i\}$, where $\tau_i \in \mathbb{R}^2$ is the location and $\theta_i \in \mathbb{R}^1$ and $s_i \in \mathbb{R}^1$ are orientation and scale of the part i (represented as a rectangular patch) in the image plane. As a result, the inference is over a set of discrete part configurations $\mathbf{l}_i \in \mathbb{Z}$ (for part i), where $\mathbb{Z}$ is the enumeration of poses for a part in an image ($\mathbf{l}_i$ is a discrete version of $\mathbf{x}_i$). With an additional assumption of pairwise potentials that account for kinematic constraints, the model forms a tree-structured graph known as the *tree-structured pictorial structures* (PS) model. An approximate inference with continuous variables is also possible [20, 21].

Inference in the tree-structured PS model first proceeds by sending recursively defined messages of the form:

$$m_{i\to j}(\mathbf{l}_j) = \sum_{\mathbf{l}_i} p(\mathbf{l}_i, \mathbf{l}_j)p(\mathbf{z}|\mathbf{l}_i) \prod_{k\in A(i)\setminus j} m_{k\to i}(\mathbf{l}_i), \quad (7)$$

where $m_{i\to j}$ is the message from part i to part j, with $p(\mathbf{l}_i, \mathbf{l}_j)$ measuring the compatibility of poses for the two parts and $p(\mathbf{z}|\mathbf{l}_i)$ the likelihood, and $A(i) \setminus j$ is the set of parts in the graph adjacent to i except for j. Compatibility, $p(\mathbf{l}_i, \mathbf{l}_j)$, is often measured by the physical consistency of two parts at the joint, or by their statistical (e.g., angular) co-occurrence with respect to one another. In a tree-structured PS graph, these messages are sent from the outermost extremities inward and then back outward.

Once all of the message updates are complete, the marginal posteriors for all of the parts can be estimated as

$$p(\mathbf{l}_i|\mathbf{z}) \propto p(\mathbf{z}|\mathbf{l}_i) \prod_{j\in A(i)} m_{j\to i}(\mathbf{l}_i). \quad (8)$$

Similarly, the most likely configuration can be obtained as a MAP estimate:

$$\mathbf{l}_{i,MAP} = \arg\max_{\mathbf{l}_i} p(\mathbf{l}_i|\mathbf{z}). \quad (9)$$

One of the key benefits of the pictorial structures (PS) paradigm is its simplicity and efficiency. In PS exact inference is possible in the time linear to the number of discrete configurations a given part can assume. Because of this property, recent implementations [2] can handle the pixel-dense configurations of parts that result in millions of potential discrete states for each body part. The linear complexity comes from the observation that a generally complex non-Gaussian prior over neighboring parts, $p(\mathbf{x}_i, \mathbf{x}_j)$, can be expressed as a Gaussian prior over the transformed locations corresponding to joints, mainly $p(\mathbf{x}_i, \mathbf{x}_j) = \mathcal{N}(T_{ij}(\mathbf{x}_i); T_{ji}(\mathbf{x}_j), \Sigma_{ij})$. This is done by defining a transformation, $T_{ij}(\mathbf{x}_i)$, that maps a common joint between parts i and j, defined in the part i's coordinate frame, to its location in the image space. Similarly, $T_{ji}(\mathbf{x}_j)$ defines the transformation from the same common joint defined in the j's coordinate frame to the location in the image plane. This transformation allows the inference to use an efficient solution that involves convolution (see [8] for more details).

Performance: Recently, it has been shown that the effectiveness of a PS model is closely tied to the quality of the part likelihoods [2]. Discriminatively trained models [16] and more complex appearance models [2] tend to outperform models defined by hand [8]. Methods that learn likelihood cascades, corresponding to better and better features tuned to a particular image, have also been explored for both superior speed and performance [16]. Most recent discriminative formulation of PS model allows joint learning of part appearances and model structure [26] using structural Support Vector Machine (SVM).

Speed: Cascades of part detectors serve not only to improve performance but also to speed up the inference (e.g., [23]). Fast likelihoods can be used to prune away large parts of the search space before applying more complex and computationally expensive likelihood models. Other approaches to speed up performance include data-driven methods (e.g., data-driven Belief Propagation). These methods look for the parts in an image first and then assemble a small set of the part candidates into the body (akin to the methods of [14, 17]). The problem with such approaches is that any occluded parts are missed altogether because they cannot be detected by the initial part detectors. Inference can also be sped up by using progressive search refinement methods [9]. For example, some methods use upper body detectors to restrict the search to promising parts of the image instead of searching the whole image.

Non-tree-structured extensions: Although tree-structured PS models are computationally efficient and exact, they generally are not sufficient to model all the necessary constraints imposed by the body. More complex relationships among the parts that fall outside of the realm of these models include nonpenetration constraints and occlusion constraints [20]. Incorporating such relationships into the model adds loops corresponding to long-range dependencies between body parts. These loops complicate inference because (1) no optimal solutions can be found efficiently (message-passing algorithms, like BP, are not guaranteed to converge in loopy graphs) and (2) even approximate inference is typically computationally expensive. Despite these challenges, it has been argued that adding such constraints is necessary to improve performance [4]. To alleviate some of the inference complexities with these non-tree-structured models, a number of competing methods have been introduced. Early attempts used sampling techniques from the tree-structured posterior as proposals for evaluation of a more complex non-tree-structured model [7, 8]. To obtain optimality guarantees, branch-and-bound search was recently proposed by Tian et al. [24], with the tree-structured solutions as a lower bound on the more complex loopy model energy.

Open Problems

Despite much progress in the field, pose estimation remains a challenging and still largely unsolved task. Progress has been made in estimating the configurations of mostly unoccluded and isolated subjects. Open problems include dealing with multiple, potentially interacting people (e.g., [7]) and tolerance to unexpected occlusions. Future research is also likely to expand on the types of postures and imaging conditions that the current algorithms can handle.

To date, most successful pose estimation approaches have been bottom-up. This observation applies to both discriminative approaches and part-based approaches. However, it seems shortsighted to assume that the general pose estimation problem can be solved purely in a bottom-up fashion. Top-down information may be useful for enforcing global pose consistency, and a combination of top-down and bottom-up inference is likely to lead to success faster. The recent success of combining bottom-up part-based models with 3D top-down priors [3] is encouraging and should be built upon to produce models that can deal with more complex postures and motions. Earlier attempts at building hierarchical models [27] may also be worth revisiting with the newfound insights.

Finally, there is significant evidence suggesting that successfully estimating pose independently at every frame is a very ill-posed problem. Spatio-temporal models that aggregate information over time [3] are emerging as a way to regularize performance obtained in individual frames and smooth out the noise in the estimates. Leveraging all sources of generic prior knowledge, such as spatial layout of the body and temporal consistency of poses, and rich image observation models is critical in advancing the state of the art.

References

1. Agarwal A, Triggs B (2006) Recovering 3d human pose from monocular images. IEEE Trans Pattern Anal Mach Intell 28(1):44–58
2. Andriluka M, Roth S, Schiele B (2009) Pictorial structures revisited: people detection and articulated pose estimation. In: IEEE conference on computer vision and pattern recognition (CVPR), Miami
3. Andriluka M, Roth S, Schiele B (2010) Monocular 3d pose estimation and tracking by detection. In: IEEE conference on computer vision and pattern recognition (CVPR), San Francisco
4. Bergtholdt M, Kappes J, Schmidt S, Schnorr C (2010) A study of parts-based object class detection using complete graphs. Int J Comput Vis 87(1–2):93–117
5. Bo L, Sminchisescu C (2010) Twin gaussian processes for structured prediction. Int J Comput Vis 87(1–2):28–52
6. Bo L, Sminchisescu C, Kanaujia A, Metaxas D (2008) Fast algorithms for large scale conditional 3d prediction. In: IEEE conference on computer vision and pattern recognition (CVPR), Anchorage
7. Eichner M, Ferrari V (2010) We are family: joint pose estimation of multiple persons. In: European conference on computer vision (ECCV), Heraklion
8. Felzenszwalb PF, Huttenlocher DP (2005) Pictorial structures for object recognition. Int J Comput Vis 61(1): 55–79
9. Ferrari V, Marn-Jimnez MJ, Zisserman A (2008) Progressive search space reduction for human pose estimation. In: IEEE conference on computer vision and pattern recognition (CVPR), Anchorage
10. Gall J, Rosenhahn B, Brox T, Seidel H-P (2010) Optimization and filtering for human motion capture. Int J Comput Vis 87(1–2):75–92
11. Jiang H (2009) Human pose estimation using consistent max-covering. In: IEEE international conference on computer vision, Kyoto
12. Kanaujia A, Sminchisescu C, Metaxas D (2007) Semi-supervised hierarchical models for 3d human pose reconstruction. In: IEEE conference on computer vision and pattern recognition (CVPR), Minneapolis
13. Lee MW, Cohen I (2004) Proposal maps driven mcmc for estimating human body pose in static images. In: IEEE conference on computer vision and pattern recognition (CVPR), Washington, DC
14. Mori G, Ren X, Efros A, Malik J (2004) Recovering human body configurations: Combining segmentation and recognition. In: IEEE conference on computer vision and pattern recognition (CVPR), Washington, DC
15. Navaratnam R, Fitzgibbon A, Cipolla R (2007) The joint manifold model for semi-supervised multi-valued regression. In: IEEE international conference on computer vision, Rio de Janeiro
16. Ramanan D (2006) Learning to parse images of articulated bodies. In: Neural information and processing systems, Vancouver
17. Ren X, Berg AC, Malik J (2005) Recovering human body configurations using pair-wise constraints between parts. In: International conference on computer vision, Beijing
18. Shakhnarovich G, Viola P, Darrell T (2003) Fast pose estimation with parameter sensitive hashing. In: International conference on computer vision, Nice
19. Sigal L, Balan A, Black MJ (2007) Combined discriminative and generative articulated pose and non-rigid shape estimation. In: Neural information and processing systems, Vancouver
20. Sigal L, Black MJ (2006) Measure locally, reason globally: occlusion-sensitive articulated pose estimation. In: IEEE conference on computer vision and pattern recognition (CVPR), New York
21. Sigal L, Isard M, Sigelman BH, Black MJ (2003) Attractive people: Assembling loose-limbed models using non-parametric belief propagation. In: Advances in neural information processing systems, Vancouver
22. Sigal L, Memisevic R, Fleet DJ (2009) Shared kernel information embedding for discriminative inference. In: IEEE conference on computer vision and pattern recognition, Miami

23. Singh VK, Nevatia R, Huang C (2010) Efficient inference with multiple heterogeneous part detectors for human pose estimation. In: European conference on computer vision (ECCV), Heraklion, pp 314–327
24. Tian T-P, Sclaroff S (2010) Fast globally optimal 2d human detection with loopy graph models. In: IEEE conference on computer vision and pattern recognition (CVPR), San Francisco
25. Urtasun R, Darrell T (2008) Sparse probabilistic regression for activity-independent human pose inference. In: IEEE conference on computer vision and pattern recognition (CVPR), Anchorage
26. Yang Y, Ramanan D (2011) Articulated pose estimation with flexible mixture-of-parts. In: IEEE conference on computer vision and pattern recognition (CVPR), Colorado Springs
27. Zhang J, Luo J, Collins R, Liu Y (2006) Body localization in still images using hierarchical models and hybrid search. In: IEEE conference on computer vision and pattern recognition (CVPR), New York

I

ICP

►Iterative Closest Point (ICP)

Illumination Estimation, Illuminant Estimation

Stephen Lin
Microsoft Research Asia, Beijing Sigma Center,
Beijing, China

Related Concepts

►Color Constancy; ►Incident Light Measurement

Definition

The purpose of illumination estimation is to determine the direction, intensity, and/or color of the lighting in a scene. In contrast to direct measurement of lighting, the illumination information is inferred from cues within the scene, without use of a special probe or color calibration chart.

Background

The appearance of objects and scenes can vary considerably with respect to illumination conditions. In [1], differences in face appearance due to lighting were found to be greater than those due to identity. Since such appearance variations can affect the performance of certain computer vision algorithms, much research has focused on illumination estimation, so that lighting can be accounted for in image understanding.

To simplify inference, methods for illumination estimation typically assume that the illumination originates from distant light sources. With this assumption, the illumination can be considered to be uniform across the scene, such that only a single lighting condition needs to be estimated. Most techniques perform this estimation on a single input image, as this allows for wider applicability.

Methods

Different image cues have been utilized to estimate illumination. Several techniques categorized by cue are described in the following.

Shading

Many methods for illumination estimation are based on an analysis of shading over the surface of an object. They typically utilize the relationship between shading and lighting described by the Lambertian reflectance model:

$$I(x) = \rho(x)N(x) \cdot L$$

where x indexes the shaded image pixels, I denotes image intensity (shading), ρ is the albedo, $N(x)$ is the surface normal, and L is the light vector that encodes the direction and magnitude of illumination. To solve for L, shading-based techniques for illumination estimation generally assume the surface of interest to have a uniform albedo and a known geometry. If the absolute albedo value is unknown, then L can be estimated up to an unknown scale factor.

K. Ikeuchi (ed.), *Computer Vision*, DOI 10.1007/978-0-387-31439-6,

While some methods focus on recovering only the direction of a single illuminant [2, 3], most address the more common scenario of multiple illumination sources. Hougen and Ahuja [4] solve a set of linear equations to determine light intensities from a set of sampled directions. Yang and Yuille [5] use image intensities and known surface normals at occluding boundaries to constrain illuminant directions. Ramamoorthi and Hanrahan [6] compute a low-frequency illumination distribution from a deconvolution of reflectance and lighting. Zhang and Yang [7] estimate lighting directions from critical points which have surface normals perpendicular to an illuminant direction. Based on this, Wang and Samaras [8] segmented images into regions of uniform lighting and then performed estimation by recursive least-squares fitting of the Lambertian reflectance model to these regions.

Illumination may alternatively be estimated by uncalibrated photometric stereo [9], without the need for known albedos and surface normals. This approach requires a set of images taken under different lighting conditions as input.

Cast Shadows

Several techniques analyze cast shadows for illumination estimation. For an object of known shape, the shadows that it casts provide constraints on the lighting directions and their corresponding intensities. Sato et al. [10–13] formulated these constraints as a system of equations in terms of observed brightness values within shadows and a set of sampled lighting directions at which source intensities are to be solved. These methods require a single input image for objects that cast shadows onto a uniform-colored surface; two images are needed to cancel out the effects of color variation for surfaces with texture. This approach was extended by Okabe et al. [14] to a lighting representation of Haar wavelets. Kim and Hong [15] later proposed a single-image method that handles surface texture by incorporating regularization and some user-specified information.

Specular Reflections

Some methods consider specular reflections in estimating illumination. From the locations of specular reflections on an object of known shape, these techniques compute the corresponding light source directions according to the mirror reflection property. This approach was used by Nishino et al. [16] to obtain an initial approximation of the illumination distribution, which is then refined using a more sophisticated model of reflectance. Illumination cues from specular reflections are combined with those from shading and shadows by Li et al. [17] to minimize the effects of scene texture on lighting estimation. Without needing explicit object shape recovery, Nishino et al. [18] and Wang et al. [19] proposed to estimate lighting from specular reflections on human eyes, which are highly reflective and have a similar shape from person to person.

Color

Much research focuses on estimating the color of illumination, rather than the directional distribution. Early methods solve for light color based on assumed properties of the imaged scene, such as the average color being achromatic [20] or that the scene contains a maximally reflective white patch [21]. More recent techniques are guided by more detailed knowledge about illumination and surface colors in the natural world. A statistical model of lights and surfaces from training data is used by Brainard and Freeman [22] and Finlayson et al. [23] to obtain a solution. Knowledge about illuminations and surfaces is instead used by Forsyth [24] and Finlayson et al. [25] to constrain the range of illuminant colors that could possibly result in the observed image. A comprehensive review of illumination color estimation methods is provided in [26].

Application

Illumination estimation has been employed in various applications based on appearance modeling. In [18], lighting estimates from eye reflections are used for robust face recognition under varying illumination conditions. Estimates of scene illumination have also been used to realistically composite virtual objects into an image in an illumination-consistent manner [27, 28]. In digital cameras, methods for estimating illumination color are incorporated into automatic white balance algorithms. Recent methods for estimating light color, however, have been found to be inadequate in improving color-based object recognition [29].

References

1. Moses Y, Adini Y, Ullman S (1994) Face recognition: the problem of compensating for changes in illumination direction. In: Proceedings of European conference on computer vision (ECCV). Springer, Heidelberg/Berlin, pp 286–296
2. Zheng Q, Chellappa R (1991) Estimation of illuminant direction, albedo, and shape from shading. IEEE Trans Pattern Anal Mach Intell 13:680–702
3. Samaras D, Metaxas D (1999) Coupled lighting direction and shape estimation from single images. In: Proceedings of the international conference on computer vision. IEEE Computer Society, Washington, DC, pp 868–874
4. Hougen DR, Ahuja N (1993) Estimation of the light source distribution and its use in integrated shape recovery from stereo and shading. In: Proceedings of the international conference on computer vision. IEEE Computer Society, Washington, DC, pp 148–155
5. Yang Y, Yuille AL (1991) Sources from shading. In: Proceedings of the IEEE conference on computer vision and pattern recognition (CVPR). IEEE Computer Society, Washington, DC, pp 534–539
6. Ramamoorthi R, Hanrahan P (2001) A signal-processing framework for inverse rendering. In: Proceedings of ACM SIGGRAPH. ACM, New York, pp 117–128
7. Zhang Y, Yang YH (2001) Multiple illuminant direction detection with application to image synthesis. IEEE Trans Pattern Anal Mach Intell 23:915–920
8. Wang Y, Samaras D (2002) Estimation of multiple illuminants from a single image of arbitrary known geometry. In: Proceedings of European conference on computer vision (ECCV). Lecture notes in computer science, vol 2352. Springer, Berlin/Heidelberg, pp 272–288
9. Basri R, Jacobs D, Kemelmacher I (2007) Photometric stereo with general, unknown lighting. Int J Comput Vis 72:239–257
10. Sato I, Sato Y, Ikeuchi K (1999) Illumination distribution from brightness in shadows: adaptive estimation of illumination distribution with unknown reflectance properties in shadow regions. In: Proceedings of the international conference on computer vision. IEEE Computer Society, Washington, DC, pp 875–883
11. Sato I, Sato Y, Ikeuchi K (1999) Illumination distribution from shadows. In: Proceeding of the IEEE Conference on computer vision and pattern recognition (CVPR). IEEE Computer Society, Washington, DC, pp 306–312
12. Sato I, Sato Y, Ikeuchi K (2001) Stability issues in recovering illumination distribution from brightness in shadows. Proc IEEE Conf Comput Vis Pattern Recognit (CVPR) II:400–407
13. Sato I, Sato Y, Ikeuchi K (2003) Illumination from shadows. IEEE Trans Pattern Anal Mach Intell 25:290–300
14. Okabe T, Sato I, Sato Y (2004) Spherical harmonics vs. haar wavelets: basis for recovering illumination from cast shadows. Proc IEEE Conf Comput Vis Pattern Recognit (CVPR) I:50–57
15. Kim T, Hong K (2005) A practical single image based approach for estimating illumination distribution from shadows. In: Proceedings of the international conference on computer vision. IEEE Computer Society, Washington, DC, pp 266–271
16. Nishino K, Zhang Z, Ikeuchi K (2001) Determining reflectance parameters and illumination distribution from sparse set of images for viewdependent image synthesis. In: Proceedings of the international conference on computer vision. IEEE Computer Society, Washington, DC, pp 599–606
17. Li Y, Lin S, Lu H, Shum HY (2003) Multiple-cue illumination estimation in textured scenes. In: Proceedings of the international conference on computer vision. IEEE Computer Society, Washington, DC, pp 1366–1373
18. Nishino K, Belhumeur P, Nayar S (2005) Using eye reflections for face recognition under varying illumination. Proc Int Conf Comput Vis I:519–526
19. Wang H, Lin S, Liu X, Kang SB (2005) Separating reflections in human iris images for illumination estimation. In: Proceedings of the international conference on computer vision. IEEE Computer Society, Washington, DC, pp 1691–1698
20. Buchsbaum G (1980) A spatial processor model for object colour perception. J Franklin Inst 310:1–26
21. Land EH (1977) The retinex theory of color vision. Sci Am 237:108–128
22. Brainard DH, Freeman WT (1997) Bayesian color constancy. J Opt Soc Am A 14:1393–1411
23. Finlayson GD, Hordley S, Hubel PM (2001) Color by correlation: a simple, unifying framework for color constancy. IEEE Trans Pattern Anal Mach Intell 23:1209–1221
24. Forsyth DA (1990) A novel algorithm for colour constancy. Int J Comput Vis 5:5–36
25. Finlayson GD, Hordley S, Tastl I (2006) Gamut constrained illuminant estimation. Int J Comput Vis 67:93–109
26. Ebner M (2007) Color constancy. Wiley, Chichester
27. Sato I, Sato Y, Ikeuchi K (1999) Acquiring a radiance distribution to superimpose virtual objects onto a real scene. IEEE Trans Vis Comput Graph 5:1–12
28. Lalonde JF, Efros AA, Narasimhan SG (2009) Estimating natural illumination from a single outdoor image. In: Proceedings of the international conference on computer Vision. IEEE Computer Society, Washington, DC
29. Funt B, Barnard K, Martin L (1998) Is colour constancy good enough? In: Proceedings of the European conference on computer vision (ECCV). Springer, London, pp 445–459

Image Alignment

►Image Registration

Image Decompositions

Marshall F. Tappen
University of Central Florida, Orlando, FL, USA

Definition

An image decomposition is the result of a mathematical transformation of an image into a new set of images that represent different aspects of the input image or scene pictured in that image. The original image can typically be reconstructed from these new images.

Background

While images are primarily stored as an array of pixel values, an image can be represented in a number of different ways. For instance, an image can be easily transformed into two images, one containing the high-frequency variation in the input image and a second containing the low-frequency variation. This process decomposes the input image into two images, each of which expresses different information about the original image.

This process is useful when further processing will treat these two images differently. If the decomposition is chosen correctly, the image is decomposed into a set of images that can each be processed uniformly. Thus, the decomposition facilitates adaptive processing of the content of an image.

Theory

Image decompositions can be roughly divided into two different types of decompositions, image-based decompositions and intrinsic image decompositions. Image-based decompositions represent the image itself using new images, while the intrinsic image decompositions reflect the content of the scene pictured in the image itself.

Image-Based Decompositions

Similar to the background example above, many image-based decompositions focus on representing multi-scale frequency content in the scene. The Gaussian pyramid is one of the most basic decompositions representing multi-scale content. The decomposition consists of a set of images of progressively smaller resolution, with each image being one level of the pyramid. Each level is created by filtering the image at the level below, then downsampling the result. This creates a multi-resolution set of images.

Depending on the application, the usefulness of the Gaussian pyramid may be limited because each level contains redundant information. This can be eliminated by modifying the pyramid creation process to create a Laplacian pyramid [1]. In the Laplacian pyramid, the input image is progressively downsampled. The image at level i in the Laplacian pyramid is computed by taking the difference between the ith level of the Gaussian pyramid and the upsampled version of level $i + 1$, which has been downsampled from the ith level of the Gaussian pyramid. Effectively, each level of the Laplacian pyramid expresses the image information at a particular scale. Figure 1 shows an example of the Laplacian decomposition of an image.

In [2], Simoncelli et al. extended this decomposition process to also separate orientation into different images, creating the steerable pyramid decomposition. Similar decompositions can also be generated by using a different process to separate the images. In [3], the bilateral filter is used to generate a two-image decomposition.

These decompositions are also connected to other image transformations, particularly wavelets. The connections are discussed in [2].

Intrinsic Image Decompositions

While image-based decompositions are focused on the pixel values themselves, intrinsic image decompositions create images that are based on the content of the scene. The intrinsic image decomposition is based on the intrinsic image approach for representing scene characteristics. In this approach, each intrinsic characteristic of the scene is represented by a distinct image. In the intrinsic image decomposition, these images are chosen to both represent intrinsic characteristics and image content.

In [4], Weiss uses video data to separate an image into illumination, or shading, and albedo components. In this decomposition, an input image pixel at location n, $I(n)$, is equal to the product of a shading image and an albedo image, or $I(n) = S(n) \times A(n)$. In [5] and [6], Tappen et al. show how the intrinsic image decomposition can be computed from a single image. Figure 2 shows an example of an intrinsic image decomposition for the image on the left.

Application

Image decompositions are frequently used to generate images that processed separately. In [7], Portilla et al. use the steerable pyramid to denoise images. Heeger and Bergen showed that texture can be generated by

Image Decompositions, Fig. 1 These images are a Laplacian pyramid created from the well-known *Lena* image. Each image captures the variation at a specific scale

Image Decompositions, Fig. 2 An example of an intrinsic image decomposition. The image on the *left* is decomposed into shading and albedo components

forcing the marginal histograms of the levels of a steerable pyramid to match those of a pyramid generated from a reference image [8]. More complete measures of statistical similarity are used in [9], leading to improved synthesis results.

As mentioned earlier, the bilateral filter is used in [3] to separate the image into large- and fine-scale variations to combine images taken under different illumination. In [10], Bousseau et al. describe how user input can improve intrinsic image decompositions and demonstrate how they can be applied for graphics applications.

References

1. Burt P, Adelson E (1983) The Laplacian pyramid as a compact image code. IEEE Trans Commun 31(4):532–540
2. Simoncelli EP, Freeman WT, Adelson EH, Heeger DJ (1992) Shiftable multiscale transforms. IEEE Trans Inf Theory 38(2):587–607
3. Eisemann E, Durand F (2004) Flash photography enhancement via intrinsic relighting. ACM Trans Graph 23:673–678
4. Weiss Y (2001) Deriving intrinsic images from image sequences. In: IEEE international conference on computer vision, Vancouver, vol 2, p 68
5. Tappen MF, Freeman WT, Adelson EH (2005) Recovering intrinsic images from a single image. IEEE Trans Pattern Anal Mach Intell 27(9):1459–1472

6. Tappen MF, Adelson EH, Freeman WT (2006) Estimating intrinsic component images using non-linear regression. In: The proceedings of the 2006 IEEE computer society conference on computer vision and pattern recognition (CVPR), New York, vol 2, pp 1992–1999
7. Portilla J, Strela V, Wainwright M, Simoncelli E (2003) Image denoising using scale mixtures of gaussians in the wavelet domain. IEEE Trans Image Process 12(11): 1338–1351
8. Heeger DJ, Bergen JR (1995) Pyramid-based texture analysis/synthesis. In: Proceedings of ACM SIGGRAPH 1995. ACM, New York, pp 229–238
9. Portilla J, Simoncelli EP (2000) A parametric texture model based on joint statistics of complex wavelet coefficients. Int J Comput Vis 40(1):49–70
10. Bousseau A, Paris S, Durand F (2009) User assisted intrinsic images. ACM Trans Graph (Proceedings of SIGGRAPH Asia 2009) 28(5)

Image Enhancement and Restoration

Guoshen Yu[1] and Guillermo Sapiro[2]
[1]Electrical and Computer Engineering, University of Minnesota, Minneapolis, MN, USA
[2]Electrical and Computer Engineering, Computer Science, and Biomedical Engineering, Duke University, Durham, NC, USA

Synonyms

Image inverse problems

Related Concepts

▶Denoising; ▶Image-Based Modeling; ▶Inpainting

Definition

Image enhancement and restoration is a procedure that attempts to improve the image quality by removing the degradation while preserving the underlying image characteristics.

Background

Image quality is often deteriorated during acquisition, compression, and transmission. Typical degradations include image blur introduced by lens out-of-focus, resolution downgrade due to acquisition equipment pixel limitation, noise spots introduced at high ISO, and JPEG block artifact, as illustrated in Fig. 1. Image enhancement and restoration is a procedure that attempts to improve the image quality by removing the degradation while preserving the underlying image characteristics. For some specific degradations as mentioned above, image enhancement and restoration is also known as deblurring, super-resolution zooming, denoising, and deblocking. While jointly addressed here and in most of the literature, *restoration* often refers to the case where one attempts to mathematically invert the degradation (e.g., invert the blurring filter), and *enhancement* refers to the improvement of the overall image quality without explicit mathematical inversion of the degradation process.

Theory

The problems of image enhancement and restoration are ill posed since they amount to recovering some image information that has been eliminated during the degradation. Solving these problems must therefore rely on some prior knowledge of the image, or in mathematical terms *image models*, to regularize the solution. Mathematically, let f denote an ideal image, U a linear degradation operator, w an additive noise, and

$$y = Uf + w \tag{1}$$

the degraded (observed) image. While this model does not cover all possible degradation scenarios, it is very popular and useful, and serves to illustrate the underlying image enhancement and restoration key concepts. Modern image enhancement and restoration estimates the underlying image f from the degraded observation y by, for example, minimizing a functional of the form

$$\hat{f} = \arg\min_h \left(\|y - Uh\|^2 + \varphi(h) \right), \tag{2}$$

where the first term ensures that the restored image $\hat{f}$ and the degraded image y agree with the image degradation (Eq. 1), and the second term $\varphi(h)$ regularizes the solution via a certain image model. The technology of image enhancement and restoration thus has been developed hand in hand with a better understanding of image modeling.

Image Enhancement and Restoration, Fig. 1 From *left* to *right*. Ideal image, image degraded by out-of-focus, resolution downgrade, noise spots, and JPEG block artifact

Image Enhancement and Restoration, Fig. 2 *Left*: noisy image. *Right*: image denoised by Gaussian smoothing

The most classic image model that dated from the 1960s assumes that image content is uniformly smooth [10]. This model results in a number of well-known image enhancement and restoration algorithms, including Gaussian smoothing for denoising, bicubic interpolation for zooming, and Wiener filter for deblurring [12]. All these algorithms are implemented with *linear* filtering uniformly applied over the image, typical isotropic local filters smoothing out the image. While the uniformly smooth assumption holds on regular image regions such as sky or a blackboard, that typically dominates a natural image, it is obviously oversimplified on other important types of image transition structures, such as contours, that are smooth along one direction but not the other, and textures that are oscillatory patterns. As shown in Fig. 2, although image noise is attenuated, image contours become blurred at the end of restoration when this simple uniformly smooth model is assumed.

Anisotropic image models attempting to address this problem came into the scene in the early 1990s (with some works dating to the 1960s as well, by Gabor). As opposed to the uniformly smooth assumption, the anisotropic models assume that an image is piecewise smooth, in other words, smooth inside each sub-region, and that at a contour or boundary of the regions where the image intensity sharply changes, the smoothness holds only along the contour direction but not in the perpendicular direction. These models give clearly a better image description and have been elegantly formulated in some partial differential equation frameworks such as anisotropic diffusion [7, 15] and total variation [16]. The resulting algorithms implement *nonlinear* filtering adaptive to the image content, uniformly smoothing inside each image sub-region, and smoothing only along the contour direction on the region boundaries. Therefore, image contours are better preserved.

Since the boom of wavelets in the early 1990s, multi-resolution harmonic analysis has lead to considerable efforts and improvements on image modeling and restoration [4, 12]. Wavelet analysis models an image from multiple resolutions; at each resolution, translating local wavelet atoms oscillating at the corresponding scale are used. The wavelet response is typically high on image transition structures, such as contours and textures, and negligible on regular regions. As a result, it does not only implement *nonlinear* adaptive filtering, but also reveals the important concept of "sparse modeling": wavelet analysis represents an image with only a few large wavelet coefficients that absorb most of the image energy, while the majority of wavelet coefficients quickly decay to zero. The wavelet's sparsity as well as its performance in image enhancement and restoration have been later improved by geometric adaptive harmonic analysis such as curvelets [5] that include local directional atoms to catch the image contours.

In order to further promote the resulting sparsity relative to prefixed harmonic analysis dictionaries (Dictionary here means an ensemble of harmonic analysis atoms), such as wavelets or curvelets, *sparsifying learned dictionaries*, i.e., dictionaries that are learned

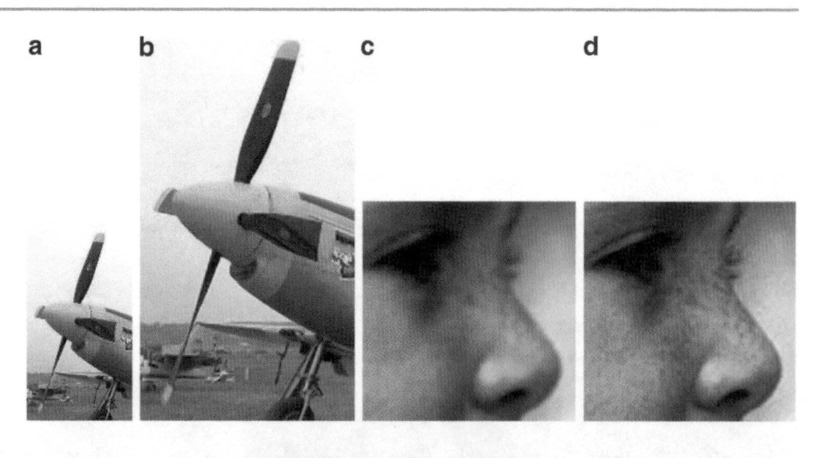

Image Enhancement and Restoration, Fig. 3 Image enhancement and restoration examples. (**a**) and (**b**) Super-resolution zooming: Low-resolution and zoomed images. (**c**) and (**d**) Deblurring: blurred and deblurred images (Figures reproduced from [17])

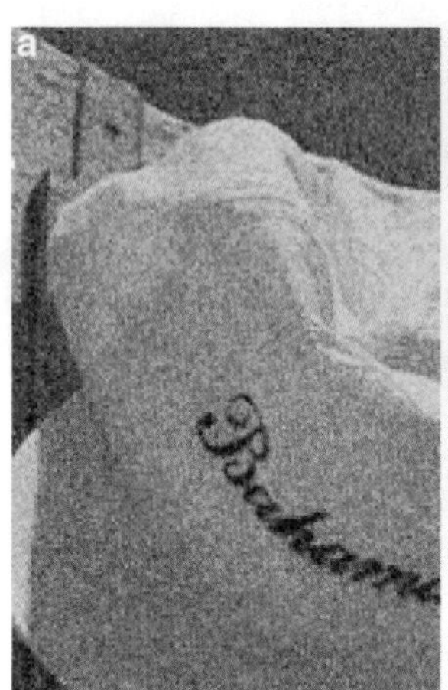

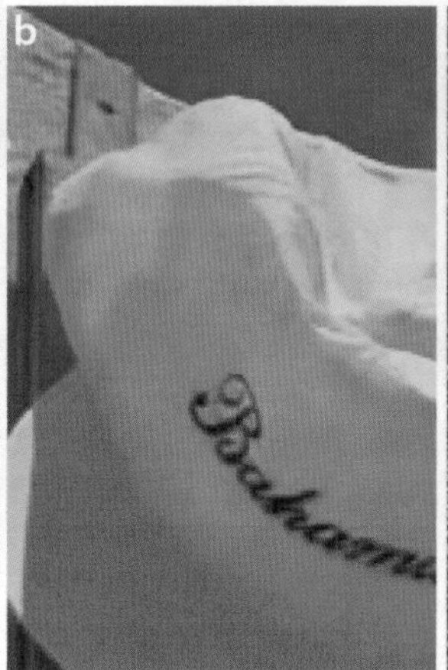

Image Enhancement and Restoration, Fig. 4 Image enhancement and restoration examples. (**a**) and (**b**) Denoising: Noisy and denoised images. (**c**) and (**d**) Inpainting: image with 80 % random missing pixels and restore image. (This problem is related to the task of reconstructing a color image from undersampled color channels, as present in most low/mid, end digital cameras.) The results are obtained following the technique in [17]

from images of interest to yield sparse representations for that class of data, have emerged [1, 13], leading to further improved image enhancement and restoration performance [11].

Non-local image modeling is based on the observation that images typically contain repetitive local patterns (self-similarity). Since the pioneering work of the nonlocal means denoising algorithm [3] in 2005 (see also [2, 14]), non-local modeling has been extensively studied in image enhancement and restoration [8, 9].

Gaussian mixture models, a statistical model widely applied in machine learning, have been shown particularly effective for image enhancement and restoration [17]. The models assume that local image patches follow a mixture of Gaussian distributions. The resulting *piecewise linear* algorithm is not only extremely fast, but also reveals some connections to sparse modeling and non-local modeling.

State-of-the-art image enhancement and restoration results are obtained with algorithms derived from the last three image models, namely, sparse modeling with learned dictionaries, non-local modeling, and Gaussian mixture models. Figure 3 illustrates some examples.

Open Problems

For image enhancement and restoration problems such as removing Gaussian white noise from an image and filling small holes at random positions in an image, it seems that the current performance is already acceptable, as illustrated in Fig. 4, and has arguably reached a quality boundary uneasy to go beyond. For other more difficult problems such as deblurring and zooming, although substantial visual quality improvement has been achieved with respect to classic algorithms such as Wiener filter and bicubic interpolation, objective performance improvement is relatively limited despite considerable efforts that have been devoted. Theoretical performance bounds of image enhancement and restoration remains to be understood. The recent very exciting compressive sensing theory [6] reveals the performance bounds of the sparse modeling approaches given some random degradation operations, but is inapplicable to typical degradations such as blurring and subsampling and to the most successful learned dictionaries. The extension of these results to more realistic image degradation scenarios and image models is among the current challenges of image restoration and enhancement.

References

1. Aharon M, Elad M, Bruckstein A (2006) K-SVD: an algorithm for designing overcomplete dictionaries for sparse representation. IEEE Trans Signal Process 54(11):4311
2. Awate SP, Whitaker RT (2005) Higher-order image statistics for unsupervised, information-theoretic, adaptive, image filtering. In: Proceedings of conference on computer vision and pattern recognition (CVPR), vol 2, San Diego, pp 44–51
3. Buades A, Coll B, Morel JM (2006) A review of image denoising algorithms, with a new one. Multiscale Modeling Simul 4(2):490–530
4. Burt P, Adelson E (1983) The Laplacian pyramid as a compact image code. IEEE Trans Commun 31(4):532–540
5. Candes EJ, Donoho DL (2004) New tight frames of curvelets and optimal representations of objects with C2 singularities. Commun Pure Appl Math 56:219–266
6. Candès EJ, Tao T (2006) Near-optimal signal recovery from random projections: Universal encoding strategies? IEEE Trans Inf Theory 52(12):5406–5425
7. Catté F, Lions PL, Morel JM, Coll T (1992) Image selective smoothing and edge detection by nonlinear diffusion. SIAM J Numer Anal 29(1):182–193
8. Dabov K, Foi A, Katkovnik V, Egiazarian K (2007) Image denoising by sparse 3-D transform-domain collaborative filtering. IEEE Trans Image Process 16(8):2080–2095
9. Gilboa G, Osher S (2008) Nonlocal operators with applications to image processing. Multiscale Modeling Simul 7(3):1005–1028
10. Lindenbaum M, Fischer M, Bruckstein A (1994) On Gabor's contribution to image enhancement. Pattern Recognit 27(1):1–8
11. Mairal J, Elad M, Sapiro G (2007) Sparse representation for color image restoration. IEEE Trans Image Process 17(1):53–69
12. Mallat S (2009) A wavelet tour of signal processing: the sparse way, 3rd edn. Academic, Burlington
13. Olshausen BA, Field DJ (1996) Natural image statistics and efficient coding*. Netw Comput Neural Syst 7(2):333–339
14. Ordentlich E, Seroussi G, Verdu S, Weinberger M, Weissman T (2003) A discrete universal denoiser and its application to binary images. In: Proceedings of international conference on image processing, vol 1, Barcelona
15. Perona P, Malik J (1990) Scale-space and edge detection using anisotropic diffusion. IEEE Trans Pattern Anal Mach Intell 12(7):629–639
16. Rudin LI, Osher S, Fatemi E (1992) Nonlinear total variation based noise removal algorithms. Phys D 60(1–4): 259–268
17. Yu G, Sapiro G, Mallat S (2010) Solving inverse problems with piecewise linear estimators: from Gaussian mixture models to structured sparsity. Image Processing, IEEE Transactions 2481–2499

Image Inverse Problems

▶Image Enhancement and Restoration

Image Mosaicing

▶Image Stitching

Image Plane

Peter Sturm
INRIA Grenoble Rhône-Alpes, St Ismier Cedex, France

Synonyms

Retina

Related Concepts

▶Pinhole Camera Model

Definition

The image plane is the planar surface on which the image is generated in an image formation process or a model thereof.

Background

In most cameras, the photosensitive elements are arranged on a planar support. In image formation models, the image plane is the (mathematical) plane where the image is formed and within which pixels or film are supposed to be located.

There exist cameras where the photosensitive area is not flat. For instance, in most early panoramic image acquisition systems that proceeded by scanning a scene with a rotating slit camera, the film was wrapped onto the inside of a cylindrical surface [1, 2]. In that case, one may still devise an equivalent theoretical image formation model that has a planar image support surface.

References

1. McBride B (2011) A timeline of panoramic cameras. http://www.panoramicphoto.com/timeline.htm Accessed 3 August 2011
2. Benosman R, Kang S (2001) A brief historical perspective on panorama. In: Benosman R, Kang S (eds): Panoramic vision: sensors, theory, and applications. Springer, Verlag, pp 5–20

Image Registration

Daniel C. Alexander
Centre for Medical Image Computing, Department of Computer Science, University College London, London, UK

Synonyms

Image alignment

Definition

Image registration aligns corresponding features of images via spatial transformations.

Background

Computer vision or image processing systems often need to align multiple images of the same or similar scenes. In medical imaging, for example, radiologists routinely compare images of a patient acquired at different times to monitor changes. The intensity difference between two images highlights such changes but only if the corresponding features are in the same location. However, patients' positions in imaging devices vary between visits, so raw images never have perfect alignment. Image registration transforms or warps one image so that the important objects and regions are in the same position as in the other image. The difference image then reveals intrinsic physical changes. Figure 1 illustrates the idea. The problem becomes more challenging when the images come from different devices (inter-modality registration) or from different subjects (intersubject registration).

The same problem arises in nonmedical imaging applications. Surveillance systems, for example, often need to look for differences between images at different times, for example, to subtract the background and highlight activity in a scene viewed by a security camera. Fixed cameras can wobble in the wind and produce misaligned images that require registration before the difference image provides a meaningful result. Stitching images together to create panoramas [1–3] also requires image registration to align the overlapping parts of the images being stitched together; Fig. 2 illustrates this application. Similarly, super-resolution techniques [4] align multiple images of the same scene and infer subpixel detail.

Theory

The process of automatic image registration involves optimizing a cost function, which expresses the similarity of the two images, with respect to the parameters of a transformation of one of the images. Mathematically, the optimization problem is

$$\{T^\star, g^\star\} = \underset{\{T,g\}}{\operatorname{argmin}}(f(I_1, T(g(I_2)))), \qquad (1)$$

where I_1 is the target image, which is fixed; I_2 is the source image, which the transformations T and g act upon; T is a spatial transformation, or warp; and g affects only the image intensity at each pixel position; the optimization seeks the transformations $T^\star$ and $g^\star$ that minimize the cost function f.

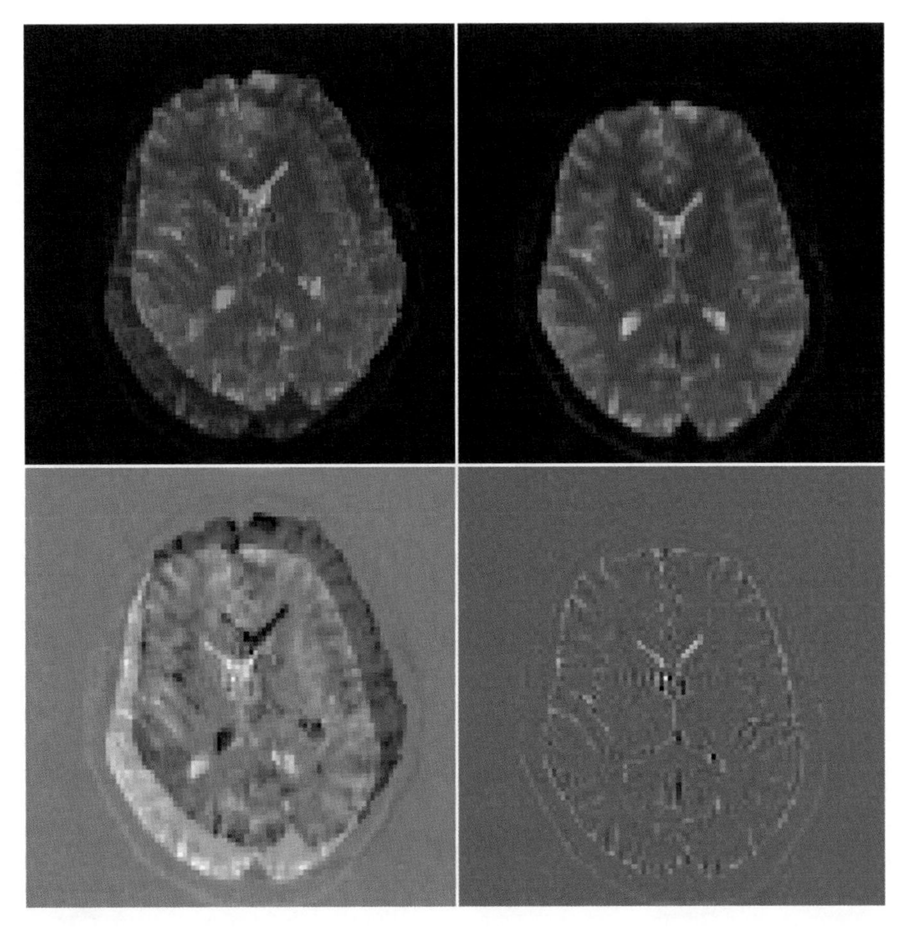

Image Registration, Fig. 1 Intrasubject brain image registration. *Top left*: overlaid images of the same brain from different acquisitions, one in *red* and one in *green*; *bottom left*: difference image of the two unaligned images; *top right*: overlaid images after registration via a rigid transformation; *bottom right*: difference image after alignment

The process decomposes into four key components, which the following subsections discuss one by one. The literature contains many review papers, for example, [5–8], that discuss each component in more detail.

Features

Various image features can drive the registration process. Broadly, the feature set is either sparse or dense.

Sparse feature sets consist of geometric features identified in the image through some preprocessing step. These features might be salient points identified by a user or by an automatic detector; SIFT features [9] or variants thereof are a common choice. Features may also be more complex geometric objects, such as salient lines, curves, surfaces, or regions.

Dense features are typically pixel-by-pixel image intensities. Each feature may be a single scalar intensity or may have multiple components, as in multispectral images. Dense feature sets may not include every pixel in the image and often exclude pixels that lie outside the salient region of the image, such as the brain region in Fig. 1.

Cost Function

The cost function provides a measurement of similarity between two images. The definition of similarity depends on the set of features.

For sparse feature sets, the cost function typically uses a measure of distance between matched features in the two images. For example, if the feature set is a list of salient points in each image, the Euclidean distance between each corresponding pair of points

Image Registration, Fig. 2 Image registration for stitching. The panoramic image at the *top* comes from stitching together various images including the two at the *bottom*. Image registration provides the spatial transformation that associates corresponding salient points in the two images, such as those marked by the *green arrows*

provides a measure of similarity. This requires a preceding step to establish correspondence between pairs of points, in a similar way to various other computer vision tasks, such as stereo matching.

Registration based on dense features, that is, pixel intensities, typically uses statistical measures of similarity between pixel intensities in corresponding locations. The most direct measure of similarity uses the average intensity difference

$$f_1(I_1, I_2) = -\sum_{x \in X} |I_1(x) - I_2(x)|, \tag{2}$$

where X is the salient set of pixels. Equation 2 uses the L_1 norm, but other norms are equally possible.

Direct intensity comparisons, as in f_1, assume that the pixel intensity at corresponding locations is the same subject to some noise perturbation. However, that assumption often does not hold. For example, differences in intensity scale between images arise frequently. Where such intensity differences are likely, similarity measures based on the correlation of pixel intensities between the two images are more appropriate. For example, the normalized cross correlation

$$f_2(I_1, I_2) = \frac{\sum_{x \in X}(I_1(x) - \bar{I}_1)(I_2(x) - \bar{I}_2)}{\sqrt{\sum_{x \in X}(I_1(x) - \bar{I}_1)^2 \sum_{x \in X}(I_2(x) - \bar{I}_2)^2}}, \tag{3}$$

where $\bar{I}$ is the mean intensity of image I over region X.

The mapping between intensities of corresponding pixels is sometimes more complex than a simple scale change. It may be nonlinear and non-monotonic. For example, in inter-modality medical image registration, two images of the same object may have the same regional structure but different regional contrast: image 1 has higher intensity than image 2 in some regions, vice versa in others, and intensity correlation at alignment remains low. Entropy-based similarity measures [10–13] provide a useful alternative. A common choice is the normalized mutual information

$$f_3(I_1, I_2) = \frac{H(I_1) + H(I_2)}{H(I_1, I_2)}, \tag{4}$$

where

$$H(I) = -\sum_{x \in X} p(I(x)) \log p(I(x)) \quad (5)$$

is the entropy of image I, with p the distribution of image intensities in I, and

$$H(I_1, I_2) = -\sum_{x \in X} p(I_1(x), I_2(x)) \log p(I_1(x), I_2(x)) \quad (6)$$

is the joint entropy of images I_1 and I_2, with p now the joint intensity distribution. The cost function f_3 is minimum when one image predicts the other most parsimoniously, that is, when the intensity mapping from I_1 to I_2 requires the least information to describe. Mutual-information-based cost functions prove remarkably effective and robust. In practice, they are often preferred to direct comparison or correlation-based cost functions, such as f_1 and f_2, even for intrasubject intra-modality registration.

Transformation Model

A variety of models are available for the spatial transformation, T. Simple transformations, such as rigid, affine, or polynomial transformations, are global in the sense that even well-separated pixels undergo highly correlated displacements. More complex models, such as spline [14], radial-basis-function [15], elastic [16], or fluid [17] transformations, can have more local properties, so that the displacement of one pixel under the transformation correlates only with that of proximal pixels.

In some applications, simple global transformations are sufficient. For example, in brain imaging, rigid transformations are often sufficient to align two images from the same subject. Since a rigid skull encases the brain, it deforms very little between image acquisitions. The registration needs to correct only for the difference in position and orientation of the subject in the imaging device. Higher-order global transformations, such as full affine or polynomial transformations, can improve alignment significantly even when the physical transformation is rigid, because they can capture artifactual distortions introduced by the image device. Image stitching often uses a homography, which is a global transformation that accounts for changes in perspective.

Local transformations can capture more subtle changes between images. They are essential, for example, for detecting and quantifying local atrophy (shrinkage) of brain tissue that occurs over time in various neurological conditions [18]. In general, in medical imaging, local transformations are usually necessary for good alignment in intersubject image registration, where local variations in size and shape of organs and body structure arise.

In practice, the intensity transformation, g in Eq. 1, is often the identity. However, g becomes important in images that contain more complex information at each pixel than single or multiple scalar values. For example, vector or tensor images are common in remote sensing and medical imaging. In such images, each pixel has an associated orientation. Nontrivial g is essential to ensure that local orientations remain consistent with the image structure through the spatial transformation; see, for example, [19, 20].

Optimization

The wide range of optimization algorithms available today, from simple line search or gradient descent to stochastic and genetic optimization procedures, provides many candidates for driving the minimization of the cost function that solves the registration problem. The choice of optimization procedure depends on the feature set. Registration via sparse feature matching often relies on algorithms like RANSAC [21], which are robust to errors in point correspondence, whereas most image registration algorithms with dense features use some form of gradient descent. The cost function is almost always non-convex, and an effective optimization procedure for reliable image registration cannot ignore local minima. Even for simple rigid transformations, local minima often arise and reliable rigid registration with gradient descent requires repeated runs with multiple starting points [22]. The optimization problem tends to become harder the more complex the transformation model. In local registration, the optimization has a much larger number of parameters, so takes longer, and repeated runs can be impractical. Hierarchical approaches, which start with a simple global registration to get a good starting point and gradually add parameters and reoptimize, are common to obtain a good local registration. Multi-resolution strategies, which start with low-resolution images and gradually increase resolution, also help.

Application

The medical imaging community is a large consumer and developer of image registration techniques. Intrasubject image registration enables fusion of information in images from different devices. Intrasubject registration also enables tracking of changes over time during development or disease. In drug trials, for example, imaging offers the potential to observe the effects of a prospective treatment and establish its efficacy noninvasively; image registration is essential for monitoring such effects.

Another major application is spatial normalization for group studies, which study the variation in the size, shape, and internal organization of a particular organ or object. A common application is human brain mapping where morphological variability is well studied in a range of conditions. Intersubject image registration ensures that a collection of similar images are in the same spatial frame of reference so that studies of variation are meaningful. This spatial normalization allows, for example, medical imaging researchers to characterize differences in organ size, shape, and structure between different populations, such as normal healthy adults and patients with a certain condition.

Image registration for image stitching enables day-to-day image processing for digital camera users, as standard packages like Photoshop include such operations. Google Maps is a large-scale application of the same technology.

Many implementations of image registration are freely available. Tried and tested global registration software includes the FLIRT package [22] and Nifty Reg [23]. The popular b-spline registration algorithm [14] has implementations with some variations in FNIRT [24] and Nifty Reg [25], which also offers a GPU implementation. The DARTEL package [26] is designed specifically for spatial normalization of large brain image ensembles. The recent ANTS package [27] combines several state-of-the-art ideas and performs well in a head-to-head evaluation with other standard packages [28].

Open Problems

Consistency remains an open problem in image registration. Basic algorithms do not ensure that the transformation from registering images A to B is the perfect inverse of that from registering images B to A. Symmetry constraints on the cost function to ensure binary consistency are straightforward to enforce. However, the problem becomes more complex as the number of images to align increases: ensuring consistency of A to B to C with C to A is more challenging. Groupwise registration, as in [26], goes some way towards ameliorating this problem.

Topological differences or changes present a further open challenge. Most transformation models do not accommodate differences in topology naturally between images. In fact, significant effort has gone into developing diffeomorphic transformation models that cannot fold or tear. However, topological differences arise frequently. In intersubject medical image registration, for example, it is not uncommon for an anatomical feature in one person to be entirely missing in another. The same problem can arise even intrasubject, say, before and after surgery to remove a tumor.

References

1. Szeliski R, Shum HY (1997) Creating full view panoramic image mosaics and environment maps. In: Proceedings of the 24th annual conference on computer graphics and interactive techniques. SIGGRAPH '97, New York, NY, USA. ACM/Addison-Wesley, pp 251–258
2. Szeliski R (2006) Image alignment and stitching: a tutorial. Found Trends Comput Graph Vis 2:1–104
3. Brown M, Lowe D (2007) Automatic panoramic image stitching using invariant features. Int J Comput Vis 74: 59–73. doi:10.1007/s11263-006-0002-3
4. Irani M, Peleg S (1991) Improving resolution by image registration. CVGIP 53(3):231–239
5. Maintz J, Viergever MA (1998) A survey of medical image registration. Med Image Anal 2(1):1–36
6. Lester H, Arridge SR (1999) A survey of hierarchical non-linear medical image registration. Pattern Recognit 32(1):129–149
7. Hajnal JV, Hill DLG, Hawkes DJ (2001) Medical image registration. CRC Press, Boca Raton. ISBN 0-8493-0064-9
8. Zitov B, Flusser J (2003) Image registration methods: a survey. Image Vis Comput 21(11):977–1000
9. Lowe DG (2004) Distinctive Image Features from Scale-Invariant Keypoints. Int J Comput Vision 60(2): 91–110
10. Studholme C, Hill DLG, Hawkes DJ (1995) Multiresolution voxel similarity measures for mr-pet registration. In: Proceedings of the information processing in medical imaging, Kluwer, pp 287–298
11. Collignon A, Maes F, Delaere D, Vandermeulen D, Suetens P, Marchal G (1995) Automated multi-modality image registration based on information theory. In: Proceedings of the information processing in medical imaging, Kluwer, pp 263–274

12. Viola P, Wells WM (1995) Alignment by maximization of mutual information. In: Proceedings of the international conference on computer vision, IEEE, pp 16–23
13. Pluim JPW, Maintz JBA, Viergever MA (2003) Mutual information based registration of medical images: a survey. IEEE Trans Med Imaging 22:986–1004
14. Rueckert D, Sonoda LI, Hayes C, Hill DLG, Leach MO, Hawkes DJ (1999) Non-rigid registration using free-form deformations: application to breast mr images. IEEE Trans Med Imaging 18:712–721
15. Fornefett M, Rohr K, Stiehl HS (2001) Radial basis functions with compact support for elastic registration of medical images. Image Vis Comput 19:87–96
16. Bajcsy R, Kovacic S (1989) Multiresolution elastic matching. Comput Vis Graph Image Process 46(1):1–21
17. Crum WR, Scahill RI, Fox NC (2001) Automated hippocampal segmentation by regional fluid registration of serial MRI: validation and application in alzheimer's disease. NeuroImage 13(5):847–855
18. Scahill RI, Schott JM, Stevens JM, Rossor MN, Fox NC (2002) Mapping the evolution of regional atrophy in alzheimer's disease: unbiased analysis of fluid-registered serial mri. Proc Natl Acad Sci 99:4703–4707
19. Alexander DC, Pierpaoli C, Basser PJ, Gee JC (2001) Spatial transformations of diffusion tensor magnetic resonance images. IEEE Trans Med Imaging 20:1131–1139
20. Zhang H, Yushkevich PA, Alexander DC, Gee JC (2006) Deformable registration of diffusion tensor mr images with explicit orientation optimization. Med Image Anal 10: 764–785
21. Fischler MA, Bolles RC (1981) Random sample consensus: a paradigm for model fitting with applications to image analysis and automated cartography. Commun ACM 24:381–395
22. Jenkinson M, Smith S (2001) A global optimisation method for robust affine registration of brain images. Med Image Anal 5(2):143–156
23. Ourselin S, Roche A, Subsol G, Pennec X, Ayache N (2001) Reconstructing a 3d structure from serial histological sections. Image Vis Comput 19:25–31
24. Andersson J, Smith S, Jenkinson M (2008) FNIRT: FMRIB's non-linear image registration tool. In: 14th annual meeting of the organization for human brain mapping, OHBM
25. Modat M, Ridgway GR, Taylor ZA, Lehmann M, Barnes J, Fox NC, Hawkes DJ, Ourselin S (2009) Fast free-form deformation using graphics processing units. Comput Methods Programs Biomed 98(3):278–284
26. Ashburner J (2007) A fast diffeomorphic image registration algorithm. NeuroImage 38(1):95–113
27. Avants B, Epstein C, Grossman M, Gee J (2008) Symmetric diffeomorphic image registration with cross-correlation: evaluating automated labeling of elderly and neurodegenerative brain. Med Image Anal 12(1):26–41. Special issue on the third international workshop on biomedical image registration – WBIR 2006
28. Klein A, Andersson J, Ardekani BA, Ashburner J, Avants B, Chiang MC, Christensen GE, Collins DL, Gee J, Hellier P, Song JH, Jenkinson M, Lepage C, Rueckert D, Thompson P, Vercauteren T, Woods RP, Mann JJ, Parsey RV (2009) Evaluation of 14 nonlinear deformation algorithms applied to human brain mri registration. NeuroImage 46(3):786–802

Image Stitching

Matthew Brown
Dept of Computer Science, University of Bath, Bath, UK

Synonyms

Image Mosaicing; Panoramic stitching

Related Concepts

▸Environment Mapping

Definition

Image stitching is the process of combining multiple overlapping images to generate a new image with a larger field of view than the originals.

Background

Image stitching can enhance the capabilities of an ordinary camera, enabling the capture of larger field-of-view, higher-resolution images. A popular example is the construction of panorama images by seamlessly combining several images of a scene taken from the same point (This process is known as panoramic stitching, which refers to the special case of image stitching for rotational motion). By capturing images with variable exposure settings, it can also be used to generate images with a higher dynamic range than the originals.

Stitching techniques were originally used in photogrammetry to produce maps from aerial and satellite images. Early techniques involved manual specification of matching images and control points (correspondences) between them [1]. Later methods used automated image alignment [2, 3] and interactive viewers to visualize the results [4]. Modern stitching pipelines offer fully automated operation [3], seam selection [5], and photometric, as well as geometric alignment [6].

Half of the images aligned

All images aligned using bundle adjustment

After photometric alignment, seam selection and blending

Image Stitching, Fig. 1 Panoramic stitching. Images are first geometrically aligned (using a rotational motion model in this case). Photometric alignment is used to compensate for brightness variations between the images, and the final panorama is rendered using seam selection and pyramid blending. (**a**) Half of the images aligned. (**b**) All images aligned using bundle adjustment. (**c**) After photometric alignment, seam selection and blending

A typical pipeline for image stitching consists of the following stages (see Fig. 1):

1. Estimating two-frame motion and discovering overlaps between the images
2. Global alignment (e.g., using bundle adjustment [7])
3. Photometric alignment and seam selection/deghosting
4. Rendering the final panorama with blending and/or tone mapping

Theory and Applications

Image stitching is possible when a one-to-one mapping exists between the source image coordinates. Two commonly occurring examples are: (1) a camera rotating about it's optical center and (2) cameras viewing a planar scene. If the cameras are assumed to be rectilinear, the image coordinates are related by a homography

$$\tilde{\mathbf{u}}_2 = \mathbf{H}_{12}\tilde{\mathbf{u}}_1 , \quad (1)$$

where $\tilde{\mathbf{u}}_1, \tilde{\mathbf{u}}_2$ are the homogeneous coordinates in image 1 and 2 and $\mathbf{H}_{12}$ is a 3×3 matrix that encodes the relative camera positions. For example, in the rotational case, $\mathbf{H}_{12}$ is given by

$$\mathbf{H}_{12} = \mathbf{K}_2\mathbf{R}_2\mathbf{R}_1^T\mathbf{K}_1^{-1} , \quad (2)$$

where $\mathbf{R}_1, \mathbf{R}_2$ are the rotation matrices of cameras 1 and 2 and $\mathbf{K}_1, \mathbf{K}_2$ contain the intrinsic parameters.

A typical image stitching approach begins by robustly estimating $\mathbf{H}_{12}$ from correspondences of local image features [8]. A standard method is to use the RANSAC algorithm [9] to sample the space of transformation hypotheses, for all images with a sufficiently large number of feature matches. One can then reason about the adjacency relationships and recognize panoramas by making a match/no-match decision for each pair and finding connected components in the resulting graph of image matches [3].

After pairwise alignment, gaps and inconsistencies can still exist. Bundle adjustment [7] can be used to minimize projection errors between feature matches in all images and generate globally consistent results. Best results are achieved by parameterizing in terms of the intrinsic and extrinsic parameters of the cameras (e.g., rotation, focal length, radial distortion) [2]. Direct methods [10] (using all of the pixel data instead of only feature points) may optionally be used for accurate final registration.

Once the images are geometrically aligned, the remaining task is to render a seamless output view. The appropriate render surface may depend on the images being aligned: rectilinear renderings (preserving straight lines) might be best for stitching planar surfaces such as whiteboards, spherical or cylindrical render surfaces are popular for wide-angle panoramas. Multiperspective renderings can be used to preserve important geometric properties in the output [11].

Ideally, one can capture or estimate irradiance values per pixel, and given perfect alignment, these would be equal in all images overlapping a given ray. In practice, however, several sources of error contribute toward differences in the recorded radiances. Some common examples are parallax due to motion of the camera center, errors or unmodeled parameters in the camera pose estimate, and moving objects in the scene. Several algorithms have been developed to eliminate the visual seams that result. The best approaches find seam lines which minimize differences between image intensities or radiances [12], and smoothly interpolate between images using pyramid blending [13] or gradient domain fusion [5]. The final results can be tone mapped for display.

An example of an automated capture system capable of stitching gigapixel panoramas with feature-based alignment, seam selection, and dynamic tone mapping is given in [14].

References

1. Slama CC (ed) (1980) Manual of photogrammetry, 4th edn. American Society of Photogrammetry, Falls Church, Virginia
2. Szeliski R, Shum H (1997) Creating full view panoramic image mosaics and environment maps. Comput Graph (SIGGRAPH'97) 31(Annual Conference Series):251–258
3. Brown M, Lowe D (2007) Automatic panoramic image stitching using invariant features. Int J Comput Vis 74(1):59–73
4. Chen S (1995) QuickTime VR – an image-based approach to virtual environment navigation. In: ACM transactions on graphics (SIGGRAPH'95), vol 29, pp 29–38
5. Agarwala A, Dontcheva M, Agarwala M, Drucker S, Colburn A, Curless B, Salesin D, Cohen M (2004) Interactive digital photomontage. In: ACM transactions on graphics (SIGGRAPH'04)
6. Eden A, Uyttendaele M, Szeliski R (2006) Seamless image stitching of scenes with large motions and exposure differences. In: IEEE computer society conference on computer vision and pattern recognition (CVPR'06)
7. Triggs W, McLauchlan P, Hartley R, Fitzgibbon A (1999) Bundle adjustment: a modern synthesis. In: Vision algorithms: theory and practice, number 1883 in LNCS. Springer, Corfu, September 1999, pp 298–373
8. Szeliski R (2010) Computer vision: algorithms and applications, Chapter 9. Springer, http://www.springer.com/computer/image+processing/book/978-1-84882-934-3
9. Fischler M, Bolles R (1981) Random sample consensus: a paradigm for model fitting with application to image analysis and automated cartography. Commun ACM 24: 381–395
10. Irani M, Anandan P (1999) About direct methods. In: Triggs B, Zisserman A, Szeliski R (eds) Vision algorithms: theory and practice, number 1883 in LNCS. Springer, Corfu, September 1999, pp 267–277
11. Zelnik-Manor L, Peters G, Perona P (2005) Squaring the circle in panoramas. In: Tenth IEEE international conference on computer vision (ICCV'05), Beijing, pp 1292–1299
12. Davis J (1998) Mosaics of scenes with moving objects. In: IEEE computer society conference on computer vision and pattern recognition (CVPR'98), pp 354–360
13. Burt P, Adelson E (1983) A multiresolution spline with application to image mosaics. ACM Trans Graph 2(4): 217–236
14. Kopf J, Uyttendaele M, Deussen O, Cohen M (2007) Capturing and viewing gigapixel images. In: ACM transactions on graphics (SIGGRAPH'07), vol 26

Image-Based Lighting

Tien-Tsin Wong
Department of Computer Science and Engineering, The Chinese University of Hong Kong, Hong Kong SAR, China

Synonyms

Environment mapping; Reflection mapping

Related Concepts

▶Plenoptic Function

Definition

Image-based lighting [1, 2] is a rendering technique to compute the reflection from a 3D object lit in a distant environment, represented as an image, typically in the form of a cubemap (Fig. 1).

Background

Due to the computational expense of global illumination (e.g., radiosity and Monte Carlo ray tracing), most real-time graphics systems are depth-buffering based and only support local illumination. This hurts the realism of the rendered images. Environment mapping [1, 3] is proposed to simulate the reflection of the surrounding environment on an object surface (Fig. 1). The enclosing environment is assumed to be infinitely far away (distant environment) because all surface points on the object are assumed to be lit by the same environment. Due to the high computational expense, early implementations of environment mapping account only for the light contribution along the mirror reflection direction. Hence, most surfaces rendered by the environment mapping are over-shiny.

Image-based lighting can be regarded as a more comprehensive realization of the environment mapping, by accounting not only the light contribution along the mirror reflection but also the whole enclosing sphere. The surface reflectance property (bidirectional reflectance distribution function, BRDF) is also considered so as to render not only shiny or glossy surfaces but also most kinds of surfaces. Moreover, the environment maps are usually captured as high dynamic range (HDR) images to further increase the photorealism. Note that the image-based lighting remains work even low-dynamic range (LDR) environment maps are used instead.

Theory

Extending from accounting only the mirror reflection direction to the whole enclosing sphere drastically increases the computational expense. Hence, the challenge is how to efficiently compute the following integration for each surface point:

$$I(x,s) = \int_{\Omega} L_{\text{in}}(\omega)\rho(x,\omega,u)v(x,\omega)(\omega \cdot n)d\omega \quad (1)$$

where x is the current surface point of interest; Ω is the surrounding environment (the distant environment map); ω is the incoming light direction; u is the viewing direction from x towards the eye; I is the reflected light; $L_{\text{in}}(\omega)$ is the incoming light contribution along direction ω, in other words, a point in the environment map; v is the visibility function; ρ is the BRDF; and n is the surface normal at x. Note that L_{in}, ρ, and v are spherical functions.

One way to evaluate the above integration is to approximate the environment map by a much smaller number (say m) of point light sources. The position and color of the point lights are obtained by importance sampling of the environment map [4]. In other words, the above integration is approximated by a summation of light contribution of m point lights. The rendered image can simply be generated by adding m images, each rendered by illuminating the scene with a point light source.

By adopting the image-based relighting techniques [5], the above integration can be evaluated more efficiently. The idea is to first encode the spherical function with basis functions. This effectively converts a huge spherical function (table) into a coefficient vector s_i as follows. Since the basis functions B_i are known, they need not be stored:

$$S(\omega) \approx \sum_{i}^{k} s_i B_i(\omega) \quad (2)$$

where S is a spherical function and k is the total number of basis functions, which is much smaller than the number of entries in the original spherical table.

By embedding $\rho(x,\omega,u)v(x,\omega)(\omega \cdot n)$ into a spherical function C, both C and L_{in} can be encoded with the same basis and stored as two coefficient vectors, c_i and l_i, respectively. If the selected basis functions are orthonormal, the above integration (Eq. 1) can be simply evaluated as a dot product between two coefficient vectors [1],

$$I \approx \sum_{i}^{k} c_i l_i \quad (3)$$

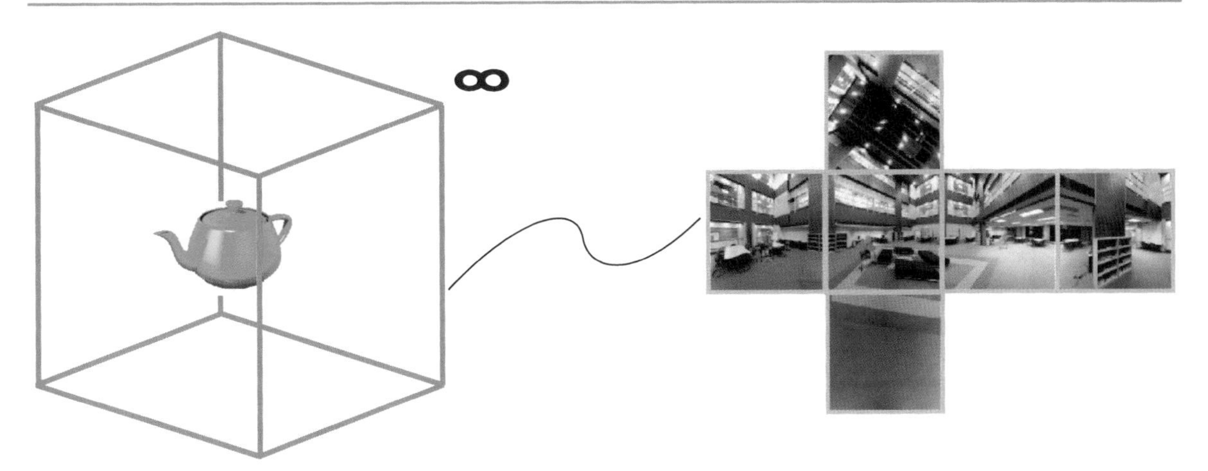

Image-Based Lighting, Fig. 1 Image-based lighting assumes the object being rendered is enclosed by an image-based environment positioned infinitely far away, or equivalently, the object is infinitesimally small. The environment map is typically represented as a cubemap

Basis Functions

The key to efficient image-based lighting is to select an appropriate basis for representing the spherical functions. Various bases have been proposed for image-based relighting [5]. They are directly applicable to image-based lighting.

A pioneer work is proposed by Nimeroff et al. [6]. They efficiently relit the scene under various natural illumination (overcast or clear skylight). The illumination function is decomposed into a linear combination of *steerable functions*.

Principal component analysis is naturally a potential choice for basis function [7]. Singular value decomposition can be used to extract a set of eigenimages from the input reference images. The desired image can then be synthesized by a linear combination of these basis images given a set of coefficients [8, 9] if all surfaces are Lambertian.

Earlier works do not consider the spherical nature of the illumination computation. Wong et al. [5, 10] chose the *spherical harmonic basis*, which is commonly used for compressing BRDF. Pleasant rendering results are obtained with 16–25 basis functions. However, spherical harmonic is also well known in oversmoothing the high-frequency signal (e.g., shadow) in the original spherical function, leading to low-frequency results.

To achieve all-frequency rendering, Haar *wavelet basis* is proposed [11]. It may introduce visual artifact when the distant environment contains a dominant but small-size spot. The cause of such artifact is due to the digitization of the spherical function and the limited reconstruction involving only important wavelet coefficients.

Spherical radial basis function (SRBF) [12–14] is another approach to capture all-frequency signal. The local support nature of SRBF allows its implementation to be very efficient and simple. The multiscale spherical radial basis function [15] avoids the visual artifact of the Haar wavelet basis while remains able to achieve all-frequency rendering.

Application

Image-based lighting can be applied to produce realistic rendering for both off-line movie production or real-time computer games. The parallel nature of image-based lighting (all surface points have to evaluate Eq. 1 independently) facilitates its real-time realization on modern graphics processing unit (GPU).

References

1. Cabral B, Max N, Springmeyer R (1987) Bidirectional reflection functions from surface bump maps. In: Proceedings of the 14th annual conference on computer graphics and interactive techniques. ACM, New York, pp 273–281

2. Debevec P (2002) Image-based lighting. IEEE Comput Graphics Appl 22(2):26–34
3. Blinn JF, Newell ME (1976) Texture and reflection in computer generated images. Commun ACM 19(10):542–547
4. Agarwal S, Ramamoorthi R, Belongie S, Jensen HW (2003) Structured importance sampling of environment maps. ACM Trans Graphics 22(3):605–612
5. Wong TT, Heng PA, Or SH, Ng WY (1997) Image-based rendering with controllable illumination. In: Proceedings of the 8th Eurographics workshop on rendering (Rendering techniques'97), St. Etienne, pp 13–22
6. Nimeroff JS, Simoncelli E, Dorsey J (1994) Efficient re-rendering of naturally illuminated environments. In: Fifth Eurographics workshop on rendering, Darmstadt, Germany, pp 359–373
7. Ho PM, Wong TT, Choy KH, Leung CS (2003) PCA-based compression for image-based relighting. In: Proceedings of IEEE international conference on multimedia and expo 2003 (ICME 2003), vol I, Baltimore, Maryland, USA,pp 473–476
8. Belhumeur PN, Kriegman DJ (1996) What is the set of images of an object under all possible lighting conditions. In: Proceedings of IEEE conference on computer vision and pattern recognition (CVPR), San Francisco
9. Zhang Z (1998) Modeling geometric structure and illumination variation of a scene from real images. In: Proceedings of the international conference on computer vision (ICCV'98), Bombay, India
10. Wong TT, Fu CW, Heng PA, Leung CS (2002) The plenoptic illumination function. IEEE Trans Multimedia 4(3): 361–371
11. Ng R, Ramamoorthi R, Hanrahan P (2003) All-frequency shadows using non-linear wavelet lighting approximation. ACM Trans Graphics 22(3):376–381
12. Wong TT, Leung CS, Choy KH (2005) Lighting precomputation using the relighting map. In: Engel W (ed) ShaderX^3: advanced rendering with DirectX and OpenGL. Charles Rivers Media, Hingham, pp 379–392
13. Leung CS, Wong TT, Lam PM, Choy KH (2006) An RBF-based image compression method for image-based rendering. IEEE Trans Image Process 15(4):1031–1041
14. Tsai YT, Shih ZC (2006) All-frequency precomputed radiance transfer using spherical radial basis functions and clustered tensor approximation. In: Proceedings of ACM SIGGRAPH 2006, Boston, pp 967–976
15. Lam PM, Ho TY, Leung CS, Wong TT (2010) All-frequency lighting with multiscale spherical radial basis functions. IEEE Trans Visual Comput Graphics 16(1):43–56

Image-Based Modeling

Ping Tan
Department of Electrical and Computer Engineering, National University of Singapore, Singapore, Singapore

Synonyms

Three dimensional Modeling from Images

Definition

Image-based modeling refers to the process of using two-dimensional images to create three-dimensional models. These models often consist of a geometric shape and a texture map defined over this shape.

Background

Three dimensional models are mathematical representations of three dimensional objects or scenes. These models are useful for various applications such as simulation, robotics, virtual reality, and digital entertainment. Automatically creating these models has been an important research topic since the early days of computer vision.

Broadly speaking, there are two schools of image-based modeling methods. One group employs range sensor for shape modeling, while the other group uses pure images with binocular or multi-view stereo. One of the origins of the first group is [1], which developed a technique to automatically generate a virtual three dimensional model by observing actual objects along the line of physics-based paradigm. Recently, this direction of research has been accelerated by the development of handy range sensor, such as the Microsoft Kinect and other consumer depth cameras. A representative work from the other group of pure image-based method is [2], which introduced an interactive method to model architectural scenes by fitting geometric primitives to the input images. This direction of research is later generalized to model more general objects. Typically, a cloud of 3D points is first recovered from the input images. A detailed and precise shape representation is then derived from these points, and a texture map is created to represent the color of each point on the shape. This process is illustrated in Figure 1.

The process of obtaining 3D points from input images is known as 3D reconstruction. It is a well-studied problem, and most of the relevant theoretic results are summarized in this handbook [4]. There are also a number of well-established 3D reconstruction software systems such as [5–7]. Though the reconstructed 3D points can be used directly for certain measurements, most of applications require polygonal meshes, NURBS surfaces, or solid shape models. 3D modeling is the process of creating these shape models

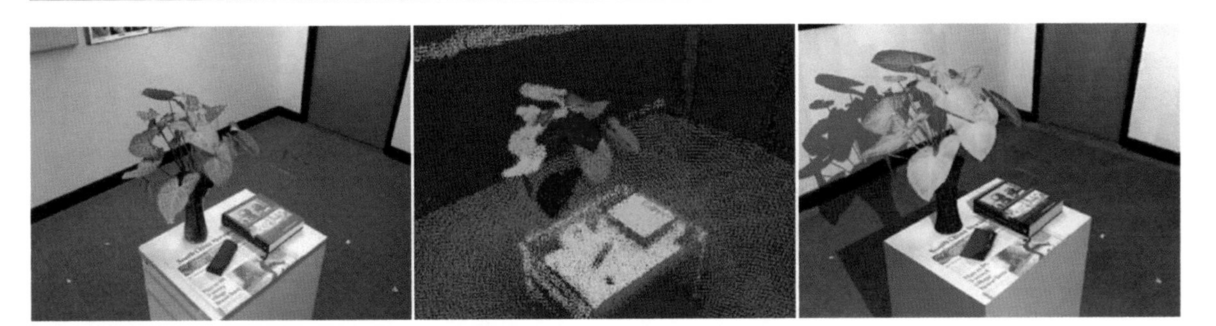

Image-Based Modeling, Fig. 1 A typical pipeline of image-based modeling. In the *left* is one of the input image. In the *middle* is the set of reconstructed 3D points. In the *right* is a rendering of the recovered 3D model of the scene (These pictures are from [3])

from 3D points and 2D images. This modeling process can be automatic or interactive. The prior knowledge of the shape to be modeled often plays an important role in this process. A comprehensive review of recent modeling techniques can be found in [8].

In image-based modeling, the appearance of a shape is often modeled by specifying a color for each point on the shape. These colors can be stored in a so-called texture map. Essentially, this representation assumes the surface is Lambertian and its Bidirectional Reflectance Distribution Function (BRDF) is a constant, that is, the color. In comparison, appearance modeling dedicates to model both the shape and precise surface reflectance properties. More details of appearance modeling can be found in [9].

Theory

In this short chapter, we only focus on the process of creating shape models from 3D points and 2D images. One way to generate these shape models is to obtain a minimal surface automatically from the input data. The minimal surface is a surface that minimizes a functional of the following form

$$\int\int w ds.$$

Here, ds is the infinitesimal surface element and w is the consistency of the surface according to the input 3D points and 2D images. This consistency can be simply the Euclidean distance between the surface and the set of 3D points. Given a set of points $\mathcal{P}$, Zhao et al. [10] defined w as $d(\mathbf{x}, \mathcal{P})$ which is the smallest distance between a surface point $\mathbf{x}$ and other points in the set $\mathcal{P}$. This functional is then minimized by the level set method [11]. Faugeras and Keriven [12] defined w as a function of both the surface position $\mathbf{x}$ and its normal direction $\mathbf{n}$ to facilitate the surface modeling. Lhuillier and Quan [13] further incorporated image reprojection errors and silhouettes in this function w.

The minimal surface-based approach works well when the points are dense and the surface is smooth. However, it has difficulties to model discontinuous surfaces such as hair fibers (linear structure), clothes (open surface patches), tree branches (tree and fractal structure), or buildings (regular axis aligned boxes). There are many existing methods which are designed to exploit the prior knowledge of a shape to facilitate modeling. Wei et al. [14] modeled hair by "growing" 3D smooth curves guided by 3D points and images. Bhat et al. [15] used videos to obtain the parameters of a cloth simulation system. Tan et al. [16] recovered some basic branch elements from the 3D points and used them to generate a fractal branch structure. Xiao et al. [17, 18] identified building facades and repetitive structures on these facades to model streets. Furukawa et al. [19, 20] assumed the scene consists of mutual orthogonal planes at different depth to model buildings. Though generating good results, these methods are limited to model the type of surface that matches their underlying prior shape assumption. A general modeling method is still missing to handle all these different data in a unified framework.

Application

Image-based modeling can be applied in autonomous robotics to generate a three-dimensional map of their environment for path/action planning. It can also be used in industry vision for product quality inspection. The three-dimensional models can also be applied in digital entertainment such as games and movies. Recently, Google Earth and Microsoft Virtual Earth start to provide 3D map services, which can be a very good test bed of large-scale image-based modeling techniques.

References

1. Sato Y, Wheeler MD, Ikeuchi K (1997) Object shape and reflectance modeling from observation. Proceedings of the 24th Annual Conference on Computer Graphics and Interactive Techniques. SIGGRAPH '97, New York, NY, USA, ACM Press/Addison-Wesley Publishing Co, pp 379–387
2. Debevec PE, Taylor CJ, Malik J (1996) Modeling and rendering architecture from photographs: A hybrid geometry- and image-based approach. Proceedings of the 23rd Annual Conference on Computer Graphics and Interactive Techniques. SIGGRAPH '96, New York, NY, USA, ACM, pp 11–20
3. Quan L, Wang J, Tan P, Yuan L (2007) Image-based modeling by joint segmentation. Int J Comput Vis 75:135–150
4. Hartley R, Zisserman A (2003) Multiple view geometry in computer vision, 2 edn. Cambridge University Press, New York
5. Lhuillier M, Quan L (2005) A quasi-dense approach to surface reconstruction from uncalibrated images. IEEE Trans Pattern Anal Mach Intell 27:418–433
6. Furukawa Y, Ponce J (2010) Accurate, dense, and robust multiview stereopsis. IEEE Trans Pattern Anal Mach Intell 32:1362–1376
7. Snavely N, Seitz SM, Szeliski R (2006) Photo tourism: exploring photo collections in 3d. ACM Trans Graph 25: 835–846
8. Quan L (2010) Image-based modeling, 1 edn. Springer, New York
9. Weyrich T, Lawrence J, Lensch HPA, Rusinkiewicz S, Zickler T (2009) Principles of appearance acquisition and representation. Found Trends Comput Graph Vis 4:75–191
10. Zhao HK, Osher S, Fedkiw R (2001) Fast surface reconstruction using the level set method. In: Proceedings of the IEEE workshop on variational and level set methods (VLSM'01). VLSM '01, Washington, DC, USA. IEEE Computer Society, p 194
11. Osher S, Sethian JA (1988) Fronts propagating with curvature-dependent speed: algorithms based on hamilton-jacobi formulations. J Comput Phys 79:12–49
12. Faugeras OD, Keriven R (1998) Complete dense stereovision using level set methods. In: Proceedings of the 5th European conference on computer vision-volume i – volume i (ECCV '98). London, UK. Springer, pp 379–393
13. Lhuillier M, Quan L (2003) Surface reconstruction by integrating 3d and 2d data of multiple views. In: Proceedings of the 9th IEEE international conference on computer vision – volume 2. ICCV '03, Washington, DC, USA. IEEE Computer Society, p 1313
14. Wei Y, Ofek E, Quan L, Shum HY (2005) Modeling hair from multiple views. ACM Trans Graph 24:816–820
15. Bhat KS, Twigg CD, Hodgins JK, Khosla PK, Popović Z, Seitz SM (2003) Estimating cloth simulation parameters from video. In: Proceedings of the 2003 ACM SIGGRAPH/Eurographics symposium on computer animation. SCA '03, Aire-la-Ville, Switzerland. Eurographics Association, pp 37–51
16. Tan P, Zeng G, Wang J, Kang SB, Quan L (2007) Image-based tree modeling. ACM Trans Graph 26
17. Xiao J, Fang T, Tan P, Zhao P, Ofek E, Quan L (2008) Image-based facade modeling. ACM Trans Graph 27: 161:1–161:10
18. Xiao J, Fang T, Zhao P, Lhuillier M, Quan L (2009) Image-based street-side city modeling. ACM Trans Graph 28:114: 1–114:12
19. Furukawa Y, Curless B, Seitz S, Szeliski R (2009) Manhattan-world stereo. In: Proceedings of the IEEE international conference on computer vision and pattern recognition (CVPR '09). IEEE Computer Society
20. Furukawa Y, Curless B, Seitz S, Szeliski R (2009) Reconstructing building interiors from images. In: Proceedings of the IEEE international conference on computer vision. ICCV '09. IEEE Computer Society

Image-Based Rendering

Shing Chow Chan
Department of Electrical and Electronic Engineering, The University of Hong Kong, Hong Kong, China

Synonyms

Image-based rendering (IBR)

Related Concepts

►Light Field; ►Lumigraph; ►Plenoptic Function

Definition

Image-based rendering (IBR) refers to a collection of techniques and representations that allows 3D scenes and objects to be visualized and manipulated in a realistic way without full 3D model reconstruction.

Background

One of the primary goals in computer graphics is photorealistic rendering. Motivated by the difficulties in achieving full photorealism with conventional 3D and model-based graphics, image-based rendering which works directly with real images has proposed as an alternative approach to reduce the rendering and capturing complexity. Depending on how the images are being taken and the auxiliary information, such as depths, etc., required, a number of image-based representations supporting different viewing freedom and functionalities are available. These range from the familiar two-dimension (2D) panoramas to more sophisticated representations such as the four-dimension (4D) light fields [9], lumigraphs [8], and variants, which are special cases of the radiance received at every viewing position, visual angle, wavelength, and time, called the plenoptic function.

The rendering of novel views can therefore be viewed as the reconstruction of the plenoptic function from its samples. Image-based representations are usually densely sampled high-dimensional data with large data sizes, but their samples are highly correlated. Because of the multidimensional nature of image-based representations and scene geometry, much research has been devoted to the efficient capturing, sampling, rendering, and compression of IBR.

Theory

Representation

In IBR, new views of scenes are reconstructed from a collection of densely sampled images or videos. Examples include the well-known panoramas [5], light fields [9], lumigraph [8], layered depth images [13], concentric mosaics (CM) [14], etc. Figure summarizes the concept of CM and light field (see the sections on light field, lumigraph, and plenoptic function for more illustration). The reconstruction problem (i.e., rendering) is treated as a multidimensional sampling problem, where new views are generated from densely sampled images and depth maps instead of building accurate 3D model of the scenes.

Depending on the functionality required, there is a spectrum of IBR as shown in Fig. . They differ from each other in the amount of geometry information of the scenes/objects being used. At one end of the spectrum, like traditional texture mapping, very accurate geometric models of the scenes and objects say generated by animation techniques is used, but only a few images are required to generate the textures. Given the 3D models and the lighting conditions, novel views can be rendered using conventional graphic techniques. Moreover, interactive rendering with movable objects and light sources can be supported using advanced graphic hardware.

At the other extreme, light field or lumigraph rendering relies on dense sampling (by capturing more image/videos) with no or very little geometry information for rendering without recovering the exact 3D models. An important advantage of the latter is its superior image quality, compared with 3D model building for complicated real world scenes. Another important advantage is that it requires much less computational resources for rendering regardless of the scene complexity because most of the quantities involved are precomputed or recorded. This has attracted considerable attention in the computer graphic community in developing fast and efficient rendering algorithms for real-time relighting and soft-shadow generation [2, 12, 19, 22].

Broadly speaking, image-based representations can be classified according to the geometry information used into three main categories: (1) representations with no geometry, (2) representations with implicit geometry, and (3) representations with explicit geometry. 2-D panoramas, McMillan and Bishop's plenoptic modeling [11], and 3D concentric mosaics and light fields/lumigraph belong to the first category, and they can be viewed as the direct interpolation of the plenoptic function. Layere-based, object-based representations [4], pop-up light [16] using depth maps fall into the second. Finally, conventional 3D computer graphic models and other more sophisticated representations [7, 21, 22] belong to the last category. Although these representations also sample the plenoptic function, further processing of the plenoptic function has been performed to infer the scene geometry or surface property such as bidirectional reflectance distribution function (BRDF) of objects. Such image-based modeling approach has emerged as a more promising approach to enrich the photorealism and user interactivity of IBR. Moreover, since 3D models of the scenes are unavailable, conventional image-based representations are limited to the change of viewpoints and sometimes limited amount of relighting. Recently, it was found

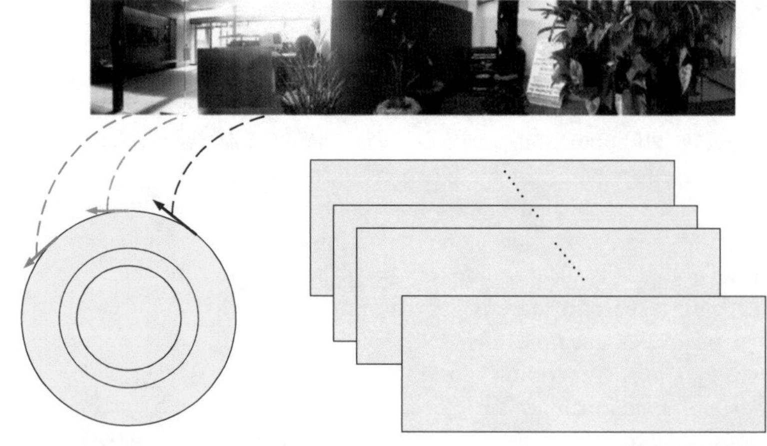

Concentric mosaic
By constraining camera motion to planar concentric circles, concentric mosaic can be created by compositing slit images taken at different locations of each circle.

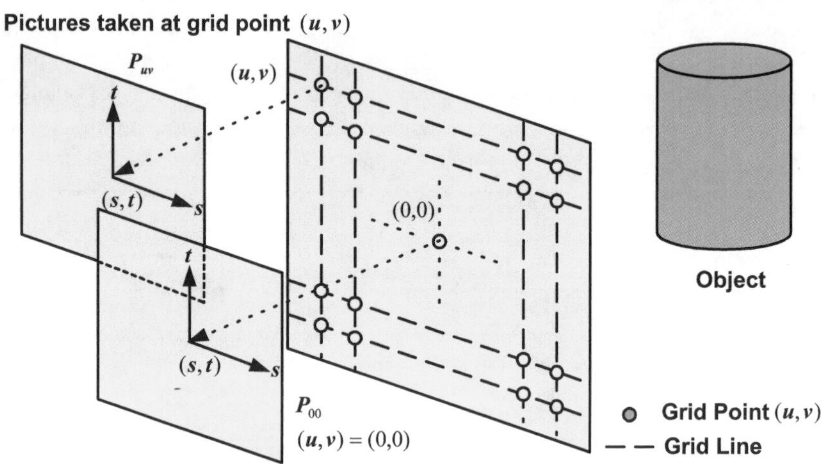

Image-Based Rendering, Fig. 1 Concentric mosaic and light field [3]

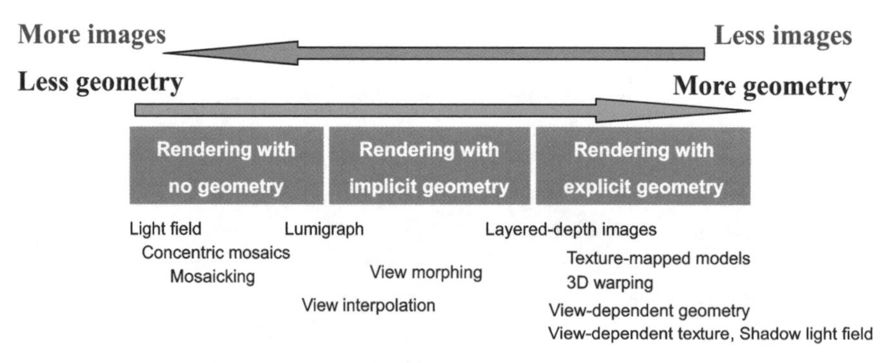

Image-Based Rendering, Fig. 2 Spectrum of IBR representations

that real-time relighting and soft-shadow computation are feasible using the IBR concepts and the associated 3D models using precomputed radiance transfer (PRT) [19] and precomputed shadow fields [22].

Earlier image-based representations are usually static, and their extension usually requires multiple camera arrays. Much research has been devoted to the capturing, compression, transmission, and processing of these dynamic representations. For a review as of 2007, see [17].

Rendering

Rendering refers to the process of rendering of new views from the images and possibly other auxiliary information captured in the representations. For early image-based representations which do not employ any geometry information, rendering can be done simply by image blending as in panoramas [5] and ray-space interpolation in light field [9]. In ray-space interpolation, each ray that corresponds to a target screen pixel is mapped to nearby sampled rays. Figure 4a shows the example renderings of a simplified light field using ray-space interpolation [17]. For more sophisticated representations which use more geometry information such as layered depth images [13], surface light field [21], and pop-up light field [16], graphics hardware has been exploited to accelerate the rendering process. The geometry information can either be implicit that relies on positional correspondences or explicit in the form of depth along known lines-of-sight or 3D coordinates. Representations of the former usually involve weakly calibrated cameras and rely on image correspondences to render new views, say by triangulating two reference images into patches according to the correspondences as in joint view triangulation (JVT) [10]. These include view interpolation, view morphing, JVT, and transfer methods with fundamental matrices and trifocal tensors. Representations employing explicit geometry include sprites, relief textures, layered depth images (LDIs), view-dependent texture, surface light field, pop-up light field, shadow light field, etc.

In general, the rendering methods can be broadly classified into three groups [17]: (1) point-based, (2) layer-based, and (3) monolithic.

Point-Based Rendering works on 3D point clouds or point correspondences, and typically each point is rendered independently. Points are mapped to the target screen through forward mapping and variants. For the 3D point X in Fig. 3, the mapping can be written as

$$X = C_r + \rho_r P_r x_r = C_t + \rho_t P_t x_t \quad (1)$$

where x_t and x_r are homogeneous coordinates of the projection of X on target screen and reference images, respectively. C and P are camera center and projection matrix, respectively, and ρ is a scale factor. Since C_t, P_t, and the focus length f_t are known for the target view, ρ_t can be computed using the depth of X. Given x_r and ρ_r, one can compute the exact position of x_t on the target screen and transfer the color accordingly. Gaps or holes may exist due to magnification and disocclusion, and splatting techniques have been proposed to alleviate this problem. The painter's algorithm is frequently used to avoid the problem of the mapping of multiple pixels from the reference view to the same pixel in the target view.

Layered Techniques usually discretize the scene into a collection of planar layers with each layer consisting of a 3D plane with texture and optionally a transparency map. The layers can be thought of as a continuous set of polygonal models, which is amenable to conventional texture mapping and view-dependent texture mapping. Usually, each layer is rendered using either point-based or polygon meshes as in monolithic rendering techniques before being composed in the back-to-front order using the painter's algorithm to produce the final view. Layer-based rendering is also easier to implement using graphic processing unit (GPU). Since the rendering of IBR requires very low complexity, it is even possible to perform the calculation using CPU by working on individual layer or object [4].

Monolithic Rendering usually represents the geometry as continuous polygon meshes with textures, which can be readily rendered using graphics hardware. The 3D model normally consists of vertices, normals of vertices, faces, and texture mapping coordinates. The data can be stored in a variety of data formats. The most popular formats are .obj, .3ds, .max, .stl, .ply, .wrl, .dxf, etc.

Relighting, shadow generation, and interactivity have played an increasingly important role in 3D interactive rendering. The most popular algorithms are shadow mapping, shadow volume, ray tracing, precomputed radiance transfer, precomputed shadow field, etc. Some of them have better rendering quality, while others are more efficient for real-time rendering. Thanks to the development of GPU, basic lighting, and shading algorithms like shadow mapping and shadow volume have been realized on the fly. Modern GPUs can even offer programmable rendering pipelines for customized rendering effects and "shader" is a set of software instructions running on

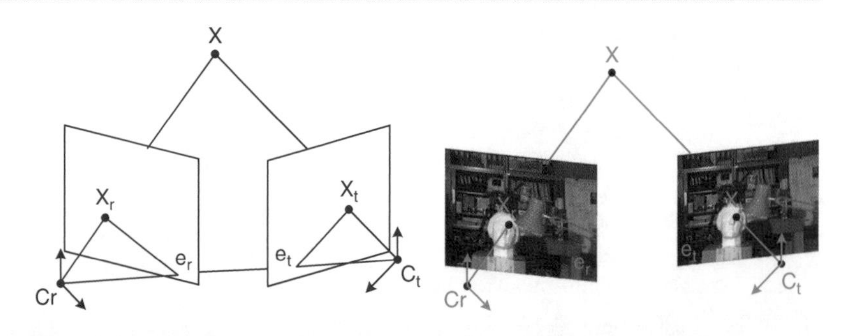

Image-Based Rendering, Fig. 3 Forward mapping

Image-Based Rendering, Fig. 4 Example renderings using (**a**) ray-space interpolation [17], (**b**) forward mapping in layered representation (with two layers – dancer and background) [4], (**c**) monolithic rendering using 3D polygonal mesh (*left*) estimated by multiview stereo and real-time rendering with shadow light field technique on GPU [23]

these GPUs to control the pipelines. Using shader programming, high-quality shadow rendering algorithms like precomputed shadow field can be done in real time. Figure 4 shows example renderings of the three techniques, and Fig. 5 summarizes the types of representations and rendering in IBR called the geometry-rendering matrix.

Compression

In general, there are two approaches to reduce the data size of image-based representations. The first one is to reduce their dimensionality, often by limiting viewpoints or scarifying some realism. Panoramas and concentric mosaics are such examples. The second approach is to exploit the high correlation

Image-Based Rendering, Fig. 5 Geometry-rendering matrix

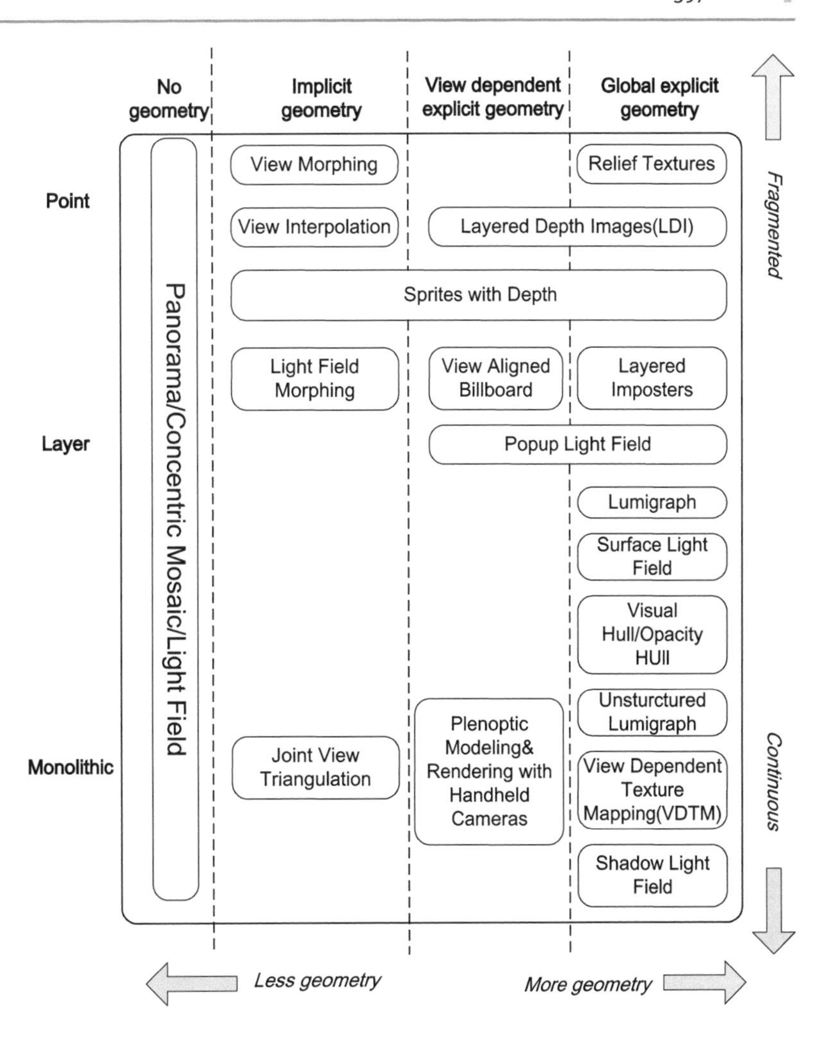

(i.e., redundancy) within the representation using waveform coding or model-based techniques. The scene geometry may be used explicitly or implicitly. The second approach can be further classified into four broad categories: pixel-based methods, disparity compensation/prediction (DCP) methods, model-based/model-aided methods, and object-based approach.

In pixel-based methods, the correlation between adjacent image pixels is exploited using conventional techniques such as vector quantization and transform coding. In the DCP methods, scene geometry is utilized implicitly by exploiting the disparity of image pixels, resulting in better compression performance. (Disparity refers to the relative displacement of pixels in images taken at adjacent physical locations.) Model-based/model-aided approaches recover the geometry of the objects or scene in coding the observed images. The models and other information such as prediction residuals or view-dependent texture maps are then encoded. In the object-based approach, the representations are segmented into IBR objects, each with its image sequences, depth maps, and other relevant information such as shape information. The main advantage is that it helps to reduce the rendering artifacts and hence the required sampling rate. For additional references, see the section on light fields.

Unlike conventional video coding, higher dimensional image-based representations such as 3D concentric mosaics (CMs) require random access at the line level, whereas the 4D light field and lumigraph require random access at the pixel level. It is usually time-consuming to retrieve and decode a single line or pixel from the compressed which is of variable length due to entropy coding. This is referred to as the "random access problem" of IBR and is usually tackled

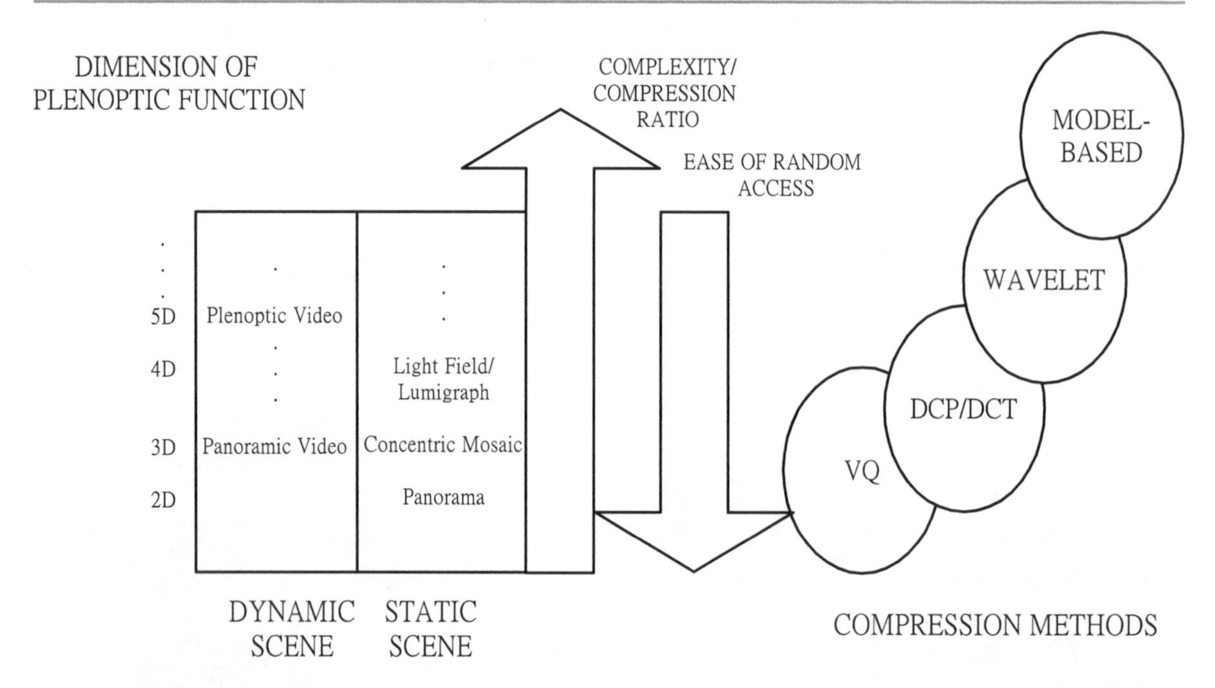

Image-Based Rendering, Fig. 6 Comparison of different image-based representation and compression methods in terms of their complexity. The ease of random access increases as the dimension of plenoptic function decreases, while the complexity and potential for compression both increase with the dimension. *DCP* disparity compensation/prediction, *VQ* vector quantization

by grouping the compressed data of several basic units for rendering (such as lines in CMs or image blocks in light fields) together and employ pointers to locate them efficiently. Moreover, interdependence of decoding resulting from DCP should be reduced to avoid decoding excessively unnecessary intermediate data. This is also required for selective transmission or decoding of the compressed representations due to their large bandwidth and storage requirement. A simple comparison of difficult image-based representations and compression methods in terms of their complexities, compression ratios, and ease of random access is shown in Fig. 6. For more information, see [15, 17] and references in the light field section.

Application

The potential for photorealistic visualization and simplicity in rendering of IBR has tremendous appeal. They have already found applications in architectural modeling [6], cultural heritage preservation [23], virtual tour, and digital museum [18], multiview TV [3, 4], etc. Other potential applications include digital edutainment, E-commerce and photorealistic modeling, and real-time rendering in computer graphics and mobile devices. Another emerging application is view synthesis in 3D and multiview videos and display.

Open Problems

Though there has been substantial progress in capturing, representing, rendering, and modeling of scenes, the ability to handle general complex scenes remains challenging for IBR. A substantial amount of work is still required to ensure robustness in handling reflection translucency, highlights, depth estimation, capturing complexity, object manipulation, etc. Since IBR uses images for rendering, interacting with IBR representations remains challenging. Recent approaches have focused on using advanced computer vision techniques, such as stereo/multiview vision and photometric stereo, and depth sensing devices to extract more geometry information from the scene so as to enhance the functionalities of IBR representations. While there has been considerable progress in relighting and interactive rendering of individual real static objects, such operations are still difficult for real and complicated scenes. For dynamic scenes, the huge amount of data and vast amount of viewpoints to be provided present one of the major challenges to IBR. Advanced algorithms for processing and manipulation of the high-dimensional representation to achieve such function as object extraction, model completion, scene inpainting, etc., are all major challenges to be addressed. Finally, the efficient transmission, compression, and display of dynamic IBR and models are also urgent issues awaiting for satisfactory solution in order for IBR to establish itself as an essential media for communication and presentation.

References

1. Adelson EH, Bergen JR (1991) The plenoptic function and the elements of early vision. In: Landy M, Movshon JA (eds) Computational models of visual processing. MIT, Cambridge, MA
2. Agrawala M, Ramamoorthi R, Heirich A, Moll L (2000) Efficient image-based methods for rendering soft shadows. In: Proceedings of ACM SIGGRAPH, New Orleans, 372–384
3. Chan SC, Shum HY, Ng KT (2007) Image-based rendering and synthesis: technological advances and challenges. IEEE Signal Process Mag 24(6):22–33
4. Chan SC, Gan ZF, Ng KT, Ho KL, Shum HY (2009) An object-based approach to image/video-based synthesis and processing for 3-D and multiview televisions. IEEE Trans. Circuits Syst Video Technol 19(6):821–831
5. Chen SE (1995) QuickTime VR – an image-based approach to virtual environment navigation. In: Proceedings of ACM SIGGRAPH, Los Angeles, 29–38
6. Debevec PE, Taylor CJ, Malik J (1996) Modeling and rendering architecture from photographs: a hybrid geometry- and image-based approach. In: Proceedings of ACM SIGGRAPH, New Orleans, 11–20
7. Debevec PE, Yu Y, Borshukov G (1998) Efficient view-dependent image-based rendering with projective texture-mapping. In: Eurographics workshop on rendering, Vienna, 150–116
8. Gortler SJ, Grzeszezuk R, Szeliski R, Cohen MF (1996) The lumigraph. In: Proceedings of ACM SIGGRAPH, New Orleans, 43–54
9. Levoy M, Hanrahan P (1996) Light field rendering. In: Proceedings of ACM SIGGRAPH, New Orleans, 31–42
10. Lhuillier M, Quan L (2003) Image-based rendering by joint view triangulation. IEEE Trans Circuits Syst Video Technol 13(11):1051–1063
11. McMillan L, Bishop G (1995) Plenoptic modeling: an image-based rendering system. In: Proceedings of ACM SIGGRAPH, Los Angeles, 39–46
12. Ng R, Ramamoorthi R, Hanrahan P (2004) Triple product wavelet integrals for all-frequency relighting. In: Proceedings of ACM SIGGRAPH, Los Angeles, 477–487
13. Shade J, Gortler S, He LW, Szeliski R (1998) Layered depth images. In: Proceedings of ACM SIGGRAPH, Orlando, 231–242
14. Shum HY, He LW (1999) Rendering with concentric mosaics. In: Proceedings of ACM SIGGRAPH, Los Angeles, 299–306
15. Shum HY, Kang SB, Chan SC (2003) Survey of image-based representations and compression techniques. IEEE Trans Circuits Syst Video Technol 13(11):1020–1037
16. Shum HY, Sun J, Yamazaki S, Li Y, Tang CK (2004) Pop-up light field: an interactive image-based modeling and rendering system. ACM Trans Graph 23(2):143–162
17. Shum HY, Chan SC, Kang SB (2007) Image-based rendering. Springer, New York
18. Snavely N, Simon I, Goesele M, Szeliski R, Seitz SM (2010) Scene reconstruction and visualization from community photo collections. Proc IEEE Internet Vis 98(8): 1370–1390
19. Sloan P, Kautz J, Snyder J (2002) Precomputed radiance transfer for real-time rendering in dynamic, low-frequency lighting environment. In: Proceedings of ACM SIGGRAPH, San Antonio, 527–536
20. Szeliski R, Shum HY (1997) Creating full view panoramic image mosaics and environment maps. In: Proceedings of ACM SIGGRAPH, Los Angeles, 251–258
21. Wood DN, Azuma DI, Aldinger K, Curless B, Duchamp T, Salesin DH, Stuetzle W (2000) Surface light fields for 3D photography. In: Proceedings of ACM SIGGRAPH, New Orleans, 287–296
22. Zhou K, Hu Y, Lin S, Guo B, Shum HY (2005) Precomputed shadow fields for dynamic scenes. In: Proceedings of ACM SIGGRAPH, Los Angeles, 1196–1201
23. Zu ZY, Ng KT, Chan SC, Shum HY (2010) Image-based rendering of ancient Chinese artifacts for multi-viewdisplays – a multi-camera approach. In: Proceedings of 2010 IEEE international symposium on circuits and systems (ISCAS), Paris, 3252–3255

Image-Based Rendering (IBR)

►Image-Based Rendering
►Light Field
►Plenoptic Function

Implicit Polynomial Curve

►Algebraic Curve

Implicit Polynomial Surface

►Algebraic Surface

Incident Light Measurement

Stephen Lin
Microsoft Research Asia, Beijing Sigma Center,
Beijing, China

Related Concepts

▶Illumination Estimation, Illuminant Estimation

Definition

Incident light measurement is the recording of incoming illumination at a given scene point or in a given scene.

Background

The distribution and intensity of light incident upon a surface point or in a scene affects the amount of reflected radiance to the camera and more generally the appearance of objects. Knowledge of the incident light can aid in shape recovery through photometric analysis techniques such as shape-from-shading and photometric stereo, or may be used in reducing appearance variation caused by lighting. Various methods have been used for measurement of incident light. Different from algorithms for illumination estimation, incident light measurement does not infer lighting from indirect scene cues such as shading, but rather obtains direct observations of the light sources.

Methods for incident light measurement typically introduce a probe or a sensor into the scene to view the incoming radiance. In general, the probes are mirrored spheres that allow for precise readings of light from a broad range of incident directions. Besides lighting distribution, the color of incident illumination may be measured using a color calibration target such as a white reference standard. Unlike illumination estimation methods, light measurement with such devices often allows for accurate recovery of both direct illumination from light sources and more subtle indirect illumination from reflected light within the scene.

Some techniques are intended to measure incident light at a certain scene point. Excluding the effects of light occluders, these methods equivalently measure far lighting that originates from distant light sources and is considered to be uniform throughout the scene. Other methods are more general in that their measurements also determine the location and brightness of near light sources, whose illumination varies within the scene. Such methods utilize triangulation, usually from two or more probes or sensors placed in the scene, to locate the positions of local light sources.

Methods

Several methods for incident light measurement are described in the following.

Spherical Probes

To measure distant illumination or the light incident at a given scene point, a common approach is to place a mirrored spherical probe at the scene point. From the reflections on the sphere, the corresponding directions of the incident light are computed from the known surface orientation of each sphere point and the mirror reflection property, which states that the incident angle of light is equal to the reflected light angle. Incident lighting environments of various scenes were measured in this manner by Debevec [1]. High dynamic range imaging was used to obtain accurate measurements of relative light source brightness.

To recover spatially variant incident lighting due to local light sources, Powell et al. [2] used three mirrored spheres at known relative positions to triangulate light source locations. For triangulation, correspondences need to be computed among the mirrored reflections of the spheres. Illumination color is also measured from the color of diffuse reflections on the spheres. Zhou and Kambhamettu [3] also employed triangulation, but instead computed correspondences in a stereo image pair of a single sphere. Shifts in specular reflections as seen from the two stereo viewpoints indicate the distance of light sources. Here, the spheres also exhibit diffuse reflection, which provides information on light intensities. Using this setup, they later proposed a method [4] based on ray tracing and convex hull computation to measure a more general light source model [5].

Hemispherical Imaging

An alternative to lighting probes is to directly place sensors within the scene. Drettakis et al. [6] employed image mosaicing of several snapshots captured within the scene to form a panoramic image of the incident lighting. Sato et al. [7] instead used a pair of omnidirectional cameras, each outfitted with a fish-eye lens. Correspondences in the omnidirectional images are computed with an omnidirectional stereo algorithm to obtain a 3D model of the incident lighting, and high dynamic range imaging is used to measure the intensity of radiance.

Color Calibration Target

The incident light color may be measured by inserting a white reference standard into the scene. Deviations from white of the reflected light indicate the color of illumination. This approach to measuring incident lighting color is described by Barnard et al. [8] for their construction of an image dataset for computational color constancy. Directional variations in illumination color may be measured by imaging the white reference standard at different orientations.

Application

Incident light measurement is often employed for augmented reality [1, 7], to ensure that inserted virtual objects exhibit an appearance consistent with the scene's illumination environment. Measurements of real-world lighting have also been utilized in computer graphics applications to give rendered objects a more natural appearance. Applications of light color measurement include evaluation of color constancy algorithms [8] and spectral reflectance recovery using multiple illumination colors [9].

References

1. Debevec P (1998) Rendering synthetic objects into real scenes: bridging traditional and image-based graphics with global illumination and high dynamic range photography. In: Proceedings of ACM SIGGRAPH. ACM, New York, pp 189–198
2. Powell MW, Sarkar S, Goldgof D (2001) A simple strategy for calibrating the geometry of light sources. IEEE Trans Pattern Anal Mach Intell 23:1022–1027
3. Zhou W, Kambhamettu C (2002) Estimation of illuminant direction and intensity of multiple light sources. In: Proceedings of European conference on computer vision (ECCV). Lecture notes in computer science, vol 2353. Springer, Berlin/Heidelberg, pp 206–220
4. Zhou W, Kambhamettu C (2008) A unified framework for scene illuminant estimation. Image Vis Comput 26:415–429
5. Langer MS, Zucker SW (1997) What is a light source? In: Proceedings of the IEEE conference on computer vision and pattern recognition (CVPR). IEEE Computer Society, Washington, DC, pp 172–178
6. Drettakis G, Robert L, Bougnoux S (1997) Interactive common illumination for computer augmented reality. In: Proceedings of the Eurographics workshop on rendering. Springer, London, pp 45–57
7. Sato I, Sato Y, Ikeuchi K (1999) Acquiring a radiance distribution to superimpose virtual objects onto a real scene. IEEE Trans Vis Comput Graph 5:1–12
8. Barnard K, Martin L, Funt B, Coath A (2002) A data set for colour research. Color Res Appl 27:147–151
9. Han S, Sato I, Okabe T, Sato Y (2010) Fast spectral reflectance recovery using dlp projector. In: Proceedings of Asian conference on computer vision. Springer, Berlin

Inexact Matching

▶Many-to-Many Graph Matching

Information Fusion

▶Data Fusion

Inherent Optical Properties

▶Underwater Effects

Inpainting

Marcelo Bertalmío[1], Vicent Caselles[1], Simon Masnou[2] and Guillermo Sapiro[3]
[1]Universitat Pompeu Fabra, Barcelona, Spain
[2]Institut Camille Jordan, Universitè Lyon 1, Villeurbanne, France
[3]Electrical and Computer Engineering, Computer Science, and Biomedical Engineering, Duke University, Durham, NC, USA

Synonyms

Disocclusion; Error concealment; Filling in

This contribution is dedicated to the memory of Vicent Caselles, outstanding researcher, exceptional friend.

Definition

Given an image and a region Ω inside it, the inpainting problem consists in modifying the image values of the pixels in Ω so that this region does not stand out with respect to its surroundings. The purpose of inpainting might be to restore damaged portions of an image (e.g., an old photograph where folds and scratches have left image gaps) or to remove unwanted elements present in the image (e.g., a microphone appearing in a film frame). See Fig. 1. The region Ω is always given by the user, so the localization of Ω is not part of the inpainting problem. Almost all inpainting algorithms treat Ω as a hard constraint, whereas some methods allow some relaxing of the boundaries of Ω.

This definition, given for a single-image problem, extends naturally to the multi-image case; therefore, this entry covers both image and video inpainting. What is not however considered in this text is *surface* inpainting (e.g., how to fill holes in 3D scans), although this problem has been addressed in the literature.

Background

The term *inpainting* comes from art restoration, where it is also called *retouching*. Medieval artwork started to be restored as early as the Renaissance, the motives being often as much to bring medieval pictures "up to date" as to fill in any gaps. The need to retouch the image in an unobtrusive way extended naturally from paintings to photography and film. The purposes remained the same: to revert deterioration (e.g., scratches and dust spots in film) or to add or remove elements (e.g., the infamous "airbrushing" of political enemies in Stalin era USSR). In the digital domain, the inpainting problem first appeared under the name "error concealment" in telecommunications, where the need was to fill in image blocks that had been lost during data transmission. One of the first works to address automatic inpainting in a general setting dubbed it "image disocclusion" since it treated the image gap as an occluding object that had to be removed, and the image underneath would be the restoration result. Popular terms used to denote inpainting algorithms are also "image completion" and "image fill-in."

Application

The extensive literature on digital image inpainting may be roughly grouped into three categories: patch-based, sparse, and PDEs/variational methods.

From Texture Synthesis to Patch-Based Inpainting

Efros and Leung [14] proposed a method that, although initially intended for texture synthesis, has proven most effective for the inpainting problem. The image gap is filled in recursively, inwards from the gap boundary: each "empty" pixel P at the boundary is filled with the value of the pixel Q (lying outside the image gap, that is, Q is a pixel with valid information) such that the neighborhood $\Psi(Q)$ of Q (a square patch centered in Q) is most similar to the (available) neighborhood $\Psi(P)$ of P. Formally, this can be expressed as an optimization problem:

$$\begin{aligned} \text{Output}(P) &= \text{Value}(Q),\ P \in \Omega,\ Q \notin \Omega, \\ Q &= \arg\min d(\Psi(P), \Psi(Q)), \end{aligned} \tag{1}$$

where $d(\Psi(P), \Psi(Q))$ is the sum of squared differences (SSD) among the patches $\Psi(P)$ and $\Psi(Q)$ (considering only available pixels):

$$d(\Psi_1, \Psi_2) = \sum_i \sum_j |\Psi_1(i,j) - \Psi_2(i,j)|^2, \tag{2}$$

and the indices i, j span the extent of the patches (e.g., if Ψ is an 11×11 patch, then $0 \le i, j \le 10$). Once P is filled in, the algorithm marches on to the next pixel at the boundary of the gap, never going back to P (whose value is, therefore, not altered again). See Fig. 2 for an overview of the algorithm and Fig. 3 for an example of the outputs it can achieve. The results are really impressive for a wide range of images. The main shortcomings of this algorithm are its computational cost, the selection of the neighborhood size (which in the original paper is a global user-selected parameter but which should change locally, depending on image content), the filling order (which may create unconnected boundaries for some objects), and the fact that it cannot deal well with image perspective (it was intended to synthesize frontal textures; hence, neighborhoods are compared always with the same size and orientation). Also, results are poor if the image gap is very large and

Inpainting, Fig. 1 The inpainting problem. *Left*: original image. *Middle*: inpainting mask Ω, in *black*. *Right*: an inpainting result (Figure taken from [20])

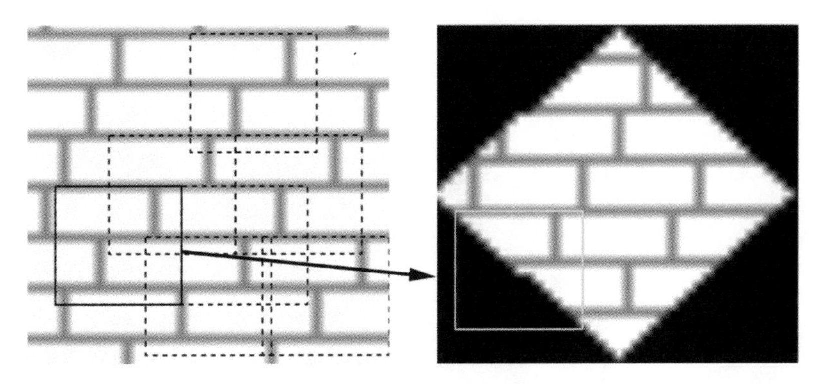

Inpainting, Fig. 2 Efros and Leung's algorithm overview (figure taken from [14]). Given a sample texture image (*left*), a new image is being synthesized one pixel at a time (*right*). To synthesize a pixel, the algorithm first finds all neighborhoods in the sample image (boxes on the *left*) that are similar to the pixels neighborhood (box on the *right*) and then randomly chooses one neighborhood and takes its *center* to be the newly synthesized pixel

disperse (e.g., an image where 80 % of the pixels have been lost due to random *salt and pepper noise*).

Criminisi et al. [12] improved on this work in two aspects. Firstly, they changed the filling order from the original "onion-peel" fashion to a priority scheme where empty pixels at the edge of an image object have higher priority than empty pixels on flat regions. Thus, they are able to correctly inpaint straight object boundaries which could have otherwise ended up disconnected with the original formulation. See Fig. 4. Secondly, they copy entire patches instead of single pixels, so this method is considerably faster. Several shortcomings remain, though, like the inability to deal with perspective and the need to manually select the neighborhood size (here, there are two sizes to set, one for the patch to compare with and another for the patch to copy from). Also, objects with curved boundaries may not be inpainted correctly.

Ashikhmin [2] contributed as well to improve on the original method of Efros and Leung [14]. With the idea of reducing the computational cost of the procedure, he proposed to look for the best candidate Q to copy its value to the empty pixel P not searching the whole image but only searching among the candidates of the neighbors of P which have already been inpainted. See Fig. 5. The speedup achieved with this simple technique is considerable, and also there is a very positive effect regarding the visual quality of the output. Other methods reduce the search space and computational cost involved in the candidate patch search by organizing image patches in tree structures, reducing the dimensionality of the patches with techniques like principal component analysis (PCA), or using randomized approaches.

While most image inpainting methods attempt to be fully automatic (aside from the manual setting of some parameters), there are user-assisted methods that provide remarkable results with just a little input from the user. In the work by Sun et al. [27], the user must specify curves in the unknown region, curves corresponding to relevant object boundaries. Patch synthesis is performed along these curves inside the image gap,

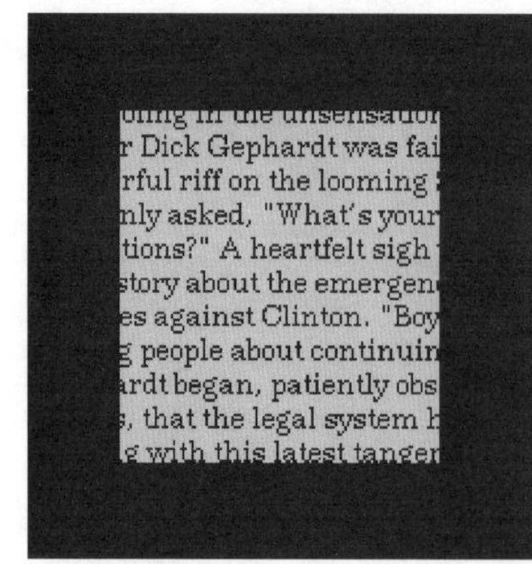

Inpainting, Fig. 3 *Left*: original image, inpainting mask Ω in *black*. *Right*: inpainting result obtained with Efros and Leung's algorithm, images taken from their paper [14]

Inpainting, Fig. 4 *Left*: original image. *Right*: inpainting result obtained with the algorithm of Criminisi et al. [12], images taken from their paper

by copying from patches that lie on the segments of these curves which are outside the gap, in the "known" region. Once these curves are completed, in a process which the authors call *structure propagation*, the remaining empty pixels are inpainted using a technique like the one by Ashikhmin [2] with priorities as in Criminisi et al. [12]. Barnes et al. [5] accelerate this method and make it interactive, by employing randomized searches and combining into one step the structure propagation and texture synthesis processes of Sun et al. [27].

The Role of Sparsity

After the introduction of patch-based methods for texture synthesis by Efros and Leung [14], and image inpainting by Criminisi et al. [12], it became clear that the patches of an image provide a good dictionary to express other parts of the image. This idea has been successfully applied to other areas of image processing, for example, denoising and segmentation.

More general sparse image representations using dictionaries have proven their efficiency in the context of inpainting. For instance, using overcomplete dictionaries adapted to the representation of image geometry and texture, Elad et al. [15] proposed an image decomposition model with sparse coefficients for the geometry and texture components of the image and showed that the model can be easily adapted for image inpainting. A further description of this model follows.

Let u be an image represented as a vector in $\mathbb{R}^N$. Let the matrices D_g and D_t of sizes $N \times k_g$ and

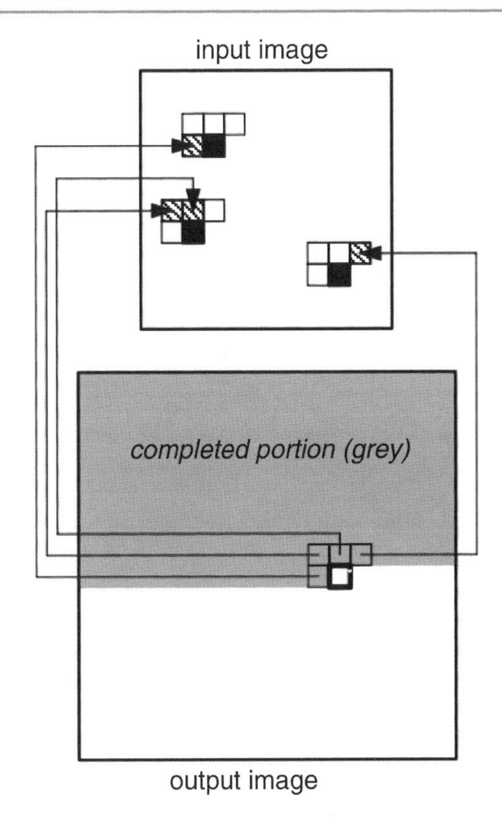

Inpainting, Fig. 5 Ashikhmin's texture synthesis method (figure taken from [2]). Each pixel in the current L-shaped neighborhood generates a shifted candidate pixel (*black*) according to its original position (hatched) in the input texture. The best pixel is chosen among these candidates only. Several different pixels in the current neighborhood can generate the same candidate

$N \times k_t$ represent two dictionaries adapted to geometry and texture, respectively. If $\alpha_g \in \mathbb{R}^{k_g}$ and $\alpha_t \in \mathbb{R}^{k_g}$ represent the geometry and texture coefficients, then $u = D_g\alpha_g + D_t\alpha_t$ represents the image decomposition using the dictionaries collected in D_g and D_t. A sparse image representation is obtained by minimizing

$$\min_{(\alpha_g,\alpha_t):u=D_g\alpha_g+D_t\alpha_t} \|\alpha_g\|_p + \|\alpha_t\|_p, \tag{3}$$

where $p = 0, 1$. Although the case $p = 0$ represents the sparseness measure (i.e., the number of nonzero coordinates), it leads to a nonconvex optimization problem whose minimization is more complex. The case $p = 1$ yields a convex and tractable optimization problem leading also to sparseness. Introducing the constraint by penalization (thus, in practice, relaxing it) and regularizing the geometric part of the decomposition with a total variation semi-norm penalization, Elad et al. [15] propose the variational model:

$$\min_{(\alpha_g,\alpha_t)} \|\alpha_g\|_1 + \|\alpha_t\|_1 + \lambda\|u - D_g\alpha_g - D_t\alpha_t\|_2^2 + \gamma TV(D_g\alpha_g), \tag{4}$$

where TV denotes the total variation, $\lambda, \gamma > 0$. This model can be easily adapted to a model for image inpainting. Observe that $u - D_g\alpha_g - D_t\alpha_t$ can be interpreted as the noise component of the image and λ is a penalization parameter that depends inversely on the noise power. Then the inpainting mask can be interpreted as a region where the noise is very large (infinite). Thus, if $M = 0$ and $= 1$ identify the inpainting mask and the known part of the image, respectively, then the extension of (4) to inpainting can be written as

$$\min_{(\alpha_g,\alpha_t)} \|\alpha_g\|_1 + \|\alpha_t\|_1 + \lambda\|M(u - D_g\alpha_g - D_t\alpha_t)\|_2^2 + \gamma TV(D_g\alpha_g). \tag{5}$$

Writing the energy in (5) using $u_g := D_g u$, $u_t := D_t u$ as unknown variables, it can be observed that $\alpha_g = D_g^+ u_g + r_g$, $\alpha_t = D_t^+ u_t + r_t$, where D_g^+, D_t^+ denote the corresponding pseudoinverse matrices and r_g, r_t are in the null spaces of D_g and D_t, respectively. Assuming for simplicity, as in Elad et al. [15], that $r_g = 0$, $r_t = 0$, the model (5) can be written as

$$\min_{(\alpha_g,\alpha_t)} \|D_g^+ u_g\|_1 + \|D_t^+ u_t\|_1 + \lambda\|M(u - u_g - u_t)\|_2^2 + \gamma TV(u_g). \tag{6}$$

This simplified model is justified in Elad et al. [15] by several reasons: it is an upper bound for (5), it is easier to solve, it provides good results, it has a Bayesian interpretation, and it is equivalent to (5) if D_g and D_t are non-singular or when using the ℓ^2 norm in place of the ℓ^1 norm. The model has nice features since it permits to use adapted dictionaries for geometry and texture and treats the inpainting as missing samples, and the sparsity model is included with ℓ^1 norms that are easy to solve.

This framework has been adapted to the use of dictionaries of patches and has been extended in several directions like image denoising, filling in missing pixels (Aharon et al. [1]), color image denoising, demosaicing, and inpainting of small holes (Mairal et al. [21]

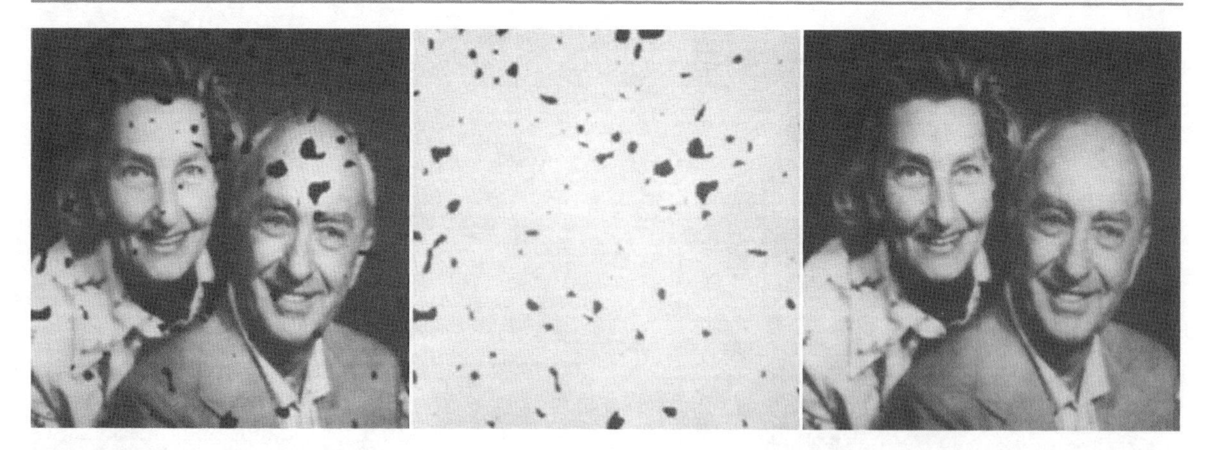

Inpainting, Fig. 6 An inpainting experiment taken from Ogden et al. [24]. The method uses a Gaussian pyramid and a series of linear interpolations, downsampling, and upsampling

and further extended to deal with multiscale dictionaries and to cover the case of video sequences in Mairal et al. [22]. To give a brief review of this model, some notation is required. Image patches are squares of size $n = \sqrt{n} \times \sqrt{n}$. Let D be a dictionary of patches represented by a matrix of size $n \times k$, where the elements of the dictionary are the columns of D. If $\alpha \in \mathbb{R}^k$ is a vector of coefficients, then $D\alpha$ represents the patch obtained by linear combination of the columns of D. Given an image $v(i, j), i, j \in \{1, \ldots, N\}$, the purpose is to find a dictionary $\hat{D}$, an image $\hat{u}$, and coefficients $\hat{\alpha} = \{\hat{\alpha}_{i,j} \in \mathbb{R}^k : i, j \in \{1, \ldots, N\}\}$ which minimize the energy

$$\min_{(\alpha,D,u)} \lambda\|v - u\|_2 + \sum_{i,j=1}^{N} \mu_{i,j}\|\alpha_{i,j}\|_0 + \sum_{i,j=1}^{N} \|D\alpha_{i,j} - R_{i,j}u\|_2, \tag{7}$$

where $R_{i,j}u$ denotes the patch of u centered at (i, j) (dismissing boundary effects), and $\mu_{i,j}$ are positive weights. The solution of the nonconvex problem (7) is obtained using an alternate minimization: a sparse coding step where one computes $\alpha_{i,j}$ knowing the dictionary D for all i, j, a dictionary update using a sequence of one rank approximation problem to update each column of D (Aharon et al. [1]), and a final reconstruction step given by the solution of

$$\min_{u} \lambda\|v - u\|_2 + \sum_{i,j=1}^{N} \|\hat{D}\alpha_{i,j} - R_{i,j}u\|_2. \tag{8}$$

Again, the inpainting problem can be considered as a case of nonhomogeneous noise. Defining for each pixel (i, j) a coefficient $\beta_{i,j}$ inversely proportional to the noise variance, a value of $\beta_{i,j} = 0$ may be taken for each pixel in the inpainting mask. Then the inpainting problem can be formulated as

$$\min_{(\alpha,D,u)} \lambda\|\beta \otimes (v - u)\|_2 + \sum_{i,j=1}^{N} \mu_{i,j}\|\alpha_{i,j}\|_0 + \sum_{i,j=1}^{N} \|(R_{i,j}\beta) \otimes (D\alpha_{i,j} - R_{i,j}u)\|_2, \tag{9}$$

where $\beta = (\beta_{i,j})_{i,j=1}^{N}$ and $\otimes$ denotes the elementwise multiplication between two vectors.

With suitable adaptations, this model has been applied to inpainting (of relatively small holes), to interpolation from sparse irregular samples and super-resolution, to image denoising, to demosaicing of color images, and to video denoising and inpainting, obtaining excellent results; see Mairal et al. [22].

PDEs and Variational Approaches

All the methods mentioned so far are based on the same principle: a missing/corrupted part of an image

Inpainting, Fig. 7 Amodal completion: the visual system automatically completes the broken edge in the *left* figure. The *middle* figure illustrates that, here, no global symmetry process is involved: in both figures, the same edge is synthesized. In such simple situation, the interpolated curve can be modeled as Euler's elastica, that is, a curve with clamped points and tangents at its extremities and with minimal oscillations

can be well synthetized by suitably sampling and copying uncorrupted patches (taken either from the image itself or built from a dictionary). A very different point of view underlies many contributions involving either a variational principle, through a minimization process, or a (non necessarily variational) partial differential equation (PDE).

An early interpolation method that applies for inpainting is due to Ogden et al. [24]. Starting from an initial image, a Gaussian filtering is built by iterated convolution and subsampling. Then, a given inpainting domain can be filled in by successive linear interpolations, downsampling, and upsampling at different levels of the Gaussian pyramid. The efficiency of such approach is illustrated in Fig. 6.

Masnou and Morel proposed in [23] to interpolate a gray-valued image by extending its isophotes (the lines of constant intensity) in the inpainting domain. This approach is very much in the spirit of early works by Kanizsa, Ullman, Horn, Mumford, and Nitzberg to model the ability of the visual system to complete edges in an occlusion or visual illusion context. This is illustrated in Fig. 7. The general completion process involves complicated phenomena that cannot be easily and univocally modeled. However, experimental results show that, in simple occlusion situations, it is reasonable to argue that the brain extrapolates broken edges using elastica-type curves, that is, curves that join two given points with prescribed tangents at these points, a total length lower than a given L, and minimize the Euler elastica energy $\int |\kappa(s)|^2 ds$, with s the curve arc length and κ the curvature.

The model by Masnou and Morel [23] generalizes this principle to the isophotes of a gray-valued image. More precisely, denoting $\tilde{\Omega}$ a domain slightly larger than Ω, it is proposed in [23] to extrapolate the isophotes of an image u, known outside Ω and valued in $[m, M]$, by a collection of curves $\{\gamma_t\}_{t \in [m,M]}$ with no mutual crossings, that coincide with the isophotes of u on $\tilde{\Omega} \setminus \Omega$ and that minimize the energy

$$\int_m^M \int_{\gamma_t} (\alpha + \beta |\kappa_{\gamma_t}|^p) ds\, dt. \tag{10}$$

Here α, β are two context-dependent parameters. This energy penalizes a generalized Euler's elastica energy, with curvature to the power $p > 1$ instead of 2, of all extrapolation curves γ_t, $t \in [m, M]$.

An inpainting algorithm, based on the minimization of (10) in the case $p = 1$, is proposed by Masnou and Morel in [23]. A globally minimal solution is computed using a dynamic programming approach that reduces the algorithmical complexity. The algorithm handles only simply connected domains, that is, those with no holes. In order to deal with color images, RGB images are turned into a luma/chrominance representation, for example, YCrCb, or Lab, and each channel is processed independently. The reconstruction process is illustrated in Fig. 8.

The word *inpainting*, in the image processing context, has been coined first by Bertalmío, Sapiro, Caselles, and Ballester in [6], where a PDE model is proposed in the very spirit of real paintings restoration. More precisely, u being a gray-valued image to be inpainted in Ω, a time-stepping method for the transport-like equation

$$u_t = \nabla^{\perp} u \cdot \nabla \Delta u \quad \text{in } \Omega, \qquad u \quad \text{given in } \Omega^c \tag{11}$$

is combined with anisotropic diffusion steps that are interleaved for stabilization, using the following diffusion model:

$$u_t = \varphi_\epsilon(x)\, |\nabla u|\, \nabla \cdot \frac{\nabla u}{|\nabla u|}, \tag{12}$$

where φ_ϵ is a smooth cutoff function that forces the equation to act only in Ω, and $\nabla \cdot (\nabla u / |\nabla u|)$ is the curvature along isophotes. This diffusion equation, which has been widely used for denoising an image while preserving its edges, compensates any shock possibly created by the transport-like equation. What is the meaning of Eq. (11)? Following Bertalmío et al. [6], Δu is a measure of image smoothness, and stationary points for the equation are images for which Δu

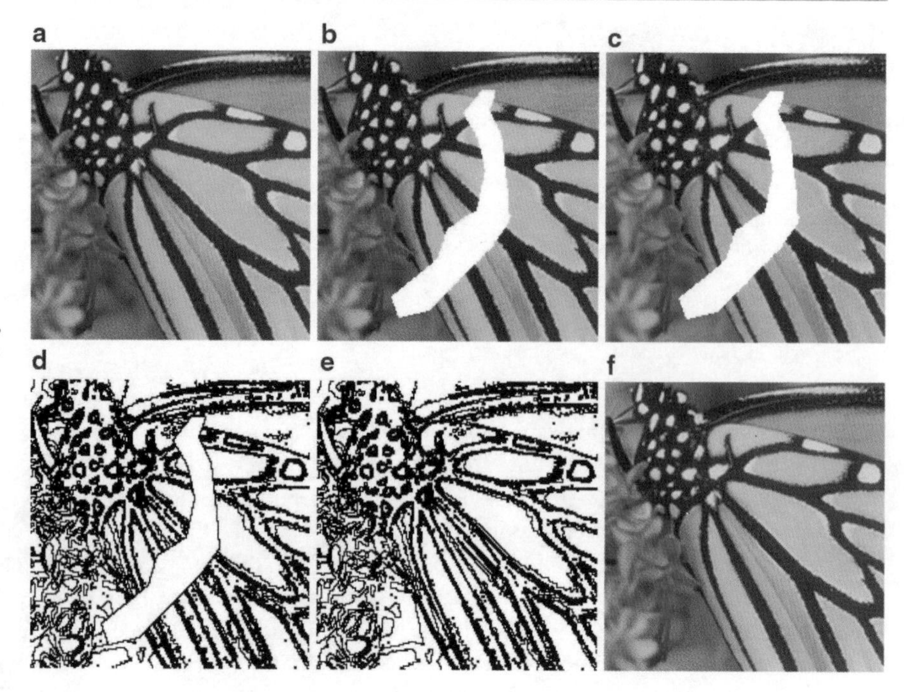

Inpainting, Fig. 8 (**a**) is the original image and (**b**) the image with occlusions in *white*. The luminance channel is shown in figure (**c**). A few isophotes are drawn in figure (**d**), and their reconstruction by the algorithm of Masnou and Morel [23] is given in figure (**e**). Applying the same method to the luminance, hue, and saturation channels yields the final result of figure (**f**)

is constant along the isophotes induced by the vector field $\nabla^{\perp}u$. Equation (11) is not explicitly a transport equation for Δu, but, in the equivalent form,

$$u_t = -\nabla^{\perp}\Delta u \cdot \nabla u, \tag{13}$$

it is a transport equation for u being convected by the field $\nabla^{\perp}\Delta u$. Following Bornemann and März [9], this field is in the direction of the level lines of Δu, which are related to the Marr-Hildreth edges. Indeed, the zero crossings of (a convoluted version of) Δu are the classical characterization of edges in the celebrated model of Marr and Hildreth. In other words, as in the real paintings restoration, the approach of Bertalmío et al. [6] consists in conveying the image intensities along the direction of the edges, from the boundary of the inpainting domain Ω toward the interior. The efficiency of such approach is illustrated in Fig. 9. From a numerical viewpoint, the transport equation and the anisotropic diffusion can be implemented with classical finite difference schemes. For color images, the coupled system can be applied independently to each channel of any classical luma/chrominance representation. There is no restriction on the topology of the inpainting domain.

Another perspective on this model is provided by Bertalmío, Bertozzi, and Sapiro in [7], where connections with the classical Navier-Stokes equation of fluid dynamics are shown. Indeed, the steady-state equation of Bertalmío et al. [6],

$$\nabla^{\perp}u \cdot \nabla\Delta u = 0,$$

is exactly the equation satisfied by steady-state inviscid flows in the two-dimensional incompressible Navier-Stokes model. Although the anisotropic diffusion equation (12) is not the exact counterpart of the viscous diffusion term used in the Navier-Stokes model for incompressible and Newtonian flows, a lot of the numerical knowledge on fluid mechanics seems to be adaptable to design stable and efficient schemes for inpainting. Results in this direction are shown in Bertalmío et al. [7].

Chan and Shen propose in [10] a denoising/inpainting first-order model based on the joint minimization of a quadratic fidelity term outside Ω and a total variation criterion in Ω, that is, the joint energy

$$\int_A |\nabla u| dx + \frac{\lambda}{2}\int_\Omega |u - u_0|^2 dx,$$

with $A \supset\supset \Omega$ the image domain and λ a Lagrange multiplier. The existence of solutions to this problem follows easily from the properties of functions of bounded variation. As for the implementation, Chan and Shen look for critical points of the energy

Inpainting, Fig. 9 An experiment taken from Bertalmío et al. [6]. *Left*: original image. *Middle*: a user-defined mask. *Right*: the result with the algorithm of [6]

Inpainting, Fig. 10 An experiment taken from Chan and Shen [10]. *Left*: original image. *Right*: after denoising and removal of text

using a Gauss-Jacobi iteration scheme for the linear system associated to an approximation of the Euler-Lagrange equation by finite differences. More recent approaches to the minimization of total variation with subpixel accuracy should nowadays be preferred. From the phenomenological point of view, the model of Chan and Shen [10] yields inpainting candidates with the smallest possible isophotes. It is therefore more suitable for thin or sparse domains. An illustration of the model's performances is given in Fig. 10.

Turning back to the criterion (10), a similar penalization on $\tilde{\Omega}$ of both the length and the curvature of all isophotes of an image u yields two equivalent forms, in the case where u is smooth enough (see Masnou and Morel [23]):

$$\int_{-\infty}^{+\infty} \int_{\{u=t\}\cap\tilde{\Omega}} (\alpha + \beta|\kappa|^p) ds\, dt = \int_{\tilde{\Omega}} |\nabla u| \left(\alpha + \beta \left| \nabla \cdot \frac{\nabla u}{|\nabla u|} \right|^p \right) dx. \quad (14)$$

There have been various contributions to the numerical approximation of critical points for this criterion. A fourth-order time-stepping method is proposed by Chan et al. in [11] based on the approximation of the Euler-Lagrange equation, for the case $p = 2$, using upwind finite differences and a min-mod formula for estimating the curvature. Such high-order evolution method suffers from well-known stability and convergence issues that are difficult to handle.

A model, slightly different from (14), is tackled by Ballester et al. in [4] using a relaxation approach. The key idea is to replace the second-order term $\nabla \cdot \frac{\nabla u}{|\nabla u|}$ with a first-order term, depending on an auxiliary variable. More precisely, Ballester et al. study in [4] the minimization of

$$\int_{\tilde{\Omega}} |\nabla \cdot \theta|^p (a + b|\nabla G * u|) dx + \alpha \int_{\tilde{\Omega}} (|\nabla u| - \theta \cdot \nabla u) dx,$$

under the constraint that θ is a vector field with subunit modulus and prescribed normal component on the boundary of $\tilde{\Omega}$, and u takes values in the same range as in Ω^c. Clearly, θ plays the role of $\nabla u / |\nabla u|$, but the new criterion is much less singular. As for G, it is a regularizing kernel introduced for technical reasons in order to ensure the existence of a minimizing couple (u, θ).

The main difference between the new relaxed criterion and (14), besides singularity, is the term $\int_{\tilde{\Omega}} |\nabla \cdot \theta|^p$ which is more restrictive, despite the relaxation, than $\int_{\tilde{\Omega}} |\nabla u| \left|\nabla \cdot \frac{\nabla u}{|\nabla u|}\right|^p dx$. However, the new model has a nice property: a gradient descent with respect to (u, θ) can be easily computed and yields two coupled second-order equations whose numerical approximation is standard. Results obtained with this model are shown in Fig. 11.

The Mumford-Shah-Euler model by Esedoglu and Shen [16] is also variational. It combines the celebrated Mumford-Shah segmentation model for images and the Euler's elastica model for curves. Being u_0 the original image defined on a domain A, and $\Omega \subset A$ the inpainting domain, Esedoglu and Shen propose to find a piecewise weakly smooth function u, that is a function with integrable squared gradient out of a discontinuity set $K \subset A$, that minimizes the criterion

$$\int_{A\setminus\Omega} \lambda \|u-u_0\|^2 dx + \int_{A\setminus K} \gamma |\nabla u|^2 dx + \int_K (\alpha + \beta\,\kappa^2) ds,$$

where $\alpha, \beta, \gamma, \lambda$ are positive parameters. The resulting image is not only reconstructed in the inpainting domain Ω, but also segmented all over A since the original image is not imposed as a hard constraint.

Two numerical approaches to the minimization of this model are discussed in Esedoglu and Shen [16]: first, a level set approach based on the representation of K as the zero-level set of a sequence of smooth functions that concentrate, and the explicit derivation, using finite differences, of the Euler-Lagrange equations associated with the criterion; second, a Γ-convergence approach based on a result originally conjectured by De Giorgi and recently proved by Röger and Schätzle in dimensions 2,3. In both cases, the final system of discrete equations is of order four, facing again difficult issues of convergence and stability.

More recently, following the work of Grzibovskis and Heintz on the Willmore flow, Esedoglu et al. [17] have addressed the numerical flow associated with the Mumford-Shah-Euler model using a promising convolution/thresholding method that is much easier to handle than the previous approaches.

Tschumperlé proposes in [28] an efficient second-order anisotropic diffusion model for multivalued image regularization and inpainting. Given a $\mathbb{R}^N$-valued image u known outside Ω, and starting from an initial rough inpainting obtained by straightforward advection of boundary values, the pixels in the inpainting domain are iteratively updated according to a finite difference approximation to the equations

$$\frac{\partial u_i}{\partial t} = \mathrm{trace}(T\,\nabla^2 u_i), \qquad i \in \{1, \cdots, N\}.$$

Here, T is the tensor field defined as

$$T = \frac{1}{(1+\lambda_{\min}+\lambda_{\max})^{\alpha_1}} v_{\min} \otimes v_{\min} + \frac{1}{(1+\lambda_{\min}+\lambda_{\max})^{\alpha_2}} v_{\max} \otimes v_{\max},$$

with $0 < \alpha_1 \ll \alpha_2$, and $\lambda_{\min}, \lambda_{\max}, v_{\min}, v_{\max}$ are the eigenvalues and eigenvectors, respectively, of $G_\sigma * \sum_{i=1}^N \nabla u_i \otimes \nabla u_i$, being G_σ a smoothing kernel and $\sum_{i=1}^N \nabla u_i \otimes \nabla u_i$ the classical structure tensor, which is known for representing well the local geometry of u. Figure 12 reproduces an experiment taken from Tschumperlé [28].

The approach of Auroux and Masmoudi in [3] uses the PDE techniques that have been developed for the inverse conductivity problem in the context of crack detection. The link with inpainting is the following: missing edges are modeled as cracks, and the image is assumed to be smooth out of these cracks. Given a crack, two inpainting candidates can be obtained as the solutions of the Laplace equation with Neumann condition along the crack and either a Dirichlet or a Neumann condition on the domain's boundary. The optimal cracks are those for which the two candidates are the most similar in quadratic norm, and they can be found through topological analysis, that is, they correspond to the set of points where putting a crack mostly decreases the quadratic difference. Both the localization of the cracks and the associated piecewise smooth inpainting solutions can be found using fast and simple finite difference schemes.

Finally, Bornemann and März propose in [9] a first-order model to advect the image information along the integral curves of a coherence vector field that extends in Ω the dominant directions of the image gradient. This coherence field is explicitly defined, at every point, as the normalized eigenvector to the minimal eigenvalue of a smoothed structure tensor whose computation carefully avoids boundary biases in the vicinity of $\partial\Omega$. Denoting c the coherence field,

Inpainting, Fig. 11 Two inpainting results obtained with the model proposed by Ballester et al. [4]. Observe in particular how curved edges are restored

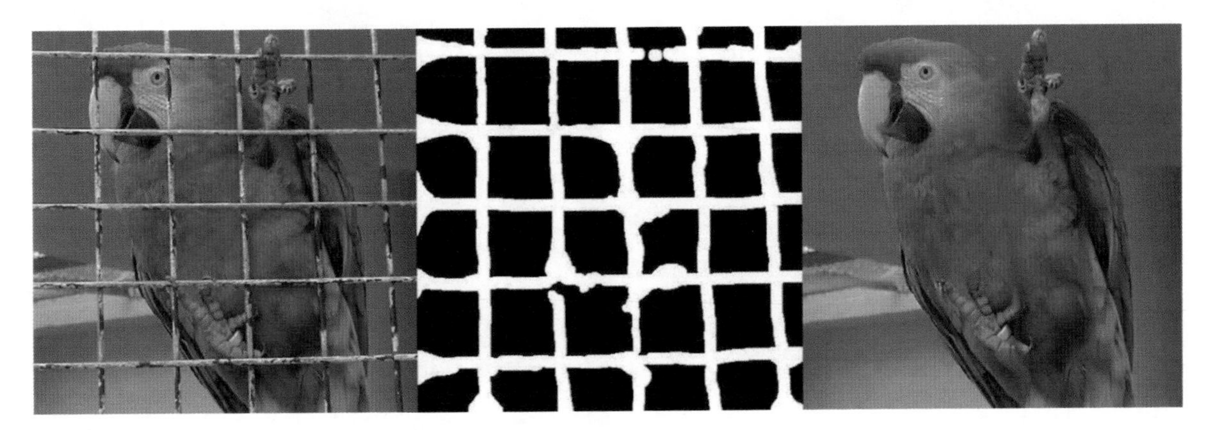

Inpainting, Fig. 12 An inpainting experiment (the *middle* image is the mask defined by the user) taken from Tschumperlé [28]

Bornemann and März show that the equation $c \cdot \nabla u = 0$ with Dirichlet boundary constraint can be obtained as the vanishing viscosity limit of an efficient fast-marching scheme: the pixels in Ω are synthesized one at a time, according to their distance to the boundary. The new value at a pixel p is a linear combination of both known and previously generated values in a neighborhood of p. The key ingredient of the method is the explicit definition of the linear weights according to the coherence field c. Although the Bornemann-März model requires a careful tune of four parameters, it is much faster than the PDE approaches mentioned so far and performs very well, as illustrated in Fig. 13.

Combining and Extending PDEs and Patch Models

In general, most PDE/variational methods that have been presented so far perform well for inpainting either thin or sparsely distributed domains. However, there is a common drawback to all these methods: they are unable to restore texture properly, and this is particularly visible on large inpainting domains, for instance, in the inpainting result of Fig. 12 where the diffusion method is not able to recover the parrot's texture. On the other hand, patch-based methods are not able to handle sparse inpainting domains like in Fig. 14, where no valid squared patch can be found that does not reduce to a point. On the contrary, most PDE/variational methods remain applicable in such situation, like in Fig. 14 where the model proposed by Masnou and Morel [23] yields the inpainting result. Obviously, some geometric information can be recovered, but no texture.

There have been several attempts to explicitly combine PDEs and patch-based methods in order to handle properly both texture and geometric structures. The contribution of Criminisi et al. [12] was mentioned

Inpainting, Fig. 13 An inpainting experiment taken from Bornemann and März [9], with a reported computation time of 0.4 s

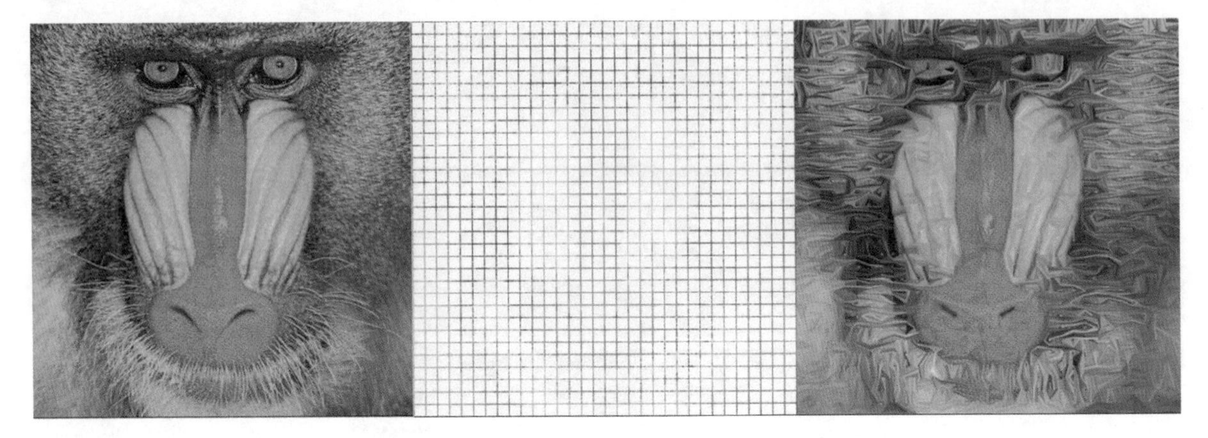

Inpainting, Fig. 14 A picture of a mandrill, the same picture after removal of 15 × 15 squares (more than 87 % of the pixels are removed), and the reconstruction with the method introduced by Masnou and Morel [23] using only the one-pixel-wide information at the squares' boundaries

already. The work of Bertalmío et al. [8] uses an additive decomposition of the image to be inpainted into a geometric component that contains all edges information, and a texture component. Then the texture image is restored using the Efros and Leung's algorithm of [14], while the geometric image is inpainted following the method proposed in Bertalmío et al. [6] (several subsequent works have proposed other methods for the individual reconstruction of each component). The final image is obtained by addition of the restored texture and geometric components. In a few situations where the additive decomposition makes sense, this approach does indeed improve the result and extends the applications domain of inpainting.

In Komodakis and Tziritas [20], the authors combine variational and patch-based strategies by defining an inpainting energy over a graph whose nodes are the centers of patches over the image. The inpainting energy has two terms, one being a texture synthesis term and the other measuring the similarity of the overlapping area of two neighboring patches (centered on nodes which are neighbors in the graph). By minimizing this energy with belief propagation, a label is assigned to each node, which amounts to copying the patch corresponding to the label over the position of the node. The results are very good on a variety of different images (e.g., Fig. 1), and the method is fast. Some potential issues are the following: there is no assurance that the iterative process converges to a global minimum, and visual artifacts may appear since the method uses a fixed grid and entire patches are copied for each pixel of the mask.

The work by Drori et al. in [13] does not involve any explicit geometry/texture decomposition, but the search for similar neighborhoods is guided by a prior rough estimate of the inpainted values using a multiscale sampling and convolution strategy, in the very spirit of Ogden et al. [24]. In addition, in contrast with many patch-based methods, the dictionary of valid patches is enriched using rotations, rescalings, and reflections. An example extracted from Drori et al. [13] is shown in Fig. 15.

Inpainting, Fig. 15 An experiment from Drori et al. [13] illustrating the proposed multiscale diffusion/patch-based inpainting method. The *upper-left* image is the original, the *upper-right* image contains the mask defined by the user, the *bottom-left* image is the result, and the *bottom-right* image shows what has been synthesized in place of the elephant

Beyond Single-Image Inpainting

All the methods mentioned above involve just a single image. For the multi-image case, there are two possible scenarios: video inpainting and inpainting a single image using information from several images.

Basic methods for video inpainting for data transmission (where the problem is known as "error concealment" and involves restoring missing image blocks) and for film restoration applications (dealing with image gaps produced by dust, scratches, or the abrasion of the material) assume that the missing data changes location in correlative frames and therefore use motion estimation to copy information along pixel trajectories. A particular difficulty in video inpainting for film restoration is that, for good visual quality of the outputs, the detection of the gap and its filling in are to be tackled jointly and in a way which is robust to noise, usually employing probabilistic models in a Bayesian framework; see, for example, the book by Kokaram [19].

Wexler et al. [29] propose a video inpainting algorithm that extends to space-time the technique of Efros and Leung [14] and combines it with the idea of coherence among neighbors developed by Ashikhmin [2]. First, for each empty pixel P, they consider a space-time cube centered in P, compare it with all possible cubes in the video, find the most similar, and keep its center pixel Q, which will be the correspondent of P. For each cube the information considered and compared is not only color but also motion vectors. Then, instead of copying the value of Q to P, they copy to P the average of all the values of the shifted correspondents of the neighbors of P: for instance, if R is at the *right* of P, and S is the correspondent of R, then the pixel to the *left* of S will be involved in the average to fill in P. This is based on the idea by Ashikhmin [2], see Fig. 5. The shortcomings of this video inpainting method are that the results present significant blur (due to the averaging), it seems to be limited only to static-camera scenarios (probably due to the simple motion estimation procedure involved) and periodic motion without change of scale, and the computational cost is quite high (due to the comparison of 3D blocks).

Shiratori et al. [26] perform video inpainting by firstly inpainting the motion field with a patch-based technique like that of Efros and Leung [14] and then propagating the colors along the (inpainted) motion trajectories. The method assumes that motion information is sufficient to fill in holes in videos, which is not always the case (e.g., with a static hole over a

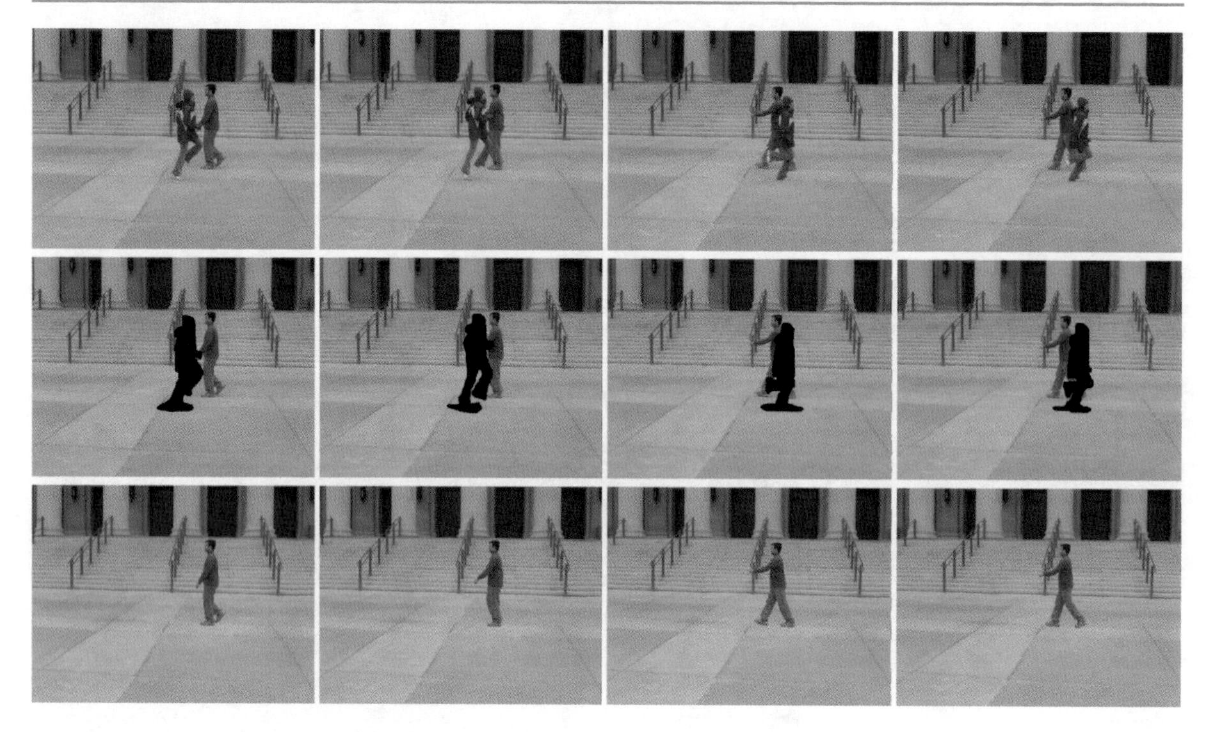

Inpainting, Fig. 16 *Top row*: some frames from a video. *Middle row*: inpainting mask Ω in black. *Bottom row*: video inpainting results obtained with the algorithm of Patwardhan et al. [25]

static region). The results present some blurring, due to the bilinear interpolation in the color propagation step.

Patwardhan et al. [25] propose a video inpainting method consisting of three steps. In the first step they decompose the video sequence into binary motion layers (foreground and background), which are used to build three image *mosaics* (a mosaic is the equivalent of a panorama image created by stitching together several images): one mosaic for the foreground, another for the background, and a third for the motion information. The other two steps of the algorithm perform inpainting, first from the foreground and then from the background: these inpainting processes are aided and sped up by using the mosaics computed in the first step. See Fig. 16 for some results. The algorithm is limited to sequences where the camera motion is approximately parallel to the image plane and foreground objects move in a repetitive fashion and do not change size: these restrictions are imposed so that a patch-synthesis algorithm like that of Efros and Leung [14] can be used.

Hays and Efros [18] perform inpainting of a single image using information from a database with several millions of photographs. They use a scene descriptor to reduce the search space from two million to two hundred images, those images from the database which are *semantically* closer to the image the user wants to inpaint. Using template matching, they align the 200 best matching scenes to the local image around the region to inpaint. Then they composite each matching scene into the target image using seam finding and image blending. Several outputs are generated so the user may select among them, and the results can be outstanding; see Fig. 17. The main shortcoming of this method is that it relies on managing and operating a huge image database. When the algorithm fails, it can be due to a lack of good scene matches (if the target image is atypical), or because of *semantic violations* (e.g., failure to recognize people, hence copying only part of them), or in the case of uniformly textured backgrounds (where this algorithm might not find the precise same texture in another picture of the database).

Open Problems

Inpainting is a very challenging problem, and it is far from being solved; see Fig. 18. Patch-based methods

Inpainting, Fig. 17 *Left*: original image. *Middle*: inpainting mask Ω, in white. *Right*: inpainting result obtained with the method by Hays and Efros [18], images taken from their paper

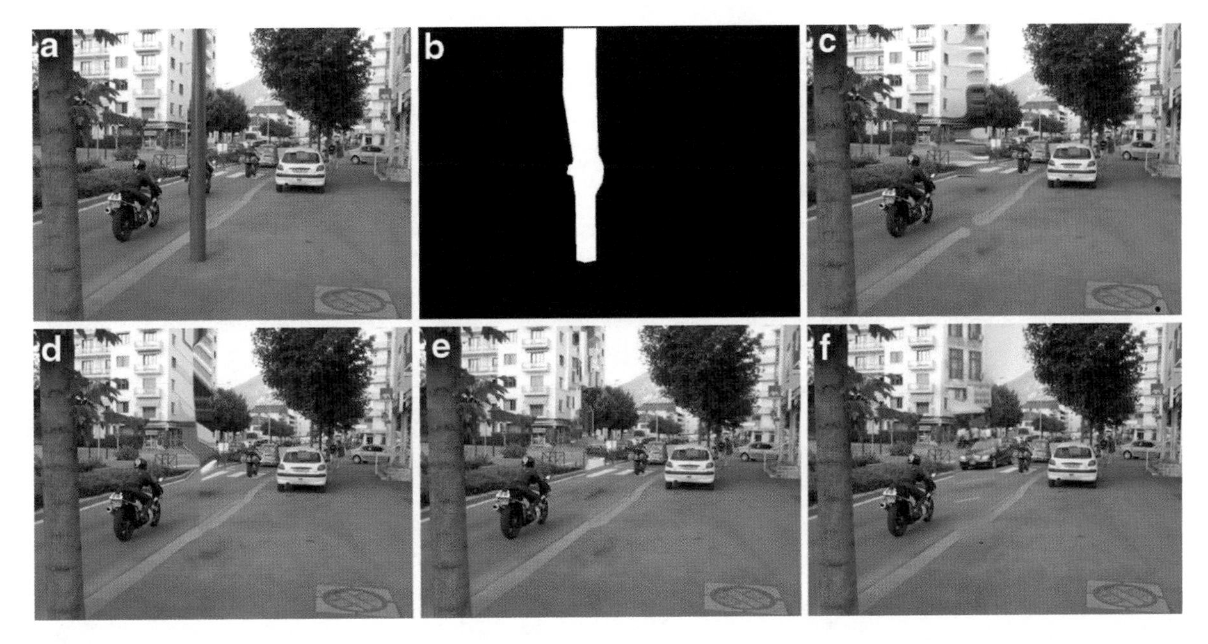

Inpainting, Fig. 18 An example where no inpainting method seems to work. (**a**) Original image, from the database provided by Hays and Efros [18]. (**b**) In *white*, the mask to be inpainted, which is not the initial mask proposed by Hayes and Efros but derives from the fuzzy mask actually used by their algorithm. (**c**) Result courtesy of D. Tschumperlé using the algorithm from [28]. (**d**) Result courtesy of T. März and F. Bornemann using the algorithm from [9]. (**e**) Result using a variant of the algorithm from Criminisi et al. [12]. (**f**) Result from Hays and Efros [18]

work best in general, although for some applications (e.g., very spread, sparsely distributed gap Ω) geometry-based methods might be better suited. And when the image gap lies on a singular location, with surroundings that cannot be found anywhere else, then all patch-based methods give poor results, regardless if they consider or not geometry. For video inpainting the situation is worse; the existing algorithms are few and with very constraining limitations on camera and object motion. Because video inpainting is very relevant in cinema postproduction, in order to replace the current typical labor intensive systems, important developments are expected in the near future.

References

1. Aharon M, Elad M, Bruckstein A (2006) K-SVD: an algorithm for designing overcomplete dictionaries for sparse representation. IEEE Trans Signal Process 54(11):4311
2. Ashikhmin M (2001) Synthesizing natural textures. In: Proceedings of the ACM symposium on interactive 3D graphics, Chapel Hill. ACM, pp 217–226

3. Auroux D, Masmoudi M (2006) A one-shot inpainting algorithm based on the topological asymptotic analysis. Comput Appl Math 25:1–17
4. Ballester C, Bertalmío, M, Caselles V, Sapiro G, Verdera J (2001) Filling-in by joint interpolation of vector fields and gray levels. IEEE Trans Image Process 10(8):1200–1211
5. Barnes C, Shechtman E, Finkelstein A, Goldman DB (2009) Patchmatch: a randomized correspondence algorithm for structural image editing. ACM Trans Graph 28(3):2
6. Bertalmío M, Sapiro G, Caselles V, Ballester C (2000) Image inpainting. In: Proceedings of SIGGRAPH'00, New Orleans, USA, pp 417–424
7. Bertalmío M, Bertozzi A, Sapiro G (2001) Navier-Stokes, fluid dynamics, and image and video inpainting. In: Proceedings of the IEEE international conference on computer vision and pattern recognition (CVPR), Hawaï
8. Bertalmío M, Vese L, Sapiro G, Osher S (2003) Simultaneous structure and texture image inpainting. IEEE Trans Image Process 12(8):882–889
9. Bornemann F, März T (2007) Fast image inpainting based on coherence transport. J Math Imaging Vis 28(3): 259–278
10. Chan TF, Shen J (2001) Mathematical models for local deterministic inpaintings. SIAM J Appl Math 62(3): 1019–1043
11. Chan TF, Kang SH, Shen J (2002) Euler's elastica and curvature based inpainting. SIAM J Appl Math 63(2): 564–592
12. Criminisi A, Pérez P, Toyama K (2004) Region filling and object removal by exemplar-based inpainting. IEEE Trans Image Process 13(9):1200–1212
13. Drori I, Cohen-Or D, Yeshurun H (2003) Fragment-based image completion. In: Proceedings of SIGGRAPH'03, vol 22(3), pp 303–312
14. Efros AA, Leung TK (1999) Texture synthesis by non-parametric sampling. In: Proceedings of the international conference on computer vision, Kerkyra, vol 2, pp 1033
15. Elad M, Starck JL, Querre P, Donoho DL (2005) Simultaneous cartoon and texture image inpainting using morphological component analysis (MCA). Appl Comput Harmon Anal 19(3):340–358
16. Esedoglu S, Shen J (2002) Digital image inpainting by the Mumford-Shah-Euler image model. Eur J Appl Math 13:353–370
17. Esedoglu S, Ruuth S, Tsai R (2008) Threshold dynamics for high order geometric motions. Interfaces Free Boundaries 10(3):263–282
18. Hays J, Efros AA (2008) Scene completion using millions of photographs. Commun ACM 51(10):87–94
19. Kokaram AC (1998) Motion picture restoration: digital algorithms for artefact suppression in degraded motion picture film and video. Springer, London
20. Komodakis N, Tziritas G (2007) Image completion using efficient belief propagation via priority scheduling and dynamic pruning. IEEE Trans Image Process 16(11):2649
21. Mairal J, Elad M, Sapiro G (2008) Sparse representation for color image restoration. IEEE Trans Image Process 17(1):53
22. Mairal J, Sapiro G, Elad M (2008) Learning multiscale sparse representations for image and video restoration. SIAM Multiscale Model Simul 7(1):214–241
23. Masnou S, Morel J.-M. (1998) Level lines based disocclusion. In: 5th IEEE international conference on image processing, Chicago, IL, Oct 4–7
24. Ogden JM, Adelson EH, Bergen JR, Burt PJ (1985) Pyramid-based computer graphics. RCA Eng 30(5):4–15
25. Patwardhan KA, Sapiro G, Bertalmío M (2007) Video inpainting under constrained camera motion. IEEE Trans Image Process 16(2):545–553
26. Shiratori T, Matsushita Y, Tang X, Kang SB (2006) Video completion by motion field transfer. In: 2006 IEEE computer society conference on computer vision and pattern recognition (CVPR), New York, vol 1
27. Sun J, Yuan L, Jia J, Shum HY (2005) Image completion with structure propagation. In: ACM SIGGRAPH 2005 papers. ACM, New York, p 868
28. Tschumperlé D (2006) Fast anisotropic smoothing of multi-valued images using curvature-preserving PDE's. Int J Comput Vis 68(1):65–82
29. Wexler Y, Shechtman E, Irani M (2004) Space-time video completion. In: IEEE computer society conference on computer vision and pattern recognition (CVPR), Washington, DC, vol 1

Interactive Segmentation

Yuri Boykov
Department of Computer Science, University of Western Ontario, London, ON, Canada

Synonyms

Labeling; Object extraction; Partitioning; Segmentation; Semiautomatic; User-assisted; User-guided

Related Concepts

▸Dynamic Programming

Definition

Interactive image segmentation is a (near) real-time mechanism for accurately marking/labeling an object of interest based on visual user interface (VUI) specifying seeds, rough delineation, partial labeling, bounding box, or other constraints. Semiautomatic interactive segmentation methods incorporate various generic image cues and/or object-specific feature detectors in order to facilitate acceptable results with minimum user efforts.

Background

The most basic object extraction techniques like *thresholding* (Fig. 1) and *region growing* are based on simple but very fast heuristics. The spectrum of applications for such techniques is limited as they are prone to many problems, most notably *leaking* as in Fig. 2. Despite significant problems, thresholding and region growing are widely known due to their simplicity and speed. For example, they could be easily run on personal computers available 15–20 years ago. More recent generations of commodity PCs allow much more robust segmentation techniques, which rely on optimization of some segmentation cost function, or an energy. An energy functional should define an explicit measure of goodness for evaluating any specific segmentation result. The main goal of optimization is to find the best segmentation with respect to the specified criteria. In the context of interactive segmentation, the energy can embed some *soft* and *hard* constraints specified by the user.

Discrete Segmentation Functionals

Many discrete optimization methods for interactive segmentation are based on classical combinatorial optimization techniques: *dynamic programming* (DP) or *s/t graph cuts*. In general, these approaches are guaranteed to find the exact global minimum solution in finite (low-order polynomial) number of steps. There are no numerical convergence issues (e.g., oscillations), and they work in near real time even on a single CPU. For efficiency, these methods are often implemented using the simplest 4-neighbor grids. In theory, this basic approach may generate some discrete metrication artifacts, but they are rarely observable on real images. Increasing the neighborhood size (e.g., to eight neighbors) adequately addresses the problem [8, 9].

Graph-path segmentation models are designed for 2D image segmentation. *Intelligent scissors* [2], also known as *live wire* in the medical imaging community [3], requires user to place seeds on the desired object boundary; see Fig. 3. The algorithm connects these seeds by computing the shortest path on a graph where edges (or nodes) are image weighted according to local contrast (intensity gradient). Such weighting makes paths "stick" to image boundaries. The shortest paths from each new seed to all other image pixels can be pre-computed in $O(n \log n)$ time. Then, an optimal path from any mouse position to the seed can be previewed in real time.

This method evaluates segmentation boundary as a path between two seeds (see Fig. 4a) using energy functionals like

$$E(\mathcal{P}_{s,t}) = \sum_{\{p,q\}\in\mathcal{P}_{s,t}} w_{pq} \tag{1}$$

where $\mathcal{P}_{s,t}$ is a set of adjacent edges from source seed s to terminal seed t and edge weights $w_{pq} \geq 0$ are segmentation boundary costs based on some local measure of intensity contrast across edge $\{p,q\}$. One example of weights w_{pq} is

$$w_{pq} \propto \frac{1}{1+|\nabla I \cdot n_{pq}|^2} \cdot ||p-q||$$

where ∇I is image gradient, vector n_{pq} is a normal to edge $\{p,q\}$, and $||p-q||$ is the geometric length of edge $\{p,q\}$. Factor $||p-q||$ differentiates diagonal edges from horizontal and vertical edges on grids with higher connectivity, which can reduce the grid bias.

There is a number of other interactive segmentation methods based on efficient DP-based optimization algorithms. For example, the methods in [10, 11] compute globally optimal cycles (closed contours), minimizing ratios of different measures of segment's boundary and region. For example, [11] can find a segment with the largest average contrast on its boundary. (Ratio of some cumulative contrast measure and the boundary length.) Optimization of ratio functionals evaluating boundary's curvature was addressed in [12].

Graph-cut segmentation models: Boykov and Jolly [5] and Boykov and Funka-Lea [6] proposed an object extraction functional for N-dimensional images that evaluates boundary and region properties of segments as

$$E(x|\theta) = -\sum_{p:x_p=0} \ln \Pr(I_p|\theta_0) - \sum_{p:x_p=1} \ln \Pr(I_p|\theta_1) + \sum_{\{p,q\}\in\mathcal{N}} w_{pq} \cdot [x_p \neq x_q] \tag{2}$$

where $[\cdot]$ are *Iverson* brackets, variables x_p are binary object/background labels at pixels p, parameters

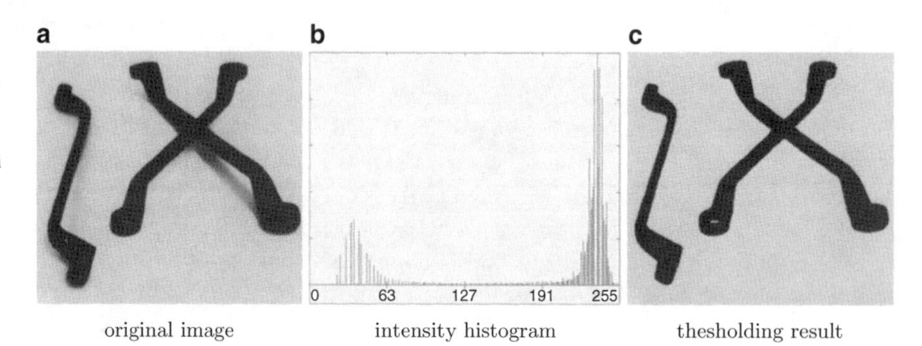

Interactive Segmentation, Fig. 1 Image *thresholding* segments a subset of pixels with intensities in a certain range, for example, $\{p : I_p < T\}$ in (**c**). Some threshold T works only if there is no overlap between the object and background intensities. The images above are from [1]

Interactive Segmentation, Fig. 2 *Region growing* is a greedy heuristic often associated with the *leaking* problem. Segment S is initialized by some *seed* in the object of interest (lake). Adjacent pixels q are iteratively added to S as long as some "growing" criteria are met, for example, $||I_q - I_p|| < T$ for some neighbor $p \in S$. A single low-contrast spot on the object boundary (horizon) will make the lake *leak* into the sky (**b**)

$\theta = \{\theta_0, \theta_1\}$ define object and background intensity distributions, and edge weights w_{pq} are a cost of discontinuity between a pair of neighboring pixels. For example,

$$w_{pq} \propto \exp\left(-\frac{|\nabla I \cdot e_{pq}|^2}{\sigma^2}\right) \cdot \frac{1}{||p-q||}$$

where e_{pq} is a unit vector collinear to edge $\{p, q\}$ and σ is a parameter controlling sensitivity to intensity contrast that is often set according to the level of noise in the image. Similarly to energy (1), the last term in (2) evaluates the image-weighted length of the segmentation boundary. In contrast to edge weights in (1), the weights above are based on intensity contrast along the edge $\{p, q\}$; see Fig. 4b. Normalization by edge length $||p-q||$ is required for grids with higher connectivity, reducing the grid bias [13].

The first two terms in (2) evaluate how well pixel intensities inside the object and background segments fit the corresponding distributions. In general, image intensity/color distributions could be extended by more sophisticated features and appearance models, for example, texture. The appearance models could be estimated from seeds or from prior data. The *grab-cut* method in [7] also uses an iterative EM-style scheme for additionally optimizing functional $E(x|\theta)$ in (2) with respect to parameters θ. In this case, sufficiently good initial appearance models can be often estimated from a user-placed box around the object. The graph-cut segmentation model also extends to video; for example, see the *snap-cut* method [14].

Functional (2) can be globally minimized over binary variables x by low-order polynomial algorithms from combinatorial optimization [15] that are fast even on a single CPU. Also, there are efficient

Interactive Segmentation, Fig. 3 *Intelligent scissors* or *live-wire* methods connect seeds placed on object boundaries. Optimal segmentation boundary (*orange curve*) can be previewed for any mouse position in real time

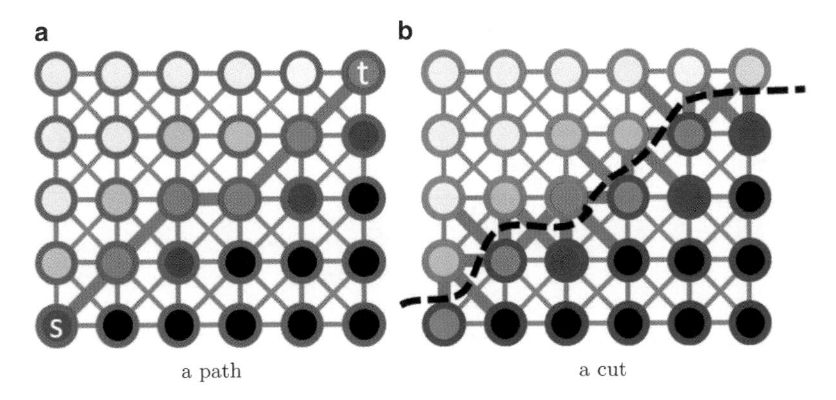

Interactive Segmentation, Fig. 4 Segmentation on graphs. The path-based methods [2–4] represent segmentation boundary as a sequence of adjacent (*green*) edges (**a**). A path could connect two seeds or form a closed cycle. The graph-cut methods [5–7] assign to pixels different labels, for example, *red* and *blue* in (**b**). Any such labeling implicitly defines a segmentation boundary, which is a *cut*, as a collection of (*green*) edges between differently labeled pixels

techniques [5, 6] for integrating interactive hard constraints (seeds) as in Fig. 5. Instead of segmentation energy (2), graph-cut framework can also use various ratio functionals [16].

Segmentation energy (2) works for N objects (labels). In general, its optimization is NP-hard for $N > 2$. An approximate solution with a factor of 2 optimality guarantee can be found via α-expansion optimization algorithm [17]. Interestingly, imposing some additional geometric constraints between object boundaries (e.g., inclusion, exclusion, minimum margin) may lead to exact polynomial optimization algorithms [18–20].

Continuous Segmentation Functionals

Many popular interactive segmentation methods use continuous representation of segments where boundaries are contours in $\mathcal{R}^2$ or surfaces in $\mathcal{R}^3$. Such methods use either physics-based or geometric functionals to evaluate such continuous segments. Traditionally, variational calculus and different forms of *gradient descent* were used to converge to a local minima from a given initial contour; see Fig. 6. This motivates the general term, *active contours*, commonly used for such methods. Recent convex formulations for standard continuous regularization functionals [21, 22]

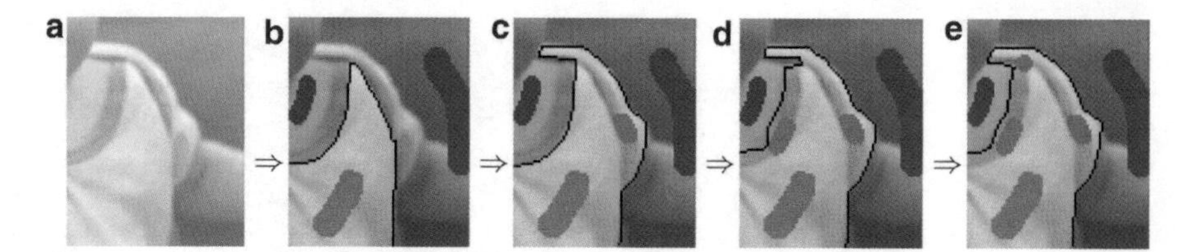

Interactive Segmentation, Fig. 5 Interactive editing of segments via hard constraints (seeds) based on graph cuts [5, 6]. A fragment of an original photo is shown in (**a**). Initial seeds and segmentation are shown in (**b**). The results in (**c**)–(**e**) illustrate changes in optimal segmentation as new hard constraints are successively added. The computation time for consecutive corrections in (**c**)–(**e**) is marginal compared to time for initial results in (**b**)

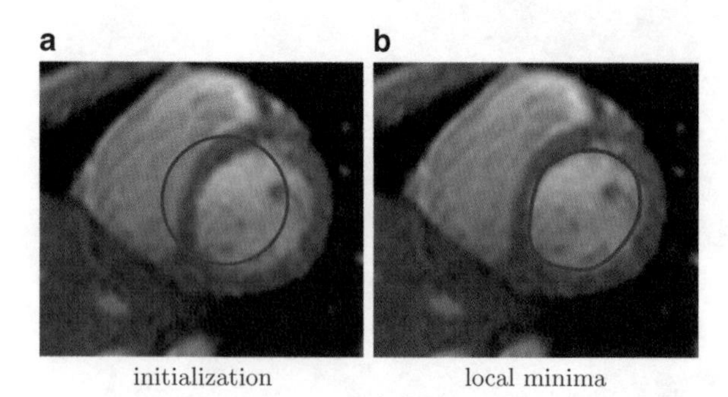

Interactive Segmentation, Fig. 6 *Snakes* and other *active contour* methods are initialized by rough delineation of the desired object (**a**). Minimization of an energy associated with the contour leads to a local minimum (**b**) with better alignment to image boundaries

and development of continuous max-flow techniques [23, 24] now also allow good quality approximations of the global minima.

Physics-based segmentation models: Snakes [25, 26], balloons [27], spline snakes [28], and other methods explicitly represent object boundaries as an elastic band or a balloon. The band is normally assigned an internal energy (elasticity and stiffness) and a potential energy with respect to some external field of predefined *image forces* attracting the band to image boundaries, that is, locations with large intensity gradients. A user can also place seeds defining additional attraction or repulsion potentials.

Geometry-based segmentation models: Note that two visually identical *snakes* appearing in the same image position may have different internal elastic energies. In many cases this may contradict a natural assumption that a segmentation result can be evaluated only by its visible appearance (In some video applications the goal is to track specific points on a moving segment, e.g. muscles of a beating heart. Physics-based (e.g. elastic) segmentation energy is well motivated is such cases). Based on this criticism of the physics-based approach, [29, 30] proposed *geodesic active contour* model evaluating contour C on a (bounded) domain Ω via geometric functionals like

$$E(C) = \int_{int(C)} f_1(p)\, dp + \int_{\Omega/int(C)} f_0(p)\, dp + \int_{\partial C} g(s) ds \qquad (3)$$

which is similar to discrete model (2). The first two integrals in (3) are over the interior and exterior regions of C, and the third integral is the Riemannian length of C under metric g. Image-based *density* function g, for example,

$$g(p) = \frac{1}{1 + ||\nabla I(p)||^2}, \quad \forall p \in \Omega$$

shortens the length of contour C if it follows image boundaries where the density is small. The geometric length term in (3) is a continuous analogue of (1) and the spatial smoothness term in (2); see [13].

Scalar functions f_1 and f_0 on Ω are interior and exterior potentials based on some known appearance models for the object and background regions. For example, one can use $f_i(p) = -\ln \Pr(I_p|\theta_i)$ as in (2). Similarly, these potentials could also enforce user-placed hard constraints (seeds).

Geometric contours C can be represented as *level sets* of some scalar embedding function $u : \Omega \rightarrow \mathcal{R}^1$, for example, $C = \{p \in \Omega : u(p) = const\}$. This approach avoids some numerical issues, for example, the need for *reparameterization*, often associated with explicit representation of contour points needed for physics-based bands. The level-set framework does not pose any topological constraints on contours and yields easy-to-implement gradient descent equations for geometric energies like (3). More recently, geometric energies like (3) are addressed with various continuous convex formulations [24, 31, 32] that are shown to converge to a good approximation of the global minimum.

The continuous geometric models are very closely related to discrete segmentation energies in the graph-cut framework [8, 13]. One theoretical advantage of the continuous formulations is absence of the grid bias. Continuous numerical schemes guarantee certain convergence rate, but some stopping threshold often needs to be chosen. Current fast implementations of continuous optimization methods, for example, [32], require GPU acceleration.

Distance-Based Segmentation Methods

Many interactive segmentation techniques optimize objective functions that are only indirectly related to the visual appearance of the segments and their boundaries. In particular, a large number of methods compute optimal (image-weighted) distance functions computed from seeds. For example, *fast-marching method* [33] extracts the boundary reached at time T by a front expanding with an image-weighted speed. This can be seen as a generalization of *region growing*. These ideas were extended in [34] where the segments are Voronoi cells w.r.t. geodesic distance $d(p,s)$ from the object and background seeds $s \in S_O \cup S_B$

$$x_p^* = \arg \min_{l \in \{O,B\}} \min_{s \in S_l} d(p,s)$$

Their image-based metric is based on gradients of the appearance models likelihoods instead of intensity gradients. Distance transforms can also work as the unary potentials in the segmentation models (2) and (3); for example, see [35].

Instead of the standard *min-sum* geodesic distance $d(p,s)$, many segmentation methods use other measures to compute Voronoi cells from the seeds. For example, *fuzzy-connectedness* methods [36, 37] compute Voronoi cells with respect to some *max–min* affinity measure. *Random walker* [38] outputs Voronoi cells for probabilistic distance function $d(p,s)$, measuring the expected time of arrival for a random walk from p to s. *Watershed* method, for example, [39], connect points to seeds using *water-drop paths* instead of geodesics. *Power watershed* algorithm [40] unifies the ideas of *watershed* and *random walker*.

Some Open Problems

Energy functionals like (2) and (3) represent only the most standard ideas for evaluating segments. Accurate evaluation of the higher-order geometric properties of the boundary, for example, curvature [12, 41], remains a difficult optimization issue. Shape priors for globally optimal segmentation [42, 43] as well as enforcement of topological constraints [44, 45] are largely open problems.

References

1. Gonzalez RC, Woods RE (2007) Digital image processing, 3rd edn. Prentice Hall, Harlow
2. Mortensen EN, Barrett WA (1998) Interactive segmentation with intelligent scissors. Graph. Models Image Process 60:349–384
3. Falcão AX, Udupa JK, Samarasekera S, Sharma S (1998) User-steered image segmentation paradigms: live wire and live lane. Graph Models Image Process 60:233–260
4. Jermyn IH, Ishikawa H (1999) Globally optimal regions and boundaries. In: International conference on computer vision, Kerkyra, vol II, pp 904–910
5. Boykov Y, Jolly MP (2001) Interactive graph cuts for optimal boundary and region segmentation of objects in N-D images. In: International conference on computer vision, Vancouver, vol II, pp 105–112
6. Boykov Y, Funka-Lea G (2006) Graph cuts and efficient N-D image segmentation. Int J Comput Vis 70(2):109–131
7. Rother C, Kolmogorov V, Blake A (2004) Grabcut – interactive foreground extraction using iterated graph cuts. ACM Trans Graph 23:307–331
8. Kolmogorov V, Boykov Y (2005) What metrics can be approximated by geo-cuts, or global optimization of length/area and flux. In: International conference on computer vision, Beijing
9. Boykov Y, Kolmogorov V, Cremers D, Delong A (2006) An integral solution to surface evolution PDEs via geo-cuts. In: European conference on computer vision (ECCV), Graz, Austria
10. Cox IJ, Rao SB, Zhong Y (1996) Ratio regions: a technique for image segmentation. In: International conference on pattern recognition, Vienna, vol II, pp 557–564

11. Jermyn IH, Ishikawa H (2001) Globally optimal regions and boundaries as minimum ratio weight cycles. PAMI 23(10):1075–1088
12. Schoenemann T, Cremers D (2007) Introducing curvature into globally optimal image segmentation: minimum ratio cycles on product graphs. In: International conference on computer vision (ICCV), Rio de Janeiro
13. Boykov Y, Kolmogorov V (2003) Computing geodesics and minimal surfaces via graph cuts. In: International conference on computer vision, Nice, vol I, pp 26–33
14. Bai X, Wang J, Simons D, Sapiro G (2009) Video SnapCut: Robust video object cutout using localized classifiers. In: ACM transactions on graphics (SIGGRAPH), Yokohama
15. Boykov Y, Kolmogorov V (2004) An experimental comparison of min-cut/max-flow algorithms for energy minimization in vision. IEEE Trans Pattern Anal Mach Intell 26(9):1124–1137
16. Kolmogorov V, Boykov Y, Rother C (2007) Applications of parametric maxflow in computer vision. In: International conference on computer vision (ICCV), Rio de Janeiro
17. Boykov Y, Veksler O, Zabih R (2001) Fast approximate energy minimization via graph cuts. IEEE Trans Pattern Anal Mach Intell 23(11):1222–1239
18. Li K, Wu X, Chen DZ, Sonka M (2006) Optimal surface segmentation in volumetric images-a graph-theoretic approach. IEEE Trans Pattern Anal Pattern Recognit (PAMI) 28(1):119–134
19. Delong A, Boykov Y (2009) Globally optimal segmentation of multi-region objects. In: International conference on computer vision (ICCV), Kyoto
20. Felzenszwalb PF, Veksler O (2010) Tiered scene labeling with dynamic programming. In: IEEE conference on computer vision and pattern recognition (CVPR), San Francisco
21. Chan T, Esedoglu S, Nikolova M (2006) Algorithms for finding global minimizers of image segmentation and denoising models. SIAM J Appl Math 66(5):1632–1648
22. Pock T, Chambolle A, Cremers D, Bischof H (2009) A convex relaxation approach for computing minimal partitions. In: IEEE conference on computer vision and pattern recognition (CVPR), Miami
23. Appleton B, Talbot H (2006) Globally minimal surfaces by continuous maximal flows. IEEE Trans Pattern Anal Pattern Recognit 28(1):106–118
24. Yuan J, Bae E, Tai XC (2010) A study on continuous max-flow and min-cut approaches. In: IEEE conference on computer vision and pattern recognition (CVPR), San Francisco
25. Kass M, Witkin A, Terzolpoulos D (1988) Snakes: active contour models. Int J Comput Vis 1(4):321–331
26. Amini AA, Weymouth TE, Jain RC (1990) Using dynamic programming for solving variational problems in vision. IEEE Trans Pattern Anal Mach Intell 12(9):855–867
27. Cohen LD, Cohen I (1993) Finite element methods for active contour models and balloons for 2-d and 3-d images. IEEE Trans Pattern Anal Mach Intell 15(11):1131–1147
28. Isard M, Blake A (1998) Active contours. Springer, London
29. Caselles V, Kimmel R, Sapiro G (1997) Geodesic active contours. Int J Comput Vis 22(1):61–79
30. Yezzi A Jr, Kichenassamy S, Kumar A, Olver P, Tannenbaum A (1997) A geometric snake model for segmentation of medical imagery. IEEE Trans Med Imaging 16(2):199–209
31. Unger M, Pock T, Cremers D, Bischof H (2008) Tvseg - interactive total variation based image segmentation. In: British machine vision conference (BMVC), Leeds, UK
32. Santner J, Pock T, Bischof H (2010) Interactive multi-label segmentation. In: Asian conference on computer vision (ACCV), Queenstown
33. Malladi R, Sethian J (1998) A real-time algorithm for medical shape recovery. In: International conference on computer vision (ICCV), Bombay, pp 304–310
34. Bai X, Sapiro G (2007) A geodesic framework for fast interactive image and video segmentation and matting. In: IEEE international conference on computer vision (ICCV), Rio de Janeiro
35. Criminisi A, Sharp T, Blake A (2008) Geos: geodesic image segmentation. In: European conference on computer vision (ECCV), Marseille
36. Udupa JK, Samarasekera S (1996) Fuzzy connectedness and object definition: theory, algorithms, and applications in image segmentation. Graph Models Image Process 58(3):246–261
37. Herman GT, Carvalho BM (2001) Multiseeded segmentation using fuzzy connectedness. IEEE Trans Pattern Anal Mach Intell (PAMI) 23(5):460–474
38. Grady L (2006) Random walks for image segmentation. IEEE Trans Pattern Anal Pattern Recognit (PAMI) 28(11):1768–1783
39. Cousty J, Bertrand G, Najman L, Couprie M (2009) Watershed cuts: minimum spanning forests and the drop of water principle. IEEE Trans Pattern Anal Mach Intell (PAMI) 31(8):1362–1374
40. Couprie C, Grady L, Najman L, Talbot H (2011). Power watersheds: a unifying graph based optimization framework. IEEE transactions on Pattern Analysis and Machine Intelligence (PAMI) 33(7):1384–1399
41. Williams DJ, Shah M (1992) A fast algorithm for active contours and curvature estimation. Comput Vis Graph Image Process 55(1):14–26
42. Veksler O (2008) Star shape prior for graph-cut image segmentation. In: European conference on computer vision (ECCV), Marseille
43. Felzenszwalb P, Veksler O (2010) Tiered scene labelling with dynamic programming. In: IEEE conference on computer vision and pattern recognition (CVPR), San Francisco
44. Vicente S, Kolmogorov V, Rother C (2008) Graph cut based image segmentation with connectivity priors. In: IEEE conference on computer vision and pattern recognition (CVPR), Anchorage
45. Nowozin S, Lampert CH (2009) Global connectivity potentials for random field models. In: IEEE conference on computer vision and pattern recognition (CVPR), Miami

Interface Reflection

►Specularity, Specular Reflectance

Interpretation of Line Drawings

▶Line Drawing Labeling

Interreflections

Michael S. Langer
School of Computer Science, McGill University, Montreal, QC, Canada

Synonyms

Mutual illumination

Related Concepts

▶Bas-Relief Ambiguity; ▶Diffuse Reflectance; ▶Radiance

Definition

Interreflections are reflections of light from one surface to another surface.

Background

Surfaces are illuminated not just by light sources but also by each other. These interreflections can provide a significant component of surface illumination, especially in concavities or enclosures. Numerical methods for computing interreflections were developed in the early twentieth century to solve problems in heat transfer such as in furnace design. The methods were developed further by the computer graphics community in the 1980s to render global illumination for scenes with Lambertian surfaces and later for scenes with specular components [7, 8].

Theory

Interreflections can be described mathematically in several equivalent ways. One way is to write the reflected light as a sum of the light that is due to the illumination that arrives at a surface directly from the light source, plus the light that arrives from other surfaces in the scene via interreflections. Suppose the scene is composed of Lambertian surfaces with albedo $\rho(\mathbf{x})$ varying across surfaces. Let $L_s(\mathbf{x})$ be the component of $\mathbf{x}$'s outgoing radiance that is due to direct illumination from the source. Then the total radiance $L(\mathbf{x})$ leaving $\mathbf{x}$ is the sum of $L_s(\mathbf{x})$ and the radiance that is due to interreflections:

$$L(\mathbf{x}) = L_s(\mathbf{x}) + \rho(\mathbf{x}) \int_S L(\mathbf{y}) K(\mathbf{x}, \mathbf{y}) d\mathbf{y}. \quad (1)$$

Here the integral is taken over all surface points $\mathbf{y} \in \mathcal{S}$ in the scene, and the function $K(\mathbf{x}, \mathbf{y})$ is a symmetric weighting function that depends on the surface normals at $\mathbf{x}$ and $\mathbf{y}$ and on the distances between $\mathbf{x}$ and $\mathbf{y}$. $K(\mathbf{x}, \mathbf{y})$ is zero if the $\mathbf{x}$ and $\mathbf{y}$ are not visible to each other.

It is common to approximate Eq. (1) by using a polygonal mesh surface whose facets have constant radiance:

$$\mathbf{r} = \mathbf{r}_s + \mathbf{PKr} \quad (2)$$

where $\mathbf{r}$ and $\mathbf{r}_s$ are the vectors of total and direct radiance, respectively, $\mathbf{P}$ is a diagonal matrix of albedos, and $\mathbf{K}$ is called the *form factor* matrix. The above equations can be generalized to non-Lambertian surfaces as well [7, 8].

A second approach is to consider the eigenfunctions of $\mathbf{K}$ which are radiance functions that are invariant to interreflections [9, 11]. These eigenfunctions are concentrated in surface concavities and at points of contact between surfaces [10].

A third approach is to use ray tracing to follow the light emitted from the source through successive reflections or bounces in the scene. The nth reflection serves as the source for the $n + 1$st reflection, and the sum of all reflections gives an infinite series. For any scene geometry and reflectance, it is possible to construct a linear operator that can be applied iteratively to decompose the interreflections into their n bounce components [15]. Understanding the various bounces is especially important for making finite approximations. For example, a two-bounce model has been used to model how surface microfacets can account for non-Lambertian reflection [14] and how color bleeding occurs between surfaces in a concavity [5].

Application

Standard shape from shading and photometric stereo methods consider only the direct illumination component [6]. When interreflections are present, these methods produce erroneous results [4]. It is possible to extend these methods to account for interreflections by first ignoring interreflections to obtain an approximate solution and then iteratively updating the solution to account for interreflections. This idea has been applied to photometric stereo [12] and to shape from shading for the special case that the surface is an unfolded book in a photocopier [16].

The above applications assume the light sources are known. But what if they are unknown? An important fact that applies in this situation is the *bas-relief ambiguity* [1]. When a Lambertian surface is illuminated by direct illumination only, there exists a family of scenes (shape, albedo, lighting) that all produce the same image. With interreflections present, the bas-relief ambiguity no longer exists [3]. One can estimate the surface shape and reflectance similarly to above, namely, by applying a photometric stereo method that is designed for unknown lighting [17] and then iteratively updating the shape and reflectance to account for interreflections [3].

Interreflections also arise in projector-camera systems. An image that is projected on a concave screen will suffer from interreflections that will lower the image contrast. This contrast reduction can be compensated for, to some extent, by modifying the originally projected pattern [2]. A related example which involves active illumination of a 3D scene is to obtain a small number of images of the scene by illuminating it with a set of high-frequency projection patterns such as checkerboards [13]. The interreflection components of the scene will have relatively low spatial frequency and will be similar in the images. This property allows one to decompose the image of a fully illuminated scene into its direct component and its interreflection component. Unlike most methods for interreflections which assume Lambertian scenes, this method allows for non-Lambertian surfaces and other forms of reflections such as volume scattering.

References

1. Belhumeur PN, Kriegman DJ, Yuille AL (1999) The bas-relief ambiguity. Int J Comput Vis 35(1):33–44
2. Bimber O, Grundhofer A, Zeidler T, Danch D, Kapakos P (2006) Compensating indirect scattering for immersive and semi-immersive projection displays. In: Virtual reality conference 2006, Alexandria, pp 151–158, 25–29
3. Chandraker MK, Kahl F, Kriegman DJ (2005) Reflections on the generalized bas-relief ambiguity. In: CVPR '05: proceedings of the 2005 IEEE computer society conference on computer vision and pattern recognition (CVPR), Washington, DC, USA. IEEE Computer Society, pp 788–795
4. Forsyth D, Zisserman A (1991) Reflections on shading. IEEE Trans Pattern Anal Mach Intell 13:671–679
5. Funt BV, Drew MS, Ho J (1991) Color constancy from mutual reflection. Int J Comput Vis 6(1):5–24
6. Horn BKP (1977) Understanding image intensities. Artif Intell 8(2):201–231
7. Immel DS, Cohen MF, Greenberg DP (1986) A radiosity method for non-diffuse environments. ACM Trans Graph 20(4):133–142
8. Kajiya JT (1986) The rendering equation. ACM Trans Graph 20(4):143–150
9. Kœnderink JJ, van Doorn AJ (1983) Geometrical modes as a general method to treat diffuse interreflections in radiometry. J Opt Soc Am 73(6):843–850
10. Langer MS (1999) When shadows become interreflections. Int J Comput Vis 34(2/3):1–12
11. Moon P (1940) On interreflections. J Opt Soc Am 30: 195–205
12. Nayar SK, Ikeuchi K, Kanade T (1991) Shape from interreflections. Int J Comput Vis 6:173–195
13. Nayar SK, Krishnan A, Grossberg MD, Raskar R (2006) Fast separation of direct and global components of a scene using high frequency illumination. ACM Trans Graph 25:935–944
14. Nayar SK, Oren M (1995) Generalization of the Lambertian model and implications for machine vision. Int J Comput Vis 14(3):227–251
15. Seitz SM, Matsushita Y, Kutulakos KN (2005) A theory of inverse light transport. In: ICCV '05: proceedings of the tenth IEEE international conference on computer vision, Washington, DC, USA. IEEE Computer Society, pp 1440–1447
16. Wada T, Ukida H, Matsuyama T (1997) Shape from shading with interreflections under a proximal light-source: distortion-free copying of an unfolded book. Int J Comput Vis 24(2):125–135
17. Yuille AL, Snow D, Epstein R, Belhumeur PN (1999) Determining generative models of objects under varying illumination: shape and albedo from multiple images using svd and integrability. Int J Comput Vis 35(3):203–222

Intrinsic Images

Marshall F. Tappen
University of Central Florida, Orlando, FL, USA

Related Concepts

▸Image Decompositions

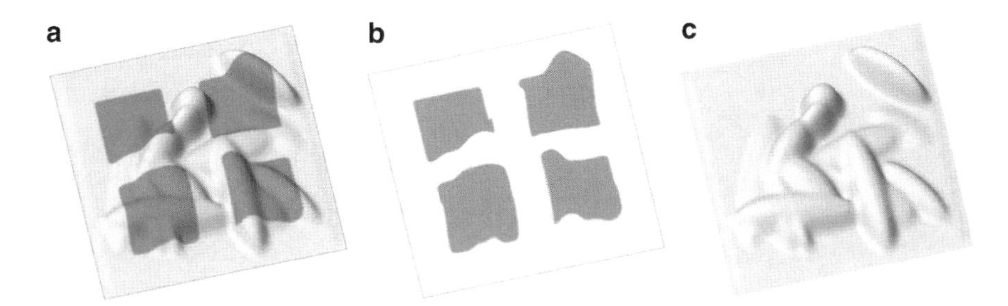

Intrinsic Images, Fig. 1 These images show an example of an intrinsic image decomposition. In this decomposition, the intrinsic images can be how the image in (**a**) can be decomposed into the albedo and shading images shown in (**b**) and (**c**), respectively

Definition

A set of images used to represent characteristics of a scene pictured in an image, with each image representing one particular characteristic of the scene.

Background

Vision systems have been categorized into low- and high-level processing, with high-level processing taking an object-centered approach [1]. In this categorization, the role of low-level processing is to extract basic characteristics at all locations in the image. These characteristics are then used to find objects.

Intrinsic images are a method for representing the low-level characteristics extracted from images. In the intrinsic image representation, proposed by Barrow and Tenenbaum in [2], one image represents each of the characteristics being used in the system. The value of each pixel represents the value of the characteristic at each point in the scene.

The types of characteristics that are conveniently expressed as intrinsic images include the illumination of each point in the scene, the motion at each point, the orientation of each point, the albedo, and the distance from the camera.

Application

Starting with [3], the term intrinsic images have also been used to refer to an image decomposition that decomposes an observed image into intrinsic images that can be recombined to recreate the observed image. The most common decomposition is, into images, representing the shading, or illumination, and albedo of each point. Figure 1 shows an example of how the image in Fig. 1a into shading and albedo images. Mathematically, the decomposition is modeled as

$$O_p = I_p \times R_p$$

where O_p is the the value of the observed image at pixel p, I is the illumination image, and R is the reflectance image.

In [3], Weiss recovers these intrinsic images from a sequence of images where the illumination varies in the scene. In [4] and [5], Tappen et al. use color and gray-scale features to estimate the decomposition from a single image.

Besides image decompositions, [6] proposes using intrinsic images that represent properties like occlusion boundaries and object depth.

References

1. Szeliski R (1990) Bayesian modeling of uncertainty in low-level vision. Int J Comput Vis 5(3):271–301
2. Barrow HG, Tenenbaum JM (1978) Recovering intrinsic scene characteristics from images. In: Hanson A, Riseman E (eds) Computer vision systems. Academic, New York, pp 3–26
3. Weiss Y (2001) Deriving intrinsic images from image sequences. In: The proceedings of the IEEE international conference on computer vision, Vancouver, pp 68–75
4. Tappen MF, Freeman WT, Adelson EH (2005) Recovering intrinsic images from a single image. IEEE Trans Pattern Anal Mach Intell 27(9):1459–1472
5. Tappen MF, Adelson EH, Freeman WT (2006) Estimating intrinsic component images using non-linear regression. In: The Proceedings of the 2006 IEEE computer society conference on computer vision and pattern recognition (CVPR), vol 2. IEEE Computer Society, Los Alamitos, pp 1992–1999
6. Hoiem D, Efros AA, Hebert M (2008) Closing the loop on scene interpretation. in: Proceedings the IEEE conference on computer vision and pattern recognition (CVPR). IEEE, Piscataway

Intrinsic Parameters

▸Intrinsics

Intrinsics

Zhengyou Zhang
Microsoft Research, Redmond, WA, USA

Synonyms

Intrinsic parameters

Related Concepts

▸Camera Parameters (Intrinsic, Extrinsic)

Definition

Intrinsics, short for *intrinsic parameters*, refer to the parameters belonging to the essential nature of a thing, which is usually a camera in computer vision. The intrinsic parameters of a camera include its focal length, the aspect ratio of a pixel, the coordinates of its principal point, and the lens distortion parameters.

See entry "▸Camera Parameters" for more details.

Inverse Compositional Algorithm

Simon Baker
Microsoft Research, Redmond, WA, USA

Definition

The inverse compositional algorithm is a reformulation of the classic Lucas-Kanade algorithm to make the steepest-descent images and Hessian constant.

Background: Lucas-Kanade

The goal of the Lucas-Kanade algorithm is to minimize the sum of squared error between a template image $T(\mathbf{x})$ and a warped input image $I(\mathbf{x})$:

$$\sum_{\mathbf{x}} [\, T(\mathbf{x}) - I(\mathbf{W}(\mathbf{x};\mathbf{p}))\,]^2 , \tag{1}$$

where $\mathbf{x} = (x, y)^{\mathrm{T}}$ are the pixel coordinates, $\mathbf{W}(\mathbf{x};\mathbf{p})$ is a parameterized set of warps, and $\mathbf{p} = (p_1, \dots p_n)^{\mathrm{T}}$ is a vector of parameters. The Lucas-Kanade algorithm assumes that a current estimate of $\mathbf{p}$ is known and then iteratively solves for increments to the parameters $\Delta\mathbf{p}$, i.e., approximately minimize

$$\sum_{\mathbf{x}} [\, T(\mathbf{x}) - I(\mathbf{W}(\mathbf{x};\mathbf{p}+\Delta\mathbf{p}))\,]^2 , \tag{2}$$

with respect to $\Delta\mathbf{p}$ and update the parameters

$$\mathbf{p} \leftarrow \mathbf{p} + \Delta\mathbf{p}. \tag{3}$$

Equation (2) is linearized by performing a first-order Taylor expansion:

$$\sum_{\mathbf{x}} \left[T(\mathbf{x}) - I(\mathbf{W}(\mathbf{x};\mathbf{p})) - \nabla I \frac{\partial \mathbf{W}}{\partial \mathbf{p}} \Delta\mathbf{p} \right]^2 . \tag{4}$$

In this expression, $\nabla I = \left(\frac{\partial I}{\partial x}, \frac{\partial I}{\partial y}\right)$ is the *gradient* of image I and $\frac{\partial \mathbf{W}}{\partial \mathbf{p}}$ is the *Jacobian* of the warp. Equation (4) has a closed-form solution as follows. The partial derivative of the expression in Eq. (4) with respect to $\Delta\mathbf{p}$ is

$$-2 \sum_{\mathbf{x}} \left[\nabla I \frac{\partial \mathbf{W}}{\partial \mathbf{p}} \right]^{\mathrm{T}} \left[T(\mathbf{x}) - I(\mathbf{W}(\mathbf{x};\mathbf{p})) - \nabla I \frac{\partial \mathbf{W}}{\partial \mathbf{p}} \Delta\mathbf{p} \right] . \tag{5}$$

Then denote

$$\mathbf{SD}_{\mathrm{lk}}(\mathbf{x}) = \nabla I \frac{\partial \mathbf{W}}{\partial \mathbf{p}} , \tag{6}$$

the *steepest-descent* images. Setting the expression in Eq. (5) to equal zero and solving give

$$\Delta\mathbf{p} = H_{\mathrm{lk}}^{-1} \sum_{\mathbf{x}} \mathbf{SD}_{\mathrm{lk}}^{\mathrm{T}}(\mathbf{x})\, E(\mathbf{x}) \tag{7}$$

where H_{lk} is the $n \times n$ (Gauss-Newton approximation to the) *Hessian* matrix

$$H_{\mathrm{lk}} = \sum_{\mathbf{x}} \mathbf{SD}_{\mathrm{lk}}^{\mathrm{T}}(\mathbf{x})\, \mathbf{SD}_{\mathrm{lk}}(\mathbf{x}) \tag{8}$$

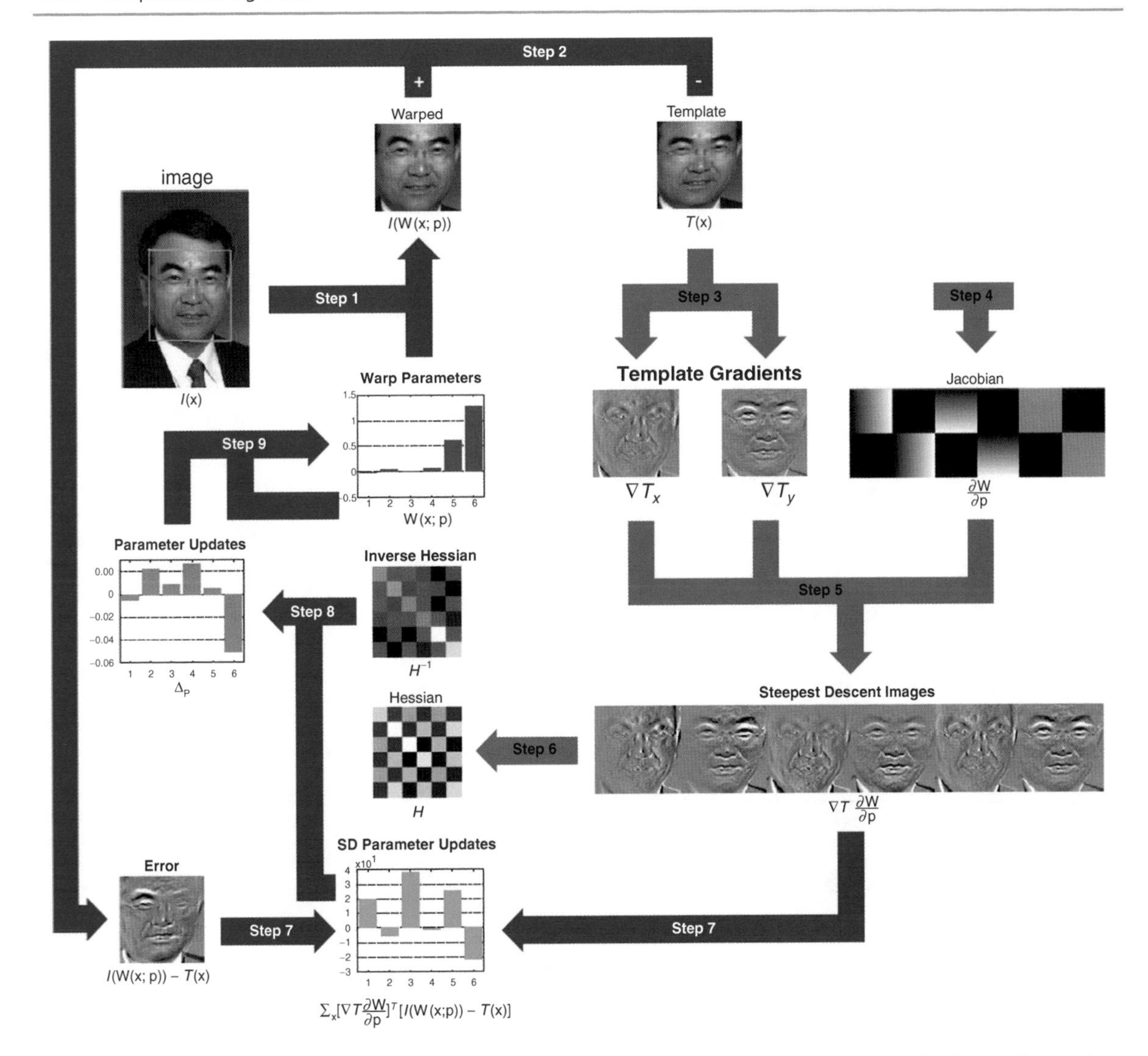

Inverse Compositional Algorithm, Fig. 1 A schematic overview of the inverse compositional algorithm. Steps 3–6 (*light-color arrows*) are performed once as a precomputation. The main algorithm simply consists of iterating image warping (Step 1), image differencing (Step 2), image dot products (Step 7), multiplication with the inverse of the Hessian (Step 8), and the update to the warp (Step 9). All of these steps can be performed efficiently

and

$$E(\mathbf{x}) = T(\mathbf{x}) - I\left(\mathbf{W}(\mathbf{x};\mathbf{p})\right) \tag{9}$$

is the *error image*. The Lucas-Kanade algorithm consists of iteratively applying Eqs. (7) and (3). Because the gradient ∇I must be evaluated at $\mathbf{W}(\mathbf{x};\mathbf{p})$ and the Jacobian $\frac{\partial \mathbf{W}}{\partial \mathbf{p}}$ at $\mathbf{p}$, they both depend on $\mathbf{p}$. Both the steepest-descent images and the Hessian must therefore be recomputed in every iteration [1, 2].

The Inverse Compositional Algorithm

Baker and Matthews [3] proposed the inverse compositional algorithm as a way of reformulating image alignment so that the steepest descent images and Hessian are constant. Although the goal of the inverse compositional algorithm is the same as the Lucas-Kanade algorithm (e.g., minimizing Eq. (1)), the inverse compositional algorithm iteratively minimizes

$$\sum_{\mathbf{x}} \left[T(\mathbf{W}(\mathbf{x};\Delta\mathbf{p})) - I(\mathbf{W}(\mathbf{x};\mathbf{p})) \right]^2, \tag{10}$$

with respect to $\Delta\mathbf{p}$ and then updates the warp

$$\mathbf{W}(\mathbf{x};\mathbf{p}) \leftarrow \mathbf{W}(\mathbf{x};\mathbf{p}) \circ \mathbf{W}(\mathbf{x};\Delta\mathbf{p})^{-1}. \quad (11)$$

The expression

$$\mathbf{W}(\mathbf{x};\mathbf{p}) \circ \mathbf{W}(\mathbf{x};\Delta\mathbf{p}) \equiv \mathbf{W}(\mathbf{W}(\mathbf{x};\Delta\mathbf{p});\mathbf{p}) \quad (12)$$

is the composition of 2 warps, and $\mathbf{W}(\mathbf{x};\Delta\mathbf{p})^{-1}$ is the inverse of $\mathbf{W}(\mathbf{x};\Delta\mathbf{p})$. The inverse compositional algorithm iterates Eq. (10) and (11) and can be shown to be equivalent to the Lucas-Kanade algorithm to first order in $\Delta\mathbf{p}$ [3].

Performing a first-order Taylor expansion on Eq. (10) gives

$$\sum_{\mathbf{x}} \left[T(\mathbf{W}(\mathbf{x};\mathbf{0})) + \nabla T \frac{\partial \mathbf{W}}{\partial \mathbf{p}} \Delta\mathbf{p} - I(\mathbf{W}(\mathbf{x};\mathbf{p})) \right]^2. \quad (13)$$

Assuming that $\mathbf{W}(\mathbf{x};\mathbf{0})$ is the identity warp, the minimum of this expression is

$$\Delta\mathbf{p} = -H_{\mathrm{ic}}^{-1} \sum_{\mathbf{x}} \mathbf{SD}_{\mathrm{ic}}^{\mathrm{T}}(\mathbf{x})\, E(\mathbf{x}), \quad (14)$$

where $\mathbf{SD}_{\mathrm{ic}}^{\mathrm{T}}(\mathbf{x})$ are the steepest-descent images with I replaced by T:

$$\mathbf{SD}_{\mathrm{ic}}(\mathbf{x}) = \nabla T \frac{\partial \mathbf{W}}{\partial \mathbf{p}}, \quad (15)$$

H_{ic} is the Hessian matrix computed using the new steepest-descent images:

$$H_{\mathrm{ic}} = \sum_{\mathbf{x}} \mathbf{SD}_{\mathrm{ic}}^{\mathrm{T}}(\mathbf{x})\, \mathbf{SD}_{\mathrm{ic}}(\mathbf{x}), \quad (16)$$

and the Jacobian $\frac{\partial \mathbf{W}}{\partial \mathbf{p}}$ is evaluated at $(\mathbf{x};\mathbf{0})$. Since there is nothing in either the steepest-descent images or the Hessian that depends on $\mathbf{p}$, they can both be precomputed. The inverse composition algorithm is illustrated in Fig. 1.

Application

The inverse compositional algorithm can be used almost anywhere the Lucas-Kanade can be. In can be applied to anything from simple translational motion to dense optical flow. Perhaps the most significant application is its use to speed-up the fitting or tracking of active appearance models [4, 5].

References

1. Hager G, Belhumeur P (1998) Efficient region tracking with parametric models of geometry and illumination. IEEE Trans Pattern Anal Mach Intell 20(10):1025–1039
2. Shum HY, Szeliski R (2000) Construction of panoramic image mosaics with global and local alignment. Int J Comput Vis 16(1):63–84
3. Baker S, Matthews I (2004) Lucas-Kanade 20 years on: a unifying framework. Int J Comput Vision 56(3):221–255
4. Cootes T, Edwards G, Taylor C (2001) Active appearance models. IEEE Trans Pattern Anal Mach Intell 23(6):681–685
5. Matthews I, Baker S (2004) Active appearance models revisited. Int J Comput Vis 60(2):135–164

IP Camera

▸Pan-Tilt-Zoom (PTZ) Camera

Irradiance

Fabian Langguth and Michael Goesele
GCC - Graphics, Capture and Massively Parallel Computing, TU Darmstadt, Darmstadt, Germany

Related Concepts

▸Radiance

Definition

Irradiance E is defined as the incident power of electromagnetic radiation on a surface per unit surface area. It is expressed in watt per square meter $(\mathrm{W} \cdot \mathrm{m}^{-2})$.

Background

Irradiance is a concept from radiometry, the science of measuring radiant energy transfer [1]. The equivalent

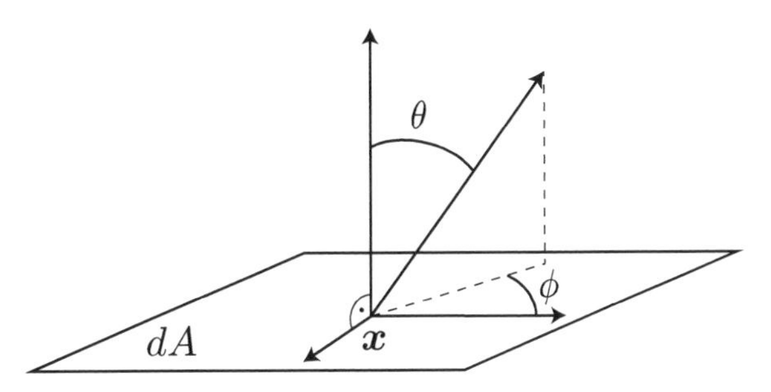

Irradiance, Fig. 1 Geometric setting

concept in photometry is illuminance, with the key difference being that illuminance is adjusted to account for the varying sensitivity of the human eye to different wavelengths of light.

Theory

The irradiance at a surface point $\boldsymbol{x}$ is proportional to the radiance $L(\boldsymbol{x}, \theta, \phi)$ arriving at $\boldsymbol{x}$ from direction (θ, ϕ) with a geometric foreshortening factor $\cos\theta$. Taking into account the whole hemisphere above the surface point, the irradiance is the integral over all incoming directions

$$E(\boldsymbol{x}) = \int_{\theta,\phi} L(\boldsymbol{x}, \theta, \phi) \cos\theta \; d\theta d\phi. \qquad (1)$$

θ denotes the angle between the surface normal and the incident direction (θ, ϕ) (see Fig. 1).

Application

For a camera with an optical lens and an aperture, the image irradiance at a camera sensor is proportional to the radiance L emitted from a small scene patch in the form that

$$E = L\frac{\pi}{4}\left(\frac{d}{f}\right)^2 \cos\alpha^4 \qquad (2)$$

where d is the aperture and f the focal length of the lens. α is the angle between the direction to the observed patch and the principal ray of the camera. For wide-angle lenses, the influence of α often results in a reduction of an image's brightness at the corners compared to the image center. This effect is also called vignetting.

The pixel values of digital images are directly related to the irradiance at the sensor of the camera via the camera's response curve [2, 3]. Many computer vision techniques such as photometric stereo use this fact to recover information about the scene from the irradiance. Early works in this field include [4] and [5].

References

1. Dutré P, Bala K, Bekaert P (2006) Advanced global illumination. AK Peters, Wellesley
2. Debevec PE, Malik J (1997) Recovering high dynamic range radiance maps from photographs. In: Proceedings of the 24th annual conference on computer graphics and interactive techniques, SIGGRAPH '97, Los Angeles. ACM/Addison-Wesley, Los Angeles, pp 369–378
3. Robertson MA, Borman S, Stevenson RL (2003) Estimation-theoretic approach to dynamic range enhancement using multiple exposures. J Electron Imaging 12(2):219
4. Bruss AR (1981) The image irradiance equation: its solution and application (AITR-623). Artificial Intelligence Laboratory, Massachusetts Institute of Technology
5. Woodham R (1980) Photometric method for determining surface orientation from multiple images. Opt Eng 19(1):139–144

Isotropic Differential Geometry in Graph Spaces

Jan J. Koenderink
Faculty of EEMSC, Delft University of Technology, Delft, The Netherlands
The Flemish Academic Centre for Science and the Arts (VLAC), Brussels, Belgium
Laboratory of Experimental Psychology, University of Leuven (K.U. Leuven), Leuven, Belgium

Synonyms

Differential Geometry of Graph Spaces; Dual Differential Geometry

Related Concepts

►Curvature; ►Curves in Euclidean Three-Space; ►Isotropic Differential Geometry in Graph Spaces

Definition

In many settings the conventional Euclidean differential geometry is not appropriate. A common case involves "graphs," an instance being images where the carrier may be modeled as the Euclidean plane, but the intensity domain is incommensurate. Isotropic differential geometry allows one to deal with such cases.

Background

Isotropic differential geometry became highly developed during the first half of the twentieth century, mainly in German-speaking countries. The bulk of the literature is still in German.

Theory

There are frequent cases in computer vision and image processing in which the Euclidean $\mathbb{E}^3$ setting from classical differential geometry is not appropriate. A simple example is an image, which may be thought of as the Euclidean plane $\mathbb{E}^2$, augmented with some "intensity domain." Intensities are nonnegative quantities that somehow reflect photon-catches in, e.g., CCD devices. Usually the physical dimension is unclear and considered irrelevant to the problem. Then the structure of the intensity domain is most appropriately modeled by the affine line $\mathbb{A}^1$, by considering the logarithm of the intensity modulo some arbitrary constant. But the $\mathbb{E}^2 \times \mathbb{A}^1$-space is quite unlike $\mathbb{E}^3$ as becomes evident when one considers Euclidean rotations about some axis in the image plane. Such rotations make no sense because photon catches and lengths are incommensurable physical quantities. The correct way to proceed is to consider "image space" to be a fiber bundle with base space $\mathbb{E}^2$ and fibers $\mathbb{A}^1$. Permissible transformations do not "mix" fibers, and Euclidean rotations about axes in the image plane are not among them.

This situation is typical in many contexts. The simplest example is perhaps a graph $y = f(x)$, where x and y are incommensurable physical quantities. Although the graph is evidently a curve in the xy-plane, it makes no sense to compute its Euclidean curvature as the result will depend on irrelevant transformations of the y-domain (Fig. 1).

A formal way to deal with such problems is to treat the y-axis as an isotropic dimension. Then the metric in the plane is essentially the separation in the x-dimension, the y-separation being treated as isotropic, i.e., nil. Thus the distance of points $\{x_1, y_1\}$ and $\{x_2, y_2\}$ is taken to be $x_2 - x_1$. Notice that this implies that points $\{x, y_1\}$ and $\{x, y_2\}$ with $y_1 \neq y_2$ are at mutually zero distance, *yet different*. One denotes such points "parallel" and assigns them the "special distance" $y_2 - y_1$. Only parallel points have a special distance, generic points only a proper distance. The group of "isotropic motions":

$$x' = x + t_x, \tag{1}$$

$$y' = \alpha x + y + t_y, \tag{2}$$

conserves proper distance and, in the case of parallel points, special distance. This group is fit to replace the group of Euclidean movements:

$$x' = x \cos\alpha - y \sin\alpha + t_x, \tag{3}$$

$$y' = x \sin\alpha + y \cos\alpha + t_y, \tag{4}$$

and indeed has a somewhat similar (with important differences!) structure.

One obtains this group if the xy-plane is interpreted as the dual number plane. Dual numbers are complex numbers $z = x + \varepsilon y$, where the imaginary unit ε is defined as a nontrivial (i.e., not equal zero) solution of the quadratic equation $\varepsilon^2 = 0$. Thus $\varepsilon \neq 0$ whereas $\varepsilon^2 = 0$, from which one derives that neither $\varepsilon > 0$, nor $\varepsilon < 0$. Thus one is forced to use intuitionistic logic, for instance dropping the law of the excluded middle. A concrete representation is by way of matrices:

$$z = x + \varepsilon y = \begin{pmatrix} x & y \\ 0 & x \end{pmatrix}, \tag{5}$$

then addition and multiplication of dual numbers may be done by matrix algebra, similar to the conventional complex numbers $x + \mathrm{i}y$ with imaginary unit $\mathrm{i}^2 = -1$,

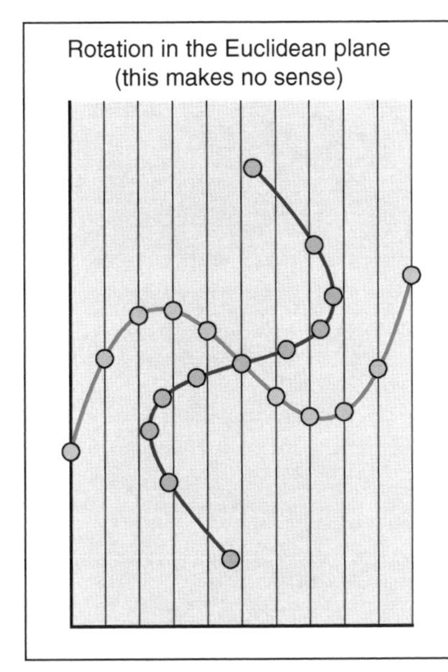

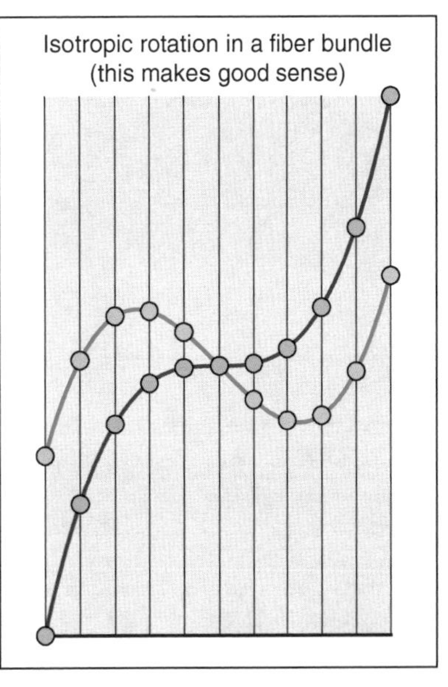

Isotropic Differential Geometry in Graph Spaces, Fig. 1 At *left* a graph in *red*, and the graph after a Euclidean rotation (*in blue*). This obviously makes no sense at all: the *blue curve* is not even a graph anymore! This is not a fiber bundle. At *right* the *red graph* has been subjected to an isotropic rotation. This makes perfect sense, one obtains another graph. Notice that the points move up and down along the fibers of the fiber space, they never leave their fiber, fibers "do not mix"

which are modeled through matrix algebra with the matrices:

$$z = x + \mathrm{i}y = \begin{pmatrix} x & -y \\ y & x \end{pmatrix}. \tag{6}$$

However, although perhaps less scary, these matrix models are an unnecessary pain in hand calculations.

A linear transformation $z' = az+b$ with $a = 1+\varepsilon\alpha$ and $b = t_x + \varepsilon t_y$ becomes:

$$\begin{aligned} z' &= (1+\varepsilon\alpha)(x+\varepsilon y) + (t_x + \varepsilon t_y) \\ &= (x + t_x) + \varepsilon(\alpha x + y + t_y), \end{aligned} \tag{7}$$

(using $\varepsilon^2 = 0$), i.e., exactly the transformation given above. Apparently the dual imaginary unit is an "infinitesimal." Indeed, the full Taylor expansion of a function F about x is:

$$F(x + \varepsilon h) = F(x) + \varepsilon h F'(x). \tag{8}$$

Specifically, one has:

$$\sin \varepsilon\xi = \varepsilon\xi, \tag{9}$$

$$\cos \varepsilon\xi = 1, \tag{10}$$

$$\mathrm{e}^{\varepsilon\xi} = 1 + \varepsilon\xi, \tag{11}$$

thus trigonometry becomes really convenient. The polar representation of a dual number becomes:

$$z = x + \varepsilon y = x(1 + \varepsilon\frac{y}{x}) = |z|\mathrm{e}^{\varepsilon \arg z}. \tag{12}$$

The dual angle is $\arctan y/x = y/x$ (notice that an isotropic angle equals its tangent!), thus the angle measure is parabolic instead of elliptic. Angles do not repeat with period 2π as in the Euclidean plane, but run between $\pm\infty$. Rotating the point 1 (that is $1+\varepsilon\,0$) about the origin over an angle α yields $1+\varepsilon\,\alpha$, thus the line $x = 1$ is (part of) a unit circle. This brings one back to the original construction, and the rotations do not "mix" the x and y dimensions in a way that would be nonsense from the perspective of physics.

The group of proper motions (translations and rotations) of the dual plane leads to a differential geometry of curves that differs from that of the Euclidean plane. Consider the curve $z(x) = x + \varepsilon y(x)$. It is evidently parameterized by arc length, for $|z_x|^2 = 1$. The tangent is $t(x) = z_x = 1 + \varepsilon y_x(x)$, and is a unit vector, for $|t(x)| = 1$. The unit normal is ε, for $t_x(x) = \varepsilon y_{xx}(x)$ with (special) length $y_{xx}(x)$. Thus the normal is constant along the curve and useless for the purposes of differential geometry. The slope of the tangent is well defined though, the tangent subtends an angle $y_x(x)$ with the x-axis. The derivative of this angle with arc-length is $y_{xx}(x)$, thus one concludes that the curve has

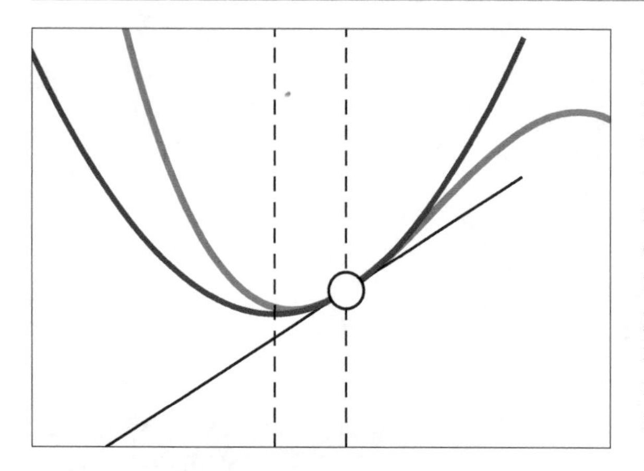

Isotropic Differential Geometry in Graph Spaces, Fig. 2 Consider the local Taylor expansion of the *red curve* at the point indicated by the *white dot*. The first order is the drawn *black line*, the second-order approximation is the *blue parabola*. The center of this "circle of the second kind" is indicated by the leftmost *dotted line*. The other *dotted line* is the normal direction

curvature $\kappa(x) = y_{xx}(x)$. Notice that it is a much simpler expression than one has in the Euclidean plane, which is:

$$k(x) = \frac{y_{xx}(x)}{(1 + y_x(x)^2)^{3/2}}. \tag{13}$$

As expected, the Euclidean and the dual curvatures agree to first order, and for very shallow curves (infinitesimally near the x-axis) the Euclidean curvature degenerates to the dual curvature.

A curve:

$$z(x) = \varepsilon \frac{(x - c)^2}{2R}, \tag{14}$$

has curvature $1/R$, thus a radius of curvature R and is centered on $x = c$. It is evidently a circle in some sense, though different from the circle encountered above. One denotes $x = \pm 1$ a unit circle of the first kind, with center at the origin, $\varepsilon x^2/2$ a unit circle of the second kind, centered at the origin. The local second-order Taylor expansion of a curve is illustrated in Fig. 2. It is a parabola with isotropic axis, thus a "circle of the second kind." It is the osculating circle to the curve in the isotropic geometry. The radius of the osculating circle is evidently the reciprocal of the second derivative, thus a curve $x + \varepsilon y(x)$ has curvature $y_{xx}(x)$ as argued above.

The differential geometry of curves and surfaces in a fiber bundle $\mathbb{E}^2 \times \mathbb{A}^1$ can be handled in a similar manner. All expressions are much simpler than those in Euclidean differential geometry, which is a very useful property, apart from the advantage that they make sense for a change. (Inappropriate applications of expressions taken from Euclidean differential geometry occur very frequently in computer vision and image processing. Although they certainly yield numerical results, they strictly speaking make no sense.) Thus the mean curvature of a surface $\{x, y, z(x, y)\}$ in Monge form becomes $2H = z_{xx} + z_{yy}$, the Gaussian curvature $K = z_{xx} z_{yy} - z_{xy}^2$, and so forth. Like in the planar case discussed above, the normal is constant, and thus useless. One uses the spatial attitude of the tangent plane instead. Any point of the surface may be mapped on the unit sphere of the second kind $z(x, y) = (x^2 + y^2)/2$ through parallelity of tangent planes. Even more conveniently, one notices that the xy-plane $\{x, y, 0\}$ is the stereographic projection of this sphere, thus conformal, but most remarkably – because different from the Euclidean case – also isometric. Thus the Gauss map becomes

$$\begin{aligned}\{x, y, z(x, y)\} \in \mathbb{E}^2 \times \mathbb{A}^1 \mapsto \{&-z_x(x, y),\\ &- z_y(x, y)\} \in \mathbb{R}^2,\end{aligned} \tag{15}$$

a map that is familiar in computer vision as "gradient space." Gradient space is often used by way of a "linear approximation," but it is really the exact Gauss map (or "spherical image") in terms of the appropriate differential geometry.

The geometry of single isotropic space is well understood, although almost all of the literature is in German. The paper by Pottmann is the only reference in English on the general (space) setting, the book by Yaglom (translated into English from Russian) is an excellent introduction to the geometry of the dual plane.

Open Problems

This section introduced a very simple setting. In general one may have to deal with a graph over a curved surface. The paper by Pottmann gives some leads as how to handle such more general cases.

References

1. Koenderink JJ, van Doorn AJ (2002) Image processing done right. In: Proceedings of the European conference on computer vision (ECCV 2002). Lecture notes in computer science, vol 2350. Springer, Heidelberg, pp 158–172
2. Pottmann H, Opitz K (1994) Curvature analysis and visualization for functions defined on Euclidean spaces or surfaces. Comput Aided Geom Des 11: 655–674
3. Yaglom IM (1979) A simple non–Euclidean geometry and its physical basis: an elementary account of Galilean geometry and the Galilean principle of relativity. (trans: Shenitzer A). Springer, New York (translated from Russian)

Iterative Closest Point (ICP)

Zhengyou Zhang
Microsoft Research, Redmond, WA, USA

Synonyms

ICP

Definition

Iterative closest point (ICP) is a popular algorithm employed to register two sets of curves, two sets of surfaces, or two clouds of points.

Background

The ICP technique was proposed independently by Besl and McKay [1] and Zhang [2] in two different contexts. Besl and McKay [1] developed the ICP algorithm to register partially sensed data from rigid objects with an ideal geometric model, prior to shape inspection. So this is a subset–set matching problem because each sensed point has a correspondence in the ideal model. Zhang [2] developed the ICP algorithm in the context of autonomous vehicle navigation in rugged terrain based on vision. His algorithm is used to register a sequence of sensed data in order to build a complete model of the scene and to plan a free path for navigation. So this is a subset–subset matching problem because a fraction of data in one set does not have any correspondence in the other set. To address this issue, Zhang's ICP algorithm has integrated a statistical method based on the distance distribution to deal with outliers, occlusion, appearance, and disappearance. However, both algorithms share the same idea: iteratively match points in one set to the closest points in another set and refine the transformation between the two sets, with the goal of minimizing the distance between the two sets of point clouds.

Theory

The ICP algorithm is very simple and can be summarized as follows:

- *Input:* two point sets, initial estimation of the transformation
- *Output:* optimal transformation between the two point sets
- *Procedure:* Iterate the following steps:
  - (i) Apply the current estimate of the transformation to the first set of points.
  - (ii) Find the closest point in the second set for each point in the first transformed point set.
  - (iii) Update the point matches by discarding outliers.
  - (iv) Compute the transformation using the updated point matches, until convergence of the estimated transformation.

Here are a few comments on this general algorithm:

- Depending on the nature of the point sets, various pose estimation techniques described in the earlier sections can be used to compute the transformation between the two sets.
- The step of finding the closest point to a given point is generally the most time-expensive one. However, this step can be easily parallelized.
- Many data structures can be used to accelerate the finding of the closest point. They include k-D tree and octree.
- Instead of using all points from the first set, a selected subset of points (such as high curvature points) can be used to speed up the process, with only moderate sacrifice of the final accuracy.
- The above algorithm is not symmetric. Let point $\hat{p}_i'$ in the second set be the closest point to a point p_i in the first set. In the other direction, point p_i is, in general, not necessarily the closest point to $\hat{p}_i'$. In

order to make the algorithm symmetric, we can find the closest point in the first transformed point set for each point in the second set and add these point matches to the overall set of matches. Better results can then be obtained at the expense of additional computational cost.

- When the ICP algorithm is applied to register curves or surfaces, they need to be sampled. The final accuracy depends on the density of sampling. The denser the sampling is, the higher the registration quality will be, but the more the computation will be required.

For more detailed and extensive discussions on ICP, the interested reader is referred to Sects. 7 and 8 of Zhang's paper [2].

There are several variants to the ICP algorithm. A useful variation is to substitute the point-to-point distance with point-to-plane distance [3]. The point-to-plane distance allows one surface to slide tangentially along the other surface, making it less likely get stuck in local minima. Consider a point p_i in the first set. Let point $\hat{p}_i'$ in the second set be its closest point. Let the surface normal at point p_i be n_i (a unit vector). Then, the point-to-plane distance measure is given by

$$d_i = n_i^T(\hat{p}_i' - p_i).$$

Surface normals can be precomputed to save computation.

References

1. Besl PJ, McKay ND (1992) A method for registration of 3-d shapes. IEEE Trans Pattern Anal Mach Intell 14(2):239–256
2. Zhang Z (1994) Iterative point matching for registration of free-form curves and surfaces. Int J Comput Vis 13(2):119–152. Also Research Report No.1658, INRIA Sophia-Antipolis, 1992
3. Chen Y, Medioni G (1992) Object modelling by registration of multiple range images. Image Vis Comput 10(3):145–155

Ives Transform

▸von Kries Hypothesis